P9-CKV-871

AAPG TREATISE OF PETROLEUM GEOLOGY

The American Association of Petroleum Geologists
gratefully acknowledges and appreciates the leadership and support
of the AAPG Foundation in the development of the
Treatise of Petroleum Geology.

STRUCTURAL TRAPS III
TECTONIC FOLD AND FAULT TRAPS

COMPILED BY
EDWARD A. BEAUMONT
AND
NORMAN H. FOSTER

TREATISE OF PETROLEUM GEOLOGY
ATLAS OF OIL AND GAS FIELDS

PUBLISHED BY
THE AMERICAN ASSOCIATION OF PETROLEUM GEOLOGISTS
TULSA, OKLAHOMA 74101, U.S.A.

ISBN 0-89181-583-X
ISSN 1043-6103

Available from:
The AAPG Bookstore
P.O. Box 979
Tulsa, OK 74101-0979

Phone: (918) 584-2555
Telex: 49-9432
FAX: (918) 584-0469

Association Editor: Susan Longacre
Science Director: Gary D. Howell
Publications Manager: Cathleen P. Williams
Special Projects Editor: Anne H. Thomas
Science Staff: William G. Brownfield
Project Production: Custom Editorial Productions

Table of Contents

TREATISE OF PETROLEUM GEOLOGY
ADVISORY BOARD

Eric A. Rudd
Floyd F. Sabins, Jr.
Nahum Schneidermann
Peter A. Scholle
George L. Scott, Jr.
Robert T. Sellars, Jr.
Faroog A. Sharief
John W. Shelton
Phillip W. Shoemaker
Synthia E. Smith
Robert M. Sneider
Stephen A. Sonnenberg
William E. Speer
Ernest J. Spradlin
Bill St. John
Philip H. Stark
Richard Steinmetz
Per R. Stokke
Denise M. Stone
Donald S. Stone
Doug Strickland
James V. Taranik
Harry Ter Best, Jr.
Bruce K. Thatcher, Jr.
M. Ray Thomasson
Jack C. Threet
Bernard Tissot
Donald F. Todd
M. O. Turner
Peter R. Vail
B. van Hoorn
Arthur M. Van Tyne
Ian R. Vann
Harry K. Veal*
Steven L. Veal
Richard R. Vincelette
Cecil von Hagen
Fred J. Wagner, Jr.
William A. Walker, Jr.
Anthony Walton
Douglas W. Waples
Harry W. Wassall, III
W. Lynn Watney
N. L. Watts
Koenradd J. Weber
Robert J. Weimer
Dietrich H. Welte
Alun H. Whittaker
James E. Wilson, Jr.
John R. Wingert
Martha O. Withjack
P. W. J. Wood
Homer O. Woodbury
Walter W. Wornardt
Marcelo R. Yrigoyen
Mehmet A. Yukler
Zhai Guangming
Robert Zinke

* Deceased

American Association of Petroleum Geologists Foundation Treatise of Petroleum Geology Fund*

Major Corporate Contributors
($25,000 or more)

BP Exploration Company Limited
Chevron Corporation
Exxon Company, U.S.A.
Mobil Oil Corporation
Oryx Energy Company
Pennzoil Exploration and Production Company
Shell Oil Company
Union Pacific Foundation
Unocal Corporation

Other Corporate Contributors
($5,000 to $25,000)

Cabot Oil & Gas Corporation
Canadian Hunter Exploration Ltd.
Conoco Inc.
Marathon Oil Company
The McGee Foundation, Inc.
Phillips Petroleum Company
Texaco Philanthropic Foundation Inc.
Transco Energy Company

Major Individual Contributors
($1,000 or more)

John J. Amoruso
C. Hayden Atchison
Richard R. Bloomer
A. S. Bonner, Jr.
David G. Campbell
Herbert G. Davis
Paul H. Dudley, Jr.
Lewis G. Fearing
James A. Gibbs
George R. Gibson
William E. Gipson
Robert D. Gunn
Merrill W. Haas
Cecil V. Hagen
Frank W. Harrison
William A. Heck
Roy M. Huffington
Harrison C. Jamison
Thomas N. Jordan, Jr.
Hugh M. Looney
Jack P. Martin
John W. Mason
George B. McBride
Dean A. McGee
John R. McMillan
Grover E. Murray
Rudolf B. Siegert
Robert M. Sneider
Jack C. Threet
Charles Weiner
Harry Westmoreland
James E. Wilson, Jr.
P. W. J. Wood

The Foundation also gratefully acknowledges the many who have supported this endeavor with additional contributions.

*Based on contributions received as of August 15, 1990.

PREFACE

The Atlas of Oil and Gas Fields and the Treatise of Petroleum Geology

The *Treatise of Petroleum Geology* was conceived during a discussion held at the annual AAPG meeting in 1984 in San Antonio, Texas. This discussion led to the conviction that AAPG should publish a state-of-the-art textbook in petroleum geology, aimed not at the student, but at the practicing petroleum geologist. The textbook gradually evolved into a series of three different publications: the Reprint Series, the Atlas of Oil and Gas Fields, and the Handbook of Petroleum Geology. Collectively these publications are known as the *Treatise of Petroleum Geology*, AAPG's Diamond Jubilee project commemorating the Association's 75th anniversary in 1991.

With input from the Advisory Board of the Treatise of Petroleum Geology, we designed this set of publications to represent, to the degree possible, the cutting edge in petroleum exploration knowledge and application: the Reprint Series to provide useful and important published literature; the Atlas to comprise a collection of detailed field studies that illustrate the many ways oil and gas are trapped and to serve as a guide to the petroleum geology of basins where these fields are found; and the Handbook as a professional explorationist's guide to the latest knowledge in the various areas of petroleum geology and related disciplines.

The Treatise Atlas is part of AAPG's long tradition of publishing field studies. Notable AAPG field study compilations include *Structure of Typical American Fields*, published in 1929 and edited by Sidney Powers; and Memoir 30, *Giant Fields of 1968-1978*, published in 1981 and edited by Michel T. Halbouty. The Treatise Atlas continues that tradition but introduces a format designed for easier access to data.

Hundreds of geologists participated in this first compilation of the Atlas. Authors are from all parts of the industry and numerous countries. We gratefully acknowledge the generous contribution of their knowledge, resources, and time.

Purpose of the Atlas

The purpose of the Atlas is twofold: (1) to help exploration and development geologists become more efficient by increasing their awareness of the ways oil and gas are trapped, and (2) to serve as a reference for both the petroleum geology of the fields described and the basins in which they occur.

Imagination is the primary tool of the explorationist. Wallace E. Pratt once said that the unfound field must first be sought in the mind. In part, what is imagined is based on what is remembered; memory is the direct link to what is created in the mind. To create ideas that lead to the discovery of new fields, the mind of the geologist builds from its knowledge of petroleum geology. To that end, the Atlas of field studies will be a primary source for locating much of the information necessary for creating prospects and will provide a connection to the phenomenon of oil and gas traps.

Next to the firsthand experience of having prospects tested with the drill bit, studying the many facets and concepts of developed fields is perhaps the best way for the geologist to develop the ability to create plays and prospects. Also, familiarity with the many ways oil and gas are trapped allows the geologist to see through the noise inherent to exploration data and to close gaps in that data.

Format of the Atlas

To facilitate data access, all field studies in the Atlas follow the same format. Once users become familiar with this format, they will know where to look for the information they seek. Different fields from different parts of the world can be easily compared and contrasted.

The following is a generalized format outline for field studies in the Atlas:

- Location
- History
  - Pre-Discovery
  - Discovery
  - Post-Discovery
- Discovery Method
- Structure
  - Tectonic History
  - Regional Structure
  - Local Structure
- Stratigraphy
- Trap
  - General Description
  - Reservoir(s)
  - Source(s)
- Exploration Concepts

Criteria for Inclusion of a Field

Fields described in the Atlas are selected using two main criteria: (1) trap type, and (2) geographic distribution. Our ultimate goal for the Atlas is to include a field study from each major petroleum-producing province and to include an example of each known trap type. Size or economic importance are not, of themselves, criteria. Many fields that are not giants are included because they are geologically unique, because they are significant examples of geological investigation and original thinking, or because they are historically important, having led to the discovery of many other fields.

Grouping of Fields into Separate Volumes

We considered several ways to group fields in these volumes. We chose trap type because the purpose of the Atlas is to make exploration geologists more effective oil and gas trap finders, regardless of where they search for traps.

Grouping oil and gas field studies into separate volumes by trap type is a difficult exercise. We decided to group the fields into volumes by designating them as structural or stratigraphic traps. Most traps are a combination of both structure and stratigraphy. Some traps are obviously more a consequence of one than the other, but many are not. The continuum that exists between purely stratigraphic and purely structural traps is what makes grouping difficult. A further complication is that many fields contain more than one trap type.

Papers Selected for *Structural Traps III: Tectonic Fold and Fault Traps*

This volume in the Atlas of Oil and Gas Fields series contains studies of fields that have traps related to tectonic folding, tectonic faulting, and/or traps that are both tectonically folded and faulted. Tectonic faulting is faulting due to the release of crustal stresses. Nontectonic faulting is faulting due to other factors, such as salt solution and collapse or detachment from sedimentary loading and slumping. Tectonic folding is folding caused by crustal stresses, as opposed to nontectonic folding, which is caused by other phenomena such as salt dome piercement or compactional drape over a preexisting high.

The first three fields in this volume, Whitney Canyon/Carter Creek, Anschutz Ranch East, and Wilson Creek, are located in the Rocky Mountains in Wyoming and Colorado. All three are anticlines associated with thrusting. The traps of Eugene Island 330 (offshore Gulf Coast) and Hibernia (offshore Canadian east coast) are rollover anticlines formed in association with growth faults. Unity field's (Sudan, Africa) trap is an anticline with complex normal faulting located in a rift. The source of the oil found at Unity is lacustrine shale. Saticoy field (Ventura basin, California) is a footwall fault trap associated with a wrench fault system. Midway-Sunset field (Ventura basin) contains several trap types: anticlinal, updip porosity pinchout, unconformity truncation, and unconformity onlap. Rio Vista (Sacramento basin, California) is a giant gas field whose trap is a faulted dome; the lateral extent of the reservoir is controlled stratigraphically. Badak (Kutai basin, Borneo) is a giant oil and gas field with several pay zones in an anticline. Stratigraphy influences the lateral position of pay zones. Urengoy (Western Siberian basin, USSR) is a huge gas and gas condensate field (estimated ultimate recovery is 288 trillion cubic feet of gas and 4.1 billion barrels of condensate). The trap is a regional dome with multiple pays. Bowdoin field (Williston basin, Montana) is an anticlinal trap with lateral stratigraphic controls. The gas produced from this field is postulated to be microbial in origin.

A unique feature of the Atlas is the exploration history of each field in the study. What seems obvious today usually was not obvious when these fields were discovered. Many times what was expected was not what was found. Geologists discovered these fields by creating concepts based on information limited by the technology available at the time. Drilling and discovery show how closely concept matches reality. Knowing the history of discovery may help explorationists realize that problems, seemingly insoluble at one time, were eventually solved. It is also instructive to learn of the sequence of thinking that solved these problems.

Careful study of these fields will enhance the prospect generator's knowledge base and, consequently, his or her ability to apply that knowledge toward future prospecting.

Edward A. Beaumont
Norman H. Foster, Editors

Whitney Canyon–Carter Creek Field—U.S.A.
Western Wyoming Thrust Belt, Wyoming

JAYNE L. SIEVERDING
Chevron USA, Inc.
Houston, Texas

FRANK ROYSE, JR.
Chevron USA, Inc.
Arvada, Colorado

FIELD CLASSIFICATION

BASIN: Fossil
BASIN TYPE: Thrust Belt
RESERVOIR ROCK TYPE: Dolomite
RESERVOIR ENVIRONMENT OF DEPOSITION: Subtidal/Intertidal Marine Carbonate Shelf
RESERVOIR AGE: Mississippian
PETROLEUM TYPE: Gas
TRAP TYPE: Anticline with Multiple Pays

LOCATION

Whitney Canyon–Carter Creek field is in the Fossil basin area of the southwestern Wyoming thrust belt province (Figure 1). It is approximately 13 mi (21 km) long by 2 mi (3 km) wide and extends from southern Lincoln County into northern Uinta County. The field is divided into the Carter Creek area to the north and Whitney Canyon area to the south. The field extends northward along the Bear River drainage divide. Surface elevations range from 6600 to 8100 ft ASL (2012–2469 m). Significant oil and gas fields nearby include Ryckman Creek and Clear Creek fields to the east and Road Hollow field to the north. At Whitney Canyon–Carter Creek, estimated in-place reserves from all potential pay zones are approximately 4.5 TCFG, 125 MMBO (condensate), and 24 MMLTS (long tons sulfur). The estimated recoverable reserves are 2.2 TCFG, 60 MMBO, and 20 MMLTS. Carmalt and St. John (1986) rank the field as No. 246 on their list of giant fields. According to recent estimates, the field has total recoverable reserves of over 500 MMBOE products, including gas, condensate, sulfur, and NGL.

HISTORY

Pre-Discovery

High-gravity oil seeps were the first indication of hydrocarbon potential in the thrust belt area. Oil seeps out of Cretaceous rocks that crop out along the toe of the Absaroka thrust sheet approximately 24 mi (19 km) to the southeast of Whitney Canyon–Carter Creek field. These seeps were recognized in the mid-1800s when they were used by pioneers to grease their wagon axles. Several nearby locations were drilled in the early 1900s establishing minor shallow oil production from Cretaceous rocks (Veatch, 1906). In 1918, La Barge field was discovered in west central Wyoming and produced oil from folded Tertiary sandstones structurally below the eastern edge of the thrust belt. Until the early 1970s, exploration was sparse and essentially unsuccessful. Hodgden and McDonald (1977) provide a complete summary of early hydrocarbon exploration history for the Wyoming, Utah, and Idaho thrust belt.

Several oil companies became involved in Whitney Canyon–Carter Creek and the associated thrust belt

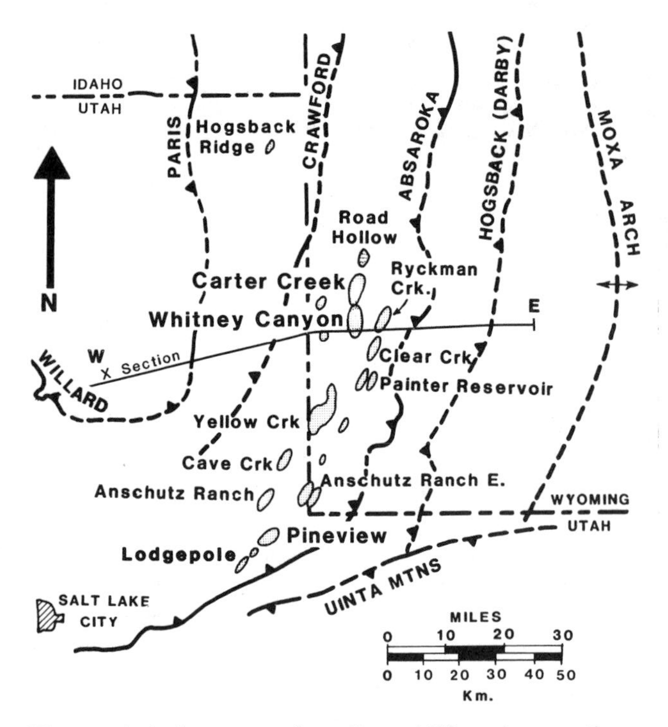

Figure 1. Index map of southwest Wyoming-northern Utah part of the Cordilleran thrust belt, showing traces of major thrust faults, position of oil and gas fields, and location of regional structure section (Figure 3).

areas because of encouraging regional geological, geophysical, and source rock studies conducted during the early 1970s. In 1969, Amoco made an exploration agreement with the Union Pacific Railroad (UPRR) that involved 6 to 7 million acres of land throughout the Rocky Mountains, including parts of the Wyoming and northern Utah thrust belt. In 1972, Chevron leased approximately 180,000 federal and minor fee acres within an area defined by the areal extent of the Absaroka thrust plate. Other companies that leased in the Whitney Canyon–Carter Creek area prior to discovery included the Kewanee Oil Co. (later acquired by Gulf) and Gulf Oil Co. (later merged with Chevron).

Amoco and Chevron joined in several ventures because Amoco's UPRR acreage was in a checkerboard pattern with Chevron's federal lands. Amoco and Chevron undertook a joint seismic program during 1974–1976 to evaluate their mutual interests. They identified several large structures in the hanging wall of the Absaroka thrust, two of which are Whitney Canyon and Carter Creek.

The discovery of Pineview field by American Quasar in 1974 was the first major success in this part of the thrust belt (Figure 1). This was the first discovery of structurally trapped hydrocarbons within the hanging wall of the Absaroka thrust. Ryckman Creek field was the next important discovery. In December 1975, hydrocarbons were recovered on a DST from the Amoco Champlin No. 224, a 50/50 Chevron-Amoco joint venture. Ryckman Creek field produces oil, condensate, and gas primarily from the Jurassic Nugget Sandstone and is located about 6 mi (10 km) east of Whitney Canyon (Figure 1). The success at Ryckman Creek prompted Amoco and Chevron to propose tests of the large structures at Carter Creek and Whitney Canyon that had been mapped previously with seismic data. The Chevron Kewanee Federal No. 1 (Sec. 24, T18N, R120W) was the first well in the Whitney Canyon–Carter Creek area. Spudded in September 1976, the well is located on the southwest end of the Carter Creek structure (Figure 2). The well's main objective was the Nugget Sandstone; however, a DST proved the formation to be tight and wet and the well was abandoned in the Triassic Ankareh in December 1976. The well location is shown on Figure 2 as a gas producer because it was later deepened and completed.

Discovery

The Whitney Canyon–Carter Creek discovery well was spudded in October 1976 in Sec. 18, T17N, R119W, 2 mi (3 km) northeast of the Whitney Canyon surface erosional feature (Figure 2). The Amoco operated ACG No. 1 well was a joint venture between Amoco (37.5%), Chevron (37.5%), Gulf (12.5%), and Champlin (UPRR) (12.5%). It was proposed as a 13,400 ft (4084 m) test of the Jurassic Nugget Sandstone with secondary objectives in the Paleozoic section. A DST established that the Nugget is porous, but wet; however, sweet gas was tested in the Triassic Thaynes Formation and a strong show of sour gas (125 ppm) was encountered in the Permian Phosphoria Formation. Drill pipe parted while coring at 10,685 ft (3257 m) in the Phosphoria. Fishing attempts were unsuccessful, and in August 1977, the well was completed through drill pipe in the Thaynes Formation from 9178 to 9500 ft (2797-2895 m) for a flow rate of 4.7 MMCFGPD, 88 BCPD, and 9 BWPD on a 48/64-in. choke.

Post-Discovery

The Whitney Canyon discovery was confirmed by the Amoco operated ACG No. 2 (Sec. 18, T17N, R119W). The 16,340 ft (4980 m) well was spudded in May 1977 and reached TD in December 1978 in the Cretaceous Frontier Formation below the Absaroka thrust fault. This well established three additional sour gas pay zones above the Absaroka thrust, including the most significant reservoir in the field, the Mississippian Mission Canyon Formation. The well tested the Mission Canyon from 13,180 to 13,303 ft (4017–4054 m) at a rate of 8.5 MMCFGPD, 130 BCPD (24/64-in. choke), and the Pennsylvanian Weber Formation from 12,034 to 12,136 ft (3668–3699 m) at a rate of 0.6 MMCFGPD, 6 BCPD (19/64-in. choke, 300 psi FTP); however, casing parted during

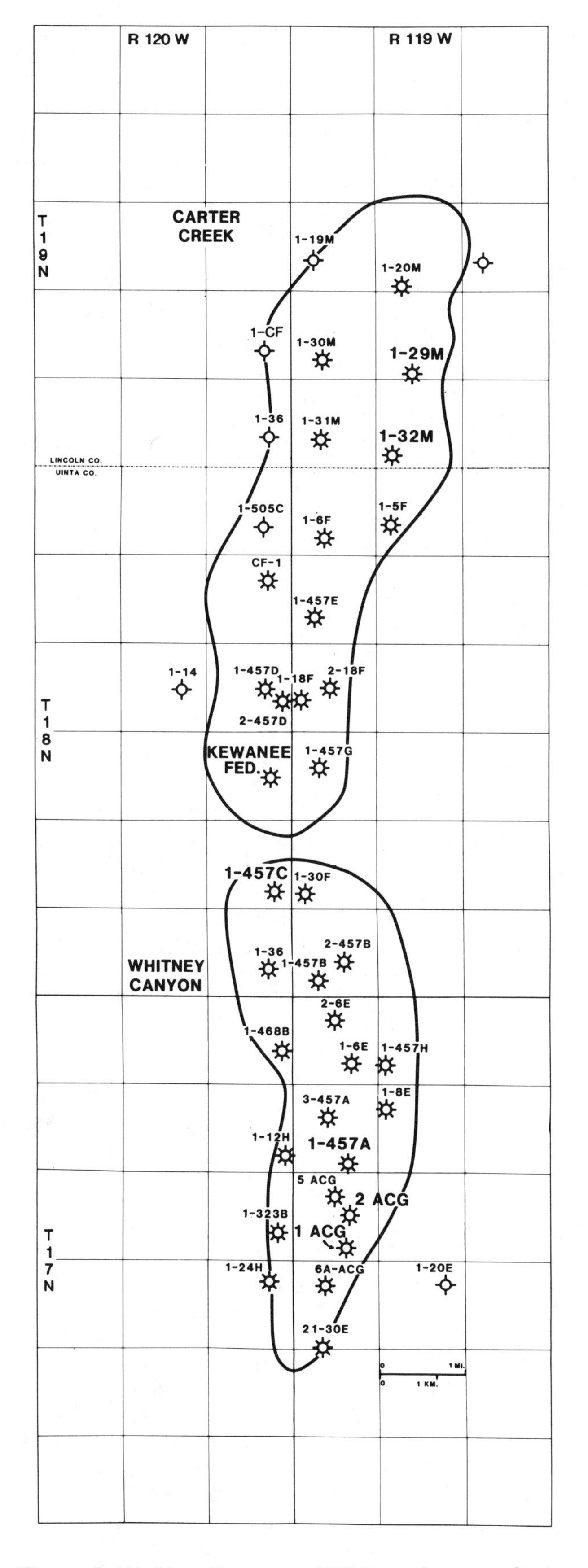

Figure 2. Well location map of Whitney Canyon-Carter Creek field. Outline is general effective limit of field production.

the Weber test, making the figures questionable. The well was completed in the Ordovician Bighorn Dolomite from 15,142 to 15,486 ft (4615-4720 m) with an IP of 5.7 MMCFGPD, 64 BCPD, and 18 BWPD (16/64-in. choke, 2850 psi FTP). This well was the first in the U.S. Rocky Mountain thrust belt to test and produce commercial amounts of gas from Paleozoic reservoirs.

The first well in the Carter Creek portion of the field was the Chevron 1-32M Federal (Sec. 32, T19N, R119W). This 16,573 ft (5051 m) well was drilled from February 1978 to June 1979. The well was completed as a Weber-Mission Canyon commingled producer from 13,019 to 13,270 ft (3968-4044 m) and 14,421 to 14,874 ft (4395-4533 m) and flowed at a rate of 11.4 MMCFGPD, 360 BCPD, and 13 BWPD (64/64-in. choke, 1423 psi FTP). The Mission Canyon was tested alone and flowed at a rate of 6.9 MMCFGPD, 216 BCPD, and 0 BWPD (64/64-in. choke, 950 psi FTP).

The Chevron 1-29M Federal (Sec. 29, T19N, R119W) established the Mississippian Lodgepole formation as a pay zone. Completed in October 1979, the well flowed at a rate of 26.0 MMCFGPD, 623 BCPD, and 0 BWPD (64/64-in. choke, 1600 psi FTP) from the Mission Canyon and Lodgepole commingled at 14,522-14,826 ft (4426-4519 m) and 15,116-15,556 ft (4577-4741 m). An isolated lower Lodgepole test of the interval from 15,492 to 15,556 ft (4721-4741 m) initially flowed at a rate of 4 MMCFGPD, 96 BCPD, and 15 BWPD (24/64-in. choke), and later flowed 17.5 MMCFGPD with no liquids. After commingling with the Mission Canyon, a production log analysis indicated a 10% Lodgepole contribution of 2.6 MMCFGPD.

Whitney Canyon and Carter Creek were originally considered as two separate fields because they occupied separate structural closures, but additional drilling and testing indicated that they were close to sharing a common gas-water contact in the main reservoir. The deepest tested Mission Canyon water-free gas occurred at -7382 ft (-2250 m) in the Amoco Champlin 457C-1 well (Sec. 25, T18N, R120W). This well tested formation water in the Mission Canyon from 14,222 to 14,230 ft (4335-4337 m), set a retainer, then moved up 28 ft (8.5 m) and perforated 14,154-14,194 ft (4314-4326 m) (-7382 ft at lowest perforation). The well then flowed 15.3 MMCFGPD, 124 BCPD, and 51 BWPD (27/64-in. choke, 2250 psi FTP). A water-free completion was eventually attained. No definitive test was ever done in a Whitney Canyon well, but testing has shown that the gas-water contact is within 200 ft (60 m) of -7382 ft (-2250 m). Communication is restricted across the structural saddle within the Mission Canyon main porosity zone and can be demonstrated through condensate chromatography. However, communication between the two structures may still exist in the very upper part of the Mission Canyon.

The Amoco Champlin 457A-1 (Sec. 7, T17N, R119W) is the field's deepest well and is also one of the most prolific gas wells in the Rocky Mountains.

This well was drilled during 1977-1979 and reached a TD of 18,494 ft (5637 m) in the Cretaceous Frontier Formation below the Absaroka thrust fault. The well flowed 2.5 MMCFGPD of sweet gas from the Triassic Thaynes Formation on a DST at 8672–8759 ft (2643–2670 m); 32.8 MMCFGPD, 600 BCPD, 62/64-in. choke, 12% H_2S from the Mission Canyon at 12,420–12,810 ft (3785–3904 m); 17.2 MMCFGPD, 197 BCPD, 54/64-in. choke, 5% H_2S from the Lodgepole at 13,410–13,470 ft (4087–4105 m); 12.2 MMCFGPD, 43 BCPD, 56/64-in. choke, 12% H_2S from the Devonian Darby at 13,550–13,590 ft (4130–4142 m) and 12.8 MMCFGPD, 224 BCPD, 54/64-in. choke, 1% H_2S from the Ordovician Bighorn at 14,130–14,230 ft (4307–4337 m) for a total productive capacity of over 77 MMCFGPD (Bishop, 1982). The well was completed in the Mission Canyon at an IP rate of 33 MMCFGPD, 720 BCPD (54/64-in. choke, 1970 psi FTP). Additional well work was done in 1981–1982 and current production decline curves indicate sustained flow of approximately 65 MMCFGPD from the completed Mission Canyon alone.

The field was developed rapidly from 1980 to 1983 under a 640 ac spacing per formation order set by the State of Wyoming. Reviews of early field development were done by Hoffman and Kelly (1981), Hoffman and Balcells-Baldwin (1982), and Bishop (1982). The field was never unitized, and two separate sour gas processing plants were built. Amoco's Whitney Canyon plant is located in Sec. 17, T17N, R119W (Webber et al., 1984), and Chevron's Carter Creek plant is in Sec. 6, T18N, R119W.

As of September 1988 there are 35 natural gas and condensate wells completed within the field. Sixteen produce from the Mission Canyon, ten are Mission Canyon and Lodgepole commingled producers, three produce from the Bighorn, two are Thaynes producers, one well is testing in the Mission Canyon, one well has Mission Canyon and Weber commingled production, one well is a Mission Canyon completion that is shut-in owing to mechanical problems, and one well is a Weber completion that is shut-in because of low flow rates and no nearby pipeline facilities. At least seven dry holes were drilled by various operators around the perimeter of the field. These were nonproductive because of low structural position, faulted out reservoir section, or extremely poor reservoir quality.

Discovery Method

Discovery of Whitney Canyon–Carter Creek field began by identifying a large, potentially productive area within the Fossil basin using regional structural cross sections based on surface geology and sparse well control, aeromagnetic data, and a regional 100% seismic line. This area was where the Absaroka thrust plate overlies Cretaceous beds. Geochemical and palynological work combined with structural studies indicated that the Cretaceous source rocks beneath the thrust sheet were within the oil-generating "window." In 1972, Chevron took leases on the prospective area, which was limited to the west by the footwall cutoff of Mesozoic beds below the Absaroka fault and to the east by the exposed toe of the Absaroka thrust sheet where active light oil seeps were known.

Smaller scale structures within the prospective area were identified mainly because of two major developments in exploration technology during the 1960s and 1970s: (1) evolution of thrust belt structural models and (2) the CDP stacking of seismic data. Useful thrust belt structural models evolved from work in other areas such as the Alberta and Appalachian thrust belts (i.e., Bally et al., 1966; Dahlstrom, 1970; Wilson and Stearns, 1958). These models were applied to the Cordilleran thrust belt and helped solve the complex structural, geometric, and stratigraphic relationships found there. Advances in digital CDP seismic recording and processing techniques enabled the delineation of previously indistinct tightly folded and faulted structures. In 1973, a joint Amoco-Chevron regional seismic program was shot that defined the general regional structural geometry and structural leads.

Cretaceous sandstones beneath the Absaroka thrust plate were thought to be the primary reservoir objectives until 1974 when American Quasar discovered oil within a hanging wall anticlinal structure at Pineview, Utah. Anticlinal closures within the thrust plate were initially considered less attractive prospects because of several previous dry holes drilled in the hanging wall by other operators that indicated flushing. The Pineview discovery shifted the exploratory emphasis in the Fossil basin area of Wyoming from footwall Cretaceous objectives to hanging-wall anticlinal closures similar to Pineview. The Whitney Canyon–Carter Creek structures were two of many such features.

STRUCTURE

Tectonic History

Whitney Canyon–Carter Creek field is in a part of the Cordilleran orogenic belt (Armstrong, 1968) characterized by large-scale imbricated thrust plates. A westward-thickening wedge of mostly shallow-water, platform-deposited, sedimentary rocks of Precambrian through Jurassic age was thrust eastward during Early Cretaceous through early Eocene time on several major thrust faults (Royse et al., 1975). Individual displacements on major thrusts were as much as 25 mi (40 km), and cumulative horizontal displacement of all thrusts was about 75 mi (121 km). Four major thrust fault systems are recognized in this region; they are, from west to east, the Paris-Willard, the Crawford, the

Absaroka, and the Hogsback (Darby-Prospect) (Figure 3). These major thrust fault systems are oldest in the west and become progressively younger to the east, and each contains more than one significant thrust fault. Locally, minor thrust fault imbrications of major thrust plates are younger than the age of major displacement.

A general reconstruction of the tectonic history is possible through paleontologic dating of synorogenic clastic deposits that correspond to times of motion on specific thrust systems (Armstrong and Oriel, 1965; Royse et al., 1975). During thrusting, thick sequences of Cretaceous and lower Tertiary clastics were shed eastward from source areas on the rising thrust plates into local adjacent subsiding basins. These synorogenic deposits, which include coarse conglomerates, thick coal-bearing fluvial and deltaic strata, and organically rich marine shale, accumulated progressively eastward through time, reflecting thrust deformation timing. Some synorogenic deposits were uplifted and recycled into younger deposits.

Regional Structure

In a regional sense, there are two, nearly parallel, eastwardly asymmetric, anticlinal fold trends within the Absaroka thrust plate in the Fossil basin. They contain several oil and gas fields in addition to Whitney Canyon-Carter Creek (Figure 1). The alignment of the Ryckman Creek, Clear Creek, Painter Reservoir, Anschutz Ranch East, and Pineview fields marks the eastern line of folding, whereas the western fold trend contains the Whitney Canyon-Carter Creek, Yellow Creek, Cave Creek, and Anschutz Ranch fields. Mid-Paleocene and younger beds lie unconformably on the folds. Rocks in the eastern fold trend are no older than Mesozoic and contain the hanging-wall cutoffs of Triassic, Jurassic, and Lower Cretaceous beds. Rocks in the western line of folds are as old as Cambrian and contain the hanging-wall cutoffs of Paleozoic beds (Figure 3).

These fold trends were created when the Absaroka plate was thrust eastward over ramps in the Absaroka thrust fault plane as indicated on Figure 4. Evolution of this type of fold, commonly called a ramp anticline, was described by Rich (1934) in his paper on the Cumberland thrust in Virginia and Tennessee. Recently, the concept has been developed in detail by Suppe (1983), who calls structures such as these fault-bend folds. Briefly stated, fault bend folds (ramp anticlines) such as Whitney Canyon-Carter Creek form when a fault cuts and transports nearly planar strata over a bend (ramp) on an otherwise planar fault surface. The size and shape of the folds depend mostly on the ramp's geometry and orientation and the amount of thrust fault displacement.

Local Structure

Whitney Canyon-Carter Creek proper is part of a large reverse-faulted, complex anticlinal closure that is contained completely within the Absaroka thrust plate. The structure formed during and as a result of Absaroka thrusting in late Santonian through early Paleocene time (85 Ma-62 Ma). Two separate anticlinal closures divide the area into the Carter Creek and Whitney Canyon fields (Figure 5). Each contains several smaller closures related to minor reverse faulting that imbricate the structures. There are approximately 1470 ft (450 m) of structural relief and 8400 ac under structural closure at Carter Creek, and 2500 ft (770 m) of structural relief and 7700 ac under closure at Whitney Canyon.

Low relief ramping and the imbrication of the Absaroka fault plane directly below the Whitney Canyon-Carter Creek structure is critical to closure development (Figure 6). Eastward movement of strata over the ramp imparted a west dip to the thrust sheet, which is the west flank of the Whitney Canyon-Carter Creek fold. Apparently, no distinct structural closure would exist without the footwall ramp. The Whitney Canyon portion has greater structural closure (greater west dip) than Carter Creek because the ramp under Whitney Canyon is steeper and has greater vertical relief. The north and south plunges of the folds are probably caused by lateral (along-strike) changes in the configuration and stratigraphic position of the Absaroka fault plane (i.e., lateral ramps).

A major structural discontinuity exists between strata above and below the Jurassic Stump-Preuss salt section, especially at Whitney Canyon. The salt acted as a zone of structural detachment. Folds and faults in upper Jurassic and Cretaceous rocks above the salt do not extend into the rocks below. Extensional structures related to post-Eocene normal faulting, as well as compressional features caused by older Absaroka thrusting, deform the post-salt strata. Seismic and drill-hole data indicate that the relatively young extensional faults do not extend below the Jurassic salt horizon and do not affect productive horizons.

Because of the structural detachment, the relatively simple fold form of the productive Paleozoic horizons is not seen at the surface and must be detected using seismic or well data. Nearly flat lying Maestrichtian and lower Tertiary fluvial and lacustrine beds cover the eastern part while the structurally disconformable post-salt Jurassic beds crop out over the western flank. The Whitney Canyon-Carter Creek structure was first defined by reconnaissance seismic because it has no surface expression.

Details of the structure became known as development drilling and 3D seismic survey data were obtained and interpreted. Seismic data are adequate to map general structural form (Figure 7); however, well data are necessary to define faults and dip changes.

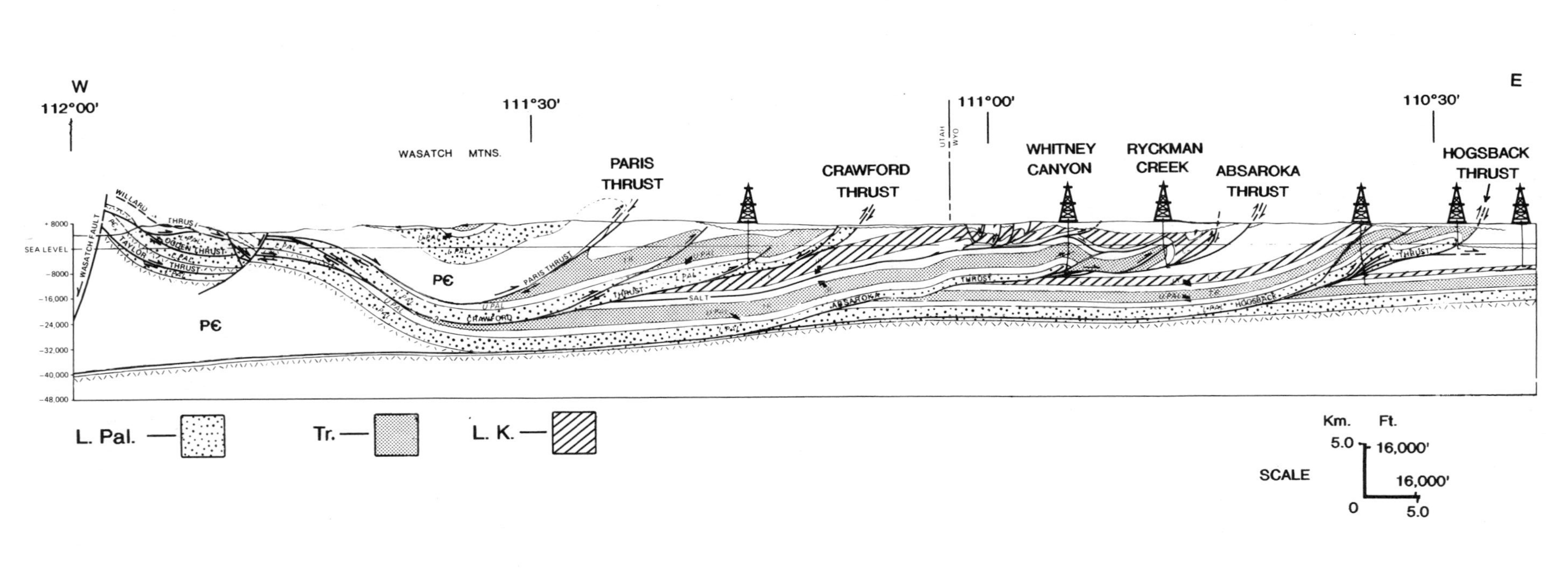

Figure 3. Generalized structural cross section of the southwest Wyoming-northern Utah thrust belt showing the position of Whitney Canyon field on the Absaroka thrust fault plane. Location of section is indicated on index map (Figure 1). Modified from Warner and Royse (1987).

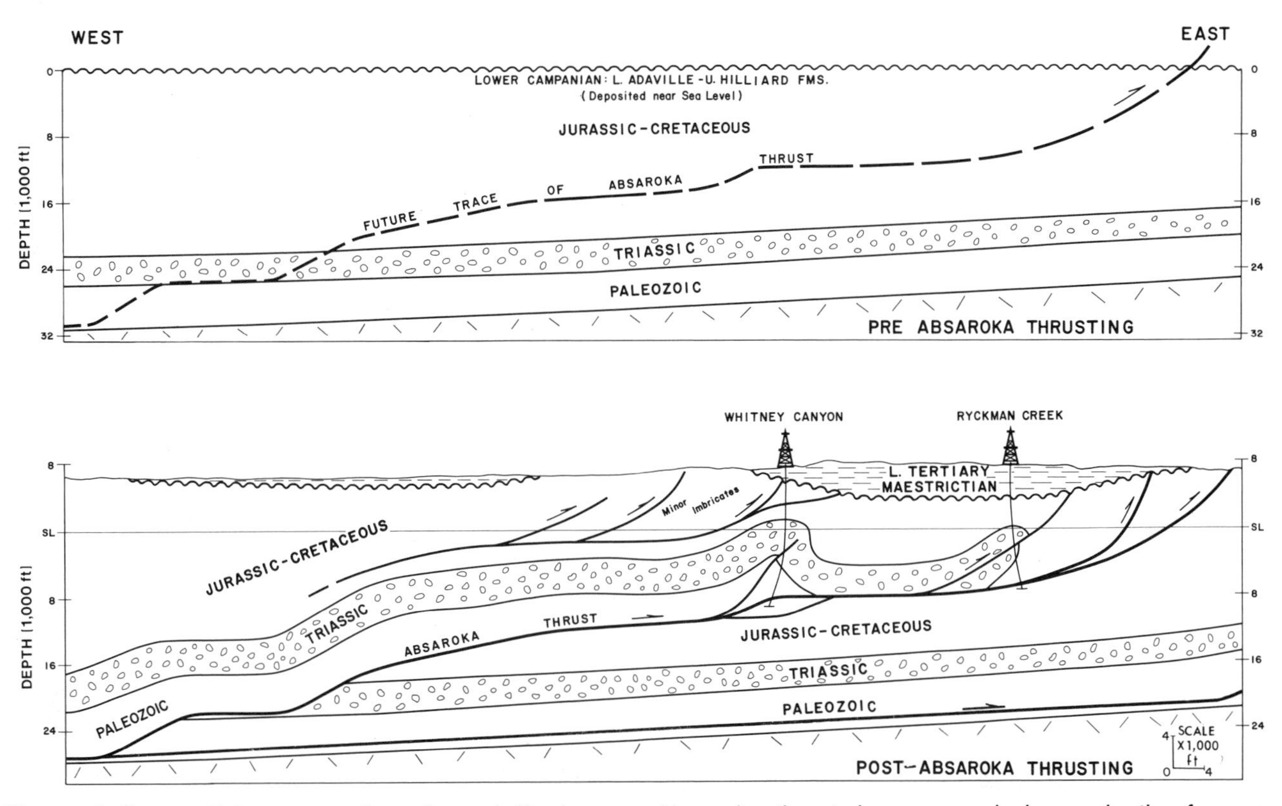

Figure 4. Sequential cross sections through Ryckman Creek and Whitney Canyon fields in Fossil basin area of Wyoming thrust belt. Upper panel shows prefault depth of stratigraphic units now preserved beneath Absaroka thrust. Lower panel shows depth of some units after Absaroka thrust was emplaced and Late Cretaceous and Tertiary sediments were deposited on eroded thrust plate (from Warner and Royse, 1987).

STRATIGRAPHY

General

The drilled stratigraphic column in the Whitney Canyon–Carter Creek area is nearly 20,000 ft (6100 m) thick and contains some units with highly variable thickness (Figure 8). Although most formations have a fairly constant stratigraphic thickness over the area, a major exception is the Jurassic Stump-Preuss formation. The Stump-Preuss contains massive salt beds that experienced flow caused by deformation and may thicken from 100 ft (30 m) to thousands of feet within a short distance. The Tertiary Wasatch and Evanston formations also have variable thickness because they were deposited syntectonically.

Several major unconformities occur throughout the section. At the top of the Bighorn Dolomite is a post-Silurian regional unconformity that marks the erosion of all the Silurian and some of the Ordovician section (Foster, 1972; Hintze, 1973). Another major but local unconformity occurs at the top of the Jurassic–Early Cretaceous Gannett formation. Uplift along the Absaroka thrust sheet caused erosion of over 10,000 ft (3050 m) of sedimentary section that completely removed the Early Cretaceous Bear River and Aspen formations and the Late Cretaceous Frontier and Hilliard formations. Unconformities also exist at the top of the Devonian Darby Formation and tops of the Jurassic Nugget Sandstone and Stump-Preuss formation.

The Tertiary and Mesozoic section is approximately 10,000 ft (3057 m) thick and contains mainly clastic sedimentary rocks and some early Mesozoic carbonate rocks. The Green River, Wasatch, and Evanston formations are the surface rocks of the Whitney Canyon–Carter Creek area and may have a combined thickness of up to 2100 ft (640 m). A generalized section consists of the:

- Tertiary to upper Cretaceous Green River, Wasatch, and Evanston formations (nonmarine clastic deposits)
- Early Cretaceous to Jurassic Gannett formation (siltstones, sandstones, and conglomerates)
- Jurassic Stump-Preuss formation (fine-grained sandstone to siltstone with massive salt beds at the base)
- Jurassic Twin Creek Limestone (massive limestone interbedded with calcareous shale and basal anhydrites)
- Jurassic–Triassic Nugget Sandstone (eolian crossbedded quartzose sandstone)

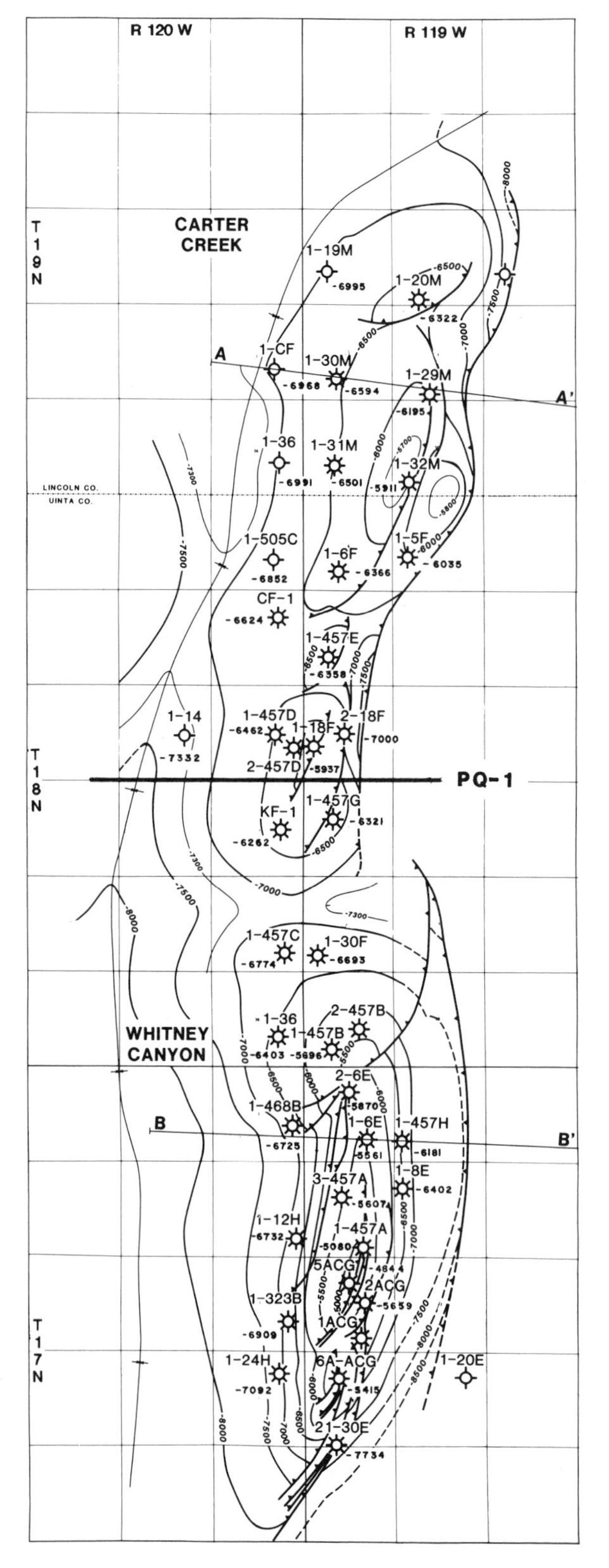

Figure 5. Generalized structure map of the Whitney Canyon-Carter Creek field contoured on the top of the Mississippian Mission Canyon Formation. Contours have been smoothed and do not show Chevron's structural interpretation of the field. Contour interval is 500 ft (150 m). Lines of structural cross sections A-A' and B-B' (Figure 6) and seismic line PQ-1 (Figure 7) are indicated.

- Triassic Ankareh Formation (interbedded shales, siltstones, sandstones, and conglomerates)
- Triassic Thaynes Formation (massive limestones and dolomitic limestones with interbedded calcareous shales)
- Triassic Woodside Formation (red interbedded shales and siltstones)
- Triassic Dinwoody Formation (interbedded shales, siltstones, and massive dolostones)

The Paleozoic section is approximately 4200 ft (1280 m) thick and contains both carbonate and clastic sedimentary rocks. The Absaroka fault cuts the section within the Cambrian; therefore, below the Cambrian are sandstones, shales, and coals of the Cretaceous (Campanian) Frontier Formation. From west to east across the field, the Absaroka fault cuts higher into the lower Paleozoic section. In the eastern half of the field the Cambrian section is missing. A generalized section consists of the:

- Permian Phosphoria Formation (interbedded carbonates and highly organic shales)
- Pennsylvanian Weber Formation (massive to cross-bedded quartzose sandstone)
- Pennsylvanian Morgan (or Amsden) Formation (interbedded shales, siltstones, sandstones, and massive dolostones)
- Mississippian Mission Canyon Formation (massive anhydrites, dolostones, and limestones)
- Mississippian Lodgepole formation (alternating dolostones and limestones)
- Devonian Darby Formation (massive dolostones and calcareous shales)
- Ordovician Bighorn Dolomite (massive dolostones)
- Cambrian Gallatin Limestone (oolitic limestones)
- Cambrian Gros Ventre shale

Whitney Canyon-Carter Creek field lacks production from three reservoirs that are commercial producers elsewhere in the Fossil basin: the Jurassic Twin Creek Limestone, the Jurassic-Triassic Nugget Sandstone, and the Permian Phosphoria Formation. The Twin Creek and Nugget are commercial hydrocarbon reservoirs only along the eastern and southern parts of the Absaroka thrust plate. Reasons for different productive horizons across the region are structural and are discussed in the section on hydrocarbon source and migration. The Nugget Sandstone is the most prolific producer within the thrust belt province but is wet and used for water disposal at Whitney Canyon-Carter Creek field. The Permian Phosphoria Formation's massive dolostones produce gas and condensate elsewhere in the western part of the Absaroka thrust plate, but tests at Whitney Canyon-Carter Creek have shown that it does not have adequate matrix or fracture porosity for economic production.

Four other formations have shows within the field area (Figure 8) but are not producing reservoirs at

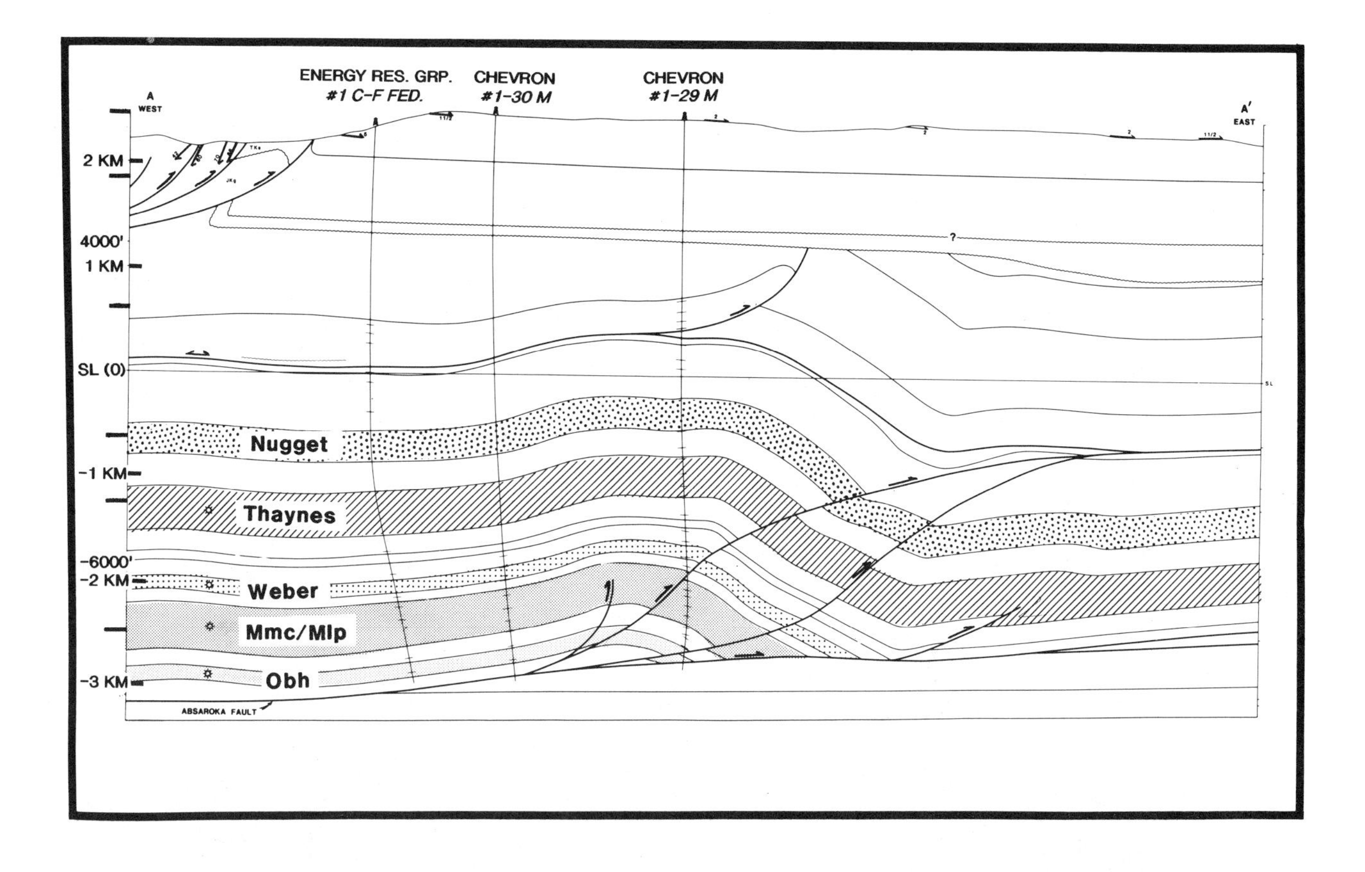

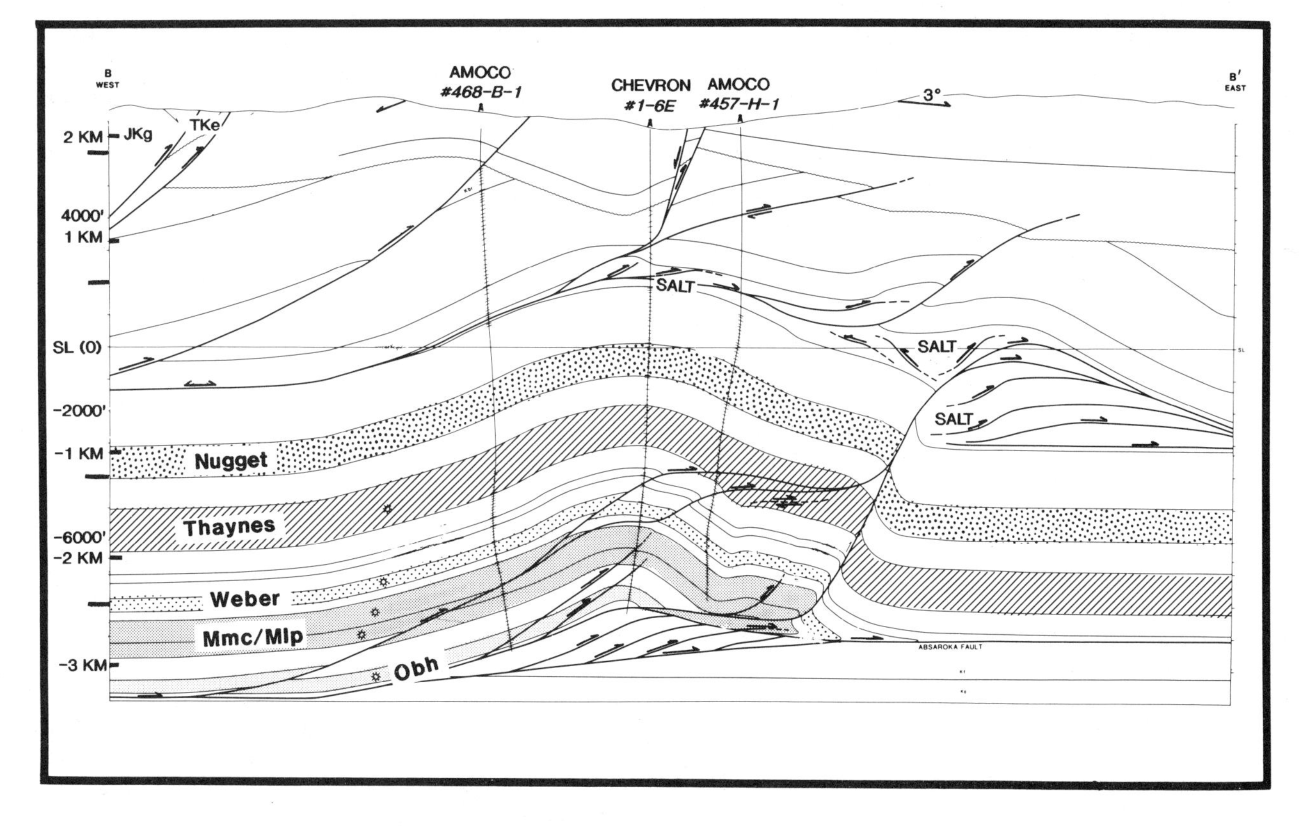

Figure 6. Structural cross sections of the Carter Creek (A–A′) and Whitney Canyon (B–B′) areas. Locations shown on Figure 5. Sections drawn with horizontal scale equal to vertical scale. A–A′ is modified from Lamerson (1982). Mmc, Mississippian Mission Canyon; Mlp, Mississippian Lodgepole; Obh, Ordovician Bighorn.

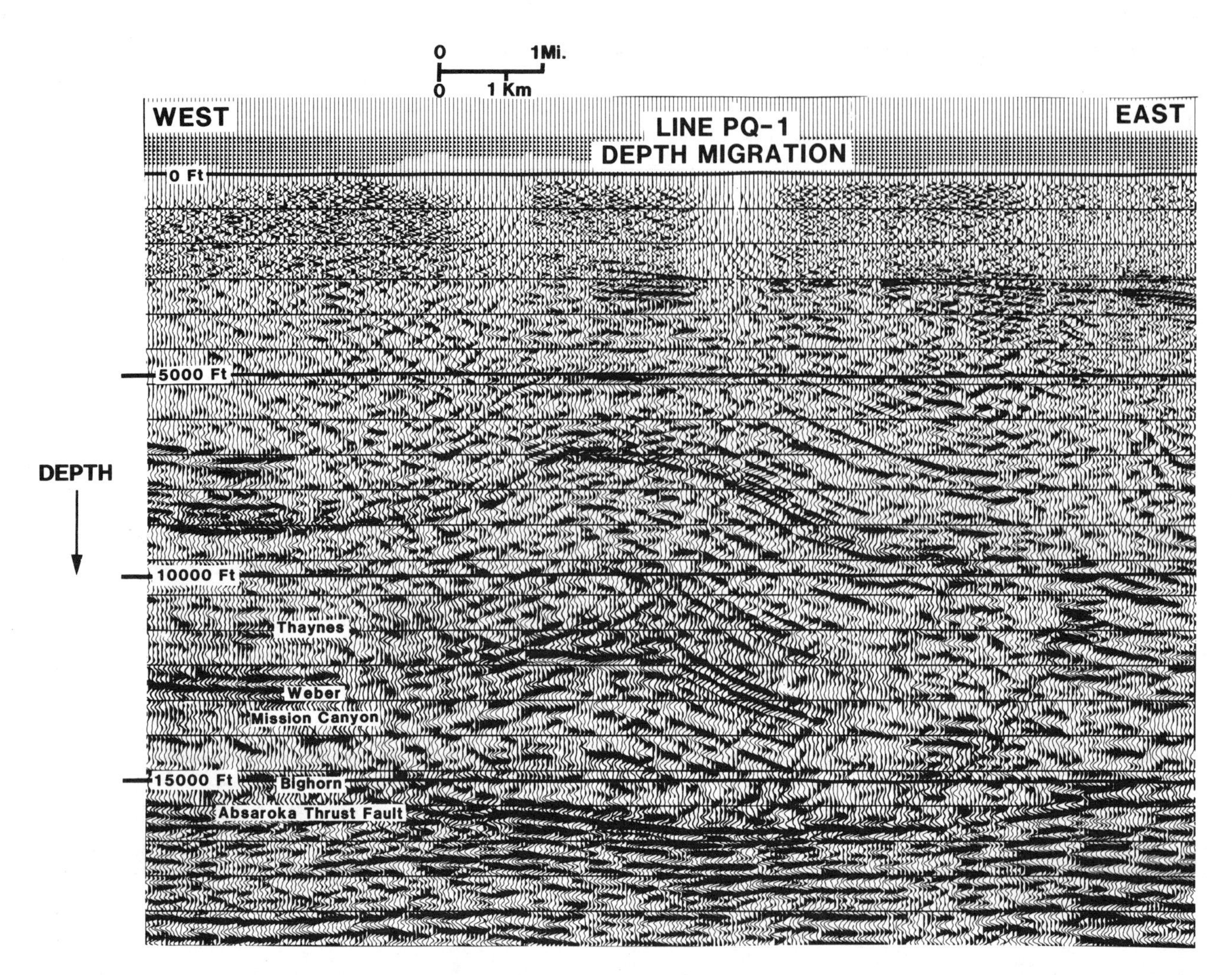

Figure 7A. Uninterpreted version of seismic line PQ-1, depth migration of 800% dynamite data. Section is oriented west-east across the southern end of Carter Creek. Location of section shown on Figure 5.

Whitney Canyon–Carter Creek or in any other discovered thrust belt field. Shows occur in the Triassic Dinwoody, the Pennsylvanian Morgan, the Devonian Darby, and from sandstones in the subthrust Cretaceous Frontier Formation. The Darby was isolated and tested in the Amoco Champlin 457A-1 well at Whitney Canyon for 12.2 MMCFGPD, but a Darby completion has not been attempted.

Source rocks occur in the subthrust Cretaceous section and are the Upper Cretaceous (Campanian) Frontier Formation and the Lower Cretaceous (Albian) Aspen Shale and Bear River Formation. The Frontier is a marginal marine deposit consisting of marine shale and sandstones, interbedded with nonmarine shales, sandstones, and coals (Hale, 1960). In outcrop to the east, the Aspen Shale is an organically rich, open-marine shale (equivalent to the Mowry Shale further to the east) that overlies the mixed marginal marine and nonmarine sandstone and shale sequence of the Bear River Formation (Haun and Barlow, 1962). The Aspen and Bear River have not been penetrated within the field area, but seismic data indicate that they underlie the Frontier Formation in a normal stratigraphic sequence. Source rock geochemistry is discussed in the section on hydrocarbon source and migration.

Trap

At Whitney Canyon–Carter Creek field, hydrocarbons are trapped structurally within two major doubly plunging anticlinal folds. These folds are discussed in detail in the section on local structure.

Reservoirs

There are five commercial reservoirs at Whitney Canyon–Carter Creek: the Mississippian Mission Canyon formation, the Mississippian Lodgepole formation, the Pennsylvanian Weber Formation, the Ordovician Bighorn Dolomite, and the Triassic Thaynes Formation. Each reservoir has a unique gas-water contact, and some have different contacts over

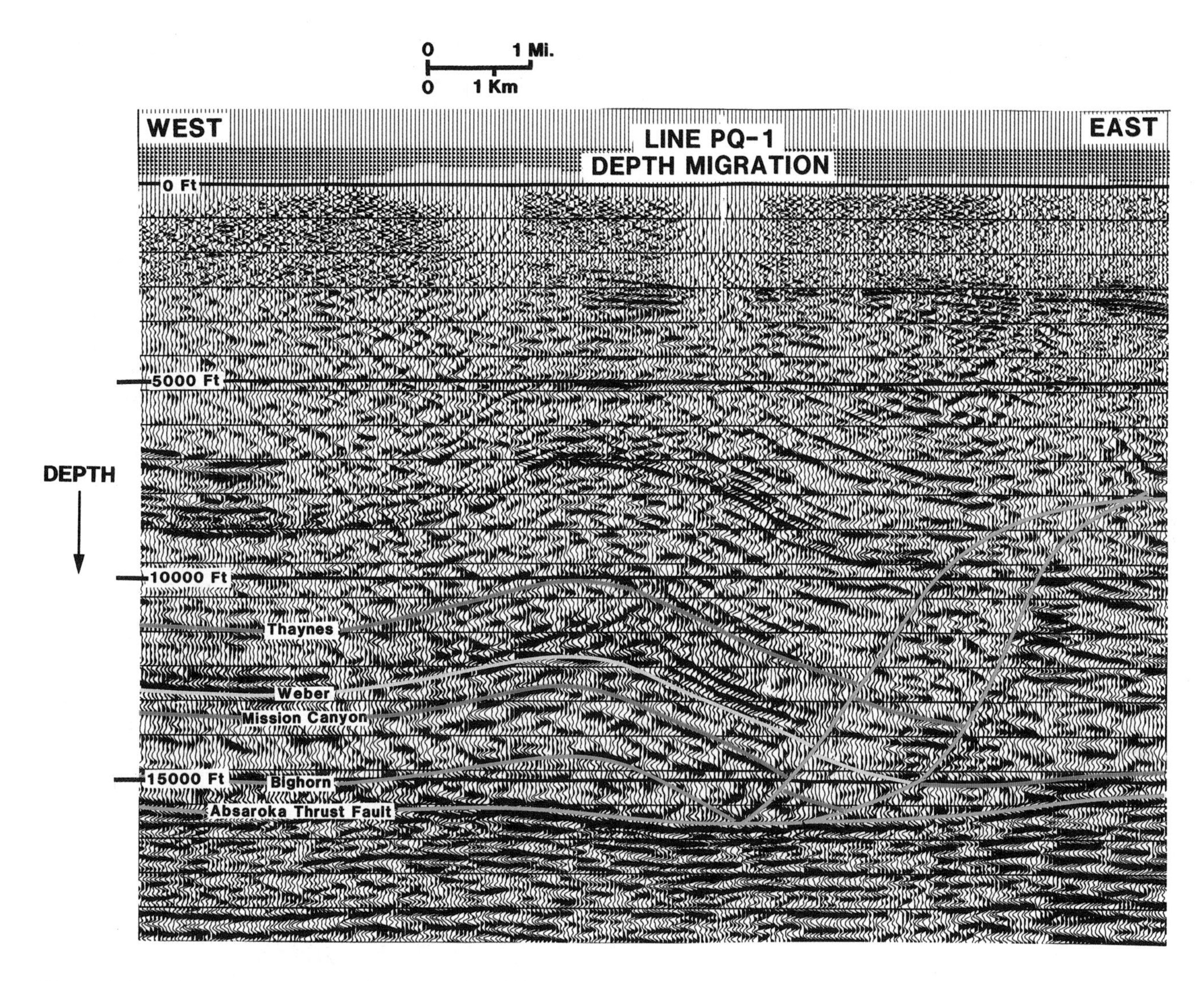

Figure 7B. Interpreted version of seismic line PQ-1.

different structurally defined parts of the field. Only in the Mission Canyon has a contact level been established by borehole well testing.

Mission Canyon Formation

The Mission Canyon is economically the most important reservoir in the field and contains 79% TGIP (percent total gas in place calculated from all producing reservoirs). The Mission Canyon is conformably above the Mississippian Lodgepole formation and unconformably below the Pennsylvanian Morgan (or Amsden) Formation. It is at an average depth of 13,300 ft (4054 m), but owing to structural complexity producing intervals may be found from 12,196 to 14,441 ft (3717-4401 m). The average Mission Canyon thickness is 750 ft (229 m); however, the average actual net pay is only 261 ft (80 m). The Mission Canyon is the dominant reservoir, and much rock and reservoir data are available. Nearly 3700 ft (1128 m) of Mission Canyon core has been recovered throughout the field. As of January 1988, 30 (out of a total of 35) producing wells are completed in the Mission Canyon interval.

During Kinderhookian to middle Meramecian time, carbonate sedimentation occurred on a large cratonic platform that encompassed most of the northern Great Plains, Wyoming, western Colorado, and southeast Utah (see Sando, 1976; Rose, 1977; Gutschick et al., 1980; Sandberg and Gutschick, 1980; Sandberg et al., 1982; and Gutschick and Sandberg, 1983). Figure 9 is a schematic block diagram that illustrates Early to middle Mississippian paleogeography. Whitney Canyon–Carter Creek field was just east (landward) of a north-to-south-trending carbonate platform margin that dipped gradually west toward a deeper water foreland basin. The Mission Canyon is a regressive unit, deposited in a shallow-marine environment becoming more restricted through time.

The Mission Canyon of western Wyoming has traditionally been subdivided into three stratigraphic and production-related units (Rose, 1977): (1) a lower 150 ft (46 m) section of very fine grained, black to

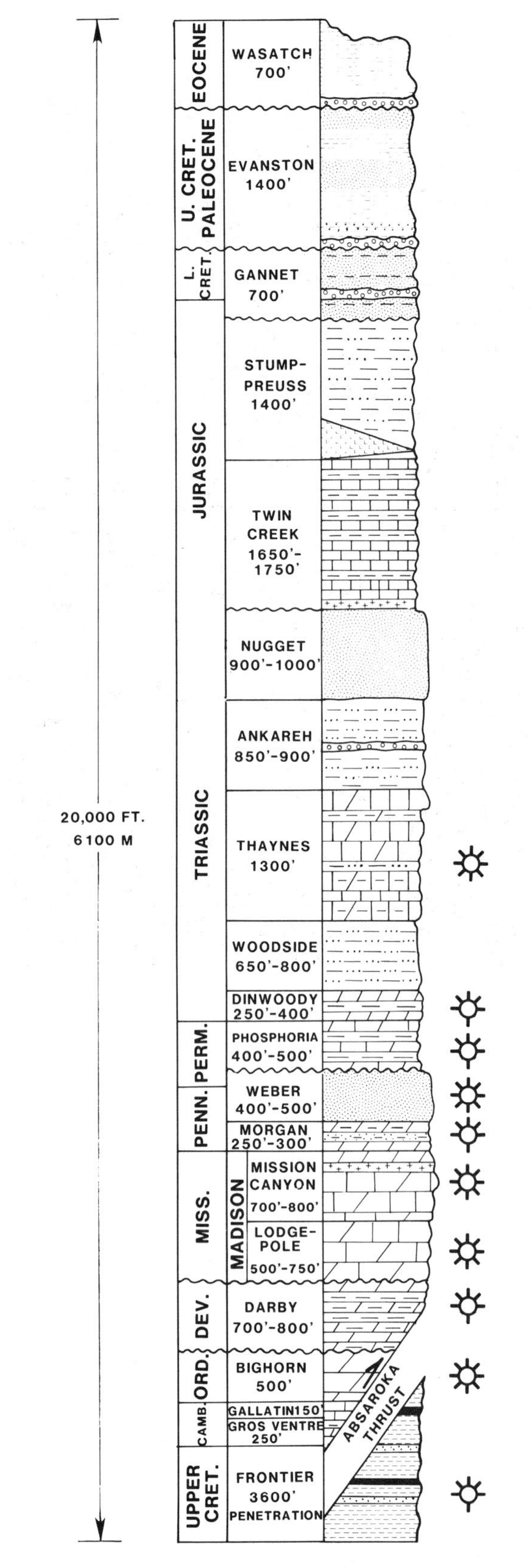

Figure 8. Generalized stratigraphic column of the Whitney Canyon-Carter Creek field area. Productive and potentially productive zones are indicated by gas and gas show symbols. Modified from Hoffman and Kelly (1981).

dark gray limestone and dolostone, (2) a middle 300 ft (91 m) section of interbedded porous dolostone and tight limestone, and (3) an upper 250 ft (76 m) fractured and brecciated section of interbedded very fine grained dolostones and massive to nodular anhydrites. The middle section is often referred to as the main porosity zone because it contains most of the porosity and contributes 90–100% of the production within Whitney Canyon-Carter Creek field. Late Mississippian solution extensively modified the upper section, especially the top 100 ft (30 m).

From core studies, the Mission Canyon may also be subdivided into seven distinct, correlatable facies based on sedimentary textures and structures that relate to depositional environment and diagenesis (Harris et al., 1988). These facies are listed below in ascending order through the section.

- Facies 1: Dark gray to black, finely crystalline lime mudstone and dolomudstone; commonly well-laminated, but may be contorted, wavy, or structureless; rare skeletal debris; common healed fractures; scattered anhydrite nodules; abundant pyrobitumen and other opaque material; tight; deposited in a low-energy environment, probably below wave base where laminations and mud-rich textures could be preserved.
- Facies 2: Brownish-gray to olive gray skeletal dolomudstone and skeletal dolowackestone; varies from faintly laminated to bioturbated and massive; locally common skeletal debris; abundant healed fractures; good intercrystalline and moldic porosity may be developed, especially within the upper part of the facies; gradational change from facies 1 to facies 2 where conditions were still low energy, but more oxygenated as indicated by burrowing organisms; an increase of probable storm-related skeletal debris accumulations implies water-level shallowing (Figure 10).
- Facies 3: Medium olive green, peloidal-lime grainstones and packstones; faintly laminated to massive; abundant filled fractures and stylolites; peloids are very fine to fine sand-sized grains; moderately well-sorted; tight; deposited in high-energy conditions within a shallow, subtidal to intertidal shoal environment facies belt.
- Facies 4: Medium gray to tan dolomudstone; variably laminated, but rarely burrowed; rare skeletal grains; common healed fractures and horizontal stylolites; variable intercrystalline porosity; porosity may exceed 12% where enhanced by fine moldic porosity or may occur in trace amounts where calcite and anhydrite plugging occur; shoaling conditions of facies 3 progressed into a shallow subtidal to intertidal mudflat environment; laminated muds accumulated in the lee of the seaward grainstone shoal (Figure 11).
- Facies 5: Interbedded olive green skeletal lime grainstones and medium gray to greenish-gray

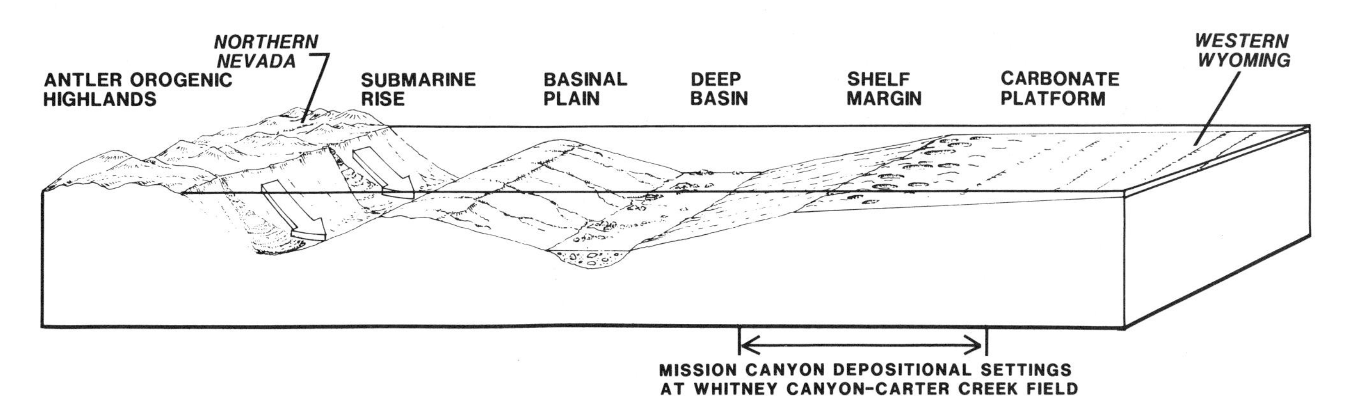

Figure 9. Schematic block diagram of interpreted geologic settings during Early to middle Mississippian time along a west-east section from northern Nevada to western Wyoming. Mission Canyon depositional settings at Whitney Canyon-Carter Creek field are indicated. Modified from Sandberg and Gutschick (1980).

dolowackestones and packstones; grainstones are thin-bedded, occasionally cross-bedded, are composed almost exclusively of crinoid debris, contain abundant healed fractures, and are tight; dolowackestones and dolopackstones are faintly laminated to burrowed and massive and have well-developed intercrystalline and moldic porosities of up to 16%; typified by frequent alternation between skeletal-lime grainstones and dolowackestones and packstones; from the lack of subaerial exposure features and alternating depositional conditions a dynamic depositional environment on a shallow-marine subtidal shelf is inferred; grainstones accumulated on mobile bars or shoals and the dolowackestones and packstones formed from the intervening carbonate muds (Figure 12).

- Facies 6: Tan dolomudstone; burrowed with scattered skeletal debris rapidly decreasing in abundance upward; common nodules and thin anhydrite beds; common open and healed fractures and stylolites; intercrystalline and moldic porosity of up to 15%; characterized by more anhydrite and less fossil debris, which implies deposition at the landward edge or inner shelf of a shallow-marine subtidal shelf (Figure 13).
- Facies 7: White bedded anhydrite and brecciated tan dolomudstone; anhydrite has a massive to chicken-wire nodular texture; dolomudstone is massive with occasional laminations and algal structures; upper part is highly chaotic; deposited on an evaporitic supratidal mudflat where bedded anhydrite and algal laminations were deposited and preserved; chaotic appearance is probably due to regional karstification and solution at the culmination of the regressive cycle (Figure 14).

The Mission Canyon was deposited on a shallowing-upward, carbonate ramp system, possibly as shown in the composite schematic block diagram in Figure 15. Variable energy, water depths, and circulation conditions existed across the ramp, creating environments from a quiet water deeper marine shelf of facies 1, to alternating higher energy shoals and lower energy subtidal marine deposition of facies 5, and to an evaporitic upper intertidal to supratidal mudflat of facies 7 (see Harris et al., 1988, for a complete discussion of depositional environments). Figure 15 shows all environments occurring in one cycle; however, another interpretation has been presented by Bartberger (1984) who, based on regional outcrop and core work, interpreted the Mission Canyon at Whitney Canyon-Carter Creek to have at least nine regressive cycles. A complete cycle would contain sedimentary rocks deposited in open-marine, restricted-marine, and intertidal/supratidal environments. According to the model, porosity developed primarily within the upper supratidal mud-supported units through very early dolomitization soon after deposition but before lithification and the onset of the next regressive cycle. Harris et al. (1988) cites evidence that, in core, the mud-supported rocks show no evidence of exposure and no consistent pattern of sands grading into muds. They believe that complex variations over a shallow-marine shelf setting could produce apparent cyclic changes in which grainstones were deposited in mobile sand bars and marine subtidal muds in the intervening lows, such as that observed in modern examples described by Halley et al. (1983).

Three major stages of diagenesis controlled Mission Canyon porosity development and modification: (1) early depositional, (2) burial, and (3) structural (Harris et al., 1988). A Mission Canyon burial-history curve (Figure 16), based on regional surface geology and structural palinspastic reconstruction, illustrates these stages. Comprehensive articles by Budai (1985) and Budai et al. (1984, 1987) analyze the Mission Canyon (Madison) regional diagenetic history in the western Wyoming thrust belt.

Most of the sucrosic intercrystalline porosity found in the present reservoir was created by dolomitization of mud-supported sediments during the early depositional diagenetic stage. Isotope data substan-

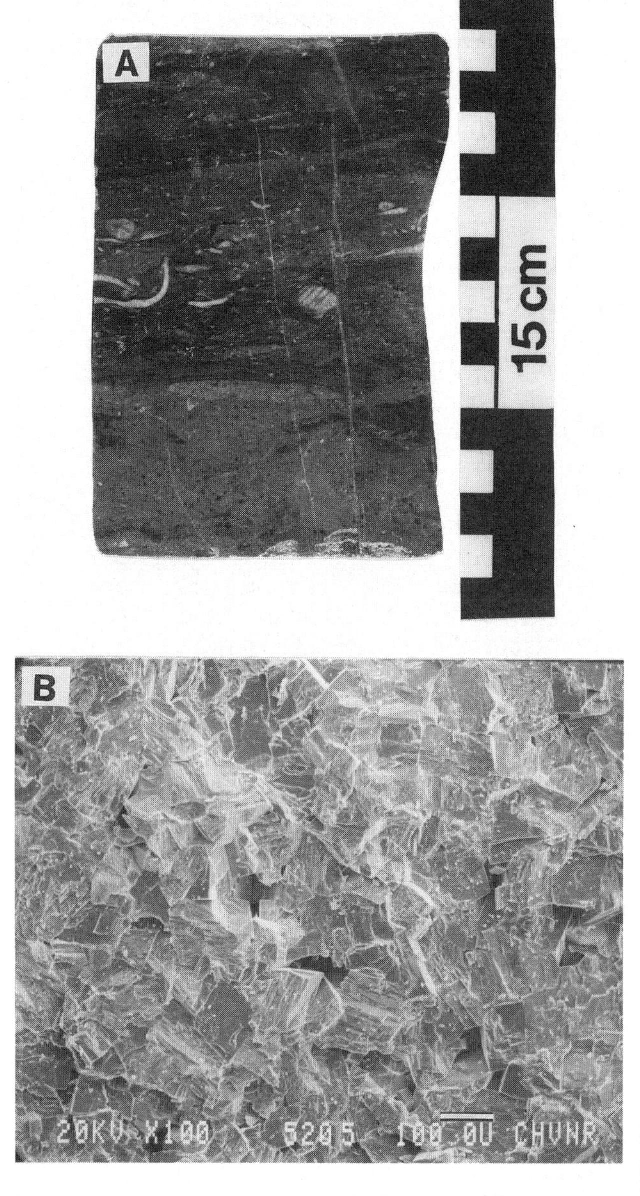

Figure 10. Core photo (A) and SEM photo (B) illustrating Mission Canyon facies 2. Core photo shows a dolomitized mudstone to wackestone. SEM photo shows 2–3% intercrystalline porosity; white bar at bottom of photo is 100 micrometers.

Figure 11. Core photo (A) and SEM photo (B) illustrating Mission Canyon facies 4. Core photo shows a laminated dolomudstone; laminations enhanced by dead oil stain from an earlier migration. SEM photo shows 10–12% sucrosic intercrystalline porosity; white bar at bottom of photo is 10 micrometers.

tiates relatively low-temperature matrix dolomite formation (Budai et al., 1984). Porosity development was facies controlled. Mud-rich, laterally correlatable facies were dolomitized preferentially over grain-supported facies. Moldic porosity was formed by dissolution of skeletal fragments after lithification and dolomitization.

Early stage porosity was retained during the burial diagenetic stage and modified during the structural diagenetic stage. Retention of porosity is substantiated by evidence of an early, prethrusting oil migration event that coated intercrystalline and moldic pore throats, but rarely fracture pores. Oil migration most likely occurred when upper Paleozoic organic source rocks were in the oil generation window during Jurassic to early Late Cretaceous time. This suggests that intercrystalline and moldic pores remained open during deep burial and oil migration; most fracture porosity was formed later by emplacement of the Absaroka thrust sheet during the structural diagenetic stage. Structural deformation also created new fluid migration pathways that

Figure 12. Core photo (A) and SEM photo (B) illustrating Mission Canyon facies 5. Core photo shows an interbedded dolomudstone to dolowackestone and skeletal lime grainstone. SEM photo shows 8-10% moldic and sucrosic intercrystalline porosity; white bar at bottom of photo is 100 micrometers.

Figure 13. Core photo (A) and SEM photo (B) illustrating Mission Canyon facies 6. Core photo shows a skeletal poor dolomudstone with anhydrite filled fractures. SEM photo shows 8-10% moldic and sucrosic intercrystalline porosity; white bar at bottom of photo is 100 micrometers.

allowed calcite, anhydrite, and dolomite cements to significantly plug all porosity types. Isotope data indicate that most of these cements were formed from relatively high temperature fluids (Budai et al., 1984).

The Mission Canyon Formation is a heterogeneous reservoir because of a complex depositional, diagenetic, and structural history that controlled porosity and permeability variability across the field. Porosity distribution across the field is variable mostly because of late diagenetic calcite, anhydrite, and dolomite cement plugging. This diagenetic porosity modification occasionally causes vast differences in average porosity within the same facies type and stratigraphic interval between two proximal wells (1.0 mi; 1.6 km). Vertical variations in porosity also occur within individual facies, but these variations are more related to original depositional textures as they are correlatable between wells. Dedolomitization

Figure 14. Core photo illustrating Mission Canyon facies 7. Photo shows a brecciated dolomudstone cemented by anhydrite.

may also have reduced porosity locally, as it was associated with late calcite fracture cementation (Budai et al., 1984).

Vertical, horizontal, and structural permeability barriers may exist within the reservoir. Some of the thicker, laterally extensive limestone beds may be vertical permeability barriers within the main porosity zone of the Mission Canyon. These limestones were originally grainstones, deposited in mobile bars or shoals, cemented tightly, and now interbedded with porous dolostones that were originally mudstones, wackestones, and packstones (Figure 17). Many limestone beds in the section are thin and discontinuous but may be sufficiently thick and extensive enough to act as a hindrance to vertical flow within the reservoir. Because there is no evidence of multiple gas-water contacts, some vertical communication may exist within the main porosity zone. Permeability barriers also may occur in areas of late stage cement porosity plugging with sharp lateral diagenetic boundaries. Reservoir pressure data cannot substantiate this, but lateral permeability changes may account for anomalous discontinuities in production and water saturations between neighboring wells. Permeability barriers may also occur in structurally deformed areas. Reverse faults with significant displacement may act as a seal and obstruct reservoir communication. Also, in the structural saddle between the Whitney Canyon and Carter Creek structures, the top of the Mission Canyon main porosity zone is projected to dip below the estimated gas-water contact.

Mission Canyon reservoir quality is controlled by matrix porosity, not fractures. Porous intervals average 6–8% porosity, with maximum porosities of 16%. Average matrix permeabilities are 0.7 to 1.5 md but may increase significantly because of micro and macro fractures. Dolostones with intercrystalline or a combination of intercrystalline and moldic porosity types are the highest quality reservoirs. Flow effective matrix permeability occurs only when average matrix porosities exceed 2 to 4%. Rock with matrix porosities below 2% has mainly fracture permeability that by itself does not promote sustained or volumetrically significant flow. Tight, fractured intervals may have significant gas shows when drilling; however, production tests indicate that these intervals become rapidly depleted and do not add significantly to production.

Lodgepole Formation

The second most important reservoir is the Mississippian Lodgepole formation with 9% TGIP. The Lodgepole is conformably below the Mission Canyon Formation and unconformably above the Devonian Darby Formation. The Lodgepole at Whitney Canyon-Carter Creek is at an average depth of 14,050 ft (4282 m) and averages 750 ft (229 m) thick. The average net pay zone is 123 ft (37 m).

The Lodgepole was deposited in the same regional carbonate depositional system as the Mission Canyon, just east (landward) of a regional north–south-trending shelf margin. In outcrop, the Lodgepole has been described as a dark gray, thin-bedded, cherty, fossiliferous, carbonaceous, silty limestone that may contain scattered bioherms (Rose, 1977).

Rau (1982) described Lodgepole core and cuttings from the Chevron 1-18F and Chevron 1-30F wells and divided the section into four shallowing-upward cycles. These units are listed below in ascending order.

- Cycle 1: siliceous, dolomitized mudstone overlain by thick (over 100 ft; 30 m) oolitic-peloidal grainstone; thick grainstone accumulation indicates that deposition and subsidence were balanced; lower half of grainstone has been completely dolomitized and has good (10–12%) intercrystalline and oomoldic porosity.
- Cycle 2: fossiliferous, dolomitized wackestones and mudstones, overlain by oolitic and skeletal grainstones, capped by anhydritic dolomitized mudstone; original dolomite porosity cemented with anhydrite and dolomite.
- Cycle 3: oolitic and skeletal grainstones overlain by dolomitized wackestones and mudstones, capped by peloidal mudstones and massive, nodular anhydrite; wackestones and mudstones

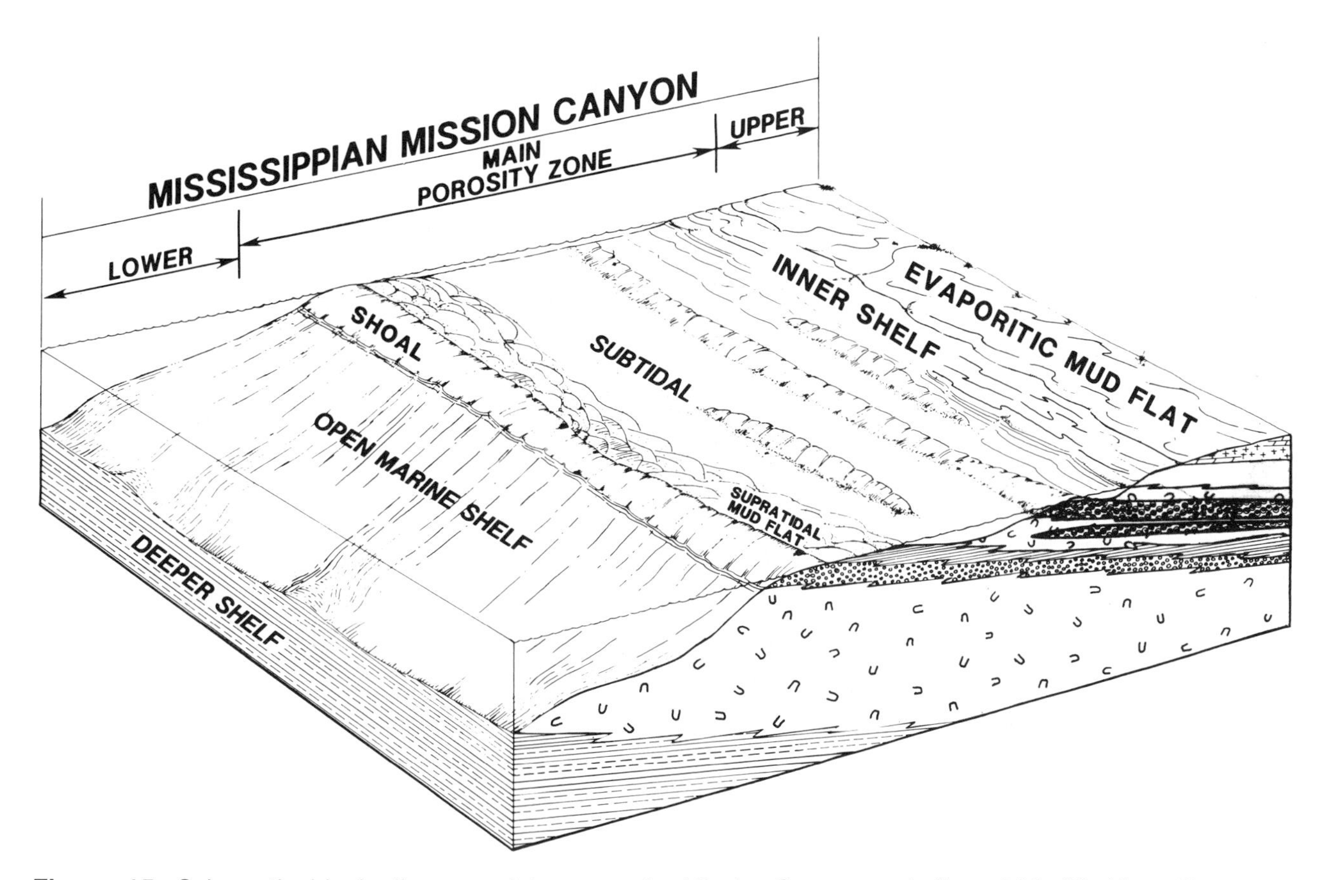

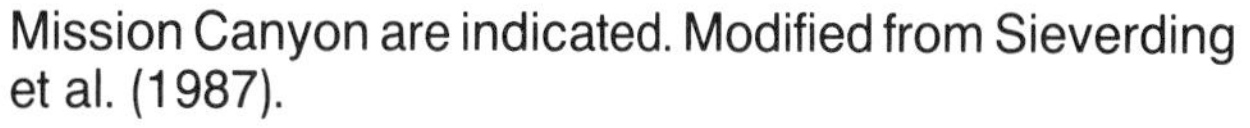

Figure 15. Schematic block diagram of interpreted Mission Canyon depositional environments. Relative locations of lower, main porosity zone, and upper Mission Canyon are indicated. Modified from Sieverding et al. (1987).

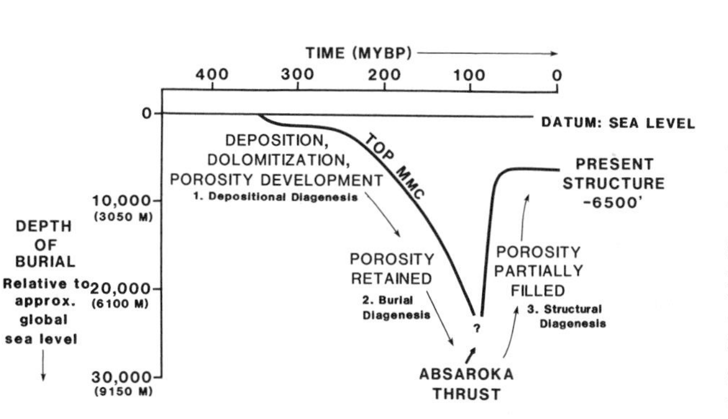

Figure 16. Burial-history curve for the Mission Canyon Formation illustrating three stages of diagenesis: (1) early depositional diagenesis, (2) burial diagenesis, and (3) late structural diagenesis. Modified from Sieverding et al. (1987).

may have 8-10% intercrystalline and moldic porosity (Figure 18).

- Cycle 4: dolomitized wackestones overlain by oolitic grainstones and dolomitized wackestones with abundant anhydrite cements, capped by a thin layer of massive, nodular anhydrite at top; basal wackestones may have 8-10% intercrystalline and moldic porosity.

Based on Rau's (1982) descriptions, the Lodgepole at Whitney Canyon-Carter Creek was deposited in overall increasingly restricted conditions through time. Each individual cycle shows shallowing upward; however, the entire section also shows a general trend from more open-marine to more restricted-marine environments. Sediment accumulated and created shallowing-upward regressive conditions. This was interspersed with minor, sporadic, and perhaps local transgressions. A major transgression occurred early in Lodgepole deposition and then again during latest Lodgepole time, also the beginning of the Mission Canyon depositional system. Rose (1977) interpreted the Lodgepole and Mission Canyon to have been deposited as a large transgressive-regressive cycle; however, at Whitney Canyon-Carter Creek both units are apparently

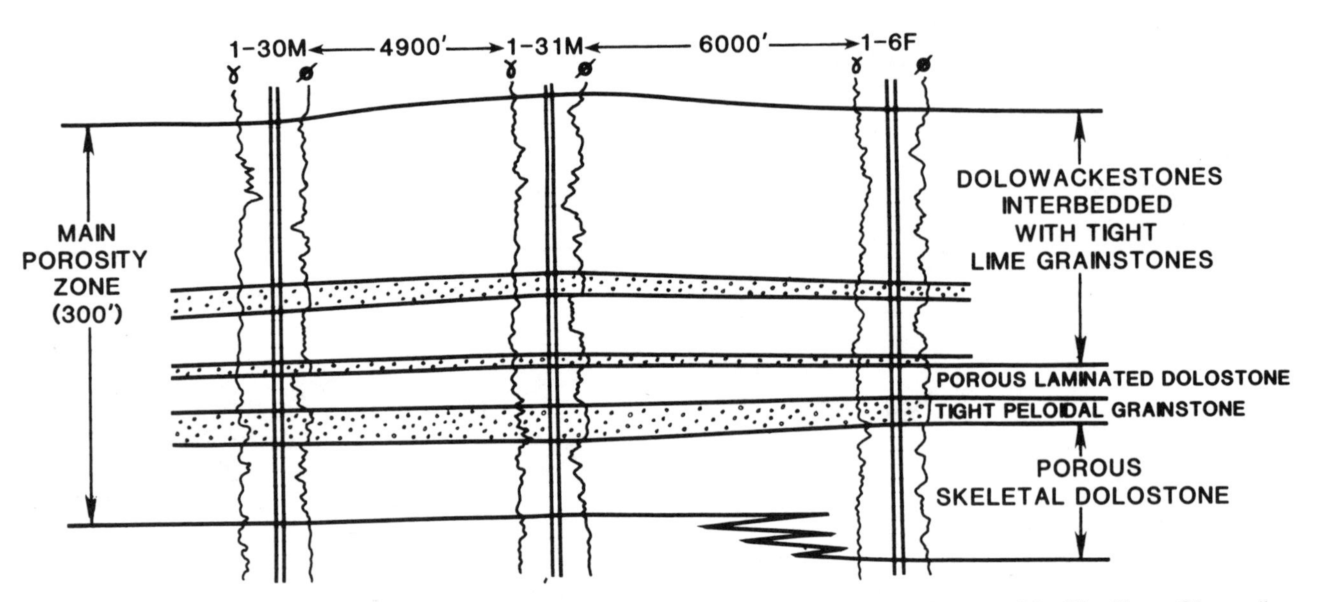

Figure 17. True stratigraphic cross section through Carter Creek field showing porous dolomites separated by laterally extensive, tight limestones that may restrict reservoir vertical permeability. Modified from Sieverding et al. (1987). Well locations are seen in Figure 2.

overall regressive deposits with very little transgressive sedimentary record preserved.

Effective Lodgepole porosity exists in two correlatable zones, an upper zone, approximately 120 ft (36 m) below the Lodgepole top, and a lower porosity zone, approximately 150 ft (46 m) above the base. Porosity zones are correlatable across the field and vary between 2% and 12% porosity. The lower zone is approximately 70 ft (21 m) thick and is a dolomitized grainstone with oomoldic and intercrystalline porosity types. The upper zone is approximately 90 ft (27 m) thick and is a dolomitized skeletal wackestone with sucrosic intercrystalline and skeletal moldic porosity types. Lodgepole diagenesis is probably similar to Mission Canyon diagenesis because of their similar genetic relationship. As in the Mission Canyon, most porosity was formed through dolomitization relatively soon after deposition. This is supported by petrographic evidence as cited by Rau (1982) and isotopic evidence by Budai et al. (1984). Late stage cementation subsequently modified porosity.

The Lodgepole gas-water contact has been estimated to be -7610, based on testing of the lower porosity zone in the Carter Creek Chevron 1-29M Fed. At Whitney Canyon, the Lodgepole gas-water contact has not yet been confirmed through production testing; therefore, Whitney Canyon and Carter Creek could be separate fields at Lodgepole level with different gas-water contacts. Communication between the upper and lower porosity zones is probably limited because they are separated by over 350 ft (100 m) of tight dolostones and limestones. Lateral permeability baffles may also exist because of diagenetic changes and structural complications.

As of September 1988, the field had ten Lodgepole completions. The Lodgepole is a viable, economic reservoir by itself but is commingled with the Mission Canyon to improve economics. In the Whitney Canyon ACG No. 5 well, the Lodgepole was completed before Mission Canyon perforation additions and flowed at rates greater than 20 MMCFGPD. Commingling is possible because the two reservoirs have approximately the same virgin reservoir pressure and gas composition, deplete in a similar manner, and are stratigraphically adjacent.

Weber Formation

The third reservoir is the mainly quartzose sandstone Pennsylvanian Weber Formation with 8% TGIP. The Weber is gradational with the underlying Pennsylvanian Morgan/Amsden Formation and is unconformably overlain by the Permian Phosphoria Formation. At Whitney Canyon-Carter Creek, the Weber is at an average depth of 12,300 ft (3749 m) and averages 700 ft (213 m) thick. The average net pay zone is 120 ft (37 m) thick.

Based on log character and lithology, the Weber may be divided into lower and upper stratigraphic units. The lower part contains thinly bedded sandstones, shales, and occasional limestones, whereas the upper part includes thicker bedded (50 ft; 15 m) massive sandstones separated by thin shaley beds. All sandstones are well sorted, are fine grained, and have abundant quartz overgrowths. Based on

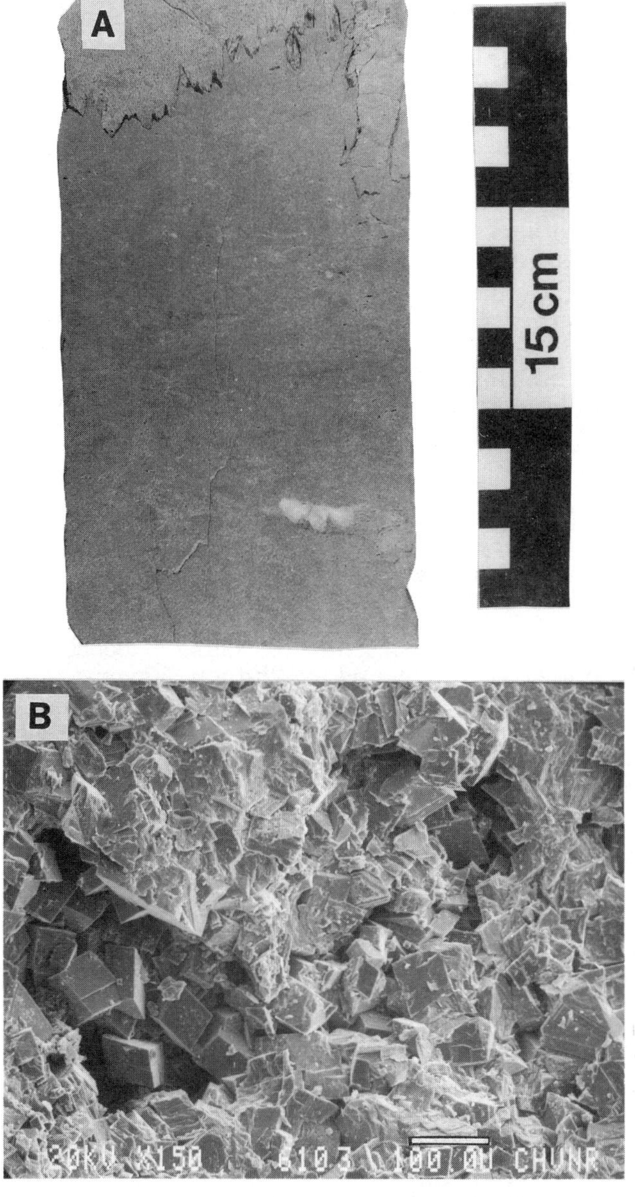

Figure 18. Core photo (A) and SEM photo (B) illustrating Lodgepole cycle 3 reservoir facies. Core photo shows a dolomitized skeletal packstone and wackestone; stylolite near top is facies boundary. SEM photo shows 6-8% intercrystalline and moldic porosity; white bar at bottom of photo is 100 micrometers.

Weber exposures in northeastern Utah and northwestern Colorado, Bissell and Childs (1958) interpreted the Weber Formation to have been deposited in shallow, near-shore marine to eolian environments, with more marine sedimentation to the west and more eolian sedimentation to the east and north. They correlated the Weber with the eolian Tensleep of Wyoming. Although Weber core data are sparse and cores lack distinctive features, core descriptions suggest that the lower Weber was deposited in a shallow-marine environment based on its thin sandstone beds and presence of shales and limestones. The upper Weber possibly was deposited from southward-advancing dunes, producing massive sands with thin interdune deposits.

Weber porosity consists of isolated, dissolution moldic pores and minor fracture porosity averaging 3-5%, which creates fairly poor reservoir quality. Quartz overgrowths are extensive and limit core measured permeabilities to less than 0.5 md; however, fractures locally increase permeabilities significantly. The Weber has not proven to be an economic reservoir by itself, but it has been commingled successfully with the Mission Canyon in the Carter Creek Chevron 1-32M Federal well. The only well completed in the Weber is the Whitney Canyon Chevron 21-30E Federal; it flowed at a rate of 1.06 MMCFGPD, 16 bbl of condensate, and 43 bbl of water but is currently shut-in. The Weber in the Whitney Canyon Amoco ACG 5 well was tested before stimulation at 4.5 MMCFGPD but was never completed.

Bighorn Dolomite

The fourth reservoir is the Ordovician Bighorn Dolomite, a massive carbonate unit with 3% TGIP. The Bighorn is conformably above the Cambrian Gallatin Limestone and is unconformably below the Devonian Darby Formation. At Whitney Canyon-Carter Creek, the Bighorn is at an average depth of 15,500 ft (4724 m) and averages 500 ft (152 m) thick. The average net pay zone is 81 ft (25 m).

From core data, the Bighorn at Whitney Canyon-Carter Creek may be described as a gray, very fine to finely crystalline massive dolomite containing a diverse open-marine fauna of crinoids, bryozoans, brachiopods, algae, and trilobites. The Bighorn Dolomite at Whitney Canyon-Carter Creek is correlative with the Fish Haven Dolomite of southeastern Idaho, which in outcrop is a medium-bedded, locally cherty dolostone (Ross, 1953; Hintze, 1973). In outcrop, the correlative Bighorn Dolomite of northwestern Wyoming is a mottled, yellowish-gray to very pale orange microcrystalline to coarsely crystalline, occasionally argillaceous dolomite containing common crinoid fragments and sparse brachiopods and corals (Richards and Nieschmidt, 1957).

The Bighorn was part of a Middle to Upper Ordovician open-marine carbonate sequence that was deposited across an extensive platform in the western United States (Foster, 1972). The Red River Formation of the Williston basin is correlative with the Bighorn (Ross, 1953). Lithologically similar Ordovician sections are found in Nevada, Utah, Colorado, and Montana; these sections may be stratigraphic syndepositional equivalents (Ross, 1953; Richards and Nieschmidt, 1957; Hintze, 1973). In Wyoming, a major unconformity is at the top of the Bighorn where nearly all Silurian rocks and much of the Ordovician section has been eroded.

A combination of intercrystalline, moldic, vugular, and fracture porosity types make the Bighorn an economic reservoir. Pay zone average wireline log porosity is 3-5%, but may be as great as 8-10%. Thin sections show that dissolution moldic, vugular, and intercrystalline porosity types are most common, especially in the upper Bighorn. Large (1 in.; 2.5 cm) solution vugs are common in core from the reservoir section. Dead oil from an earlier migration phase coats pore throats, including vugs, suggesting porosity development through early dolomitization and subsequent solution. The regional post-Silurian unconformity may have created dissolution porosity in the upper Bighorn over a large regional area.

Eight wells have produced from the Bighorn since field discovery; most production has been from the uppermost 250 ft (76 m) where porosity is best developed (Hoffman and Kelly, 1981; Hoffman and Balcells-Baldwin, 1982). The best Bighorn wells to date were the Chevron 1-6E Federal, which had an initial flow potential of 19.6 MMCFGPD, 265 BCPD, and 23 BWPD (28/64-in. choke) from gross perforations between 14,341 and 14,590 ft (4371-4447 m), and the Carter Creek Amoco 457E-1, which had an initial flow potential of 25.15 MMCFGPD and 220 BCPD (64/64-in. choke) from perforations between 15,602 and 16,108 ft (4755-4909 m). As of May 1988, only two wells in the field produce from the Bighorn: the Whitney Canyon Amoco 457A-3 and the Carter Creek Amoco 457D-2. The rest have been recompleted in the Mission Canyon/Lodgepole because of declining flow rates and greater uphole potential.

Thaynes Formation

The fifth and least significant reservoir is the Lower Triassic Thaynes Formation with 1% TGIP. The Thaynes is conformably below the Triassic Ankareh Formation and conformably above the Triassic Woodside Formation. At Whitney Canyon-Carter Creek, the Thaynes is at an average depth of 9200 ft (2804 m) and averages 1300 ft (396 m) thick. The average actual pay zone thickness is 118 ft (36 m).

The Thaynes Formation consists of silty and sandy limestones and calcareous siltstones that are interbedded and gradational with gray crystalline limestones (Kummel, 1955). During Lower Triassic time, a large shallow-marine shelf existed in western Wyoming, adjacent to a deeper water, occasionally euxinic, basin in southeastern Idaho. The shelf margin trended north-to-south just east of the Wyoming-Idaho border. The Thaynes is a transgressive marine unit that grades into deeper water silty carbonates and euxinic shales to the west and redbed facies to the east (Kummel, 1957). From work in northwestern Wyoming, Picard et al. (1969) suggest that the Thaynes is an open-marine facies in which organic mounds and oolite shoals developed along the western Wyoming shelf margin and created restricted marine conditions of redbed deposition further east.

Fracture and microsucrosic intercrystalline porosities are developed in gray, silty limestones and dolomitic limestones. Pay zone average log porosities are only 1-2% and average core permeabilities are 0.1-0.2 md; however, fractures may increase permeabilities significantly. The reservoir is considered poor, and by itself is not economic in this area because of relatively low initial potentials and steep decline curves. Thaynes completions are limited to two wells located on the crest of the tightly folded and probably fractured Whitney Canyon structure. As of January 1988, one well is shut in and the other is producing at a rate of 190 MCFGPD. Thaynes gas and condensate is sweet and does not require processing for sulfur removal. The Thaynes was probably self-sourced since it is not in direct contact with mature, subthrust Cretaceous source rocks or any other potential source (Figure 6).

HYDROCARBON SOURCE AND MIGRATION

Warner (1982) presents compelling evidence that the hydrocarbons found at Whitney Canyon-Carter Creek were derived from footwall Cretaceous source beds that were overridden by the Absaroka thrust plate, and not by phosphatic shales in the Permian Phosphoria Formation. This conclusion is based upon results from a variety of analytical techniques that include oil-to-oil and oil-to-source rock correlation (Rosenfeld et al., 1980; Siefert and Moldowan, 1981), correlation of gas chromatograms of oils and condensates whose source is known, stratigraphic and structural studies, measurements of thermal alteration of source beds, and others such as Lopatin diagrams (Waples, 1980). Gas chromatograms of condensates from Whitney Canyon-Carter Creek correlate with those of known Cretaceous sources (e.g., Bridger Lake field, Wyoming) and not with those from known Permian Phosphoria sources (e.g., Brady field, Wyoming).

Thermal maturity measurements indicate that the Cretaceous beds are currently in a hydrocarbon-generating stage, whereas Permian shales are overmature. Figure 19 shows that there is an inversion of vitrinite reflectance values on the R_o versus depth plot below the Absaroka thrust, with high values for the hanging-wall rocks and lower values for the deeper footwall section. The higher reflectance values within the shallower hanging-wall section result from deep, prethrusting burial of the Paleozoic section. Lopatin diagrams (Figure 20) show that Cretaceous source bed hydrocarbon generation began coincident with Absaroka thrust sheet emplacement (about 75 Ma) and is continuing today. Phosphoria source rock hydrocarbon generation occurred prior to the time of Absaroka thrust sheet emplacement and the creation of the Whitney Canyon-Carter Creek structural trap. Therefore, the

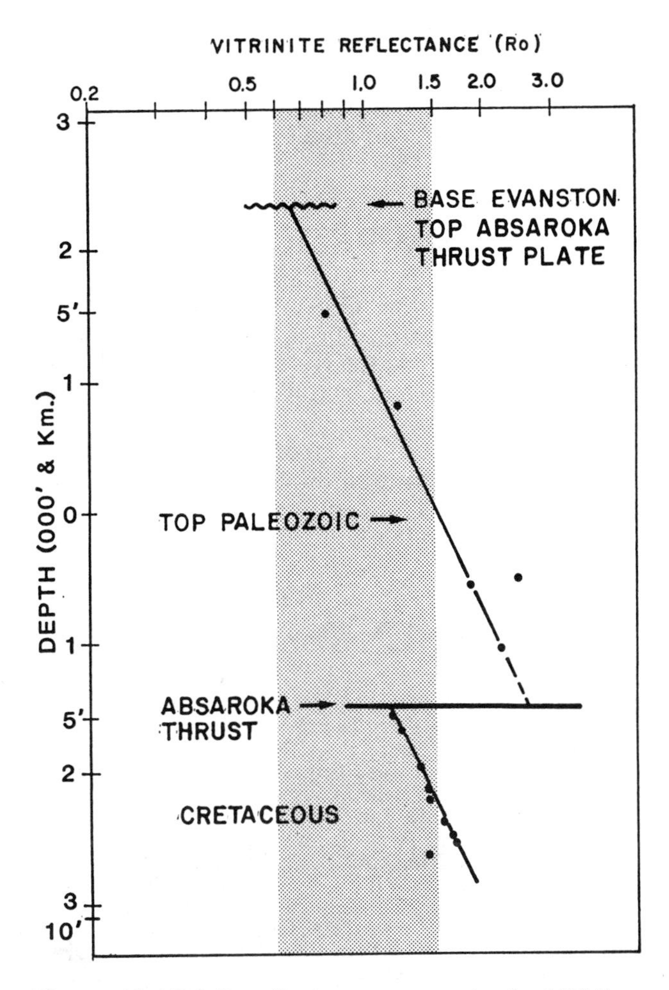

Figure 19. Vitrinite reflectance versus depth at Whitney Canyon-Carter Creek field. Higher values in the hanging-wall Paleozoic resulted from deep prefault burial. Shaded area shows hydrocarbon generation window. Modified from Warner and Royse (1987).

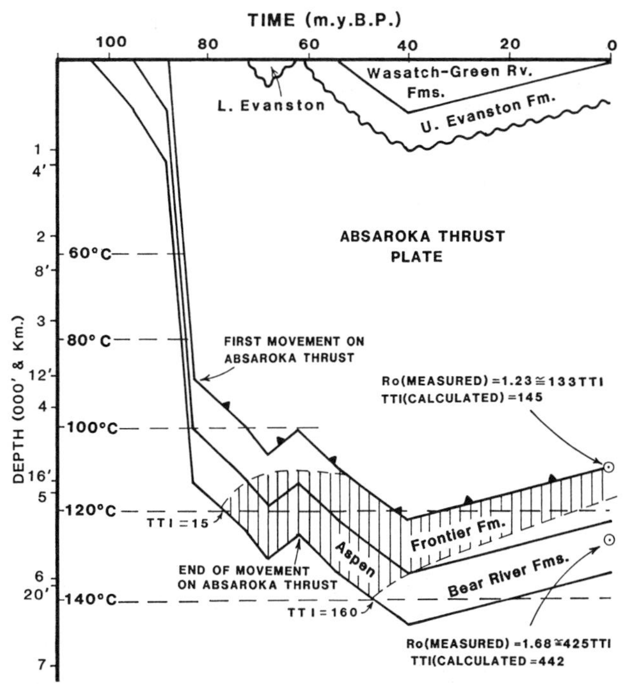

Figure 20. Lopatin diagram for subthrust Cretaceous source rocks at Whitney Canyon-Carter Creek field showing calculated iso-maturity lines. TTI, time-temperature index; R_o, vitrinite reflectance. Temperature is calculated from a constant geothermal gradient of 20°C/km (1.1°F/100 ft) and a surface temperature of 15°C (59°F). Hachured area is the oil and wet gas generative window. Level of thermal maturation calculated by Lopatin's method is similar to R_o measurements in samples from Frontier and Aspen formations. Modified from Warner and Royse (1987).

Phosphoria could not have served as an important source rock at Whitney Canyon-Carter Creek.

Footwall Cretaceous source beds are marine and nonmarine sedimentary rocks and contain a mixture of oil- and gas-prone organic matter. Proprietary company analyses indicate the Upper Cretaceous Frontier Formation and Lower Cretaceous Bear River and Aspen formations average about 1% TOC through 4000 ft (1200 m) of section, with values as high as 2.5% in the Lower Cretaceous section. The Permian Phosphoria marine shales produce higher TOC values (up to 13%) but are much thinner than the Cretaceous source beds (40 ft; 12 m).

Hydrocarbons migrated directly along the Absaroka thrust into adjacent porous Paleozoic reservoir rocks. Although Whitney Canyon-Carter Creek hanging-wall structural closure exists from the Jurassic Twin Creek to the Cambrian, not all porous units have been charged with hydrocarbons. The hydrocarbon distribution is related to the position of hanging-wall cutoffs of reservoir beds on the Absaroka thrust fault plane. Paleozoic hydrocarbon reservoirs (e.g., Bighorn, Lodgepole, Mission Canyon, and Weber) along the east flank of the structure are in direct contact with Cretaceous source rocks across the Absaroka fault, whereas water-wet porous units higher in the rock column such as the Nugget sandstones are not (Figure 6). Direct contact of reservoir and source across the Absaroka thrust migration path is apparently a requisite to hydrocarbon charging. The Nugget is a major producer at Ryckman Creek field some 6 mi (10 km) to the east because there it is in direct contact with the subthrust source rocks.

The H_2S at Whitney Canyon-Carter Creek is believed to originate from thermocatalytic reaction of oil and anhydrite (Orr, 1975) in Paleozoic strata. In other fields throughout Wyoming, H_2S is typically associated with Phosphoria-derived hydrocarbons and is not of those derived from Cretaceous sources. The significant amount of H_2S gas in the Whitney Canyon-Carter Creek Paleozoic reservoirs has been cited as evidence for a Permian Phosphoria shale source. However, H_2S, calcite, and water may be produced when hydrocarbons contact anhydrite under moderate temperature anaerobic conditions (Davis and Kirkland, 1979). Anhydrite within the reservoirs may have reacted with migrating hydrocarbons to produce H_2S. The concentration of H_2S varies within individual reservoirs, and a correlation exists between the anhydrite in the various Paleozoic units and percent H_2S (Hoffman and Kelly, 1981; Hoffman and Balcells-Baldwin, 1982). Sulfur is extracted from the gas and constitutes a significant monetary asset.

EXPLORATION AND DEVELOPMENT CONCEPTS

Discovery of hydrocarbons at Whitney Canyon-Carter Creek resulted from exploration technology advances in the early 1970s combined with a fortunate series of events throughout geologic time. As mentioned in the section *Discovery Method*, advances in structural modeling and the refinement of CDP seismic data were crucial to identifying specific prospects within the thrust belt. Also, the time of hydrocarbon generation, type of source rock, time of deformation, structural geometry, and stratigraphy all must have had a unique geologic relationship in order for hydrocarbons to trap in this area. Lower Paleozoic porous reservoir rocks were thrust over nearly flat lying Cretaceous marine and nonmarine source beds, placing the source rocks in the hydrocarbon generation window during Late Cretaceous to early Paleocene time. These reservoir beds were folded into trapping anticlinal closures and maintained direct contact with subthrust source beds across the Absaroka fault plane over a broad drainage area. The thrust zone acted as a conduit for hydrocarbons that are still being generated. Porosity was developed mainly in dolomitized intervals, whereas interbedded anhydrites and shales seal the reservoirs. Also, if the hydrocarbon phase had been liquid instead of gaseous, the relatively low matrix porosity carbonate units would not have made effective reservoirs because of restrictively high entry-level capillary pressures. Variations in this theme have resulted in many dry holes in the Wyoming, northern Utah, and eastern Idaho thrust belt area.

At the time of discovery, the field was thought to be simple and producible through conventional completion methods, but as development proceeded, it was realized that the reservoirs and structure were more complicated. In the Mission Canyon, porosity exists in layered, dolomitized units that are interbedded with tight limestones. Significant lateral porosity variations are present owing to erratic diagenesis. Structural complications may control drainage areas (e.g., in areas of large repeated sections), fracture density, and hydrocarbon migration. Known complications are now taken into consideration in reservoir management, but several wells within the field still have unexplained production and reservoir problems.

ACKNOWLEDGMENTS

We thank Chevron USA, Inc. for permitting publication of this paper. Luis F. Rivas provided most of the engineering and reservoir data. An unpublished Chevron company report by J. W. Halvorsen was used for the Weber reservoir discussion. Helpful reviews were done by P. M. Harris, J. M. Kelly, C. F. Kluth, G. A. Minnery, P. A. Schipperijn, and J. S. Vietti. Amoco Production Company granted permission to publish seismic line PQ-1.

REFERENCES CITED

Armstrong, R. C., 1968, Sevier orogenic belt in Nevada and Utah: Geological Society of America Bulletin, v. 79, p. 429-458.

Armstrong, R. C., and S. S. Oriel, 1965, Tectonic development of Idaho-Wyoming thrust belt: American Association of Petroleum Geologists Bulletin, v. 49, p. 1847-1866.

Bally, A. W., P. L. Gordy, and G. A. Stewart, 1966, Structure, seismic data, and orogenic evolution of southwestern Canadian Rocky Mountains: Canadian Petroleum Geology Bulletin, v. 14, p. 337-381.

Bartberger, C. E., 1984, Facies control of Mississippian Porosity in Whitney Canyon-Carter Creek field, Wyoming Overthrust Belt (abs.): American Association of Petroleum Geologists Bulletin, v. 69, p. 842-843.

Bishop, R. A., 1982, Whitney Canyon-Carter Creek Gas Field, Southwest Wyoming, *in* R. B. Powers, ed., Geologic studies of the Cordilleran Thrust Belt: Rocky Mountain Association of Geologists, Field Conference Guidebook, v. 2, p. 591-599.

Bissell, H. J., and O. E. Childs, 1958, The Weber Formation of Utah and Colorado, *in* Symposium on Pennsylvanian rocks of Colorado and adjacent areas: RMAG Guidebook, p. 26-30.

Budai, J. M., 1985, Evidence for rapid fluid migration during deformation, Madison Group, Wyoming and Utah Overthrust Belt, *in* M. W. Longman, K. W. Shanley, R. F. Lindsay, and D. E. Eby, Rocky Mountain carbonate reservoirs: SEPM Core Workshop No. 7, p. 377-407.

Budai, J. M., K. C. Lohmann, and R. M. Owen, 1984, Burial Dedolomite in the Mississippian Madison Limestone, Wyoming and Utah thrust belt: Journal of Sedimentary Petrology, v. 54, p. 276-288.

Budai, J. M., K. C. Lohmann, and J. L. Wilson, 1987, Dolomitization of the Madison Group, Wyoming and Utah Overthrust Belt: American Association of Petroleum Geologists Bulletin, v. 71, p. 909-924.

Carmalt, S. W., and B. St. John, 1986, Giant oil and gas fields, *in* M. T. Halbouty, ed., Future petroleum provinces of the world: American Association of Petroleum Geologists Memoir 40, p. 11-54.

Dahlstrom, C. D. A., 1970, Structural geology in the eastern margin of the Canadian Rocky Mountains: Canadian Petroleum Geology Bulletin, v. 18, p. 332-406.

Davis, J. B., and D. W. Kirkland, 1979, Bioepigenetic sulfur deposits: Economic Geology, v. 74, p. 462-468.

Foster, N. H., 1972, Ordovician system, *in* W. W. Mallory, ed., Geologic atlas of the Rocky Mountain Region: RMAG p. 76-85.

Gutschick, R. C., and C. A. Sandberg, 1983, Mississippian continental margins of the conterminous United States, *in* D. J. Stanley and G. T. Moore, eds., The shelfbreak-critical interface on continental margins: SEPM Special Publication 33, p. 79-96.

Gutschick, R. C., C. A. Sandberg, and W. J. Sando, 1980, Mississippian shelf margin and carbonate platform from Montana to Nevada, *in* T. D. Fouch and E. R. Magathan, eds., Paleozoic paleogeography of the west-central United States: SEPM, Rocky Mountain Section, Rocky Mountain Paleogeography Symposium 1, p. 111-128.

Hale, L. A., 1960, Frontier Formation-Coalville, Utah and nearby areas of Wyoming and Colorado, *in* D. P. McGookey and D. N. Miller, eds., Overthrust belt of southwestern Wyoming and adjacent areas: WGA 15th Annual Field Conference, p. 137-146.

Halley, R. B., P. M. Harris, and A. C. Hine, 1983, Bank margin environment, *in* P. A. Scholle, D. G. Bebout, and C. H. Moore, eds., Carbonate depositional environments: American Association of Petroleum Geologists Memoir 33, p. 463-506.

Harris, P. M., P. E. Flynn, and J. L. Sieverding, 1988, Mission Canyon (Mississippian) reservoir study, Whitney Canyon-Carter Creek field, southwestern Wyoming, *in* A. J. Lomando and P. M. Harris, eds., SEPM Core Workshop 12, v. 2, p. 695-740.

Haun, J. D., and J. A. Barlow, 1962, Lower Cretaceous stratigraphy of Wyoming, *in* R. L. Enyert and W. H. Curry, eds., Symposium on Early Cretaceous rocks of Wyoming and adjacent areas: WGA 17th Annual Field Conference, p. 15-22.

Hintze, L. F., 1973, Geologic history of Utah: Brigham Young University Geology Studies, v. 20, p. 3, 181 p.

Hodgden, H. J., and R. E. McDonald, 1977, History of oil and gas exploration in the Overthrust Belt of Wyoming, Idaho and Utah, *in* E. L. Heisey, et al., eds., Rocky Mountain Thrust Belt geology and resources: Joint Wyoming-Montana-Utah Geological Associations Guidebook, p. 37-69.

Hoffman, M. E., and R. N. Balcells-Baldwin, 1982, Gas giant of the Wyoming Thrust Belt: Whitney Canyon-Carter Creek field, *in* R. B. Powers, ed., Geologic studies of the Cordilleran Thrust Belt: Rocky Mountain Association of Geologists, Field Conference Guidebook, v. 2, p. 613-618.

Hoffman, M. E., and J. M. Kelly, 1981, Whitney Canyon-Carter Creek field, Uinta and Lincoln Counties, Wyoming, *in* S. G. Reid and D. D. Miller, eds., Energy resources of Wyoming: WGA 32nd Annual Field Conference, p. 99-107.

Kummel, B., 1955, Facies of Lower Triassic formations in western Wyoming, *in* Green River Basin: WGA 10th Annual Field Conference Guidebook, p. 68-74.

Kummel, B., 1957, Paleoecology of lower Triassic formations of southeastern Idaho and adjacent areas, *in* H. S. Ladd, ed., Treatise on marine ecology and paleoecology: Geological Society of America Memoir 67, v. 2, p. 437-468.

Lamerson, P. R., 1982, The Fossil Basin area and its relationship to the Absaroka Thrust Fault system: RMAG Special Publication 1982, Geologic Studies of the Cordilleran Thrust Belt, p. 279-340.

Orr, W. L., 1975, Advances in geochemistry: Proceedings of the 7th International Meeting on Organic Geochemistry: New York, Pergamon Press, p. 511-597.

Picard, M. D., R. Aadland, and L. R. High, Jr., 1969, Correlation and stratigraphy of Triassic Red Peak and Thaynes formations, Western Wyoming and Adjacent Idaho: American Association of Petroleum Geologists Bulletin, v. 53, p. 2274-2289.

Rau, R., 1982, Stratigraphy of the Mississippian Madison Group of the Whitney Canyon-Carter Creek field of southwest Wyoming: Unpublished M.S. Thesis, Kent State University, 156 p.

Rich, J. L., 1934, Mechanics of low-angle overthrust faulting illustrated by Cumberland thrust block, Virginia, Kentucky and Tennessee: American Association of Petroleum Geologists Bulletin, v. 18, p. 1584-1596.

Richards, P. W., and C. L. Nieschmidt, 1957, The Bighorn Dolomite in south-central Montana and northwestern Wyoming, *in* Crazy Mountain Basin, Billings Geological Society 8th Annual Field Conference, p. 54-62.

Rose, P. R., 1977, Mississippian carbonate shelf margins, western United States: E. L. Heisey, et al., eds., Rocky Mountain thrust belt geology and resources: Joint Wyoming-Montana-Utah Geological Associations Guidebook, p. 155-172.

Rosenfeld, J. K., T. T. Y. Hos, and H. Dembicki, Jr., 1980, Oil-to-source correlation, Pineview Field, Overthrust Belt, Utah (abs.): American Association of Petroleum Geologists Bulletin, v. 64, p. 776.

Ross, R. J., Jr., 1953, The Ordovician system in northeastern Utah and southeastern Idaho, *in* Guide to the geology of northern Utah and southeastern Idaho: IAPG 4th Annual Field Conference Guidebook, p. 22-26.

Royse, F., 1979, Structural geology of western Wyoming-northern Utah Thrust Belt and its relation of oil and gas (abs.): Oil and Gas Journal, Feb. 12, p. 155-156.

Royse, F., M. A. Warner, and D. L. Reese, 1975, Thrust belt of Wyoming, Idaho, and Northern Utah structural geometry and related stratigraphic problems: RMAG Symposium on Deep Drilling Frontiers in Central Rocky Mountains, p. 41-54.

Sandberg, C. A., and R. C. Gutschick, 1980, Sedimentation and biostratigraphy of Osagean and Meramecian starved basin and foreslope, western United States, *in* T. D. Fouch and E. R. Magathan, eds., Paleozoic paleogeography of west-central United States: SEPM, Rocky Mountain Section, West-Central U.S. Paleogeography Symposium 1, p. 129-147.

Sandberg, C. A., R. C. Gutschick, J. G. Johnson, F. G. Poole, and W. J. Sando, 1982, Middle Devonian to Late Mississippian geologic history of the Overthrust Belt region, western United States: RMAG Special Publication 1982, Geologic Studies of the Cordilleran Thrust Belt, p. 691-719.

Sando, W. J., 1976, Mississippian history of the northern Rocky Mountains region: U.S. Geological Survey, Journal of Research, v. 4, p. 317-338.

Siefert, W. K., and J. M. Moldowan, 1981, Paleoreconstruction by biological markers: Geochimica et Cosmochimica Acta, v. 45, p. 783-794.

Sieverding, J. L., P. E. Flynn, and P. M. Harris, 1987, Mission Canyon (Mississippian) Reservoir Study, Whitney Canyon-Carter Creek field, Southwestern Wyoming Thrust Belt (abs.): SEG 57th Annual Meeting Expanded Abstracts, p. 123-125.

St. John, B., A. W. Bally, and H. D. Klemme, 1984, Sedimentary provinces of the world, hydrocarbon productive and nonproductive: American Association of Petroleum Geologists Map Series, 35 p.

Suppe, J., 1983, Geometry and kinematics of fault-bend folding: American Journal of Science, v. 283, p. 684-721.

Veatch, A. C., 1906, Coal and oil in southern Uinta County, *in* Contributions to economic geology: U.S. Geological Survey Bulletin, v. 285, p. 331-353.

Waples, D. W., 1980, Time and temperature in petroleum formation: application of Lopatin's method to petroleum exploration: American Association of Petroleum Geologists Bulletin, v. 64, p. 916-926.

Warner, M. A., 1982, Source and time of generation of hydrocarbons in the Fossil basin, Western Wyoming Thrust Belt, *in* R. B. Powers, ed., Geologic studies of the Cordilleran Thrust Belt: Rocky Mountain Association of Geologists, Field Conference Guidebook, v. 2, p. 805-815.

Warner, M. A., and F. Royse, Jr., 1987, Thrust faulting and hydrocarbon generation: discussion: American Association of Petroleum Geologists Bulletin, v. 71, p. 882-889.

Webber, W. W., L. E. Petty, B. D. Ray, and J. G. Story, 1984, Whitney Canyon project is complex and successful: Oil and Gas Journal, v. 82, n. 35, p. 79-83.

Wilson, C. W., Jr., and R. G. Stearns, 1958, Structure of the Cumberland Plateau, Tennessee: Geological Society of America Bulletin, v. 69, p. 1283-1296.

Appendix 1. Field Description

Field name .... *Whitney Canyon-Carter Creek*

Ultimate recoverable reserves .... *2.2 TCFG (60 million bbl of condensate, 20 million long tons of sulfur, 57 billion m^3 of gas and condensate)*

Field location:

- **Country** .... *United States*
- **State** .... *Wyoming*
- **Basin/Province** .... *Fossil basin/Thrust Belt province*

Field discovery:

- **Year first pay discovered** .... *Triassic Thaynes Formation (rank 5 in importance) 1976*
- **Years second through fifth pays discovered** *1977–1979*

Pay zones are numbered based on relative importance of reservoir to total gas in place.

First pay (Mississippian Mission Canyon):

- **Discovery well name and general location** .... *Amoco-Chevron-Gulf WIU 2, Sec. 18, T17N, R119W*
- **Discovery well operator** .... *Amoco*
- **IP** .... *8.5 MMCFGPD, 130 BCPD (24/64-in. choke)*

Second pay (Mississippian Lodgepole):

- **Discovery well name and general location** .... *Chevron 1-29M Federal, Sec. 29, T19N, R119W*
- **Discovery well operator** .... *Chevron*
- **IP** .... *2.6 MMCFGPD (estimated from production logs)*

Third pay (Pennsylvanian Weber):

- **Discovery well name and general location** .... *Amoco-Chevron-Gulf WIU 2, Sec. 18, T17N, R119W*
- **Discovery well operator** .... *Amoco*
- **IP** .... *0.6 MMCFGPD, 6 BCPD (19/64-in. choke, 300 psi FTP)**

**Well had parted casing; test may be invalid.*

Fourth pay (Ordovician Bighorn):

- **Discovery well name and general location** .... *Amoco-Chevron-Gulf WIU 2, Sec. 18, T17N, R119W*
- **Discovery well operator** .... *Amoco*
- **IP** .... *5.7 MMCFGPD, 64 BCPD, 18 BWPD (16/64-in. choke, 2850 psi FTP)*

Fifth pay (Triassic Thaynes):

- **Discovery well name and general location** .... *Amoco-Chevron-Gulf WIU 1, Sec. 18, T17N, R119W*
- **Discovery well operator** .... *Amoco*
- **IP** .... *4.7 MMCFGPD, 196 BCPD, 9 BWPD*

All other zones with shows of oil and gas in the field:

Age	Formation	Type of Show
Permian	*Phosphoria*	*Gas*
Devonian	*Darby*	*Gas*
Subthrust Cretaceous	*Frontier*	*Gas*

Geologic concept leading to discovery and method or methods used to delineate prospect, e.g., surface geology, subsurface geology, seeps, magnetic data, gravity data, seismic data, seismic refraction, nontechnical:

Surface mapping, structural cross sections, 100% seismic data, and regional source rock studies identified a large essentially untested area where the Absaroka thrust sheet, containing potential reservoir rocks in anticlinal closures, overlay both thermally mature source rocks and potential reservoir rocks. Prospective structural closure was delineated subsequently by CDP seismic methods.

Structure:

Province/basin type .... *Western U.S. Thrust Belt; Bally 41; Klemme IIA*

Tectonic history

Early Cretaceous through early Eocene thrusting created a major multithrust system in western Wyoming. Folds were formed on the thrust plate hanging walls at fault cutoffs largely because of fault plane ramp geometries. Thrusting buried subthrust Cretaceous source rocks and placed them within the hydrocarbon generative window.

Regional structure

Located within the Absaroka thrust plate in the southwestern Wyoming thrust belt.

Local structure

Two large doubly plunging anticlinal folds that formed within the Absaroka thrust sheet; the folds are asymmetric to the east where they are bounded by reverse faults.

Trap: .... *One major anticlinal trap with multiple pays*

Basin stratigraphy (major stratigraphic intervals from surface to deepest penetration in field):

Chronostratigraphy	Formation	Depth to Top in ft (m)
Eocene	*Green River/Wasatch*	*Surface*
Paleocene/U. Cretaceous	*Evanston*	*700 (213)*
Cretaceous	*Gannett*	*2100 (640)*
Jurassic	*Stump-Preuss*	*4100 (1250)*
Jurassic	*Twin Creek*	*5500 (1676)*
Jurassic	*Nugget*	*7300 (2225)*
Triassic	*Ankareh*	*8300 (2530)*
Triassic	*Thaynes*	*9200 (2804)*
Triassic	*Woodside*	*10,500 (3200)*
Triassic	*Dinwoody*	*11,300 (3444)*
Permian	*Phosphoria*	*11,700 (3566)*
Pennsylvanian	*Weber*	*12,300 (3749)*
Pennsylvanian	*Morgan (Amsden)*	*13,000 (3962)*
Mississippian	*Mission Canyon*	*13,300 (4054)*
Mississippian	*Lodgepole*	*14,050 (4282)*
Devonian	*Darby*	*14,800 (4511)*
Ordovician	*Bighorn*	*15,500 (4724)*
Cambrian	*Gallatin/Gros Ventre*	*16,000 (4877)*
Cretaceous	*Absaroka thrust plane*	*17,000 (5181)*
Cretaceous	*Frontier*	*17,000 (5181)*

Location of well in field .... *NA*

Reservoir characteristics:

First pay zone:

Formation .... *Mission Canyon*

Age .... *Mississippian*

Average depth to top .... *13,300 ft (4054 m)*

Average gross thickness .... *750 ft (229 m)*

Net stratigraphic thickness

Average .... *261 ft (80 m)*

Maximum .... *592 ft (180 m)*

Lithology .... *Dolomitized mudstones and packstones*

Porosity type

Sucrosic intercrystalline porosity with rare to abundant moldic porosity, minor fracture porosity

Average porosity *6-8%*
Average permeability .... *0.7-1.5 md*

Second pay zone:

Formation .... *Lodgepole*
Age .... *Mississippian*
Average depth to top .... *14,050 ft (4282 m)*
Average gross thickness .... *750 ft (229 m)*
Net stratigraphic thickness
Average .... *123 ft (37 m)*
Maximum .... *219 ft (67 m)*
Lithology .... *Dolomitized mudstones and packstones*
Porosity type
Sucrosic intercrystalline porosity with rare to common moldic porosity, minor fracture porosity
Average porosity *4-6%*
Average permeability .... *0.5-1.5 md*

Third pay zone:

Formation .... *Weber*
Age .... *Pennsylvanian*
Average depth to top .... *12,300 ft (3749 m)*
Average gross thickness .... *700 ft (213 m)*
Net stratigraphic thickness
Average .... *120 ft (37 m)*
Maximum .... *180 ft (55 m)*
Lithology .... *Quartz sandstone with abundant silica overgrowths*
Porosity type .... *Unconnected, dissolution moldic porosity, also minor fracture porosity*
Average porosity *3-5%*
Average permeability .... *0.2 md*

Fourth pay zone:

Formation .... *Bighorn*
Age .... *Ordovician*
Average depth to top .... *15,500 ft (4724 m)*
Average gross thickness .... *500 ft (152 m)*
Net stratigraphic thickness
Average .... *81 ft (25 m)*
Maximum .... *108 ft (33 m)*
Lithology .... *Dolomitized mudstones, packstones, and grainstones*
Porosity type
Sucrosic intercrystalline porosity with common to abundant moldic and vugular porosity, also common to sparse fracture porosity
Average porosity *3-5%*
Average permeability .... *0.1-0.5 md*

Fifth pay zone:

Formation .... *Thaynes*
Age .... *Triassic*
Average depth to top .... *9200 ft (2804 m)*
Average gross thickness .... *1300 ft (396 m)*
Net stratigraphic thickness
Average .... *118 ft (36 m)*
Maximum .... *165 ft (50 m)*
Lithology .... *Silty and sandy limestones and limy dolomites*

Porosity type .... *Fracture porosity with rare to common microsucrosic intercrystalline porosity*
Average porosity *1-2%*
Average permeability .... *0.1-0.2 md*

Seals: .... *Each reservoir has a separate gas-water contact and several seals are present within the field*

Mission Canyon upper seal:
Massive, bedded anhydrites in the upper part of the Mission Canyon may seal the Mission Canyon main porosity zone

Lateral seal: .... *Anticlinal dip to the west, reverse fault and/or anticlinal dip to the east*

Sources:

Formation and age .... *Cretaceous Frontier*
Lithology .... *Shale*
Average total organic carbon (TOC) *1.0%*
Maximum TOC *1.2%*
Kerogen type (I, II, or III) .... *II and III*
Vitrinite reflectance (maturation) *1.1-1.7*

Formation and age .... *Cretaceous Aspen*
Lithology .... *Shale*
Average total organic carbon (TOC) *2.5%*
Maximum TOC *2.7%*
Kerogen type (I, II, or III) .... *II*
Vitrinite reflectance (maturation) .... *Aspen not penetrated within field*

Formation and age .... *Cretaceous Bear River*
Lithology .... *Shale*
Average total organic carbon (TOC) *1.5%*
Maximum TOC *1.5%*
Kerogen type (I, II, or III) .... *II and III*
Vitrinite reflectance (maturation) .... *Bear River not penetrated within field*

Time of hydrocarbon expulsion .... *Late Cretaceous (75 MYBP) to present, after the Absaroka thrust sheet was in place*
Present depth to top of source .... *17,000 ft (5182 m)*
Thickness .... *4000 ft (1200 m)*
Potential yield .... *Unknown, but substantial*

Appendix 2. Production Data

Field name .... *Whitney Canyon-Carter Creek*

Field size:

Proved acres *13,000*
Number of wells all years *39*
Current number of wells *32*
Well spacing .... *640 acre/well/formation*
Ultimate recoverable .... *2000 bcf, 60 MMBC, 20 MMLTS (57 billion m³)*
Cumulative production (9/1987) .... *600 bcf, 18 MMBC (17 billion m³)*
Annual production .... *120 bcf, 3.6 MMBC (3.4 billion m³)*
Present decline rate .... *12% per year*
Annual water production .... *180,000 bbl (28,600 m³)*

In place, total reserves .... *4.5 TCFG, 125 million bbl condensate, 24 million long tons of sulfur (130 billion m^3)*
In place, per acre-foot .... *3.5 bcf/acre*
Primary recovery .... *2000 bcf, 60 million bbl condensate, 20 million long tons of sulfur (57 billion m^3)*
Secondary recovery .... *NA*
Enhanced recovery .... *NA*
Cumulative water production (9/1987) .... *900,000 bbl (143,000 m^3)*

Drilling and casing practices:

Amount of surface casing set .... *2500 ft (762 m)*

Casing program

13⅜-in. surface casing set at 2500 ft (762 m); 9⅝-in. intermediate casing set at 12,000 ft (3657 m); 7-in. production liner and tieback set to TD

Drilling mud

A low-solids, nondispersed, salt-saturated water-based gel mud is used from surface casing to intermediate casing point; from intermediate casing point to TD either a low-solids nondispersed or highly dispersed water-based gel mud is used

Completion practices:

Interval(s) perforated

Intervals with greater than 2-3% core adjusted log matrix porosity or intervals that have high fracture potential are perforated with 2 shots per foot

Well treatment .... *28% HCl acid fracture using 200-500 gal of acid per foot of perfed interval*

Formation evaluation:

Logging suites .... *DLL-MSFL-GR, BHC-GR, CNL-LDT-GR (or CNL-FDC-GR), SHDT (or HDT)*

Testing practices

Drillstem tests are not performed in sour gas section for safety and environmental reasons; testing is done through casing after well is drilled to TD

Mud logging techniques .... *Gas detection and cuttings gas from bottom of surface casing to TD*

Mission Canyon gas characteristics:

Gravity .... *0.79 (air = 1.000)*
Gross heating value .... *1030 BTU/cubic foot*
Major components (mol%) ..... *Methane, 68%; hydrogen sulfide, 15%; ethane, 6%; carbon dioxide, 5%; propane, 2%; nitrogen, 1%*

Mission Canyon condensate characteristics:

Type .... *Condensate*
API gravity *43*
Initial GOR *30,000:1*
Sulfur, wt% *6%*

Field characteristics:

Average elevation .... *7300 ft (2225 m)*
Initial pressure .... *6050 psi (425 kg/cm^2)*
Present pressure (9/1987) .... *5000 psi (352 kg/cm^2)*
Pressure gradient .... *0.43 psi/ft (0.10 kg/cm^2/m)*
Temperature .... *210°F (99°C)*
Geothermal gradient .... *1.3°F/100 ft (2.47°C/100 m)*
Drive .... *Volumetric depletion*
Gas column thickness .... *2000 ft (610 m)*

Gas-water contacts

Mission Canyon	*-7382 ft ASL (-2250 m ASL)*
Lodgepole	*-7610 ft ASL (-2319 m ASL)*
Bighorn at Carter Creek	*-8497 ft ASL (-2590 m ASL)*
Bighorn at Whitney Canyon	*-7914 ft ASL (-2412 m ASL)*
Weber	*Unknown*
Thaynes	*Unknown*

Connate water *25%*
Water salinity, TDS .... *95,000 mg/L (Cl = 60,000 mg/L)*
Resistivity of water .... *0.12 ohm-m at 70°F (21°C)*

Transportation method and market for oil and gas:

Processed gas is generally transported on the Trailblazer pipeline to midwestern and eastern markets; condensate is pipelined to refineries in Salt Lake City, UT, or trucked to refineries in Denver; sulfur is either trucked or transported through the sulfur slurry pipeline from the Carter Creek gas processing plant to the Union Pacific Railroad sulfur terminal in Kemmerer, WY.

Anschutz Ranch East Field—U.S.A.
Utah-Wyoming Thrust Belt

R. R. WHITE and T. J. ALCOCK
Amoco Production Company
Denver, Colorado

R. A. NELSON
Amoco Production Company
Houston, Texas

FIELD CLASSIFICATION

BASIN: Utah-Wyoming Thrustbelt
BASIN TYPE: Thrustbelt
RESERVOIR ROCK TYPE: Sandstone
RESERVOIR ENVIRONMENT OF DEPOSITION: Eolian
RESERVOIR AGE: Jurassic
PETROLEUM TYPE: Oil
TRAP TYPE: Anticline

LOCATION

The Anschutz Ranch East field is located in northeastern Utah and southwestern Wyoming in the western United States (Figure 1). The field consists of two anticlinal closures; the West Lobe closure is almost entirely in Summit County, Utah, whereas the smaller East Lobe is mostly in Uinta County, Wyoming (Figure 2). The field is within the Utah-Wyoming thrust belt near the town of Evanston, Wyoming, in the central portion of the Western Overthrust province. Surface topography is relatively rugged near the field with elevations from 7000 to 9000 ft (2131 to 2743 m) above sea level. Nearby fields include Anschutz Ranch to the west, Pineview to the southwest, and Glasscock Hollow to the northeast (Figures 1 and 3).

Anschutz Ranch East is a giant gas and retrograde gas-condensate field that produces from the Jurassic Nugget Sandstone. It is by far the largest accumulation of gas and liquids in the Western Overthrust province with an estimated 800 million barrels of oil equivalent in place. Amoco Production Co. is operator of the field, which covers more than 4500 ac (1821 ha) and includes 37 producing wells and 18 injection wells. Partners in the West Lobe of the field include Amoco Production Co., The Anschutz Corp., Mobil Exploration Inc., Union Pacific Resources Co. (formerly Champlin Petroleum), Pan-Canadian Petroleum Co., BWAB Inc. (Brownley, Wallace, Armstrong, and Bander), and Chevron U.S.A. Inc. Partners in the East Lobe include Amoco Production Co., The Anschutz Corp., and Union Pacific Resources Co.

The West Lobe is the larger and more extensively studied of the two structures and contains approximately 90% of the total reserves for the Anschutz Ranch East field. Therefore, this paper will emphasize the technical aspects of the West Lobe structure unless otherwise indicated.

HISTORY

Pre-Discovery

The Western Overthrust province has been an area of active exploration interest for many years. The most recent and intensive exploration activity began in the Utah and Wyoming portions of the trend in 1969 when Amoco Production Company obtained an exclusive exploration agreement for the odd-numbered sections across the Union Pacific Railroad's land-grant acreage (Dixon, 1982). Approximately 200 wells were drilled in this thrust belt prior to the first major discovery at Pineview field in 1975 by the American Quasar Company on a farmout from Amoco, the Newton Sheep No. 1 well at Pineview providing the first commercially viable production in the trend. The well flowed 540 BOPD on test from only 8 ft (2.5 m) of perforated pay in the Nugget Sandstone at a depth of 9930 ft (3029 m).

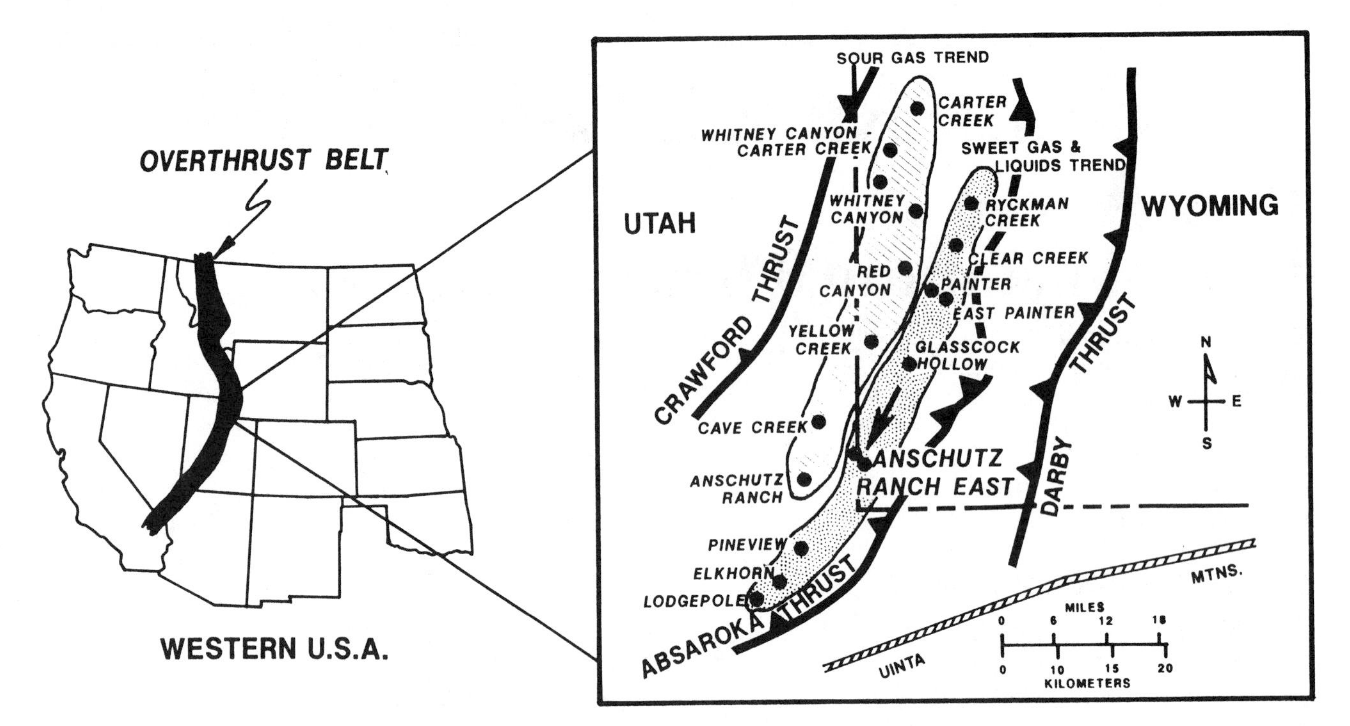

Figure 1. Location of Western Overthrust province and the Anschutz Ranch East field in the Western United States. Enlarged area shows the major producing fields and hydrocarbon trends within the Utah-Wyoming segment of the thrust belt, after Lelek (1982). Courtesy of Rocky Mountain Association of Geologists.

Nearly 20 commercial fields were discovered in the Utah-Wyoming thrust belt between 1975 and 1985 by Amoco Production Co., Chevron U.S.A Inc., Champlin Petroleum Co., The Anschutz Corp., and others. These range from the giant Whitney Canyon-Carter Creek field in the north to Lodgepole field at the south. Two productive trends soon became evident; an eastern "sweet gas and liquids trend" that produces primarily from the Nugget Sandstone and extends from Pineview field to Ryckman Creek field, and a western "sour gas trend" that produces from Paleozoic rocks and extends from the Anschutz Ranch field to Whitney Canyon-Carter Creek field (Figure 1). The Anschutz Ranch East field lies within the eastern sweet gas and liquids trend.

Discovery

The Anschutz Ranch East field was discovered in 1979 by Amoco Production Co. with the completion of the No. 1 Bountiful Livestock well (NW NW 16-4N-8E, Summit County, Utah) (Lelek, 1982). The well spudded 29 March 1979 on a farmout lease from BWAB, Inc. and Tom Brown, Inc. and was completed 31 December 1979. The discovery well penetrated 940 ft (286.5 m) of gross pay in the Jurassic Nugget Sandstone in what is now the West Lobe of the field. The East Lobe was discovered by The Anschutz Corporation, Anschutz Ranch East No. 12-26W (26-13N-121W, Uinta County, Wyoming), completed 23 December 1980. The two lobes produce gas and gas condensate from the Nugget Sandstone and have separate gas-water contacts. The West Lobe discovery well is perforated from 12,816 ft (3909 m) to 13,515 ft (4122 m) and had an initial flowing rate of 1054 BCPD, 4053 MCFD, and 28 BWPD. The East Lobe discovery well is perforated from 14,750 ft (4499 m) to 14,910 ft (4548 m) and flowed 141 BCPD, 750 MCFD, and 96 BWPD (Lelek, 1983).

The No. 1 Bountiful Livestock well was drilled to test for the possibility of an anticlinal fold in the Nugget Sandstone as interpreted from seismic data and inferred from shallower folding in the Champlin Petroleum Co. No. 404 "A" No. 1 (35-13N-121W, Uinta County, Wyoming).

Post-Discovery

Shortly after the field's discovery it was determined that the West Lobe of Anschutz Ranch East field contained a rich retrograde gas-condensate with a dew point only 230 psia (1586 kPa) below the original reservoir pressure. Pressure depletion below the dew point would have resulted in a condensate liquid drop-out of up to 40% of the hydrocarbon pore volume (HPV). Such a dropout could have led to a 75% loss of the reserves in place (Kleinsteiber et al., 1983). Because of this, a specialized plan-of-depletion was designed to optimize production while maintaining the reservoir pressure. This plan called for nitrogen

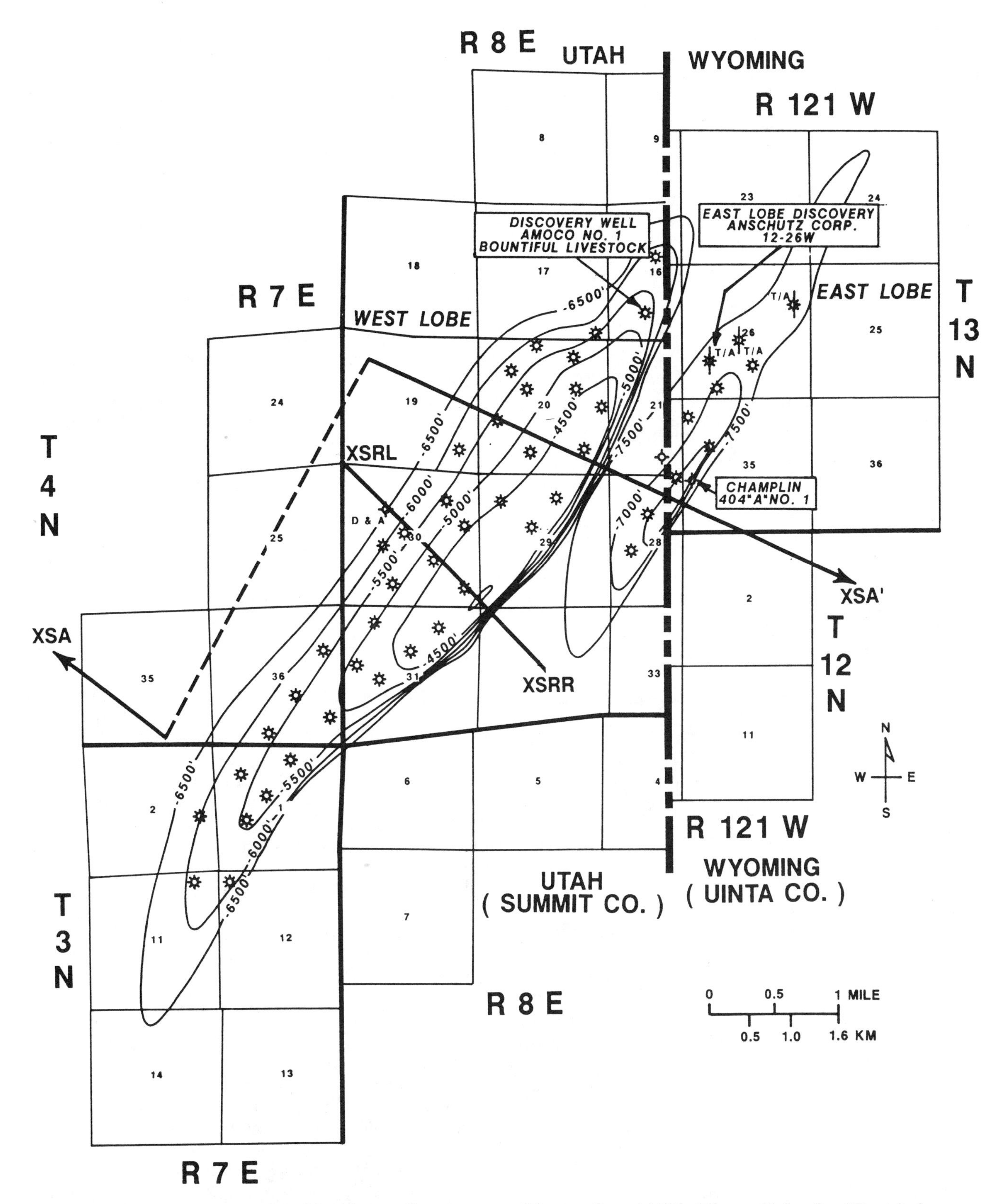

Figure 2. Structure map on top of the Nugget Sandstone reservoir at Anschutz Ranch East field, September 1987. Discovery well and structural cross sections XSR (Figure 4) and XSA (Figure 5) for the West Lobe are located.

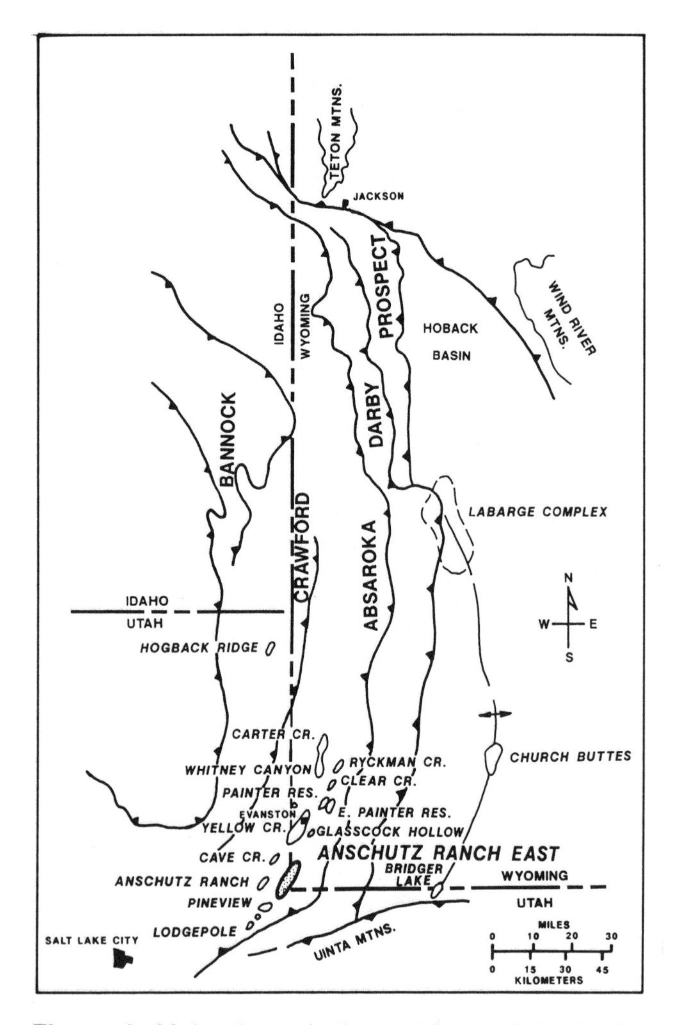

Figure 3. Major thrust faults and fields of the Utah-Wyoming thrust belt, after Frank et al. (1982). Courtesy of Rocky Mountain Association of Geologists.

generation and injection to replace the hydrocarbons produced from the reservoir.

The original injection plan in the West Lobe utilized an inverted nine-spot pattern and proposed a 10% HPV "slug" of injected buffer mix containing lean residual field gas and no greater than 35% nitrogen. Subsequent pressure maintenance has resulted in the total injection of approximately 50% of the hydrocarbon pore volume with similar nitrogen and field gas mixtures. In 1986, the injector layout was modified to an "imperfect line-drive pattern" to take advantage of permeability trends inherent in the eolian reservoir sandstones. The continuing challenge is to maintain a constant sweep of the reservoir to minimize hydrocarbon loss and to maintain field-wide pressure above the dew-point. Extensive coring (approximately 12,000 ft [3658 m] of core), core analyses, and reservoir modeling programs have been undertaken to better predict the field's response to various depletion schemes.

Currently, the West Lobe of Anschutz Ranch East field contains 30 producing wells and 18 nitrogen-gas injection wells. Production from the West Lobe is expected to continue into the early part of the twenty-first century.

In the East Lobe, sufficient differences exist between the original and dew point pressures to tolerate partial pressure maintenance (cycling) and pressure depletion alternatives. Furthermore, the amount of reserves contained in the East Lobe cannot economically justify full pressure maintenance. The East Lobe contains seven producing wells and is entering into the pressure depletion phase of production.

Significant drilling problems were encountered during the development of the field. Salt in the Jurassic Preuss Formation often caused hole problems and casing failures. Early in the field's development, a method was devised to drill and simultaneously under-ream the salt to minimize salt encroachment around the drill pipe (Holt and Johnson, 1985). More recently, oil-based drilling fluids have proved effective for controlling the thicker sections of salt. As a result of these improvements and of lower day-rate rig and subcontractor costs, drilling expenses are now half of those in the early 1980s.

Discovery Method

Approximately 25 mi (40 km) of conventional 2D seismic and shallow correlations to a nearby Cretaceous Bear River Sandstone test (Champlin Petroleum Co. No. 404, Amoco "A" No. 1) were interpreted to define the Anschutz Ranch East prospect and locate the discovery well. The Champlin No. 404 well was used to tie shallow seismic reflections, and it revealed an overturned fold in the Jurassic Preuss Formation. The presence of folding in the Preuss, stratigraphically above the Nugget-Twin Creek interval, and the seismic interpretation suggested the possibility of an anticlinal fold at the deeper Nugget level (Lelek, 1983). A 3D seismic survey including 40 mi^2 (102 km^2) of 12-fold data was acquired in 1981 to help detail the structure for further field development.

The refinement in seismic acquisition and processing characterized thrust belt exploration throughout the 1970s and early 1980s. This enabled interpreters to resolve structures below thrust detachment planes as well as below disharmonically folded salt in the Preuss Formation. However, structural resolution remains difficult for some parts of the fold even with the most advanced techniques. The discovery well at Anschutz Ranch East penetrated the Nugget interval significantly off the structural crest and close to both the eastern overturned flank and the northeastward plunging terminus of the field (Figure 2). Extensive well control and 3D seismic have led to detailed interpretations and a better definition of the structure.

STRUCTURE

Tectonic History

The Utah-Wyoming thrust belt is primarily a feature of the Laramide orogeny (Late Cretaceous to early Tertiary in age). Laramide tectonic forces were predominantly compressional and created a broad zone of relatively continuous deformation that is commonly referred to as the western North American Cordillera. This orogeny resulted in several arcuate, east-verging thrust belts from western Canada to southern California. The structural style, stratigraphy, and the amounts of crustal shortening within these thrust belts vary, and all have been areas of active interest for oil and gas exploration. The Utah-Wyoming and western Canada thrust belts are the only areas with commercial production.

A late phase of extensional deformation occurred in the Utah-Wyoming Overthrust area from the late Eocene through Pliocene epochs. Some investigators believe the extensional features were reactivated along preexisting structural trends and represent a "relaxation" of the convergent structures (Royse et al., 1975). It is also likely that some of the extentional tectonics in the Utah-Wyoming Thrust Belt are related to Tertiary extension occurring in the Basin and Range province and to the epeirogenic uplift of the western interior of North America.

Regional Structure

The western North American Cordillera is characterized by low-angle thrust faulting and high-angle uplifts from central Alaska to southern California and western Mexico. The Utah-Wyoming thrust belt extends from the Uinta Mountains at the south to just west of the southern edge of the Teton Range to the north. As much as 60 mi (96 km) of cumulative west-east shortening may have occurred across four major thrusts (Bannock, Absaroka, Darby, and Prospect, Figure 3) and numerous minor faults (Lelek, 1983). The relative age of thrusting in the Utah-Wyoming thrust belt is an extensively studied and complex subject. In general, the major thrusts are progressively younger to the east (Dixon, 1982; Wiltschko and Dorr, 1983).

Local Structure

The West and East Lobes of Anschutz Ranch East field form a leading-edge fold pair in the hanging-wall of the Absaroka thrust sheet. The larger West Lobe anticline is 7 mi (11.2 km) long by 1.5 mi (2.4 km) wide, contains more than 3500 ac (1416 ha) of hydrocarbon producing area under closure, and is approximately 2000 ft (610 m) higher than the East Lobe field. The East Lobe is substantially smaller than the West Lobe, measuring roughly 4 mi (6.4 km) long by 3/5 mi (1.0 km) wide. These two anticlines are separated by a tight, asymmetric syncline. The synclinal trough is generally deeper than -7400 ft (-2256 m) below mean sea level and is believed to control the spill point and gas-water contact of the East Lobe structure. The West Lobe of the field has its own gas-water contact at -6470 ft (-1972 m), nearly 1000 ft (305 m) higher than that of the East Lobe (Figure 4).

The geometry and relative size of the East and West Lobe fold pair is thought to be controlled by the thickness of the underlying stratigraphy carried in the hanging-wall of the Absaroka thrust. The West and East Lobe folds probably formed in response to the hanging-wall truncations of the underlying Thaynes and Ankareh formations, respectively (Figure 5). Although no wells in the West Lobe have been drilled deeper than the Ankareh Formation, the Triassic Woodside, Dinwoody, and Thaynes intervals are thought to be present in the core of the West Lobe fold (West and Lewis, 1982). The presence of these older units is based on the projection of their normal stratigraphic thicknesses to the anticipated depth of the Absaroka thrust surface. The Ankareh Formation is probably the oldest unit in the core of the East Lobe anticline owing to the lack of room beneath the Nugget and above the projected position of the Absaroka thrust.

Lelek (1983) provides an interesting historical view of several structural interpretations through the exploration and early development of the field. His summary maps representing the interpretations from early 1979 to late 1980 are reproduced here as Figure 6. Additional development drilling and geologic evaluations have extended the field southward and refined the structure throughout the West and East Lobes of the field (Figure 2).

Unmigrated and migrated seismic sections used during the early phases of exploration and development are shown in Figures 7 and 8. While the major structural culminations can be seen on these sections, details of the fold shape remain elusive. Interpretive details are known from extensive well control and 3D seismic.

STRATIGRAPHY

Stratigraphy for the southern part of the Utah-Wyoming thrust belt is illustrated in Figure 9. The major productive intervals and potential source rocks are indicated. Carbonate rocks dominate the Paleozoic section with the exception of some carbonaceous Cambrian shales and the Pennsylvanian Weber Sandstone. The Mesozoic and Tertiary sections are more varied and generally consist of sandstones, siltstones, and shales with minor intermittent carbonate and evaporitic sequences.

Formations penetrated by wells in the Anschutz Ranch East field are shown in Figure 10 with their average thicknesses. The Nugget Sandstone, which

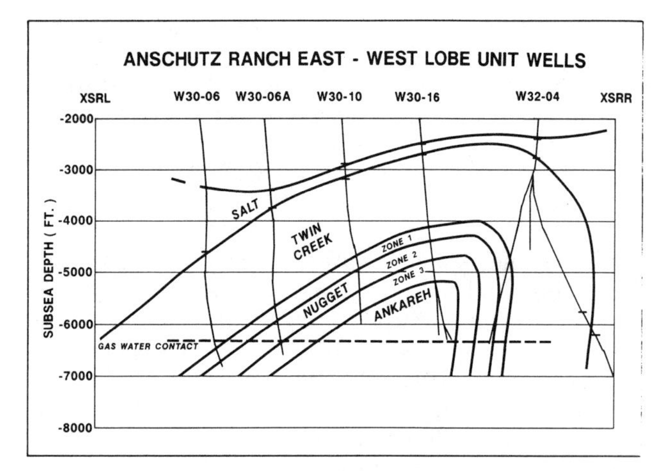

Figure 4. Structural cross section XSR of the Nugget reservoir at Anschutz Ranch East-West Lobe field.

is the primary reservoir for the eastern sweet gas and gas condensate trend, is a stratigraphic equivalent to the Navajo and Aztec sandstones of the Colorado Plateau province. At Anschutz Ranch East, the Nugget Sandstone is approximately 1050 ft (320 m) thick and is penetrated at an average depth of 12,800 ft (3900 m) in the West Lobe and at 14,700 ft (4480 m) in the East Lobe. The Nugget is overlain by approximately 100 ft (30.5 m) of Gypsum Springs anhydritic red-brown siltstone and shale, which is in turn overlain by 1000-1500 ft (305-460 m) thick sequence of Twin Creek limestones. The Nugget-Twin Creek interval generally deforms as a single mechanical package throughout the oil- and gas-producing areas of the Utah-Wyoming thrust belt.

The Upper Jurassic and Cretaceous formations are commonly detached from the Nugget-Twin Creek package in this part of the thrust belt along thick Upper Jurassic salts lying beneath the Stump-Preuss sandstone and shale section. This salt interval is highly variable in thickness, thinning to as little as 7 ft (2 m) over the crest of the West Lobe structure and thickening to as much as 1200 ft (366 m) to the northwest along the back flank of the structure. Cretaceous stratigraphy is characterized by a transition from terrestrial gray-brown-red siliciclastic rocks of the Gannet Group to the increasingly marine siltstones and shales of the Bear River, Aspen, and Frontier formations (Lelek, 1982). The Late Cretaceous to Tertiary Evanston and Wasatch formations unconformably overlie the Cretaceous section, locally with pronounced angularity (Figure 5).

The oldest rocks carried on the Absaroka thrust at Anschutz Ranch East field are probably Triassic in age. Thinly bedded Ankareh siltstones and sandstones lie conformably beneath the Nugget interval and are the oldest rocks penetrated by wells in the field. As previously discussed, the Woodside, Dinwoody, and Thaynes formations are believed to be present in the core of the West Lobe fold.

TRAP

The West and East Lobes of the field are separate structural traps (Figure 5). The Nugget and Ankareh formations in the West Lobe are folded into an elongated asymmetrical anticline with a gentle backlimb (dips approximately 30°) and a steeply dipping to overturned forelimb. The East Lobe, also asymmetrical to the southeast, has a gentle backlimb (dips approximately 15°) and a structurally complex and steeply dipping forelimb. The East Lobe is more typical of a leading edge anticline in this trend with the Nugget Sandstone truncated directly against the Absaroka thrust. Both folds are believed to overlie a relatively flat to slightly westward dipping portion of the Absaroka thrust fault surface.

The West Lobe structure contains a vertical hydrocarbon column of approximately 2100 ft (640 m) (Lelek, 1983). The gas-water contact in the West Lobe is -6470 ft (-1972 m) whereas that of the East Lobe is -7485 ft (-2281 m) below mean sea level (MSL). Nearly 80% of the Nugget section penetrated above the gas-water contact is considered "net pay" from well log analysis (D. Admire, personal communication).

The top seal for the Anschutz Ranch East field is formed by the Gypsum Springs evaporite member of the Twin Creek Formation. There are several areas within the Utah-Wyoming thrust belt where the Gypsum Springs does not form an effective seal. In some of these areas, such as in the main Anschutz Ranch and Yellow Creek fields, fractured Twin Creek limestones produce hydrocarbons beneath a seal created by the Upper Jurassic salt and evaporites in the Preuss Formation.

RESERVOIR

Rock Properties

The Nugget reservoir is anisotropic and inhomogeneous with respect to fluid flow. Many of the reservoir properties are tied to the original depositional architecture and only slightly modified by subsurface physical and chemical diagenesis (Lindquist, 1983). The Nugget reservoir in the West Lobe can be effectively divided into three petrophysical zones (Figure 11). The upper two-thirds of the formation contains large-scale eolian dune sandstones and is the best reservoir rock. The lower third of the interval is characterized by smaller-scale eolian sandstones, some with water-influenced depositional textures. This lower zone comprises the poorest Nugget reservoir rock in the field.

The average matrix porosity of the Nugget reservoir rock is 12% with the highest observed porosities approaching 22%. The average reservoir matrix permeability is 10.9 md with some permeabilities ranging as high as 400 md. Whole core analyses for the porosity and permeability are shown

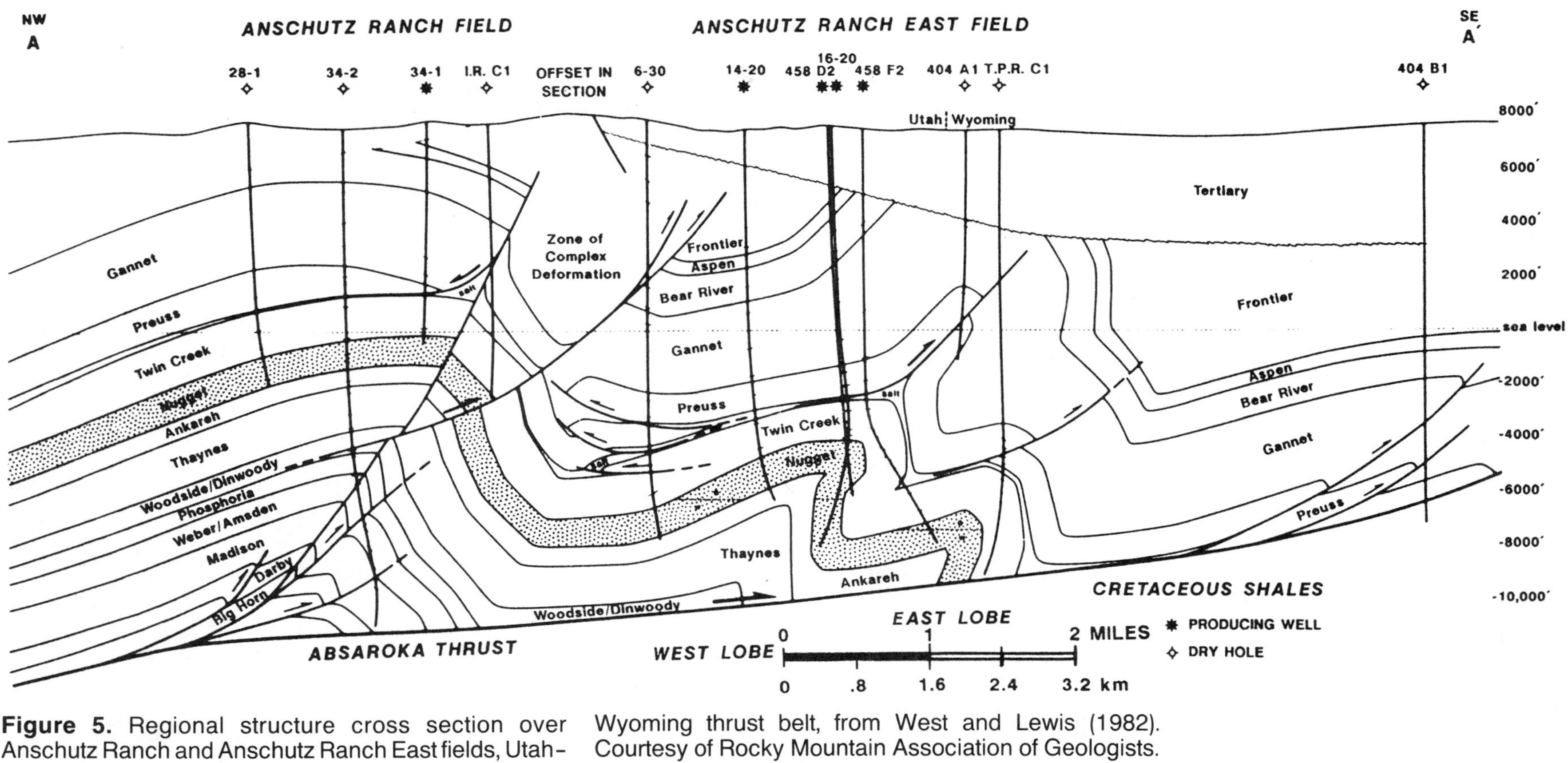

Figure 5. Regional structure cross section over Anschutz Ranch and Anschutz Ranch East fields, Utah-Wyoming thrust belt, from West and Lewis (1982). Courtesy of Rocky Mountain Association of Geologists.

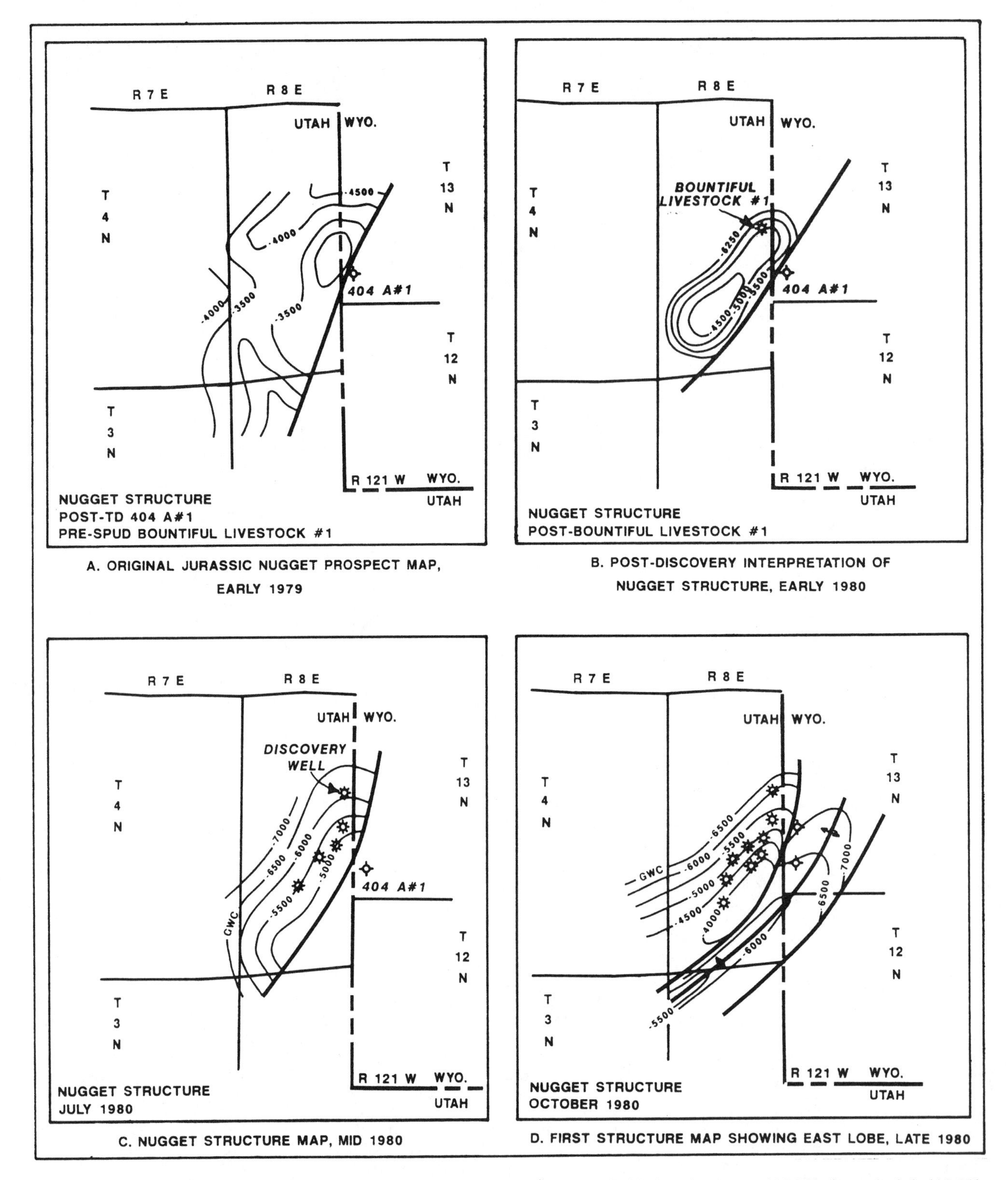

Figure 6. Historical view of major structural interpretations for the Anschutz Ranch East prospect (1979) and early field development (1980), from Lelek (1983). Courtesy of Rocky Mountain Association of Geologists.

in Figure 12. Note that the properties of this unit vary dramatically. Correlation quality of the K_{max} plot (reservoir porosity plotted against maximum permeability parallel to the bedding plane) and the K_{h90} plot (reservoir porosity vs permeability perpendicular to the bedding plane) for this data set are 31% and 51%, respectively. This high variability makes reservoir modeling predictions difficult.

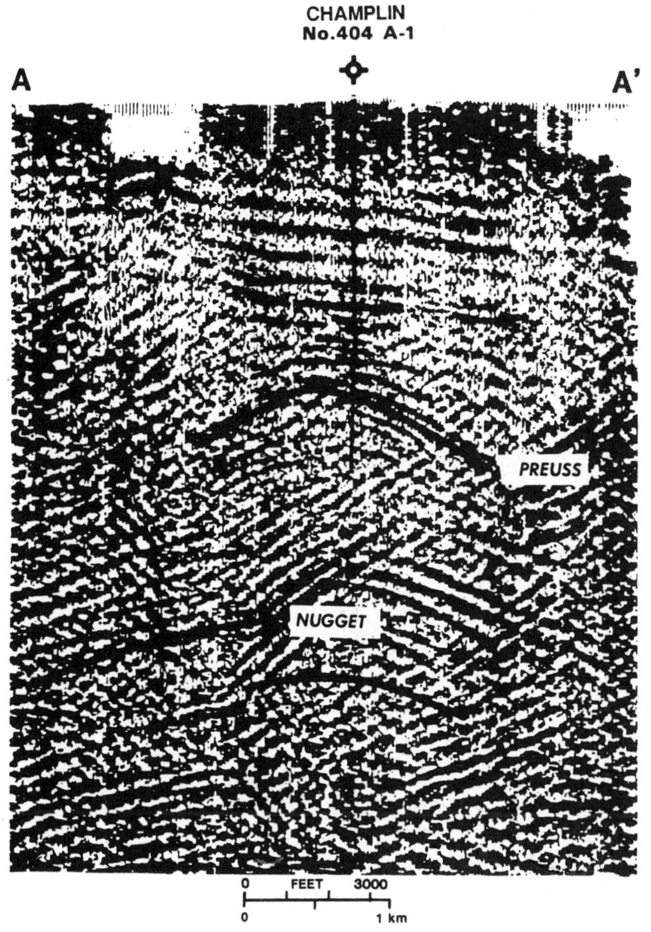

Figure 7. Unmigrated seismic section A–A′ over Anschutz Ranch East field showing the shallow Preuss structure drilled originally by the Champlin No. 404 Amoco No. 1 well, from Lelek (1982). Courtesy of Rocky Mountain Association of Geologists.

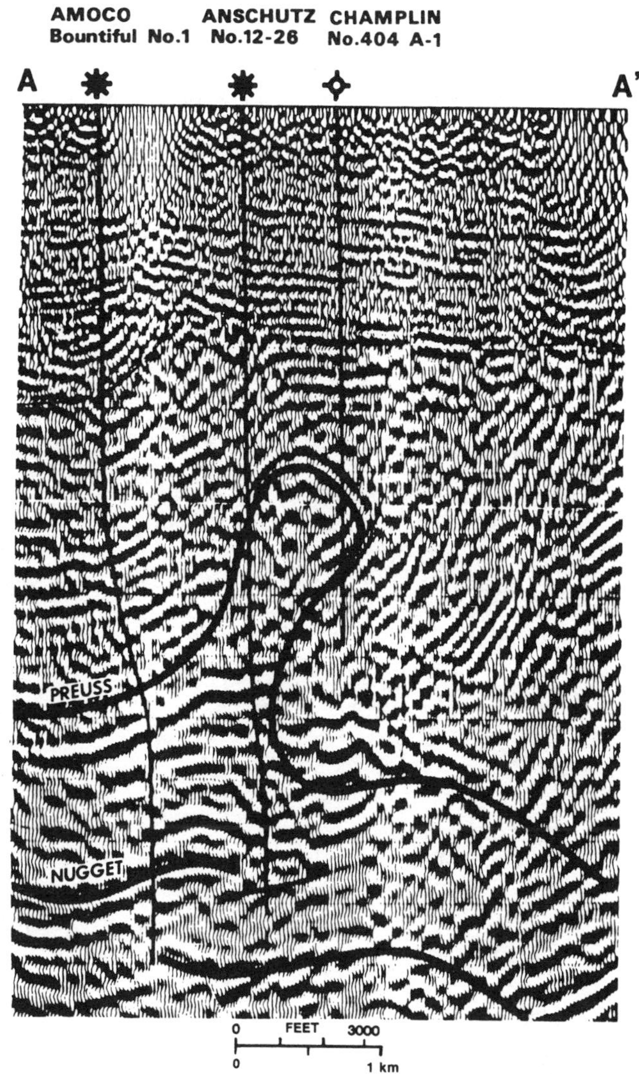

Figure 8. Depth-migrated seismic section A–A′ over Anschutz Ranch East field, generated after additional well data was obtained, from Lelek (1982). Courtesy of Rocky Mountain Association of Geologists.

Lindquist (1983) gives a detailed description of the Nugget eolian sandstones and their relationship to reservoir petrophysics. Figures 13 and 14 are reproduced from her article to show some key features of the Nugget reservoir. Typical thin sections photographed by E. D. Pittman of Amoco Research for each of the three major petrophysical zones are shown in Figures 15–17. Zone 1 (Figure 15) is characterized by laminated sandstones that are interpreted as stacked climbing wind-ripples and grainflow deposits in relatively small cross-bed sets. Zone 2 (Figure 16) is the best reservoir interval with generally higher porosities, thicker cross-bed sets, and more grainflow deposits. Zone 3 (Figure 17) is the poorest reservoir owing to its finely laminated wind-rippled and water-influenced depositional textures.

Fractures probably influence the directional permeability within the field. Core descriptions commonly identify gouge-filled fractures in Zone 1, only minimal gouge-filled fractures in Zone 2, and numerous open fractures in Zone 3. The distribution of fracture morphology is similar to that discussed by Nelson (1985), in which higher porosity sandstones commonly shear and create a low-permeability gouge. Fracturing in the lower-porosity sandstones more often creates open fractures that can enhance reservoir permeability. The fracture sets in the Nugget Sandstone at Anschutz Ranch East field are interpreted to be of dominantly tectonic, fold-related origin with generally predictable senses of motion or offset, as set forth by Nelson (1985). There is also a set of open fractures oriented at low angles to the structural axis that are thought to be related to the period of post-thrust extension. The effect of fracturing on permeability in 3D whole core measurements is shown in Figure 18.

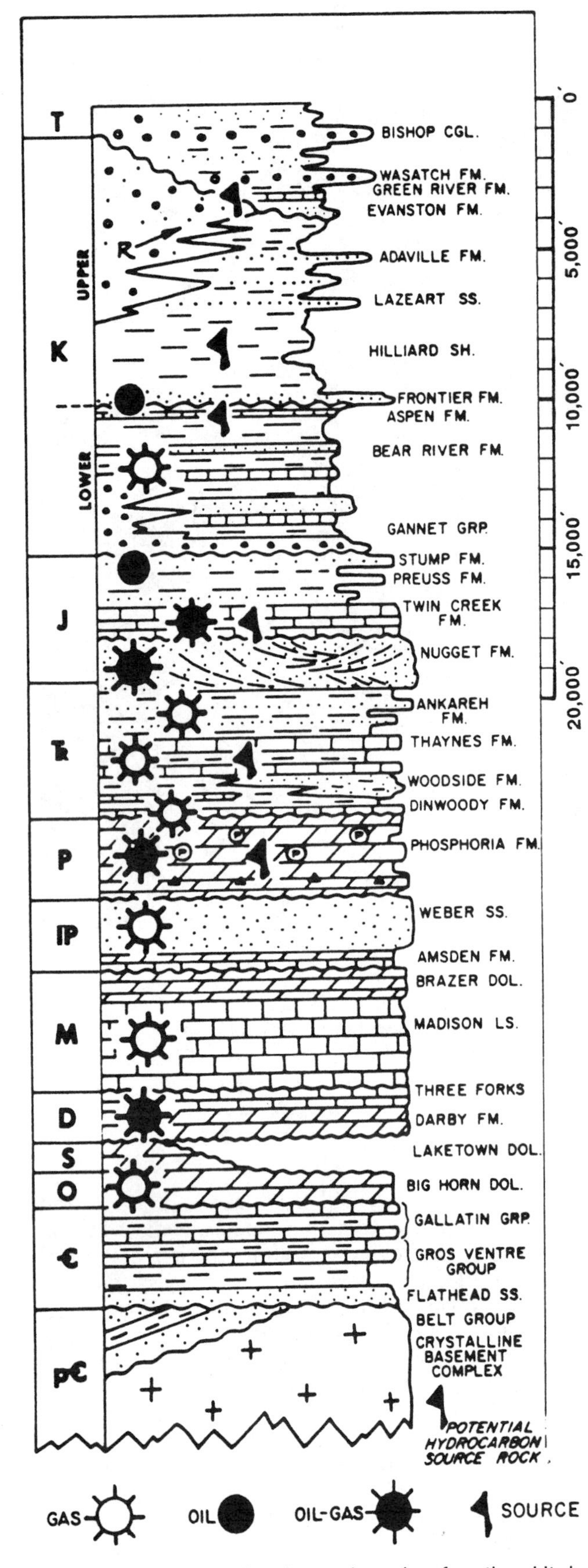

Figure 9. Generalized stratigraphy for the Utah-Wyoming thrust belt with productive and potential source rocks indicated, from Dixon (1982).

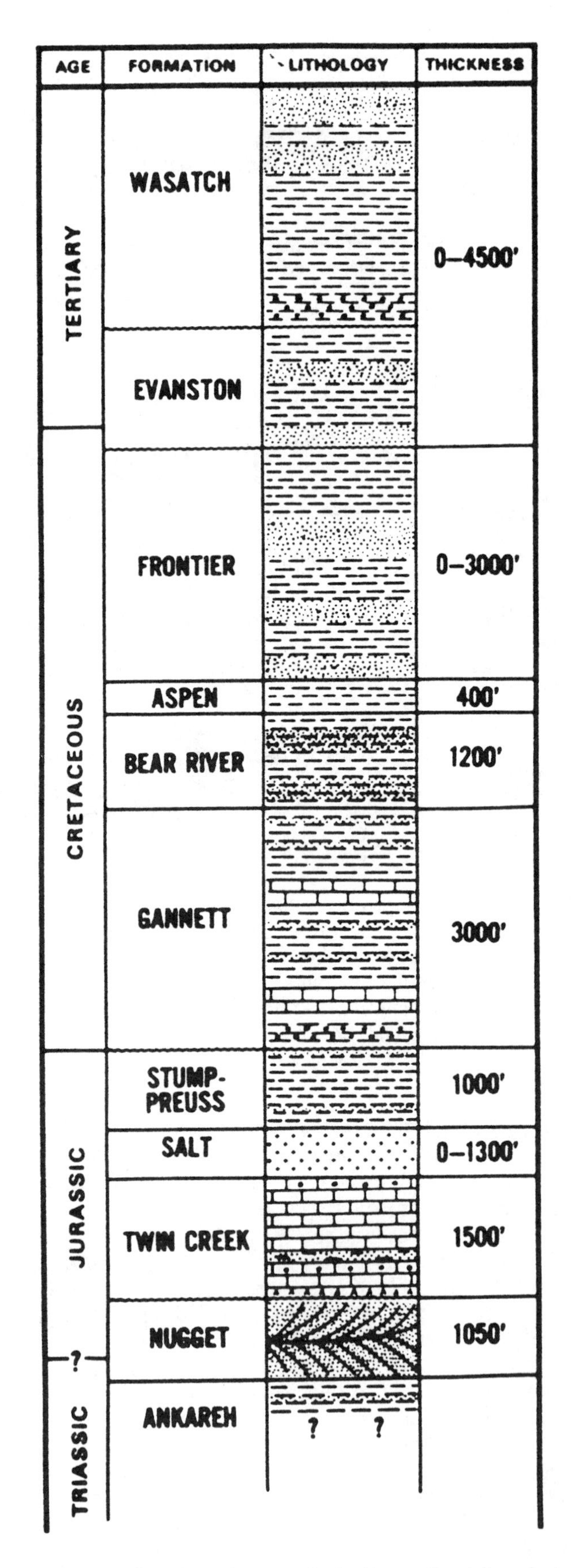

Figure 10. Stratigraphy penetrated by wells at Anschutz Ranch East field, from Lelek (1982). Courtesy of Rocky Mountain Association of Geologists.

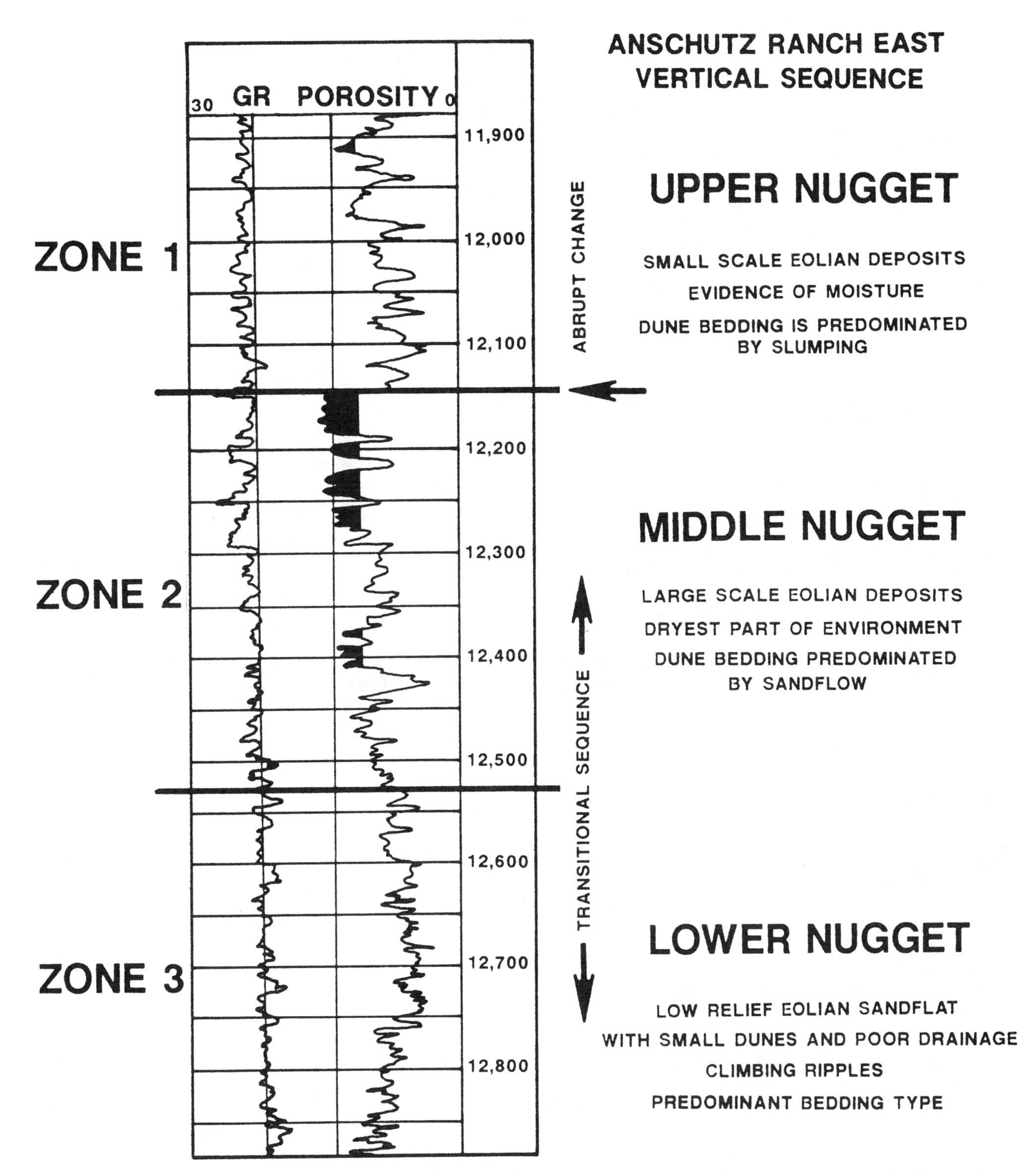

Figure 11. Typical gamma ray and calculated porosity log response showing the Three-Zone Porosity Model used at Anschutz Ranch East field. The upper two zones of the Nugget contain the best reservoir rock quality. The porosity log is shaded at values greater than 8%; the gamma ray shows the decrease in shale content, or cleanliness of Zones 1 and 2. Major depositional characteristics of each zone and the sequence transitions are shown to the right of the log. Courtesy of Norm Haskell, Amoco Research.

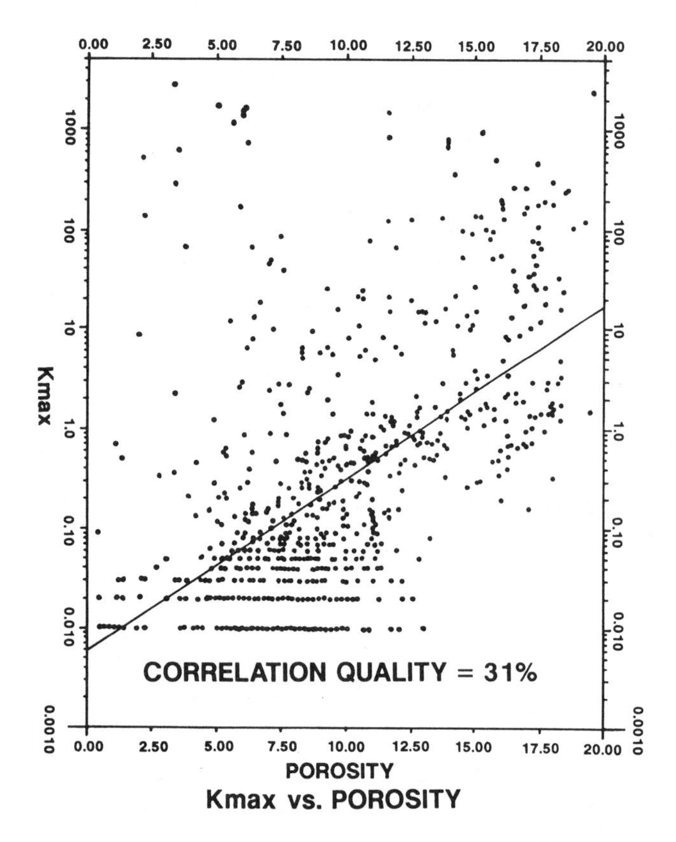

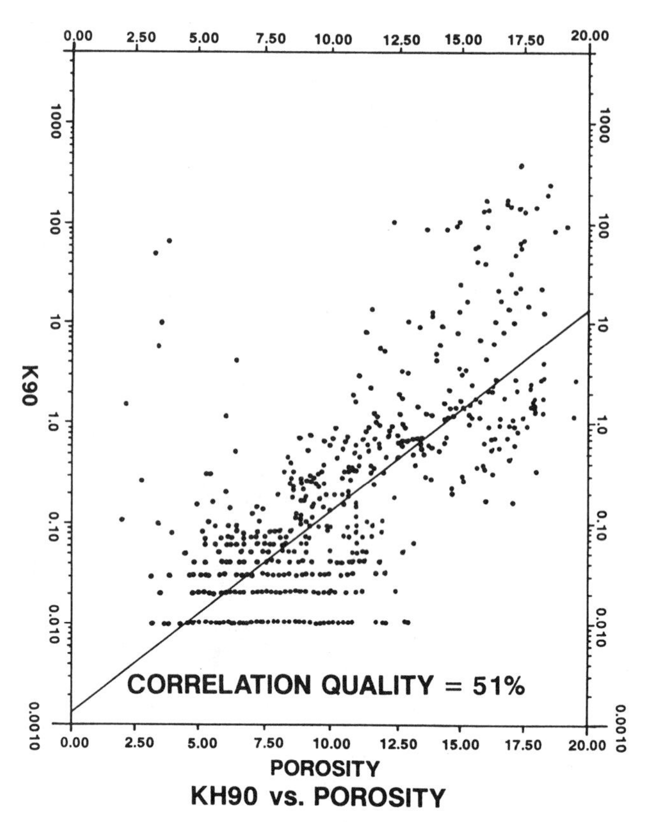

Figure 12. K_{max} and K_{h90} porosity-permeability cross plots for 3D whole core samples from two wells in the Anschutz Ranch East field, from Nelson (1985). Courtesy of Gulf Publishing Company.

Reservoir Modeling

The highly variable rock properties in the Nugget reservoir prevent the effective use of simple, single-tank models with homogeneous average properties. Therefore, early production modeling efforts at the West Lobe focused on multilayered and brick-type modeling schemes (Figures 19 and 20). Homogeneous layered modeling did not accurately match the existing well production histories, probably because of the changing directional properties of dune sandstones, reservoir fracturing, and the simplified extrapolation of well bore petrophysical properties to the interwell region. The difficulties encountered with the layered modeling schemes led to more sophisticated domain type modeling that allowed greater characterization and compartmentalization of various heterogeneous reservoir properties (Figure 20).

One particular modeling effort by Kleinsteiber et al. (1983) characterizes the reservoir as stacked "bricks" with each brick having its own set of reservoir properties. Nine categories based on sandstone bedding and fracture relationships were developed from the detailed evaluation of core samples (Table 1). The porosity and permeability cross-plots for one category (Sample Type 7) are shown in Figure 21. Note that the K_{max} and K_{h90} correlation quality has increased to 62% and 66%, respectively, compared to the lower values discussed previously for the entire sample population (see Figure 12). The calculated petrophysical properties for each of the nine categories were assigned to the bricks on a field-wide basis according to the appropriate sample type averages. From this data set and model design, 2D reservoir evaluations were run to better understand the effect of heterogeneity within the eolian Nugget reservoir (Kleinsteiber et al., 1983). Accurate reservoir modeling continues to be studied and is considered a key factor for the cost-effective production and reservoir maintenance during the remainder of the field's productive life.

Reservoir Fluid

The West Lobe of Anschutz Ranch East field contains a rich retrograde gas-condensate with a dew point only 230 psia (1586 kPa) below the initial reservoir pressure of 5310 psia (36,610 kPa) at a datum of -5324 ft (-1623 m) below MSL. The critical nature of this fluid was discovered through the analysis of recombination samples from the discovery well. Thirty additional recombination samples were gathered from 20 intervals in 8 wells, establishing a compositional gradient over the hydrocarbon column of 2100 ft (640 m). The initial gas-oil ratio

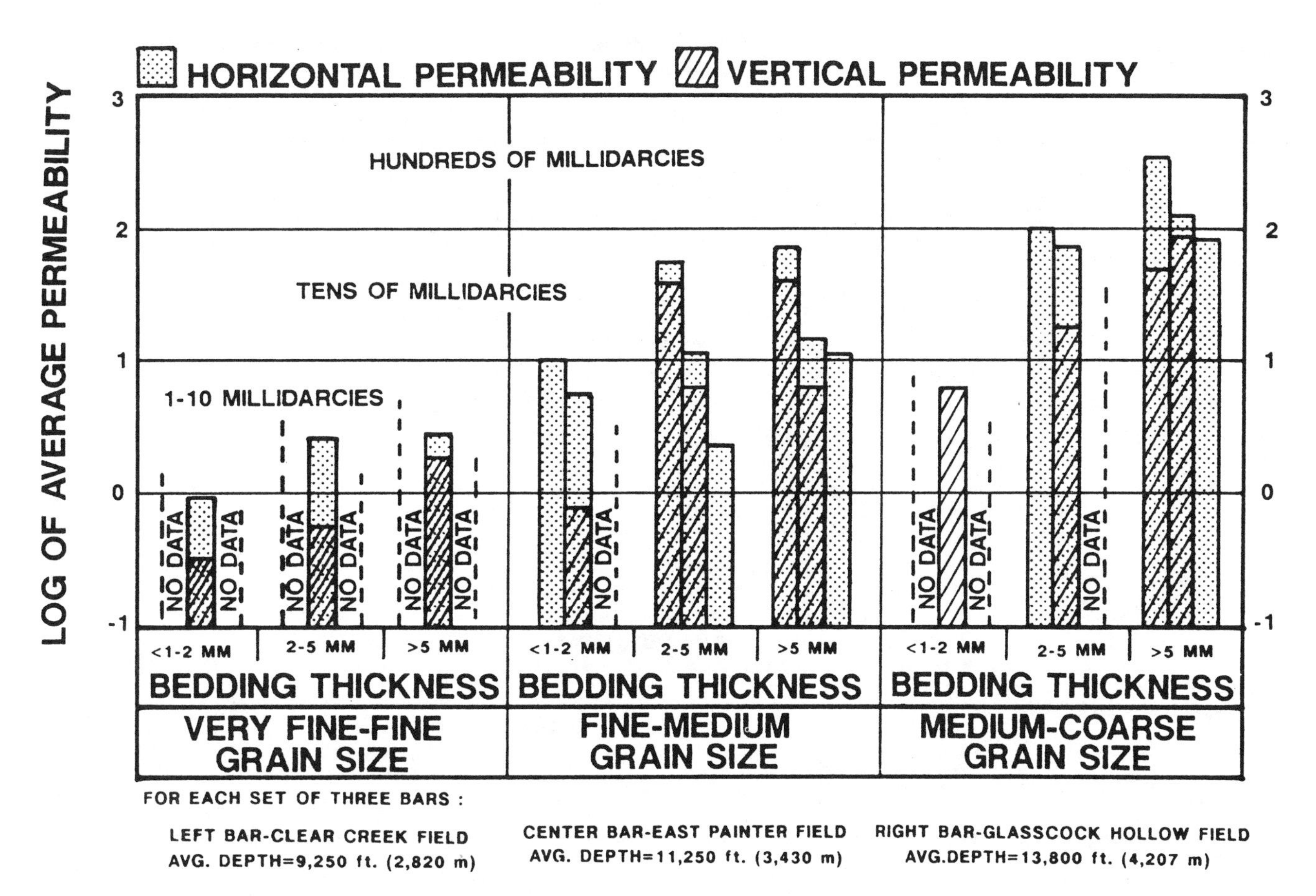

Figure 13. Generalized relationships between rock textural elements and reservoir permeability for eolian Nugget Sandstone in several Wyoming thrust belt fields in trend with the Anschutz Ranch East field. From Lindquist (1983). Copyright by SPE, 1983.

(GOR) ranges from 6,000 scf/bbl (1060 m^3/m^3) at the crest of the structure to 3500 scf/bbl (620 m^3/m^3) at the gas-water contact (Figure 22). The average initial GOR was 4500 scf/bbl (800 m^3/m^3).

Experimental constant composition volumetric expansion (CCVE) studies performed on the samples by Kleinsteiber et al. (1983) indicate up to 40% liquid dropout of the hydrocarbon pore volume and a maximum of 27% condensate saturation upon pressure depletion below the dew point pressure of 5080 psia (35,025 kPa) (Figure 23). The degradation of relative permeability and the subsequent difficulties of fluid movement caused by liquid dropout in the reservoir could result in the loss of up to 75% of the recoverable reserves in place. Full pressure maintenance, using nitrogen to make up the reservoir voidage, was therefore chosen as the most efficient and cost-effective program to best produce the field.

The pressure and dew point relationships for the East Lobe of the field are not as severe as for the West Lobe (Figure 23). The lower degree of pressure sensitivity and the smaller field volume (approximately one-tenth the volume of the West Lobe) do not justify a full pressure maintenance program at the East Lobe.

SOURCES

Potential source rocks in the area include the Cretaceous Frontier, Bear River, and Aspen shales, and the Permian Phosphoria Formation. The source of hydrocarbons for most, if not all, of the sweet gas and liquids trend is from Cretaceous shales. Hydrocarbons in the Anschutz Ranch East field appear to have come from the Cretaceous shales lying beneath the Absaroka thrust, presumably in direct contact with the Nugget reservoir at the East Lobe. It is likely that the West Lobe reservoir filled with gas spilled from the East Lobe Nugget along the highest areas of the intervening syncline (R. Mueller, personal communication, 1989). Expulsion and migration of the hydrocarbons probably began during

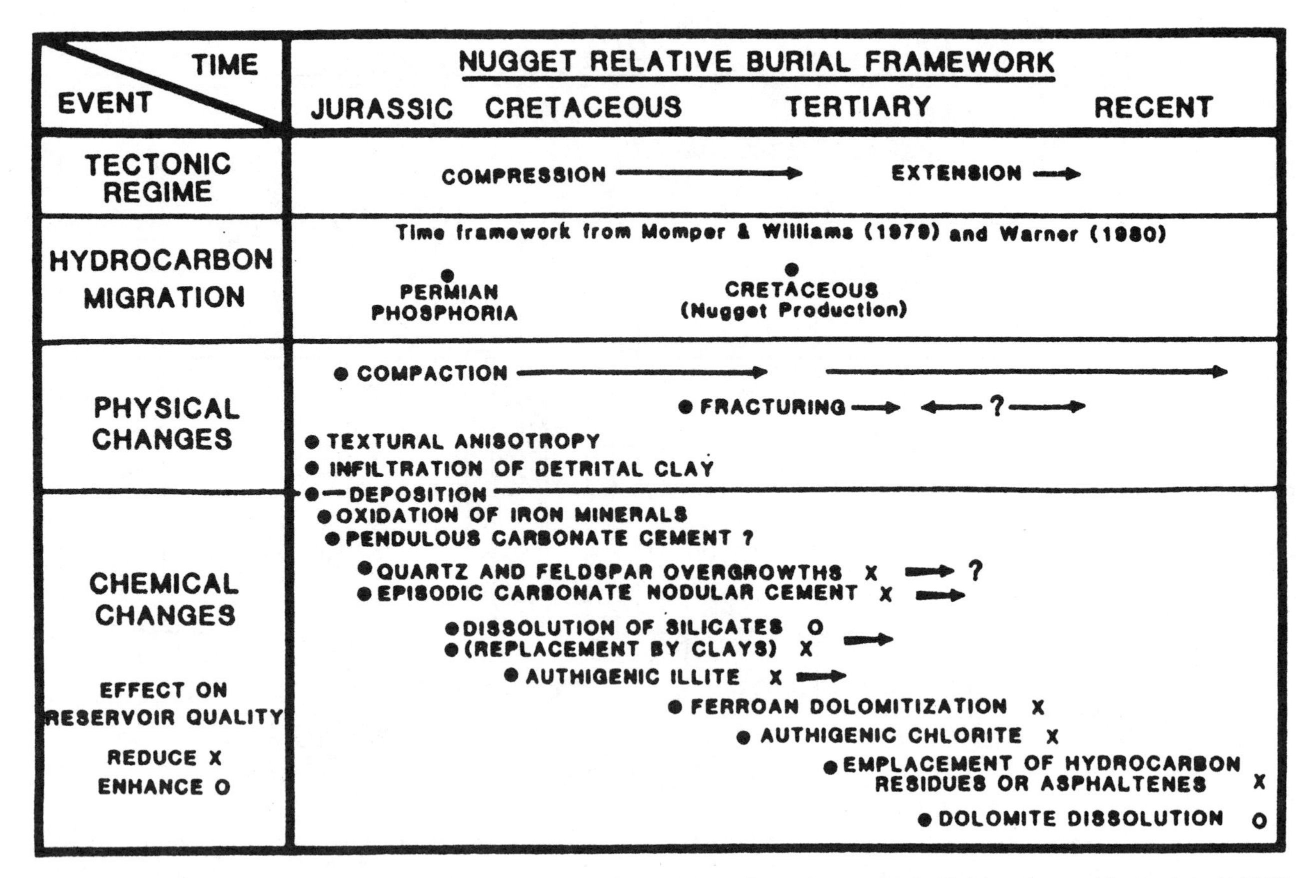

Figure 14. Generalized paragenetic sequence for Nugget Sandstone from the study of several Utah-Wyoming thrust belt fields, from Lindquist (1983). Copyright by SPE, 1983.

the Late Cretaceous and continued into the Tertiary Period (J. Williams, personal communication, 1989).

Source materials present within the Cretaceous shales are of mixed type II and type III kerogen. Current source rock analyses yield relatively low average values for the total organic carbon (TOC) content in the Frontier (0.7%) and Aspen (1.2%) formations. Values are higher for Bear River outcrop samples where the average TOC is 2.8%. Expulsion and migration of the hydrocarbons is presumably dependent on the existence of relatively rich zones of organic content. TOC values as high as 2.0%, 2.7%, and 9.3% have been measured for the Frontier, Aspen, and Bear River formations, respectively (TOC values and source rock interpretations from J. Williams, Amoco Research Center, 1989).

Early pyrobitumen staining in parts of the reservoir may indicate more than one phase of hydrocarbon migration. Two hypotheses have been proposed to explain this early staining (J. Williams, personal communication, 1989). One hypothesis, based on complex asphaltene relationships, suggests that an early Phosphoria oil migrated through the Nugget reservoir prior to the final structural configuration. An alternate hypothesis involves a two-stage migration of Cretaceous oils through and into the structure with a residual cracked or complexed product left behind as staining after the first stage of hydrocarbon migration.

EXPLORATION CONCEPTS

The Anschutz Ranch East prospect was part of the logical progression of exploration in the thrust belt area. Increased predictability of the structural geometries likely to be encountered and an increased ability to process and interpret thrust belt seismic sections helped make this play feasible. However, it is doubtful that the early seismic and geological evidence would have led to the drilling of the Anschutz Ranch East discovery well had it not been for the high success rate of the earlier (1975–1979) thrust belt exploration and the high price for oil in the 1970s. Anschutz Ranch East field was the fourteenth field discovered in the trend, approximately midway through the primary phase of exploration activity in the area.

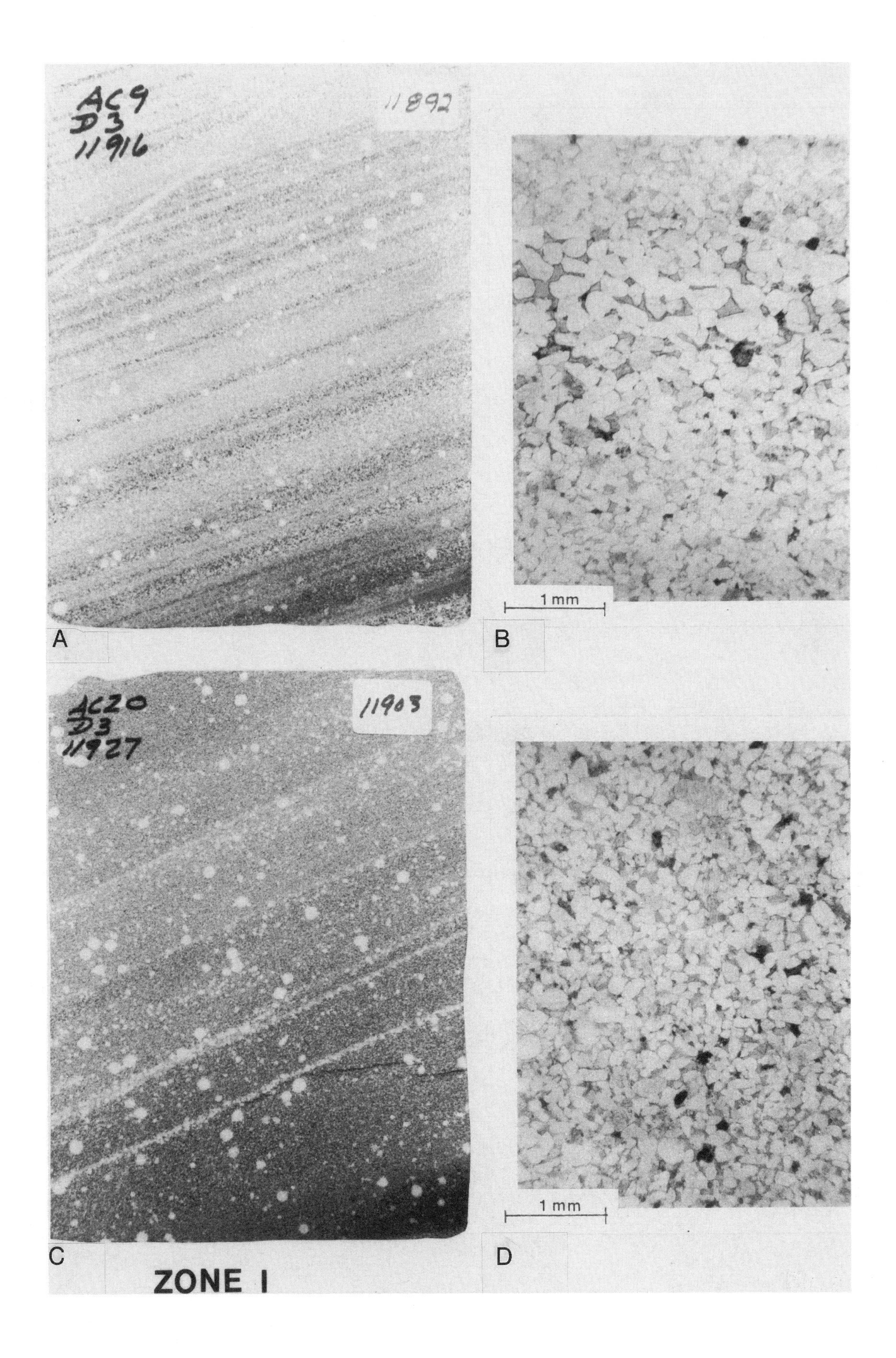

Figure 15. Typical X-radiographs and photomicrographs of thin sections for Nugget Reservoir Zone 1, Anschutz Ranch East field. Courtesy of E. D. Pittman, Amoco Research Center.

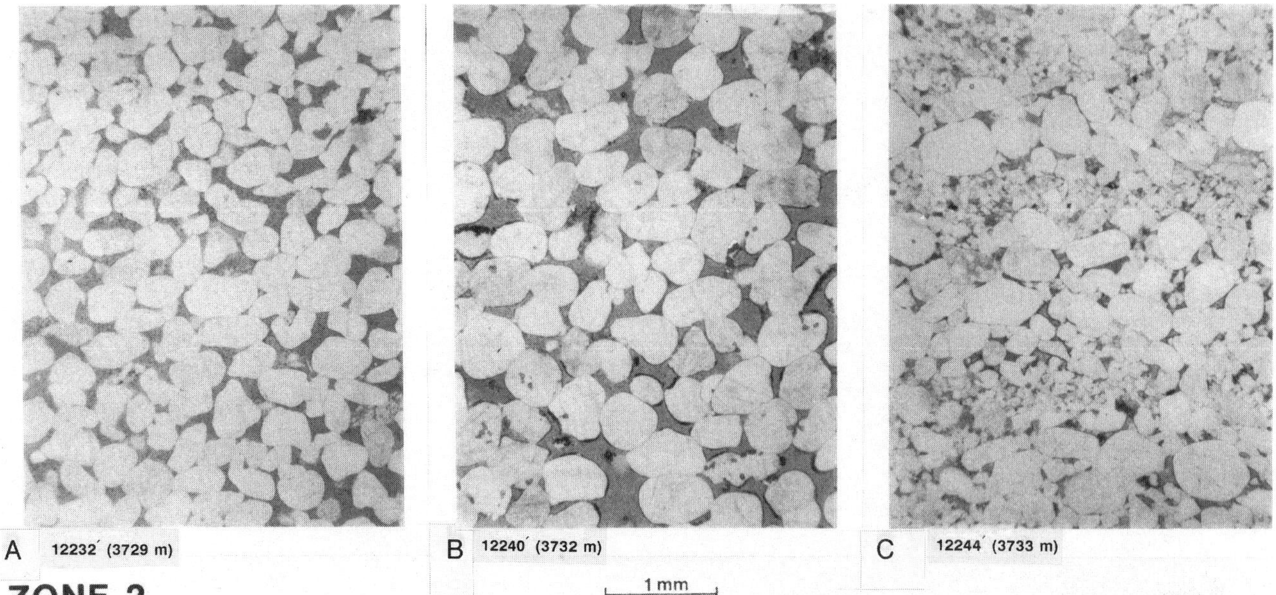

Figure 16. Typical photomicrographs of thin sections for Nugget Reservoir Zone 2, Anschutz Ranch East field. Courtesy of E. D. Pittman, Amoco Research Center.

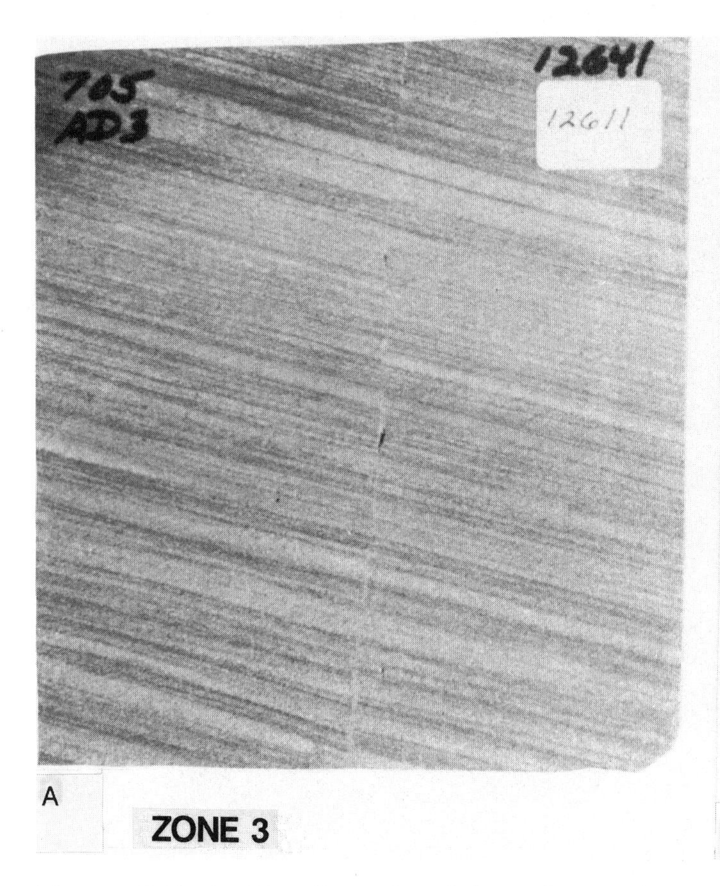

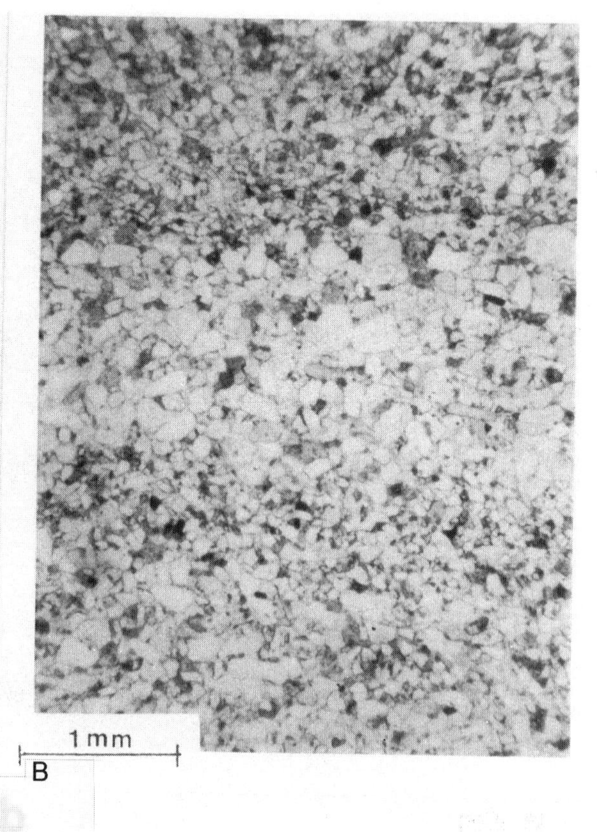

Figure 17. Typical X-radiograph and photomicrograph for Nugget Reservoir Zone 3, Anschutz Ranch East field. Courtesy of E. D. Pittman, Amoco Research Center.

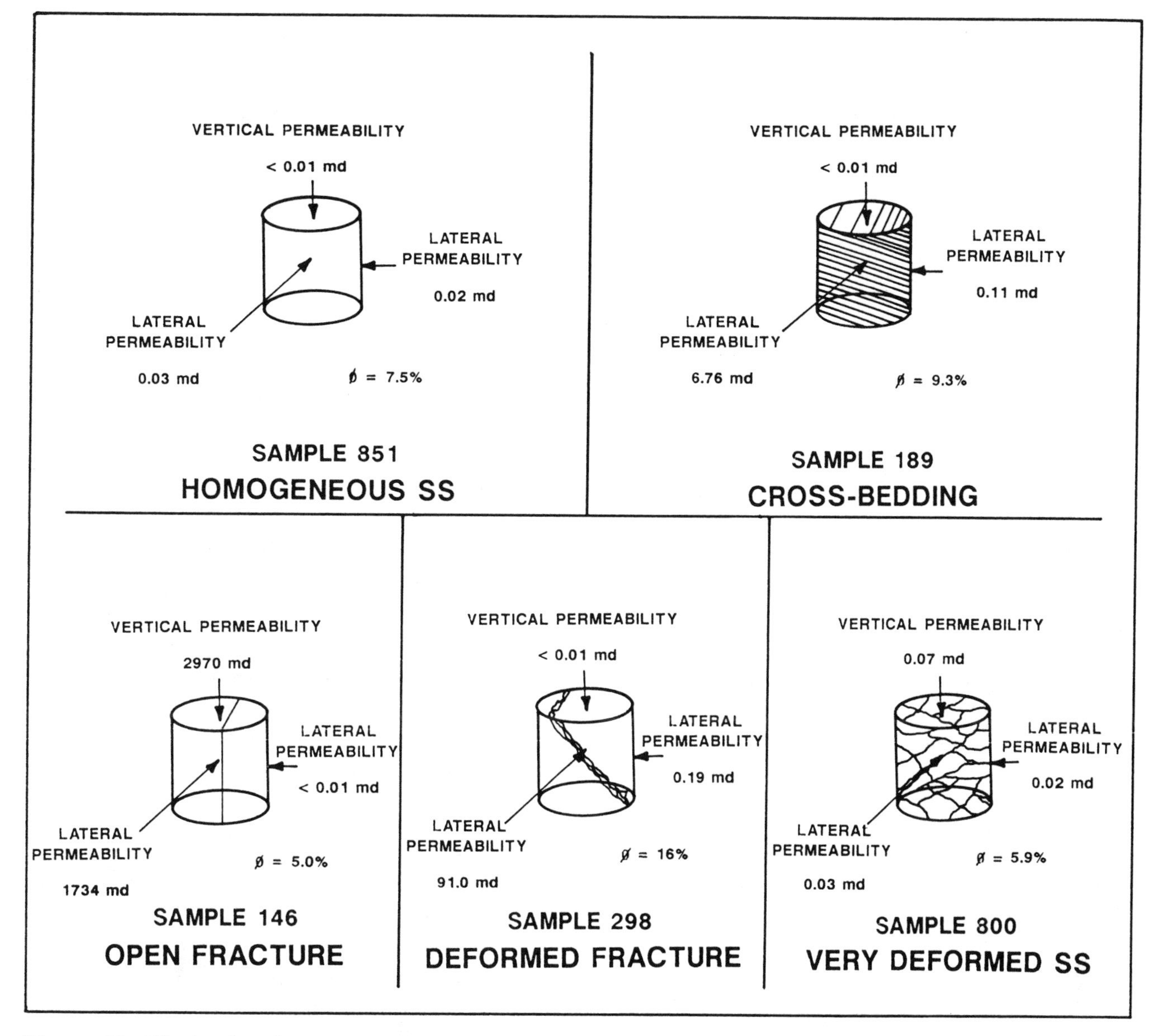

Figure 18. Effects of various natural fractures on 3D whole-core permeability in the Champlin 458, Amoco D-3 well from Anschutz Ranch East field, from Nelson (1985). Courtesy of Gulf Publishing Company.

DEVELOPMENT CONCEPTS

Fields containing retrograde gas-condensate fluids, such as Anschutz Ranch East, should be developed and managed carefully, particularly when the reservoirs are geologically complex. Detailed reservoir evaluations require extensive core, electric log, and fluid analyses. The relatively small difference between the original reservoir pressure and the condensate dew point at the West Lobe presents an additional challenge to provide an efficient and cost-effective plan of depletion. Models used to evaluate and design effective development programs are costly and time consuming. The processes required for the effective modeling needed to design an optimum recovery program for a field of this kind is economic for only the largest of fields, and the increased costs and time for such an endeavor should be anticipated, if possible, in the early stages of exploration analysis.

ACKNOWLEDGMENTS

The authors wish to thank Amoco Production Company, The Anschutz Corporation, Mobil Exploration Inc., Union Pacific Resources Company, BWAB Inc., Pan-Canadian Petroleum Co., and Chevron U.S.A. Inc. for permission to publish the material included in this paper. We gratefully acknowledge the work, ideas, and helpful revisions of numerous Amoco and Working Interest Partner employees who are not cited herein. Much of the

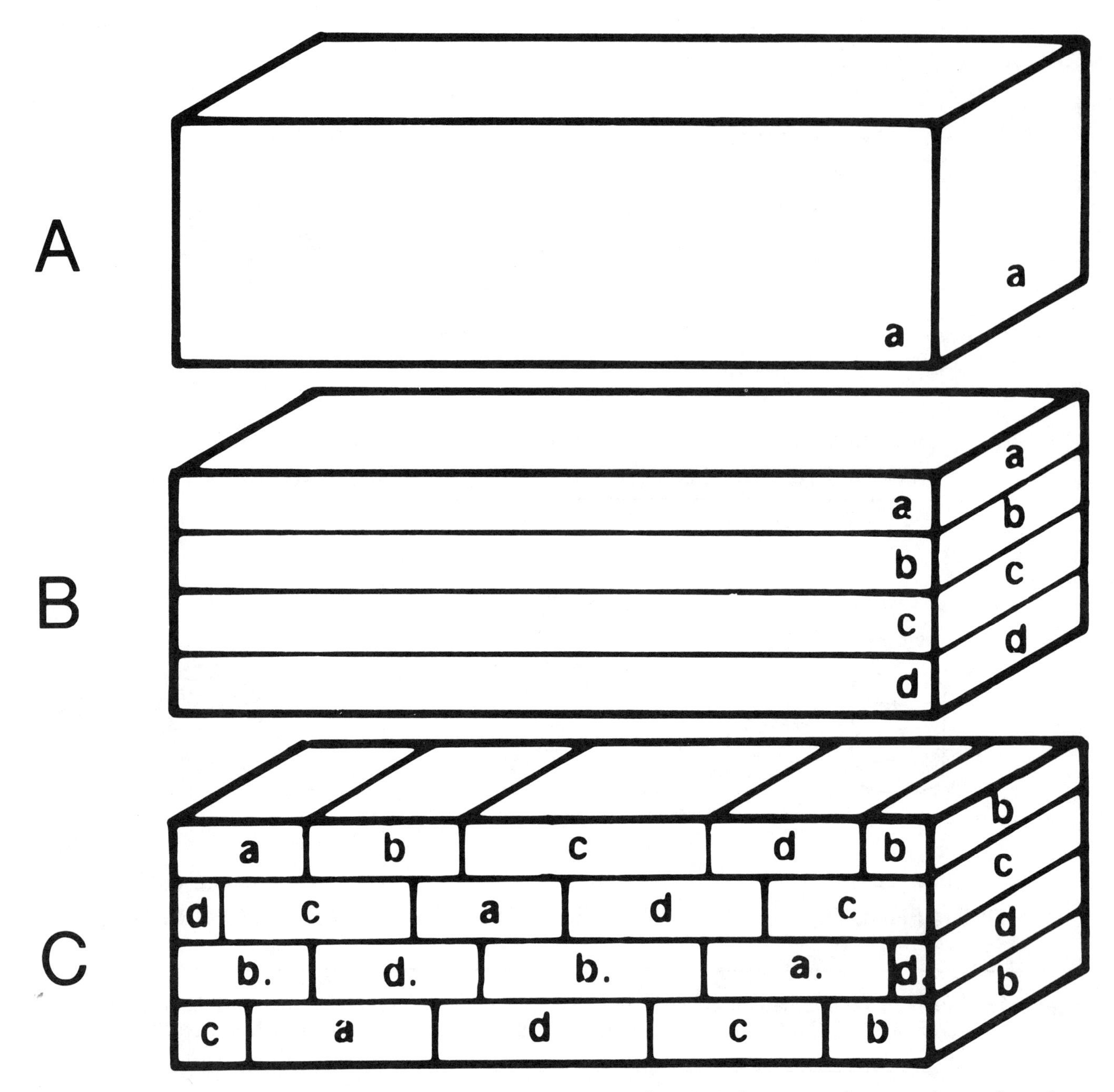

Figure 19. Typical reservoir characterization models evaluated for use at Anschutz Ranch East field. (A) Homogeneous model in which properties vary in a small and statistically random fashion. (B) Layered model in which properties vary primarily in the vertical dimension while being relatively continuous in the lateral direction. (C) Domain model in which properties are distributed in domains of similar reservoir properties and varying in both vertical and horizontal dimensions. Homogeneous- and layered-type modeling has proven to be inadequate to explain the production characteristics at Anschutz Ranch East field. From Nelson (1985). Courtesy of Gulf Publishing Company.

numerical data and illustrations presented here are from previously published sources and are carefully referenced. We thank the following publishers and professional societies for their permission to reproduce key illustrations for this article: Gulf Publishing Company, Rocky Mountain Association of Geologists, and Society of Petroleum Engineers.

REFERENCES

Dixon, J. S., 1982, Regional Structural Synthesis, Wyoming Salient of Western Overthrust Belt: American Association of Petroleum Geologists Bulletin, v. 66, n. 10, p. 1560–1580.

Frank, J. R., S. Cluff, and J. M. Bauman, 1982, Painter Reservoir, East Painter Reservoir, and Clear Creek Fields, Uinta Co., Wyo.: Rocky Mountain Association of Geologists, Geol. Stud.

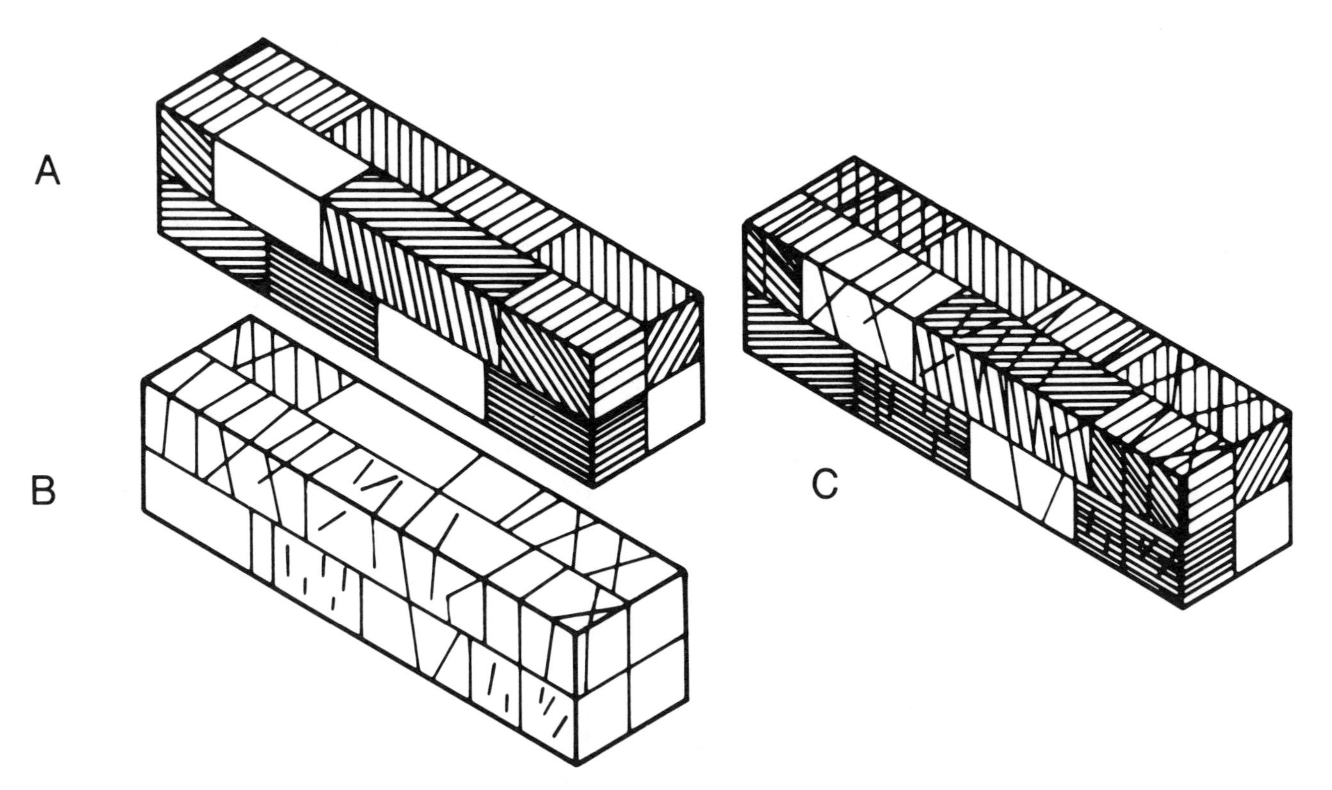

Figure 20. Generalized components of a domain type model for use at Anschutz Ranch East field. (A) Idealized slice of an eolian sandstone reservoir broken up into units of cross-bedded and nonbedded material of various cross-bed orientations. (B) Possible fracture distribution within the reservoir. (C) Complexity of a generalized reservoir model shown by the combination of bedding and fracture characteristics that result in various heterogeneous reservoir domains. Courtesy of Gulf Publishing Company.

Table 1. A sample type classification for Nugget Sandstone at Anschutz Ranch East.

1	No bedding, no fractures
2	No bedding, fractures $< 45^\circ$
3	No bedding, fractures $> 45^\circ$
4	Bedding $< 20^\circ$, no fractures
5	Bedding $< 20^\circ$, fractures $< 45^\circ$
6	Bedding $< 20^\circ$, fractures $> 45^\circ$
7	Bedding $> 20^\circ$, no fractures
8	Bedding $> 20^\circ$, fractures $< 45^\circ$
9	Bedding $> 20^\circ$, fractures $> 45^\circ$

90° is parallel to core axis
0° is perpendicular to core axis

Cordilleran Thrust Belt Field Conf., Denver, Guidebook, v. 2, p. 601-611.

Holt, C. A., and J. B. Johnson, 1985, Method for drilling moving salt formations: drilling and undereaming at the same time: SPE Paper SPE/IADC 13488, Presented at 1985 Drilling Conf., New Orleans, LA, March 6-8, 1985, 12 p.

Kleinsteiber, S. W., D. D. Wendschlag, and J. W. Calvin, 1983, Study for development of a plan of depletion in a rich gas condensate reservoir, Anschutz Ranch East Unit, Summit County, Utah, Uinta County, Wyoming: SPE Paper 12042, Presented at 58th. Annual Tech. Conf., San Francisco, Oct. 5-8, 1983, 11 p.

Lelek, J. J., 1982, Anschutz Ranch East Field, Northeast Utah and Southwest Wyoming: Rocky Mountain Association of Geologists, Geol. Stud. Cordilleran Thrust Belt Field Conf., Denver, Oct. 24-27, 1982, Guidebook, v. 2, p. 619-631.

Lelek J. J., 1983, Geologic factors affecting reservoir analysis, Anschutz Ranch East Field, Utah and Wyoming: Journal of Petroleum Technology, v. 35, n. 9, p. 1539-1545.

Lindquist, S. J., 1983, Nugget Formation reservoir characteristics affecting production in the Overthrust belt of southwestern Wyoming: Journal of Petroleum Technology, July 1983, p. 1355-1365.

Metcalfe, R. S., J. L. Vogel, and R. W. Morris, 1985, Compositional gradient in the Anschutz Ranch East Field: SPE Paper 14412, Presented at 60th Ann. Tech. Conf., Las Vegas, NV, Sept. 22-25, 1985, 11 p.

Nelson, R. A., 1985, Geologic analysis of naturally fractured reservoirs: Houston, Gulf Publishing Co., 320 p.

Pollock, C. B., and C. O. Bennet, 1983, Eight-well interference test in the Anschutz Ranch East Field: SPE Paper 11968, Presented at 58th Ann. Tech. Conf., San Francisco, CA, Oct. 5-8, 1985, 12 p.

Royse, F. Jr., M. A. Warner, and D. L. Reese, 1975, Thrust belt structural geometry and related stratigraphic problems, Wyoming-Idaho-Northern Utah: Rocky Mountain Association of Geologists Symposium, 1975, p. 41-54.

West, J., and H. Lewis, 1982, Structure and palinspastic reconstruction of the Absaroka Thrust, Anschutz Ranch Area, Utah and Wyoming: Rocky Mountain Association of Geologists, Geol. Stud. Cordilleran Thrust Belt Field Conference, Denver, Oct. 24-27, 1982, Guidebook, v. 2, p. 633-639.

Wiltschko, D. V., and J. A. Dorr, 1983, Timing of deformation in Overthrust belt and foreland of Idaho, Wyoming, and Utah: American Association of Petroleum Geologists Bulletin, v. 67, n. 8, p. 1304-1322.

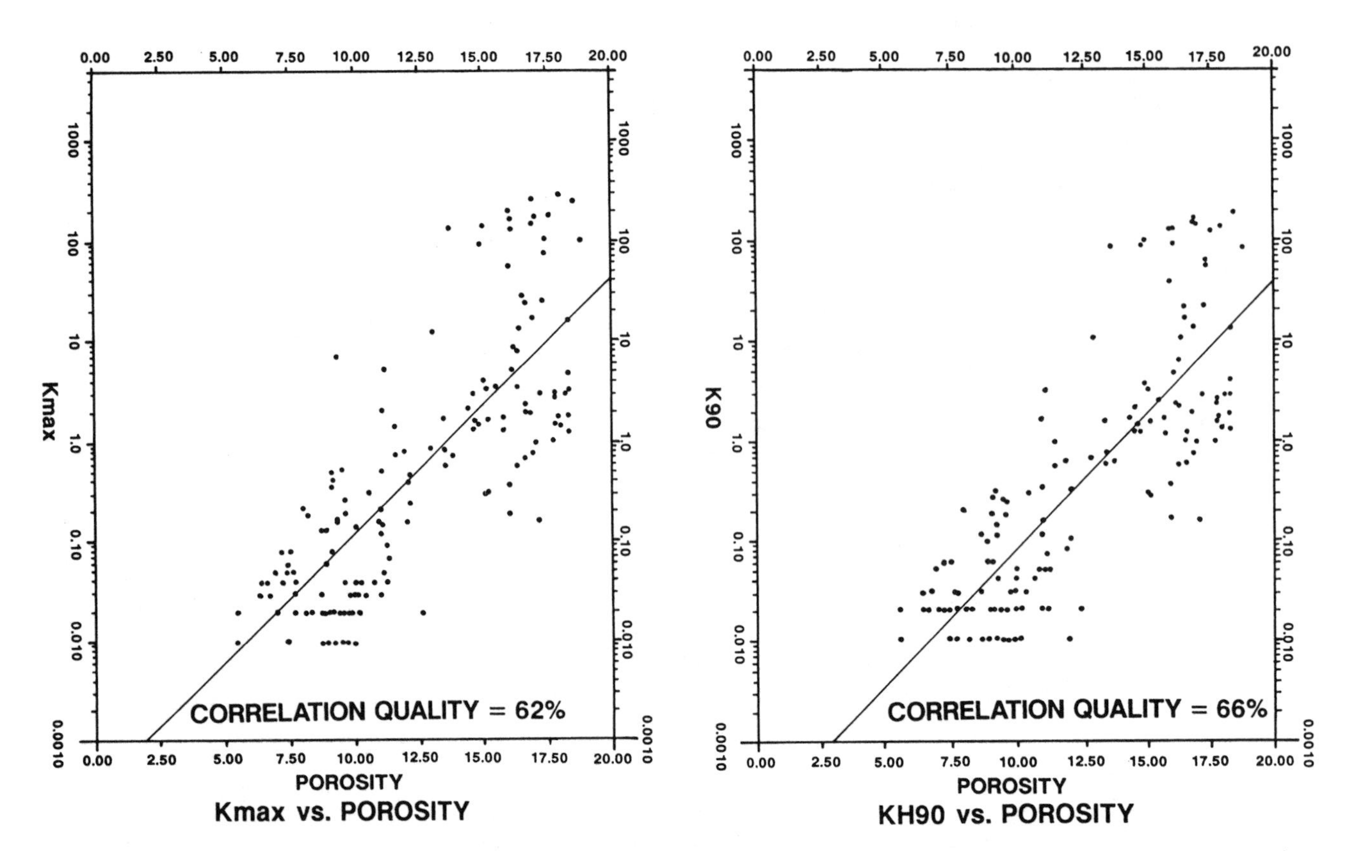

Figure 21. K_{max} and K_{h90} porosity-permeability cross plots on 3D whole core samples for sample type 7 data (bedding dip greater than 20°, unfractured). Correlation quality has risen to 62% for K_{max} and 66% for K_{h90} as compared to nonsegregated sample results shown in Figure 12. From Nelson (1985). Courtesy of Gulf Publishing Company.

Appendix 1. Field Description

Field name .... *Anschutz Ranch East (West Lobe and East Lobe)*

Ultimate recoverable reserves .... *800 MMBOE (1988)*

Field location:

- **Country** .... *United States*
- **State** .... *Utah and Wyoming*
- **Basin/Province** .... *Western Overthrust*

Field discovery:

- **Year first pay discovered** .... *West Lobe, Jurassic Nugget Sandstone 1979*
East Lobe, Jurassic Nugget Sandstone 1981
- **Year second pay discovered** .... *NA*
- **Third pay** .... *NA*

Discovery well name and general location:

First pay

West Lobe: Amoco Bountiful Livestock No. 1, NW NW Sec. 26-4N-8E, Summit Co., Utah

East Lobe: The Anschutz Corp. Anschutz Ranch East No. 12-26W, Sec. 26-13N-121W, Uinta Co., Wyoming

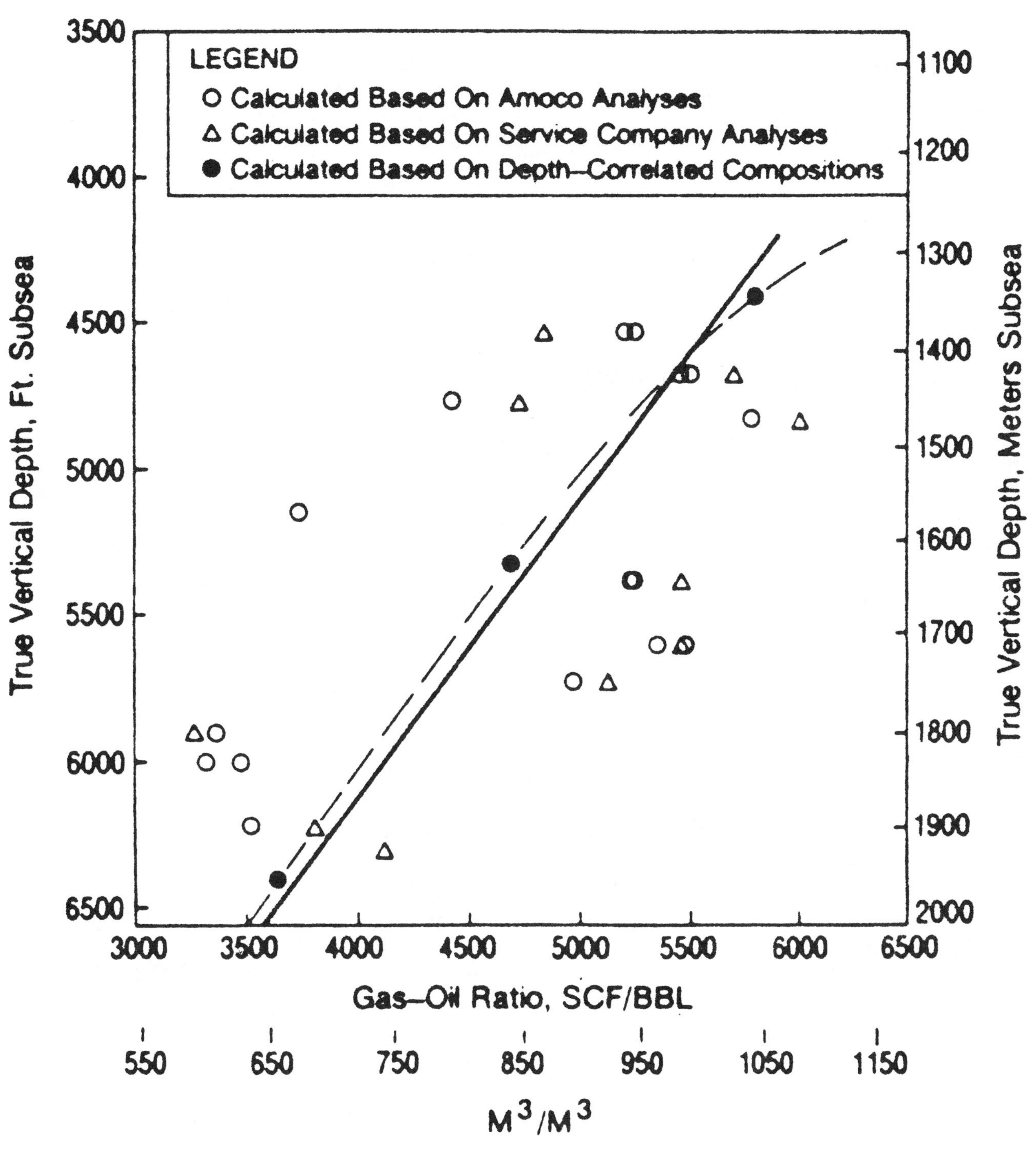

Figure 22. Reported gas oil ratio vs depth as based on Amoco and Service Company analyses, from Metcalfe et al. (1985). Courtesy of Society of Petroleum Engineers, Copyright SPE Paper 14412.

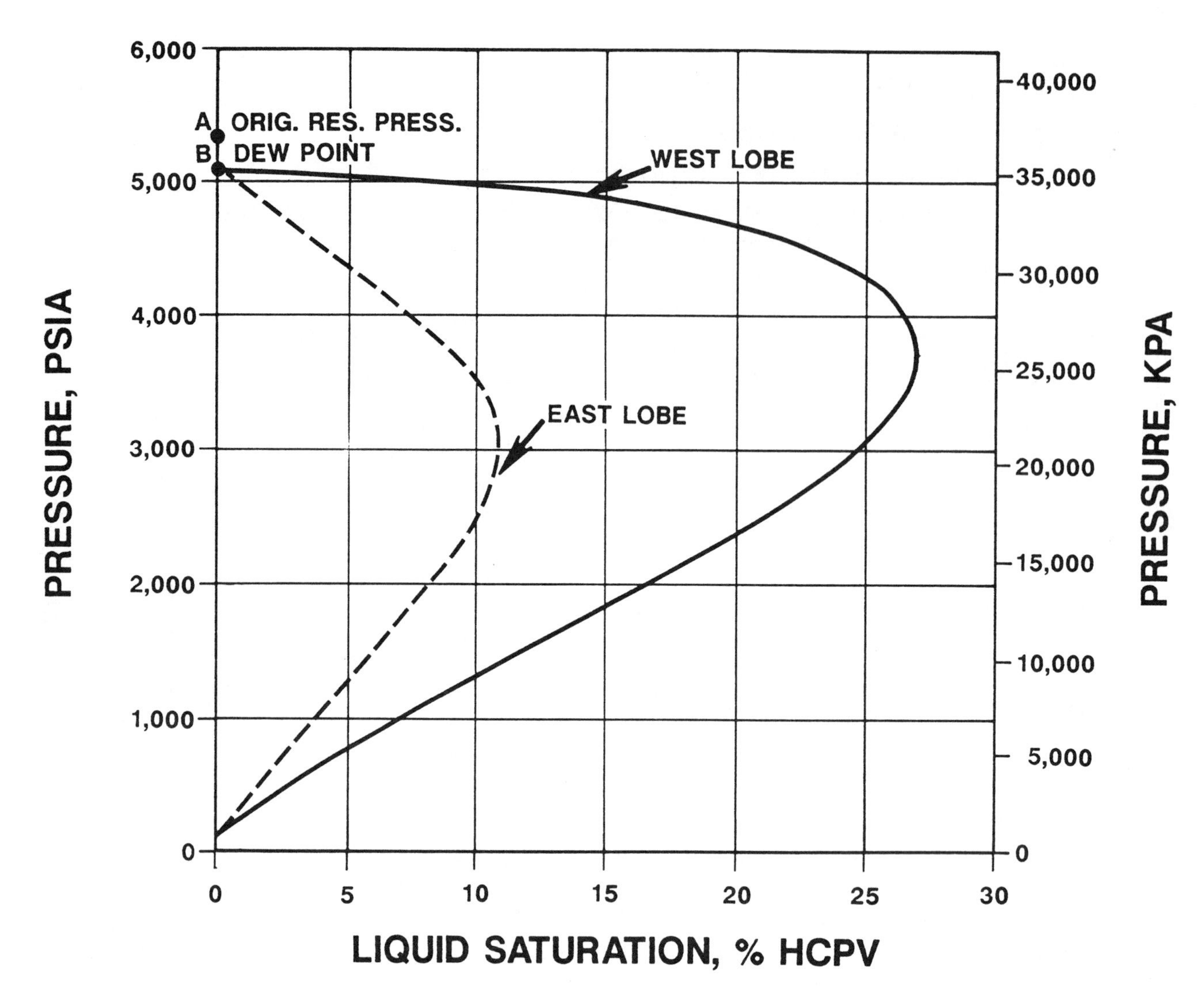

Figure 23. Typical liquid dropout curves for the depletion of Anschutz Ranch East field gas-condensates below their dew-point pressure. If pressure is allowed to decline from the original reservoir pressure at point A to the dew-point pressure at point B and below, the liquid saturation or percentage of the total liquid condensate "dropped out" is shown by the constant composition volumetric expansion (CCVE) curve measured from Anschutz Ranch East recombined hydrocarbon fluid samples (after Kleinsteiber et al., 1983). Courtesy of Society of Petroleum Engineers, Copyright SPE Paper 12042.

Discovery well operator .... *West Lobe: Amoco Production Company; East Lobe: The Anschutz Corporation*

Second pay

Third pay

IP in barrels per day and/or cubic feet or cubic meters per day:

First pay .... *West Lobe: 1054 BCPD, 4053 MCFD, 28 BWPD; East Lobe: 141 BCPD; 750 MCFD, 96 BWPD*

All other zones with shows of oil and gas in the field:

Age	Formation	Type of Show
None		

Geologic concept leading to discovery and method or methods used to delineate prospect, e.g., surface geology, subsurface geology, seeps, magnetic data, gravity data, seismic data, seismic refraction, nontechnical:

Subsurface leading-edge anticline in thrust belt delineated by 2D reflection seismic data and shallow well data, later defined by additional well data and 3D seismic.

Structure:

Province/basin type (see St. John, Bally, and Klemme, 1984)

Western Overthrust: A Bally folded belt related to Type A subduction (41).

Tectonic history

Stable shelf deposition leading to a major marine transgression in the Cretaceous. Laramide thrusting shortened the section from west to east from 30 to 60 mi (46-97 km) in the Cretaceous and early Tertiary. A secondary Tertiary extension or relaxation event followed.

Regional structure

The Utah-Wyoming thrust belt is comprised of 4 major and numerous minor north-northeast-trending thrust faults producing an imbricate fold and thrust belt. Hydrocarbon production is from leading-edge anticlines.

Local structure

Two large, en echelon, leading-edge, overturned folds associated with the Absaroka thrust. The folds are steeply dipping and overturned on the east. The shallower backlimb dips 25-30° to the west-northwest. The larger West Lobe is 2000 ft (610 m) shallower than the East Lobe.

Trap

Trap type(s) .... *Anticlinal-structural*

Basin stratigraphy (major stratigraphic intervals from surface to deepest penetration in field):

Chronostratigraphy	Formation	MSL Depth in ft (m)	Drill Depth in ft (m)
Late Cretaceous	*Aspen*	*3300 (1006)*	*4700 (1434)*
Late Cretaceous	*Bear River*	*2600 (793)*	*5400 (1647)*
Late Cretaceous	*Gannett*	*1000 (305)*	*7000 (2135)*
Jurassic	*Preuss/Stump*	*-1800 (-549)*	*9800 (2989)*
Jurassic	*Twin Creek*	*-3100 (-946)*	*11,100 (3386)*
Triassic-Jurassic	*Nugget*	*-4500 (-1373)*	*12,500 (3813)*
Triassic	*Ankareh*	*-5600 (-1708)*	*13,600 (4148)*

Location of well in field

Reservoir characteristics:

Number of reservoirs *1*

Formations .... *Nugget sandstone*

Ages .... *Triassic-Jurassic*

Depths to tops of reservoirs .... *West Lobe, 12,500 ft (3813 m); East Lobe, 14,700 ft (4484 m)*

Gross thickness (top to bottom of producing interval) .... *1050 ft (320 m) avg.*

Net thickness—total thickness of producing zones

Average .... *80% of the gross thickness*

Maximum

Average

Maximum

Lithology

Porosity type .... *Intergranular with minor fracture porosity*

Average porosity .... *Matrix 12% (2-22%)*

Average permeability .... *Matrix 10.9 md (0.01-400 md)*

Seals:

- **Upper**
  - **Formation, fault, or other feature** ...... *Gypsum Spring member of the Twin Creek*
  - **Lithology** ...... *Evaporites (anhydrite) in carbonate*
- **Lateral**
  - **Formation, fault, or other feature** ...... *Structural fold—anticlinal closure*
  - **Lithology**

Source:

- **Formation and age** ...... *Subthrust Cretaceous (Frontier, Bear River, or Aspen)*
- **Lithology** ...... *Shales*
- **Average total organic carbon (TOC)** ...... *Frontier, 0.7%; Bear River, 2.8%; Aspen, 1.2%*
- **Maximum TOC** ...... *Frontier, 2.0%; Bear River, 9.3%; Aspen, 2.7%*
- **Kerogen type (I, II, or III)** ...... *Mixed type II and type III*
- **Vitrinite reflectance (maturation)** ...... *NA*
- **Time of hydrocarbon expulsion** ...... *Late Cretaceous (L. Campanian) to Tertiary*
- **Present depth to top of source** ...... *NA*
- **Thickness** ...... *NA*
- **Potential yield** ...... *NA*

Appendix 2. Production Data

Field name ...... *Anschutz Ranch East (West Lobe and East Lobe)*

Field size:

- **Proved acres** ...... *4500 ac (1821 ha)*
- **Number of wells all years** *55*
- **Current number of wells** *55*
- **Well spacing** ...... *80 ac (198 ha)*
- **Ultimate recoverable** ...... *NA*
- **Cumulative production (to 12/31/88)** ...... *81.71 MMbbl condensate; 32.13 MMbbl NGL; gas currently reinjected*
- **Annual production (1988)** ...... *12.2 MMbbl cond. + 11.6 MMbbl NGL*
- **Present decline rate** ...... *28% per year*
  - **Initial decline rate (1983)** ...... *0.5, 25 psi (172 kPa)*
  - **Overall decline rate** ...... *28% per year*
- **Annual water production** ...... *500 Mbbl*
- **In place, total reserves** ...... *800 MMbbl*
- **In place, per acre-foot** ...... *4100 bbl/ac-ft (HCPV)*
- **Primary recovery** ...... *NA*
- **Secondary recovery** ...... *NA*
- **Enhanced recovery** ...... *Miscible flood*
- **Cumulative water production (12/31/88)** ...... *2.7 MMbbl*

Drilling and casing practices:

- **Amount of surface casing set** ...... *6000 ft (1830 m)*
- **Casing program (current)**

300 ft (91.5 m), 20-in. (50.8 cm) conductor
6000 ft (1830 m), 13⅜-in. (34 cm) surface casing
12,000 ft (3660 m), 9⅝-in. (24.4 × 27 cm) intr casing
1200 ft (366 m), 7-in. (17.8 cm) production liner

Drilling mud

Freshwater mud in surface hole, saturated mixed-salt system, intermediate hole, freshwater mud in production hole; related fluid loss inverted emulsion oil-based mud (1985 on)

Bit program .... *NA*
High pressure zones .... *None*

Completion practices:

Interval(s) perforated .... *Avg. 700 ft (214 m) perforated (full face completions)*
Well treatment .... *None (perf. underbalanced with tubing conveyed perforating—TCP guns)*

Formation evaluation:

Logging suites .... *Generally FDC-CNL, DIL-SFL, sonic, dipmeter, and occasionally FIL*
Testing practices .... *NA*
Mud logging techniques .... *NA*

Oil characteristics:

Type .... *Rich volatile retrograde gas condensate*
(Tissot and Welte Classification in "Petroleum Formation and Occurrence," 1984, Springer-Verlag, p. 419)
API gravity .... *52° but varies with depth*
Base .... *NA*
Initial GOR .... *4500:1 (3200:1 to 6000:1)*
Sulfur, wt% .... *Negligible*
Viscosity, SUS .... *NA*
Pour point .... *NA*
Gas-oil distillate .... *Gas heating value, ≈ 1200 BTU/SCr; gas composition, NA*

Field characteristics:

Average elevation .... *8000 ft (7100-9000 ft); 2440 m (2166-2745 m)*
Initial pressure .... *West Lobe, 5310 psia at -5324 ft ss (36,612 kPa at -1624 m ss)*
Present pressure .... *At or above 5310 psia (36,612 kPa)*
Pressure gradient .... *0.18 psi/ft (0.37 kPa/m) for reservoir fluid; 0.44 psi/ft (0.93 kPa/m) for formation water*
Temperature .... *210°F (98.9°C)*
Geothermal gradient .... *Normal*
Drive .... *Full pressure maintenance*
Oil column thickness .... *2100 ft (640 m)*
Gas-water contact .... *West Lobe, -6250 ft (-1906 m) ss; East Lobe, -7485 ft (-2283 m) ss*
Connate water .... *NA*
Water salinity, TDS .... *NA*
Resistivity of water .... *0.15 ohm-m at 70°F*
Bulk volume water (%) .... *NA*
Dew point .... *West Lobe, 5080 psia at -5324 ft (35,027 kPa at -1624 m) ss*

Transportation method and market for oil and gas:

Mountain Fuel gas pipeline; Amoco and Anschutz condensate pipelines; Mapco NGL pipeline.

Wilson Creek Field—U.S.A.
Piceance Basin, Northern Colorado

DONALD S. STONE
Independent Geologist
Littleton, Colorado

FIELD CLASSIFICATION

BASIN: Piceance Creek
BASIN TYPE: Foredeep
RESERVOIR ROCK TYPE: Sandstone and Limestone
RESERVOIR ENVIRONMENT OF DEPOSITION: Eolian, Shoreline, and Shallow Marine
RESERVOIR AGE: Jurassic and Pennsylvanian
PETROLEUM TYPE: Oil, Gas and Condensate
TRAP TYPE: Thrusted Anticline

LOCATION

The Wilson Creek field is located along the north edge of Rio Blanco County, about 10 mi (16 km) north of Meeker in northwestern Colorado (Figure 1), U.S.A. Wilson Creek (originally called Devil's Hole) dome forms the southeastern culmination of the northwest-trending Danforth Hills anticline, which is the northeastern boundary of the Piceance Creek basin. A strong structural saddle, associated with an east-west, oblique-slip fault at depth, separates the Wilson Creek closure from the White River Uplift (Grand Hogback) to the south.

Northwest of the Wilson Creek field, the Danforth Hills anticline contains several small, multipay oil and gas fields. Other structurally controlled oil fields occur within a 25 mi (40 km) radius; but within Colorado, Wilson Creek field is second only to the giant Rangely field (Figure 1) in the size of its oil reserves.

The topography over the field is extremely rugged (Figure 2), with elevations ranging from 7400 ft to 8500 ft (2256 m to 5844 m) and at discovery Wilson Creek had the distinction of being the highest elevation oil field in the United States. Some 240 million bbl of in-place oil is held in the Jurassic Salt Wash and Entrada sandstones. The productive area covers approximately 3900 ac (16 km^2). It is expected that 100 million bbl of oil and oil equivalent will be recovered from the Wilson Creek field if the price of oil remains near or above $20/bbl (adjusted for inflation).

HISTORY

Pre-Discovery

The Wilson Creek dome is a prominent surface feature (Figure 3), and its potential as an oil and gas trap was recognized early (Gale, 1910; Hancock and Eby, 1930). The first test well on the dome was drilled with cable tools in 1919 by Richmond Petroleum Company but was abandoned at 2200 ft (671 m) in the Cretaceous Mancos Shale after encountering only minor oil and gas shows (Sears, 1925). A second cable tool well was drilled into the Niobrara Shale Member of the Mancos Formation at 4826 ft (1471 m) in 1921 without economic success. In 1926, the Lower Cretaceous Dakota sandstone was penetrated in a Texas production Company well located in the northwest-southeast of sec. 27, T3N, R94W (Figure 4). The well was abandoned at 5913 ft (1802 m) in the Brushy Basin Member of the Morrison Formation, as the overlying, objective Dakota sandstone was found to be water bearing (Hunt, 1938, p. 192).

Discovery

The Unit 1 discovery well for the field was completed in February 1938 in the Jurassic "basal Morrison" or Salt Wash Member sandstone reservoir in the southeast-northwest of sec. 35, T3N, R94W (Figure 4), on the east edge of the productive area.

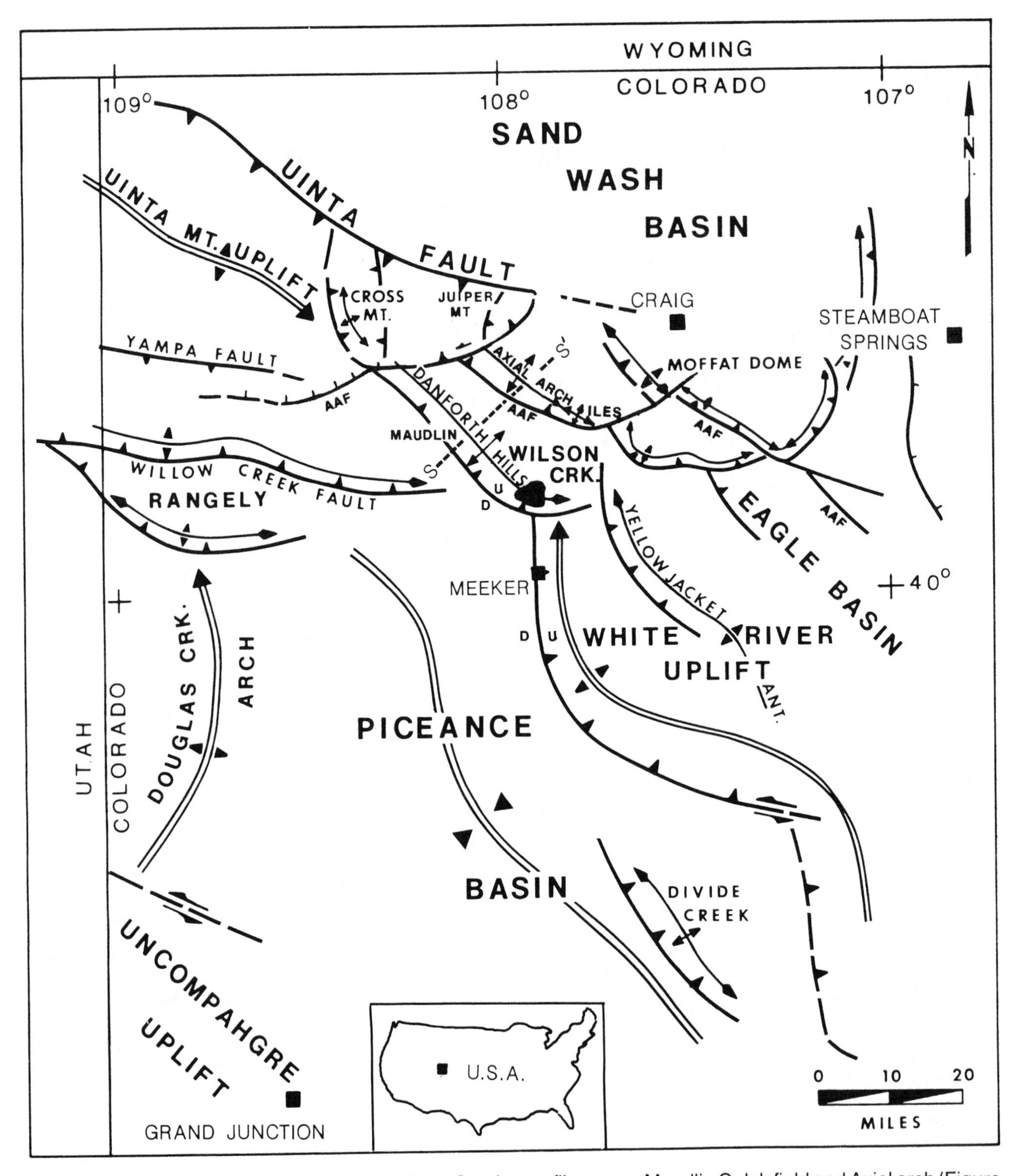

Figure 1. Index map, northwest Colorado. Wilson Creek field is black area near middle of map; S-S′ is seismic profile across Maudlin Gulch field and Axial arch (Figure 13); AAF is Ancestral Axial fault zone.

Figure 2. High-altitude photograph (USGS NHAP) of Wilson Creek field (outlined) and surrounding area. The Danforth Hills drainage divide trends northwest through Wilson Creek field.

Initial production was 490 BOPD from the interval 6664–6707 ft (2031–2044 m) but the underlying Entrada sandstone was water bearing at this location. The well was a joint venture of the Texas Company (Texaco) and the California Company (Chevron) and was drilled to 6918 ft (2109 m).

Oil was discovered in the Jurassic "Sundance" or Entrada sandstone in 1941 in the Texaco-California Company No. 5 Unit, C northeast-northeast of sec. 34, T3N, R94W. This well was completed for 300 BOPD. In 1957, there were 27 producing wells completed in the Salt Wash zone and 18 in the underlying Entrada zone. Only three dry holes were drilled. By 1964, there were 45 producing wells, but the number of active producers declined again to 27 in 1976 due primarily to water encroachment (Jones and Murray, 1976). There were only 13 wells still producing at the end of December 1988. Production for the year 1988 was 233,367 bbl including the single Minturn well (Figures 5 and 6). Gas injection was started at Wilson Creek in May 1946 and water injection in January 1959, with nine (maximum) water injection wells in the Salt Wash zone and three water disposal wells in the Entrada zone.

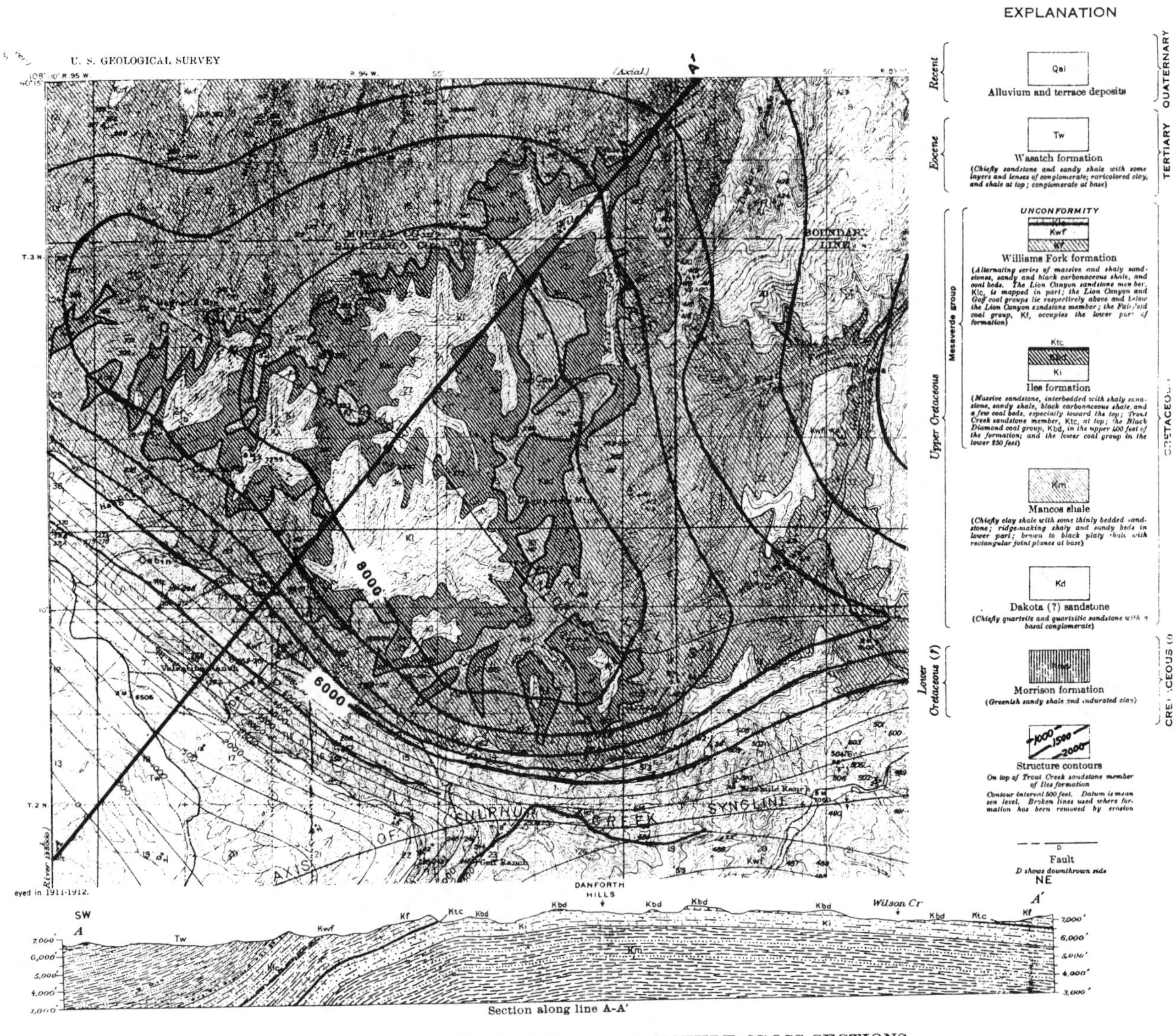

TOPOGRAPHIC AND GEOLOGIC MAP AND STRUCTURE CROSS SECTIONS OF MEEKER QUADRANGLE, COLORADO

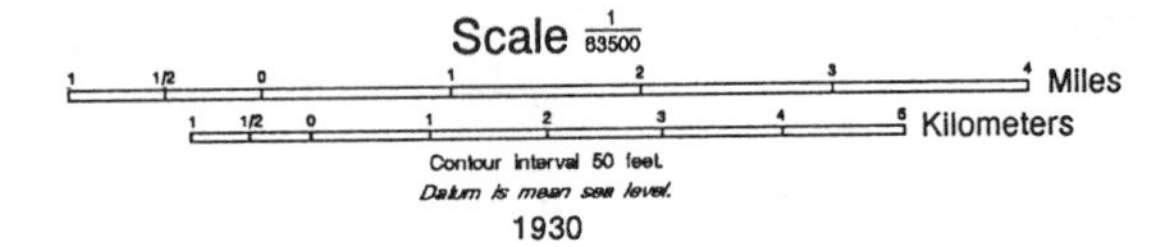

Figure 3. Early geologic and structural contour map of Wilson Creek (Devil's Hole) dome prior to discovery (after Hancock and Eby, 1930). Structural contours are on the Trout Creek sandstone of the Isles Formation; contour interval is 500 ft (152 m).

It was not until 1985 that production from the Pennsylvanian was established in the Unit 66 well, even though a deep Cambrian test had been unsuccessful 40 years earlier (1945). The Texaco Unit 66 was completed in selected intervals from 10,370 to 10,638 ft (3161 to 3242 m) in the Pennsylvanian Minturn Formation for a production rate of 18 bbl of condensate/day (58.5° API), 914 MCFD, with no water after the first few days (Figure 6). Because hydrocarbons were recovered at commercial rates from at least three discrete intervals within the Minturn Formation, it is probable that total Minturn potential is significantly greater than indicated by the current restricted production rate.

DISCOVERY METHOD

Wilson Creek dome is a prominent anticlinal closure, mappable at the surface. Upper Cretaceous Mesaverde sandstones and shales outcrop over the

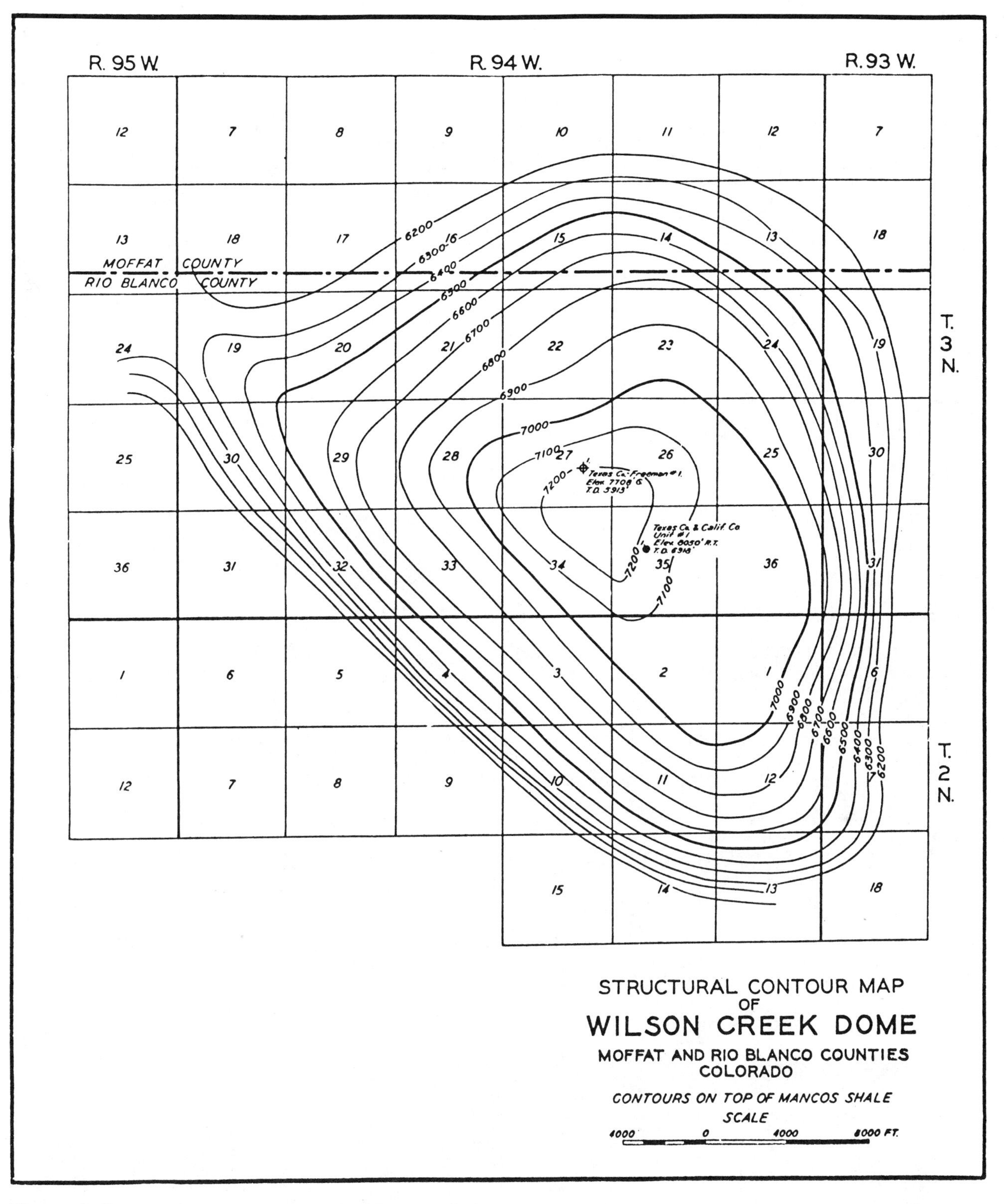

Figure 4. Structural contour map on the Mancos shale drawn at the time of discovery of Wilson Creek field (from Hunt, 1938). Contours are on the Mancos shale; contour interval is 100 ft (30.5 m).

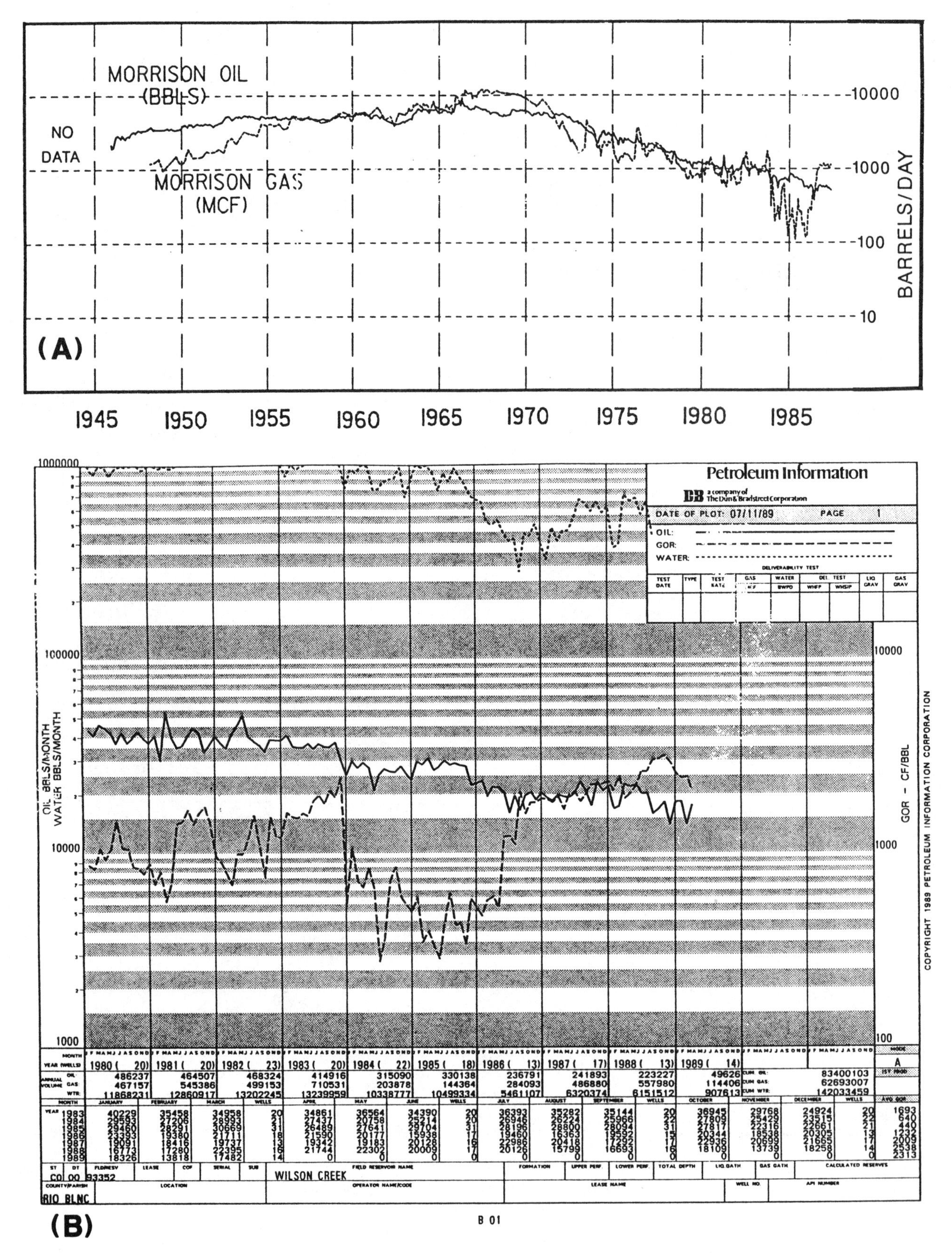

YEAR (WELLS)	1980 (20)	1981 (20)	1982 (23)	1983 (20)	1984 (22)	1985 (18)	1986 (13)	1987 (17)	1988 (13)	1989 (14)
ANNUAL VOLUME OIL:	486237	464507	468324	414916	315090	330138	236791	241893	223227	49626
GAS:	467157	545386	499153	710531	203878	144364	284093	486880	557980	114406
WTR:	11868231	12860917	13202245	13239959	10338777	10499334	5461107	6320374	6151512	907613

CUM. OIL:	83400103
CUM. GAS:	62693007
CUM. WTR:	142033459

YEAR	JANUARY	FEBRUARY	MARCH	WELLS	APRIL	MAY	JUNE	WELLS	JULY	AUGUST	SEPTEMBER	WELLS	OCTOBER	NOVEMBER	DECEMBER	WELLS	AVG GOR
1983	40229	35458	34958	20	34861	36564	34390	20	36393	35282	35144	20	36945	29768	24924	20	1693
1984	29593	27206	28993	21	27437	20758	25214	20	26946	26224	25966	22	27809	25429	23515	22	640
1985	29460	28291	30669	31	26489	27641	29704	31	28196	28800	28094	31	27817	22316	22661	21	440
1986	23393	19380	21711	18	21590	20177	15938	17	19460	16363	19592	15	20344	18538	20305	13	1232
1987	19091	18416	19737	13	19342	19183	20128	16	22986	20418	17292	17	22936	20699	21665	17	2009
1988	16773	17280	22395	16	21744	22302	20009	17	20126	15799	16693	16	18109	13739	18258	14	2538
1989	18326	13818	17482	14	0	0	0	0	0	0	0	0	0	0	0	0	2313

Figure 5. Production decline curves: (A) Salt Wash (Morrison) zone (courtesy of Texaco). (B) Salt Wash plus Entrada (Sundance) zones, 1980 to March 1989 (courtesy of Petroleum Information).

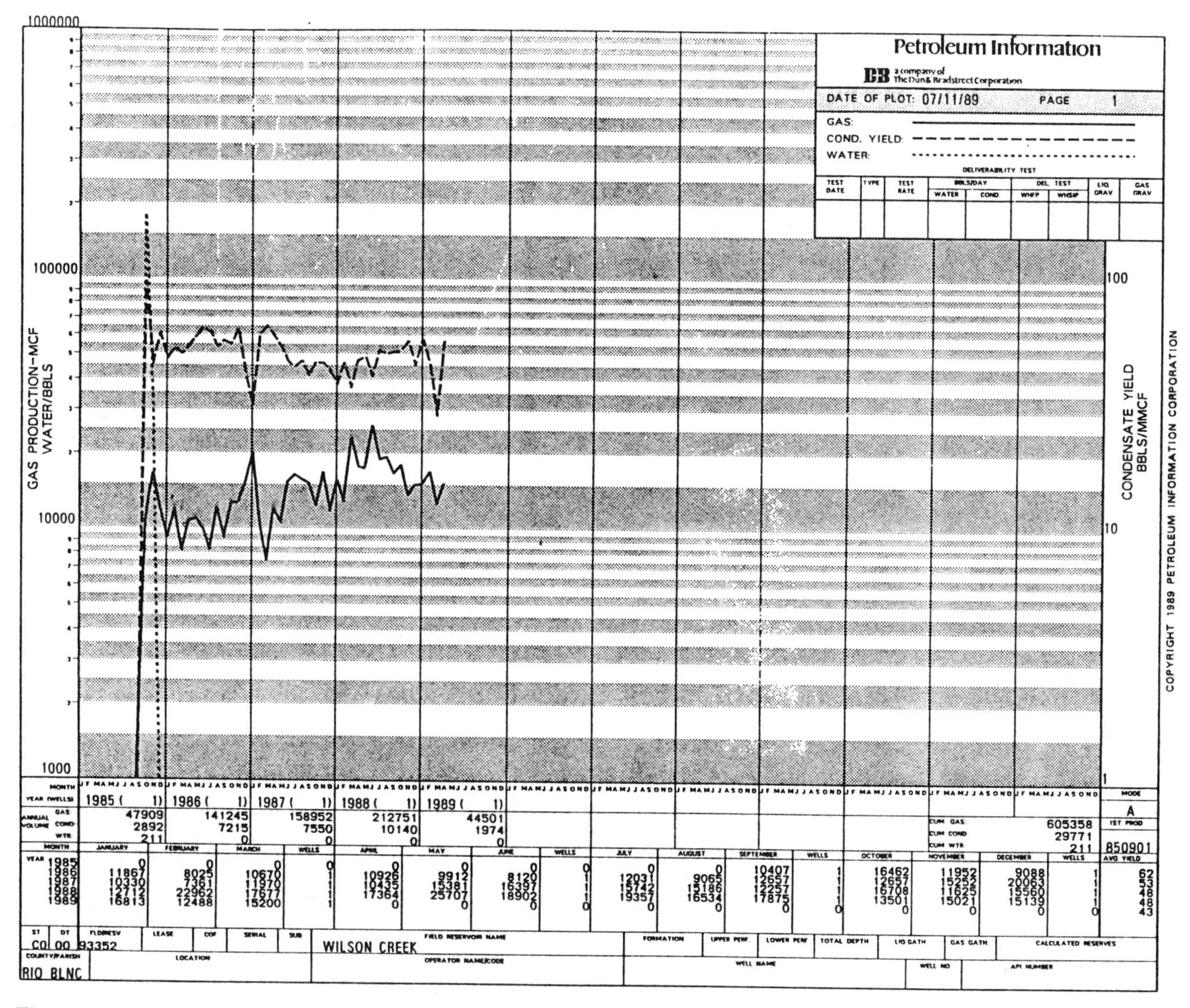

Figure 6. Production decline curve for Wilson Creek Unit 66 completed in the Pennsylvanian Minturn Formation (courtesy of Petroleum Information).

structure and are deeply incised by streams along the Danforth Hills drainage divide (Figure 2). Figure 3, reproduced from an early USGS Bulletin (Hancock and Eby, 1930), shows a domal structure with about 1000 ft (305 m) of vertical closure as mapped on the prominent Trout Creek sandstone datum. No surface oil or gas seeps have been reported over the dome. Discovery of hydrocarbons awaited only a favorable economic climate and the courage to drill below the Cretaceous, considered "deep" at that time. None of the historical field reports (e.g. Hunt, 1938; Hoffman, 1954; Ladd, 1957; Gibson, 1961) indicate that geophysical data were acquired prior to discovery.

The second Richmond Petroleum well, drilled to 4825 ft (1471 m) in 1921, was considered a "deep hole" (Hancock and Eby, 1930, p. 212) but did not reach the Dakota sandstone, a recognized objective even then (Figure 7). Further exploration on the dome was discouraged for 11 years, when the Dakota sandstone was penetrated in 1926 by a Texas production Company well and found to be water bearing. In 1926–1927, well before discovery of oil at Wilson Creek (1938), the Dakota sandstone was found productive in the nearby Moffat Dome field. The Mancos Shale was producing 1500 to 2000 bbl/month at Rangely field, and Morrison (Salt Wash) production was found at nearby Isles field on the Axial arch (Figure 1).

The discovery of oil at Wilson Creek dome taught explorationists that a closed anticlinal structure in an established oil province is never condemned as a potential oil trap until all horizons within economic depths are thoroughly tested so long as a reservoir and a hydrocarbon source can be rationalized. This

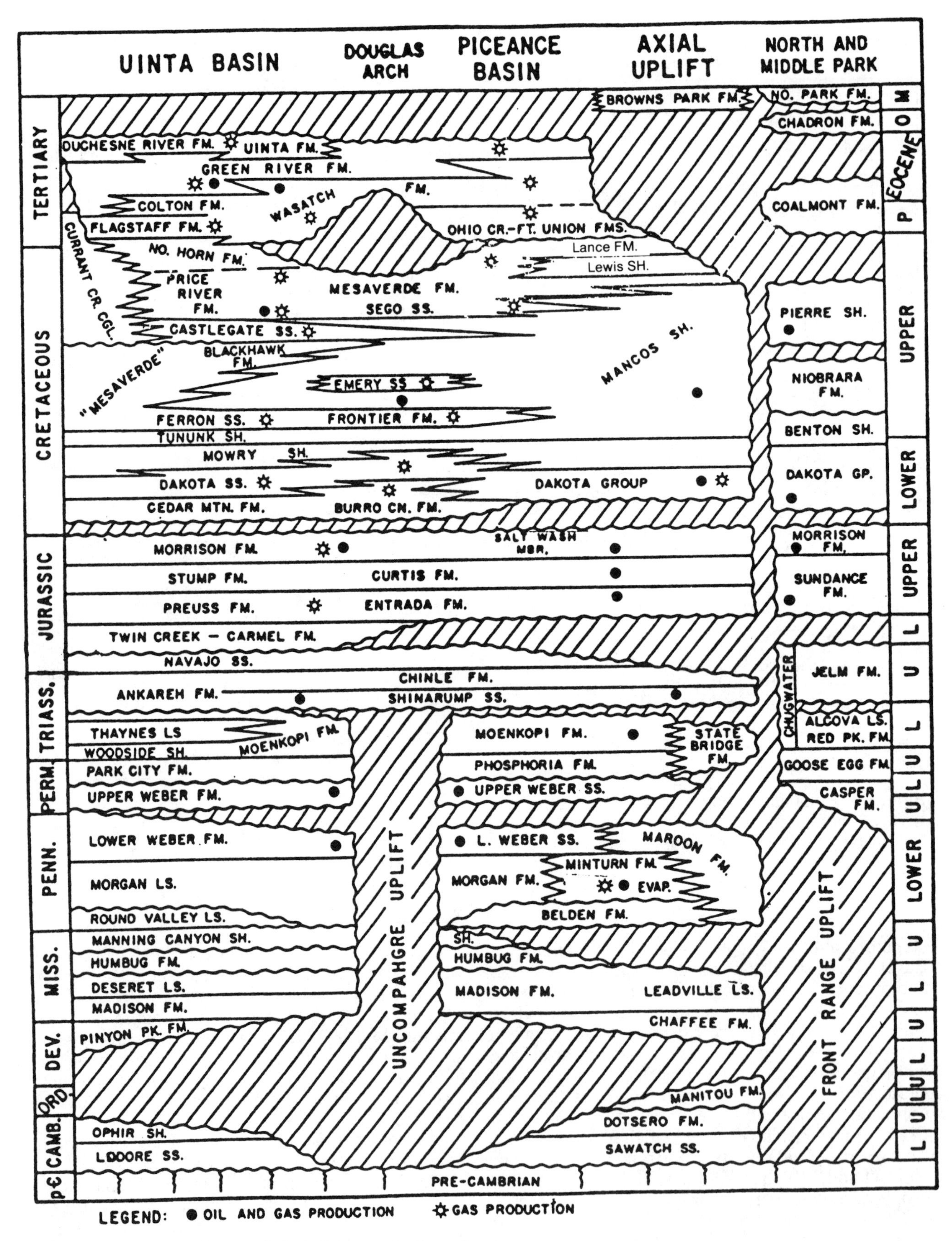

Figure 7. Stratigraphic correlation diagram, northwest Colorado and northeast Utah (modified after Sanborn, 1971).

lesson was learned with the original discovery of oil in Jurassic reservoirs below the surprisingly nonproductive Cretaceous Dakota sandstone in 1938, and again 47 years later with the discovery of hydrocarbons in the deeper Pennsylvanian Minturn zones in 1985.

STRUCTURE

Tectonic History

Wilson Creek is part of the thrusted northeastern boundary of the Piceance Creek basin, along the north edge of the Colorado Plateau. This basin is confined by the Uinta-Axial arch to the north and northeast, the White River uplift to the east, the Gunnison and Uncompahgre uplifts to the south and southwest, and the Douglas Creek arch to the west (Figure 1).

The tectonic history of this part of the Rocky Mountain foreland is particularly complex. Important structural movements along the Uinta-Axial arch began in Precambrian time. Repeated structural movements, including both extensional and compressional, occurred during Precambrian, Paleozoic, Mesozoic, and Tertiary time (Sears et al., 1982; Bryant, 1985; Hansen, 1986; Stone, 1986b; Waechter and Johnson, 1986).

Surface geologic mapping (e.g., Hancock and Eby, 1930; Hansen, 1965; Rowley et al., 1985) and subsurface seismic and borehole data (Stone, 1986b) along the Uinta uplift and southeastward into northwest Colorado show that the Uinta Mountains were formed by structural inversion of an earlier Uinta rift system or aulacogen. The Uinta aulacogen came into being probably in late Proterozoic (about 1.0 Ga) and appears to lie along the ancient continental (arc-collisional) plate boundary or suture zone (dated about 1.8 Ga) at the southeastern edge of the Wyoming Archean Province (Condie, 1982; Bryant, 1985; Duebendorfer and Houston, 1986; Reed et al., 1987). Original extensional movements occurred along a series of east-west- to west-northwest-trending, moderately dipping, normal faults with vertical separations ranging to 3 mi (5 km) or more. Major normal faults of the north and south flanks of the Uinta Mountains (which are now convergent, basin-boundary thrusts) originally formed the boundaries of the Uinta rift basin. Filling this elongate basin is the Proterozoic Uinta Mountain Group, comprised of about 4.7 miles (7.5 km) of partly synrift, fluvial, arkosic sandstones and quartzites with some coarse conglomerates in the eastern part of the trough. Paleocurrent directions were west to southwest, near-parallel to the axis of the aulacogen (Wallace, 1972; Bryant, 1985).

The Uinta aulacogen was later inverted through reversal of movement along the existing extensional fault zones, accompanied by local folding and regional uplift of the Uinta Mountain Range. Part of the Uinta fault system extended into northwest Colorado along the Axial arch, where compressional orogenic pulses occurred in Late Pennsylvanian (Ancestral Rocky Mountains orogeny) and during the Laramide orogeny (Stone, 1986b).

Lying just southwest and in front of this ancestral Axial arch, the Danforth Hills-Wilson Creek structure was formed during the Laramide orogeny in Eocene time when it came into existence as a basement-involved, thrust-generated fold (Stone, 1986c; 1987) with its axis nearly parallel to that of the Axial arch (Figure 1). There is no clear evidence for important earlier structural movements along the Danforth Hills trend.

After the deposition of the Uinta Mountain Group and a long period of erosion (400+ m.y.), Upper Cambrian clastics were deposited over a beveled Precambrian surface across both lower and middle Proterozoic crystalline rocks, and upper Proterozoic sedimentary rocks. A half-graben wedge of Uinta Mountain Group sediments is preserved beneath the Cambrian rocks along the *northeastern* side of the Ancestral axial fault zone (AAF) (Stone, 1986b). Based on the transparent seismic signature of Precambrian basement rocks all along the Danforth Hills anticline (including Wilson Creek), no Uinta Mountain Group rocks are believed to be present in the subsurface along the *southwest* side of the AAF (Figure 8).

Throughout the early Paleozoic, shelf conditions prevailed, and in northwest Colorado relatively thin sections of Ordovician and Devonian carbonates are separated by disconformities (Figure 7). The Mississippian Leadville limestone represents the last tectonic stability in the region, as the former shelf broke up into narrow basins and uplifts during the Ancestral Rocky Mountains orogeny that started in Pennsylvanian time (Mallory, 1972). As a consequence of this compressional event along the Uinta-Axial arch and southeastward into the central Colorado trough, normal displacements on the earlier extensional fault zones were reversed (i.e., became thrusts) and clastics were eroded from the upthrust blocks and shed into the adjacent, subsiding depressions in which interdigitated marine shales, carbonates, and evaporites were accumulating (DeVoto et al., 1986). At Wilson Creek field, which was downthrown to the Axial arch during the Pennsylvanian, the Weber-to-Leadville thickness is approximately 3000 ft (914 m), but across the AAF zone to the northeast this same section abruptly thins to only 2000 ft (610 m). The magnitude of fault-controlled thickness changes increases southeasterly into the Eagle basin where much of the Upper Pennsylvanian and Lower Permian is missing on relatively upthrown fault blocks by nondeposition and/or erosion (DeVoto et al., 1986; Stone, 1986b).

The overlying Upper Permian Phosphoria (Park City) marine shales and carbonates that occur to the northwest of Wilson Creek indicate a return to more stable conditions. But in the Wilson Creek area, the equivalent section is the State Bridge Formation, a transitional to nonmarine deposit.

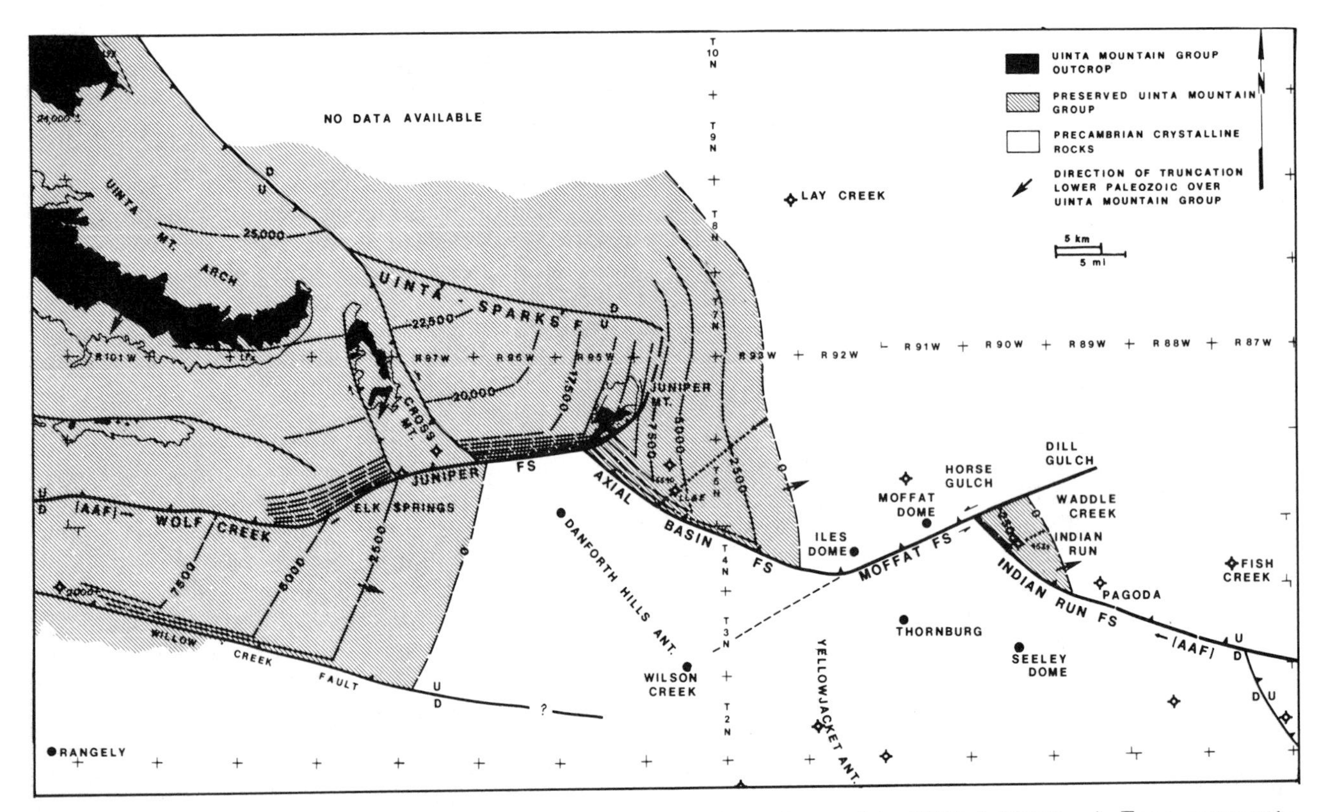

Figure 8. Distribution (hachured) and isochore map of Proterozoic Uinta Mountain Group sediments, Eastern Uinta Mountains and northwest Colorado. Isochore interval is 2500 ft (7620 m). Four segments of Ancestral Axial fault zone (AAF) are labeled individually. (After Stone, 1986b.)

There are a number of important regional unconformities within the Triassic–Jurassic section (Figure 9) reflecting regional tilts and erosional truncation (cf. Stone, 1975, Figure 3), but it was not until after the close of Cretaceous marine deposition in the Piceance Creek basin that the Wilson Creek structure was formed by northeast-southwest compression during the Tertiary Laramide orogeny.

Regional Structure

Wilson Creek dome is part of the Danforth Hills–White River uplift structural trend that forms the eastern boundary of the Piceance Creek basin for nearly 100 mi (160 km). This important structural feature is a giant basement-involved, thrust-generated fold (BITGF) with over 2 mi (3.2 km) of vertical relief along much of its length (Richard, 1986; Stone, 1986a). The anticlinal structure is asymmetric to the southwest and bordered on that side by a thrust fault zone dipping 30° or less to the northeast under the fold (Figure 10). Dips on the southwest flank (forelimb) reach 50° or more, but are more gentle (i.e., <10°) on the northeast flank (backlimb).

Local Structure

Wilson Creek dome is located near the southeastern terminus of the Danforth Hills anticline (Figure 1). At the surface, this terminus is a structural saddle between the north-trending White River uplift on the south and the northwest-trending Danforth Hills anticline on the north. At depth, however, an east-trending, north-dipping, oblique-slip fault zone (Wilson Creek fault or WCF) separates the two structures and merges with the northeast-dipping, frontal thrust common to both structures (Figures 1 and 11). Vertical drop across this composite fault zone under the crest of Wilson Creek dome is at least 14,000 ft (4267 m) on the Precambrian basement surface (Figure 10).

Seismic Profiles

The great vertical relief across the Wilson Creek structure is the result of both westerly flank dip and vertical separations across the two northeast-dipping fault zones visible on the migrated seismic profile P–P′ (Figure 12). A northeast-dipping, oblique-slip fault zone along the south flank of Wilson Creek dome is seen as a relatively low-angle fault trace on this seismic profile at about 2 sec. The weak reflections beneath this fault trace are probably "sideswipe" from the Lower Paleozoic level on the north plunge of the White River uplift. Vertical displacement on this fault at Salt Wash depth is about 0.2 sec (2-way time) or about 1300 ft (396 m). A much larger, east-dipping, frontal thrust zone underlies the steep

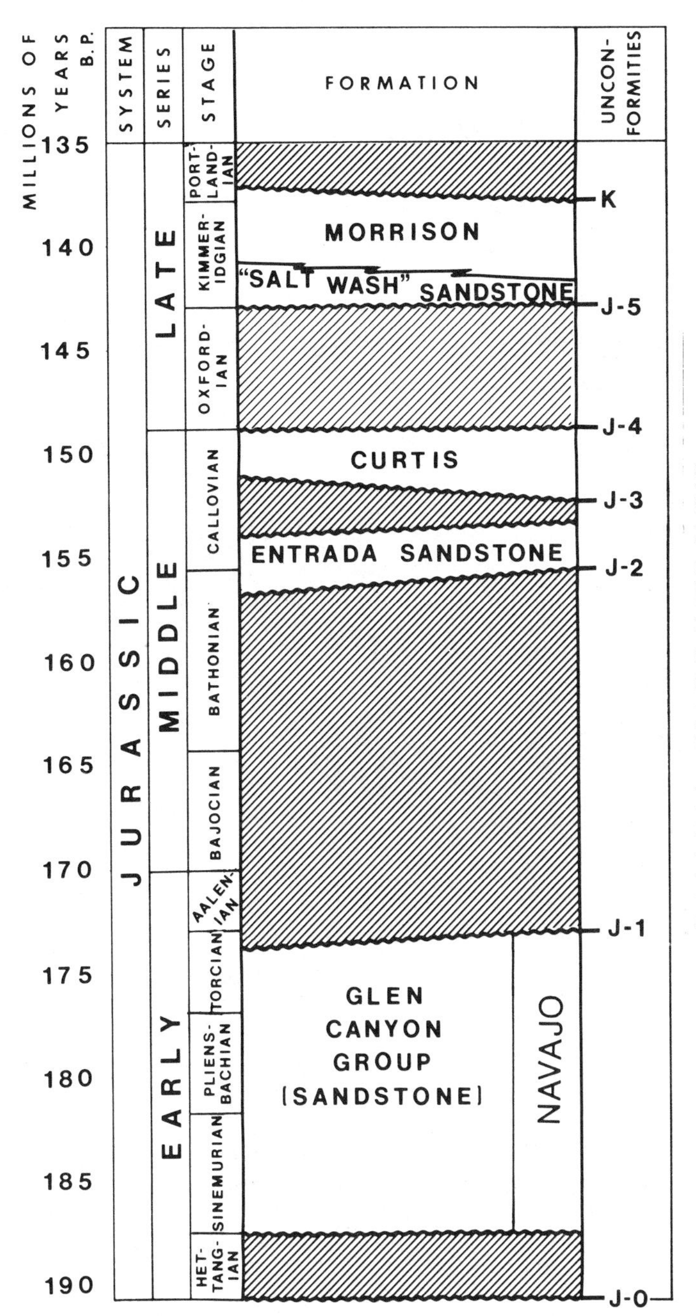

Figure 9. Time-stratigraphic chart of Jurassic section in Wilson Creek area. (After Stone, 1986a.)

west flank of the White River uplift-Danforth Hills anticlinal trend. This thrust can be seen at the far left side of the seismic profile and has an interpreted vertical separation of about 5000 ft (1524 m) at the Minturn level.

Migrated profile R-R′ (Figure 13) crosses the Wilson Creek dome about 2 mi (3.2 km) northwest of profile P-P′ but has a more northerly trend. The folding is somewhat more gentle here, while fault displacements have increased. (The two profiles intersect south of the Husky well, Figure 11.) Interpretation within the fault block between the Wilson Creek fault and the frontal Danforth Hills thrust is more difficult due to the poor quality of the data and the lower angle between line direction and structural strike.

Further north along the Danforth Hills anticline at Maudlin Gulch a vertically exaggerated seismic profile (Figure 14) shows vertical separation across a *single* thrust zone to be of the same order of magnitude as across the *two* faults at Wilson Creek field.

STRATIGRAPHY

A correlation diagram for the Piceance Creek basin is shown in Figure 7 and a type log from the Texaco-Chevron Wilson Creek Unit 66 (NW SW NE of sec. 34, T3N, R94W) deep test in Figure 15. Deposited over a beveled Precambrian surface, Upper Cambrian, Ordovician, and Devonian strata are only thin remnants of once thicker, shallow-water marine sandstones (Cambrian) and carbonates (Ordovician and Devonian). There are no apparent source rocks preserved in this lower Paleozoic section, and erosional unconformities are prominent (Ross, 1986).

About 450 ft (137 m) of Mississippian Leadville limestone was penetrated at Wilson Creek. This is a persistent transgressive carbonate unit in northwest Colorado but is not productive. A thick Pennsylvanian Minturn section may have had generative potential within the Piceance Creek basin in the interbedded Minturn and underlying Belden black shale lithofacies. However, this section is now supermature (Johnson and Nuccio, 1986). Potential reservoir rocks occur in the interbedded sandstones and carbonates. The Minturn-Belden is 2600 ft (792 m) thick at Wilson Creek and is oil and gas productive, but the contained hydrocarbons may not be indigenous. The Minturn is also productive along the Danforth Hills anticline to the northwest of Wilson Creek field (Figure 1).

The Pennsylvanian-Permian Weber sandstone, an eolian deposit, produces in several fields in northwest Colorado and northeastern Utah (e.g., Rangely and Maudlin Gulch fields), but unaccountably, not at Wilson Creek field. An overlying Permian Phosphoria (Park City) marine shale and carbonate facies that occurs some distance to the west and northwest is generally considered to be the source of Weber oil (Hoffman, 1954; Fryberger and Koelmel, 1986; Stone, 1986c), but the possibility of a contribution from the underlying Belden and Minturn shales has been pointed out recently by Waechter and Johnson (1986) and Nuccio et al. (1988).

The Pennsylvanian was a period of important tectonic activity. Deposition of Minturn-Weber rocks was accompanied by uplift along the Uinta-Axial arch (Figures 1 and 14) and also along the Uncompahgre uplift on the southwest (Stone, 1977) as reflected in the sedimentary facies (Devoto et al., 1986). The Ancestral Rocky Mountain orogeny

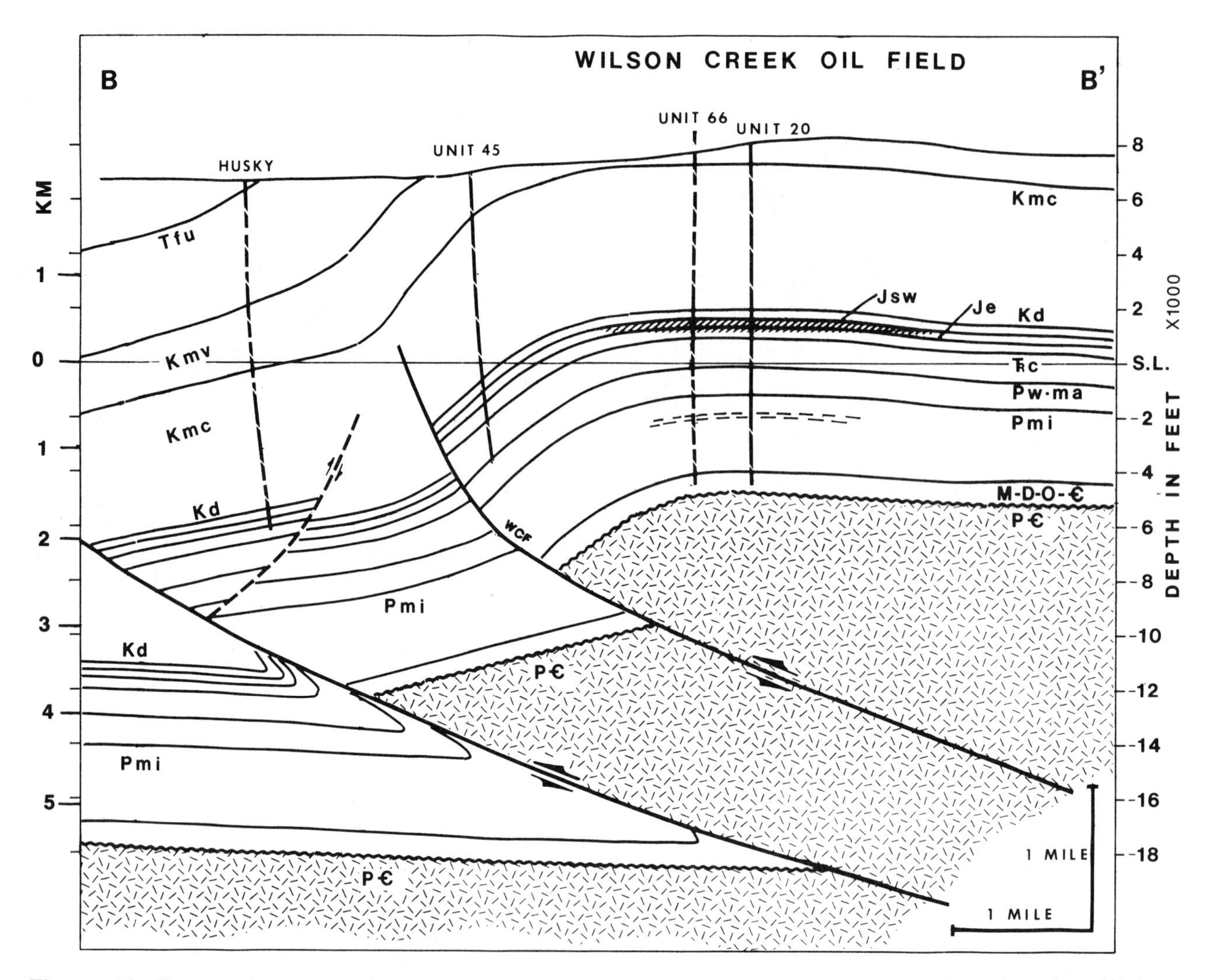

Figure 10. True-scale structural cross section B-B', Wilson Creek field (after Stone, 1986a). See Figure 11 for location. Cross-hatched area and dashed lines indicate hydrocarbon production. Formation symbols are: P€, Precambrian; M-D-O-€, Mississippian-Devonian-Ordovician-Cambrian; Pmi, Pennsylvanian Minturn; Pw·ma, Pennsylvanian-Permian Weber-Maroon; Trc, Triassic Chinle; Je, Jurassic Entrada; Jsw, Jurassic Salt Wash; Kd, Cretaceous Dakota; Kmc, Cretaceous Mancos; Kmv, Cretaceous Mesaverde; Tfu, Tertiary Fort Union. WCF, Wilson Creek fault.

probably culminated in Late Pennsylvanian (Desmoinesian) or early Missourian time, as late Missourian and Virgilian series rocks appear to be absent.

At the close of Weber deposition in the Early Permian (Wolfcampian), there was a period of erosion (Fryberger and Koelmel, 1986) before the Upper Permian Park City (Phosphoria) marine carbonates and shales and equivalent, light-colored mudstones (State Bridge Formation) were laid down. These rocks are overlain by the Triassic Moenkopi, followed by Chinle, redbeds. An unconformity occurs at the top of the Moenkopi. The overall section grades upward from the shallow-marine and eolian Weber sandstones to the nonmarine Triassic shales and into the strongly cross-stratified Navajo and Entrada eolian sandstones (Figure 16).

The Entrada sandstone, an important reservoir at Wilson Creek, was transgressed by a shallow interior sea, as represented by a basal glauconitic, coquina layer followed by the green-gray and dark gray shales of the Curtis Formation (Figures 9 and 17). The productive Salt Wash Member of the Morrison Formation, which overlies the Curtis, contains the major reservoir sandstones at Wilson Creek. A probable regressive facies, the Salt Wash grades upward into the nonmarine Brushy Basin Member of the Morrison Formation.

A major marine transgression occurred in which the Dakota sandstone at the base of the Cretaceous section is followed by a thick sequence of predominantly organic shales including the Mowry, Tununk, Frontier, and Mancos formations (Figure 7). The overlying and interfingering Upper Cretaceous

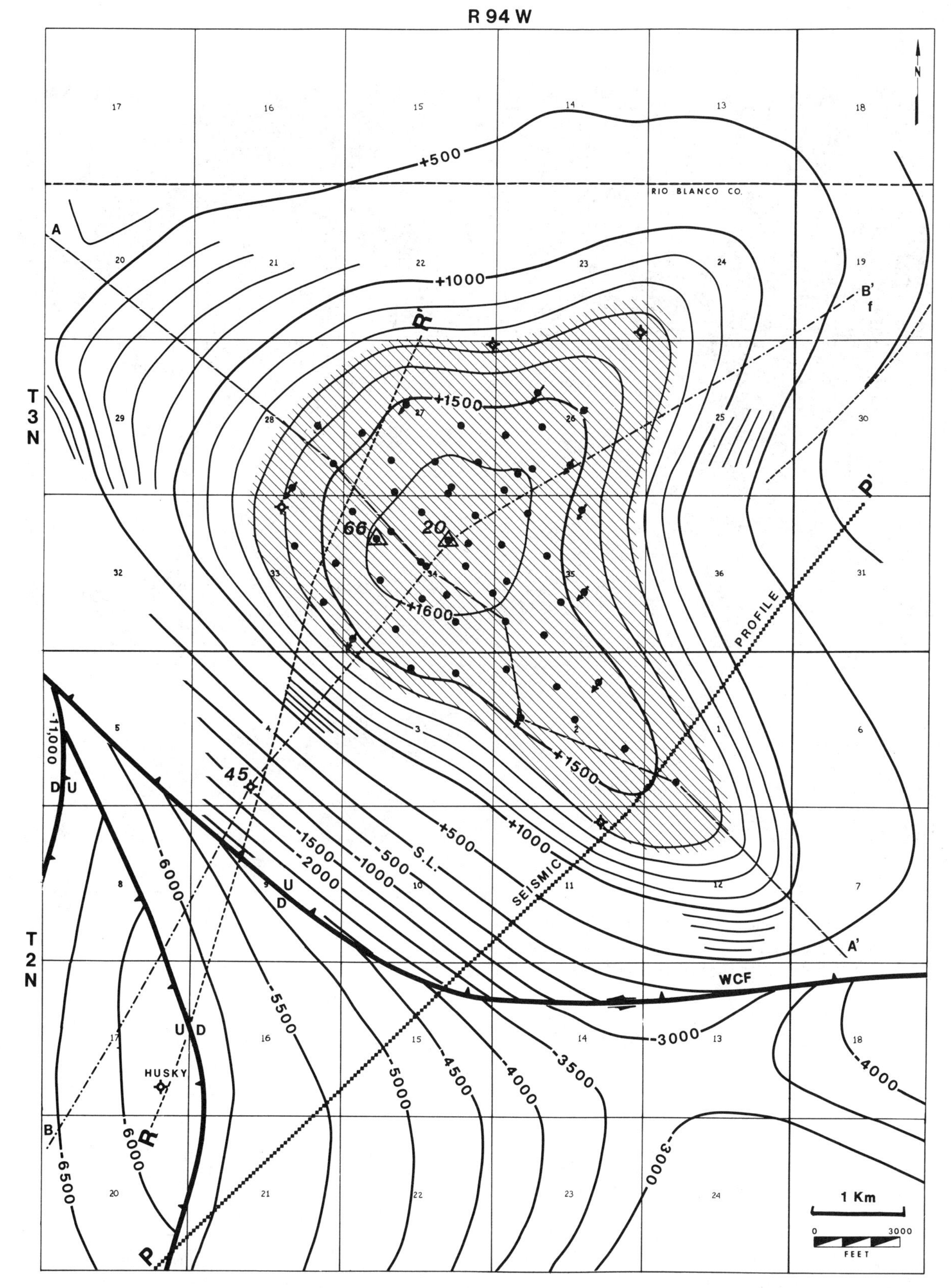

Figure 11. Structural contour map on Salt Wash Member of Morrison Formation, Wilson Creek field. Contour interval is 100 ft (30 m) in crestal area and 500 ft (152 m) elsewhere. Locations of log section A-A′ (Figure 25), structural cross-section B-B′ (Figure 10), and seismic profiles P-P′ (Figure 12) and R-R′ (Figure 13) are also shown. Salt Wash producing area is cross-hatched. (After Stone, 1986a.)

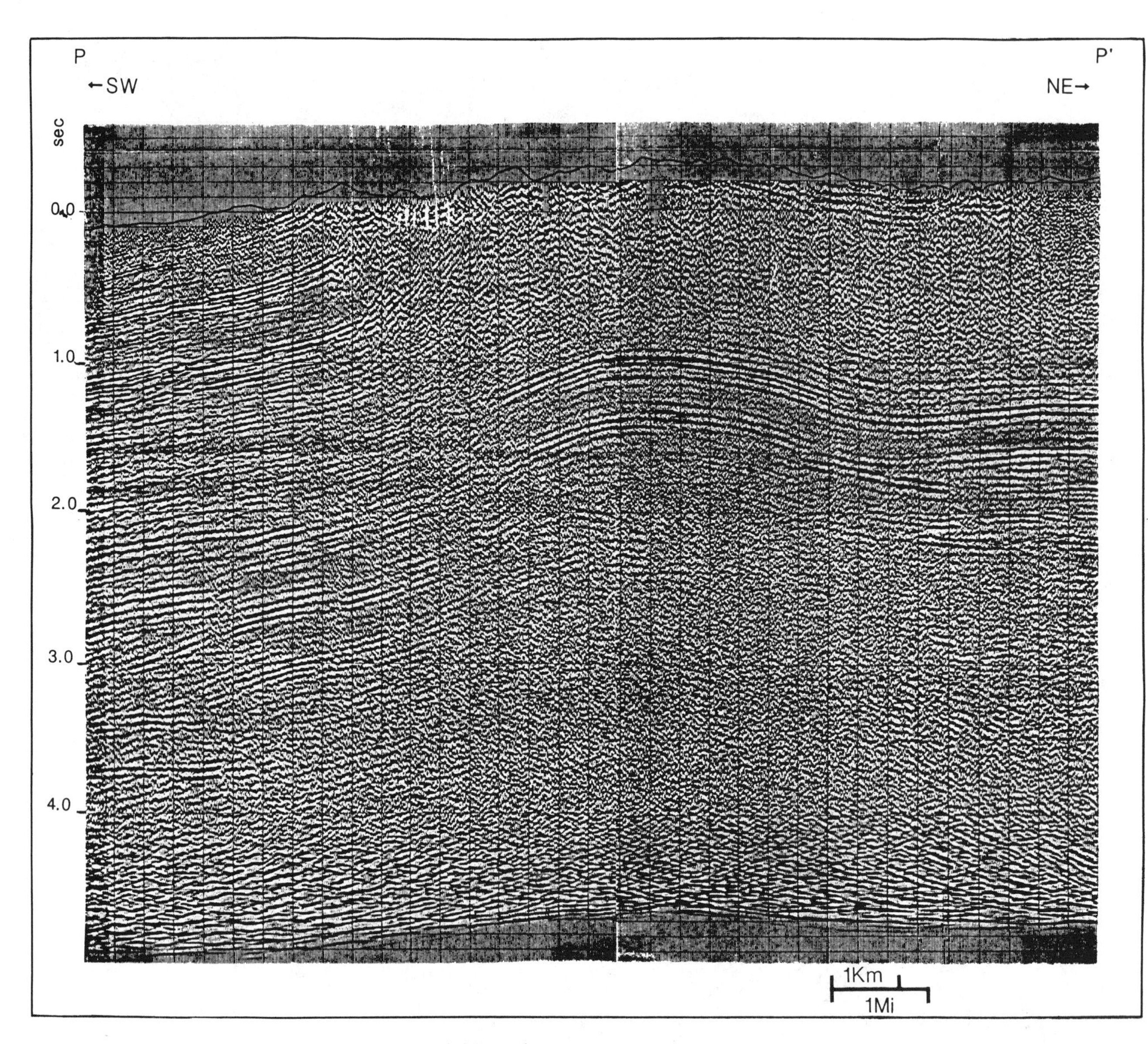

Figure 12A. Northeast-southwest, 36-fold, uninterpreted migrated seismic profile across southeast edge of Wilson Creek field (after Stone, 1986a).

Mesaverde sandstones, shales, and coals were deposited during the easterly retreat of the Cretaceous sea.

Uplift along the Danforth Hills structural trend began in Late Cretaceous-early Tertiary time as a consequence of Laramide compressional deformation; no Tertiary rocks are preserved over Wilson Creek dome. However, a thick section of nonmarine Tertiary rocks, including the lacustrine deposits of the Green River Formation, is preserved within the deeper Piceance Creek basin.

TRAP

The trapping mechanism at Wilson Creek is structural. Stratigraphic variations (lenticularity) occur within the Salt Wash pool and perhaps also within the Minturn pool, but these variations probably have little to do with the fundamental trapping mechanism—anticlinal closure. Upper and lateral seals on the Salt Wash-Entrada pool(s) are provided by the variegated shales of the Brushy Basin Member of the Morrison Formation. Interbedded

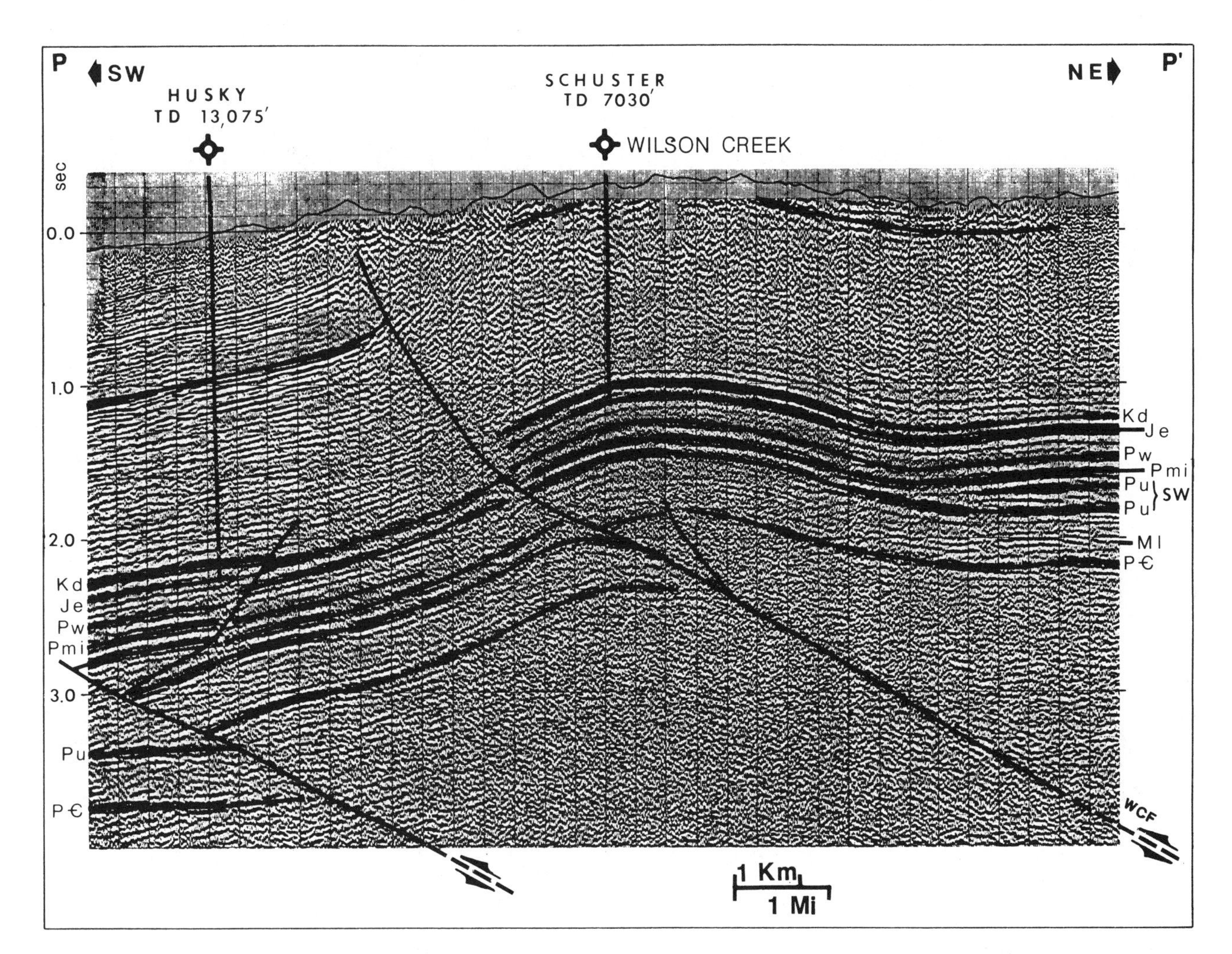

Figure 12B. Interpreted seismic profile across southeast edge of Wilson Creek field (after Stone, 1986a). Formation symbols are the same as in Figure 10 except that Pu is Paleozoic unidentified and Ml is Mississippian Leadville. SW is a zone of double reflections due to "sideswipe."

shales, tight carbonates, and/or evaporites probably are the seals for the Minturn pool(s).

Wilson Creek dome has a vertical closure of about 1000 ft (305 m), but only the top 350 ft (107 m) interval is productive.The closing contour encompasses an area of about 26,500 ac (107 km^2), 15% of which is actually productive. Its triangular or pear shape probably results from the intersection of two structural trends at nearly right angles (i.e., northeast and northwest). A structural contour map on the marker at the top of the Morrison Salt Wash Member is presented in Figure 11, and a northeast-southwest, true-scale, structural cross-section through the middle of the field is presented in Figure 10. Based largely on modern seismic control, the Wilson Creek structure may be classified as a basement-involved, thrust-generated fold (BITGF), a label that indicates fold generation in response to thrust movement initiation in a stiff, crystalline basement. Kinematic studies of foreland structures (Stone, 1986c; 1987) indicate that the causal thrust nucleates in the Precambrian basement, forming an angle of about 30° with the basement surface, and steepening as it propagates upward into the sedimentary column. So far as can be determined, the Wilson Creek BITGF did not have a pre-Laramide precursor, being produced entirely by compression during the Tertiary Laramide orogeny. The causal thrust zone along the Danforth Hills anticline is *not* a sealing fault, but has apparently acted as a conduit for migration of hydrocarbons from Cretaceous sources in the footwall.

RESERVOIRS

Jurassic Salt Wash Member

The most important reservoir interval at Wilson Creek field is in the Salt Wash Member sandstones

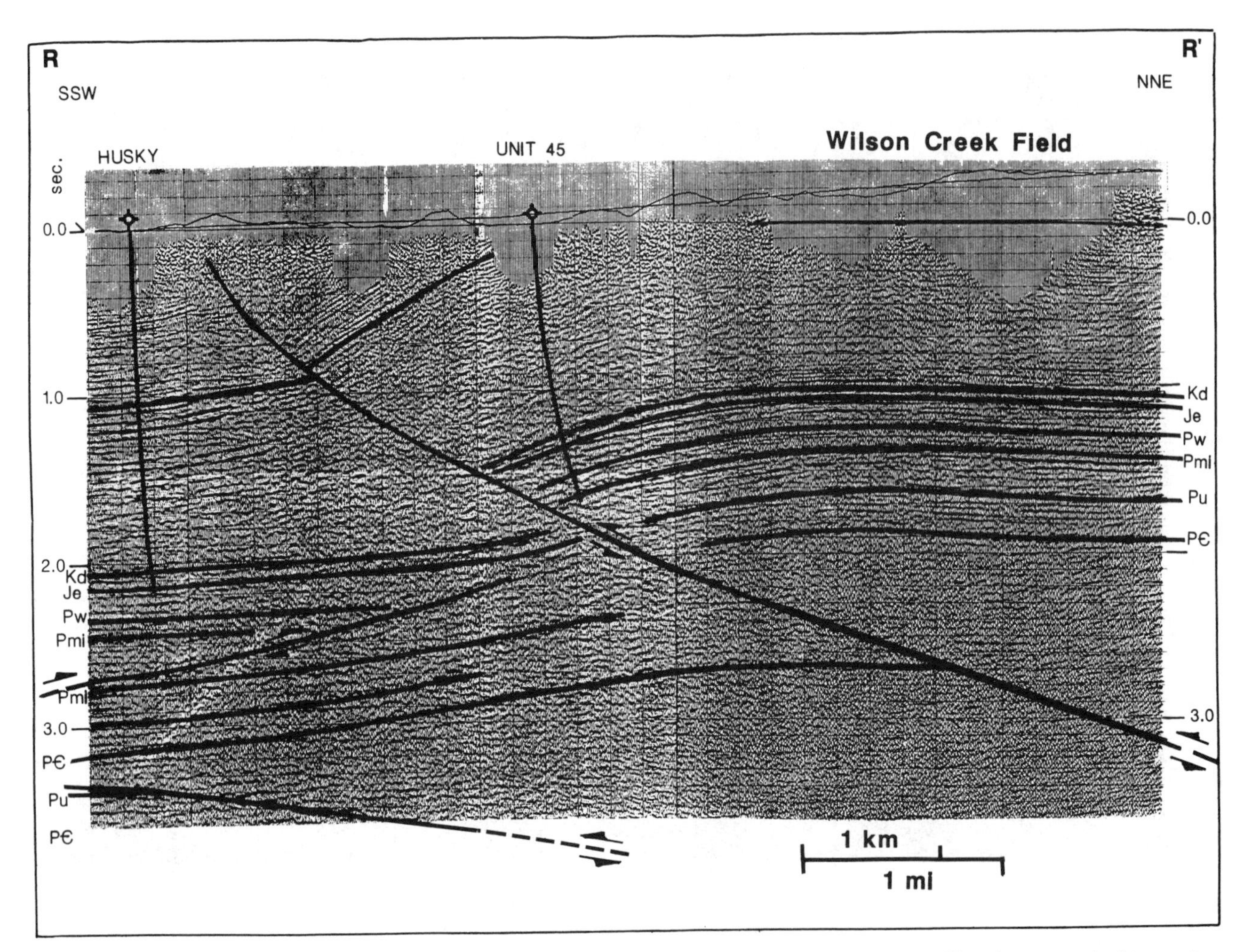

Figure 13. Interpreted north-northeast-south-southwest, 20-fold, migrated seismic profile across northwest edge of Wilson Creek field (see Figure 11 for location). Reflection identifications symbols are the same as in Figures 10 and 12.

of the lower Morrison Formation (Figures 18 and 19A, B, C, D). These sandstones are very fine to medium-grained, moderately to well-sorted quartz arenites and subarkosic arenites. Quartz is the most abundant detrital grain, with feldspar and rock fragments comprising less than 5%–7% each of the bulk mineralogy. The SEM micrographs in Figure 20A, B, C show that the detrital grains are coated with authigenic illite-smectite and less abundant chlorite clays. Illite-smectite is partially blocking the pore-throats, reducing permeability. This type of clay swells in the presence of freshwater drilling muds to produce spuriously low resistivity readings on logs (Figures 15A and 18). Trace quantities of quartz overgrowth, calcite, and anhydrite cements are also present.

The Salt Wash sandstones are white to pale green (when unstained), often cross-bedded (Figure 19B), with up to 30% frosted grains, sometimes calcareous, and usually interbedded with gray and variegated shales and siltstones. Horizontal and vertical fracturing is common throughout the reservoir (Ladd, 1957). On the outcrop at Deerlodge Park (cf. Lily Park section of Bradley, 1955), this same Salt Wash sandstone facies is stark white to light gray and forms conspicuous hummocky mounds (Figure 19A, B). Thickness of net porous sandstone (>10% porosity) within the 300 ft (91 m) Salt Wash interval at Wilson Creek ranges from 37 ft (11 m) to about 140 ft (43 m). A net pay sand isochore map of the Salt Wash zone at Wilson Creek field drawn in 1964 is shown as Figure 21. Near Elk Springs field, 30 mi (48 km) northwest of Wilson Creek, the thickness of porous Salt Wash sandstone reaches 163 ft (50 m) (Stone, 1986a, Figure 7). Porosity of the pay zone consists of well-connected primary intergranular and secondary feldspar-dissolution pores (Figures 19C, D, and 20) and ranges from 10% to 26%. Permeability is widely variable, averaging about 11 md (Ladd, 1957). Porosity vs. permeability cross plots are shown in Figure 22.

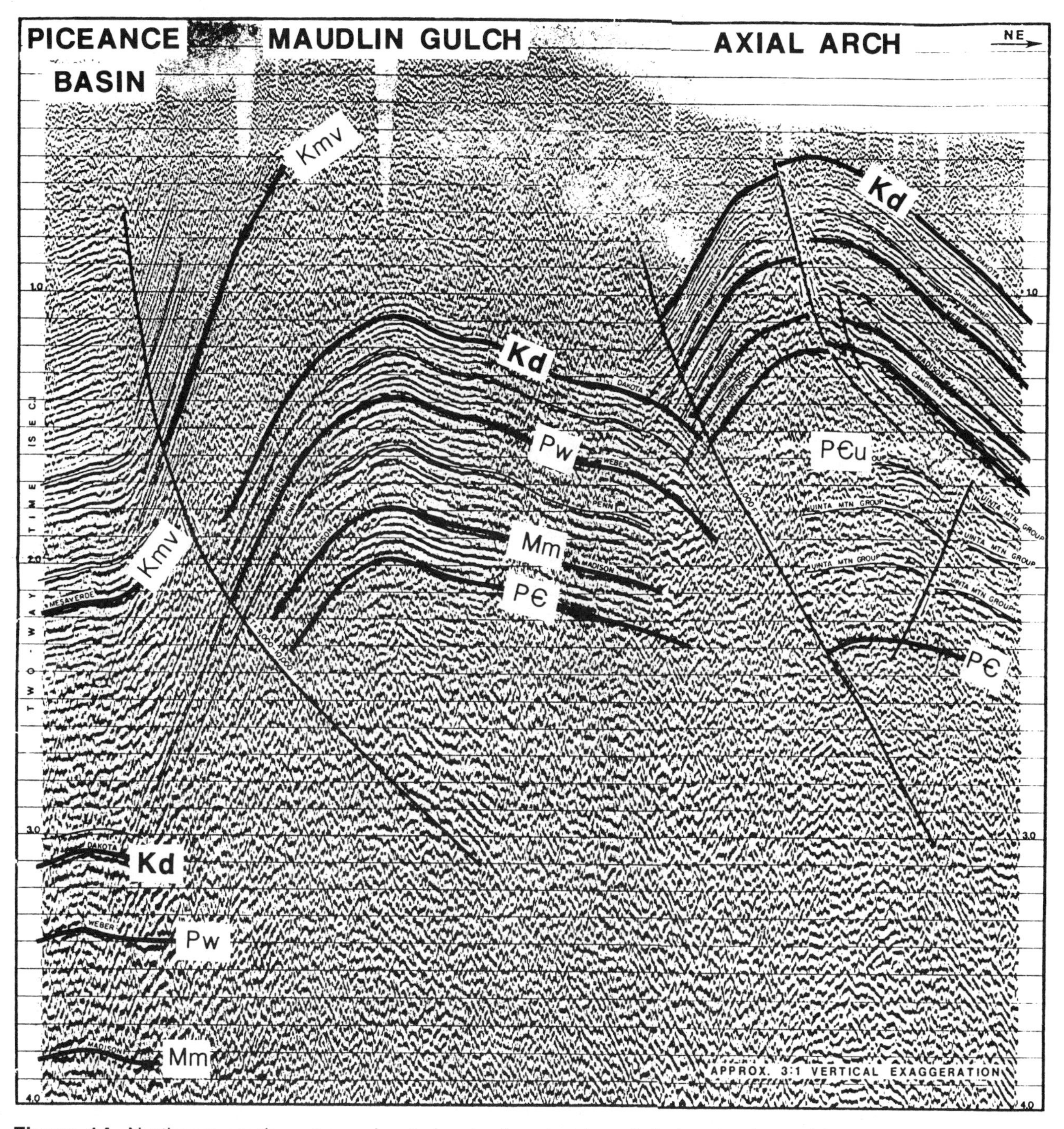

Figure 14. Northeast-southwest, unmigrated seismic profile S-S′ (see Figure 1 for location) at Maudlin Gulch field (after Richard, 1986). This field lies along the Danforth Hills anticline, 6 mi (9.7 km) northwest of Wilson Creek. Vertical exaggeration is extreme (3:1). The Danforth Hills anticline and the causal thrust zone along the west flank of Maudlin Gulch field form a major basement-involved, thrust-generated fold (BITGF) that developed during the Laramide orogeny. The profile also crosses the Axial arch (on right), which has a much more complex history of fault displacements (Stone 1986b). Formation symbols are the same as in Figures 10 and 12 except that Mm is "Madison" (Leadville equivalent) and PЄu is the upper Proterozoic Uinta Mountain Group.

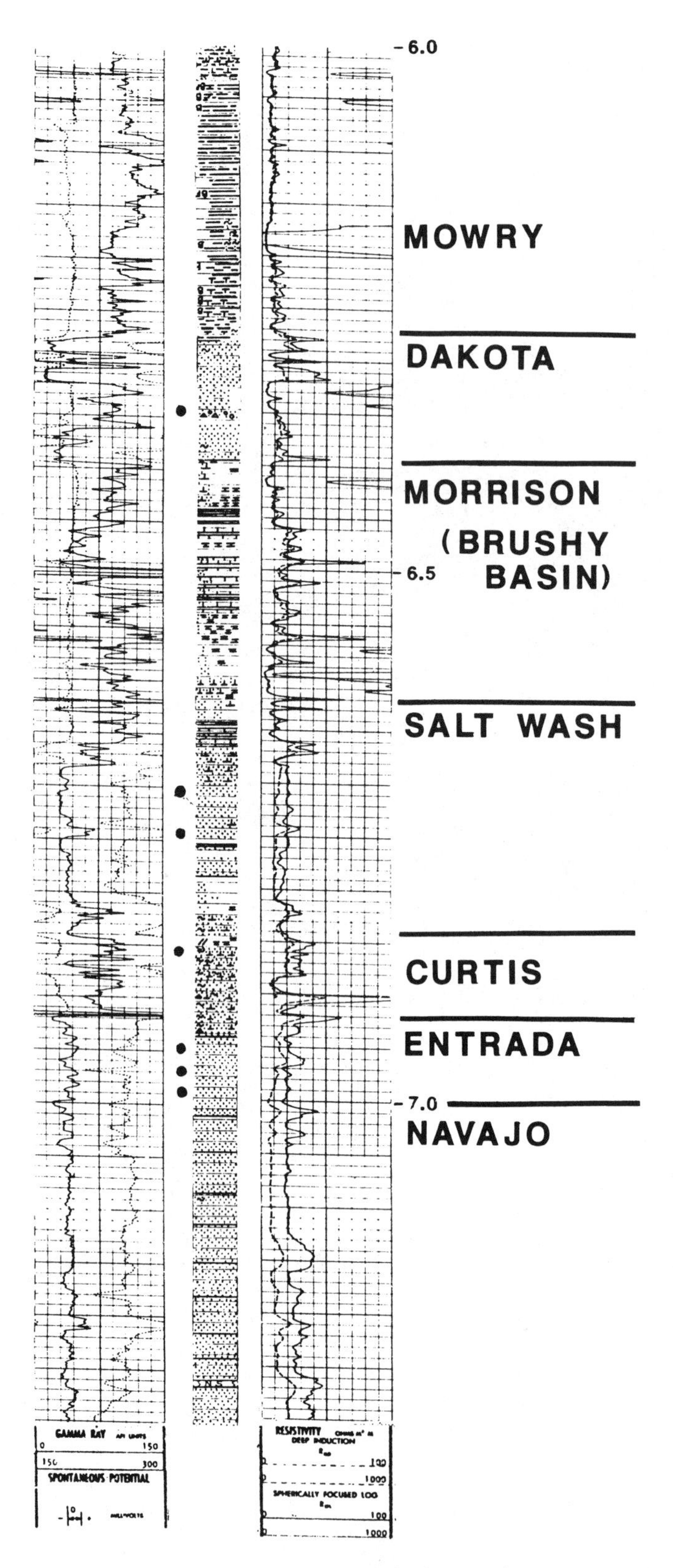

Figure 15. Gamma-ray, dual induction-SFL log of the section penetrated in the Texaco-Chevron Unit 66 (NE SW NE of sec. 34, T3N, R94W) Minturn discovery well. The Amstrat lithologic log from the nearby Unit 20 deep test is adjusted to the Unit 66 formation tops. Black circles are oil shows; black bars along middle-right vertical panels are reported test intervals (through casing); black bars along right-outside vertical panels with € symbols are cored intervals. Drill depths are shown in feet × 1000. Log scales are the same in (A), (B), and (C). Lithologies are shown with standard Amstrat symbols.

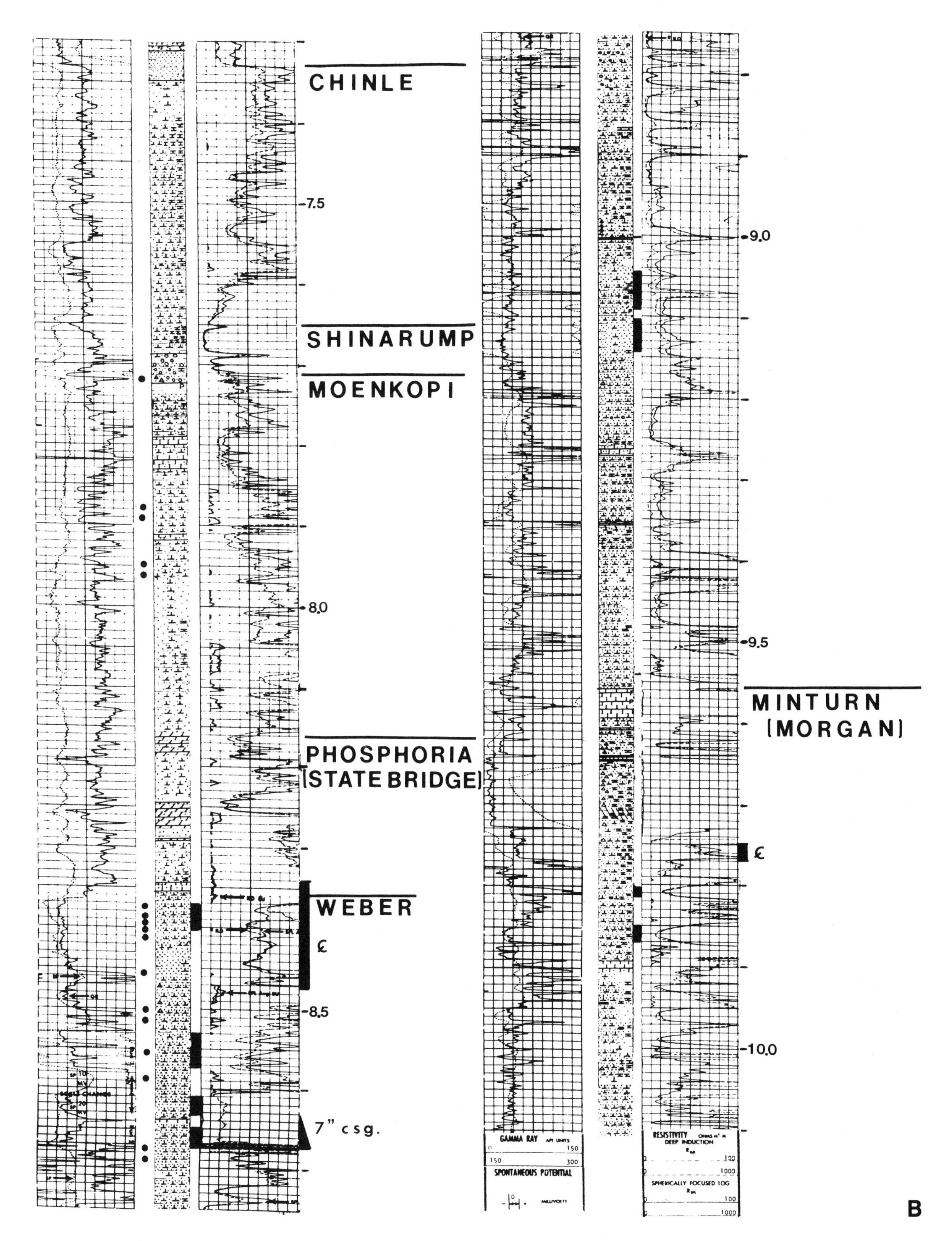

Figure 15. (Continued)

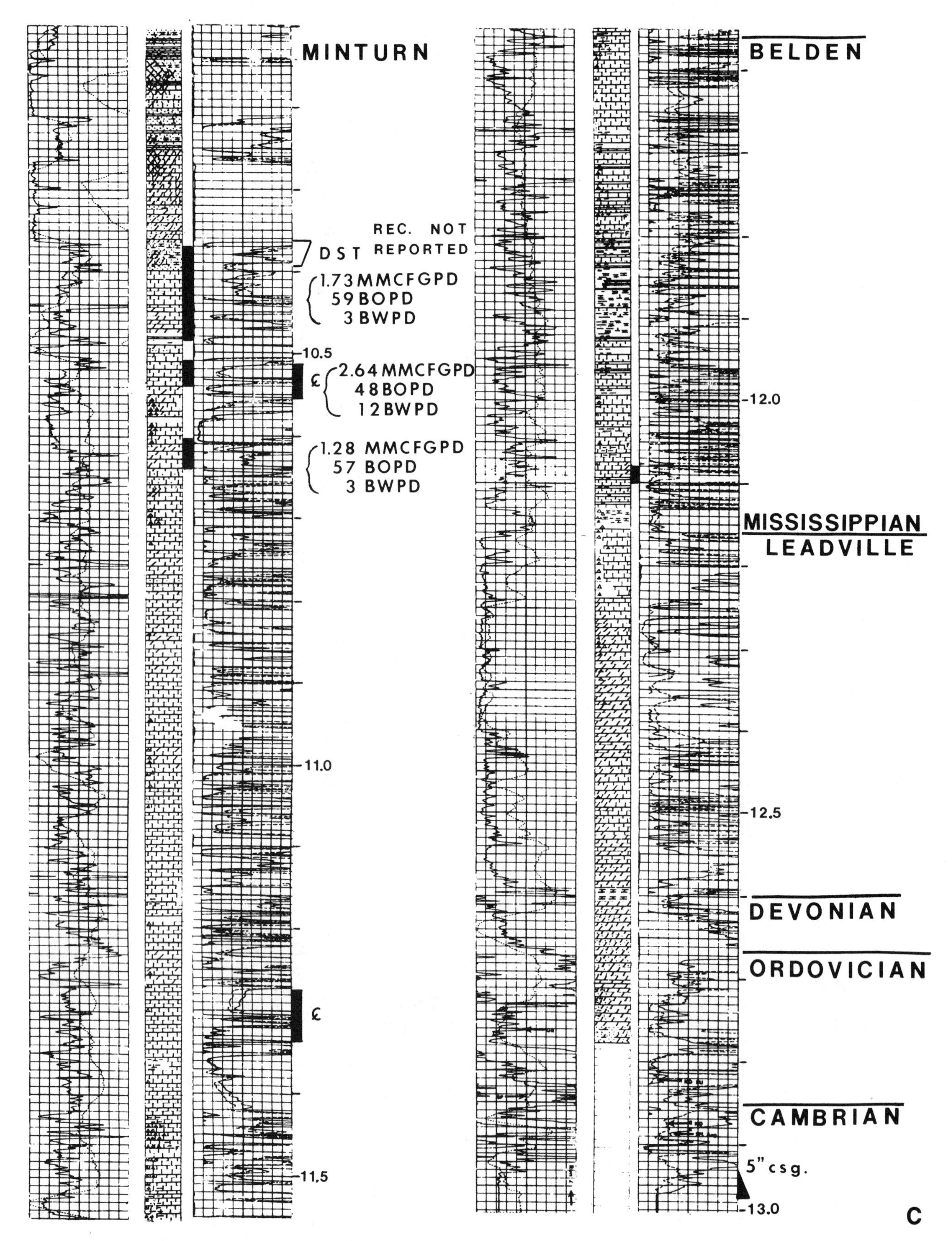

Figure 15. (Continued)

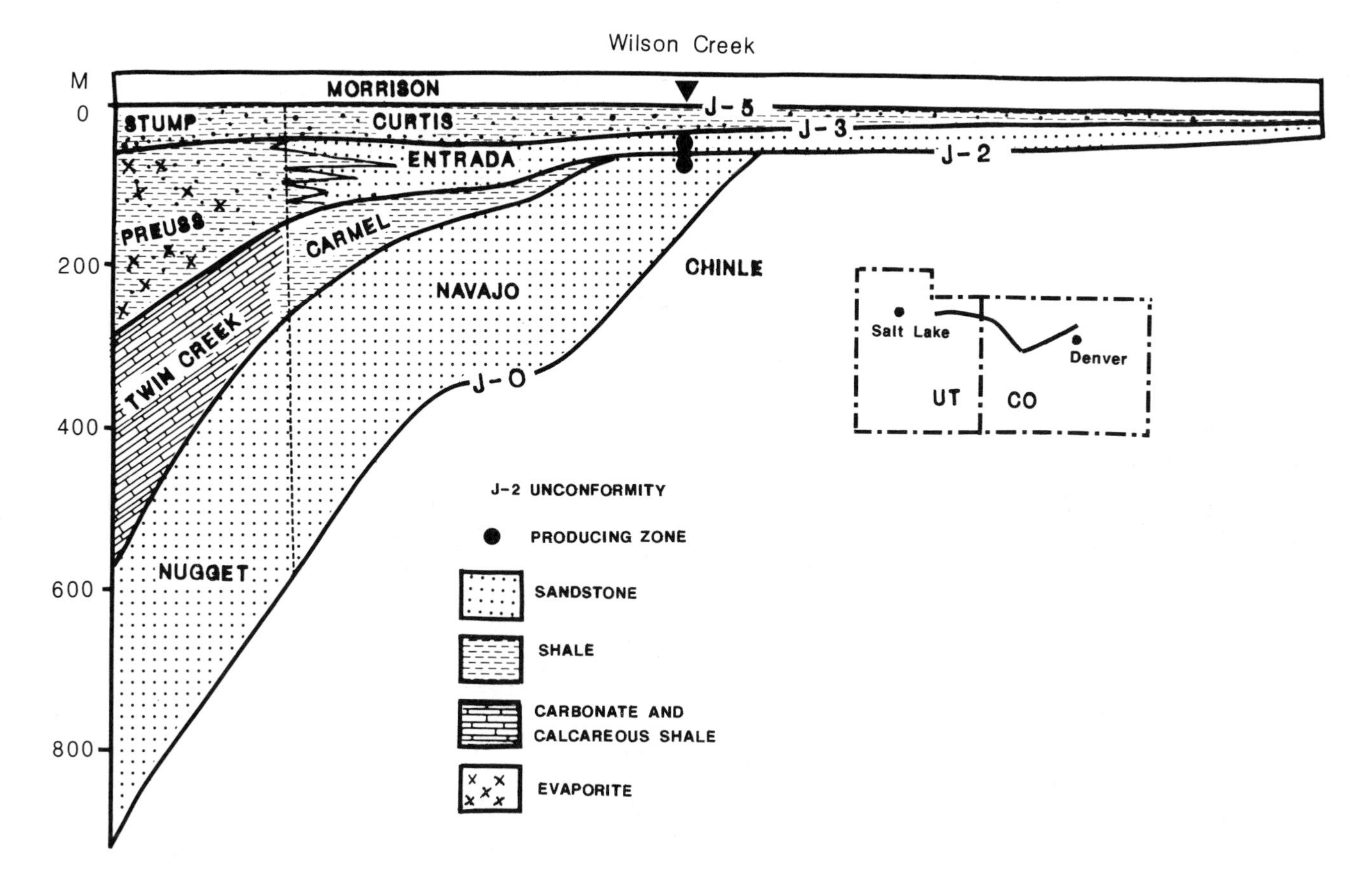

Figure 16. Regional stratigraphic section, north-central Colorado to north-central Utah. Note westward thickening into the Cordilleran geosyncline, eastward truncation of Carmel-Navajo sequence at the J-2 unconformity, and location of Wilson Creek field (black triangle). Section length is about 600 km. (Modified after Kocurek and Dott, 1983, and Pipiringos and O'Sullivan, 1978.)

Environmental Interpretation

Examination of core material from the unit 65 well (Figure 18), SW SW NE of sec. 34, T3N, R94W, recovered from 6544 to 6627 ft (1995 to 2020 m) within the Salt Wash producing interval, reveals definite marine affinities. Bioturbation and other soft-sediment deformation is evident in the calcareous gray shales interbedded with the sandstones. These conchoidal shales also contain abundant pyrite in grains or blobs indicating local reducing conditions. Small-scale cross-bedding in the sandstones is generally tangential at the base but truncated at the top, and there is no visible scour. As illustrated in Figure 19C, some sandstone laminae show reverse grading, and most of the sand grains are fairly angular, a feature not distinctive of eolian environments (T.S. Ahlbrandt, personal communication, 1987). The overall appearance of the core is subaqueous, not eolian, but the sandstone lenses could represent original eolian dunes reworked by a shallow inland sea.

The current time-stratigraphic interpretation of the Wilson Creek Salt Wash-Entrada reservoir interval illustrated in Figure 9 includes the location of regional unconformities in the Jurassic section following terminology of Pipiringos and O'Sullivan (1978). Stratigraphic studies suggest that the thick Salt Wash sandstone facies at Wilson Creek field may have developed regionally along a marine shoreline as an elongate, regressive barrier-bar or beach deposit (Figures 17 and 23). The evidence suggests that the shallow sea in which the Curtis shale was deposited transgressed southward over the eroded eolian Entrada sandstone to the latitude of Wilson Creek field. A basal Curtis, glauconitic and calcareous sandstone covers the erosion surface (J-5 unconformity) and is overlain by the gray-green and black Curtis Shale (cf. description in Dyni, 1968), the southern edge of which appears to follow approximately the trend of Salt Wash sandstone barrier bar or beach. It is proposed, therefore, that during the following period of stillstand or slow regression, the thick, porous sandstones of the Salt Wash interval were deposited parallel to the shoreline around the periphery of a south-reaching "Wilson Creek embayment" or estuary. These sandstones grade laterally and vertically upward into lighter colored red and green shales and mudstones that

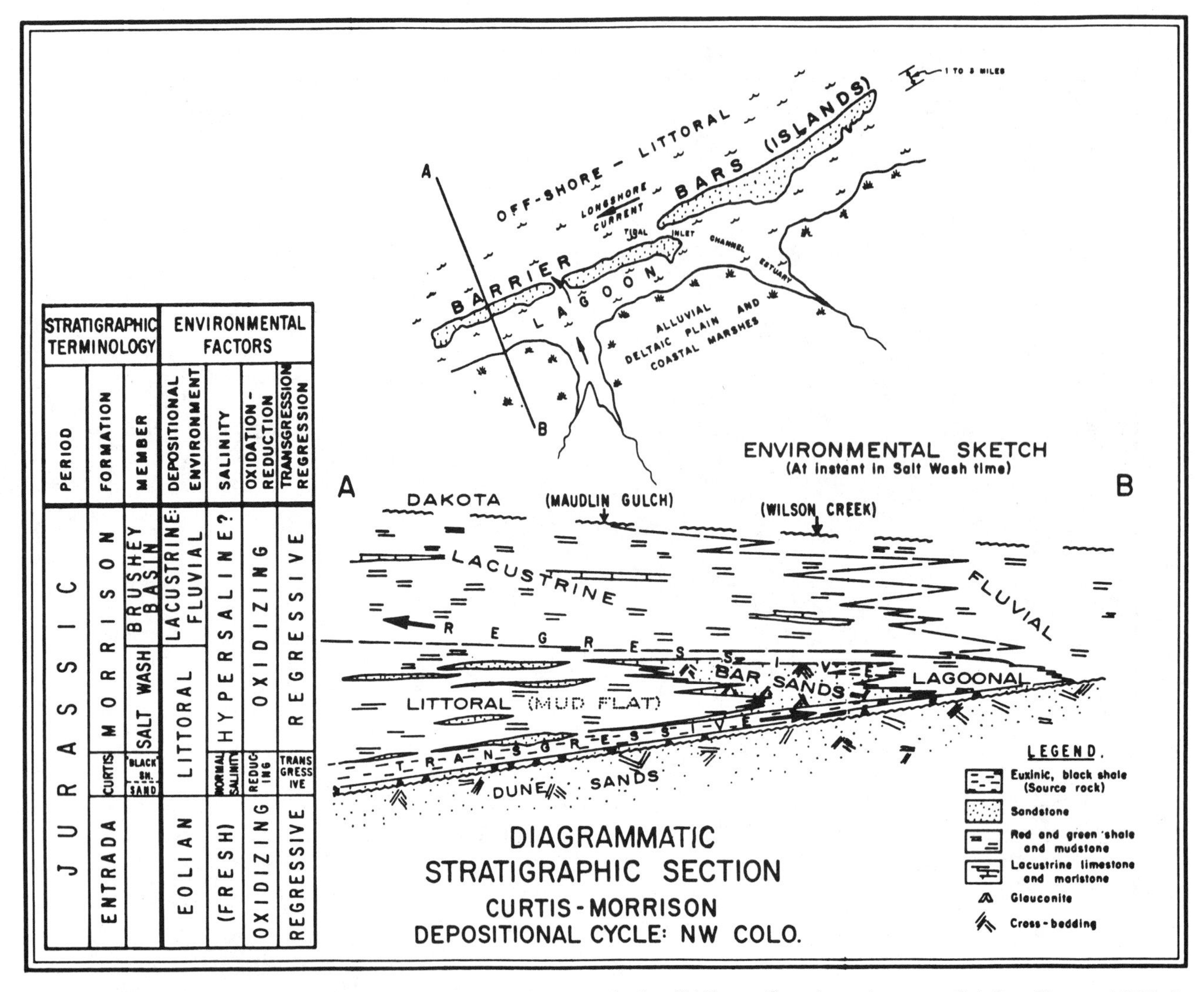

Figure 17. Diagrammatic stratigraphic section and environmental interpretation of Entrada-Morrison rocks in the "Wilson Creek embayment" (after Stone, 1986a). No scale.

probably were deposited in marsh and lacustrine oxidizing environments. The upper part of the massive sandstone build-up may represent coastal dunes.

In the Salt Wash interval, two or three discrete sandstone lenses are more common than a single, massive sandstone body. The thick sandstone porosity trend bends around the south plunge of the uplifted and eroded Cross Mountain fault block (Figure 23) and may have been influenced by the paleo Uinta-Axial arch. Also the isopach pattern may be offset by faulting. This interpretation of the depositional environment of the thick Salt Wash sandstone facies in the subsurface of northwest Colorado is quite different from the prevalent fluvial channel interpretations that are based primarily on outcrop studies on the Colorado Plateau to the south (e.g., Mullens and Freeman, 1957; Berman et al., 1980; Hartner, 1981). Certainly other environmental interpretations are possible.

Jurassic Entrada

The Entrada (Sundance) reservoir sandstones are eolian deposits (Figure 24). These sandstones have been described as white to orange-tan (unstained), quartzose, fine to medium grained, rounded to subrounded, generally well sorted, thick-bedded, cross-bedded, usually soft but with calcareous and argillaceous cement common. Climbing translatent strata formed by the migration of wind ripples under conditions of net sedimentation have been identified in the nearby Entrada outcrop (Kocurek and Dott, 1981). Fracturing may not be as conspicuous as in the Salt Wash section (Ladd, 1957), although lost circulation occurred in the Entrada zone of the Unit 66 well. The Entrada Formation is about 85 ft (26 m) thick at Wilson Creek (Figure 15A), and the pay zone occurs immediately beneath the basal Curtis coquinal layer. The Entrada Formation is separated from the Navajo sandstone below by the J-2

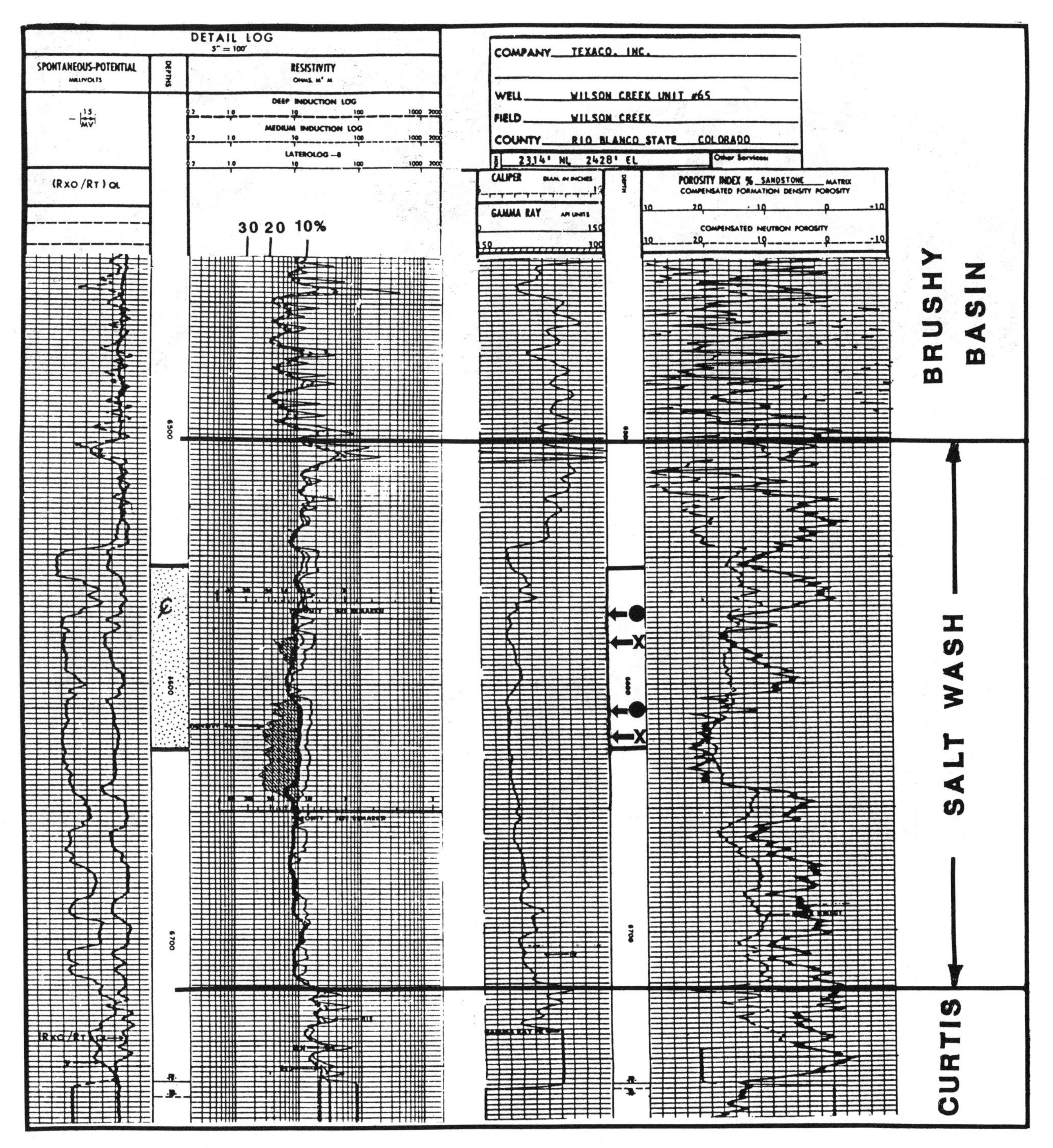

Figure 18. Induction laterolog and compensated Neutron Formation Density logs from Wilson Creek Unit 65 showing log characteristics of the Salt wash producing zone. Cored interval and level of photomicrographs (X) and SEM (●) photos of Figures 19 and 20 are marked.

unconformity, but forms a continuous (420 ft; 128 m) sandstone body with the Navajo sandstone (part of the Glenn Canyon Group) (Stone, 1975, Figure 3; Pipiringos and O'Sullivan, 1978, Plate 1; Kocurek and Dott, 1983). The Navajo lies unconformably on the Triassic Chinle red shales (Figure 9). Entrada porosity ranges from 14% to 27%, and horizontal permeability is fairly uniform at about 38 md (Ladd, 1957).

Although the Entrada and Salt Wash reservoir sandstones are separated by the thin Curtis Formation (Figure 15A), engineering and geochemical data and the physical distribution of the hydrocarbons in these two reservoirs suggest there is vertical

Figure 19. (A) and (B) Salt Wash sandstones on outcrop at Deerlodge Park. (A) Note Morrison (Brushy Basin member) shales in background and vertical fractures. (B) Note high angle, planar, tabular cross stratification, perhaps formed under the influence of reversing tidal currents. (C) and (D) are photomicrographs of Salt Wash reservoir sandstone from Wilson Creek Unit 65 core showing the angularity of detrital quartz grains and porosity (blue). Note reverse graded bedding in (C) and black residual oil staining in (D). Photo (C) is from 6623.9 ft (2019 m), and (D) is from 6582.2 ft (2006 m) of the cored interval shown in Figure 18.

fluid and pressure communication between them, so that the Salt Wash and Entrada sandstones are part of a common-pool reservoir (Figure 25).

Pennsylvanian Minturn

Pennsylvanian production was established at Wilson Creek field in 1985 (Figure 6). The gamma-ray, dual induction-SFL log of the Paleozoic section penetrated in the Texaco Wilson Creek Unit 66 (NE SW NW of sec. 34, T3N, R94W) is shown in Figure 15B and C. The lithologic log from the nearby Unit 20 (NE SW NE of sec. 34) is incorporated into the middle of the Unit 66 log and adjusted to the log tops. Both of these wells penetrated the Paleozoic section and bottomed in Cambrian. Neither well reached the Precambrian.

It is reported that the Texaco Unit 66 was completed in selected intervals from 9541 to 11,550 ft (2908 to 3520 m) in the Pennsylvanian Minturn Formation for a producing rate of 18 BOPD (58.5° API), 914 MCFD, and 109 BWPD (no water is reported in later production statistics; see Figure 6). Hydrocarbons were recovered at commercial rates from at least three intervals (Figure 15C), however, suggesting that total Minturn potential could be much greater. Production from this single Minturn well in 1988 was 10,140 BO, 221,751 MCFG, and no water. Apparently the gas from this well is used only for field operations, thus restricting producing rates. The perforated interval from 10,606 to 10,638 ft (3233 to

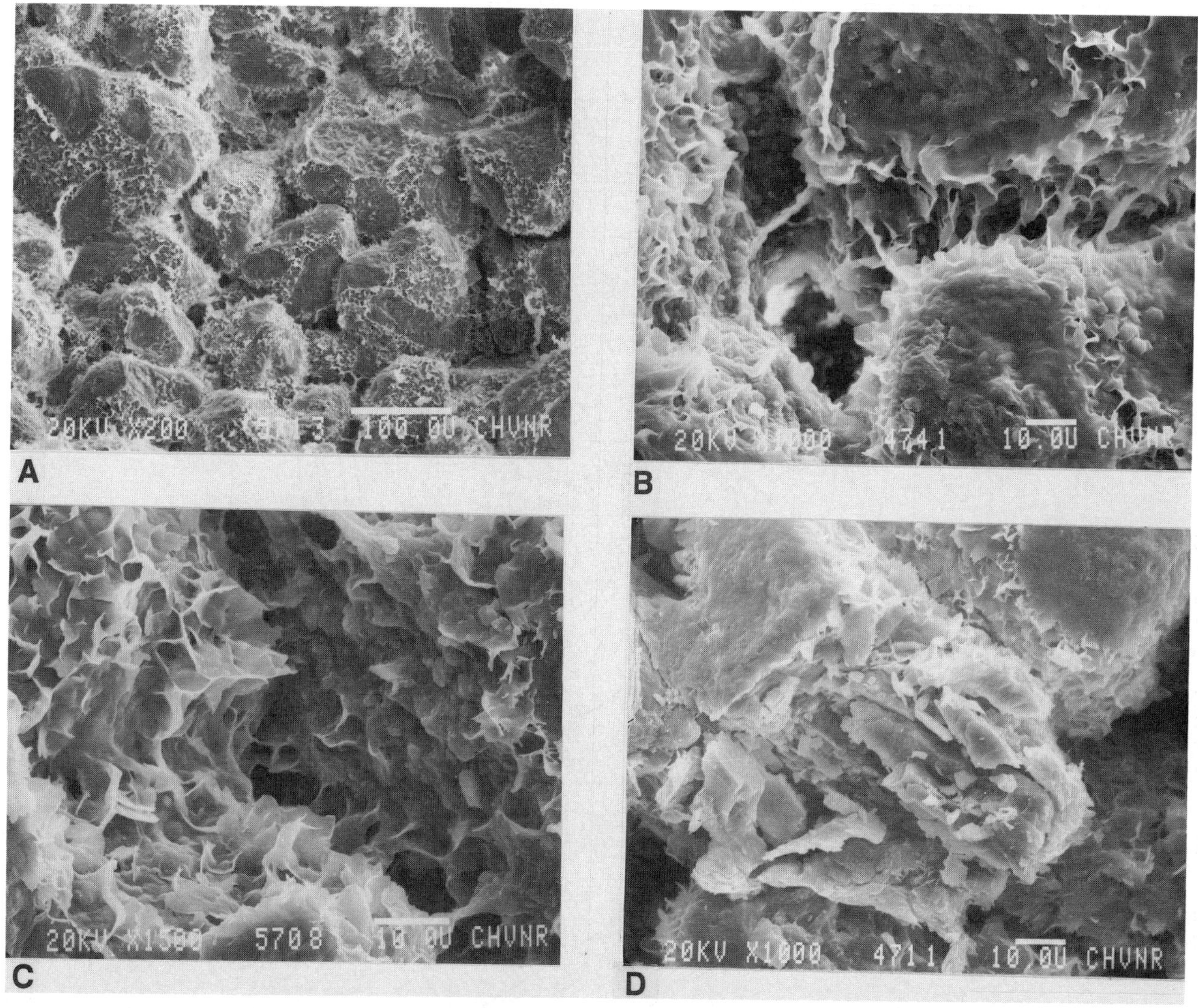

Figure 20. SEM photos of Salt Wash reservoir sandstone from Unit 65 core (Figure 18). (A), (B), and (C) show that the detrital quartz grains are coated with authigenic illite-smectite and less abundant chlorite clays. Scales are 200× in (A), 1000× in (B) and (D), and 1500× in (C). White bar, 10 micrometers (3.9×10^{-5} in.). Note in (B) and (C) that these swelling-type clays bridge pore-throats, reducing permeability. Porosity consists of well-connected primary intergranular and secondary feldspar dissolutional (D) pores.

3242 m) was reported as algal limestone; whereas the interval from 10,510 to 10,530 ft (3203 to 3210 m) was reported as very calcareous sandstone.

The regional potential of the Minturn objective seems to be confirmed by Minturn new pool discoveries at Maudlin Gulch field (8 mi [13 km] northwest of Wilson Creek; Figure 1) where the Texaco State of Colorado AD 1 (reentry; sec. 36, T4N, R95W) had a reported completion rate of 225 BOPD (of 55° API gravity), 2700 MCFD, and 215 BWPD, and at Temple Canyon field where the Texaco 2 Sweeney was completed flowing 321 BOPD (of 42° API gravity) and 547 MCFG from perforations at 9354 to 9486 ft (2851 to 2891 m) in the Minturn. Also, 2733 bbl of high-gravity oil were recovered from the Minturn in a much earlier well at Moffat Dome field (Figures 1 and 8). Texaco and Fina abandoned a 10,000 ft (3024 m) Mississippian Leadville test at Moffat Dome field in 1986 after recovering oil at noncommercial rates from the Minturn.

The Amstrat lithologic study of samples from the Wilson Creek Unit 20 (Figure 15) describes the upper 600 ft (182 m) of Minturn (Morgan) rocks as mostly sandstone, buff, red, brown, very fine to fine grained, with interbedded gray-brown and black, dense limestone, occasional gypsum and anhydrite, with a 35 ft (11 m) "fusiline" limestone at the top. The lower 1400 ft (427 m) of Minturn (to the top of the Belden Formation) is primarily limestone, gray-black or brown, with a distinctive (approximate) 200 ft (61 m) zone of interbedded evaporites at the top and some thin black shales at the base. Although this lithologic

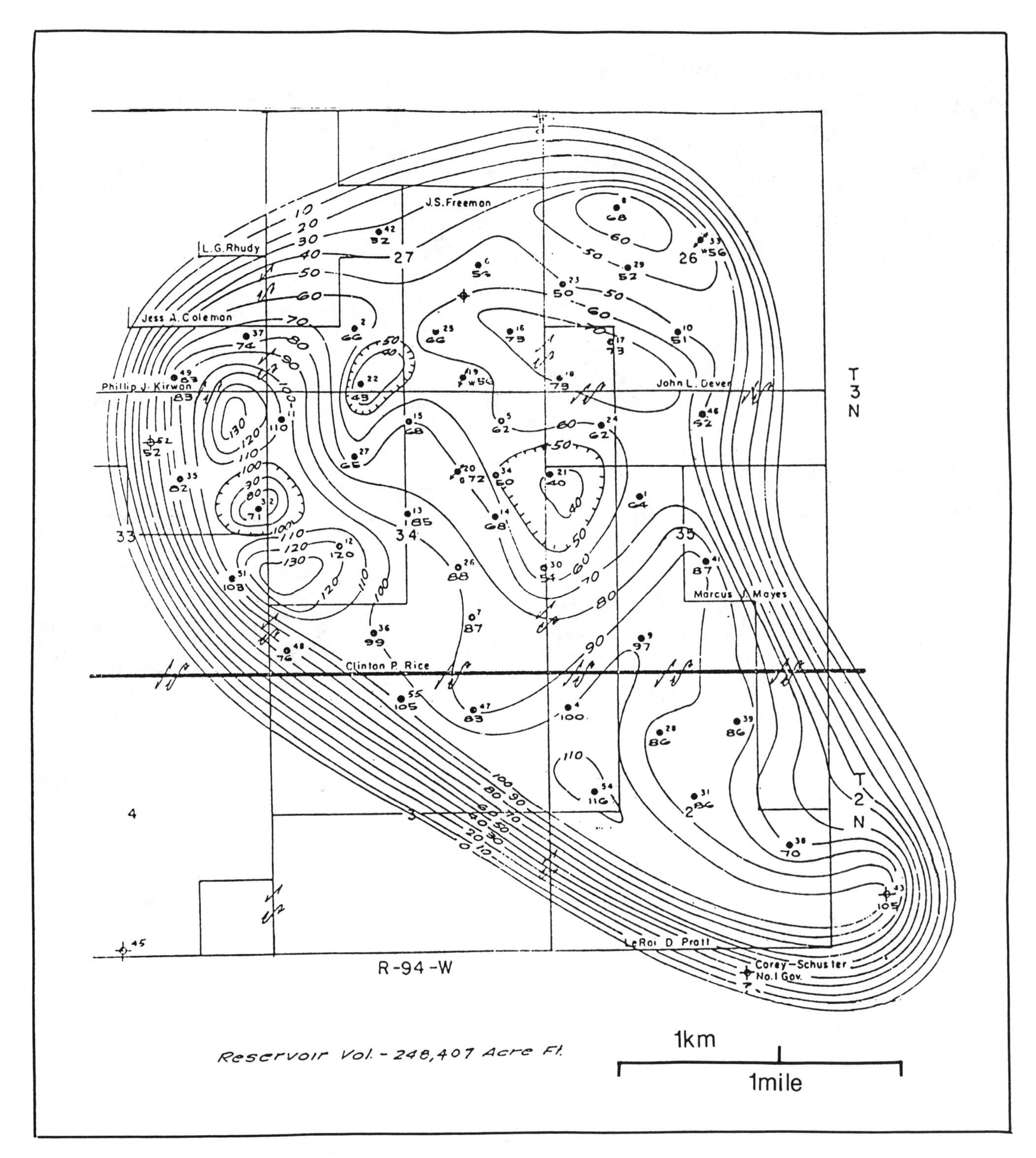

Figure 21. Net-pay sandstone isochore map of Salt Wash pool at Wilson Creek field drawn by operator in 1964. Isochore interval is 10 ft (3 m).

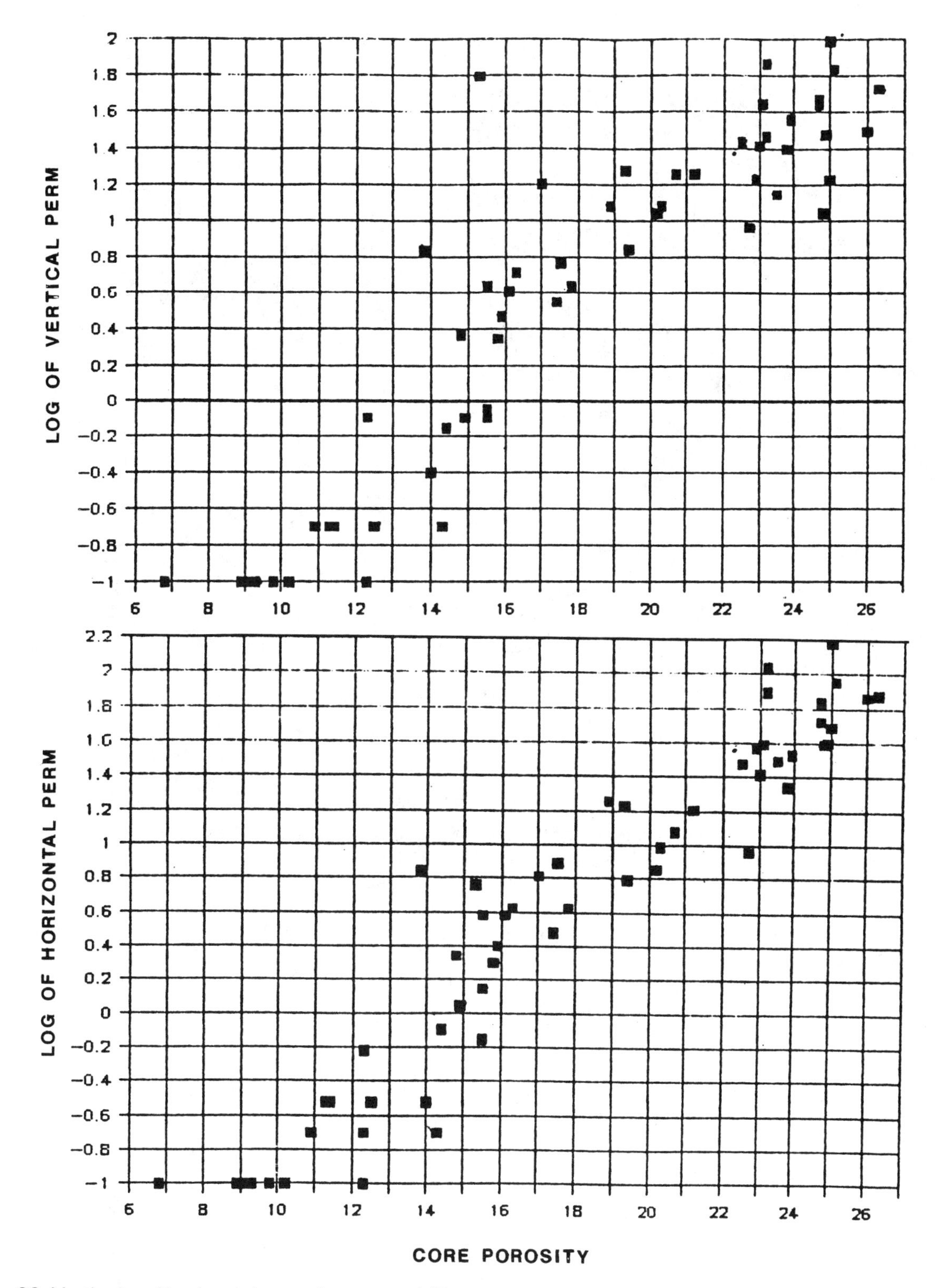

Figure 22. Vertical and horizontal porosity-permeability cross-plots of Salt Wash reservoir sandstone from Unit 65 core. Note similarity of the two plots.

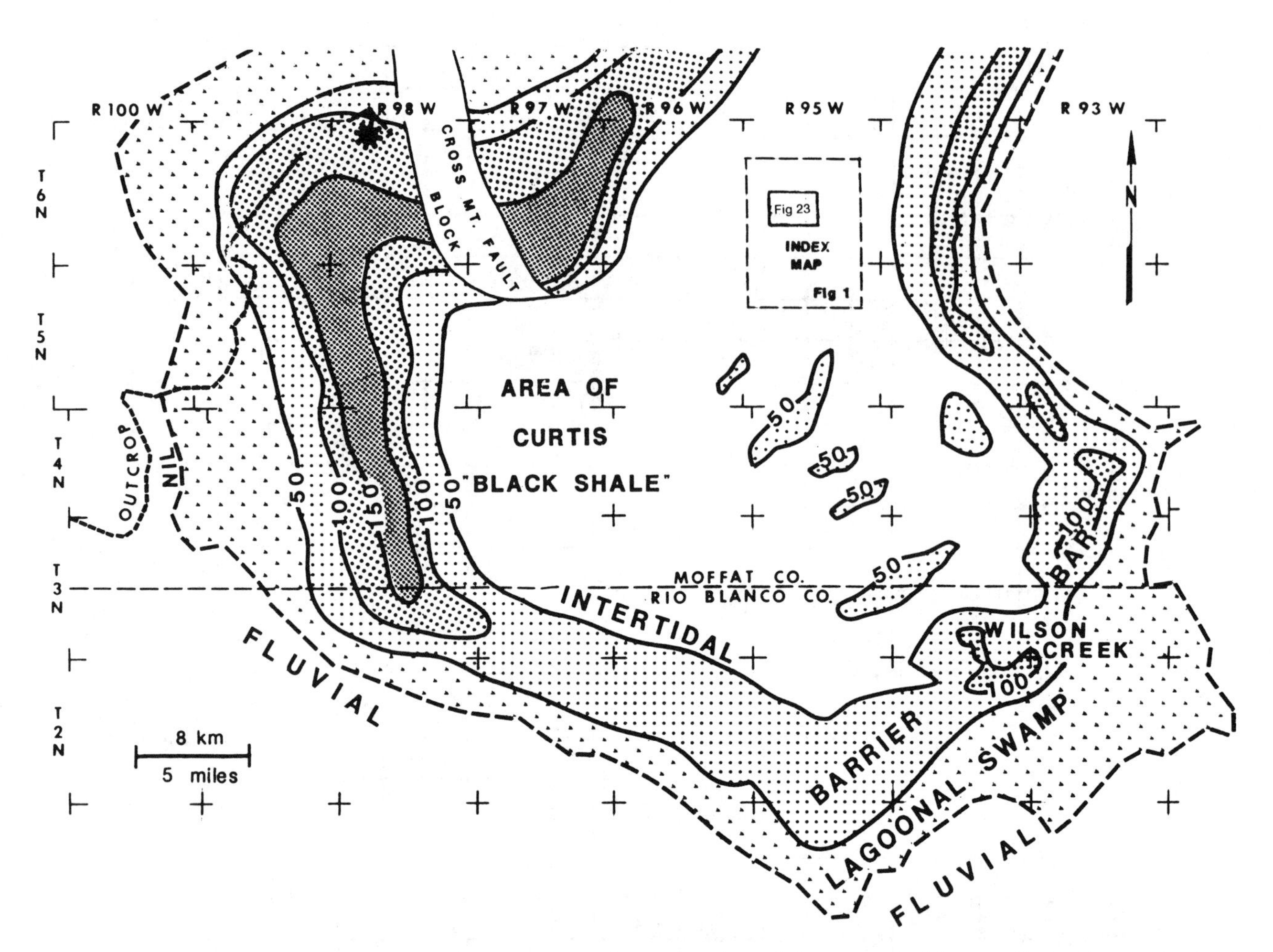

Figure 23. Environmental interpretation, net sand porosity (>10%) isopach map of Salt Wash sandstone. Isopach interval is 50 ft (15 m), and each interval is highlighted by a different stipple pattern. The arcuate porosity trend is considered a barrier bar or beach sand buildup deposited along the shoreline of the "Wilson Creek embayment" during Late Jurassic regression. Coastal sand dunes probably contributed to the stacked sandstone buildups. Cross Mountain may have had some expression at the time of deposition, and Laramide fault offsets could complicate the isopach pattern. Star is the Deerlodge Park outcrop. (After Stone, 1986a.)

log does not mention algal material, some algal limestone fragments have been reported in the cuttings of the Unit 66 well.

Boggs (1966), Tillman (1971), and Walker (1972) working in the eastern part of the Eagle basin near the town of Minturn describe shallow-water, algal bioherms and biostromes within the Minturn section. Similar algal structures occur at the Miller Creek outcrop on the Yellow Jacket anticline some 15 mi (24 km) southeast of Wilson Creek field (DeVoto, 1980; Brinton and Wray, 1986). These algal mounds are similar in many respects to those encountered in Paradox and San Juan basin Pennsylvanian carbonate reservoirs, containing typical colonies of *Chaetetes* corals and phylloid algae (*Ivanovia*). At Wilson Creek, however, the hydrocarbons recovered in tests of the Minturn section may have come from the sandstones as well as from the carbonates. In either lithology, fracturing probably is required for commercial production.

UNPRODUCTIVE RESERVOIR ROCKS

Dakota

Although the Dakota sandstones at Wilson Creek field are well developed, appear to exhibit good reservoir characteristics, and are under 1000 ft (305 m) of structural closure, oil or gas at commercial rates has not been recovered from this formation. A number of drill-stem tests of the Dakota interval have yielded only water with some oil and gas shows.

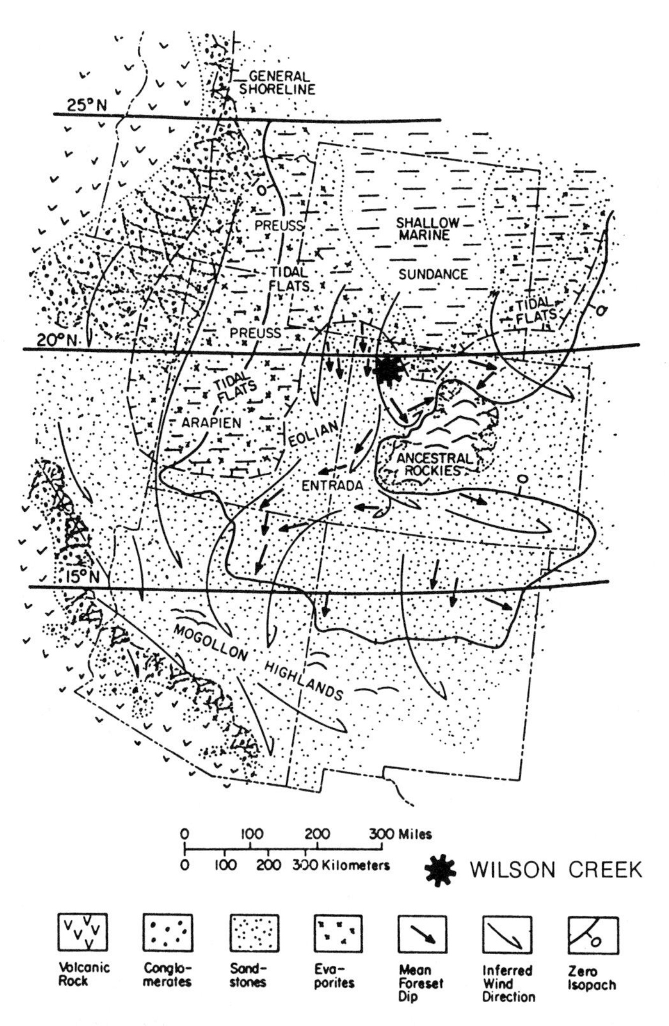

Figure 24. Middle Jurassic (early-middle Callovian) paleogeography (modified after Kocurek and Dott, 1983).

This condition is clearly anomalous as the Dakota is an important producing zone in the nearby Maudlin Gulch field (Gibbs, 1982) and other fields along the Danforth Hills anticline. Since Dakota sandstone benches are generally tested together at Wilson Creek, there is still the possibility that water recovery from one bench has masked the presence of hydrocarbons in another.

Weber

It is equally anomalous that hydrocarbons in commercial quantities have not been recovered from the Weber sandstone at Wilson Creek field. However, the two deep tests within the Wilson Creek closure, the earlier Unit 20 well and the recent Unit 66 well, both had live oil shows in the upper part of the Weber Formation. The record indicates that the Unit 20 well drilled in 1945 was left full of mud for a number of years before testing. Thus test results are questionable. The Weber in the Unit 66 well is reported to have been saturated with dark oil in cores (Figure 15B). Black, residual oil-staining was present on cross-beds.

Truncated cross-beds and routlets cut off at the top were observed in the Weber core, features considered indicative of eolian deposition. However, the upper 20–30 ft (6–9 m) of the Weber may have been reworked by marine processes (B. Huntsman, personal communication, 1987). The core was highly fractured with one natural, unhealed fracture occurring about every inch. Some 400–500 barrels of mud were lost at the top of the Weber in the Unit 66 well, and a poor cement job behind casing resulted from the loss of cement into the Weber fracture system. Two intervals were tested through pipe according to the completion record, but recoveries were not reported.

The eolian Weber sandstone (Fryberger, 1979; Fryberger and Koelmel, 1986) is described as fine grained, moderately sorted, usually well cemented, and generally of low permeability. The relatively low porosity and permeability of the Weber are not surprising considering its probable deep burial (estimated 18,000 ft; 5486 m) prior to Laramide uplift. Reservoir quality is apparently much poorer at Wilson Creek field than at Maudlin Gulch field (i.e., 5–10% vs. 10–15% porosity), where the Weber is oil productive (Gibbs, 1982). With effective fractures, however, it is still possible that the Weber may sustain commercial production at Wilson Creek in some future development test.

OIL AND FIELD CHARACTERISTICS

The oil recovered from the Salt Wash and Entrada zones has gravities of 47° to 50° API and a striking green color when fresh from the well head, turning dark brown on exposure to the air. U.S. Bureau of Mines routine distillation analyses of these oils are presented in Table 1 and other geochemical data in Table 2. These crude oils are about 70% gasoline and naphtha and only 10% residuum and contain 0.12 wt% sulfur. Viscosity is 33 Saybolt universal seconds at 100°F. The Minturn oil or condensate is about 60° API gravity and essentially colorless with perhaps a slight yellow tinge. This oil would appear to be simply a more mature, thermally altered version of the oil in the shallower Jurassic zones (cf. chromatograms in Figure 26).

A summary of engineering data from the Wilson Creek Salt Wash and Entrada reservoirs compiled from various sources is shown in Table 3. Note that the original reservoir pressure in the Salt Wash reservoir was reported at 2184 psi, while the bubble point pressure was only 1067 psi, so that the reservoir was undersaturated at discovery. Bubble point in the Entrada was even less, at 240 psi, and with only 110 cf gas per bbl of oil. Water drive is active in

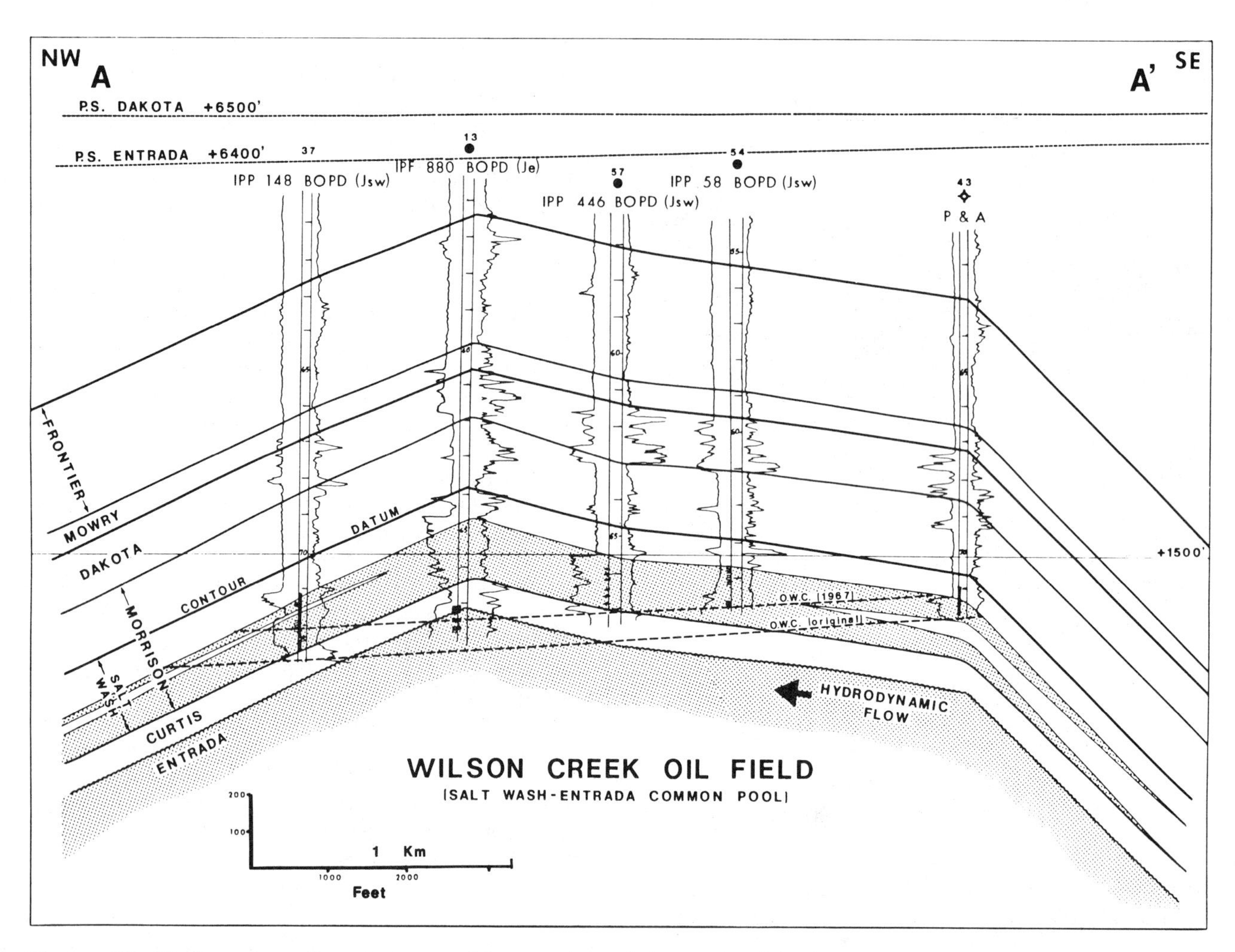

Figure 25. Northwest-southeast structural log section (pre-1970 SP and resistivity) across Wilson Creek field (see Figure 11 for location) illustrating the southeasterly pinchout of Salt Wash barrier bar sandstones, amount of structural relief, fracture communication between the two sandstone reservoirs, and hydrodynamic tilt (after Stone, 1986a). Drill depths (× 100 ft) shown in 100 ft intervals. Stipple pattern, porous sandstone. P. S., potentiometric surface. Jsw, Salt Wash completion; Je, Entrada completion. O.W.C., estimated oil-water contact (tilted) of Salt Wash-Entrada common pool at 1938 discovery ("original") and in 1967 after water encroachment.

the blanket Entrada sandstone and high recoveries are expected (i.e., 460 bbl/ac-ft), but it is considerably less active in the lenticular Salt Wash sandstones.

SOURCE(S)

Possible source rocks in the northwestern Colorado area include (1) the Cretaceous Mowry-Mancos shales, (2) the Jurassic Curtis dark shales, (3) the Permian Phosphoria marine mudstones and shales (present to the west and northwest of Wilson Creek), and (4) the Pennsylvanian Minturn and Belden dark shales. While the quality of the dark-colored Mancos, Phosphoria, and Minturn-Belden shales and mudstones as source rocks is established in the literature (see Woodward et al., eds., 1984; Nuccio et al., 1988), source potential of the Curtis Shale has not been confirmed by publication of oil-source rock correlations.

Salt Wash-Entrada Pool(s)

Analyses of three core samples of dark, calcareous Curtis Shale from the Wilson Creek Unit 47 (sec. 3, T2N, R94W) indicate low organic carbon content (i.e., average retortable carbon/organic carbon = 0.34), and its role as a possible source rock is questionable. The geochemical signature of the 47–50° API, green, paraffinic-naphthenic oils contained in both the Salt Wash and Entrada sandstone reservoirs at Wilson Creek field is very similar to, but somewhat more mature than, most oils presumed to have come from Cretaceous Mowry-Mancos shale

Table 1. U.S. Bureau of Mines routine distillation analysis of crude oils from the Salt Wash (Morrison) and Entrada (Sundance) reservoirs, Wilson Creek field (from Wenger et al., 1957).

REPORT OF CRUDE PETROLEUM ANALYSIS

Bureau of Mines Laramie Laboratory
Sample PC-49-857

IDENTIFICATION

Wilson Creek field
Morrison sandstone
6664-6681 feet

Colorado
Rio Blanco County
Sec. 35, T. 3N., R. 94W.

GENERAL CHARACTERISTICS

Specific gravity, 0.788 A. P. I. gravity 48.1° Pour point, ° F. 25
Sulfur, percent, 0.12 Color, brownish-green
Saybolt Universal viscosity at 100° F., 33 sec.; at ° F., sec.

DISTILLATION, BUREAU OF MINES ROUTINE METHOD

STAGE 1—Distillation at atmospheric pressure, 760 mm. Hg.
First drop, 26 ° C. (79 ° F.)

Fraction No.	Cut at ° C.	Cut at ° F.	Percent	Sum. Percent	Sp. Gr., 60/60° F.	° A. P. I., 60° F.	C. I.	Aniline Point, ° C.	S. U. Visc., 100° F.	Cloud Test, ° F.
1	50	122	7.5	7.5	0.636	91.0				
2	75	167	5.6	13.1	665	81.3	5.1	59.2		
3	100	212	8.1	21.2	708	68.4	16	57.0		
4	125	257	9.0	30.2	734	61.3	19	54.8		
5	150	302	8.0	38.2	755	55.9	21	53.7		
6	175	347	5.8	44.0	773	51.6	23	55.0		
7	200	392	5.9	49.9	786	48.5	23	58.2		
8	225	437	5.2	55.1	801	45.2	24	64.4		
9	250	482	5.8	60.9	814	42.3	25	68.8		
10	275	527	6.5	67.4	828	39.4	27	72.4		
STAGE 2—Distillation continued at 40 mm. Hg.										
11	200	392	2.9	70.3	0.843	36.4	30	77.5	40	20
12	225	437	6.0	76.3	847	35.6	28	82.7	45	40
13	250	482	4.1	80.4	857	33.6	30	87.2	55	55
14	275	527	3.9	84.3	871	31.0	33	91.4	74	75
15	300	572	3.9	88.2	880	29.3	35	96.6	115	90
Residuum			10.8	99.0	929	20.8		too dark		

Carbon residue of residuum, 4.8 percent; carbon residue of crude, 0.6 percent.

APPROXIMATE SUMMARY

	Percent	Sp. Gr.	° A. P. I.	Viscosity
Light gasoline	21.2	0.671	79.4	
Total gasoline and naphtha	49.9	0.721	64.8	
Kerosine distillate	11.0	808	43.6	
Gas oil	15.0	838	37.4	
Nonviscous lubricating distillate	8.9	852-.877	34.6-29.9	50-100
Medium lubricating distillate	3.4	877-.885	29.9-28.4	100-200
Viscous lubricating distillate				Above 200
Residuum	10.8	929	20.8	
Distillation loss	1.0			

REPORT OF CRUDE PETROLEUM ANALYSIS

Bureau of Mines Laramie Laboratory
Sample PC-49-254

IDENTIFICATION

Wilson Creek field
Sundance sandstone
6551-6570 feet

Colorado
Rio Blanco County
Sec. 34, T. 3 N., R. 94 W.

GENERAL CHARACTERISTICS

Specific gravity, 0.792 A. P. I. gravity 47.2° Pour point, ° F. below 5
Sulfur, percent, 0.12 Color, brownish-green
Saybolt Universal viscosity at 100° F., 33 sec.; at ° F., sec.

DISTILLATION, BUREAU OF MINES ROUTINE METHOD

STAGE 1—Distillation at atmospheric pressure, 760 mm. Hg.
First drop, 22 ° C. (72 ° F.)

Fraction No.	Cut at ° C.	Cut at ° F.	Percent	Sum. Percent	Sp. Gr., 60/60° F.	° A. P. I., 60° F.	C. I.	Aniline Point, ° C.	S. U. Visc., 100° F.	Cloud Test, ° F.
1	50	122	5.8	5.8	0.633	92.0				
2	75	167	5.4	11.2	662	82.2	3.7	55.5		
3	100	212	8.3	19.5	707	68.6	15	52.5		
4	125	257	8.7	28.2	733	61.5	18	56.2		
5	150	302	7.2	35.4	754	56.2	21	55.4		
6	175	347	6.9	42.3	774	51.3	23	55.3		
7	200	392	5.3	47.6	787	48.3	23	59.2		
8	225	437	5.7	53.3	799	45.6	24	64.1		
9	250	482	5.7	59.0	813	42.6	25	69.0		
10	275	527	6.5	65.5	827	39.6	27	72.5		
STAGE 2—Distillation continued at 40 mm. Hg.										
11	200	392	4.8	70.3	0.844	36.2	31	77.8	40	25
12	225	437	4.7	75.0	851	34.8	30	83.0	46	40
13	250	482	5.3	80.3	859	33.2	31	87.7	57	60
14	275	527	3.7	84.0	873	30.6	34	92.2	80	80
15	300	572	4.6	88.6	882	28.9	36	98.0	135	95
Residuum			9.9	98.5	933	20.2		too dark		

Carbon residue of residuum, 5.0 percent; carbon residue of crude, 0.6 percent.

APPROXIMATE SUMMARY

	Percent	Sp. Gr.	° A. P. I.	Viscosity
Light gasoline	19.5	0.673	78.8	
Total gasoline and naphtha	47.6	0.723	64.2	
Kerosine distillate	11.4	806	44.1	
Gas oil	15.5	839	37.2	
Nonviscous lubricating distillate	9.2	854-.877	34.2-29.9	50-100
Medium lubricating distillate	4.9	877-.887	29.9-28.0	100-200
Viscous lubricating distillate				Above 200
Residuum	9.9	933	20.2	
Distillation loss	1.5			

Table 2. Geochemical comparisons of oils from Salt Wash and Entrada reservoirs, Wilson Creek field (modified after Stone, 1986a).

Reservoir	API Gravity	Light Gasoline Percent	Residuum Percent	Sulfur Weight Percent	Paraffin-icity	Carbon Isotopes $^{13}C/^{12}C$ $\delta(0/00)$	Infrared @14.35μ	Porphy-rins	PR/PH
Salt Wash	48.3°	21.2	10.8	.12	2.62	+1.4	.057	none	1.57
Entrada	47.3°	19.5	9.9	.12	2.48	+1.6	.059	none	1.62

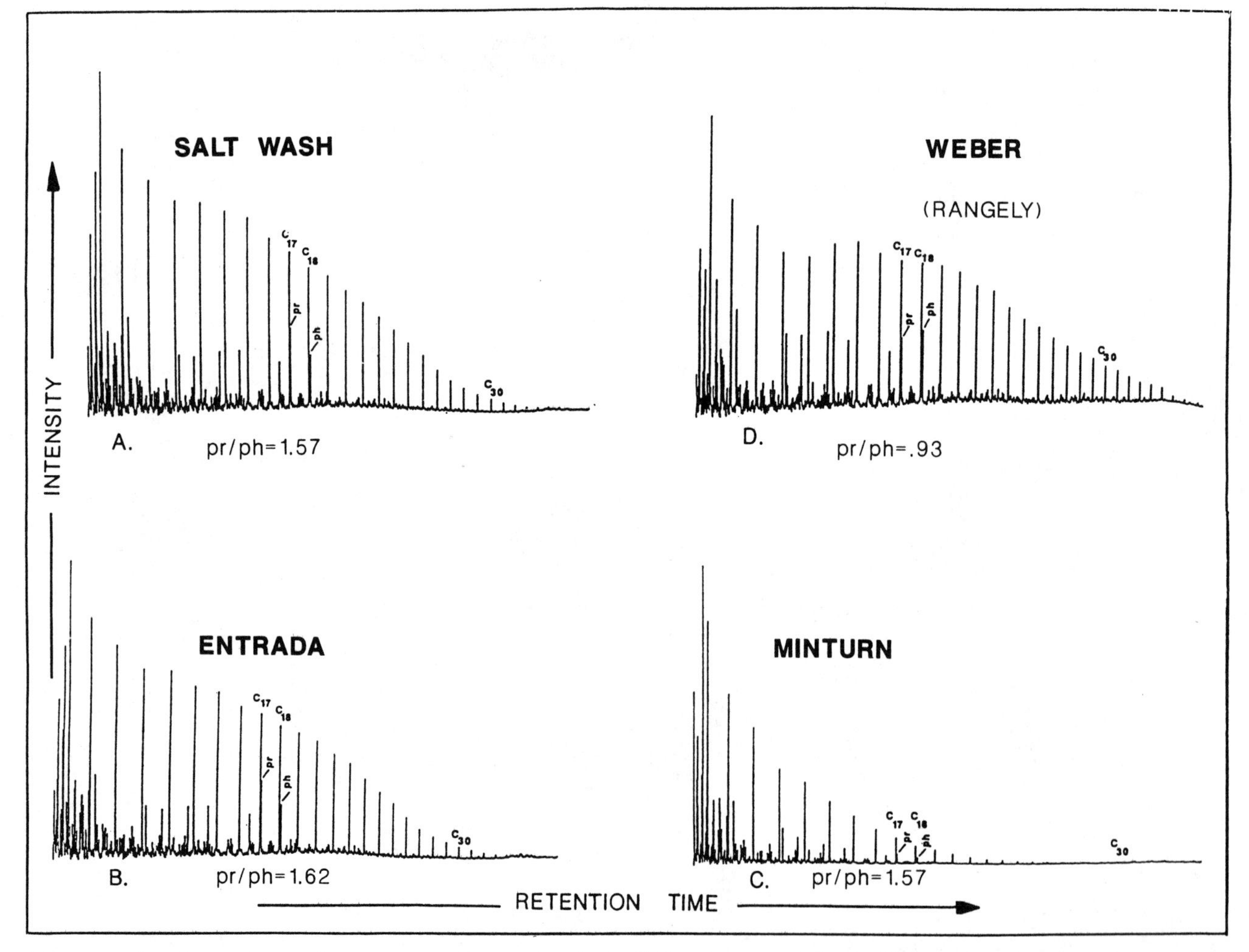

Figure 26. Gas chromatograms of C_{9+} saturated hydrocarbons in oils from Wilson Creek and Rangely fields. (A), (B), and (C) are Salt Wash, Entrada, and Minturn oils, respectively, from Wilson Creek field. (D) is Weber oil from Rangely field. Pristane/phytane (Pr/Ph) ratios of all three Wilson Creek oils are near 1.6, typical of oils from a Cretaceous Mowry-Mancos source; whereas the Rangely Weber oil has a Pr/Ph ratio of .93 typical of oils from a Paleozoic (Phosphoria) source.

Table 3. Reservoir data from engineering report by operator (1964), Wilson Creek field.

	Salt Wash	Entrada
Approx. avg. drill depth, ft:	6500	6700
Gravity (API°); oil base:	48-50; paraffin	47-50; paraffin
Avg. pay thickness, ft:	63.6	40.1
Max. pay thickness, ft:	130	156
Original productive area, ac:	3904	1297
Pay sand volume, ac-ft:	248,407	52,037
Avg. porosity, %:	16	19.6
Avg. porosity, range:	10-23	14-27
Avg. permeability, md:	66	38
Avg. connate water saturation, %:	39	33
Water characteristics, total solids:	3332	19,306
RW @ 68°F:	1.85	.41
Reservoir temp., °F:	172	171
Reservoir drive:	volumetric & partial water drive	water drive
Original reservoir pressure, psi:	2184	2202
Bubble point pressure, psi:	1067	240
Solution gas @ bubble point, CF/B:	420	110
B.T.U. of gas:	1685	—
Specific gravity of gas:	0.985	—
Fm. volume factor @ original pressure:	1.227	1.105
Oil originally in place:		
M reservoir bbl:	186,913	53,200
M stock tank bbl:	146,380	48,200
Recovery factor, %:	45	50
Est. cumm. recovery, bbl:	66,000,000	24,000,000
Approx. recovery factor, B/ac-ft:	268	460

source rocks throughout northwest Colorado. This conclusion is based on a number of geochemical characteristics (e.g., Table 2) including pre-1970 program temperature chromatography (PTC), the shape of Correlation Index (C.I.) curves (cf. Stone, 1986d) from the U.S. Bureau of Mines routine distillation analyses (see Smith, 1940; Smith and Hale, 1966) shown in Figure 27, and data from recent gas chromatography (Figure 26). The pristane/phytane ratios of 1.5 to 2.0 measured from the Salt Wash and Entrada chromatograms (Figure 26 and Table 2) are typical of Cretaceous oils throughout the Rocky Mountain area.

A maturity development plot shown in Figure 28A indicates that Cretaceous Mancos Shale source rocks reached the oil window (zone 3) in early Tertiary about the time of early Laramide deformation. All along the deep footwall of the Danforth Hills thrust, Mowry-Mancos source rocks were within the oil window as indicated by the maximum depth of burial of about 15,000 ft (4572 m) on the burial-history plot (Figure 28B). The maturity development plot (Figure 28A) is based on the present 1.6°F/100 geothermal gradient measured from well logs (corrected), which is more likely to be *lower* rather than *higher* (Nuccio and Schenk, 1987) than the paleogeothermal gradient. Lower Mancos shales in the footwall may have reached zone 4 (wet gas) if the paleogeothermal gradient was significantly above 1.6°F/100, a condition that could have resulted from footwall temperature elevation produced by the overriding Wilson Creek thrust block (cf. Furlong and Edman, 1983). The light 50° API oils in the Salt Wash and Entrada reservoirs show evidence of thermal cracking and are clearly mature (Figure 26).

Minturn Pool(s)

The dark appearance of the Minturn and Belden shales generally has been considered evidence of its potential role as a source rock (e.g., Sanborn, 1971; MacMillan, 1980, p. 194). Analysis (in 1967) of a Minturn black shale fragment from the Chevron Yellow Jacket Unit 1 well (sec. 9, T1N, R91W) indicated good source-rock characteristics based on a retortable carbon/organic carbon value of .91 (Stone, 1986a). From samples collected along the outcrop in the nearby Eagle basin (Figure 1), Nuccio and Schenk (1986) report total organic carbon (TOC) values range from 0.77 wt% to 2.10 wt% in the Minturn shales and 0.45 wt% to 5.85 wt% in the Belden Formation, indicating that these are, or were, probable source rocks. Vitrinite reflectance (R_o) values of 2.4% to 4.12% for the Belden and 0.69% to 2.62% for the Minturn indicate that these rocks are now supermature with respect to the oil window but may still be in the upper range of gas generation. Both marine (type II) and terrestrial (type III) organic material occur in the Belden Formation, which could have provided sources for both oil and gas.

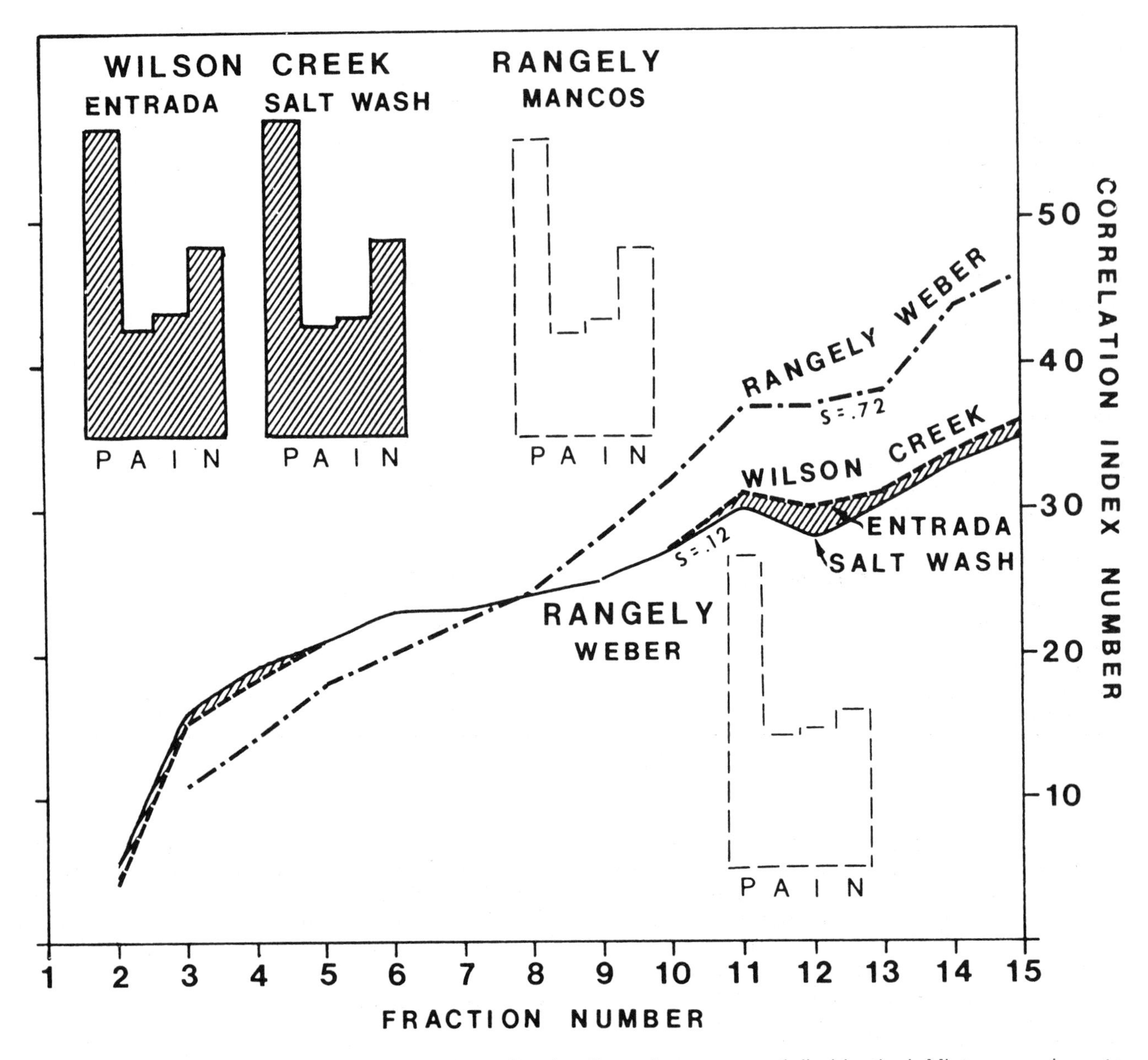

Figure 27. Geochemical comparisons: Wilson Creek and Rangely fields. In the upper half of the figure are chromatographic (C_1-C_{20}) diagrams that characterize oil through analysis of heights of 25 component peaks. The results are shown graphically in terms of paraffins (P), aromatics (A), isoparaffins (I), and naphthenes (N), or "PAIN." The diagrams of Entrada and Salt Wash oils at Wilson Creek and Cretaceous Mancos oil at Rangely are essentially identical. Minturn condensate also falls in this category. The U.S. Bureau of Mines Correlation Index (C.I.) curves of Salt Wash and Entrada oils also have a typical Cretaceous signature. But the PAIN diagram and C.I. curve of Weber oil from Rangely field are distinctly different and typical of oils from Paleozoic (Phosphoria) source rocks. "S" is sulfur content (wt%).

The Minturn-Belden source rocks passed through the oil window long before the end of Cretaceous time and the onset of the Laramide orogeny (Figure 28). Thus, if the condensate in the Minturn zone(s) at Wilson Creek field was generated from indigenous source rocks it must have been stored in a primary stratigraphic or early structural trap and later remigrated into the Laramide Wilson Creek structure. This possibility is not supported by available geological evidence. Furthermore, vitrinite reflectance measured in Belden Shale from the nearby Gilman outcrop locality (Nuccio et al., 1988) is 3.7%, corresponding to a TTI value of between 40,000 and 50,000 (Waples, 1980), which is well beyond the oil window and into the dry gas zone. The burial history and maturity development plots suggest that these Pennsylvanian shales may have passed through the oil window as early as Permian time and perhaps even earlier if a high geothermal gradient is assumed (Nuccio and Schenk, 1987), but Wilson Creek dome probably was not incipient until very latest Cretaceous.

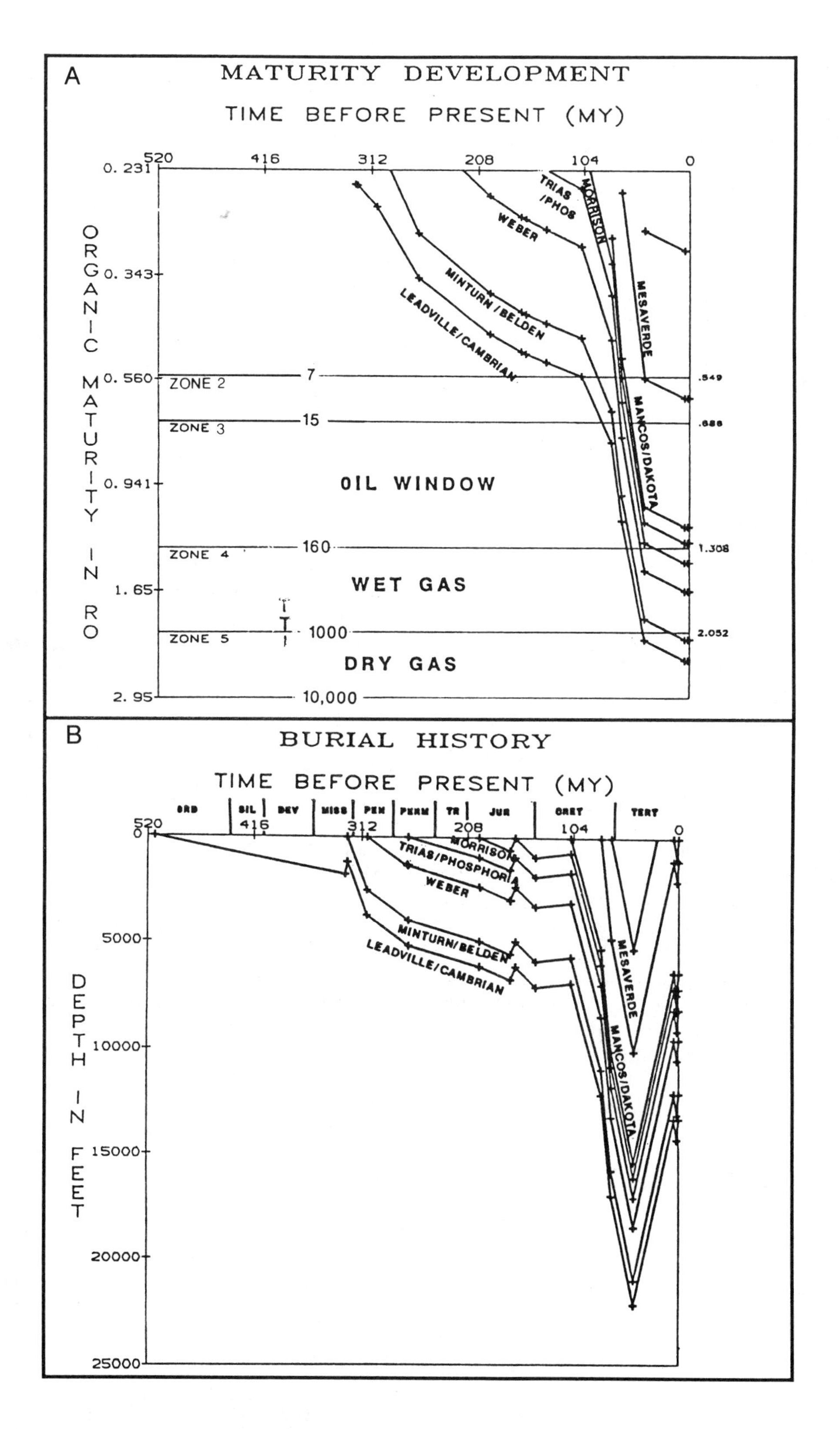

Figure 28. (A) Maturity development for Wilson Creek. Note that Pennsylvanian Minturn-Belden (Pmi) source rocks passed through the oil window during Late Cretaceous prior to Laramide uplift, while Cretaceous Mowry-Mancos source rocks entered the oil window at this time and still remain there. Within the deeper Piceance Creek basin (in the footwall of the Danforth Hills thrust), the Minturn-Belden shales are now in the dry gas zone, while Mancos shales are in the high gravity (condensate) zone. R_o, vitrinite reflectance; TTI, time-temperature index (Waples, 1980). (B) Burial history for Wilson Creek field. Deepest burial was Late Cretaceous-early Tertiary, before the Laramide orogeny.

Based on comparisons of a chromatogram of the oil from the Minturn in the Wilson Creek Unit 66 well with chromatograms of produced Mancos and Weber oils from Rangely field, it is concluded that the Minturn oil, as well as the oil in the Entrada and Salt Wash at Wilson Creek field, are most like the oil produced from the Mancos Formation at Rangely field (Stone, 1986a). This would indicate that a "PAIN" diagram (a comparison of PTC peak heights for paraffins, aromatics, isoparaffins, and naphthenes to C_{20}) of Minturn oil should appear essentially the same as those shown and described in Figure 27 for the Salt Wash and Entrada oils at Wilson Creek field and the Mancos oil from Rangely field, except that the Minturn oil has been subject to greater thermal alteration by cracking and destruction of the heavier hydrocarbons, particularly above C_{20}.

If both the Salt Wash-Entrada and the Minturn reservoirs contain high gravity "Cretaceous-type" paraffinic crudes that were derived from a Mowry-Mancos source, then these oils must have migrated across or along the Danforth Hills thrust fault zone to charge the Salt Wash-Entrada and Minturn pools at Wilson Creek field. The low hydrocarbon column-to-closure ratio (350/1000) and the undersaturated oil in the Salt Wash-Entrada pool suggest that migration occurred early in the growth of the Wilson Creek BITGF. Hydrocarbon dispersal efficiency may also have been low, and gas may have been selectively lost along the fault zone in the migration process. The presence of probable Cretaceous oils in Jurassic and Pennsylvanian reservoirs at Wilson Creek field, however, does not rule out the possibility that either Curtis or Minturn-Belden shales may have contributed hydrocarbons to some fields in the northwest Colorado area. It is worth mentioning that the Pennsylvanian oils in the Paradox basin are relatively high-gravity, paraffinic crudes (Wenger and Reid, 1960; Hite et al., 1984) derived from black shales associated with evaporites, similar to the shales found in the Minturn.

HYDRODYNAMICS

A generalized potentiometric surface map for the Entrada Formation in northwest Colorado is presented in Figure 29. This map is based on drill-stem test pressure data, reported original reservoir pressures, outcrop elevations where cut by streams, and data from flowing water wells. The map is interpretive. Datum pressure comparisons (i.e., pressure-depth plots) indicate that the Salt Wash and Entrada sandstones are generally in vertical communication and probably comprise a hydrostratigraphic unit (notwithstanding the difference in reported salinities in Table 3), so that the potentiometric map can be assumed to have general application also to the Salt Wash reservoir. As illustrated in Figure 25, the Salt Wash and Entrada reservoir intervals probably are part of a common pool at Wilson Creek field and almost always contain oils of the same chemical composition in individual oil fields in the northwest Colorado area, even though this chemical composition (and source) may vary from field to field.

The potentiometric map shows a general north to northwest flow direction within the Entrada Formation with recharge occurring at outcrops to the south along the White River uplift-Yellow Jacket anticline complex. A regional, northerly gradient of about 25 ft/mi (7.6 m/km) across Wilson Creek field is suggested by the map. This hydrodynamic flow should have produced some tilt on the oil-water contact at least within the blanket-type reservoir sandstone of the Entrada at Wilson Creek; and indeed, this is what was observed in the early stage of field development. A contour map on the Entrada oil-water contact (Figure 30) drawn by the operator in 1950 shows a general northward to northeastward tilt. Ladd (1957, p. 25) noted that the original water table level was at about +1250 ft (381 m) on the southwest, dropping to about +1150 ft (351 m) on the northeast, a distance of about 2.5 mi (4 km). This would indicate a tilt on the oil-water contact of about 40 ft/mi (7.6 m/km) to the north-northeast, which equates to a potentiometric gradient of about 13 ft/mi (2.46 m/km) using a tilt factor of about 3.4 for 50° API gravity oil (see Levorsen, 1967, Figure 13-16, or Dahlberg, 1982, Table 5-1). The hydrodynamic tilt on the Salt Wash-Entrada pool at Wilson Creek is diagrammatically shown in Figure 25, as is the rise of the oil-water contact during the period of 1938 to 1967 caused by withdrawal of oil from the reservoir and accompanying water encroachment. General northerly hydrodynamic tilts can also be seen at Maudlin Gulch and Iles fields (Figure 29).

The patterns of northward hydrodynamic flow within the overlying Dakota sandstone and underlying Weber sandstone are very similar to that of the Entrada, but datum pressures for available reservoirs within the geologic column at any given location vary considerably. At Wilson Creek field early drill-stem test data indicate that the Dakota potentiometric surface was about 100 ft (30 m) higher than in the Salt Wash-Entrada pool (i.e., 6500 ft or 1980 m vs. 6400 ft or 1950 m) as illustrated in Figure 25. Theoretically, this datum pressure differential is sufficient to cause *downward* vertical migration of oil originally contained in the Dakota reservoir into the lower datum-pressured Salt Wash-Entrada pool if there was some avenue of vertical communication between the two reservoirs (i.e., fractures and/or faults). This datum pressure differential also could have inhibited *upward* migration from the Salt Wash pool to the Dakota reservoir. However, upward migration has not occurred at nearby Maudlin Gulch field where the Dakota pool contains oil of the normal Cretaceous type, but Salt Wash reservoirs contain an oil most like the (Phosphoria) oil found in the Weber pool in the same field, and at Rangely field (Stone, 1975, symposium). The chemical similarity of the oil in the Wilson Creek Salt Wash-Entrada

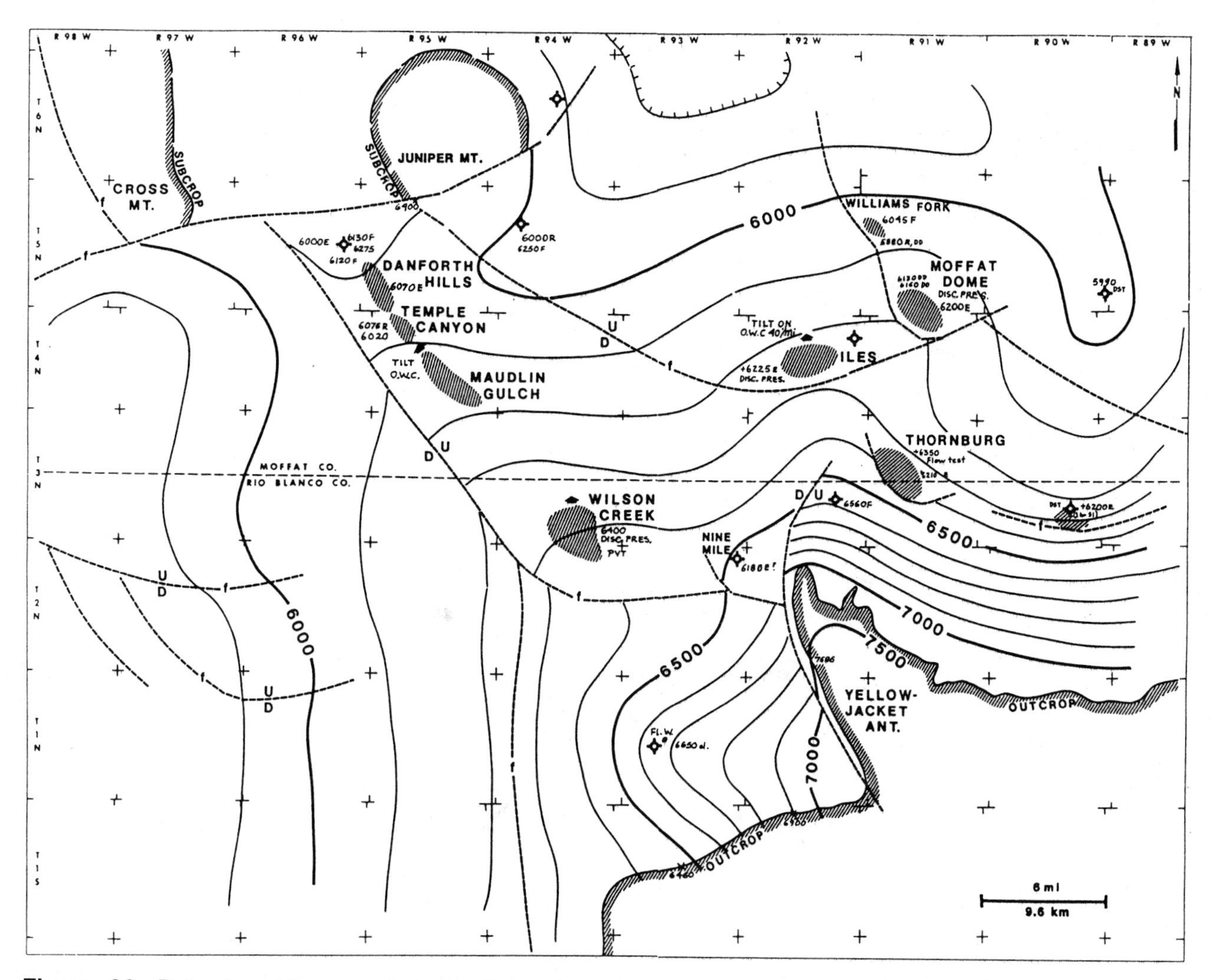

Figure 29. Potentiometric map, Entrada Formation, northwest Colorado. Small "f" denotes fault. Black arrows show direction of hydrodynamic tilt on oil-water contacts of named oil fields shown by cross-hatching (after Stone, 1986a).

pool and the oil found in Dakota pools of the area could be cited in support of the concept of thieving of Cretaceous oil by the lower datum pressure Jurassic reservoirs in the Wilson Creek field, but the relatively unaltered, mature character of this oil is a contradiction.

A potentiometric map has not been prepared for the productive Minturn Formation because there are insufficient data. However, logically, hydrodynamic flow in the Minturn also should be in a northward direction from the high outcrop elevations on the south and east. But low transmissibility is expected within the inhomogeneous, low-permeability Minturn section in northwest Colorado. In the few deep tests of the area, datum pressures in the Minturn are generally higher than in overlying zones.

EXPLORATION CONCEPTS

The discovery of oil at Wilson Creek field was probably not the result of any purposeful regional play, but confirms the lure of the closed contour. Persistence was the key to success at Wilson Creek dome, however, because the first few shallow tests here (considered "deep" when drilled) were failures and only deeper drilling in later years resulted in the field discovery.

An economic accumulation of hydrocarbons requires a reservoir rock, a hydrocarbon source, a trap, and favorable association of these factors over geologic time. At Wilson Creek, the Salt Wash and Entrada sandstone reservoirs were not the original objectives in the first or second round of exploratory drilling. These units generally had been considered products of a nonmarine (eolian and/or fluvial) environment and therefore not likely to be charged by *indigenous* hydrocarbons. Reevaluation of time-honored geologic suppositions sometimes leads to exploration success, however, as eventual deeper drilling resulted in the discovery of oil in the Jurassic section, and much later in the Pennsylvanian Minturn Formation. Fortuitously, these reservoirs received their hydrocarbon charge from migration along or through the flanking thrust fault zone from

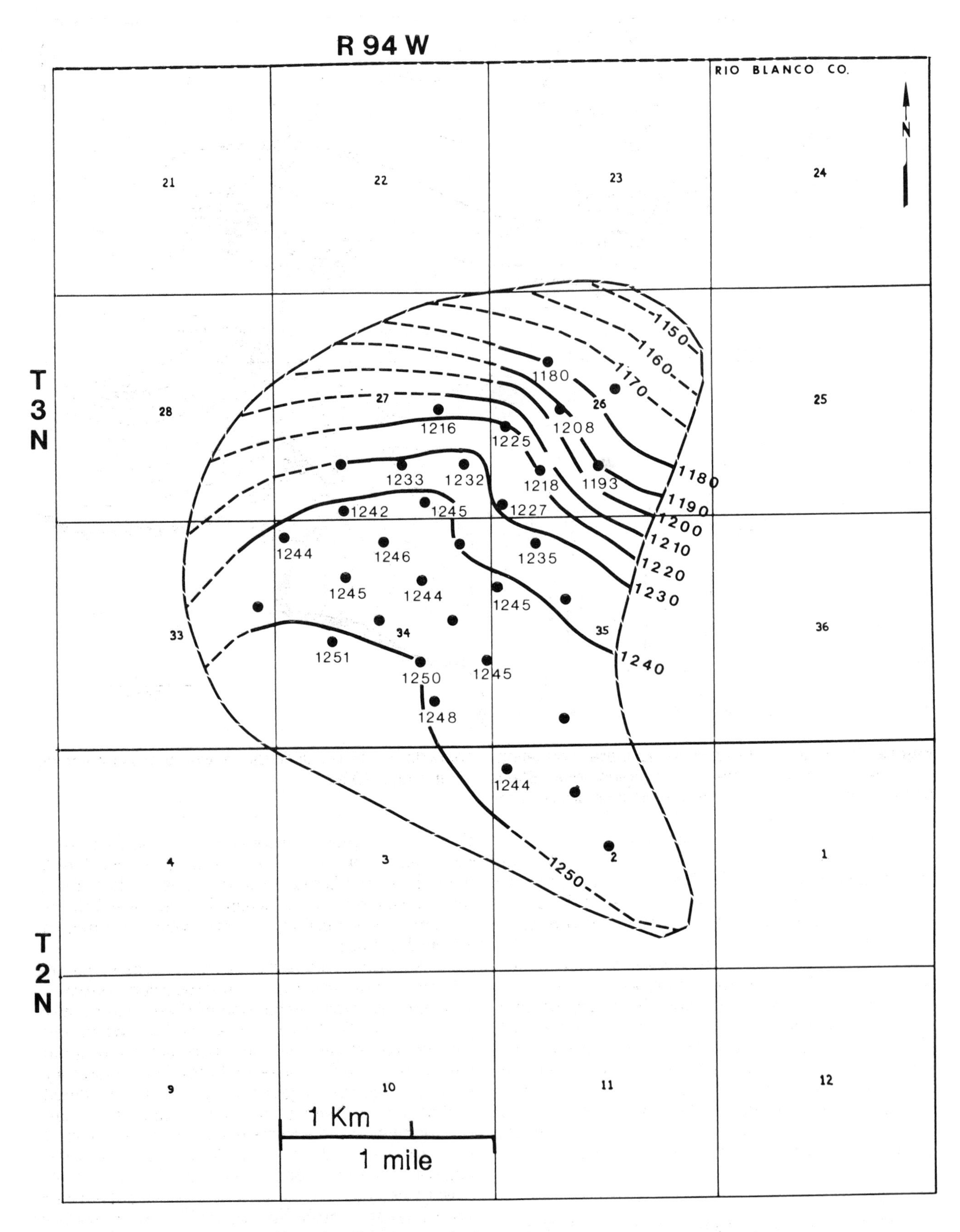

Figure 30. Contours on the Entrada oil-water contact in 1950, Wilson Creek field. Contour interval is 10 ft (3 m) (from Stone, 1986a).

(exotic) Cretaceous sources in the footwall sometime after the initiation of Laramide deformation.

Thus an important exploration concept confirmed by this study is that major hydrocarbon accumulations do not require indigenous sources, but may occur in any situation where migration paths are available between source and reservoir rocks in trapping situations, and where the timing is favorable. In the central Rocky Mountain area many important oil and gas fields have resulted from the migration of hydrocarbons derived from Cretaceous sources in the footwall, across regional Laramide thrust zones, into older reservoir rocks of hanging wall anticlinal traps. Some examples are fields of the Wyoming thrust belt (Warner, 1982), Lance Creek field (Hubbell and Wilson, 1963), and Big Medicine Bow field in the Laramie basin of southeastern Wyoming, where (as at Wilson Creek field) a high-gravity Cretaceous-type oil is produced from both Jurassic (Salt Wash and Entrada equivalents) and Pennsylvanian (Weber equivalent) sandstone reservoirs in a large BITGF structural trap. The importance of effective lateral and vertical migration through and along fault zones as an accumulation mechanism is shown not only by this study of Wilson Creek field but also by similar data compilations at Maudlin Gulch, Moffat Dome, Isles, and other smaller fields in the northwest Colorado area (Stone, 1975).

ACKNOWLEDGMENTS

Some of the critical supporting data in this report were contributed by the following people. Dana Rowan of Texaco (Englewood, Colorado) provided the production decline curve for the Salt Wash zone, burial-history and maturity development plots, and some of the data for Appendix 1. He also made arrangements for the collection of oil samples by Ron Scott at the Wilson Creek field office. Don Anders of the USGS (Lakewood, Colorado) ran analyses of the oil samples and provided the chromatograms. Brent Huntsman of Chevron (Englewood) arranged for my viewing of the Salt Wash core from the Unit 66 well; Tom Ahlbrandt of the USGS (Reston, Virginia) attended the viewing and shared his sedimentologic expertise. Brent Huntsman also provided the photomicrographs, and Pat Flynn of Chevron provided the SEM photos and descriptions of the Salt Wash core material. Ron Nordquist (ex-Tenneco) arranged for Tenneco's contribution of the seismic profile R-R′ across Wilson Creek field. Permission by the Rocky Mountain Association of Geologists to use parts of my earlier papers and the seismic profile across Maudlin Gulch field from Richard (1986) is gratefully acknowledged.

REFERENCES CITED

Berman, A. E., D. Poleschook, and T. E. Dimelow, 1980, Jurassic and Cretaceous systems of Colorado, *in* H.C. Kent and K.W. Porter, eds., Colorado geology: Rocky Mountain Association of Geologists, p. 111-128.

Boggs, S., Jr., 1966, Petrology of Minturn Formation, east-central Eagle County, Colorado: American Association of Petroleum Geologists Bulletin, v. 50, p. 1399-1422.

Bradley, W. A., 1955, Jurassic and pre-Mancos Cretaceous stratigraphy of the eastern Uinta Mountains, Utah-Colorado, *in* Geology of northwest Colorado: Intermountain Association of Geologists and Rocky Mountain Association of Geologists, p. 21-26.

Brinton, L., and J. L. Wray, 1986, Pennsylvanian (Minturn Formation) algal mound facies, Rio Blanco County, Colorado, *in* D. S. Stone, ed., New interpretations of northwest Colorado geology: Rocky Mountain Association of Geologists, p. 103-111.

Bryant, B., 1985, Structural ancestry of Uinta Mountains, *in* M. D. Picard, ed., Geology and resources of the Uinta Basin of Utah: Utah Geological Association, p. 115-119.

Condie, K. C., 1982, Plate-tectonics model for Proterozoic continental accretion in the southwest U.S.: Geology, v. 10, p. 37-42.

Dahlberg, E. C., 1982, Applied hydrodynamics in petroleum exploration: New York, Springer-Verlag, 161 p.

DeVoto, R. H., 1980, Pennsylvanian stratigraphy and history of Colorado, *in* H. C. Kent and K. W. Porter, eds., Colorado geology: Rocky Mountain Association of Geologists, p. 71-102.

DeVoto, R. H., B. L. Bartleson, C. J. Schenk, and N. B. Waechter, 1986, Late Paleozoic stratigraphy and syndepositional tectonism, northwestern Colorado, *in* D. S. Stone, ed., New interpretations of northwest Colorado geology: Rocky Mountain Association of Geologists, p. 37-49.

Duebenderfer, E. M., and R. S. Houston, 1986, Kinematic history of the Cheyenne belt, Medicine Bow Mountains, southeastern Wyoming: Geology, v. 14, p. 171-174.

Dyni, J. R., 1968, Geologic map of the Elk Springs Quadrangle, Moffat County, Colorado: USGS Geological Quadrangle Map GQ-702.

Fryberger, S. G., 1979, Eolian-fluviatile (continental) origin of ancient stratigraphic trap for petroleum in Weber Sandstone, Rangely oil field, Colorado: The Mountain Geologist, v. 16, p. 1-36.

Fryberger, S. G., and M. Koelmel, 1986, Rangely field: eolian system-boundary trap in the Permo-Pennsylvanian Weber Sandstone of northwest Colorado, *in* D. S. Stone, ed., New interpretations of northwest Colorado geology: Rocky Mountain Association of Geologists, p. 129-149.

Furlong, K., and J. D. Edman, 1983, Graphical approach to determination of hydrocarbon maturation in overthrust terrains: American Association of Petroleum Geologists Bulletin (Abs.), v. 67, p. 465.

Gale, H. S., 1910, Coal fields of northwestern Colorado and northeastern Utah: USGS Bulletin 415, 265 p.

Gibbs, J. F., 1982, Maudlin Gulch field, *in* M. C. Crouch III, ed., Oil and gas fields of Colorado-Nebraska and adjacent areas: Rocky Mountain Association of Geologists, p. 338-344.

Gibson, H. A., 1961, Wilson Creek field, *in* J. M. Parker, ed., Oil and gas field volume, Colorado-Nebraska: Rocky Mountain Association of Geologists, p. 258-300.

Hancock, E. T., and J. B. Eby, 1930, Geology and coal resources of the Meeker Quadrangle, Moffat and Rio Blanco counties, Colorado: USGS Bulletin 812-C, p. 191-242.

Hansen, W. R., 1965, Geology of the Flaming Gorge area, Utah-Colorado-Wyoming: USGS Professional Paper 490, 196 p.

Hansen, W. R., 1986, History of faulting in the eastern Uinta Mountains, Colorado and Utah, *in* D. S. Stone, ed., New interpretations of northwest Colorado geology: Rocky Mountain Association of Geologists, p. 5-17.

Hartner, J. D., 1981, Depositional environments of the Salt Wash Sandstones (Upper Jurassic) in portions of Rio Blanco and Moffat counties, Colorado: Master's thesis, Colorado School of Mines, Golden, Colorado, 110 p.

Hite, R. J., D. E. Anders, and T. G. Ging, 1984, Organic-rich source rocks of Pennsylvanian age in the Paradox Basin of Utah and Colorado, *in* J. Woodward et al., eds., Hydrocarbon source rocks of the greater Rocky Mountain region: Rocky Mountain Association of Geologists, p. 255–274.

Hoffman, F. H., 1954, Wilson Creek oil field, Rio Blanco County, Colorado, *in* F. S. Jensen, H. H. R. Sharkey, and D. S. Turner, eds., The oil and gas fields of Colorado: Rocky Mountain Association of Geologists, p. 289–290.

Hubbell, R. G., and J. M. Wilson, 1963, Lance Creek field, Niobrara County, Wyoming, *in* P. J. Katich and D. W. Bolyard, eds., Geology of northern Denver Basin and adjacent uplifts: Rocky Mountain Association of Geologists, p. 248–256.

Hunt, E. H., 1938, Geology of Wilson Creek dome, Rio Blanco and Moffat counties, Colorado: Colorado School of Mines, The Mines Magazine, v. 28, n. 5, p. 192–195.

Johnson, R. C., and N. F. Nuccio, 1986, Structural and thermal history of the Piceance Creek Basin, western Colorado, in relation to hydrocarbon occurrence in the Mesaverde Group, in C. W. Spencer and R. F. Mast, eds., Geology of tight gas reservoirs: American Association of Petroleum Geologists Studies #24, p. 165–205.

Jones, D. C., and D. K. Murray, comp., 1976, Oil and gas fields of Colorado statistical data: Colorado Geological Survey Information Series 3, 57 p.

Kocurek, G., and R. H. Dott, Jr., 1981, Distinctions and uses of stratification types in the interpretation of eolian sand: Journal of Sedimentary Petrology, v. 51, p. 579–595.

Kocurek, G., and R. H. Dott, Jr., 1983, Jurassic paleogeography and paleoclimate of the central and southern Rocky Mountains region, *in* M. W. Reynolds and E. D. Dolly, eds., Mesozoic paleogeography of the west-central United States: SEPM Rocky Mountain Paleogeography Symposium 2, p. 101–116.

Ladd, J. B., 1957, The Wilson Creek field, Rio Blanco County, Colorado: Journal of Petroleum Technology, August, p. 23–27.

Levorsen, A. I., 1967, Geology of petroleum: San Francisco, W. H. Freeman and Company, 724 p.

MacMillan, L., 1980, Oil and gas of Colorado: a conceptual view, *in* H. C. Kent and K. W. Porter, eds., Colorado geology: Rocky Mountain Association of Geologists, p. 191–197.

Mallory, W. W., 1972, Pennsylvanian arkose and the Ancestral Rocky Mountains, *in* Geologic Atlas of the Rocky Mountain region: Rocky Mountain Association of Geologists, p. 131–132.

Mullens, T. E., and V. L. Freeman, 1957, Lithofacies of the Salt Wash Member of the Morrison Formation, Colorado Plateau: Geological Society of America Bulletin, v. 68, p. 505–526.

Nuccio, V. F., and C. J. Schenk, 1986, Thermal maturity and hydrocarbon source-rock potential of the Eagle basin, northwestern Colorado, *in* D. S. Stone, ed., New interpretations of northwest Colorado geology: Rocky Mountain Association of Geologists, p. 259–264.

Nuccio, V. F, and C. J. Schenk, 1987, Burial reconstruction of early and middle Pennsylvanian Belden Formation, Gilman area, Eagle basin, northwest Colorado: USGS Bulletin 1787C, p. 31–36.

Nuccio, V. F., C. J. Schenk, and S. Y. Johnson, 1988, Estimation of paleogeothermal gradients and their relationship to the timing of petroleum generation, Eagle Basin, northwestern Colorado, *in* USGS research on energy resources—1988 program and abstracts (V. E. McKelvey forum on mineral and energy resources): USGS circular 1025, p. 38.

Pipiringos, G. N., and R. B. O'Sullivan, 1978, Principal unconformities in Triassic and Jurassic rocks, western interior United States—a preliminary survey: USGS Professional Paper 1035-A, 29 p.

Reed, J. C., Jr., M. E. Bickford, W. R. Premo, J. N. Aleinkoff, and J. S. Palister, 1987, Evolution of the Early Proterozoic Colorado province: Constraints from U-Pb geochronology: Geology, v. 15, p. 861–865.

Richard, J. J., 1986, Interpretation of a seismic section across the Danforth Hills anticline (Maudlin Gulch) and Axial arch in northwest Colorado, *in* D. S. Stone, ed., New interpretations of northwest Colorado geology: Rocky Mountain Association of Geologists, p. 191–193.

Ross, R. J., Jr., 1986, Lower Paleozoic of northwest Colorado: a summary, *in* D. S. Stone, ed., New interpretations of northwest Colorado geology: Rocky Mountain Association of Geologists, p. 99–102.

Rowley, D. R., W. R. Hansen, O. Tweto, and P. Carrara, 1985, Geologic map of the Vernal 1° × 2° Quadrangle, Colorado, Utah, and Wyoming: USGS Miscellaneous Investigations Series Map I-1526.

Sanborn, A. F., 1971, Possible future petroleum of Uinta and Piceance basins and vicinity, northeast Utah and northwest Colorado, *in* I. H. Cram, ed., Future petroleum provinces of the United States—their geology and potential: American Association of Petroleum Geologists Memoir 15, v. 1, p. 489–508.

Sears, J. D., 1925, Geology and oil and gas prospects of part of Moffat County: USGS Bulletin 751.

Sears, J. W., P. J. Graff, and G. S. Holden, 1982, Tectonic evolution of Lower Proterozoic rocks, Uinta Mountains, Utah and Colorado: Geological Society of America Bulletin, v .93, p. 990–997.

Smith, H. M., 1940, Correlation index to aid in interpreting crude-oil analyses: U.S. Bureau of Mines Technical Paper 610, 34 p.

Smith, H. M., and J. H. Hale, 1966, Crude oil characterizations based on Bureau of Mines routine analyses: U.S. Dept. of the Interior Bureau of Mines Report of Investigations 6846, 28 p.

Stone, D. S., 1975, A dynamic analysis of sub-surface structure in northwestern Colorado, *in* D. W. Bolyard, ed., Deep drilling frontiers of the central Rocky Mountain: Rocky Mountain Association of Geologists, p. 33–40. Geochemistry and hydrodynamics of northwest Colorado fields presented in oral symposium at Steamboat Springs, CO.

Stone, D. S., 1977, Tectonic history of the Uncompaghre Uplift, *in* H. K. Veal, ed., Exploration frontiers of the central and southern Rockies: Rocky Mountain Association of Geologists, p. 23–30.

Stone, D. S., 1986a, Geology of the Wilson Creek field, Rio Blanco County, Colorado, *in* D. S. Stone, ed., New interpretations of northwest Colorado geology: Rocky Mountain Association of Geologists, p. 229–246.

Stone, D. S., 1986b, Seismic and borehole evidence for important Pre-Laramide faulting along the Axial Arch in northwest Colorado *in* D.S. Stone, ed., New interpretations of northwest Colorado geology: Rocky Mountain Association of Geologists, p. 19–36.

Stone, D. S., 1986c, Geometry and kinematics of thrust-fold structures in central Rocky Mountain foreland: American Association of Petroleum Geologists Bulletin (Abs.) v. 70, p. 1057.

Stone, D. S., 1986d, Rangely field summary 2, seismic profile, structural cross-section, and geochemical comparison, *in* D. S. Stone, ed., New interpretations of northwest Colorado geology: Rocky Mountain Association of Geologists, p. 226–228.

Stone, D. S., 1987, Structural transect across the Wyoming foreland: a model for kinematic analysis of basement-involved, thrust-generated folds: The Outcrop (Abs.), v. 36, n. 12, p. 4, and presented before the Rocky Mountain Association of Geologists, Dec. 18, 1987.

Tillman, R. W., 1971, Petrology and paleoenvironments, Robinson Member, Minturn Formation (Desmoinesian), Eagle Basin, Colorado: American Association of Petroleum Geologists Bulletin, v. 55, p. 593–620.

Walker, T. R., 1972, Bioherms in the Minturn Formation (Des Moines Age), Vail-Minturn area, Eagle County, Colorado, *in* R. H. DeVoto, ed., Paleozoic stratigraphy and structural evolution of Colorado Symposium: Colorado School of Mines Quarterly, v. 67, n. 4, p. 249–278.

Waechter, N. B., and W. E. Johnson, 1986, Pennsylvanian-Permian paleostructure and stratigraphy as interpreted from seismic data in the Piceance Basin, northwest Colorado, *in* D. S. Stone, ed., New interpretations of northwest Colorado geology: Rocky Mountain Association of Geologists, p. 51–64.

Wallace, C. A., 1972, A basin analysis of the Upper Precambrian Uinta Mountain Group, Utah: Ph.D. dissertation, University of California, Santa Barbara, California, 412 p.

Waples, D. W., 1980, Time and temperature in petroleum formation: application of Lopatin's method to petroleum exploration: American Association of Petroleum Geologists Bulletin, v. 64, p. 916–926.

Warner, M. A., 1982, Source and time of generation of hydrocarbons in the Fossil Basin, western Wyoming thrust belt, *in* P. B. Powers, ed., Geologic studies of the Cordilleran thrust belt, v. 2: Rocky Mountain Association of Geologists, p. 805–815.

Wenger, W. J., and B. W. Reid, 1960, Properties of petroleum from

the Four Corners area of Arizona, Colorado, New Mexico, and Utah: U.S. Dept. of the Interior Bureau of Mines Report of Investigations 5587, 25 p.
Wenger, W. J., M. L. Whisman, W. J. Lanum, and J.S . Ball, 1957, Characteristics and analyses of ninety-two Colorado crude oils: U.S. Bureau of Mines Report of Investigations 5309, 60 p.
Woodward, J., F. F. Meissner, and J. L. Clayton, eds., 1984, Hydrocarbon source rocks of the greater Rocky Mountain region: Rocky Mountain Association of Geologists, 557 p.

SUGGESTED READINGS

Stone, D. S., ed., 1986, New interpretations of northwest Colorado geology: Rocky Mountain Association of Geologists, 308 p. Contains 29 papers on geology of this region.
Tweto, O. (compiler), 1976, Geologic map of the Craig 1° × 2° Quadrangle, northwestern Colorado: USGS Miscellaneous Investigations Series Map I-972. Includes the Danforth Hills anticline and Wilson Creek dome.
Tweto, O., 1983, Geologic sections across Colorado: USGS Miscellaneous Investigations Series Map I-1416. Sections A-A′, and B-B′ are pertinent.

Appendix 1. Field Description

Field name .... *Wilson Creek*

Ultimate recoverable reserves .... *100 million bbl oil and oil equivalents*

Field location:

- **Country** .... *United States*
- **State** .... *Colorado*
- **Basin/Province** .... *Piceance basin/White River uplift*

Field discovery:

- **Year first pay discovered** .... *Jurassic Salt Wash Sandstone Member 1938*
- **Year second pay discovered** .... *Jurassic Entrada Sandstone 1941*
- **Third pay** .... *Pennsylvanian Minturn Formation 1984*

Discovery well name and general location
(i.e., Jones No. 1, Sec. 2T12NR5E; or Smith No. 1, 5 mi west of Sheridan, Wyoming):

- **First pay** .... *SE NW Sec. 35, T3N, R99W, 12 mi NNW of Meeker, CO, Salt Wash (Morrison)*
- **Second pay** .... *Texas Co.& California Co., #5 Unit, NE NE Sec. 34, T3N, R99W, Entrada (Sundance)*
- **Third pay** .... *Wilson Creek Unit #66, NE SW NE Sec. 34, T3N, R94W, Minturn Fm.*

Discovery well operator .... *The Texas Co. & California Co.*
(if more than one pay in field, list operators of discovery well in other pays)

- **Second pay** .... *The Texas Co & California Co.*
- **Third pay** .... *Texaco and Chevron*

IP in barrels per day and/or cubic feet or cubic meters per day:

- **First pay** .... *490 BOPD*
- **Second pay** .... *300 BOPD*
- **Third pay** .... *18 BCPD, 914 MCFD, 109 BWPD*

All other zones with shows of oil and gas in the field:

Age	Formation	Type of Show
Cretaceous	*Dakota*	*Oil stain; gas on mudlog*
Triassic	*Shinarump*	*Oil stain*
Permian/Pennsylvanian	*Weber*	*Oil stain; saturation in core*

Geologic concept leading to discovery and method or methods used to delineate prospect, e.g., surface geology, subsurface geology, seeps, magnetic data, gravity data, seismic data, seismic refraction, nontechnical:

Surface geology; confirmed by drilling.

Structure:

Province/basin type (see St. John, Bally, and Klemme, 1984)
Bally 222; Klemme IIA

Tectonic history
A thick section of Phanerozoic rocks, from Paleozoic (clastics, carbonates, and evaporites), through Mesozoic and Tertiary (clastics) is present in northwest Colorado, but stratigraphic thicknesses and facies can vary significantly across regional fault zones. The field lies along a major basement-involved, thrust-generated anticline produced entirely by Laramide compression. But important Precambrian and Pennsylvanian/Permian (Ancestral Rockies orogeny) structural movements occurred along the nearby Uinta-Axial Arch.

Regional structure
Wilson Creek dome forms the southeastern culmination of the Danforth Hills anticline, which is the northeastern boundary of the Piceance Creek basin.

Local structure
Wilson Creek is a large anticlinal closure, asymmetric to the southwest and bordered on that side by a major, basement-involved thrust fault zone dipping about 30° northeasterly under the fold. Dips on the

southwestern forelimb can reach 50° or more; but on the northeastern backlimb, dips are generally less than 10°.

Trap

Trap type(s) .... *Anticlinal trap with multiple pays and stratigraphic variations*

Basin stratigraphy (major stratigraphic intervals from surface to deepest penetration in field):

Chronostratigraphy	Formation	Depth to Top in ft
Cretaceous	*Mesaverde*	*Surface*
	Mancos	*1120*
	Dakota	*6270*
Jurassic	*Morrison*	*6400*
	Entrada	*6900*
	Navajo	*7000*
Permian/Pennsylvanian	*Weber*	*8350*
Pennsylvanian	*Minturn*	*9560*
Mississippian	*Leadville*	*12,160*
Cambrian	*Sawatch*	*12,850*

(From well in central part of field.)

Location of well in field .... *NA*

Reservoir characteristics:

Number of reservoirs *3*

Formations .... *Salt Wash Member of Morrison Fm.; Entrada Fm.; Minturn Fm.*

Ages .... *Salt Wash, Jurassic; Entrada, Jurassic; Minturn, Pennsylvanian*

Depths to tops of reservoirs .... *Salt Wash, 6600 ft; Entrada, 6700 ft; Minturn, 10,370 ft*

Gross thickness (top to bottom of producing interval) .... *Salt Wash, 300 ft; Entrada, 156 ft; Minturn, 268 ft*

Net thickness—total thickness of producing zones

Average .... *Salt Wash, 64 ft; Entrada, 40 ft*

Maximum .... *Salt Wash, 105 ft; Entrada, 156 ft; Minturn, not determinable*

Average

Maximum

Lithology

Salt Wash and Entrada: fine-grained, quartzose, well-sorted sandstone

Minturn: arkosic sandstone and gray-brown, dense, fractured (?) limestone with possible algal material

Porosity type .... *Intergranular sandstone porosity in Salt Wash and Entrada; Minturn unknown (probably fractured)*

Average porosity .... *Salt Wash, 19.4%; Entrada, 19.7%; Minturn, unknown*

Average permeability .... *Salt Wash, 11 md; Entrada, 38 md*

Seals:

Upper

Formation, fault, or other feature .... *Morrison (Brushy Basin Member)*

Lithology .... *Shale and tight siltstone*

Lateral

Formation, fault, or other feature .... *Same as above*

Lithology .... *Same as above*

Source:

Formation and age .... *Mowry-Mancos Shale (Cretaceous)*

Lithology .... *Dark gray shale*

Average total organic carbon (TOC) .... *2.5 ± 1%*

Maximum TOC *4.0%*

Kerogen type (I, II, or III) .... *II (80%) and III (20%)*
Vitrinite reflectance (maturation) .... *R_o = 0.5–1.3*
Time of hydrocarbon expulsion .... *Tertiary (Eocene to present)*
Present depth to top of source .... *9000–10,000 (in footwall of thrust)*
Thickness .... *1400 ft (426 m)*
Potential yield .... *NA*

Appendix 2. Production Data

Field name .... *Wilson Creek*

Field size:

- **Proved acres** .... *Salt Wash, 3904 ac; Entrada, 1297 ac; Minturn, unknown*
- **Number of wells all years** .... *62 producing*
- **Current number of wells (as of 12/1988)** *14*
- **Well spacing** .... *60 ± ac/well (all zones)*
- **Ultimate recoverable** .... *Nearly 100 million bbl (OAOE) (all zones)*
- **Cumulative production** .... *83.4 million bbl; 63.2 bcf CSHD*
- **Annual production (1988)** .... *0.233 million bbl oil; 0.77C bcf gas*
- **Present decline rate** .... *Salt Wash nil %*
  - **Initial decline rate** .... *NA*
  - **Overall decline rate** .... *NA*
- **Annual water production (1988)** .... *151,512 bbl*
- **In place, total reserves** .... *240 million bbl*
- **In place, per acre-foot** .... *1000+ bbl/ac-ft*
- **Primary and secondary recovery** .... *90 ± million bbl and 70 BCF*
- **Enhanced recovery** .... *NA*
- **Cumulative water production** .... *141.2 million bbl*

Drilling and casing practices:

- **Amount of surface casing set** .... *500–900 ft*
- **Casing program** .... *Set 10¾-in. or 8⅝-in. surface casing, cement to surface; drill to TD with low solids, light mud 9 lb/gal, run 7-in. or 5½-in. casing*
- **Drilling mud** .... *LSND*
- **Bit program** .... *NA*
- **High pressure zones** .... *None (lost circulation zones in Mancos Shale)*

Completion practices:

- **Interval(s) perforated** .... *Salt Wash; Entrada; or Minturn*
- **Well treatment** .... *Perforate and frac if necessary to stimulate*

Formation evaluation:

- **Logging suites** .... *DIL or electric log, wells drilled since late 1950s include a porosity log*
- **Testing practices** .... *Swab to test perforated intervals; DSTs seldom used*
- **Mud logging techniques** .... *No mud logs run except on post-1950 stepout or deeper pool wildcat wells*

Oil characteristics:

- **Type** .... *NA*
  (Tissot and Welte Classification in "Petroleum Formation and Occurrence," 1984, Springer-Verlag, p. 419)
- **API gravity** .... *Salt Wash and Entrada, 47–50.5°; Minturn, 58.4–60°*
- **Base** .... *Paraffinic-naphthenic (all zones)*
- **Initial GOR** .... *Salt Wash, 420 cf/bbl; Entrada, 110 cf/bbl*
- **Sulfur, wt%** .... *Salt Wash, 0.14; Entrada, 0.03*
- **Viscosity, SUS** .... *Salt Wash, 33; Entrada, 33*

Pour point .... *Salt Wash, 25; Entrada, 5*
Gas-oil distillate .... *Salt Wash, 15%; Entrada, 15.5%*

Field characteristics:

Average elevation .... *8100 ft*
Initial pressure .... *Salt Wash, 2184 psi; Entrada, 2202 psi; Minturn, NA*
Present pressure .... *NA*
Pressure gradient .... *0.33 psi/ft*
Temperature .... *Salt Wash, 172°F; Entrada, 171°F; from log (uncorrected)*
Geothermal gradient .... *0.016°F*
Drive .... *Volumetric + water drive for Salt Wash and Entrada; Minturn probably volumetric*
Oil column thickness .... *300+ ft*
Oil-water contact .... *+1175 to +1350 ft (tilted) Entrada pool*
Connate water .... *Salt Wash, 39%; Entrada, 33%*
Water salinity, TDS .... *Salt Wash, 2342; Entrada, 16,566*
Resistivity of water .... *Salt Wash, 0.95 at 172°F; Entrada, 0.17 at 172°F*
Bulk volume water (%) .... *NA*

Transportation method and market for oil and gas:

Oil pipeline; gas is used on site for field operations.

Eugene Island Block 330 Field—U.S.A.
Offshore Louisiana

DAVID S. HOLLAND
JOHN B. LEEDY
DAVID R. LAMMLEIN
Pennzoil Exploration and Production Company
Houston, Texas

FIELD CLASSIFICATION

BASIN: Gulf of Mexico
BASIN TYPE: Passive Margin
RESERVOIR ROCK TYPE: Sandstone
RESERVOIR ENVIRONMENT OF
DEPOSITION: Deltaic (Delta Front Sands)
RESERVOIR AGE: Pleistocene
PETROLEUM TYPE: Oil and Gas
TRAP TYPE: Rollover Anticline

INTRODUCTION AND LOCATION

The Eugene Island Block 330 field is located in the Gulf of Mexico offshore Louisiana, U.S.A., approximately 272 km (170 mi) southwest of New Orleans (Figure 1). The field covers portions of seven 2024-ha (5000-ac) blocks (Figure 2) in the west-central part of the Eugene Island area, South Addition. The Eugene Island Block 330 field lies within the northern Gulf Coast basin, near the southern edge of the Louisiana Outer Continental Shelf (OCS). Water depths in this area of the shelf range from 64 to 81 m (210 to 266 ft).

The field consists of two rollover anticlines, bounded to the north and east by a large arcuate, down-to-the-basin growth fault system. More than 25 Pleistocene sandstones are productive at depths of 701 to 3658 m (2300 to 12,000 ft). Faulting and permeability barriers separate these sands into more than 100 oil and gas reservoirs.

The Eugene Island Block 330 field was the largest producing field in Federal OCS waters from 1975 to 1980 (Figure 3). The field now ranks second in annual hydrocarbon production (Figure 3) and fourth in cumulative hydrocarbon production (Figure 4). Ultimate recoverable reserves are estimated at 307 million bbl of hydrocarbon liquids and 1.65 tcf of gas. The field is ranked number 437 among the giant oil and gas fields of the world (Carmalt and St. John, 1986).

HISTORY

Our 20-year evaluation of the Eugene Island Block 330 field may be divided into three phases: pre-discovery, discovery, and post-discovery (including development, production, and continued exploration and development). Pennzoil's Block 330 history of operations (Figure 5) characterizes activities in the entire field.

Pre-Discovery

Evaluation of this part of the Louisiana OCS began in 1968 as part of the industry's preparation for the 1970 OCS lease sale offshore southwest Louisiana (Figure 5). In the 1970 sale 240,361 ha (593,458 ac) were offered for lease.

The technical evaluation for the lease sale delineated several rollover anticlines downthrown to a major, arcuate growth fault system in the Eugene Island Block 330 area. Favorable sand conditions for the area were extrapolated from lithofacies and isopachous trends defined by well data (Norwood and Holland, 1974). Prior to the lease sale, Pennzoil regarded the Block 330 area as a gas prospect, because the Pliocene–Pleistocene objective section was considered geochemically immature. Before 1970, wells drilled to the north of Block 330 in Eugene Island Blocks 266, 273, and 292 had established gas production in Pleistocene sands at depths between 915 and 1525 m (3000 and 5000 ft).

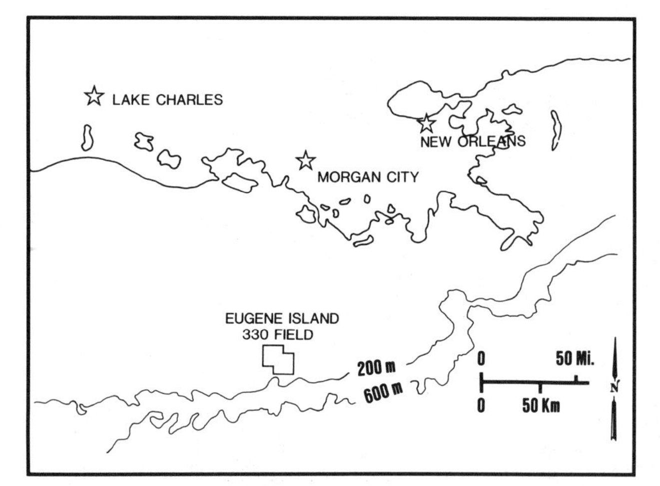

Figure 1. Eugene Island Block 330 field location map, offshore Louisiana.

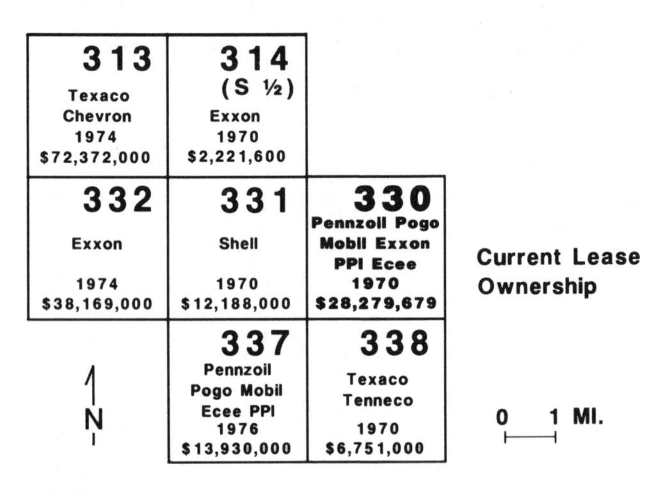

Figure 2. Acquisition history and current lease ownership of blocks comprising Eugene Island Block 330 field. Initial lease bonus and year acquired are shown. All blocks are currently held by production. Only the south half (S ½) of Block 314 is included in the field.

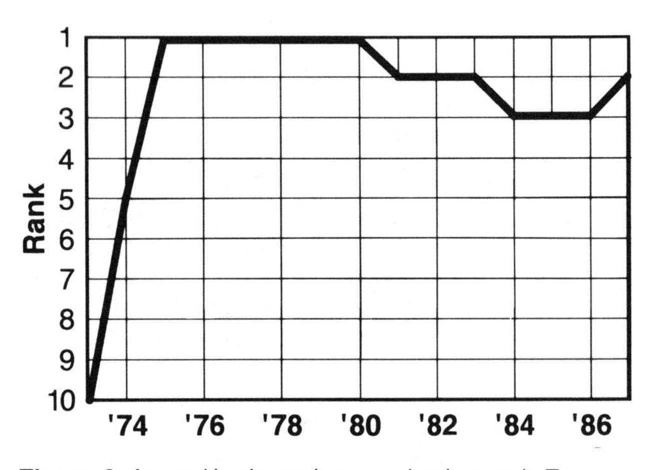

Figure 3. Annual hydrocarbon production rank, Eugene Island Block 330, U.S. Federal OCS waters. U.S. Minerals Management Service (MMS) data.

Rank	Field	Production (MMBOE)
1	West Delta 30	541
2	Bay Marchand 2	512
3	Grand Isle 43	488
4	Eugene Island 330	481

MMBOE - Million Barrels Oil Equivalent
SOURCE: Dwight's Energy Data

Figure 4. Top four cumulative oil- and gas-producing fields, U.S. Federal OCS. Cumulative production to 30 June 1987.

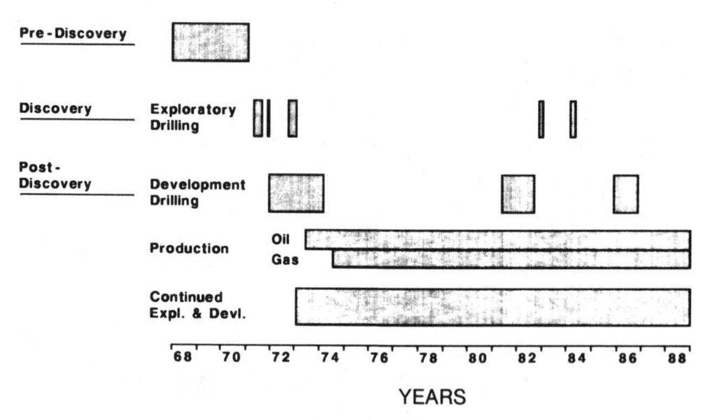

Figure 5. Eugene Island Block 330 exploration, development, and production history.

Four blocks—Eugene Island 314, 330, 331, and 338—were leased by various companies in the 1970 lease sale (Figure 2). Of the remaining blocks encompassing the field, Blocks 313 and 332 were leased in 1974, and Block 337 in 1976. Bonus bids for the seven blocks totaled $174 million, or an average of $5,355 per acre.

Discovery

Only 2 months after acquiring the leases (Figure 5), exploratory drilling was under way with the simultaneous drilling of the Pennzoil Block 330 #1 well (completed 6 March 1971) and the Shell Block 331 #1 well (completed 24 March 1971). The Pennzoil well, drilled on the crest of the structure (Figure 6), reached a total depth of 1920 m (6300 ft) and penetrated four hydrocarbon-bearing sandstones totaling 34 m (110 ft) of net oil and gas sand. The Shell well, located 100 m (330 ft) west of the Block 330/331 lease line (Figure 6), was drilled to a total

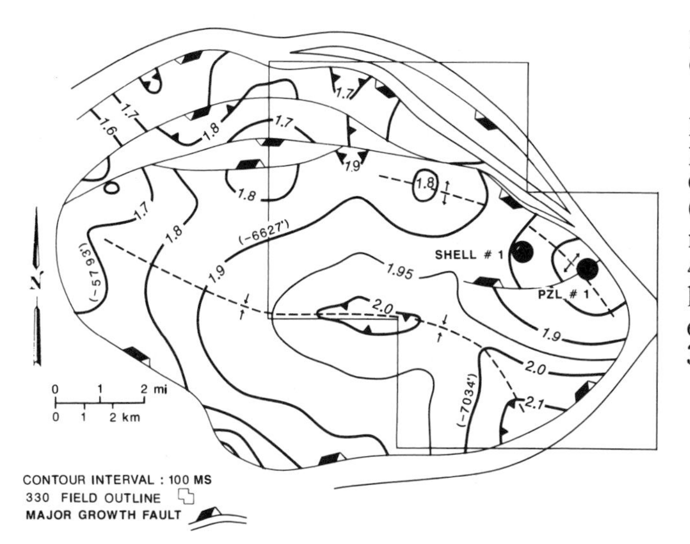

Figure 6. Initial 1970 seismic structure map of the JD sandstone, Eugene Island Block 330 field vicinity. Locations of Pennzoil #1 and Shell #1 wells shown.

depth of 3383 m (11,098 ft), and logged five hydrocarbon-bearing sandstones totaling 60 m (198 ft) of net oil and gas sand. The two wells established a cumulative hydrocarbon column of about 458 m (950 ft), identified six major Pleistocene productive zones (four oil and two gas/condensate), and discovered a major oil and gas field.

Subsequent exploratory drilling encountered additional pay zones, and confirmed and delineated the original reservoirs. To date, 31 exploratory wells have been drilled in the Block 330 field (Figure 7).

Post-Discovery

The first development well was spudded on Block 330 from Pennzoil's "A" platform in November 1971 (Figure 5). Ten production platforms have been set and 227 development wells drilled in the Eugene Island Block 330 field (Figure 7). The 258 exploratory and development wells together have identified more than 25 pay sands.

During the period 1982–1986, additional hydrocarbons were discovered and developed on the upthrown side of the large growth fault system in Eugene Island Blocks 330, 315, and 316, and also south of the field in Eugene Island Blocks 336, 337, and 354. These adjacent fields will not be addressed in this discussion.

Production in the Eugene Island Block 330 field began in September 1972, which is 1 year and 9 months after initial lease acquisition (Figure 5). Maximum fieldwide daily production of 95,290 bbl of liquids and 482,000 mmcf of gas was attained in 1977 (Figure 8). Cumulative production through September 1987 is more than 271 million bbl of oil and condensate and 1.259 tcf of gas (Figure 8). Daily production was 23,600 bbl of liquids and 111 mmcf of gas, as of September 1987.

Pennzoil's Eugene Island Block 330 has been producing since March 1973. Lewis et al. (1982) provide a petroleum engineering review of the development and production history of the field. Cumulative production through September 1987 is more than 101 million bbl of oil and condensate and 477 bcf of gas. This represents 37% of the liquid hydrocarbons and 38% of the gas produced from the entire field. The oil production decline rate (Block 330 only) has averaged 11% per year.

DISCOVERY METHOD

Pennzoil-United, a forerunner of the present-day Pennzoil Company, and its partners, in an effort to lease acreage in the newly emerging Pliocene-Pleistocene trend offshore Louisiana, began evaluation in 1968 for the December 1970 Federal OCS lease sale. Pennzoil's evaluation strategy was to give top ranking to prospects that (1) have the potential for large reserves, (2) are located on the crests of structural highs (especially four-way dip closures), and (3) are located within subbasins with alternating sand-shale sequences.

Electrical well logs and paleontological reports formed the data base for a regional stratigraphic study of the western part of the Louisiana continental shelf. Six types of stratigraphic maps were prepared for as many as 17 regional stratigraphic markers in the Pleistocene, Pliocene, and upper-middle Miocene. The maps included:

1. **Paleobathymetric maps**—showing water depth at the time of deposition. Several authors, among which are Limes and Stipe (1959), Fisher (1969), and Norwood and Holland (1974), observed and documented the relationship between rock facies, depositional environment, and hydrocarbon accumulations in the Pleistocene, Pliocene, and Miocene rock systems of the Gulf of Mexico. It was generally accepted that deposition in water depths of 91 to 457 m (300 to 500 ft) is optimum for hydrocarbon accumulations in the Gulf of Mexico. This range in paleodepth for the shale is classified as outer neritic (outer shelf, paleoecological zone 3) to upper bathyal (upper slope, paleoecological zone 4). Mapping of paleoenvironmental data defines the shape of the delta platform, proximity to the delta plain, and hence the distribution of the alternating sand-shale facies (Norwood and Holland, 1974).
2. **Sand percentage maps**—sand percentages between 10 and 35% defining the alternating sand-shale facies, the most favorable facies for hydrocarbon accumulations.

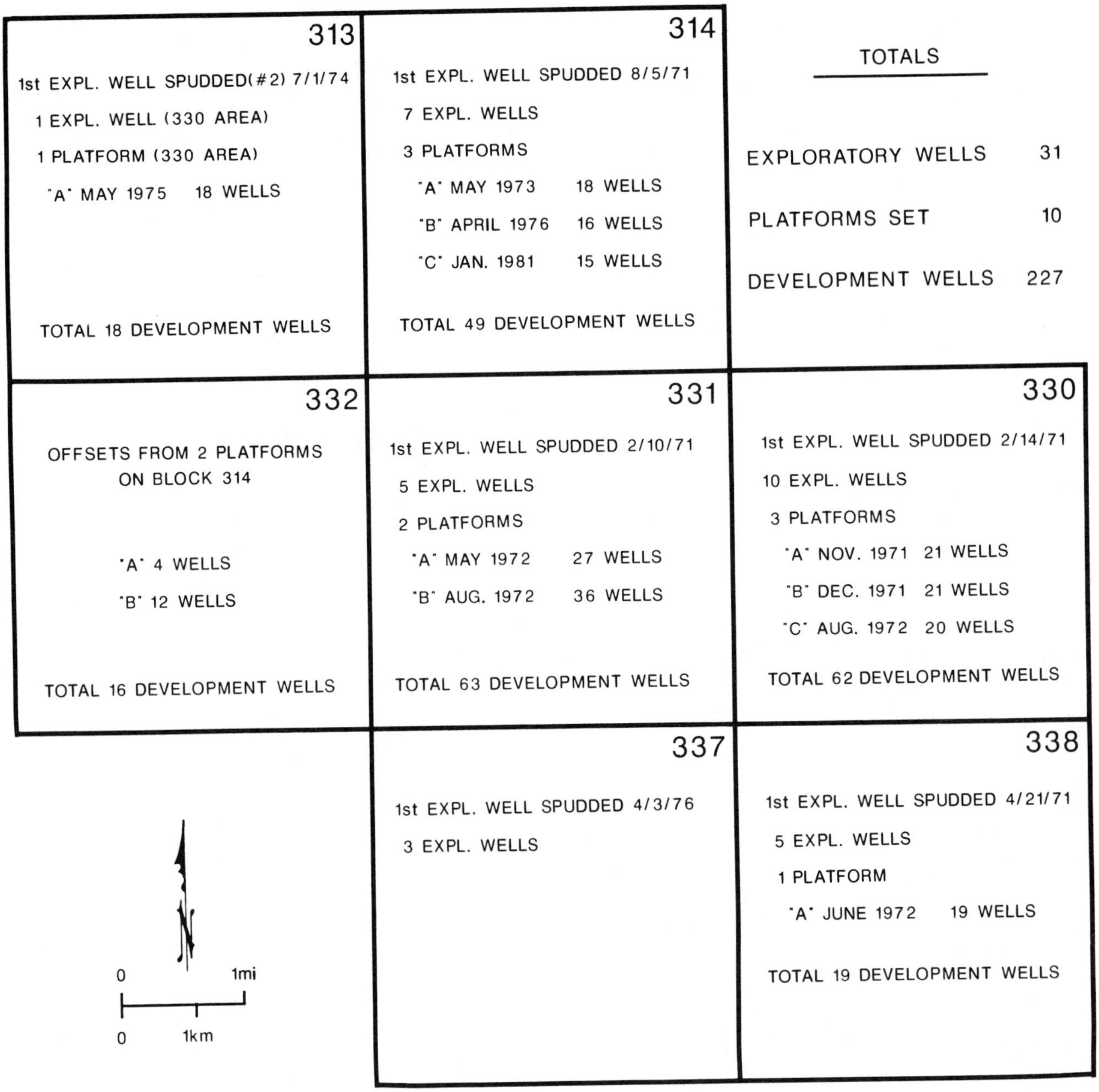

313

1st EXPL. WELL SPUDDED(#2) 7/1/74
1 EXPL. WELL (330 AREA)
1 PLATFORM (330 AREA)
"A" MAY 1975 18 WELLS
TOTAL 18 DEVELOPMENT WELLS

314

1st EXPL. WELL SPUDDED 8/5/71
7 EXPL. WELLS
3 PLATFORMS
"A" MAY 1973 18 WELLS
"B" APRIL 1976 16 WELLS
"C" JAN. 1981 15 WELLS
TOTAL 49 DEVELOPMENT WELLS

TOTALS

EXPLORATORY WELLS	31
PLATFORMS SET	10
DEVELOPMENT WELLS	227

332

OFFSETS FROM 2 PLATFORMS ON BLOCK 314
"A" 4 WELLS
"B" 12 WELLS
TOTAL 16 DEVELOPMENT WELLS

331

1st EXPL. WELL SPUDDED 2/10/71
5 EXPL. WELLS
2 PLATFORMS
"A" MAY 1972 27 WELLS
"B" AUG. 1972 36 WELLS
TOTAL 63 DEVELOPMENT WELLS

330

1st EXPL. WELL SPUDDED 2/14/71
10 EXPL. WELLS
3 PLATFORMS
"A" NOV. 1971 21 WELLS
"B" DEC. 1971 21 WELLS
"C" AUG. 1972 20 WELLS
TOTAL 62 DEVELOPMENT WELLS

N
0 1mi
0 1km

337

1st EXPL. WELL SPUDDED 4/3/76
3 EXPL. WELLS

338

1st EXPL. WELL SPUDDED 4/21/71
5 EXPL. WELLS
1 PLATFORM
"A" JUNE 1972 19 WELLS
TOTAL 19 DEVELOPMENT WELLS

Figure 7. Exploratory and development summary for the Eugene Island Block 330 field.

3. **Net sand maps**—indicating total sand and potential reservoir rock within a stratigraphic interval.
4. **Isopachous maps**—predicting stratigraphic interval thicknesses with limited control.
5. **Trend analysis maps**—combining optimum paleobathymetric and sand percentage fairways with known production to highlight areas of possible prime hydrocarbon accumulation.
6. **Fence diagrams**—visual aids for understanding stratigraphic variability.

Along with the stratigraphic study, reconnaissance seismic structural interpretation and mapping of key horizons were initiated using available contractor 6- and 12-fold seismic data. Numerous anomalies within prime depositional trends were identified. Additional 12- and 24-fold seismic data were purchased to provide infill coverage over prospects selected for more detailed mapping and evaluation. Over highly selective areas, including the Eugene Island Block 330 area, proprietary seismic data were shot and processed. By combining the subsurface information

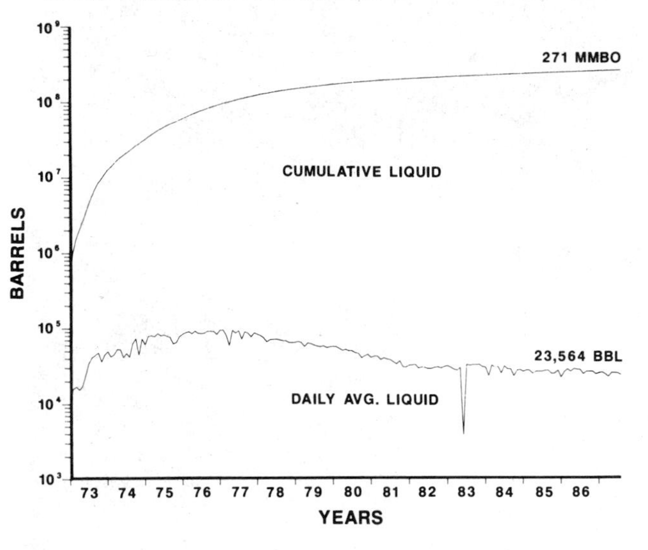

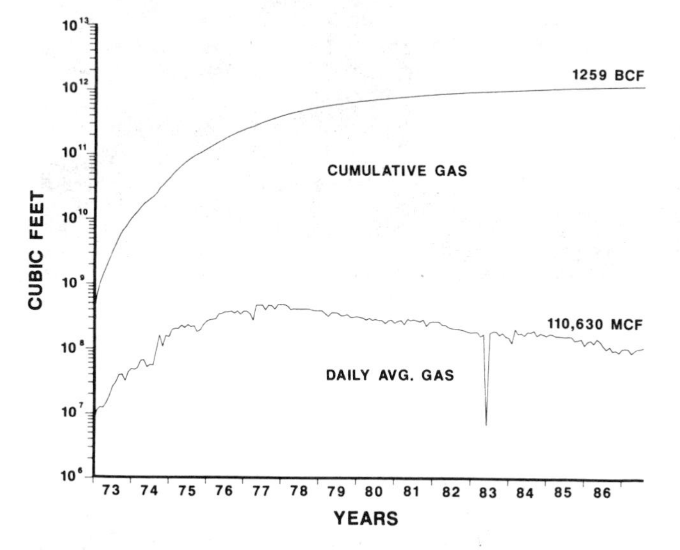

Figure 8. Cumulative and daily average hydrocarbon production for Eugene Island Block 330 field. The drop in daily average production in 1983 was the result of a pipeline rupture.

with detailed seismic-reflection and velocity data, structural interpretations were completed on three Pleistocene horizons throughout the evaluation area.

Through this regional seismic mapping procedure the Block 330 prospect was identified (Figure 6). An anticlinal structure with excellent trapping conditions was mapped on the downthrown side of a large growth fault system, with the crest on Block 330. Geological data indicated that the Block 330 anticline was located in a prime Pleistocene depositional fairway. Subsequent exploration drilling confirmed the anticline as part of a giant oil and gas field.

Current Exploration Technology

Since the conclusion of the initial exploration and development phases in 1973, the Eugene Island Block 330 field has been the subject of continuing geologic, geophysical, and engineering studies (Figure 5). Today's advancements in geophysical acquisition, processing, and interpretation have increased the importance of seismic data in reevaluating developed fields and defining new prospects. In 1986, 1637 km (1017 mi) of 3D seismic data were acquired to evaluate Block 330. These data are interpreted on an interactive workstation to identify and evaluate prospects for additional exploration and development drilling. The use of 3D data has enhanced mapping resolution by defining individual fault-block reservoirs more precisely.

Hydrocarbon indicators recognized on seismic data over the Block 330 area correlate with many of the oil and gas reservoirs in the field. Figure 9 shows a 3D seismic section over Block 330. High-amplitude reflections and flat spots are associated with the productive zones. For example, the high-amplitude flat spot at about 1.5 sec is a reflection from the oil-water contact within the HB sandstone. Petrophysical and log analyses indicate that most oil and gas reservoirs have reduced acoustic impedance values relative to those of water sands and shales (Figure 10). This explains the observed correlation between seismic-reflection amplitudes and hydrocarbon accumulations in the field.

Extracted amplitude displays (Figure 11) are generated from the 3D data set to aid in the evaluation of seismic hydrocarbon indicators and in the definition of productive limits controlled by oil-water contacts, faults, and permeability barriers.

In addition to traditional stratigraphic techniques using well logs and paleontology, seismic stratigraphy combined with dipmeter analysis is used to define stratigraphic sequences and lithologies. These data indicate that the Eugene Island Block 330 field is within the sand-rich delta-front facies of a Pleistocene delta complex (see *Stratigraphy*).

Surface Manifestations

Recent advances in petroleum geochemistry have helped establish the value of near-surface geochemical surveys (Jones and Drozd, 1983; Faber and Stahl, 1984; Brooks et al., 1986). Results from these studies demonstrate that the composition of gases in near-surface soils and marine sediments commonly shows a striking similarity to the composition of the underlying hydrocarbons.

Sediments from sea-bottom piston cores, collected on Eugene Island Block 330 along surface fault traces, contain minor anomalous concentrations of light

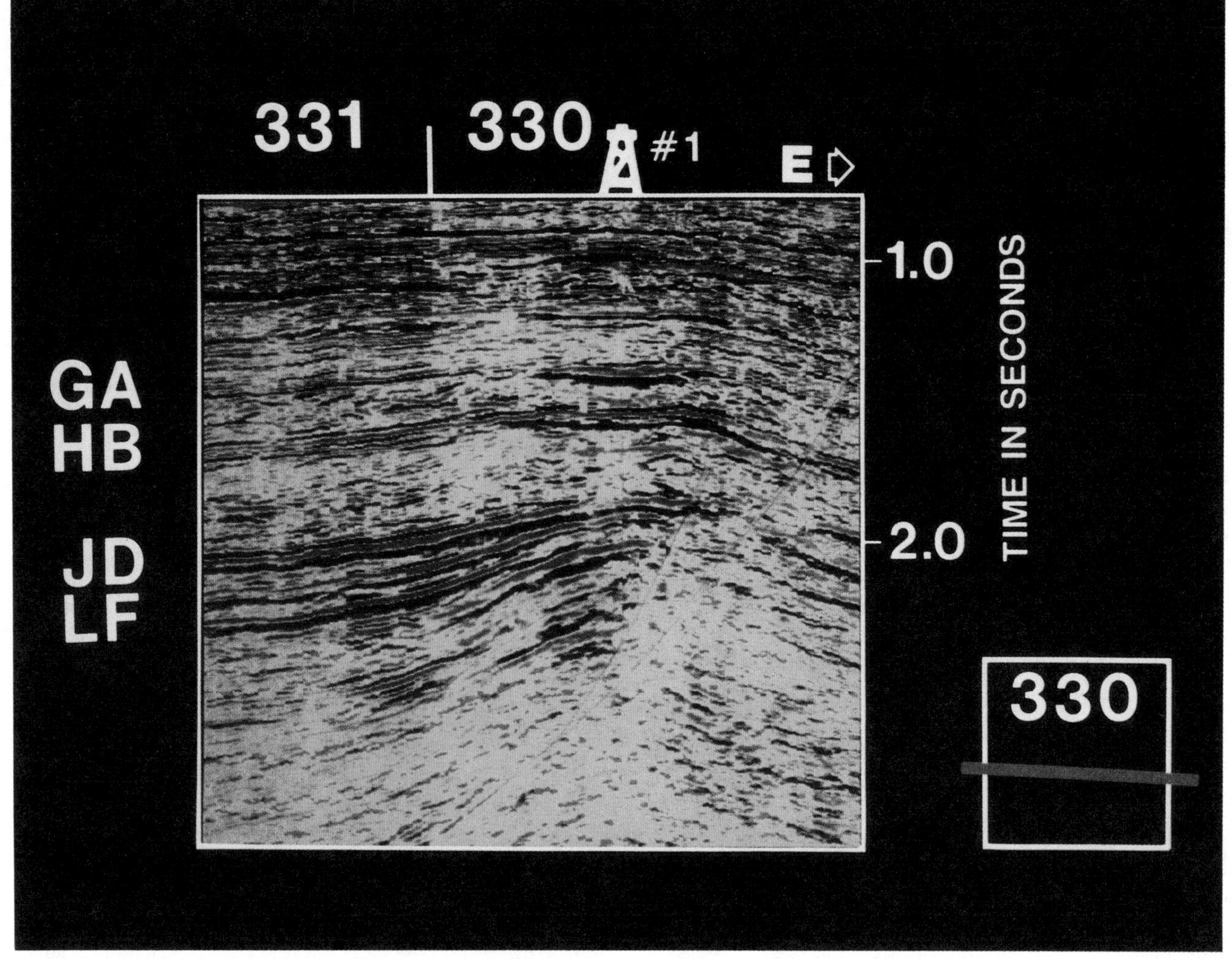

Figure 9. 3D seismic line over crest of Block 330 showing HB amplitude at approximately 1.5 seconds, and JD sandstone amplitude at approximately 1.9 seconds.

hydrocarbons and aromatic sulfur compounds. The latter are particularly significant because these dibenzothiophenes are important constituents of the Eugene Island 330 crude oil. The presence of the hydrocarbon microseepage, and a significant heat-flow anomaly identified from analyses of well and sea-floor data, suggest present-day fluid migration associated with the major growth fault bounding the field on the north and east.

STRUCTURE

Tectonic History

The Block 330 field is located in the Gulf Coast basin (Figures 12 and 13). The siliciclastic portion of the basin extends from the Sigsbee escarpment on the south to the northern pinchout of Tertiary sediments on the coastal plain, and from the Florida escarpment to the East Mexico shelf (Martin, 1978). Grabens and down-to-the-basin faulting are present along the inner coastal plain. These graben and fault systems are related to the formation of the Gulf of Mexico by early rifting and subsequent sea-floor spreading and subsidence in the Jurassic through Early Cretaceous (Hall et al., 1982). According to the compilation of hydrocarbon provinces by St. John et al. (1984), the Gulf Coast basin is classified as 1143 under the modified scheme of Bally and Snelson (1980) (i.e., basins located on the rigid lithosphere, not associated with the formation of megasutures; Atlantic-type passive margins that straddle continental and oceanic crust; overlying earlier backarc basins). Klemme (1971) classifies the Gulf Coast basin as IICc/IV (i.e., continental multicycle basin; crustal collision zone-downwarp into small ocean basin; open/delta basin).

Regional Structure

Structural features on the Louisiana OCS include salt domes, salt-withdrawal subbasins, growth

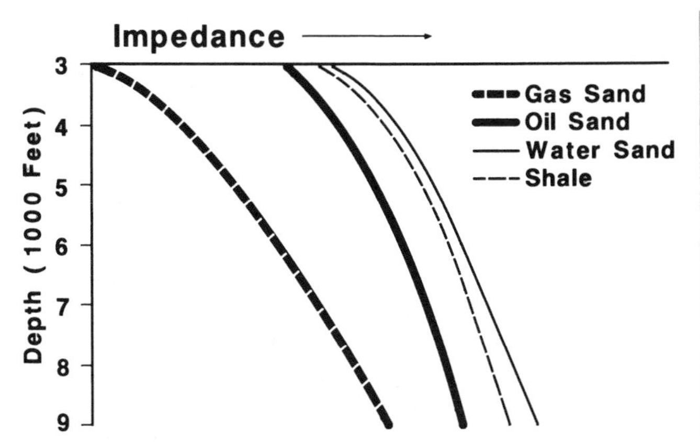

Figure 10. Characteristic petrophysically derived acoustic impedance vs. depth plot for water sands, shales, and hydrocarbon-bearing sands, illustrating the empirical basis for observed bright seismic amplitudes associated with pay zones in the field.

faults, and diapir-related faults. Rollover anticlines are common on the downthrown sides of growth faults (Bruce, 1973). The mobilization of salt into anticlines, domes, and diapirs is of particular tectonic importance in the Gulf Coast basin.

Woodbury et al. (1973) subdivided the salt structures in the Texas and Louisiana coast, shelf, and slope areas according to the shape, size, and horizontal cross-sectional area of the salt at a depth of 3658 m (12,000 ft). Based upon their system, the Eugene Island Block 330 field is near the northern edge of the salt-tectonic province characterized by semicontinuous diapiric uplifts (Figure 14).

Deformation related to salt movement has been occurring since Late Jurassic (Martin, 1978) and continues today (Humphris, 1978). Deformation was contemporaneous with sedimentation, being both a consequence of and an influence on deposition. The rapid influx of Upper Cretaceous and lower Tertiary terrigenous sediments mobilized the salt, prograded the northern Gulf margin to the south, and produced regional, down-to-the-basin growth faults.

Local Structure

The Block 330 field is located within an oval-shaped structural and depositional basin (Figure 6) on the downthrown side of a large, arcuate down-to-the-basin growth fault system (Holland et al., 1980). Antithetic faults and other permeability barriers divide the sandstone into more than 100 oil and gas reservoirs (see *Reservoirs* section).

Along the northern rim of the subbasin, hydrocarbons accumulated in two anticlines formed by rollover on the downthrown side of the growth fault system (Figure 6). The anticlines have different growth histories and are separated by a small syncline. Maximum structural growth for the eastern anticline occurred within *Angulogerina "B"* time (1.8 MYBP) (Figure 15), after which the growth rate steadily decreased. The eastern structure is sharply defined and has as much as 549 m (1800 ft) of structural closure. The western anticline has less relief, with 122 to 152 m (400 to 500 ft) of structural closure. Maximum structural growth occurred later than that of its eastern counterpart.

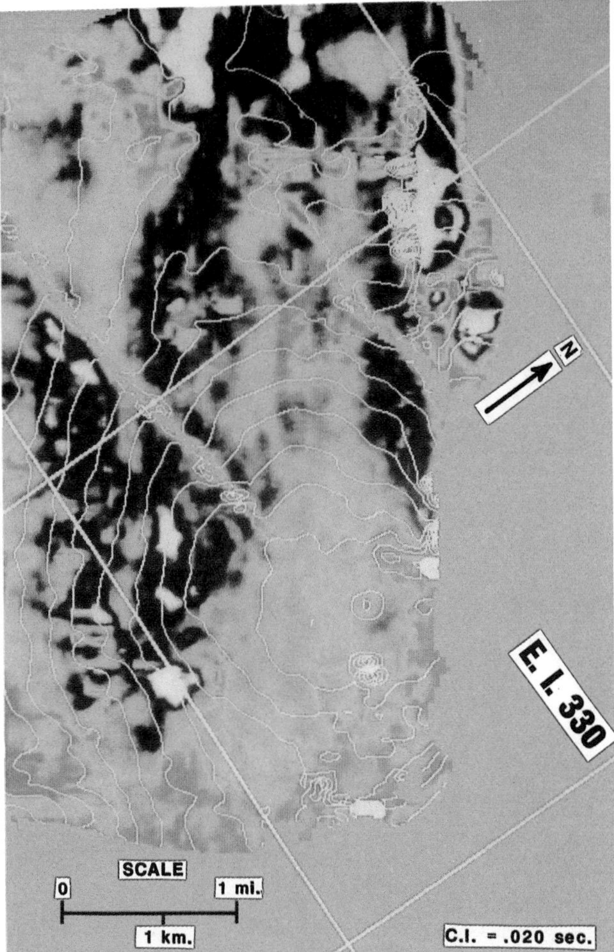

Figure 11. Extracted amplitude map of JD sandstone generated from 3D seismic data volume with superimposed structural contours.

STRATIGRAPHY

Sediments in the Gulf Coast basin are entirely Mesozoic and Cenozoic in age, occurring in offlapping sequences that become generally younger to the south (Figure 13). Beginning in the Late Cretaceous and early Tertiary, the northern Gulf received a large influx of terrigenous sediments presumably in response to the Laramide orogeny to the west. Thick accumulations of siliciclastic sediments were deposited in depocenters overlying the Jurassic Louann Salt. These depocenters shifted generally seaward and eastward throughout the Cenozoic

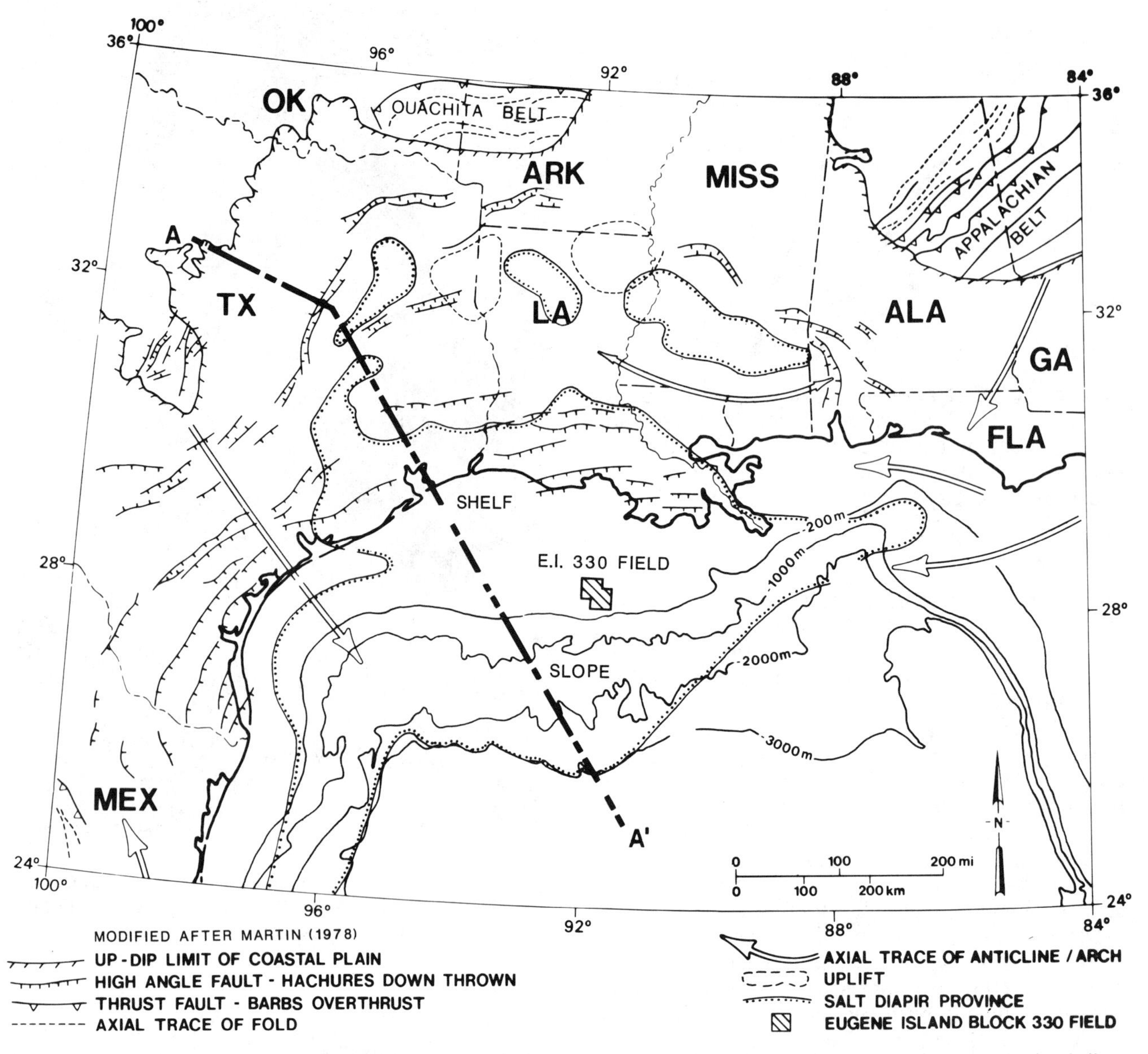

Figure 12. Structural-physiographic map of the northern Gulf of Mexico area. Location of Figure 13 cross section A–A′ shown as heavy dashed line. Modified after Martin (1978).

(Woodbury et al., 1973). The Pliocene-Pleistocene section in the Gulf Coast consists of transgressive and regressive sequences of fluvial-deltaic facies. Caughey (1975) described the Pleistocene section as a fluvial, delta-plain, delta-front, and prodelta regressive sequence.

The major producing sandstone reservoirs in the Eugene Island 330 field are shown on the type log (Figure 15). These Pleistocene sandstones are interpreted to have been deposited as delta-front sands (Figure 16). Gently sloping shingled seismic reflectors (Figure 17), classified by Mitchum et al. (1977) as oblique-progradational, are interpreted as representing delta-front depositional surfaces. Dipmeter data (Figure 18) show depositional current dips that indicate delta-front foreset bedding (Gilreath and Maricelli, 1964). These delta-front sequences typically have 15–30% sand in beds that show a range in thicknesses and have gradational bases and sharp upper contacts (Martin, 1978). Norwood and Holland (1974) recognized the relatively high percentages of Pleistocene reservoired oil in these alternating sand-shale facies.

Seismic stratigraphic-sequence analysis illustrates the numerous sequence boundaries defined by reflection terminations and distinctive reflector character (Figure 19). Synthetic seismograms are used routinely to tie well data to seismic data. Seismic stratigraphy utilizes this accurate seismic identification of reservoir facies to help extrapolate and predict sand conditions and geometries away from well control.

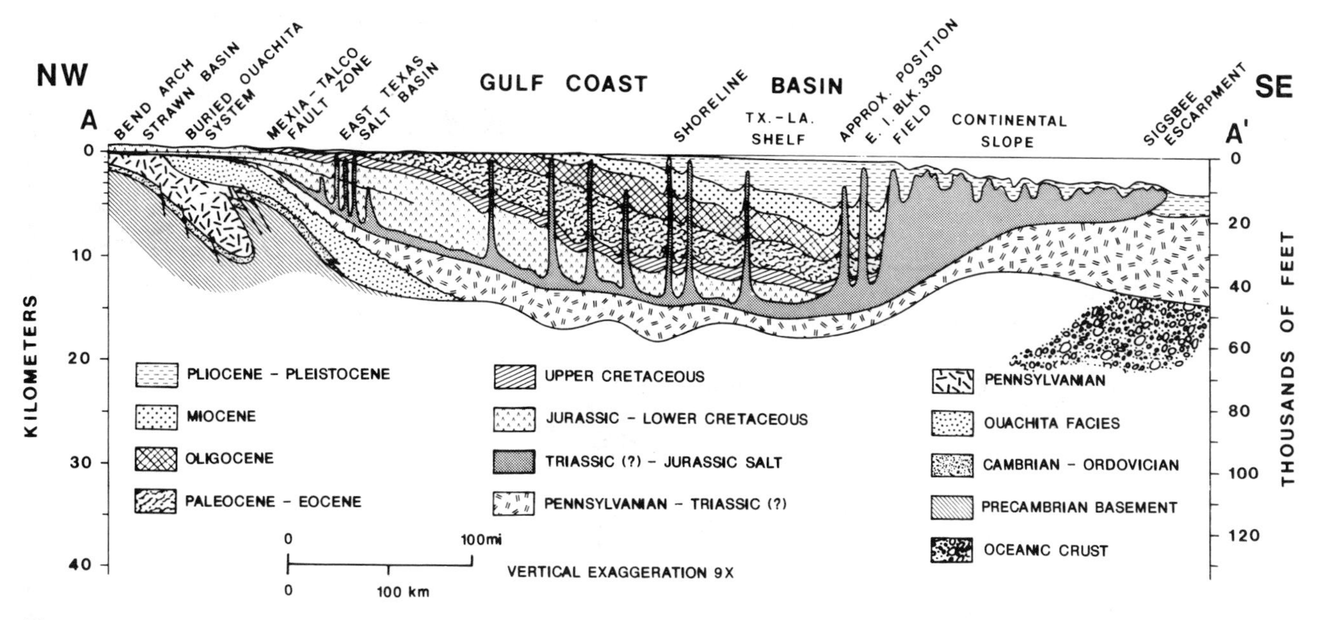

Figure 13. Diagrammatic cross section (A–A′) of northern Gulf of Mexico. Location shown on Figure 12. Modified after Martin (1978).

Seismic stratigraphic-facies analysis also indicates that the source of sediments deposited in the subbasin was from the north, predominantly through Block 331. The developing structure in Block 330 deflected sediment dispersal. During the *Angulogerina "B"* time (1.8 MYBP) of maximum growth on the Block 330 structure, the OI through JD reflectors are interpreted as onlapping the structural crest (Figure 19). Channels evident at the GA horizon (Figure 20) indicate transport away from the anticline during the younger *Trimosina "A"* time (0.5 MYBP). Erosional channels are of high exploration interest owing to their part in redistributing reservoir sediments.

TRAP

The trapping mechanism for the field is four-way dip closure on two rollover anticlines located on the downthrown side of the growth fault (i.e., JD sandstone, Figures 6 and 27). Faulting was contemporaneous with deposition. The fault has been active since at least the early Pleistocene (2.1 MYBP) and continues to be active today. Maximum thinning is seen during *Angulogerina "B"* time (1.8 MYBP). Additional porosity pinchout traps, due to facies changes or nondeposition, are found on the flanks of the Block 330 anticline, as seen on the LF sandstone structure map in Figure 29.

Impervious shales above individual reservoirs act as seals (Figure 15). Shale beds only a few feet thick, if continuous, are known to be effective vertical seals. Lateral seals are formed by major or minor faults, although permeability barriers caused by facies changes also are common (Figure 29).

RESERVOIRS

Most of the production in the field is from the seven Pleistocene sandstones shown in Figure 15. Six of these are predominantly oil sands; the JD is gas productive; and the OI produces both oil and gas. Many of these sandstones are divided into two or more units. For example, the GA sandstone is divided into five units, of which the GA-1, GA-2, and GA-5 sand units are productive. More than 25 producing sandstone units have been developed in the field.

Reservoir parameters for the major oil and gas sands on Pennzoil's Block 330 are listed in Appendix 1. Porosities average about 30% and water saturation ranges from 20 to 40%. Permeabilities range from 10 md to more than 6 darcys. Approximately 80% of the reservoirs have permeabilities of more than 100 md and about 20% have more than 1 darcy (Holland et al., 1980).

The major producing sandstones are categorized as quartzose to slightly arkosic arenites. Although the grain size ranges from coarse silt to medium sand (Appendix 1), most sandstones are fine grained and well sorted. A wide range in grain roundness has been noted.

Quartz (both monocrystalline and polycrystalline) is the predominant framework mineral, amounting to as much as 89% of the bulk rock volume. Feldspars and chert also are present, and may represent as much as 25% of the total composition. Rock fragments (igneous, metamorphic, and shale balls) account for 13% of the sandstone composition. Heavy minerals (zircon and chlorite), mica, and glauconite are present in small percentages.

The depositional shale and clay content is variable and can comprise as much as 9% of the bulk rock

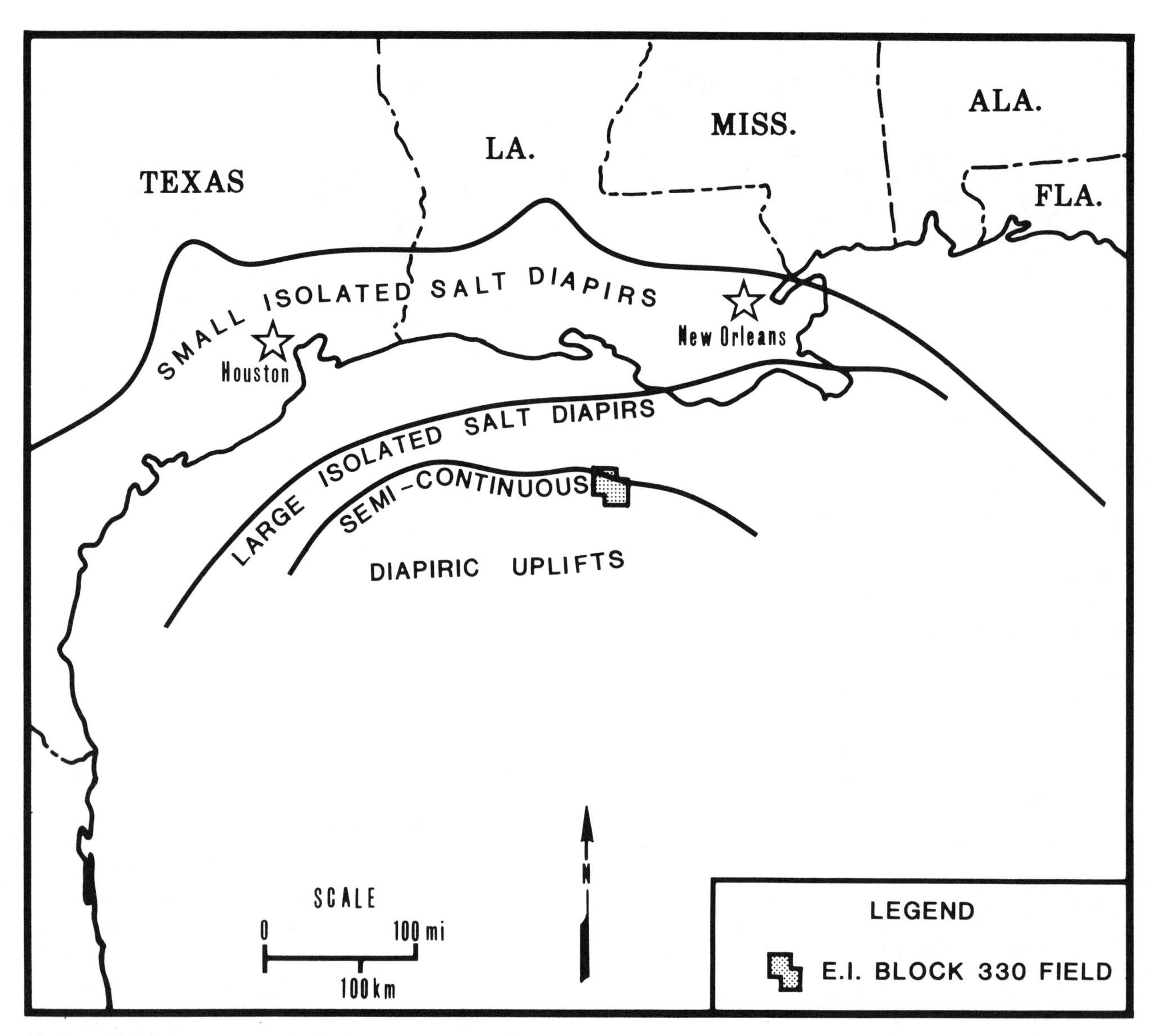

Figure 14. Major groups of salt diapirs and salt tectonic provinces in the northern Gulf of Mexico. (Holland et al., 1980, after Woodbury, et al., 1973.)

volume. This shale is structural, laminated, or dispersed as pore fillings. Various cements are present, including authigenic clay (0 to 8%), silica (0 to 9%), dolomite (0 to 12%), pyrite (0 to 2%), and feldspar (trace).

Porosity is commonly primary intergranular (Figure 21A and B) or reduced primary intergranular. Some secondary moldic porosity also formed by feldspar dissolution (Holland et al., 1980).

GA Sandstone

The GA sandstone is productive in the GA-1, GA-2, and GA-5 units. The GA-1 is the shallowest major producing unit in the field and is found at an average depth of 1311 m (4300 ft). The structural configuration at the GA horizon is an unfaulted rollover anticline (Figure 24A). The eastern structure dips 2 to 3° on all flanks except near the growth fault, where the dip increases to 6°. The productive area at the GA-2 horizon is 341 ha (842 ac), is crestal, and is located wholly within Block 330.

The GA sandstone is interpreted to be a delta-mouth bar deposit, with massive sands as much as 122 m (400 ft) thick (Figure 24B). The sand is very fine to fine grained (0.08 to 0.19 mm) and contains various amounts of detrital shale and clay. The rocks consist predominantly of monocrystalline quartz, with minor quantities of feldspar, chert, and rock fragments. Depositional shale occurs primarily as laminae (Figure 21A). The sands contain relatively little cement (0 to 4%) except the GA-5 unit, which is heavily cemented by dolomite. Silica cement is seen as overgrowths on detrital quartz grains. Pore-lining clay occurs in small quantities and consists largely

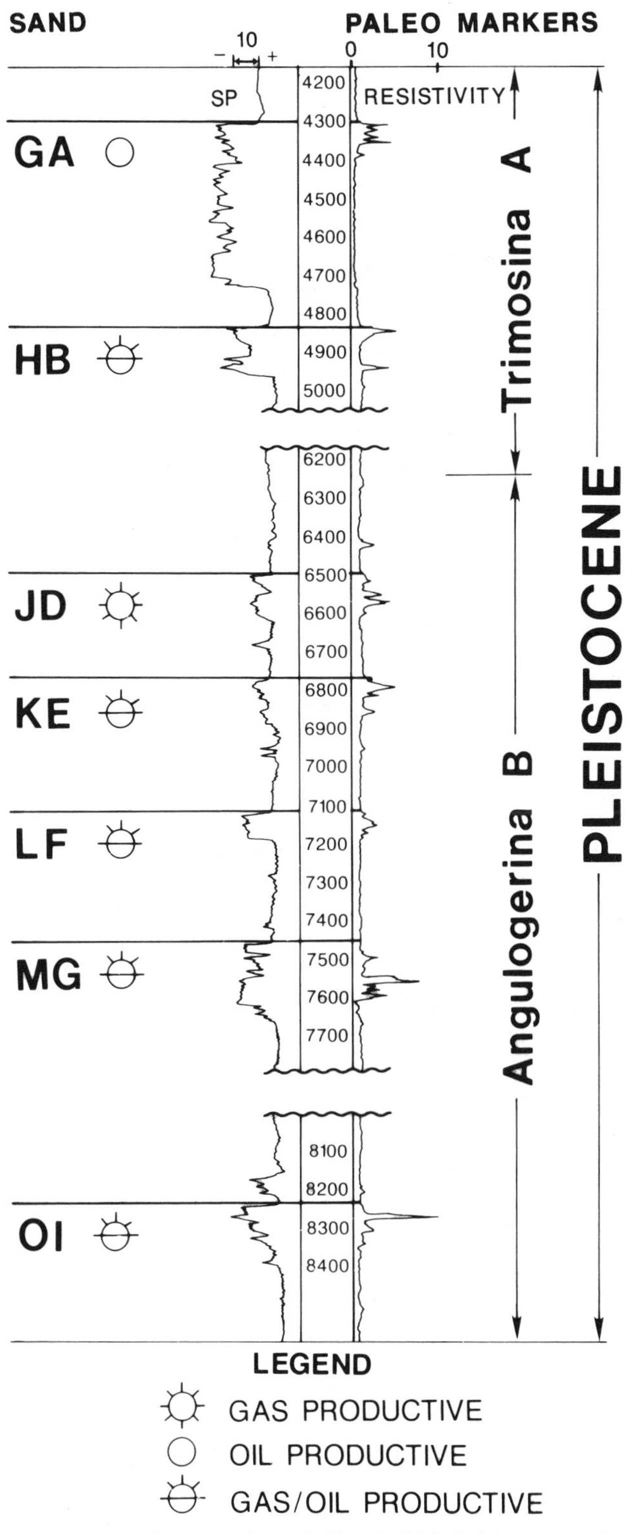

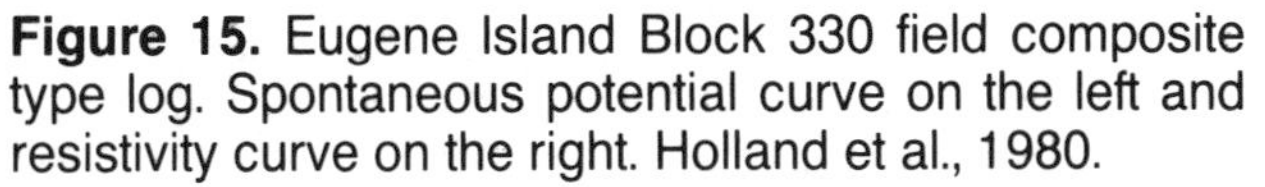

Figure 15. Eugene Island Block 330 field composite type log. Spontaneous potential curve on the left and resistivity curve on the right. Holland et al., 1980.

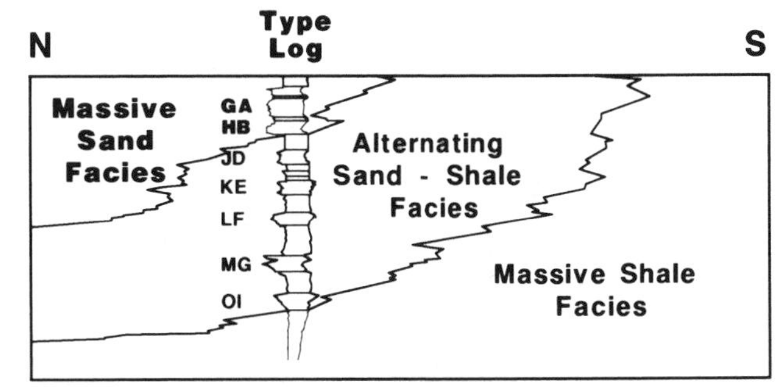

Figure 16. Depositional model for sandstones in Eugene Island Block 330 field. The delta-front facies are shown as alternating sand and shale. Modified after Norwood and Holland (1974).

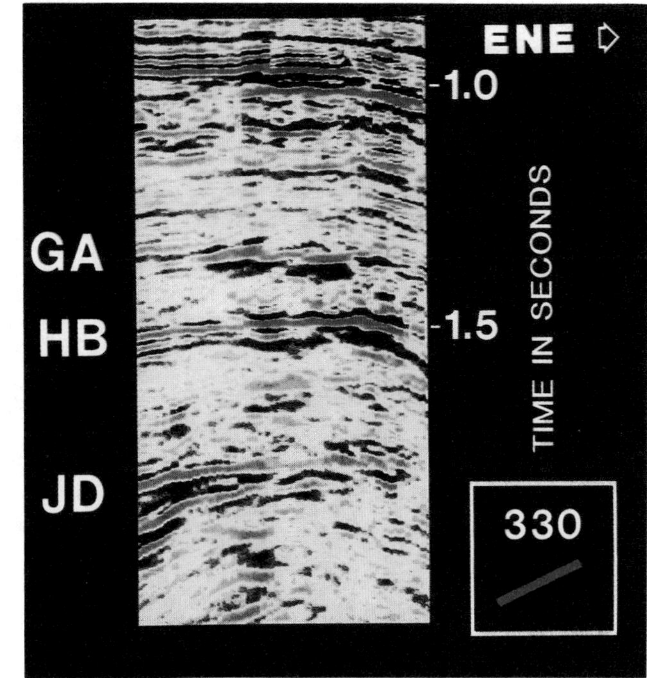

Figure 17. 3D seismic line showing shingled reflections of the GA horizon, indicating delta-front sand deposition on Block 330.

of smectite, illite, and kaolinite. Porosity is intergranular and volume is reduced by shale fill or authigenic clay cementation.

The GA-2 unit is a major reservoir having a maximum net oil pay of 30 m (100 ft) and an oil column of 38 m (126 ft). The reservoir is bounded on the southwest by a permeability barrier and on the northeast by dip closure (Figure 24A). The reservoir exhibits a strong bottom water drive (Figure 25) and has a pressure gradient of 0.49 psi/ft. Reservoir parameters for the GA-2 sandstone are listed in Appendix 1.

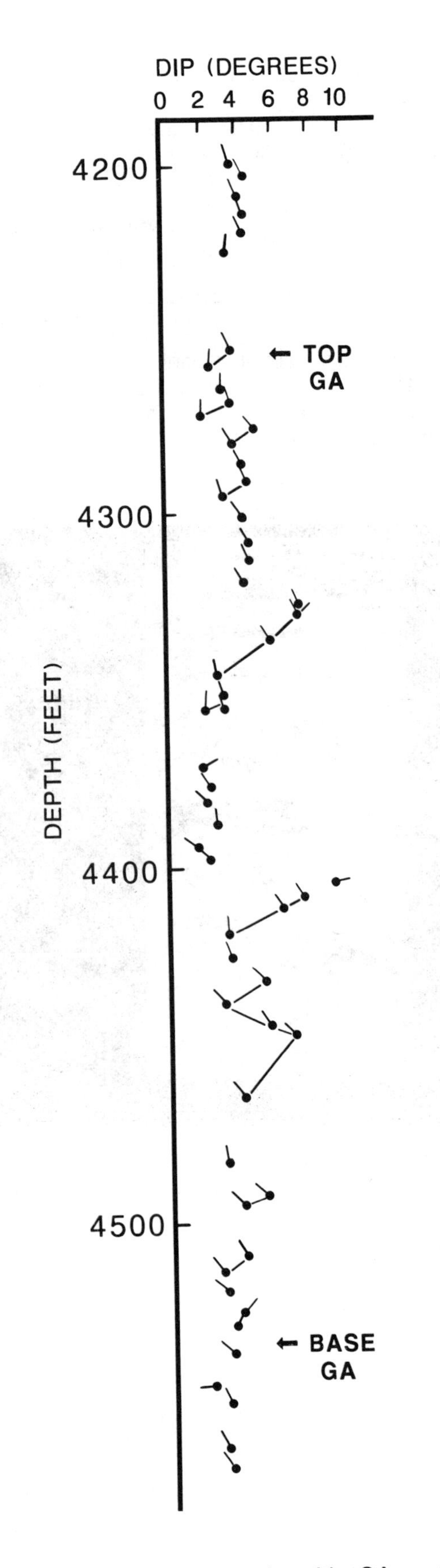

Figure 18. Dipmeter interpretation of the GA sandstone, Block 330 #B-13 well. Current patterns correspond to shingled seismic reflectors. Holland et al., 1980.

Original oil-in-place for the GA sandstones is estimated at 40 million bbl, 10 million bbl of which was initially considered recoverable. Cumulative production (March 1988) for the GA sandstone on Block 330 is 14.8 million bbl of oil and 8.8 bcf of gas. Eight wells completed in the GA-2 unit have each produced more than 1 million bbl of oil, and the #C-2 well had produced 2.9 million bbl as of March 1988. The oil gravity is 23° API and the initial gas/oil ratio was 400–500 scf/bbl.

The GA-2 reservoir has overproduced original recoverable volumetric estimates by 43%. It is now believed that the GA sandstones will ultimately produce 17.0 million bbl of oil. This could be the result of the contribution to the overall production from silty and shaly laminated sandstones that originally were not considered pay because of the low resistivity log response. This low-resistivity sandstone has porosity in the 18 to 25% range and permeability of less than 10 md. Even though these zones have little capacity for significant lateral flow, they may recharge the more permeable productive zones adjacent to them. These zones could then be classified as "feeder pay" and would increase the estimated recoverable reserves.

HB Sandstone

The HB sandstone is divided into the HB-1, HB-1B, HB-2, and HB-3 producing units. The reservoir produces predominantly oil and has a small gas cap in the HB-1 unit. As discussed earlier, high-amplitude seismic anomalies with associated flat spots correspond to the GA and HB pay zones (Figure 9). Average depth of the reservoir is 1463 m (4800 ft). Porosity averages 28% within this generally well-sorted, medium-grained sand. Average water saturation is 33%.

Structurally, the HB sandstone is mapped as rollover anticlines, cut only by minor faulting (Figure 26A). The HB sandstone is productive on the crest of the eastern anticline, and the hydrocarbons are entirely within Block 330 (Figure 26A). The oil reservoir covers 416 ha (1029 ac) and has a maximum oil column of 29 m (96 ft). The gas cap covers 140 ha (345 ac) and has a maximum thickness of 8 m (27 ft).

Original oil-in-place reserves for the HB sandstone are 38 million bbl of oil, 16.4 million of which is considered recoverable. Cumulative production from the HB sandstones on Block 330 is 14.1 million bbl of oil and 16.2 bcf of gas (March 1988). The oil gravity is 25° API, and the initial gas/oil ratio was 361 scf/bbl. The reservoir exhibits a strong water drive and has a pressure gradient of 0.49 psi/ft. Critical in the development of the HB reservoir was the placement of wells outside the gas cap but close enough to the gas/oil contact to effectively produce the oil rim.

The HB sandstones are interpreted as prograding delta-front sands, with net sand thickening generally to the north (Figure 26B). Grain size (0.05–0.25 mm)

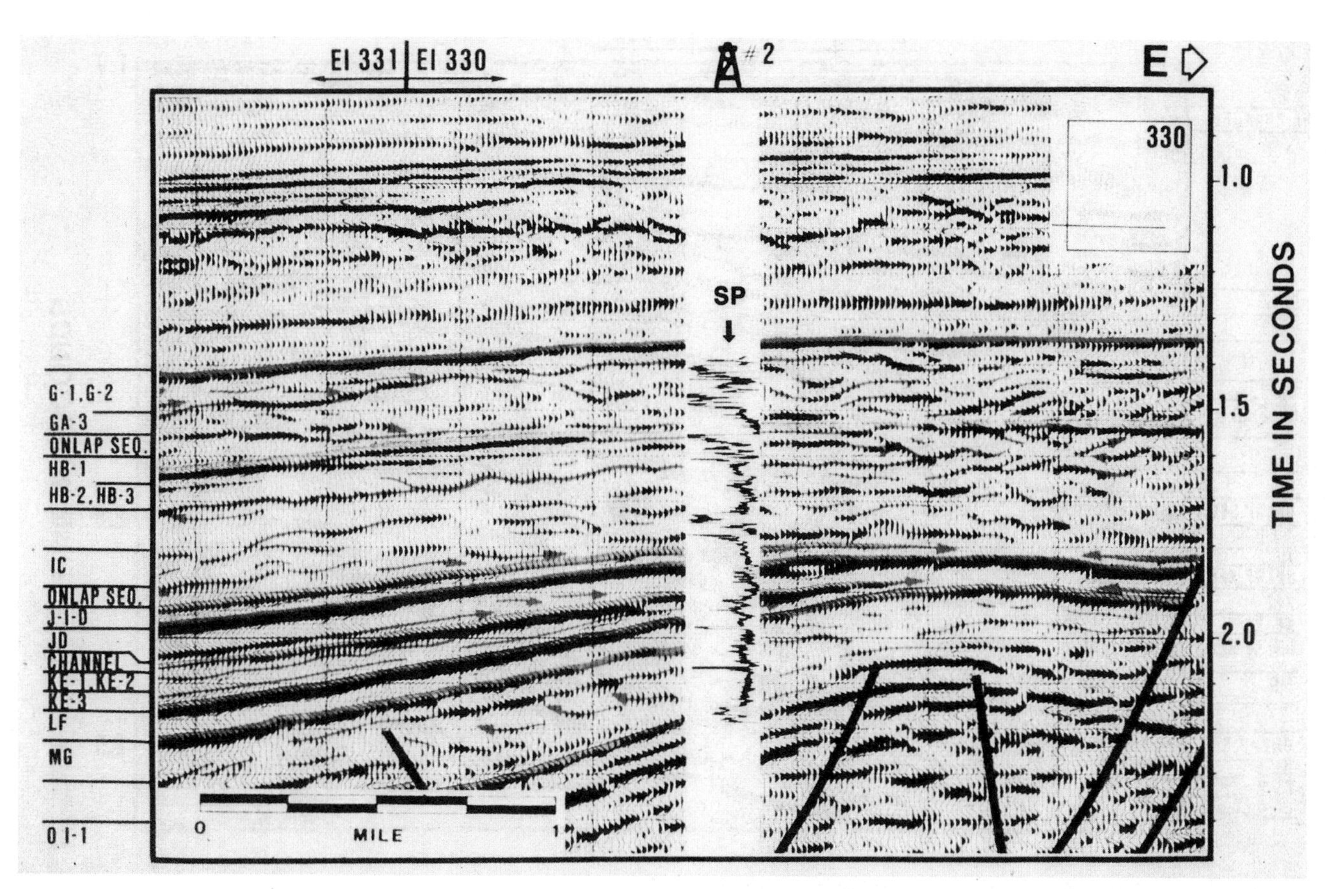

Figure 19. Seismic line across the southern part of Block 330, showing stratigraphic interpretation, sequence boundaries, and onlapping reflectors onto the 330 structure. SP log curve insert shows Eugene Island 330 #2 well tie.

increases and sorting improves from the base to the top of the sand unit. The framework grains (Appendix 1) are primarily quartz (average 68% of bulk rock volume), with minor amounts of feldspar (7 to 11%), chert (6 to 14%), and rock fragments (1 to 10%) (Figure 22A). The shale within this zone is predominantly laminar. The HB-1 sand contains local concentrations (as much as 27%) of shell fragments. Pore-filling dolomite cement occurs in the finest grained reservoir rock. Pores are also partly occluded by quartz overgrowths and authigenic clay cement (Figure 21B).

JD Sandstone

The JD sandstone has the largest productive area of any reservoir in the Block 330 field, 2611 ha (6450 ac) and parts of seven blocks (Figure 27A). It is also the only reservoir that produces predominantly gas, with more than 1 tcf of gas originally in place. An estimated 736 bcf of gas and 16 million bbl of condensate are recoverable. The average producing depth is 1981 m (6400 ft).

The structural configuration at the JD horizon is interpreted as two anticlines, separated by an intervening syncline (Figure 27A). The eastern structure is divided into several fault blocks. The JD sandstone is characterized by a relatively high-amplitude seismic reflection, as seen at approximately 1.8 to 2.1 sec on Figure 9.

The JD sandstone is interpreted to be a delta-mouth bar/delta-front deposit (Figure 27B). The distributary mouth is near the intersection of Block 313 and the northwestern corner of Block 331, where the sand is thicker, cleaner, and coarser grained. In Block 330, sand conditions are more variable and distal, with silt and shale laminations, finer grain size, and less sorting.

The JD sandstone on Block 330 is predominantly quartzose, with various, but significant, quantities of shell fragments (0 to 56%) (Figure 22B) and limited amounts of detrital feldspar, mica, and rock fragments. Dolomite and lesser amounts of authigenic clay (illite), silica, and pyrite are present as cements (Figure 21C).

Reservoir parameters for the JD sandstone on Block 330 are listed in Appendix 1. The average porosity is 29% and the average water saturation is 35%. Net pay averages about 11 m (36 ft) and the maximum thickness is 28 m (92 ft). Permeability ranges from 20 to 4100 md and averages 720 md.

The only commercial JD oil reservoir is located in the southern fault block on the eastern structure on Block 338 (Figure 27A). The oil rim has a column

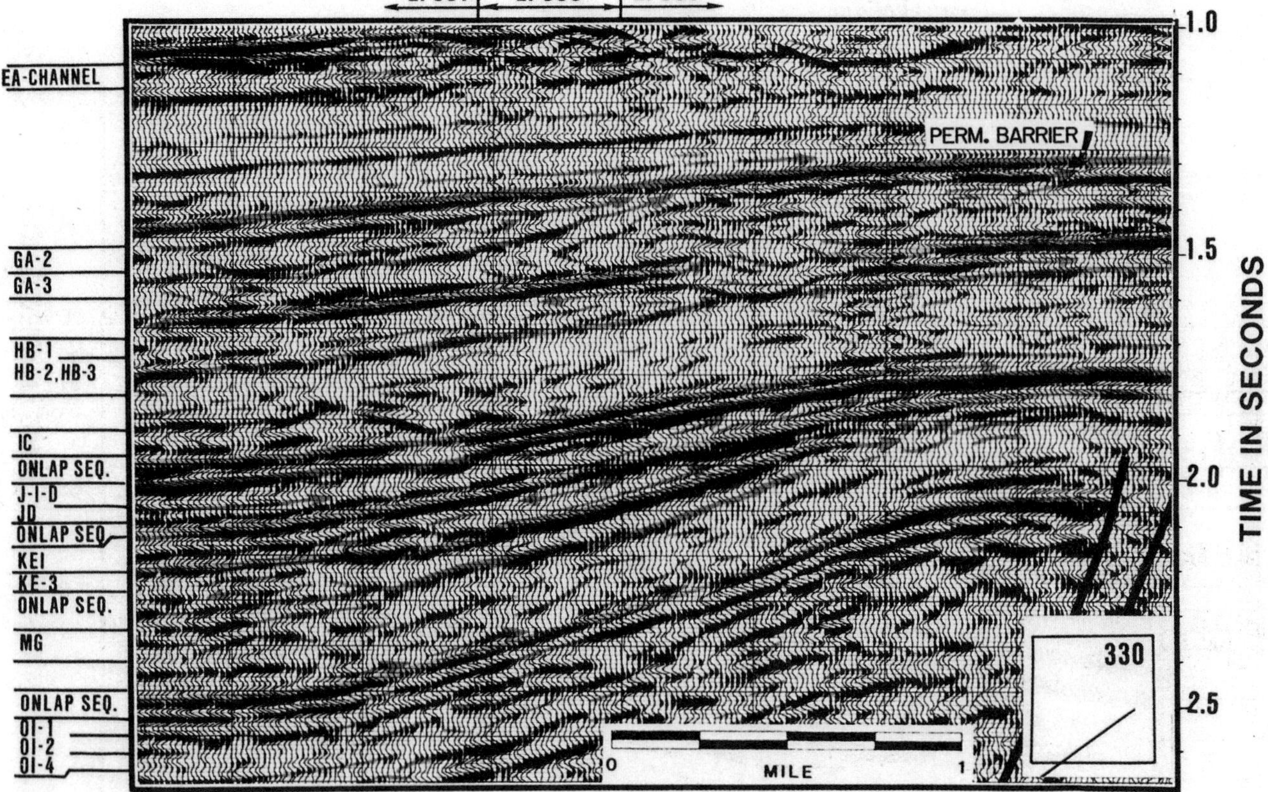

Figure 20. Seismic stratigraphic interpretation, showing GA sandstone channeling into underlying HB sandstone.

of 39 m (129 ft). The gas column above the oil rim is 257 m (842 ft). Faulting is less significant on the western structure, which has a gas column of 113 m (370 ft) and no oil rim. A directional water drive is present, watering out wells successively from west to east.

Cumulative production from the JD sandstone on Block 330 is 226 bcf of gas and 4.7 million bbl of condensate (March 1988). The reservoir has a limited water drive, with bottom-hole pressures dropping from an original 3784 psig to a current 1678 psig (July 1987).

KE Sandstone

The KE sandstone is productive on the eastern structure (Block 330) and on the crest of the western structure (Blocks 314 and 331) (Figure 28A). The sandstone consists of the KE-1 and KE-2, each approximately 12 m (40 ft) thick. Both units have gas caps and oil rims. A permeability barrier bounds both KE-1 and KE-2 on the southeastern side of the eastern structure. The average producing depth is 2017 m (6600 ft). The sandstone has an average porosity of 29% and a permeability range from 10 to 4500 md (Appendix 1).

The sand is a delta-front/distributary-mouth deposit, which is as much as 91 m (300 ft) thick in the southeast corner of Block 313 (Figure 28B). The KE sandstone is predominantly fine-grained and contains minor amounts of detrital shale matrix (Appendix 1). Framework grains consist mainly of quartz, with lesser amounts of feldspars, chert, and rock fragments (as much as 21%) (Figures 21D and 22C). Silica and dolomite are the most common cements.

The eastern structure is broken by normal down-to-the-south and antithetic faults that subdivide the reservoir into several producing fault blocks. The reservoir has a maximum oil column of 134 m (441 ft) and a gas cap of 227 m (745 ft). The estimated original recoverable reserves for the eastern structure are 16 million bbl of oil, 18 bcf solution gas, 46 bcf nonassociated gas, and 940,000 bbl of condensate. Cumulative production from the KE sandstone on Block 330 is 14.3 million bbl of oil, 26.4 bcf of gas, and 559,000 bbl of condensate (March 1988). The oil reserves are essentially depleted and production of the gas cap is continuing. The gravity of the oil is 35° API. The initial gas/oil ratio was 893 scf/bbl. The reservoir has a predominant water drive.

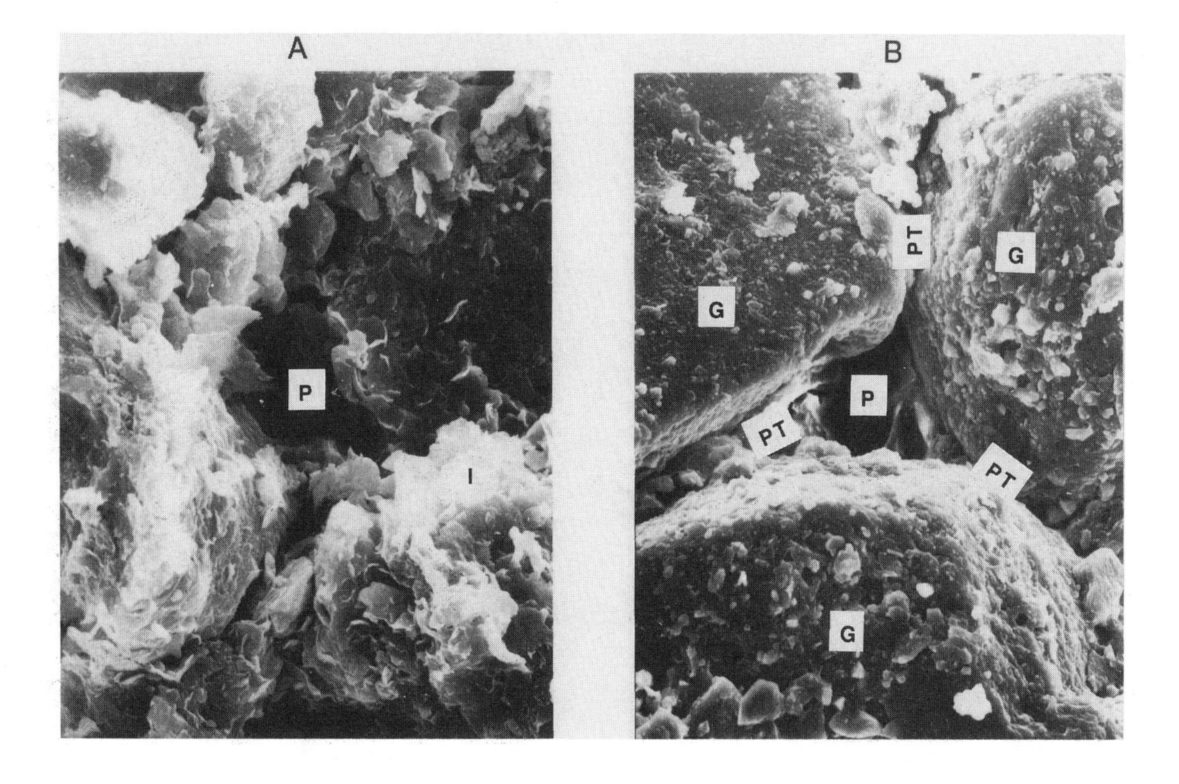

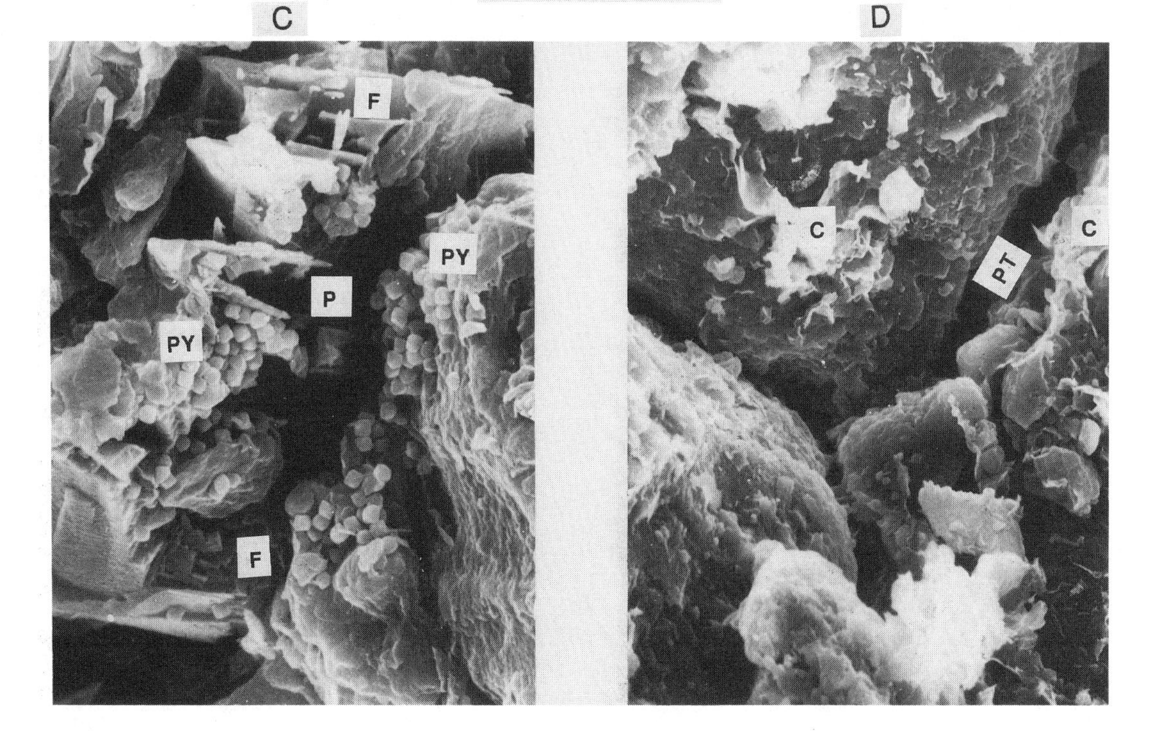

Figure 21. SEM photomicrographs of sidewall core samples. **(A)** Intergranular pore (P) lined with partially recrystallized illite (I). GA sandstone, E.I. 330 #C-3, 5039 ft (scale = 10 microns). **(B)** Sand grains (G), intergranular pore (P), and open pore throats (PT) in clean, shale-free sandstone. HB sandstone E.I. 330 #C-17 ST, 5358 ft (scale = 30 microns). **(C)** Primary, intergranular pore (P) lined with pyrite octahedrons (PY). Feldspar (F) is partially leached, resulting in secondary porosity development. JD sandstone, E.I. 330 #B-2, 7491 ft (scale = 4 microns). **(D)** Pore walls coated with partially recrystallized, discontinuous, detrital illite and smectite clays (C). Pore throats (PT) open and well-defined. KE sandstone, E.I. 330 #B-2, 7934 ft (scale = 10 microns).

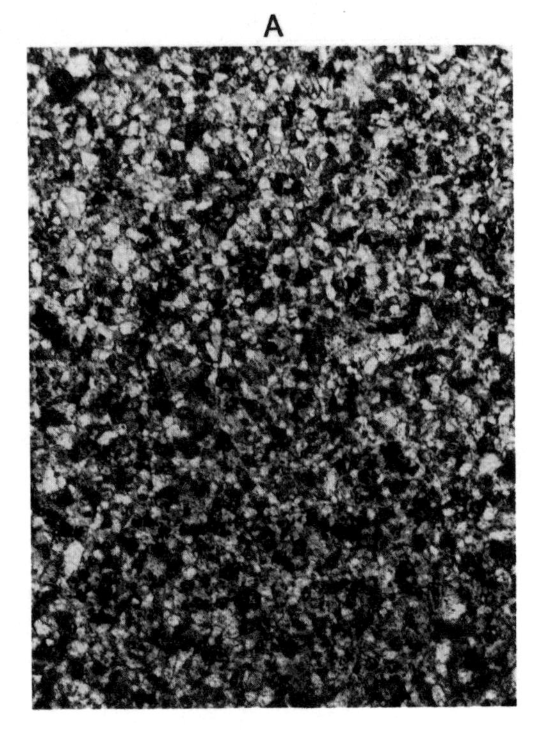

SCALE BAR

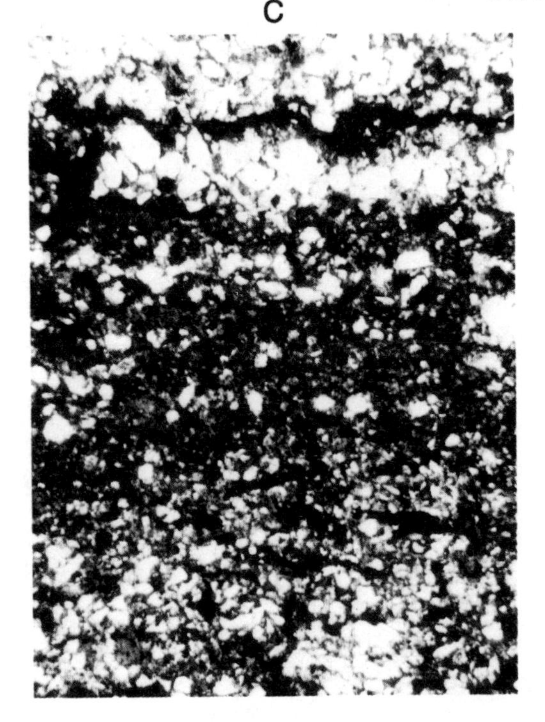

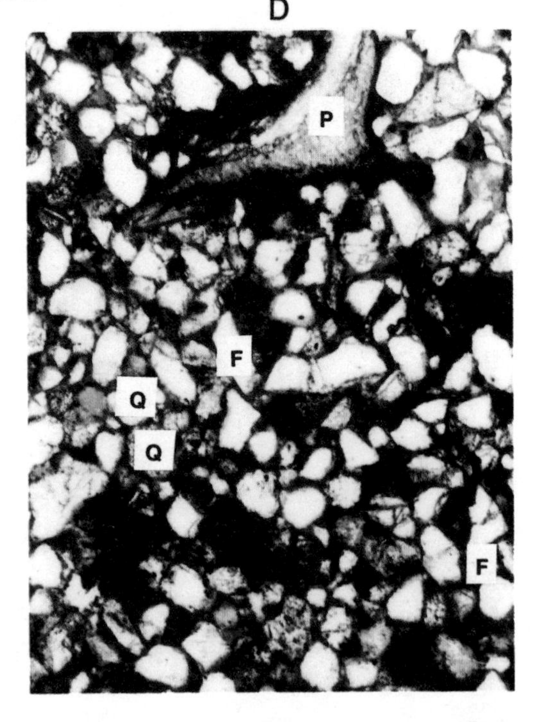

Figure 22. Thin-section photomicrographs of selected sidewall core samples from Eugene Island Block 330. (Plane polarized light: scale = 0.25 mm.) **(A)** Quartz-rich, dolomitic siltstone with intergranular porosity. HB sandstone, E.I. 330 #A-13, 6485 ft. **(B)** Fine-grained, loosely packed sandstone (subarkose) with well-developed intergranular porosity. JD sandstone, E.I. 330 #A-1, 6603 ft. **(C)** Laminated, very fine grained sandstone and interbedded silty shale. KE sandstone, E.I. 330 #B-2, 7966 ft. **(D)** Fine-grained sandstone (subarkose). C, chert; F, plagioclase feldspar; P, pelecypod fragment; Q, quartz. LF sandstone, E.I. 330 #B-2 well, 8338 ft.

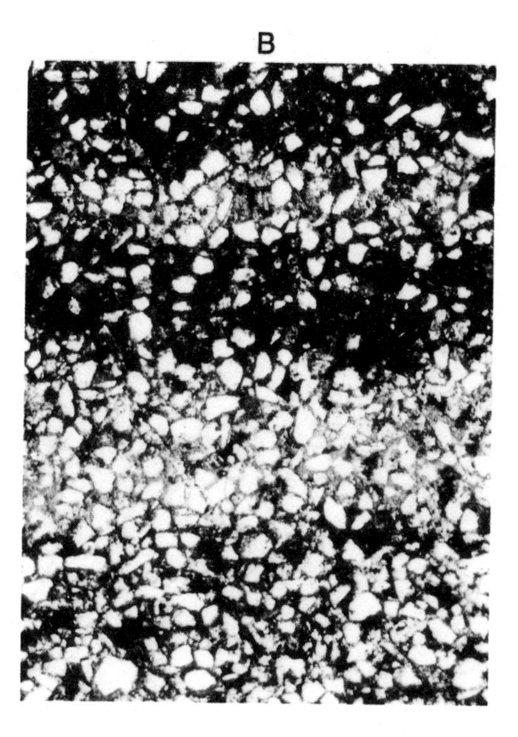

SCALE BAR

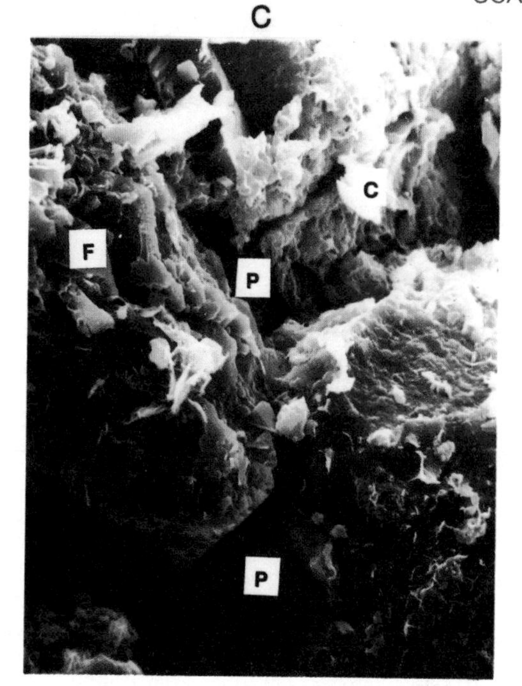

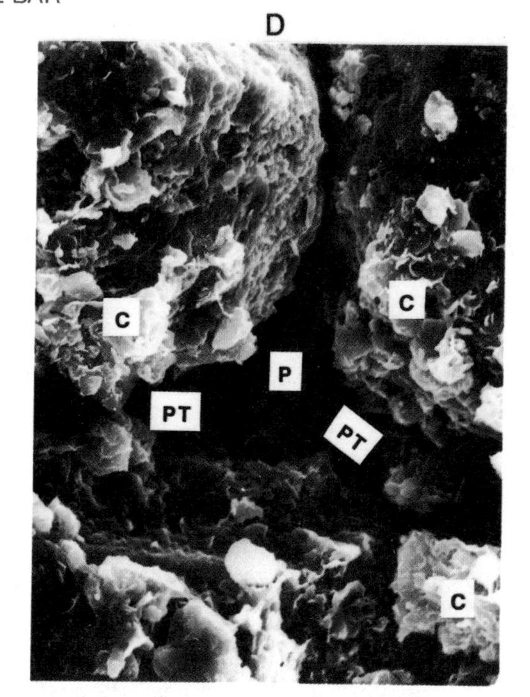

Figure 23. Thin-section photomicrographs. (Plane polarized light: scale = 0.25 mm.) **(A)** Single lamina of very fine grained, poorly sorted quartz sandstone in shale. Q, quartz; PF, plant fragment. MG sandstone, E.I. Block 330 #A-12, 7712 ft. **(B)** Two laminae of very fine grained, well-sorted quartz arenite interlaminated with sandy shale. OI sandstone, E.I. Block 330 #A-4, 8764 ft. **(C)** SEM photomicrograph (scale = 30 microns). Intergranular pores (P) devoid of pore lining minerals (bottom portion) or lined with discontinuous overgrows of sodium feldspar (F), illite and smectite clays (C). LF sandstone, E.I. Block 330 #B-6, 7788 ft. **(D)** SEM photomicrograph (scale = 8 microns). Large intergranular pore (P) with discontinuous illite and smectite clay coats (C). Pores interconnected through open pore throats (T). OI sandstone, E.I. 330 #B-15, 7815 ft.

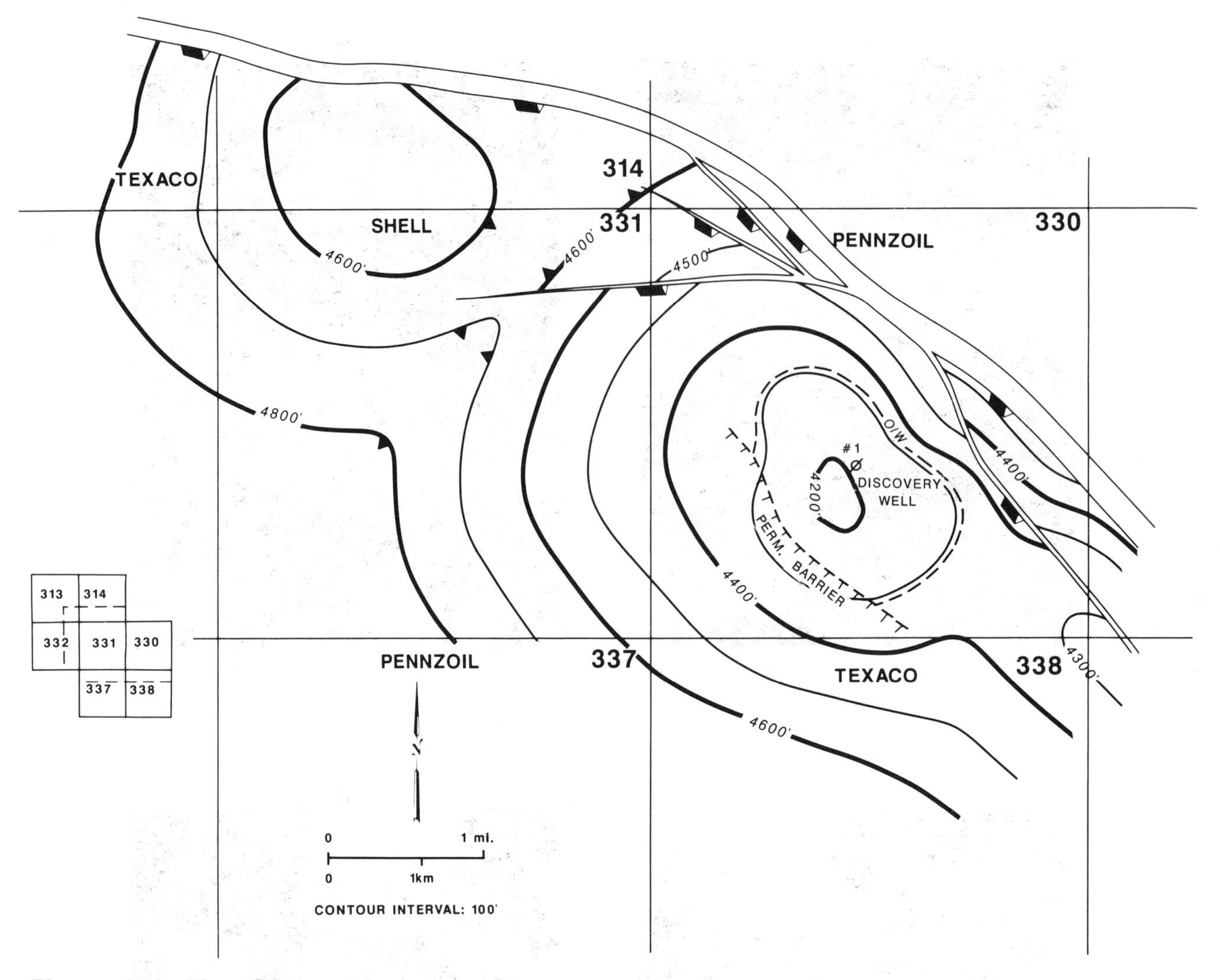

Figure 24A. Top GA sandstone structure map. Discovery well (E.I. 330 #1) indicated. Contour values subsea level.

LF Sandstone

The LF sandstone is the most prolific oil-producing reservoir in the field. It has the second largest area with 1472 productive ha (3635 ac) in the oil column and 245 productive gas cap ha (606 ac). Cumulative production is more than 63 million bbl of oil, and ultimate recoverable reserves are estimated at 69 million bbl. Original oil-in-place reserves are 180 million bbl. Average producing depth is 2164 m (7100 ft). Reserves are equally divided between the eastern and western structures. The structure in Block 330 has a gas cap containing 15 bcf gas reserves.

The structural configuration at the LF sandstone horizon is similar to that of the shallower sands (Figure 29A). The eastern structure is broken into several fault blocks by down-to-the-south and antithetic faults. A permeability barrier is present on the crest. Figure 30 illustrates the dramatic stratigraphic thinning onto the crest of Block 330 within the JD to LF interval. As suggested by mapping, the Block 330 structure was actively growing during this time, and southward sediment flow was deflected to the west off the crest.

The LF sandstone is interpreted as a distributary-mouth bar deposit, grading into a delta-front deposit to the southeast in Block 330. The gross interval isopach map (Figure 29B) shows thick deposition in the area near Blocks 313 and 331. The sand progressively thins and is shaled out on the crest in Block 330.

The LF in Block 330 is a very fine grained, clean sandstone. The small grain size allows for a high interstitial water surface area and a high (40%) irreducible water saturation. This results in an abnormally low resistivity log response (average 1.3 ohms) and masks its true productivity (Figure 31). The gross sand thickness averages 17 m (55 ft), and, because of the absence of shale, most of this thickness is filled with hydrocarbons.

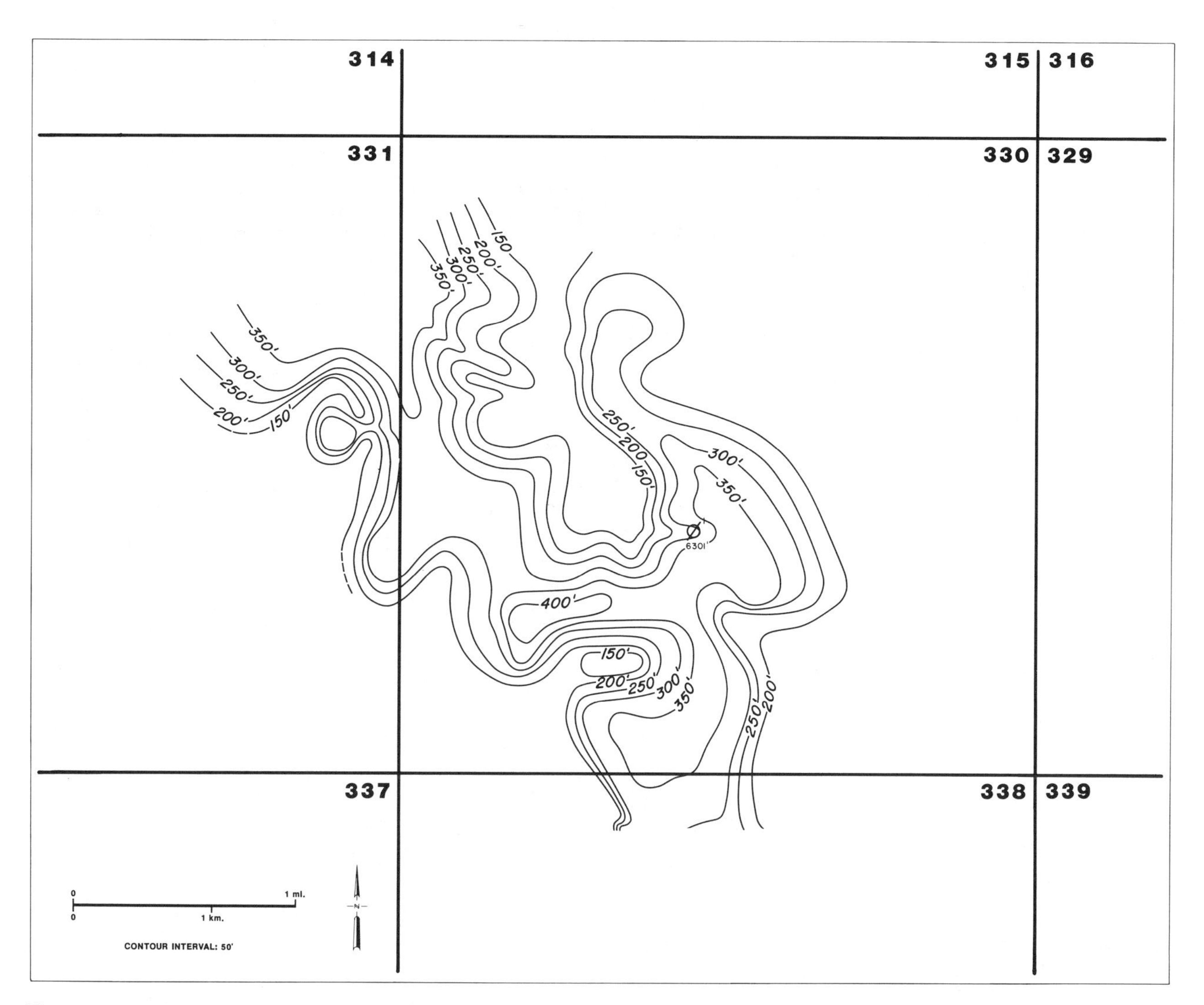

Figure 24B. GA net sandstone isopach map.

The LF sandstones are generally quartzose with various, but minor, amounts of detrital shale (Figure 22D). Feldspars, chert, and rock fragments may constitute as much as 30% of the total rock volume (Figure 23C). These sands are more uniformly cemented than the other sandstones in the field. The cements include authigenic clay, silica, dolomite, and pyrite.

The LF sandstone is a combination gas-cap expansion and partial-water-driven reservoir. The pressure gradient is 0.57 psi/ft. The original oil column was 212 m (696 ft) with a gas column of 113 m (370 ft). The oil gravity is 34° API, with an initial gas/oil ratio of 941 scf/bbl.

MG Sandstone

The MG sandstone consists of four oil- and gas-productive units in the west-central part of Block 330 and in Block 331 (Figure 32). Permeability barriers are present throughout the area. MG

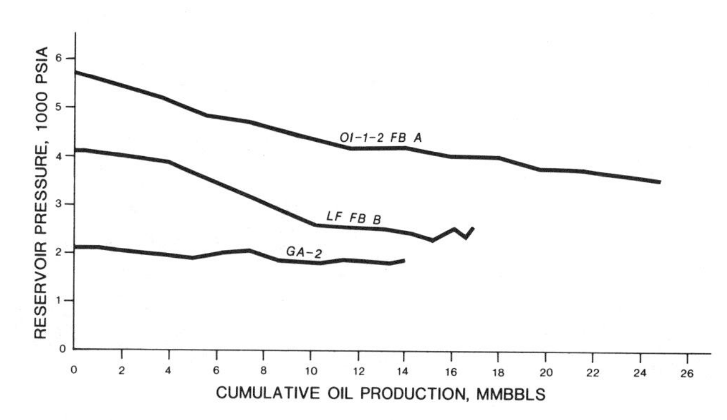

Figure 25. Pressure history for three Eugene Island Block 330 reservoirs.

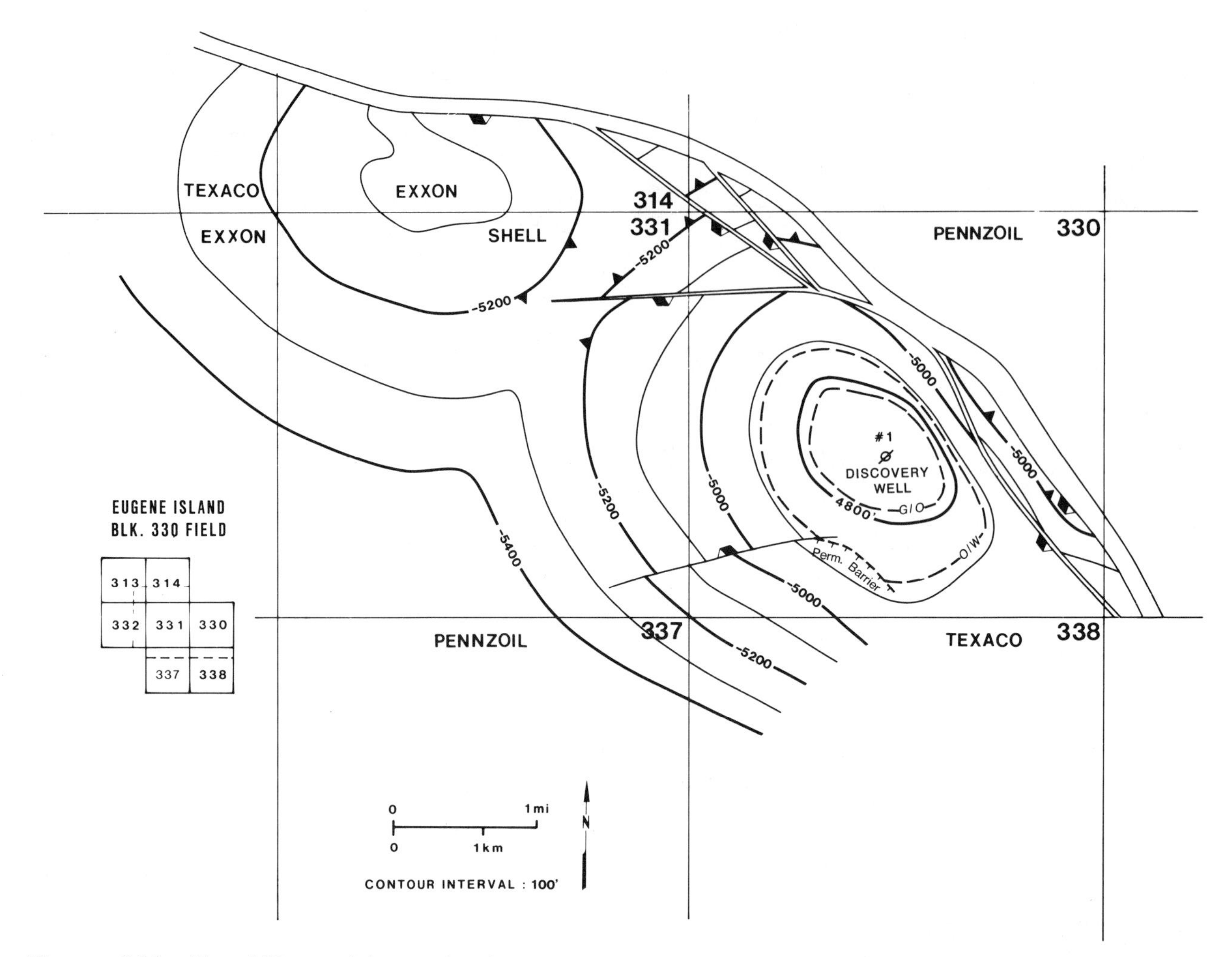

Figure 26A. Top HB sandstone structure map. Discovery well (E.I. 330 #1) shown.

sandstones have produced 4.3 million bbl of oil, 37.9 bcf of gas, and 1.3 million bbl of condensate in Block 330 (March 1988). The oil gravity is 36° API, with an initial gas/oil ratio of 1225 scf/bbl (Appendix 2). The MG sandstone has predominantly produced as a depletion-drive reservoir. The pressure gradient is 0.58 psi/ft.

These deltaic MG sandstones are well sorted and consist of various grain sizes. Mean grain size ranges from very fine (Figure 23A) to coarse (0.07 to 0.57 mm). Quartz is the dominant framework mineral, with feldspar, chert, and rock fragments present in smaller amounts (1 to 18%). Plagioclase feldspar is more abundant than potassium feldspar. The sandstones contain very little cement. Most of the clay is smectite, which is present in laminar and structural form.

OI Sandstone

The OI sandstone is composed of the OI-1, OI-2, OI-3, and OI-4 units. This sandstone is stratigraphically the deepest major oil- and gas-producing reservoir in the field. Productive depths range from 2134 to 2743 m (7000 to 9000 ft). The only known reservoirs deeper than the OI sandstone are the *Lenticulina* sands, which produce minor amounts of hydrocarbons on Block 331.

The structural configuration at the OI horizon is two complexly faulted anticlines (Figure 33A). The major down-to-the-south growth-fault system divides into several subparallel faults at the OI horizon. These faults are sealing, as evidenced by the unusual pay separation between fault blocks "A" and "B." Fault block "A" has a 159 m (521 ft) oil column with a 36 m (118 ft) gas cap. Fault block "B," immediately to the north and downthrown to fault block "A," is a gas/condensate reservoir. The gas column extends 536 m (1760 ft) from the crest to the gas-oil contact. The oil column in fault block "B" is 34 m (110 ft) and extends 414 m (1357 ft) below the oil-water contact in fault block "A" (Figure 33A).

The OI sandstones are a distributary channel deposit with a maximum net sand thickness of 67

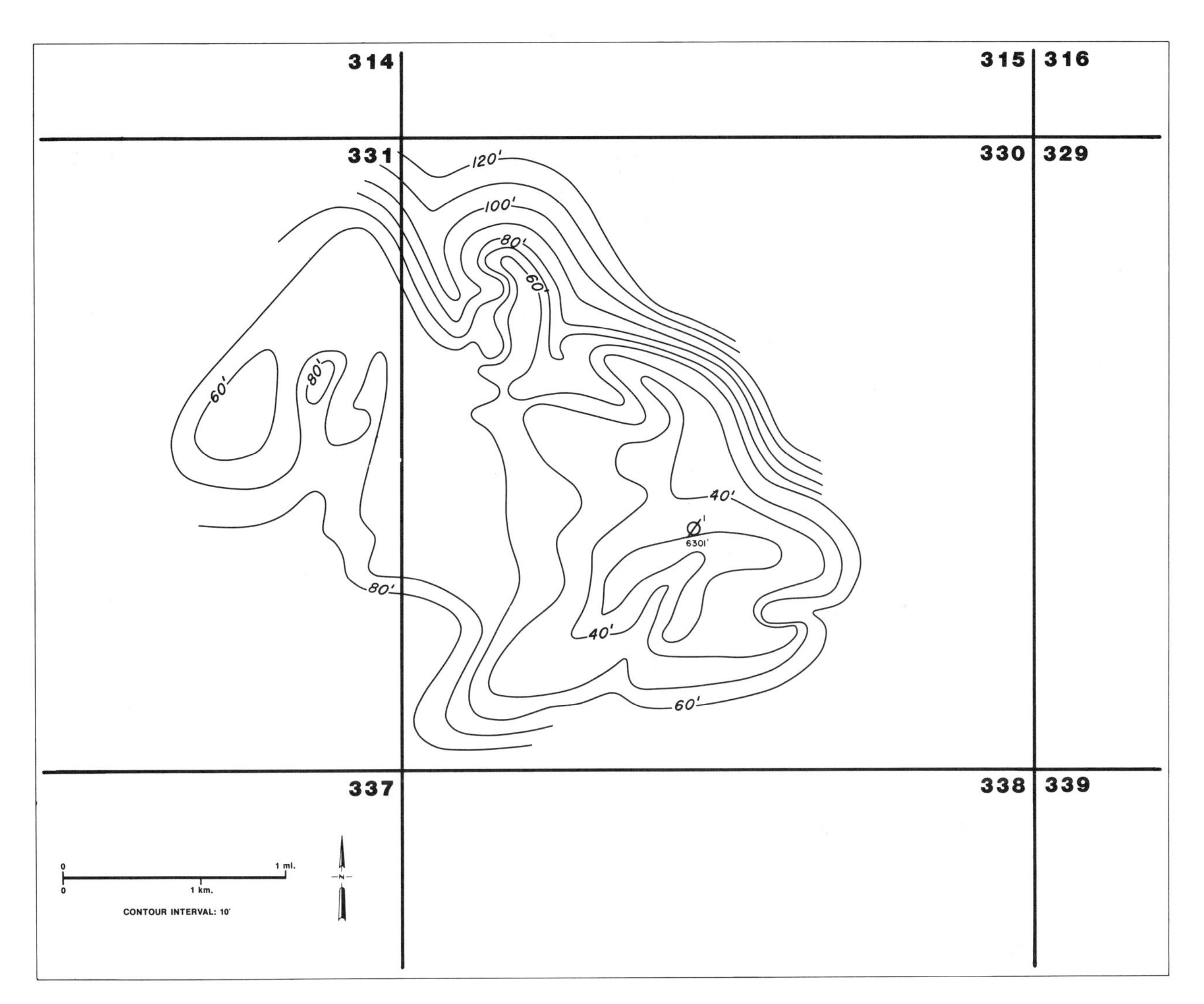

Figure 26B. HB net sandstone isopach map.

m (220 ft) (Figure 33B). Generally, the sands are well-sorted, very fine to medium-grained sandstones (Appendix 1), most of which coarsen upward. The rocks are quartzose, and locally contain significant quantities of feldspars (Figure 23B) and rock fragments. Plagioclase feldspar is more common than orthoclase feldspar. OI sandstones contain limited cementing material, with silica and pyrite being the most common. Clays occur mainly in detrital shale laminae (Figure 23D).

Reservoir parameters for the OI sandstones are listed in Appendix 1. The OI sandstone is second only to the LF sandstone in oil production. Ultimate recoverable field reserves are estimated to be 49 million bbl of oil, 60 bcf of solution gas, and 122 bcf of nonassociated gas. Cumulative production on Block 330 is 33.3 million bbl of oil plus condensate and 130.1 bcf of gas (March 1988). The sands are overpressured, with a pressure gradient of 0.76 psi/ft.

Gravity segregation is the dominant drive mechanism in the OI reservoir. PVT fluid studies indicate that normal pressure depletion increases oil viscosity in the reservoir, thereby decreasing the oil mobility and recovery efficiency. Computer reservoir studies indicate that pressure maintenance by crestal gas injection is the most efficient method of secondary recovery for the OI reservoir. Gas injection enhances the natural gravity separation of the oil and gas within the formation by displacing the oil downdip to the producing wellbores.

The gas injection, pressure-maintenance program in the OI fault block "A" reservoir was implemented in December 1979 (Figure 34). A total of 17.3 bcf of gas is planned to be injected over the duration of the program, with 14.4 bcf of gas being recoverable.

During the 7 years of production prior to gas injection, reservoir pressure dropped 1454 psi with production of 11.5 million bbl of oil and 11.75 bcf of gas. In contrast, during the 10 years of gas

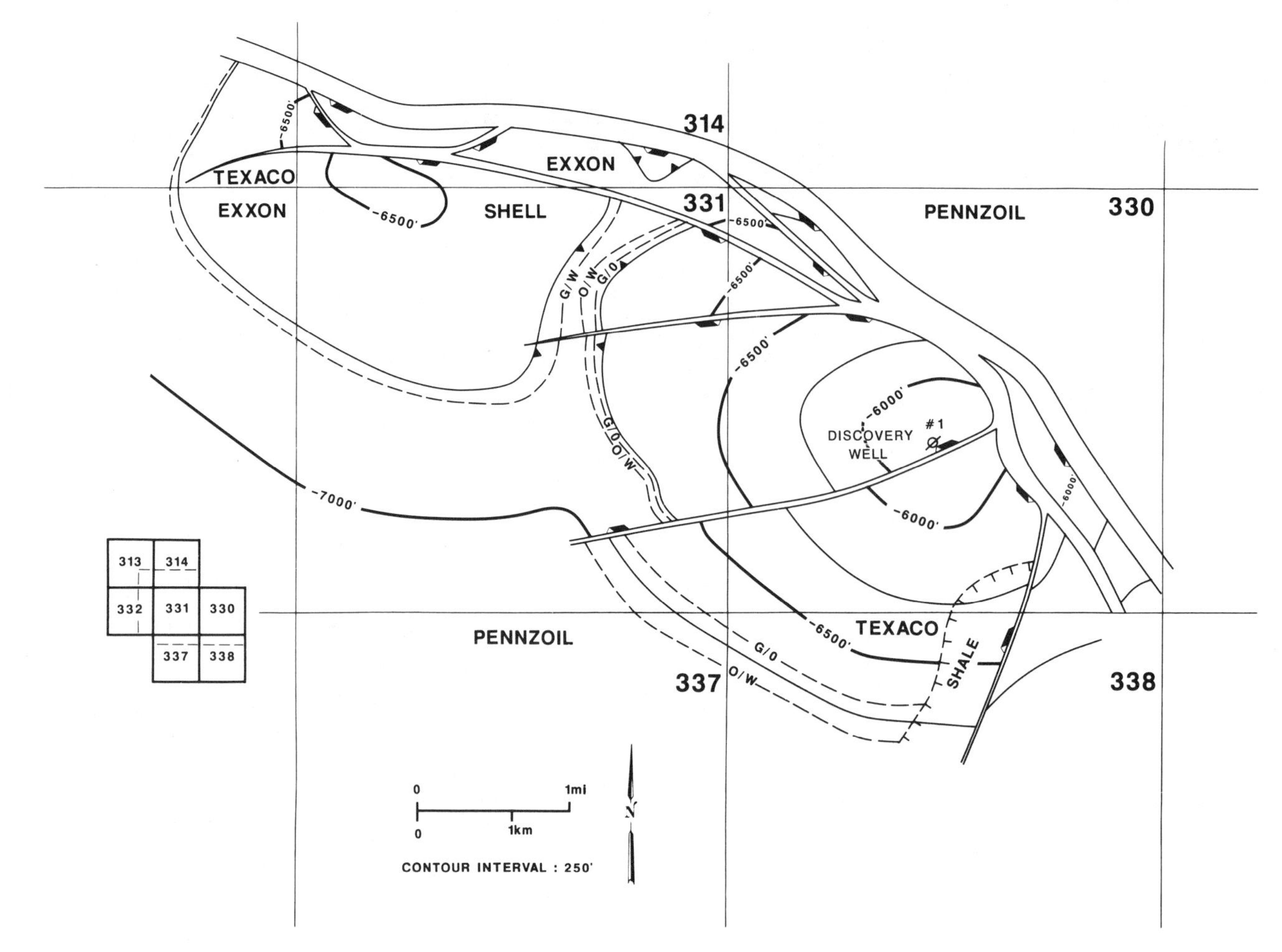

Figure 27A. Top JD sandstone structure map. Discovery well (E.I. 330 #1) indicated.

injection, reservoir pressure has dropped only 671 psi with production of 14 million barrels of oil and 12.1 bcf of gas (Figure 34).

Reservoir recovery is now 52% of the original oil-in-place, and ultimate recovery is expected to exceed 58%. An additional 30% increase to the originally estimated 28% primary-recovery efficiency can be attributed to this pressure-maintenance program. These data confirm the significant role of secondary recovery by gas injection in the OI reservoir on Eugene Island Block 330.

FAULTS

The major growth fault system bounding the field was instrumental in the formation of rollover anticlinal traps. Faulting was contemporaneous with deposition throughout the Pleistocene and continues today. Salt domes around the periphery of the subbasin were integral in the formation of growth faults, and some of the extremities of the fault system extend to these diapirs. Antithetic faults also are present and subdivide the structure into numerous producing faultblock reservoirs. Faults, especially on the eastern structure, are sealing faults, as evidenced by a down-dip gas reservoir juxtaposed against an updip oil reservoir (as discussed in the section on the OI sandstone) (also see Figure 33A).

The major down-to-the-basin growth faults were initially recognized and mapped using the pre-1970 seismic data. Additional faulting, especially at greater depths, and refined fault locations have been identified using subsequent development drilling, 2D, and now 3D, seismic data.

SOURCE

Petroleum Geochemistry

Recent studies of Gulf of Mexico oils have shown that most offshore oils can be assigned to one of two

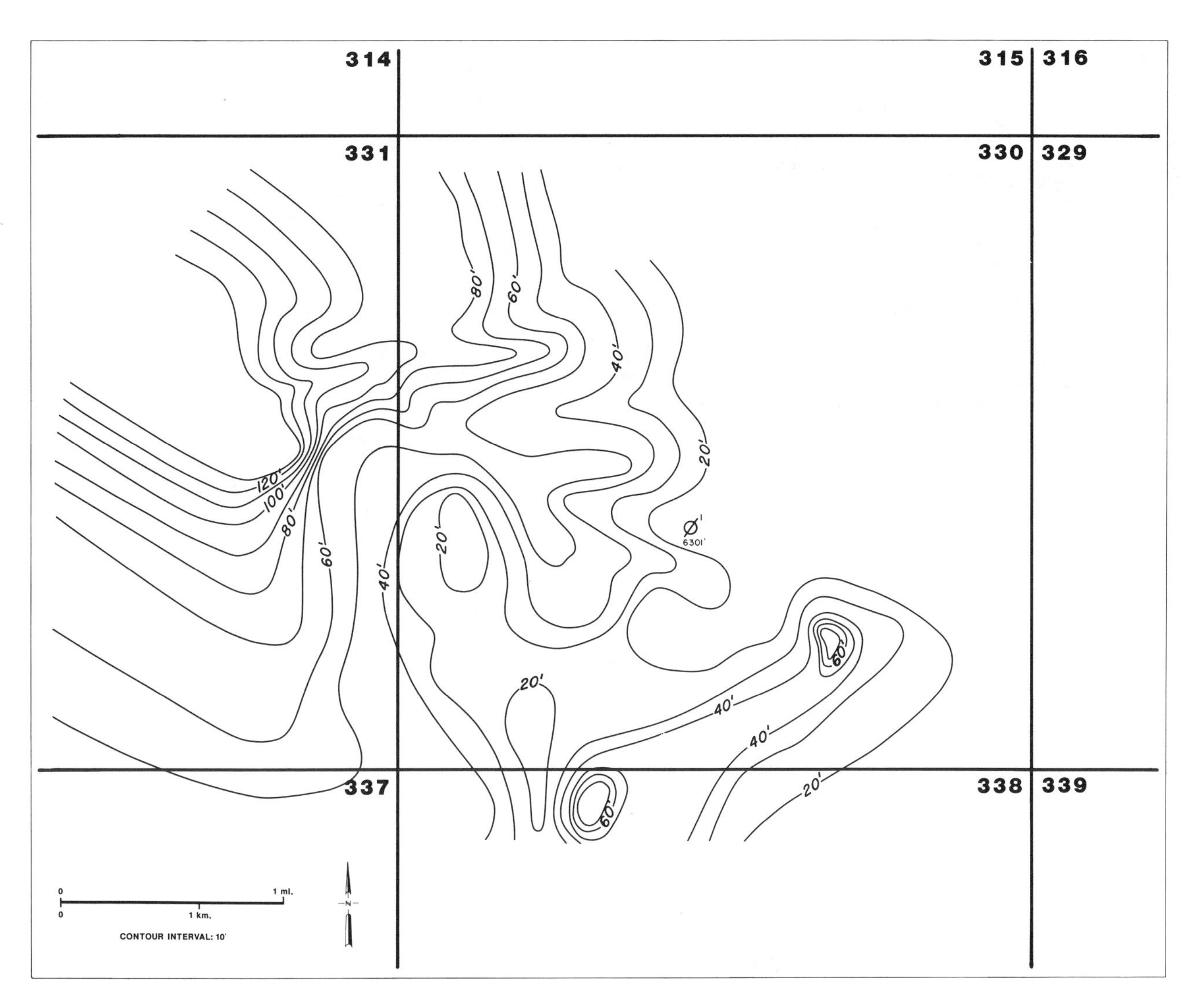

Figure 27B. JD net sandstone isopach map.

broad genetic families (Walters and Cassa, 1985; Kennicutt and Thompson, 1988). The smaller and more distinctive family is represented by oils in Pleistocene reservoirs of the Louisiana shelf-slope break region and includes the oils from the Eugene Island Block 330 field. The crude oil of this group is characterized by significantly higher sulfur and vanadium concentrations and lower pristane/phytane ratios relative to other Gulf of Mexico oils. These geochemical characteristics suggest that these oils were derived from a marine source facies deposited under more reducing (anoxic) conditions than for the bulk of the offshore oils.

Representative chromatograms of oil from Block 330 are shown on Figure 35, and the characteristics of the oil are summarized on Tables 1, 2, and 3. Oil from reservoirs above 1951 m (6400 ft) show evidence of considerable biodegradation and water washing, but more deeply reservoired oil is relatively unaltered and contains a full suite of normal alkanes. The unaltered oils are assignable to the paraffinic-naphthenic class of Tissot and Welte (1984), whereas the biodegraded oils belong to the aromatic-intermediate class. Most of the variation in physical and chemical properties of Block 330 oils can be attributed to the effects of biodegradation. Source-sensitive characteristics such as isotopic composition, V/V+Ni ratio, and selected biomarker ratios indicate that all Eugene Island Block 330 oils are genetically similar and were derived from a common source facies. Pleistocene shales associated with the Block 330 reservoirs do not qualify as the source rocks for these oils because of their thermal immaturity (R_o = 0.31-0.47%) and low organic carbon content (TOC 0.30-0.80%), and the highly oxidized nature of their predominantly terrestrial kerogen (Holland et al., 1980).

The Eugene Island Block 330 oils show abundant evidence of long-distance vertical migration. Based on a variety of biomarker and gasoline-range

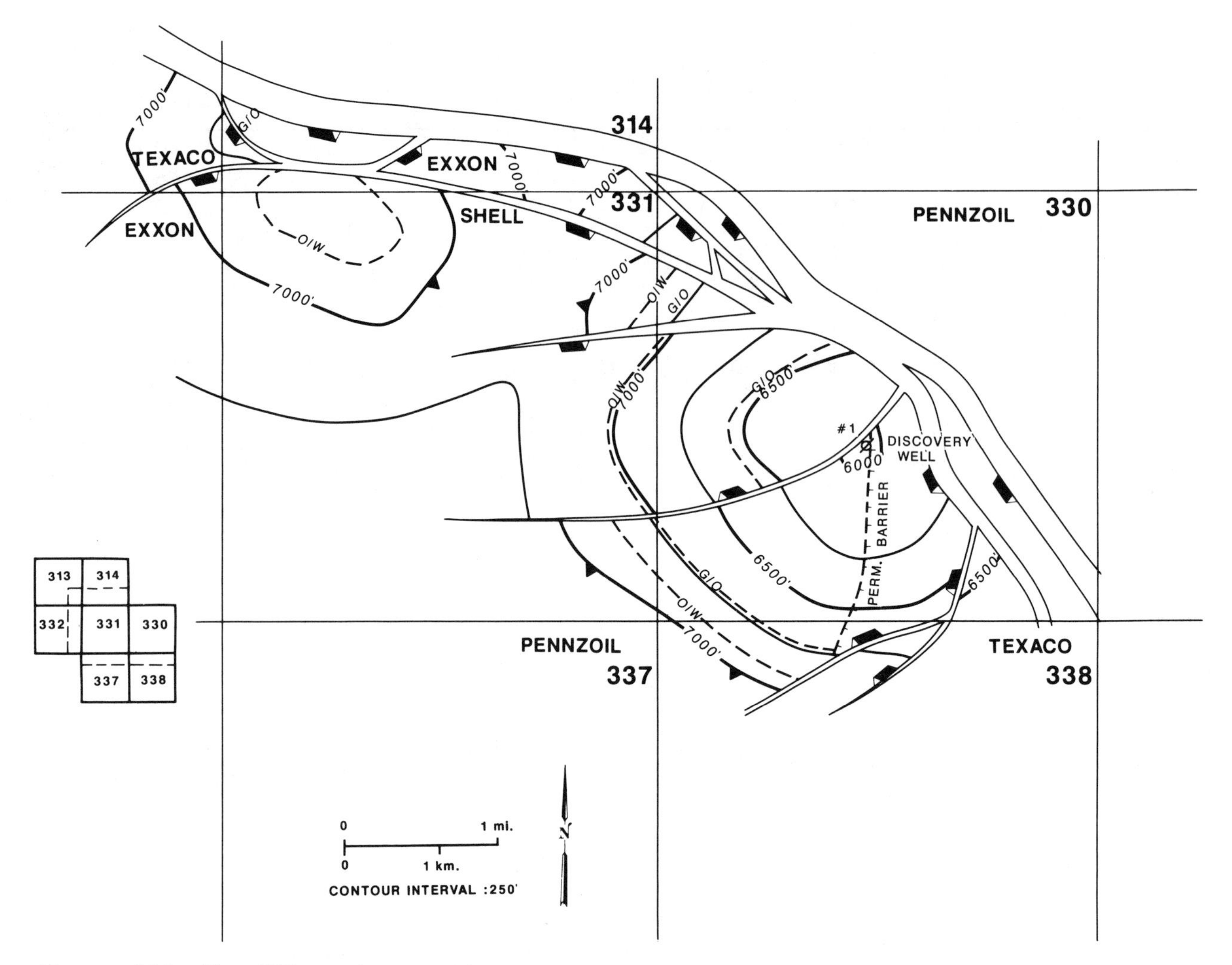

Figure 28A. Top KE sandstone structure map. Discovery well (E.I. 330 #1) shown.

maturity indicators, these oils are estimated to have been generated at depths of 4572 to 4877 m (15,000 to 16,000 ft) at vitrinite reflectance maturities of 0.08 to 1.0% and temperatures of 150 to 170°C (300 to 340°F). Their presence in shallow, thermally immature reservoirs requires significant vertical migration. This is illustrated on Figure 36, which represents a burial and maturation history for the field at the time of petroleum migration, that is, at the end of *Trimosina "A"* time approximately 500,000 years ago. A plot of the present measured maturity values versus depth is superimposed on the calculated maturity profile for *Trimosina "A"* time to illustrate the close agreement between measured and predicted maturity profiles. The clear discrepancy between reservoir maturity and oil maturity is striking and suggests that the oil migrated more than 3650 m (12,000 ft) from a deep, possibly upper Miocene, source facies. Petroleum migration along faults is indicated based on the observed temperature and hydrocarbon anomalies at the surface and the distribution of pay in the subsurface. These results are consistent with those of Young et al. (1977), who concluded that most Gulf of Mexico oils originated 2438 to 3350 m (8000 to 11,000 ft) deeper than their reservoirs, from source beds 5 to 9 million years older than the reservoirs.

EXPLORATION AND DEVELOPMENT CONCEPTS

Regional Play and General Application of Geologic Parameters

The Pliocene-Pleistocene play on the offshore Louisiana shelf, of which the Eugene Island Block 330 field is a part, is typified by the following geological characteristics:

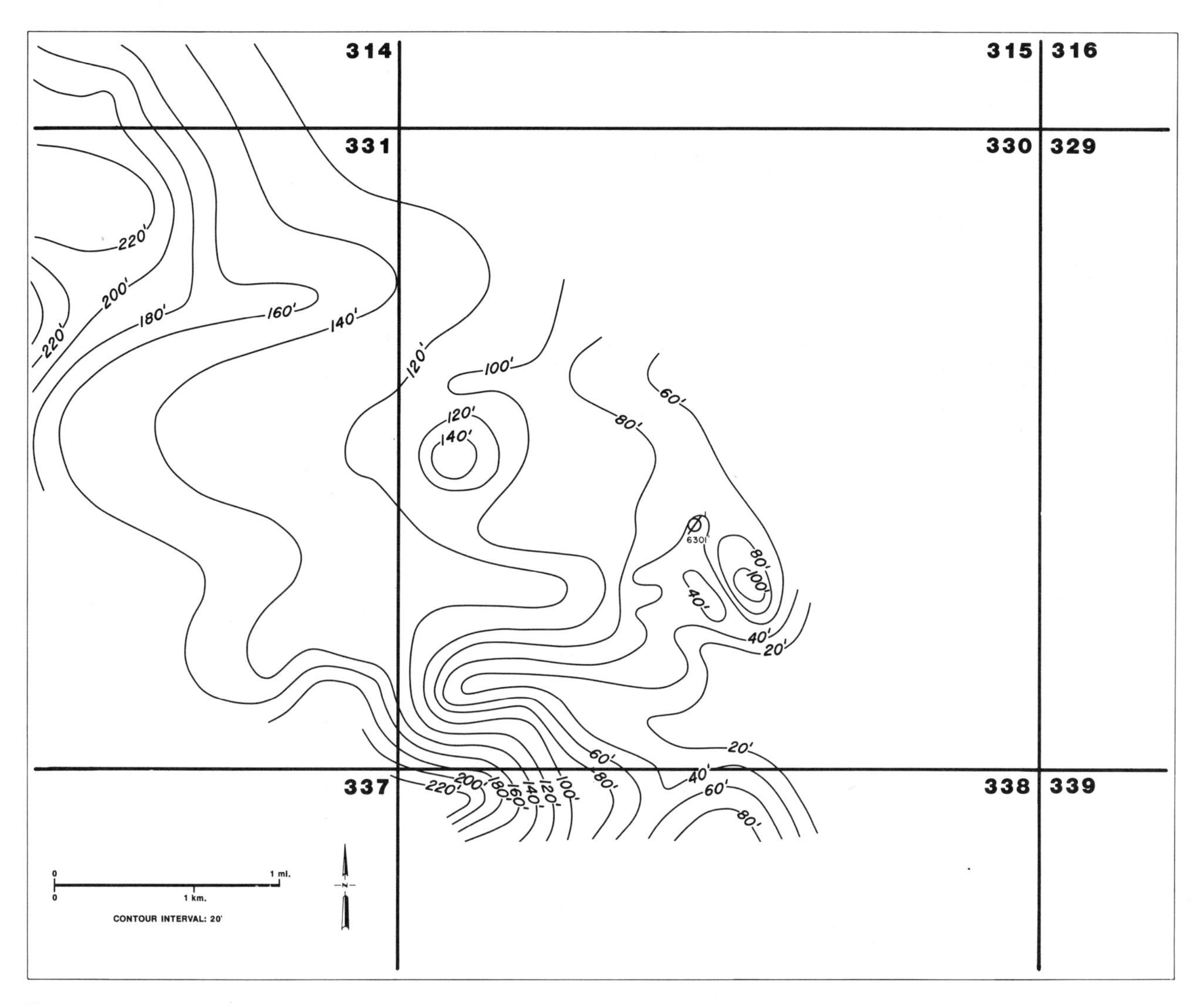

Figure 28B. KE net sandstone isopach map.

1. Structural features dominated by growth faults, salt domes, and salt-related faulting.
2. Thick accumulations of predominantly deltaic deposits of alternating sand and shale.
3. Young reservoirs (less than 2.5 m.y. old) with migrated hydrocarbons whose origins are in deeper, organic-rich marine shales.
4. Rapidly changing stratigraphy, due to deposition and subsequent reworking.
5. Numerous oil and gas fields with stacked reservoirs, long hydrocarbon columns, and high producing rates.

The exploration keys to success in this play are identifying prime stratigraphic fairways for favorable sand deposition and locating structures within these fairways that provide good avenues for migration and adequate trapping mechanisms. The timing of the trap formation is also important, and must be related to structural movement and depositional history.

Numerous fields of various sizes, trapping styles, and producing horizons surround the Block 330 field (Figure 37). Most of these fields can be classified into the following trap styles: (1) rollover anticlines, (2) faulted anticlines, (3) diapirs, or (4) upthrown fault traps. Cumulatively, the Block 330 field and its surrounding fields, located on Figure 37, have produced more than 338 million bbl of liquids and 4.748 tcf of gas.

Many of the geological characteristics outlined above can be applied to similar rifted, passive continental margins with thick deltaic deposits. The parameters most characteristic of the Gulf Coast basin are the apparent long vertical migration paths of hydrocarbons into younger reservoirs and the ubiquity of salt domes.

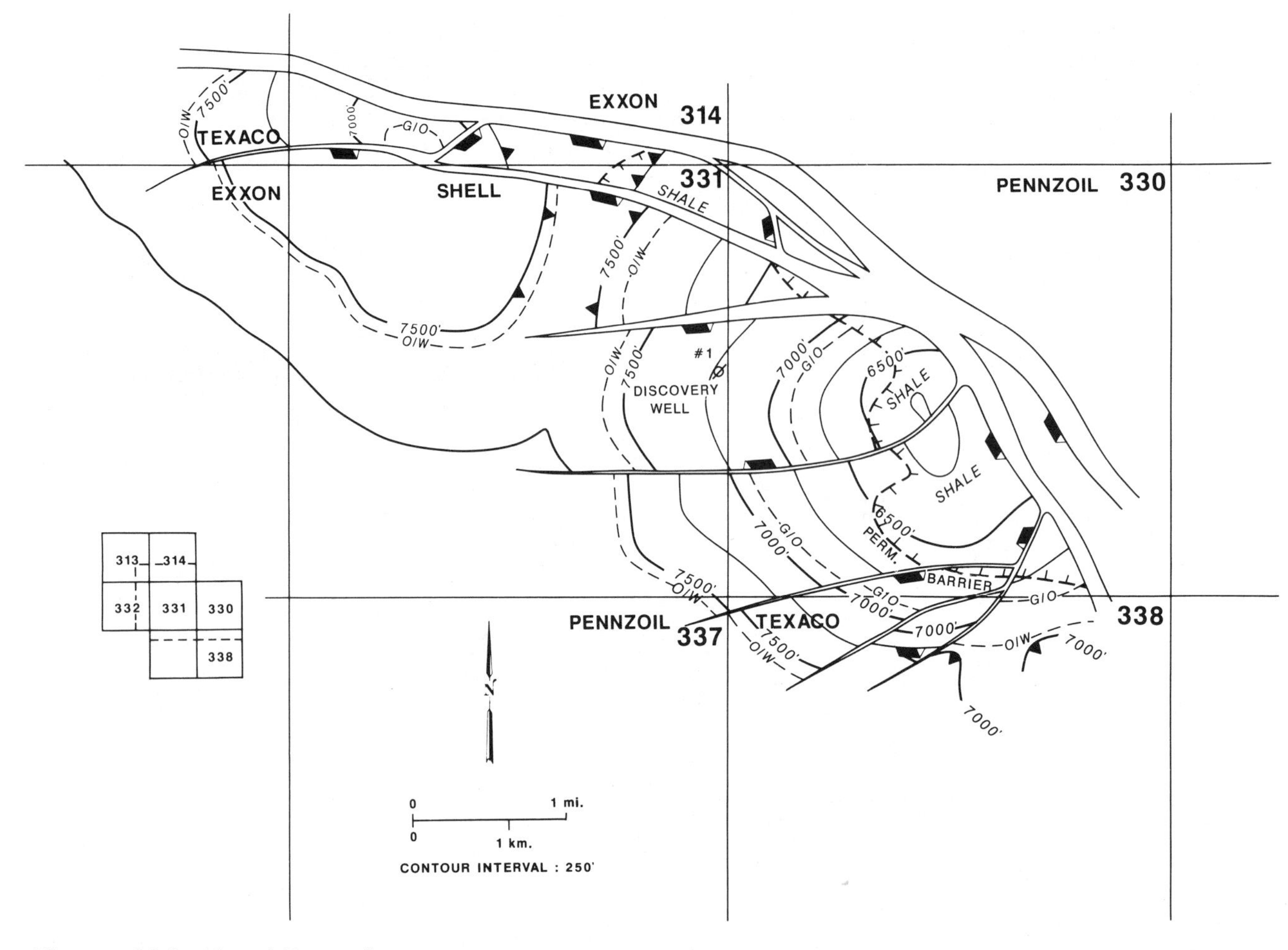

Figure 29A. Top LF sandstone structure map. Discovery well (E.I. 330 #1) indicated.

Lessons

The Eugene Island Block 330 field is a giant oil and gas field resulting from the favorable occurrence and history of many key geologic parameters and conditions. The field received the highest geologic rankings as an undrilled, presale prospect in 1970 on the basis of geologic and geophysical data and concepts that were state-of-the-art at that time. Clearly, if the Block 330 field were identified as an undrilled prospect in today's exploration environment, with modern data and concepts, it would rank as a top exploration prospect, be better understood geologically and geophysically, and generate a very high level of lease sale bidding interest.

Exploration with modern seismic data would allow us to define in more and finer detail the structural and stratigraphic elements of the field and the seismic hydrocarbon indicators associated with many of the reservoirs. Modern seismic stratigraphic and seismic hydrocarbon indicator theory and interpretation techniques were not available in the early 1970s, and their use undoubtedly would enhance the exploration and development programs. Present-day exploratory drilling would probably closely parallel the 1970s drilling history; wells would be located on the west side of Block 330, at the crest and on the flanks of the structure, to define reservoir numbers, thickness, column, and areal extent. A greater density of better quality seismic data would also help to optimize and refine exploratory well locations, drilling depths, the number of wells necessary, and postdrilling evaluations.

The acquisition of a 3D seismic survey prior to development would be the most beneficial exploitation tool available today. 3D seismic data would be used for precisely locating platforms, development wells, and later exploratory wells and for optimizing well spacing and recovery efficiency, thereby optimizing and expediting field development and production. However, it would be very difficult to improve upon the actual record of initial production 1 year and 9 months after lease acquisition.

Over the years, we have performed numerous geologic, geophysical, and engineering studies of the Block 330 field. The implementation of recommen-

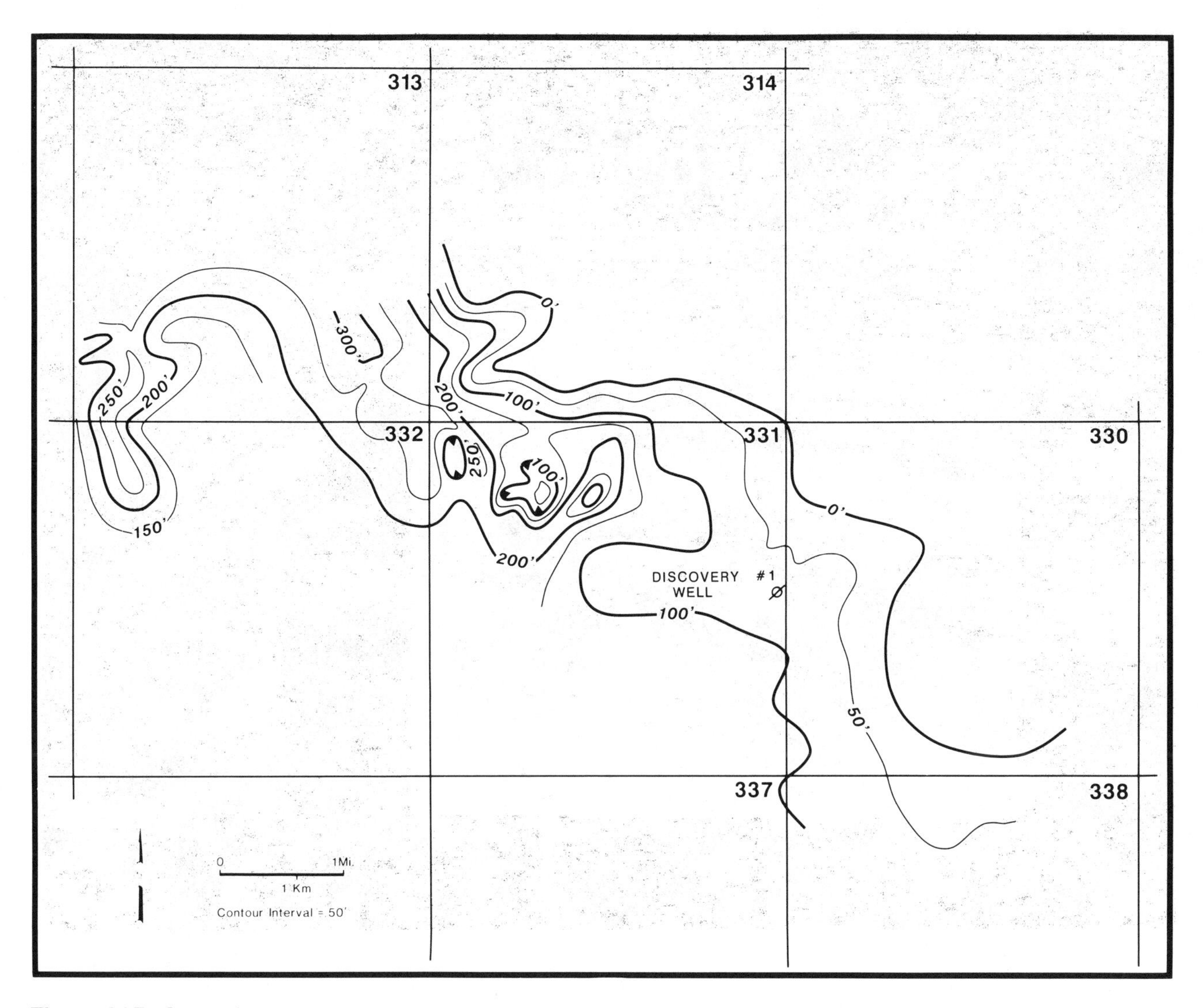

Figure 29B. Gross interval LF sandstone isopach map. Interpreted distributary-mouth bar deposits reach maximum thicknesses near the southeast corner of Block 313, and progressively thin toward Block 330.

dations from these studies has included projects such as exploration and development drilling, and pressure maintenance by gas injection. From 1978 to 1988, these operations, activities, and natural factors have increased ultimate recoverable reserves from 225 million bbl to 307 million bbl of hydrocarbon liquids and from 950 bcf to 1.65 tcf of gas.

As noted earlier (Figure 5), the field has experienced several phases of exploration and development. Exploratory drilling in 1982–1986 led to the discovery of additional hydrocarbons on the upthrown side of the main growth fault on Block 330, a fourth platform was set on the block, and eight development wells were drilled. In addition, 15 other fields have been discovered in the area over the last 18 years (Figure 37).

Currently, Block 330 is in another reevaluation phase. The most important aspect of our present evaluation is the use of 3D seismic data and techniques to identify new exploration and development prospects previously unidentified by 2D seismic methods and drilling. These highly economic prospects will utilize existing production facilities, add additional recoverable reserves, and extend the producing life of the field well into the future. One example is the HB sandstone prospect in the northwest corner of Block 330 (Figure 38). A proposed development well (Figure 39) will test a high-amplitude seismic event in an undrilled fault block recently identified using 3D data. Seismic amplitudes at the HB sandstone level are as intense as those observed in zones of known pay to the south. Also within this same fault block, two deeper levels of high-amplitude reflectors correlate to known pay zones and are targets for two additional development wells. Using 3D seismic data, we have also identified

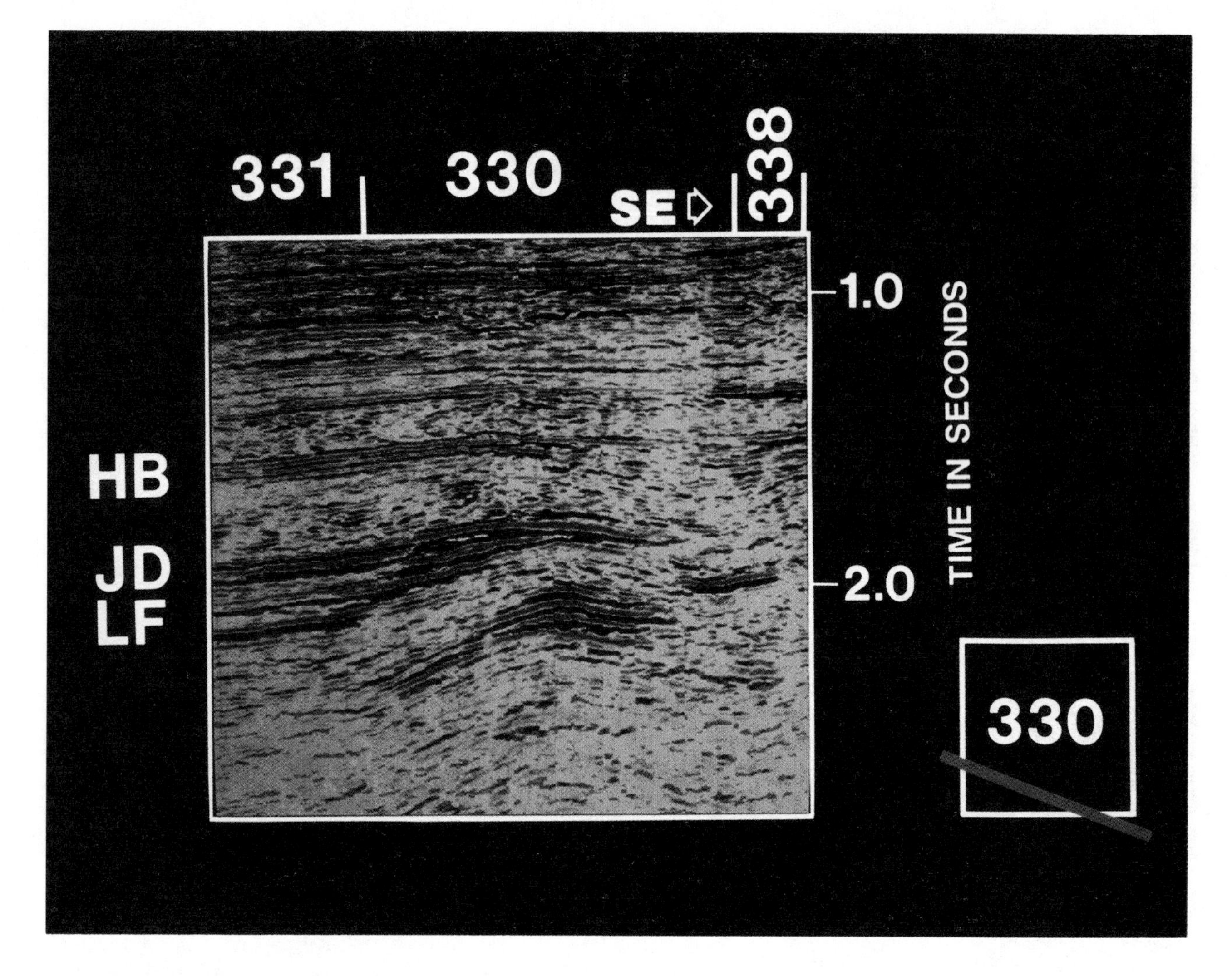

Figure 30. Seismic line from 3D seismic data volume illustrating stratigraphic thinning within the JD to LF interval over the crest of Block 330.

a deeper exploratory prospect, below present production levels on Block 330, that will be tested by a future exploratory well.

These prospects and others will be tested by drilling and should add additional economic reserves to an already giant field. Our ongoing evaluation has continually refined our geologic understanding and, although the field is mature, exploration and development continue.

ACKNOWLEDGMENTS

The authors thank Pennzoil Exploration and Production Company and the Eugene Island Block 330 partners for permission to publish this paper. We gratefully acknowledge all current and past Pennzoil U.S. Offshore Division personnel who have contributed to the current understanding of the field, including the following, who have particularly assisted in this and previous papers: V. F. Laiche, R. L. Lewis, D. C. Nester, E. M. Norwood, Jr., W. E. Nunan, C. F. Oudin, III, J. F. Rielly, D. Schumacher, B. R. Stewart, C. E. Sutley, and R. L. Woodhams. Richard Vessel of David K. Davies and Associates, Inc., assisted in the petrographic analysis.

We also thank the reviewers whose comments have significantly improved the manuscript, including J. M. Austin, B. E. Ball, F. W. Broussard, D. Curtis, R. McDonald, III, M. J. Padget, R. W. Spoelhof, J. F. van Sant, and K. R. Whaley.

We are also grateful to J. L. Howard and S. L. Winzig for typing the manuscript and to J. Rodriguez, Jr., C. Hinojosa, Jr., P. D. Maloney, C. A. Fullbright, and S. A. Shroyer for drafting the figures.

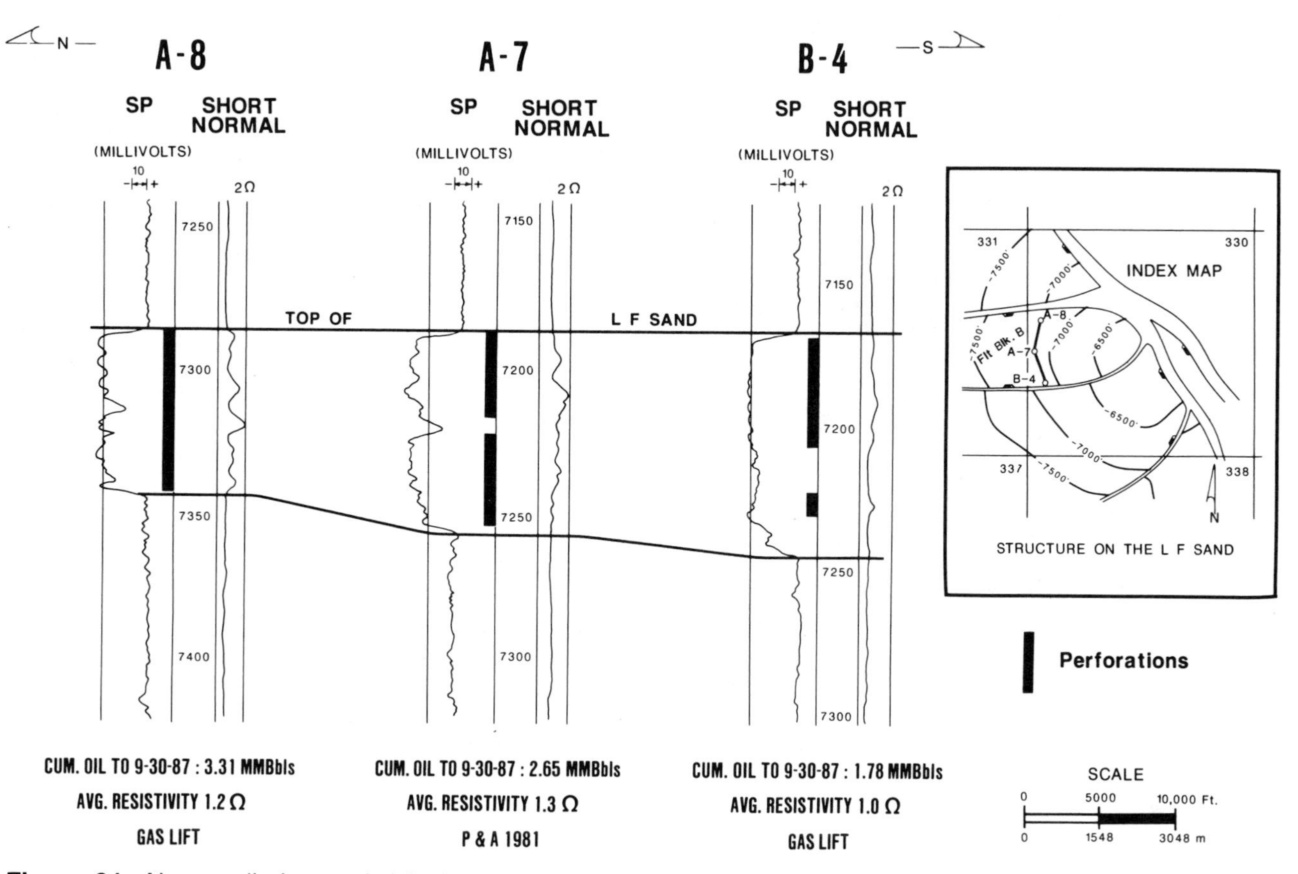

Figure 31. Abnormally low resistivity log response of LF sandstone seen in wells on Block 330.

REFERENCES

Bally, A. W., and S. Snelson, 1980, Realms of subsidence, *in* Facts and principles of world petroleum occurrence: Canadian Society of Petroleum Geologists Memoir 5, p. 9-94.

Brooks, J. M., M. C. Kennicutt, II, and B. D. Carey, 1986, Offshore surface geochemical exploration: Oil and Gas Journal, v. 84, n. 24, p. 66-72.

Bruce, C. H., 1973, Pressured shale and related sediment deformation: mechanism for development of regional contemporaneous faults: American Association of Petroleum Geologists Bulletin, v. 57, p. 878-885.

Carmalt, S. W., and B. St. John, 1986, Giant oil and gas fields, *in* M. T. Halbouty, ed., Future petroleum provinces of the world: American Association of Petroleum Geologists Memoir 40, p. 11-54.

Caughey, C. A., 1975, Pleistocene depositional trends host valuable gulf oil reserves: Oil and Gas Journal, v. 73, n. 36, 37, p. 90-94, 240-242.

Faber, E., and W. Stahl, 1984, Geochemical surface exploration for hydrocarbons in the North Sea: American Association of Petroleum Geologists Bulletin, v. 68, p. 363-386.

Fisher, W. L., 1969, Facies characterization of Gulf Coast basin delta systems, with some Holocene analogues: Gulf Coast Association of Geological Societies Transactions, v. 19, p. 239-261.

Gilreath, J. A., and J. J. Maricelli, 1964, Detailed stratigraphic control through dip computations: American Association of Petroleum Geologists Bulletin, v. 48, p. 1902-1910.

Hall, D. J., T. D. Cavanaugh, J. S. Watkins, and K. J. McMillen, 1982, The rotational origin of the Gulf of Mexico based on regional gravity data, *in* J. S. Watkins and C. L. Drake, eds., Continental margin geology: American Association of Petroleum Geologists Memoir 34, p. 115-125.

Holland, D. S., W. E. Nunan, D. R. Lammlein, and R. L. Woodhams, 1980, Eugene Island Block 330 field, offshore Louisiana, *in* M. T. Halbouty, ed., Giant oil and gas fields of the decade: 1968-1978: American Association of Petroleum Geologists Memoir 30, p. 253-280.

Humphris, C. C., Jr., 1978, Salt movements on continental slope, northern Gulf of Mexico, *in* A. H. Bouma et al., eds., Framework, facies, and oil trapping characteristics of the upper continental margin: American Association of Petroleum Geologists Studies in Geology 7, p. 69-85.

Jones, V. T., and R. J. Drozd, 1983, Prediction of oil or gas potential by near-surface geochemistry: American Association of Petroleum Geologists Bulletin, v. 67, p. 932-952.

Kennicutt, M. C., II, and K. F. M. Thompson, 1988, A suggested genetic classification of Gulf of Mexico offshore oils (abs.): American Association of Petroleum Geologists Bulletin, v. 72, p. 205.

Klemme, H. D., 1971, What giants and their basins have in common: Oil and Gas Journal, v. 69, n. 9, 10, 11; pt. 1, p. 85-90; pt. 2, p. 103-110; pt. 3, p. 96-100.

Lewis, R. L., H. J. Dupuy, Jr., and D. S. Holland, 1982, Eugene Island Block 330 field—development and production history: International Petroleum Exhibition and Technical Symposium Proceedings, Society of Petroleum Engineers, SPE 10003, p. 1-13.

Limes, L. L., and J. C. Stipe, 1959, Occurrence of Miocene oil in South Louisiana: Gulf Coast Association of Geological Societies Transactions, v. 9, p. 77-90.

Martin, R. G., 1978, Northern and eastern Gulf of Mexico continental margin: stratigraphic and structural framework, *in* A. H. Bouma et al., eds., Framework, facies, and oil trapping

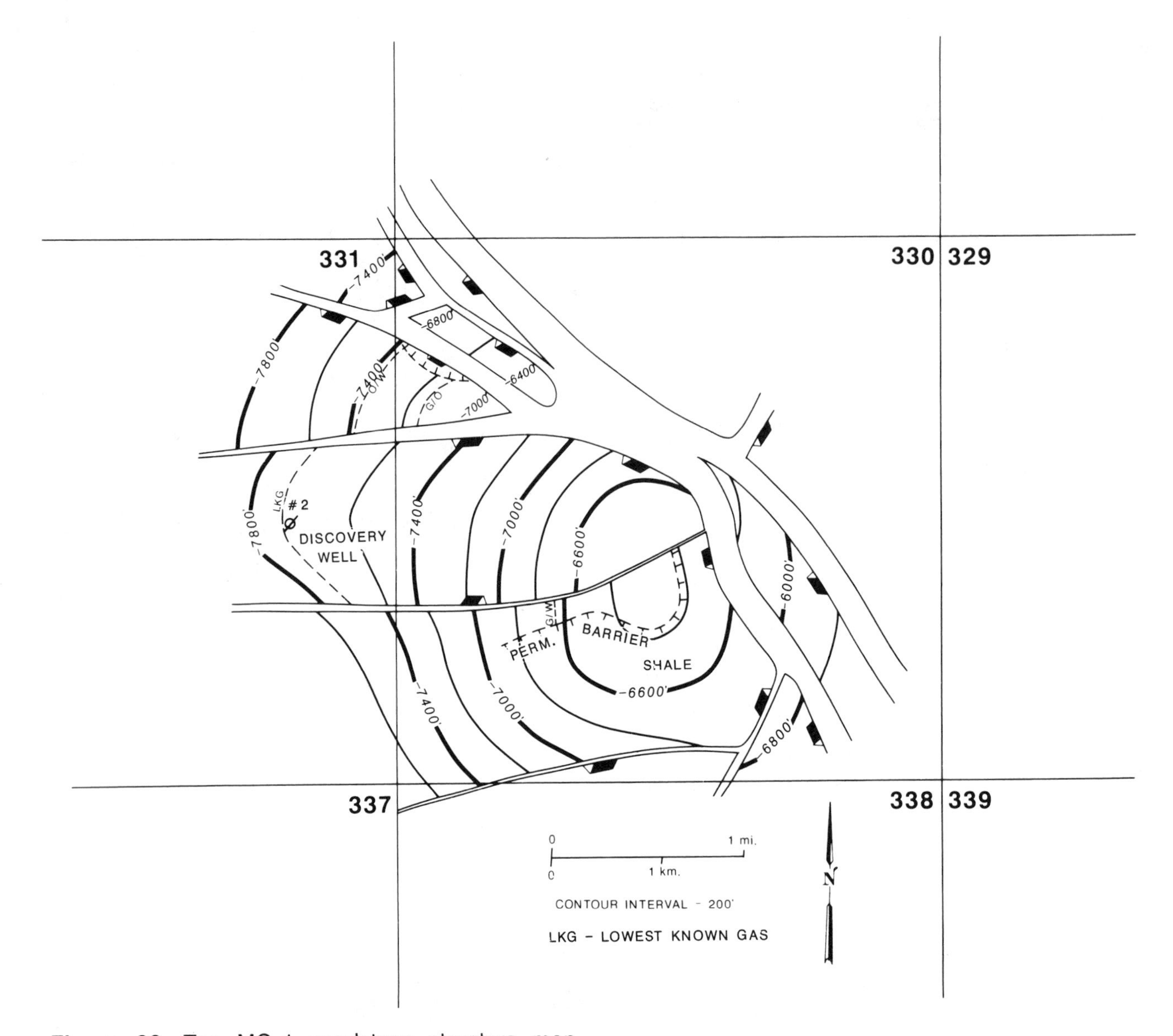

Figure 32. Top MG-1 sandstone structure map. Additional MG sandstone units are similar in structure. MG-1 discovery well (E.I. 331 #2) indicated.

characteristics of the upper continental margin: American Association of Petroleum Geologists Studies in Geology 7, p. 21-42.

Mitchum, R. M., Jr., P. R. Vail, and J. B. Sangree, 1977, Seismic stratigraphy and global changes of sea level, part 6: stratigraphic interpretation of seismic reflection patterns in depositional sequences, *in* C. E. Payton, ed., Seismic stratigraphy—applications to hydrocarbon exploration: American Association of Petroleum Geologists Memoir 26, p. 117-133.

Norwood, E. M., Jr., and D. S. Holland, 1974, Lithofacies mapping a descriptive tool for ancient delta systems of the Louisiana outer continental shelf: Gulf Coast Association of Geological Societies Transactions, v. 24, p. 175-188.

St. John, B., A. W. Bally, and H. D. Klemme, 1984, Sedimentary provinces of the world, hydrocarbon productive and nonproductive: American Association of Petroleum Geologists Map Series, 35 p.

Tissot, B., and D. Welte, 1984, Petroleum formation and occurrence: New York, Springer-Verlag, 2nd ed., 699 p.

Walters C. C., and M. R. Cassa, 1985, Regional organic geochemistry of offshore Louisiana: Gulf Coast Association of Geological Societies Transactions, v. 35, p. 277-286.

Woodbury, H. O., I. B. Murray, Jr., P. J. Pickford, and W. H. Akers, 1973, Pliocene and Pleistocene depocenters, outer continental shelf. Louisiana and Texas: American Association of Petroleum Geologists Bulletin, v. 57, p. 2428-2439.

Young, A., P. H. Monaghan, and R. T. Schweisberger, 1977, Calculation of ages of hydrocarbons in oils: physical chemistry applied to petroleum geochemistry, Part I: American Association of Petroleum Geologists Bulletin, v. 61, p. 573-600.

SUGGESTED READING

Brown, L. F., Jr., and W. L. Fisher, 1982, Seismic stratigraphic interpretation and petroleum exploration: American Association of Petroleum Geologists Continuing Education Course

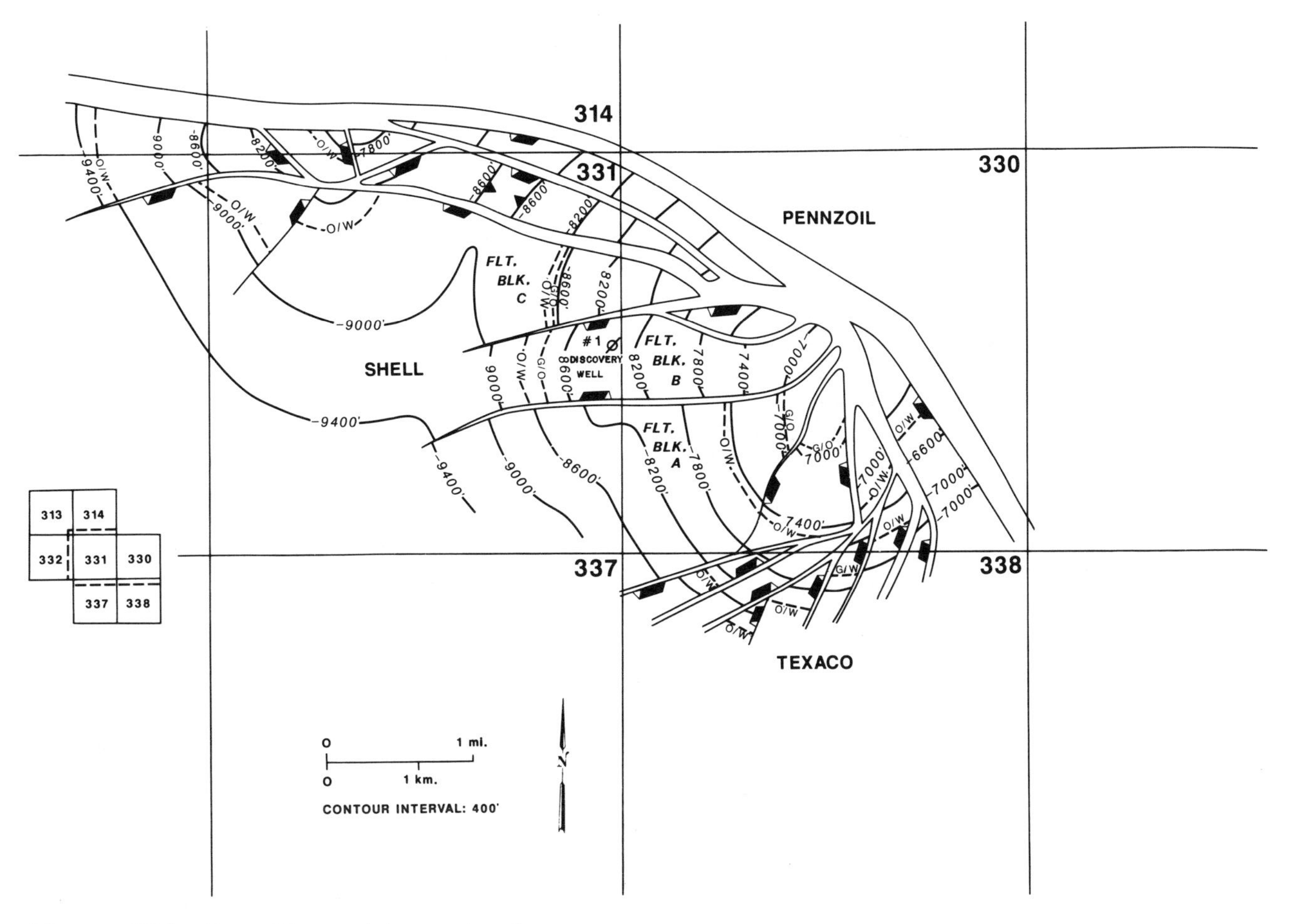

Figure 33A. Top OI sandstone structure map. Discovery well (E.I. 331 #1) shown.

Note Series #16, 181 p. Review of seismic stratigraphy and the geometry of depositional systems.

Dow, W. G., 1978, Petroleum source beds on continental slopes and rises: American Association of Petroleum Geologists Bulletin, v. 62, p. 1584–1606. Reviews conditions required for the formation of petroleum source beds, with special emphasis on the Gulf of Mexico.

Dow, W. G., 1984, Oil source beds and oil prospect definition in the Upper Tertiary of the Gulf Coast: Gulf Coast Association of Geological Societies Transactions, v. 34, p. 329–339. Proposes localized preservation of oil-prone kerogen during the Tertiary in anoxic intraslope basins.

Jackson, M. P. A., and W. E. Galloway, 1984, Structural and depositional styles of Gulf Coast Tertiary continental margins: application to hydrocarbon exploration: American Association of Petroleum Geologists Continuing Education Course Note Series #25, 226 p. Good summary with illustrations of growth faulting, rollover anticlinal traps, and deltaic depositional sequences.

Jones, R. W., 1987, Organic facies, *in* J. Brooks and D. Welte, eds., Advances in petroleum geochemistry, v. 2: London, Academic Press, p. 1–90. A thorough review of the concept of organic facies, their characteristics and petroleum potential, and the application of the organic facies concept in exploration.

Martin, R. G., and J. E. Case, 1975, Geophysical studies in the Gulf of Mexico, *in* A. E. M. Nairn and F. G. Stehli, eds., Ocean basins and margins; the Gulf of Mexico and the Caribbean: New York, Plenum Press, v. 3, p. 65–106. Gulf of Mexico physiography and origin.

Nunn, J. A., and R. Sassen, 1986, The framework of hydrocarbon generation and migration, Gulf of Mexico continental slope: Gulf Coast Association of Geological Societies Transactions, v. 36, p. 257–262. Discussion of maturation modeling results, which suggest that Cretaceous and Early Tertiary sediments are possible source rocks for oil in Pliocene-Pleistocene reservoirs.

Rice, D. D., 1980, Chemical and isotopic evidence of the origins of natural gases in offshore Gulf of Mexico: Gulf Coast Association of Geological Societies Transactions, v. 30, p. 203–213. Describes significant differences in isotopic composition of methane in reservoirs across the Gulf of Mexico basin and discusses their origin.

Williams, D. F., and I. Lerche, 1987, Salt domes, organic-rich source beds and reservoirs in intraslope basins of the Gulf Coast region, *in* I. Lerche and J. J. O'Brien, eds., Dynamical geology of salt and related structures: Orlando, Florida, Academic Press, p. 751–786.

Williams, D. F., and I. Lerche, 1987, Hydrocarbon production in the Gulf Coast region from organic-rich source beds of ancient intraslope basins: Energy Exploration and Exploitation, v. 5, p. 199–218. Discusses a model in which salt structures underlying the continental margin of the northern Gulf of Mexico are suggested to play a major role in determining the origin, distribution, and maturation history of petroleum source beds.

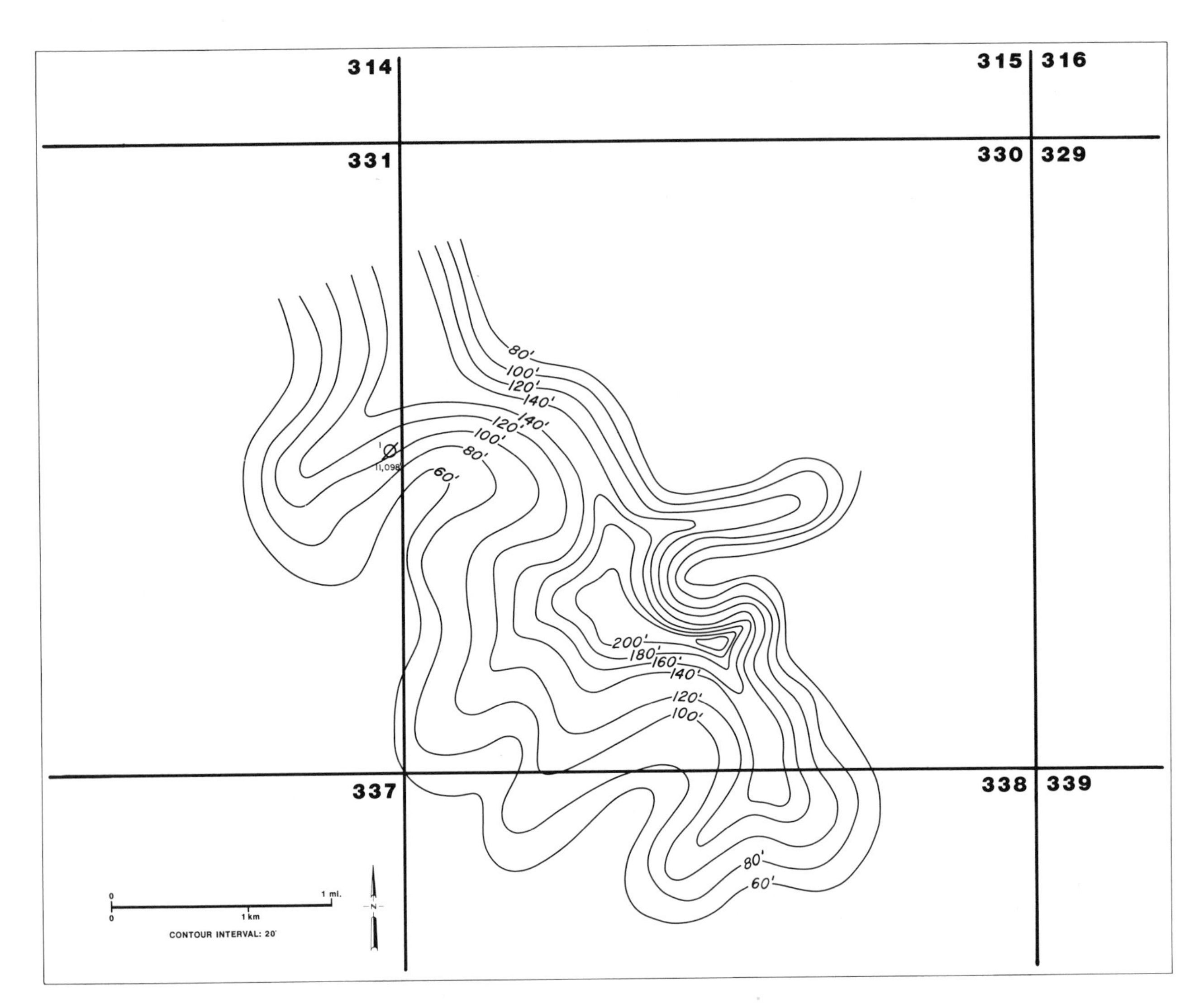

Figure 33B. OI net sandstone isopach map.

Appendix 1. Field Description

Field name .... *Eugene Island Block 330 field*

Ultimate recoverable reserves .... *307 MMbbl liquids + 1650 bcf gas*

Field location:

- **Country** .... *U.S.A.*
- **State** .... *Offshore Louisiana*
- **Basin/Province** .... *Gulf of Mexico*

Field discovery:

- **Year first pay discovered** .... *Numerous Pleistocene reservoirs 1971*
- **Year second pay discovered** *1971*
- **Third pay**

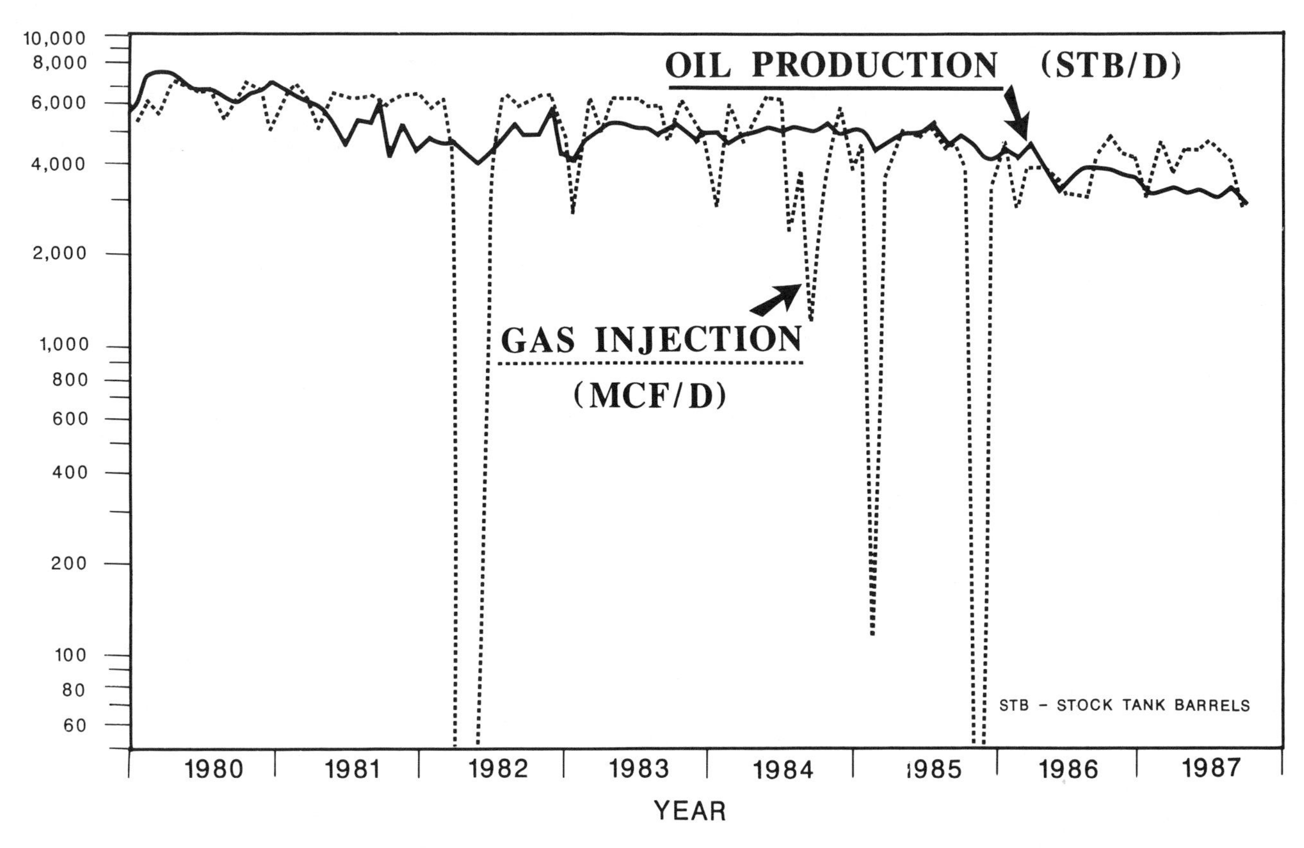

Figure 34. Gas-injection project, OI sandstone, Block 330. Oil production (solid line) and gas injection (dashed line) shown.

Discovery well name and general location:

First pay .... *Pennzoil 1 OCS G-2115 (4 pay sands) 5525 ft FSL, 6882 ft FWL Blk. 330*
Second pay .... *Shell 1 OCS G-2116 (5 pay sands) 7426 ft FSL, 301 ft FEL blk. 331*
Third pay

Discovery well operator .... *Pennzoil*
Second pay .... *Shell*
Third pay

Discovery well and initial production rate for Eugene Island Block 330 field pay sands:

Pay Sand	Discovery Well	Year Discovered	IP	Well
DA	*E.I. 330 #3*	*1971*	*NP*	
GA-1	*E.I. 330 #1*	*1971*	*109 BOPD + 22 MCFGPD*	*(330 A-18A)*
GA-2	*E.I. 330 #1*	*1971*	*211 BOPD + 35 MCFGPD*	*(330 C-1)*
GA-5	*E.I. 330 #3*	*1971*	*NP*	
HB-1	*E.I. 330 #1*	*1971*	*552 BOPD + 128 MCFGPD*	*(330 C-2)*
HB-2, 1B	*E.I. 330 #1*	*1971*	*366 BOPD + 952 MCFGPD*	*(330 C-1)*
HB-3	*E.I. 330 #1*	*1971*	*473 BOPD + 228 MCFGPD*	*(330 C-3)*
IC-2	*E.I. 330 #A-13*	*1973*	*510 BOPD + 323 MCFGPD*	*(330 A-13*
IC-4	*E.I. 330 #4*	*1971*	*864 BOPD + 389 MCFGPD*	*(330 A-18)*
IC-5	*E.I. 330 #2*	*1971*	*NP*	
IC-7	*E.I. 330 #A-13*	*1973*	*NP*	

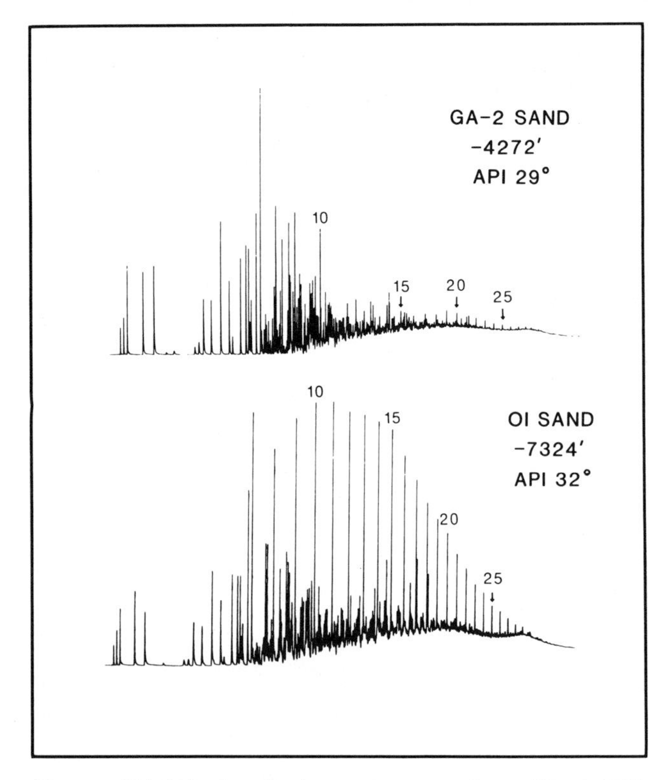

Figure 35. Whole-oil chromatograms from Block 330 GA-2 and LF reservoirs showing relatively unaltered deeply buried oil and strongly altered shallower oil. Numbers refer to the carbon numbers of the normal alkanes.

Table 1. Crude oil characteristics.

	Biodegraded Oils	Unaltered Oils
Reservoirs	GA,HB	KE,LF,OI
Age	Trim - A	Ang - B
Depth, TVD	4200-4800	6600-7400
API gravity	25-29	31-38
Sulfur, %	1.1-1.5	0.5-1.1
Vanadium, PPM	9-17	1-10
Pristane/phytane	1.50	1.20
% Saturates/% aromatics	48/40	57/32
Carbon isotopes (Sats.)	-27.3	-27.5
Sulfur isotopes	-6.0	-6.4

Table 2. Hydrocarbon characteristics—Eugene Island Block 330.

Sand	GA-2	HB-1	JD	KE-1	LF	MG-2	OI-1
API gravity	23°	25°	—	35°	34°	36°	32°
BTU (d-dry, w-wet)			1163d/1142w				1172d/1151w
Initial GOR	402	361	—	893	941	1225	1078
Sulfur (%)	1.1	1.4	0.15	0.95	0.7	—	1.05
Viscosity (cp)	2.68	3.45	—	0.54	0.37	0.65	0.48
Pour point	NA	NA	—	NA	NA	NA	NA
Gas-oil distillate (scf/bbl)	—	—	32	60	—	35	34

Table 3. Field characteristics—Eugene Island Block 330.

Sand	GA-2	HB-1	JD	KE-1	LF	MG-2	OI-1
Avg. elevation (ft)	-4300	-4800	-6400	-6600	-7100	-7300	-7600
(m)	-1311	-1463	-1951	-2017	-2164	-2225	-2316
Initial pressure (psig)	2095	2400	3784	3805	4076	4327	5652
(kg/cm^2)	147	169	266	268	287	304	397
Present pressure (psig)	1799	1997	1678	3206	2576	1617	3522
(7-87) (kg/cm^2)	126	140	118	225	118	114	248
Pressure gradient (psi/ft)	.487	.486	.579	.551	.559	.585	.764
($kg/cm^2/m$)	.112	.112	.134	.127	.129	.135	.176
Temperature °F/°C	130°/54°	140°/60°	145°/63°	155°/68°	160°/71°	166°/74°	187°/86°
Geothermal gradient °F/ft	1.3	1.2	0.9	1.1	1.5	1.3	1.5
°C/m	2.4	2.2	1.6	2.0	2.7	2.4	2.7
Drive	W	W	partial W	W	partial W	D	G
Oil column thickness (ft)	126	96	21	441	696	224	521
(m)	38	29	6	134	212	68	159
Gas column thickness (ft)	—	49	842	512	364	774	1800
(m)	—	15	257	156	111	236	549
Oil-water content (ft)	-4312	-4876	-6778	-7003	-7610	-7448	-7461
(m)	-1314	-1486	-2066	-2135	-2320	-2270	-2274
Gas-oil contact (ft)	—	-4780	-6757	-6562	-6914	-7224	-8710
(m)	—	-1457	-2060	-2000	-2107	-2202	-2655
Water salinity (ppm)	125,000	85,000	82,000	84,000	94,000	75,000	140,000
Water resistivity	0.036	0.047	0.352	0.045	0.039	0.045	0.028
($ohm\text{-}m^2/m$)	@ 135°F	@ 140°F	@ 145°F	@ 155°F	@ 160°F	@ 163°F	@ 168°F
Bulk vol. water (%)	11	9	10	11	12	10	6

W = water drive, D = Depletion drive, G = Gravity

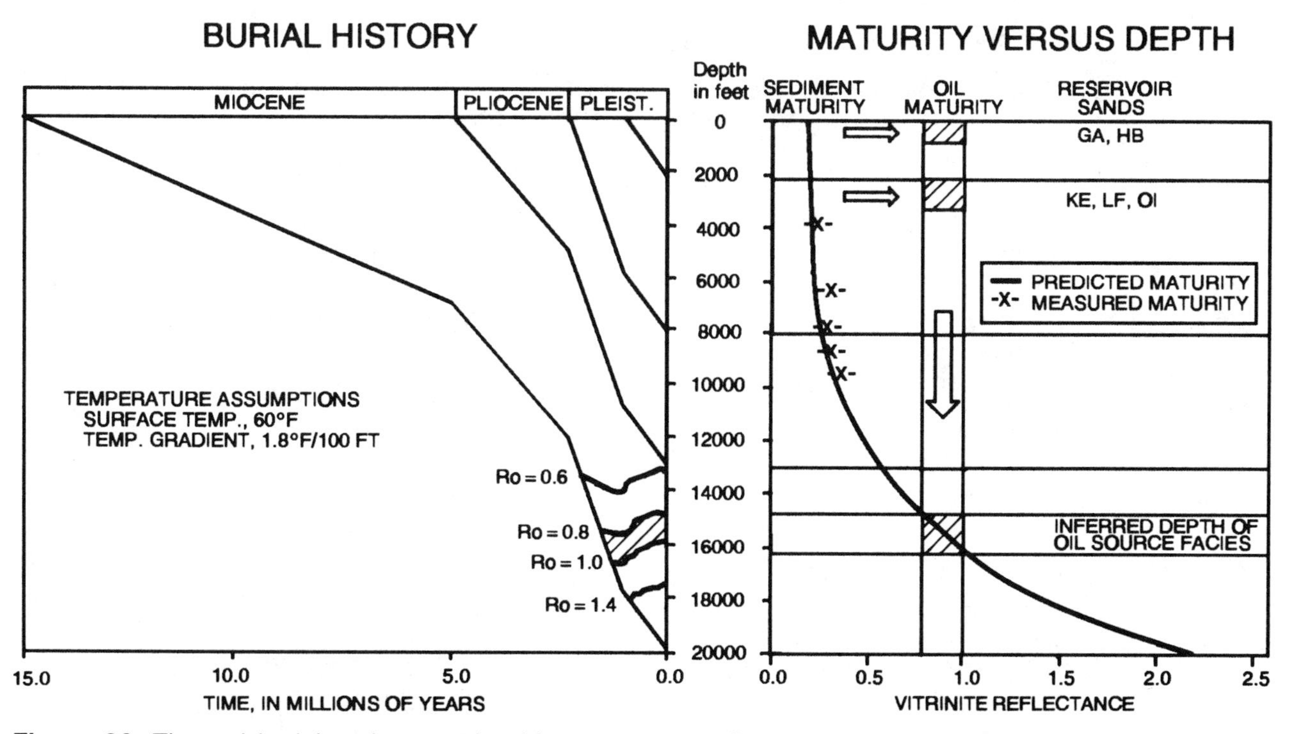

Figure 36. Thermal burial and maturation history at the end of *Trimosina "A"* time for Eugene Island Block 330 oils based on vitrinite reflectance (R_o) data. See type log (Figure 15) for ages of sandstone units.

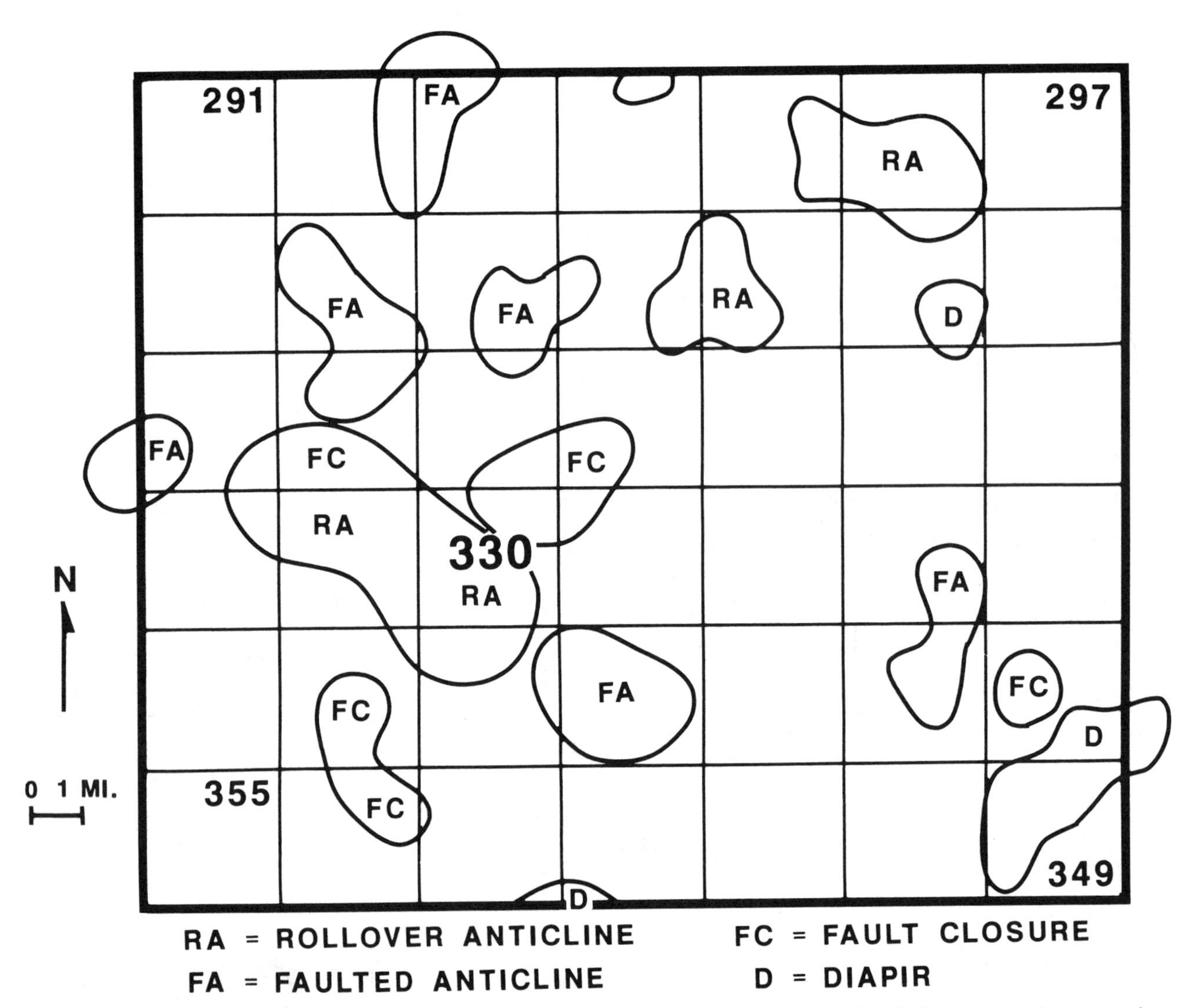

Figure 37. Field outlines and trapping styles for fields surrounding Eugene Island Block 330 field. Total cumulative production for fields shown is more than 328 million bbl of oil and 4.7 tcf of gas.

J-1-D	*E.I. 331 #1*	*1971*	*840 MCFGPD + 48 BCPD*	*(331 A-17)*
JD	*E.I. 330 #1*	*1971*	*1616 MCFGPD + 38 BCPD*	*(330 A-11)*
KD-2	*E.I. 330 #4*	*1971*	*1371 MCFGPD + 33 BCPD*	*(330 A-11)*
KE-1	*E.I. 330 #1*	*1971*	*108 BOPD + 80 MCFGPD*	*(330 A-1)*
KE-2	*E.I. 330 #1*	*1971*	*480 BOPD + 436 MCFGPD*	*(330 A-1)*
LF	*E.I. 331 #1*	*1971*	*516 BOPD + 490 MCFGPD*	*(331 A-2)*
MG-1	*E.I. 331 #2*	*1971*	*224 BOPD + 274 MCFGPD*	*(330 B-9)*
MG-2	*E.I. 331 #2*	*1971*	*831 BOPD + 816 MCFGPD*	*(330 A-12)*
MG-3	*E.I. 331 #2*	*1971*	*821 BOPD + 620 MCFGPD*	*(330 B-14)*
MG-4	*E.I. 331 #2*	*1971*	*600 BOPD + 411 MCFGPD*	*(330 B-9)*
NH	*E.I. 331 #1*	*1971*	*262 BOPD + 162 MCFGPD*	*(330 B-10ST)*
NI	*E.I. 330 #2*	*1971*	*NP*	
OI-1	*E.I. 331 #1*	*1971*	*6400 MCFGPD + 531 BCPD*	*(330 A-3)*
OI-2	*E.I. 330 #A-4*	*1972*	*1769 MCFGPD + 19 BCPD*	*(330 A-14)*
OI-3	*E.I. 330 #A-3*	*1972*	*216 BOPD + 522 MCFGPD*	*(330 C-14)*
OI-4	*E.I. 330 #4*	*1971*	*238 BOPD + 230 MCFGPD*	*(331 A-15)*
Lentic	*E.I. 331 #B-1*	*1972*	*370 BOPD + 1075 MCFGPD*	*(331 B-1)*

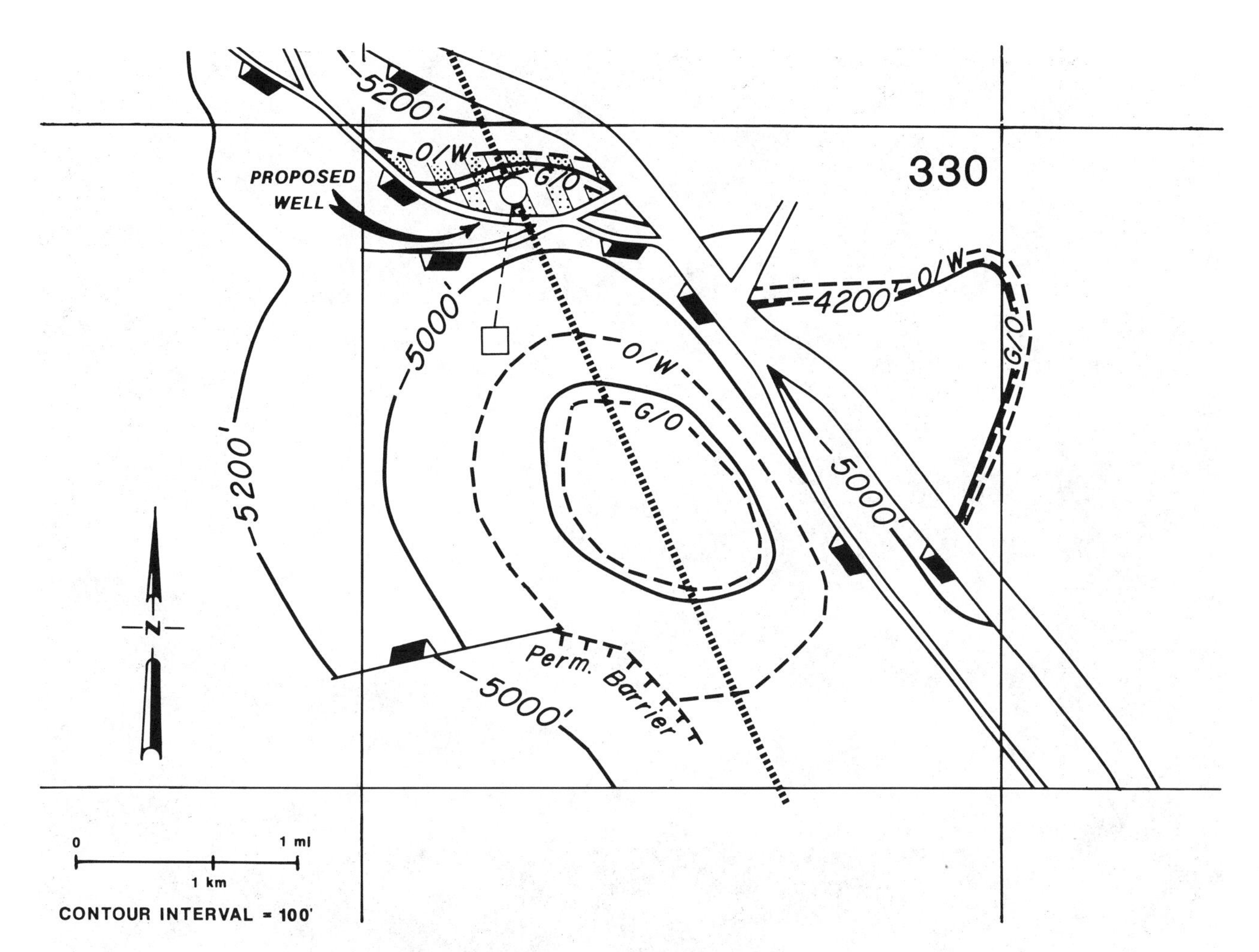

Figure 38. Top HB sandstone structure map. Potential oil and gas reserves will be tested by a deviated development well from the "A" platform (square) on Block 330.

All other zones with shows of oil and gas in the field:

Age	Formation	Type of Show
Pleistocene	*DA*	*Gas*
Pleistocene	*GA-1*	*Oil*
Pleistocene	*GA-5*	*Gas/oil*
Pleistocene	*HB-2, 1B*	*Oil*
Pleistocene	*HB-3*	*Oil*
Pleistocene	*IC-2*	*Oil*
Pleistocene	*IC-4*	*Oil*
Pleistocene	*IC-5*	*Gas/oil*
Pleistocene	*IC-6*	*Oil*
Pleistocene	*IC-7*	*Oil*
Pleistocene	*J-1-D*	*Gas*
Pleistocene	*JD-1*	*Gas*
Pleistocene	*KD-2*	*Gas/oil*
Pleistocene	*KE-1*	*Gas/oil*
Pleistocene	*MG-1*	*Gas/oil*
Pleistocene	*MG-3*	*Gas/oil*
Pleistocene	*MG-4*	*Gas/oil*

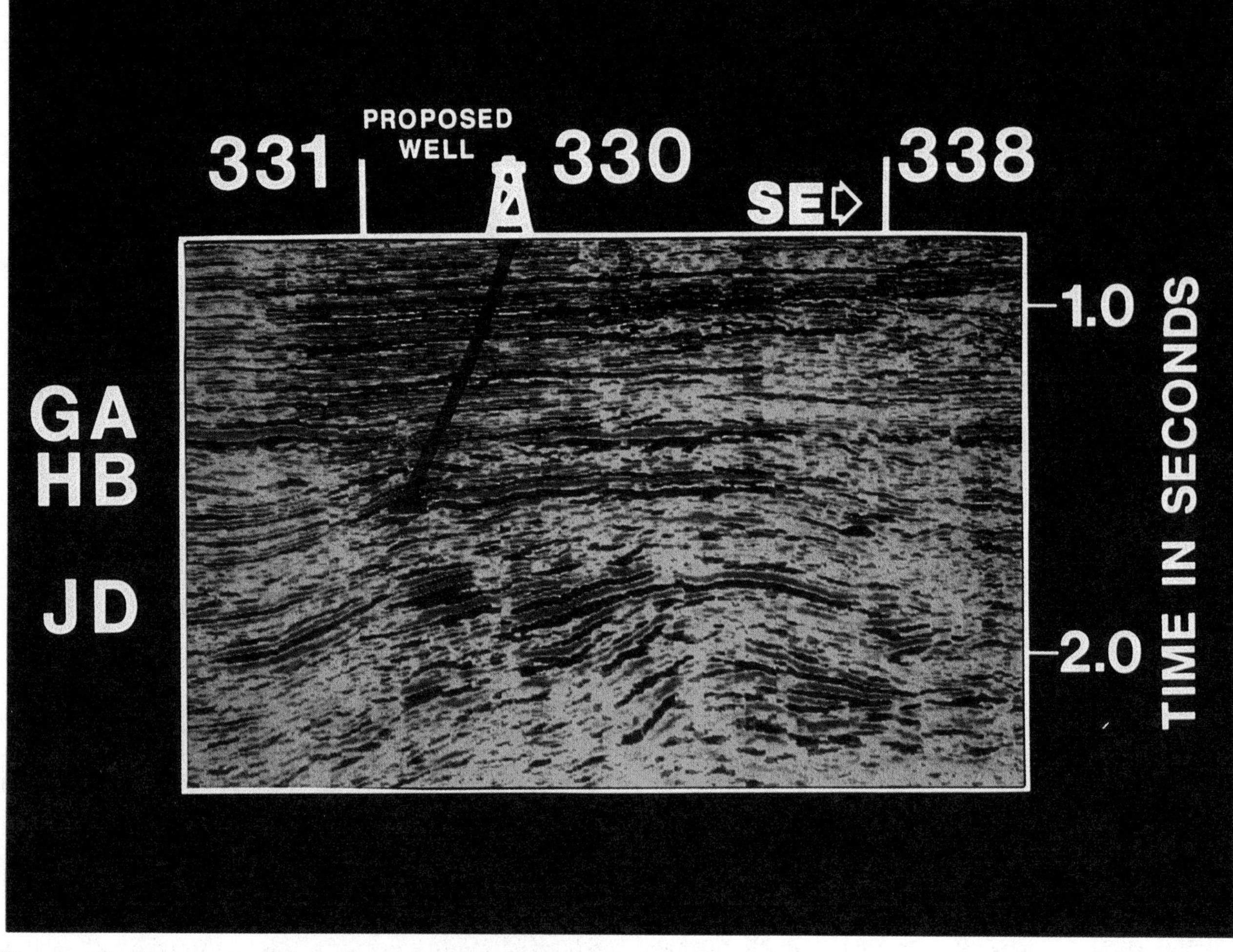

Figure 39. Seismic line from 3D seismic data volume illustrating the HB development prospect shown in Figure 38.

Pleistocene	*NH*	*Gas/oil*
Pleistocene	*NI*	*Gas/oil*
Pleistocene	*OI-2*	*Gas/oil*
Pleistocene	*OI-3*	*Gas/oil*
Pleistocene	*OI-4*	*Gas/oil*
Pleistocene	*Lentic*	*Gas/oil*

Geologic concept leading to discovery and method or methods used to delineate prospect, e.g., surface geology, subsurface geology, seeps, magnetic data, gravity data, seismic data, seismic refraction, nontechnical:

Use physical data to define structural areas of large possible reserves, on the crests of structural highs (with 4-way dip closure) and within minibasins, which lithofacies mapping identified as having alternating sand-shale sequences.

Structure:

Province/basin type .... *Bally 1143; Klemme IICc/IV*

Tectonic history

The Gulf Coast basin is typified by graben systems and high-angle growth faults occurring in a semicircular zone parallel to the present coastline. Salt tectonism, with associated doming and faulting, is the most active structural process.

Regional structure

The field is located in a small minibasin formed by an arcuate system of growth faults.

Local structure

Rollover anticlines on the downthrown side of large northwest-trending, salt-related growth fault.

Trap:

Trap type(s) .... *Rollover anticline into a large basinal growth fault with multiple pays*

Basin stratigraphy (major stratigraphic intervals from surface to deepest penetration in field):

Chronostratigraphy	Formation	Depth to Top in m (ft)
Trimosina A *0.5 m.y.*	*GA sandstone*	*-1310 (-4300)*
Trimosina A *0.5 m.y.*	*HB sandstone*	*-1478 (-4850)*
Angulogerina B *1.8 m.y.*	*JD sandstone*	*-1951 (-6400)*
Angulogerina B *1.8 m.y.*	*KE sandstone*	*-2088 (-6850)*
Angulogerina B *1.8 m.y.*	*LF sandstone*	*-2164 (-7100)*
Angulogerina B *1.8 m.y.*	*MG sandstone*	*-2286 (-7500)*
Angulogerina B *1.8 m.y.*	*NH sandstone*	*-2316 (-7600)*
Angulogerina B *1.8 m.y.*	*OI sandstone*	*-2347 (-7700)*
Lenticulina *1 2.3 m.y.*	*L sandstone*	*-3353 (-11,000)*

Location of well in field

Reservoir characteristics:

Number of reservoirs *25*

Formations .... Trimosina A-Lenticulina *1*

Ages .... *0.5 m.y.-2.5 m.y. (Pleistocene)*

Depths to tops of reservoirs .... *1311-3658 m (4300-12,000 ft)*

Gross thickness (top to bottom of producing interval) .... *1981 m (6500 ft)*

Net thickness—total thickness of producing zones

Average .... *165 m (540 ft)*

Maximum .... *336 m (1100 ft)*

Average net pay/interval .... *8-17 m (27-56 ft)*

Maximum net pay/interval .... *31 m (103 ft)*

Lithology

Very fine to medium-grained, moderately well to well-sorted, quartzose to slightly arkosic arenites

Porosity type

Primary intergranular or reduced primarily intergranular porosity. Some moldic porosity where feldspars have been diagenetically partially dissolved.

Average porosity .... *30% (25-35%)*

Average permeability .... *500-1350 md*

Seals:

Upper

Formation, fault, or other feature .... *Overlying shale*

Lithology .... *Shale*

Lateral

Formation, fault, or other feature .... *Growth fault or permeability barrier*

Lithology .... *Shale*

Source:

Formation and age .... *Unknown, probably Miocene*

Lithology .... *Marine shales*

Average total organic carbon (TOC) .... *Unknown*

Maximum TOC .... *Unknown*

Kerogen type (I, II, or III) .... *II*
Vitrinite reflectance (maturation) .... $R_o = 0.8–1.0$ *(inferred)*
Time of hydrocarbon expulsion .... *Pleistocene*
Present depth to top of source .... *4572–4877 m (15,000–16,000 ft) (inferred)*
Thickness .... *Unknown*
Potential yield .... *Unknown*

Appendix 2. Production Data

Field name .... *Eugene Island Block 330 field*

Field size:

Proved acres .... *"JD" 6450 ac (2611 ha)*
Number of wells all years *258*
Current number of wells .... *135 (producing in 1986)*
Well spacing .... *NA*
Ultimate recoverable .... *307 million bbl oil + 1650 bcf gas (48.8 million m³ oil + 46.7 billion m³ gas)*
Cumulative production .... *271 million bbl oil + 1259 bcf gas (43.1 million m³ oil + 35.7 billion m³ gas)*
Annual production (1986) .... *9.1 million bbl oil + 41 bcf gas (1.4 million m³ oil + 1.2 billion m³ gas)*
Present decline rate *13%*
Initial decline rate *32%*
Overall decline rate *11%*
Annual water production (1986) .... *9,177,801 bbl (1,459,179 m³)*
In place, total reserves .... *750 million bbl oil + 2200 bcf gas (119 million m³ + 62.3 billion m³)*
In place, per acre-foot .... *1200 bbl oil, 2.25 MMcf gas*
Primary recovery .... *300 million bbl oil + 1650 bcf gas (47.7 million m³ oil + 46.7 billion m³ gas)*
Secondary recovery .... *6.11 million bbl oil (0.97 million m³)*
Enhanced recovery *0*
Cumulative water production (12/1986) .... *136,944,560 bbl (21.78 million m³)*

Drilling and casing practices:

Amount of surface casing set .... *1097 m (3600 ft)*
Casing program
(1) Wells TD within normal pressure (through MG sands)—26-in. at 465 ft, 16-in. at 850 ft, 10¾-in. at 3600 ft, 7⅝-in. at 7700 ft; (2) wells TD within abnormal pressure (through OI sands)—26-in. at 465 ft, 20-in. at 850 ft, 13 ⅜-in. at 3600 ft, 9⅝-in. at 6800 ft, 7-in. at 7400 ft
Drilling mud .... *Seawater gel to 3600 ft, lignosulfate 3600 ft to TD*
Bit program .... *Typical bit: sealed bearing, milled-tooth rock bit*
High pressure zones .... *NH through OI sands (approx. 7500 ft down)*

Completion practices:

Interval(s) perforated .... *Avg. perforated interval per major reservoir: GA, 32 ft; HB, 19-in.; JD, 66 ft; KE, 30 ft; LF, 60 ft; MG, 19 ft; OI, 24 ft*
Well treatment .... *Retarded mud acid treatment as needed for formation damage*

Formation evaluation:

Logging suites .... *Induction electric logs (SP, GR, resistivity) + sonic, FDC-CNL, dipmeter, sidewall cores*

Testing practices

(1) Routine wireline sidewall cores; (2) drill-stem testing in major sands; (3) conventional coring in selected wells and intervals; and (4) pressure transient analysis in all reservoirs after initial completion

Mud logging techniques

Logging unit installed from surface to TD to monitor total gas, rate of penetration, pressure, chlorides, and shale densities; and to catch cutting samples and perform gas chromatography

Oil characteristics:

Type .... *Paraffinic-naphthenic to aromatic-intermediate*
API gravity .... *23-37.5°*
Base .... *Paraffinic*
Initial GOR .... *(LF) 941*
Sulfur, wt% .... *0.19-1.53 (avg. 1.0%)*
Viscosity, SUS *0.37*
Pour point .... *NA*
Gas-oil distillate .... *(JD) 32 scf/bbl*

Field characteristics:

Average elevation .... *(LF sand) -2164 m (-7100 ft)*
Initial pressure .... *(LF) 4076 psig (287 kg/cm²)*
Present pressure .... *(LF) 2576 psig (181 kg/cm²)*
Pressure gradient .... *(LF) 0.306 psi/ft (0.072 kg/cm ²/m)*
Temperature .... *(LF) 160°F (71°C)*
Geothermal gradient .... *0.9-1.5°F/100 ft (1.64-2.73°C/100 m)*
Drive .... *Water drive + gas expansion*
Oil column thickness .... *(LF) Oil, 212 m (696 ft); gas, 111 m (364 ft)*
Oil-water contact .... *(LF) -2320 m (-7610 ft)*
Connate water .... *NA*
Water salinity, TDS .... *(LF) 94,000 ppm*
Resistivity of water .... *(LF) 0.0387 ohm-m²/m at 160°F (71°C)*
Bulk volume water (%) .... *(LF) 12*

Transportation method and market for oil and gas:

Gas sales are metered on each platform and transported via Sea Robin pipeline to an onshore processing facility. Oil is metered and pumped through the Bonito pipeline system to a sales point near shore.

Hibernia Oil Field—Canada
Jeanne d'Arc Basin, Grand Banks Offshore Newfoundland

A. H. MACKAY
A. J. TANKARD
Petro-Canada Resources
Calgary, Alberta, Canada

FIELD CLASSIFICATION

BASIN: Jeanne d'Arc
BASIN TYPE: Rift
RESERVOIR ROCK TYPE: Sandstone
RESERVOIR ENVIRONMENT OF DEPOSITION: Fluvial and Nearshore Marine
RESERVOIR AGE: Jurassic and Cretaceous
PETROLEUM TYPE: Oil
TRAP TYPE: Rollover Anticline

LOCATION

In 1979, the Hibernia discovery well, Chevron et al Hibernia P-15, was drilled in 80 m (262 ft) of water on the Grand Banks, 315 km (195 mi) east of St. John's, Newfoundland (Figure 1). The P-15 well has an estimated productive capability in excess of 3200 m^3 OPD (20,000 BOPD) and was the first well capable of commercial oil production on the east coast of Canada (Arthur et al., 1982). Subsequent drilling has defined a complex field with oil reserves of 83×10^6 m^3 (522×10^6 bbl).

The Grand Banks is the broad continental shelf that extends 450 km (280 mi) seaward of Newfoundland. In the Hibernia area, the seafloor consists of 1 to 5 m (3.3 to 16.4 ft) of coarse-grained surficial sediment above semiconsolidated Tertiary bedrock. The surficial sediment has been scoured by icebergs and wave activity (Benteau and Sheppard, 1982). Extreme storm conditions also occur on the Grand Banks. On 15 February 1982, while drilling the Mobil et al Hibernia J-34 well, the Ocean Ranger semisubmersible capsized and sank with the loss of 84 lives. This tragedy emphasizes the harsh operating environment.

HISTORY

Pre-Discovery

Petroleum exploration offshore Newfoundland during the past two decades has resulted in acquisition of over 375,000 km (230,000 mi) of reflection seismic data and the drilling of more than 100 wells. Exploration drilling on the Grand Banks began in 1966 with the dry and abandoned (D & A) Pan Am Tors Cove D-52 wildcat, followed by over 30 more D & A wells (Leavitt, 1987). In 1973, exploration focused on the Jeanne d'Arc basin after a significant oil show in the Mobil et al Adolphus 2K-41 well (Figures 2 and 3). The 2K-41 well tested 43 m^3 OPD (270 BOPD) from a 1.2 m (4 ft) thick porous sandstone of Late Cretaceous age. However, subsequent drilling indicated a lack of reservoir-quality sandstones in the area. In 1973–1974, the Mobil et al Egret K-36 and N-46 wells were drilled 57 km (35 mi) southwest of the Adolphus 2K-41 well and penetrated thick, porous, but wet Lower Cretaceous sandstones overlying rich Jurassic source rock. The occurrence of oil as well as substantial reservoir and source rock in the Jeanne d'Arc basin were encouraging and emphasized the potential of the Hibernia area.

Discovery

In February 1979, Chevron Canada Resources and Columbia Gas Development of Canada concluded a farm-in arrangement on a 212,468 ha (525,005 acre) block covering the Hibernia area with the landholders Mobil Oil Canada, Gulf Canada Resources, and Petro-Canada (Table 1; Canada-Newfoundland Offshore Petroleum Board [CNOPB], 1986). On 27 May of the same year, the Chevron et al Hibernia P-15 well (lat. 46°44′58.98″N, long. 48°46′51.18″W) was spudded. This well tested a large fault-bounded rollover anticline that was mapped from reflection seismic,

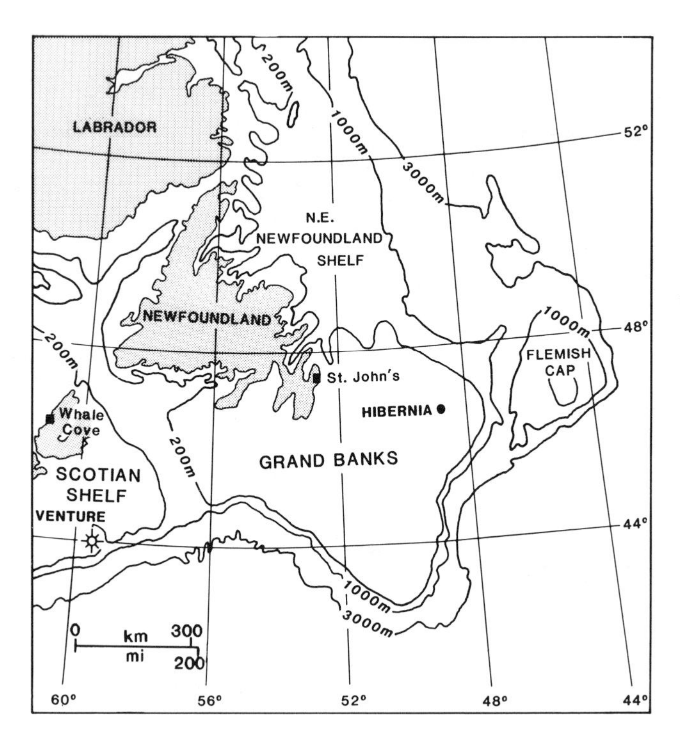

Figure 1. Location of the Hibernia oil field on the Grand Banks, the broad continental shelf offshore Newfoundland.

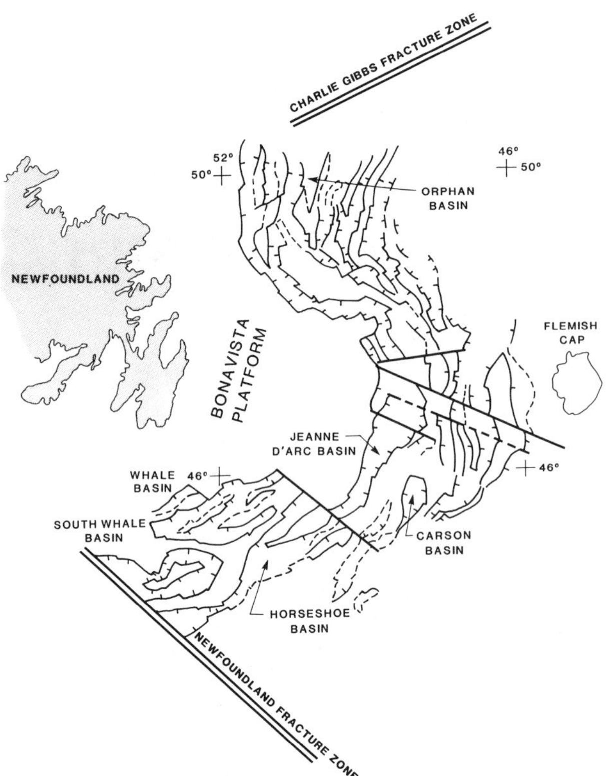

Figure 2. Grand Banks is separated from Scotian and Labrador shelves by Newfoundland and Charlie Gibbs fracture zones, respectively. Normal extensional and transfer faults dissected basement into a suite of half-grabens. Map is based on seismic structure mapping. (After Welsink et al., 1989.)

and it encountered pay zones in the Upper Jurassic Jeanne d'Arc sandstone (-4105 m; -13,468 ft) and Lower Cretaceous Hibernia (-3727 m; -12,228 ft), Catalina (-3196 m; -10,486 ft), and Avalon (-2191 m; -7,188 ft) sandstones (Figures 4 and 5). Production capabilities in excess of 3200 m^3 OPD (20,000 BOPD) were estimated for P-15 (Arthur et al., 1982).

Post-Discovery

Discovery of the Hibernia oil field initiated a new phase of drilling activity on the Grand Banks that resulted in the Ben Nevis, Hebron, South Tempest, Nautilus, Mara, Terra Nova, and Whiterose discoveries (Figure 3). In the Hibernia field, nine delineation wells and 6365 km (3955 mi) of 3D seismic outlined a complex structure with reserves of 83×10^6 m^3 (522×10^6 bbl) oil (Figure 6). The most recent delineation well, Mobil et al Hibernia C-96, was drilled in 1984. The reserves occur in the Hibernia sandstone that spans 6703 ha (16,563 acres) and contains 71×10^6 m^3 (447×10^6 bbl) of recoverable oil, and the Avalon sandstone that covers 1788 ha (4418 acres) and contains 12×10^6 m^3 (75×10^6 bbl) of recoverable oil (Figure 5). Potential exists for additional reserves in other parts of the Avalon interval and in the less prolific Jeanne d'Arc and Catalina reservoir sandstones. Approximately 1300 m (4265 ft) of core have been recovered.

The development plan prepared by Mobil Oil Canada (1986) on behalf of the joint venture partners (Table 1) envisages production in the 1990s. The proposed production system shown in Figure 7 consists of topside facilities mounted on a gravity based structure comprised of three components: a concrete base slab, a caisson reaching from the seabed to approximately 5 m (16.4 ft) above mean low water level, and four shafts extending above the caisson to support the topside facilities (Mobil Oil Canada, 1986). An estimated 48 wells will be drilled from the platform and 35 from semi-submersibles for an annual oil production rate of 17.5×10^3 m^3 OPD (110,000 BOPD). As production rates from the Hibernia reservoir decline, additional production will be initiated from the Avalon reservoir to maintain overall field performance (CNOPB, 1986). It is anticipated that reservoir pressure will be maintained by a combination of produced gas injected into the crestal areas and water injection into the downdip parts of the Hibernia reservoir (Mobil Oil Canada, 1986).

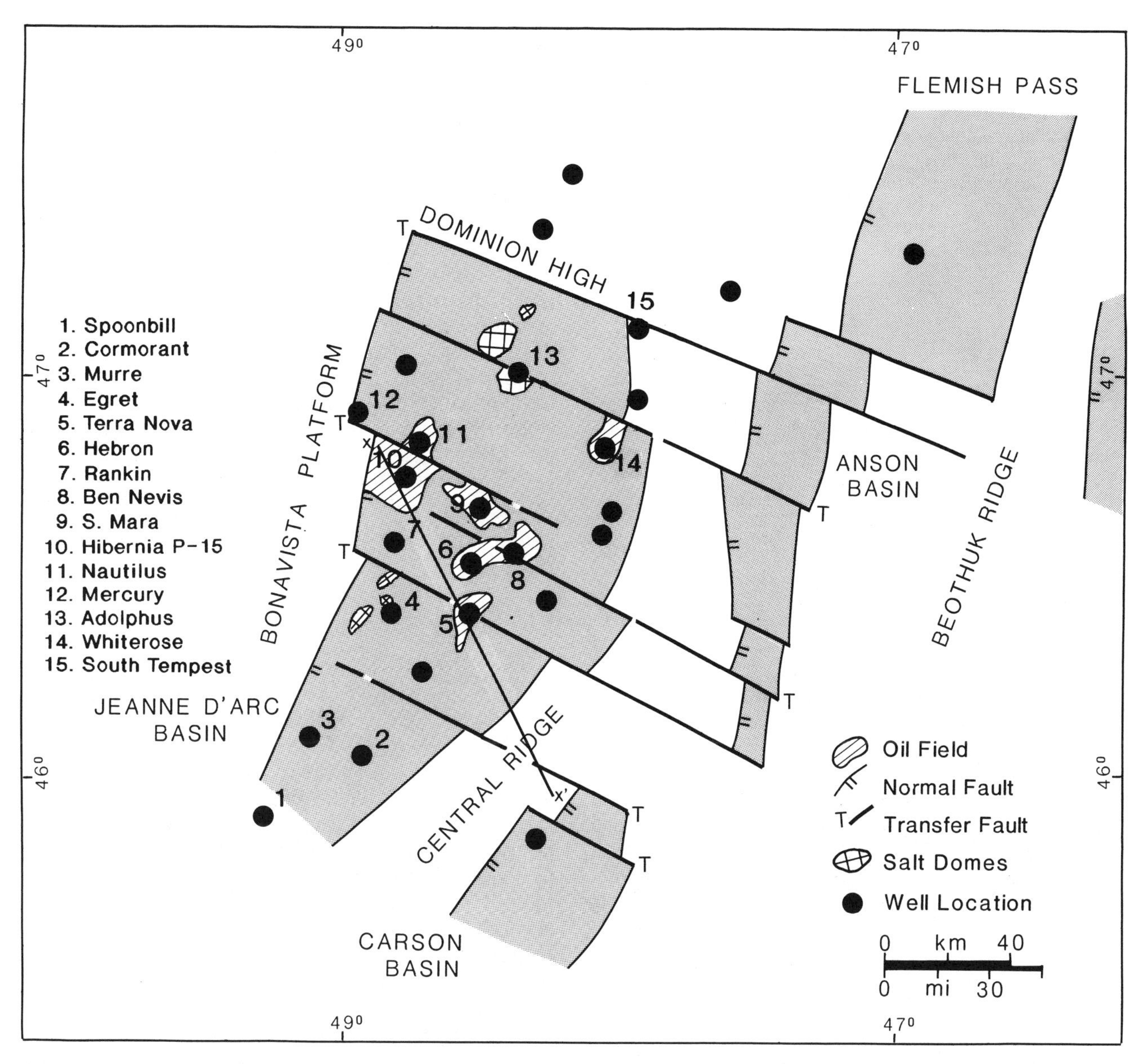

Figure 3. The Jeanne d'Arc basin was formed by listric normal faults and transfer faults that offset the margins, resulting in an irregular funnel-shaped geometry. Oil field distribution is related to linkage of these fault systems. X–X′ is location of seismic line illustrated in Figure 9. (Modified after Tankard and Welsink, 1987.)

DISCOVERY METHOD

Exploration drilling on the southern Grand Banks began in 1966 and concentrated on salt diapiric structures. Thirty D & A wells followed. By the early 1970s, regional seismic programs had defined and characterized many of the half-grabens beneath the central Grand Banks. In 1973, the Adolphus 2K-41 well was drilled on a salt pillar in the Jeanne d'Arc basin and tested oil. Although subsequent drilling indicated a lack of adequate reservoir in the area, the potential of the Jeanne d'Arc basin was established. The search for more substantial reservoirs resulted in drilling of the Egret K-36 and N-46 wells in 1973 and 1974, respectively (Figure 3). Although these wells were dry and abandoned, thick, Lower Cretaceous sandstones and rich Jurassic source rocks were encountered, providing sufficient encouragement for further drilling in this basin. Reflection seismic indicated a large structural closure associated with rollover into a listric basin-bounding fault, the Hibernia structure. The Chevron et al Hibernia P-15 well was drilled on this structure in 1979. This departure from salt-structure exploration was rewarded with a significant oil discovery.

Table 1. Land Ownership, Hibernia oil field.

Participant	Interest (%)
Mobil Oil Canada Ltd.	28.12500
Gulf Canada Resources Inc.	25.00000
Petro-Canada Inc.	25.00000
Chevron Canada Resources Ltd.	16.40625
Columbia Gas Development of Canada Ltd.	5.46875*

*Columbia Gas Development of Canada share purchased by Chevron Canada Resources Ltd. in 1988.
Source: CNOPB (1986).

STRUCTURE

Tectonic History

The Grand Banks records a long history of deformation, including Paleozoic convergence and three major episodes of extension and ocean opening in the Mesozoic (Figure 4). Avalon basement consists of crystalline and metasedimentary rocks (Williams, 1984). Deep reflection seismic, gravity, and magnetic data suggest that Paleozoic structural fabrics had a profound effect on Mesozoic extension (Tankard and Welsink, 1987, 1989). Mesozoic extension dissected the Grand Banks into a series of rift basins, one of which is the oil-prone Jeanne d'Arc basin (Figure 2; Bally III or Klemme IIIA; cf. St. John et al., 1984). Hydrocarbons accumulated in Upper Jurassic and Lower Cretaceous synrift sediments.

Extensional subsidence in the Triassic and Early Jurassic resulted in deposition of terrestrial redbeds and evaporites; examples include the Adolphus 2K-41, Cormorant N-83, Murre G-67, and Spoonbill C-30 wells (Jansa and Wade, 1975; Tankard and Welsink, 1987). This rift system was later aborted, and subsequent marine encroachment deposited a thick succession of Lower and Middle Jurassic calcareous mudstones. Extensional subsidence again dominated the Grand Banks from late Callovian to Aptian time and produced a series of deep, subparallel half-grabens that contain thick accumulations of alluvial and shallow-marine sediments that are punctuated by numerous unconformities. This history of intermittent subsidence is summarized in Figures 4 and 8. Separation of the Iberian peninsula during the Aptian terminated this episode of rifting. The locus of latest Cretaceous extension was the Orphan basin to the northeast but was reflected in accelerated post-rift subsidence in Jeanne d'Arc basin. The Tertiary section is composed of interfingering marine shales and sandstones. The Jeanne d'Arc basin contains up to 17 km (11 mi) of Mesozoic and Cenozoic sediments (Tankard et al., 1989).

Regional Structure

Late Callovian to Aptian extension formed the structural architecture of the Jeanne d'Arc basin. Deposition and preservation were controlled by multidirectional and intersecting fault systems (Figure 2). The Murre fault forms the western edge of the Jeanne d'Arc basin, separating it from the Bonavista Platform. A listric geometry and rotation of basement and the synrift succession into the fault plane characterize the Murre fault, which soles out at about 26 km (16 mi) along an intracrustal decollement (Tankard and Welsink, 1987). The eastern edge of the Jeanne d'Arc basin is terraced by a number of antithetic listric and planar faults (Figure 9). Cross-basin transfer faults were contemporaneous with normal faulting; transfer faults are very high angle basement tear faults that are nearly orthogonal to the normal faults (Figure 3). Smaller scale faulting resulted from strike-slip motion of the cross-basin transfer faults.

Local Structure

The Hibernia structure was created by extensional faulting and rollover that were accentuated by salt diapirism. The Nautilus fault, which forms the northeast boundary of the Hibernia field, is a transfer fault that offsets the listric normal Murre fault. A series of smaller scale faults dissects the Hibernia structure internally; these are attributed to the right-lateral shear couple, rollover, and diapirism (Figure 6; Tankard and Welsink, 1987). These fault styles have a profound effect on thickness variations of several stratigraphic sequences and on hydrocarbon trapping. The extensional faults form first-order seals. All the reservoir sands have down-dip oil-water contacts to the east and south (Figures 6 and 10).

STRATIGRAPHY

Deposition of Triassic Eurydice redbeds and Upper Triassic to Lower Jurassic Argo evaporites occurred during an early episode of rifting (Figure 4). The Eurydice Formation is composed of reddish shales and siltstones with rare feldspathic sandstones. The Argo Formation consists predominantly of halite with minor interbeds of anhydrite and shale. After this episode of rifting ceased, regional subsidence dominated the Grand Banks, resulting in a broad epeiric basin. Lower and Middle Jurassic stratigraphy was little affected by synsedimentary faulting or unconformities and consists of a thick layered succession of carbonates and calcareous mudstones. This succession has been sampled in the southern Jeanne d'Arc basin in Cormorant N-83, Murre G-67, and Spoonbill C-30 (Jansa and Wade, 1975; Tankard and Welsink, 1987).

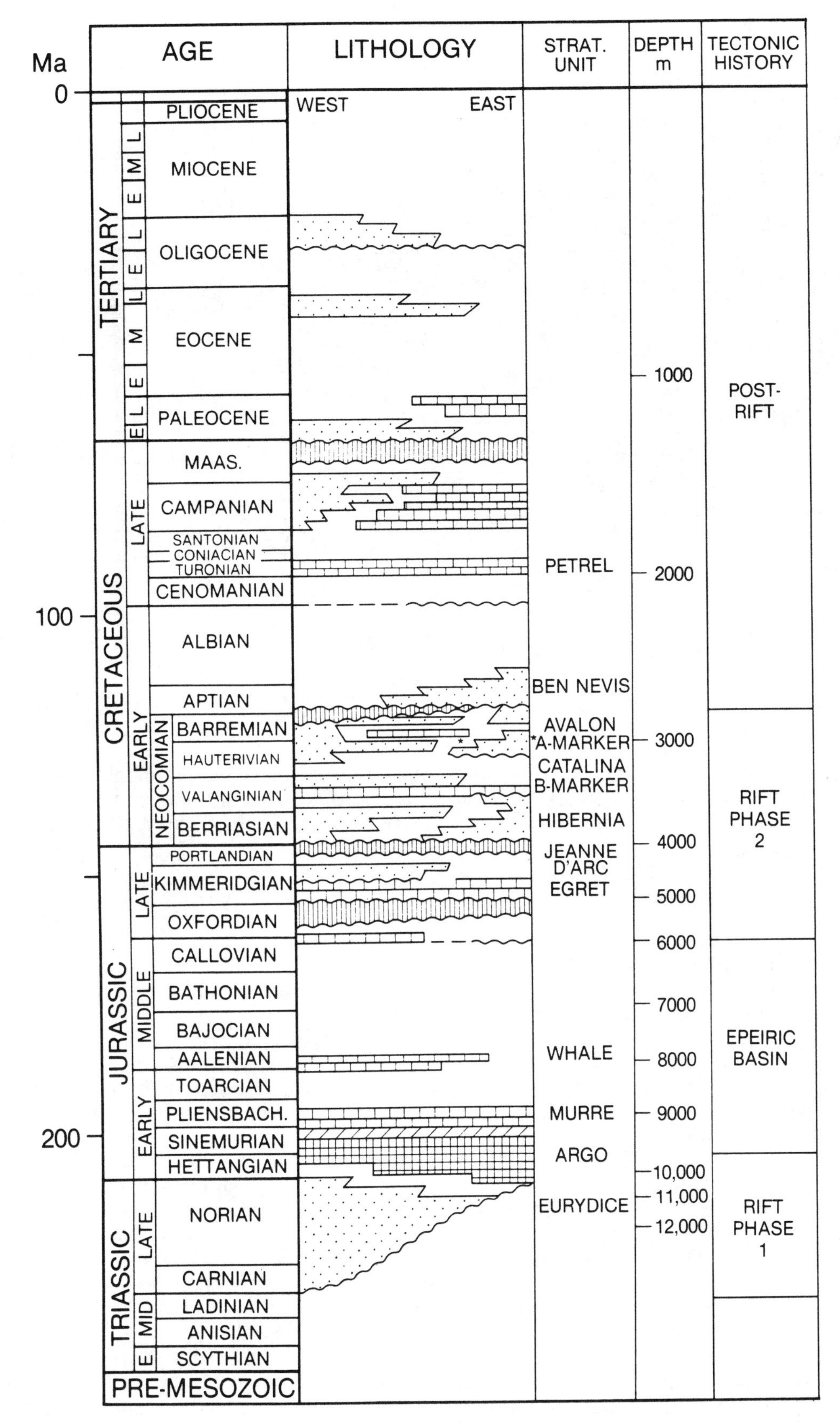

Figure 4. Tectonostratigraphic column for Jeanne d'Arc basin showing five major basin-forming stages. Unconformity-bounded sequences are based on seismic stratigraphy. Several prolific reservoir sequences were deposited during the mid-Mesozoic episode of rifting. (After Tankard et al., 1989.)

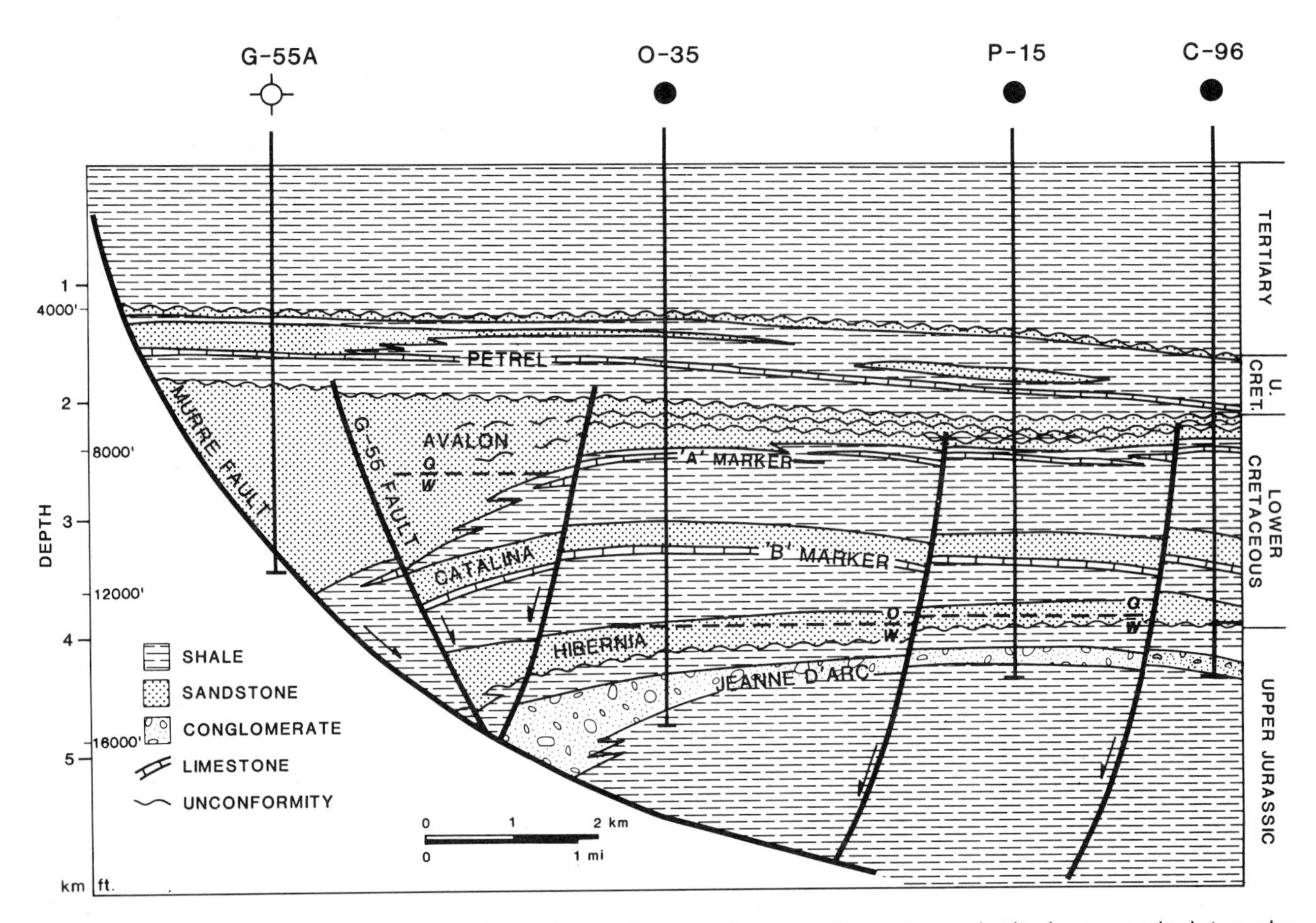

Figure 5. Stratigraphic cross section of synrift succession of Hibernia field showing rollover into the listric normal Murre fault. Hibernia and Avalon sandstones form the principal reservoir intervals; Jeanne d'Arc and Catalina reservoirs are secondary. (Modified after Tankard and Welsink, 1987.)

Late Callovian to Kimmeridgian extension and subsidence was initially gentle. At least 1300 m (4265 ft) of Egret oolitic limestones, organic-rich shales, siltstones, and very fine grained sandstones were deposited in weak marine and paralic swamp environments (Tankard and Welsink, 1987). Organic-rich shales of early-middle Kimmeridgian age are the principal source of oil in the Jeanne d'Arc basin (Powell, 1985; Tankard et al., 1989; von der Dick, 1989; von der Dick et al., in press). The climax of Kimmeridgian-Valanginian fault-controlled subsidence resulted in deposition of coarse terrestrial material. The Jeanne d'Arc Formation is a secondary reservoir in the Hibernia field and the deepest encountered (Figure 4). The sequence of interbedded fluvial sandstones and conglomerates was penetrated in the Chevron et al Hibernia P-15 well at -4105 m (13,468 ft). Portlandian transgressive onlap of shallow-marine sandstones and mudstones terminated the sequence.

During Berriasian time a suite of sandy braid deltas was stacked against the basin margin, forming the Hibernia Formation, the most prolific of the two primary reservoirs in the Hibernia field. The Hibernia sequence consists of thin, upward-coarsening delta front facies that are overlain by fluvial sandstones and capped by transgressive marine mudstones that inundated the sandy delta facies (Figure 5). Basin deepening and transgression culminated in deposition of the B-marker limestone at the end of Hibernia time.

The B-marker is a 20 to 80 m (65 to 262 ft) thick, transgressive limestone and calcareous sandstone that is regionally persistent and forms a prominent seismic reflector (Tankard et al., 1989). Subsequent regression deposited a thick sequence (180–295 m; 591–967 ft) of interbedded sandstones, siltstones, and mudstones of the Catalina Formation, forming a secondary reservoir unit.

Unlike the B-marker, the early Barremian A-marker is a diachronous unit consisting of en echelon arranged limestones and calcareous sandstones that form the base of the Avalon sequence (Figure 5). The Avalon Formation is a primary reservoir zone and the youngest reservoir unit in the Hibernia field. It consists of a southwestward-thickening wedge of

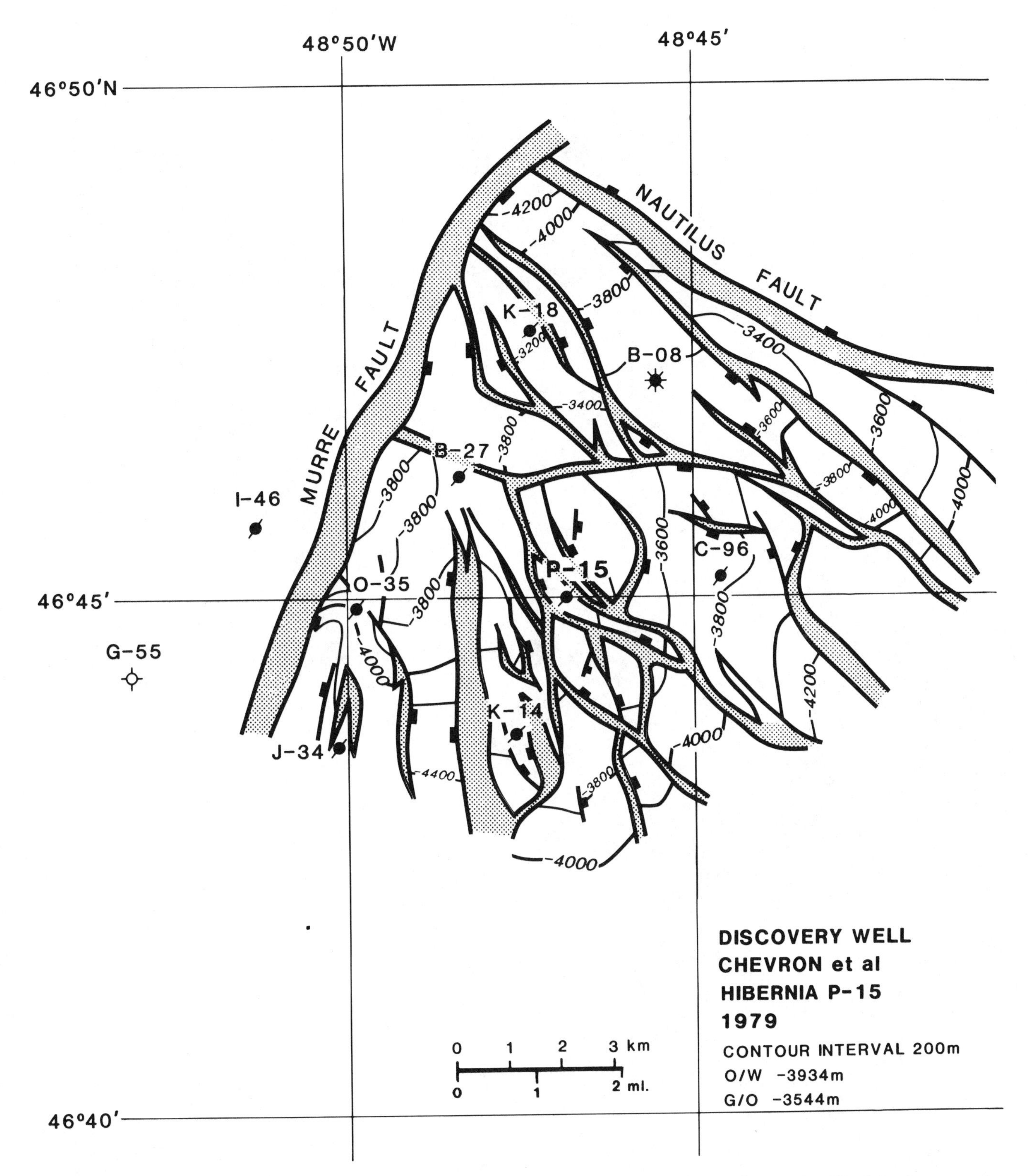

Figure 6. Depth-structure map at top of Hibernia reservoir, showing Murre basin-forming fault, Nautilus transfer fault, and smaller-scale Riedel shear and tension gashes that dissect the field.

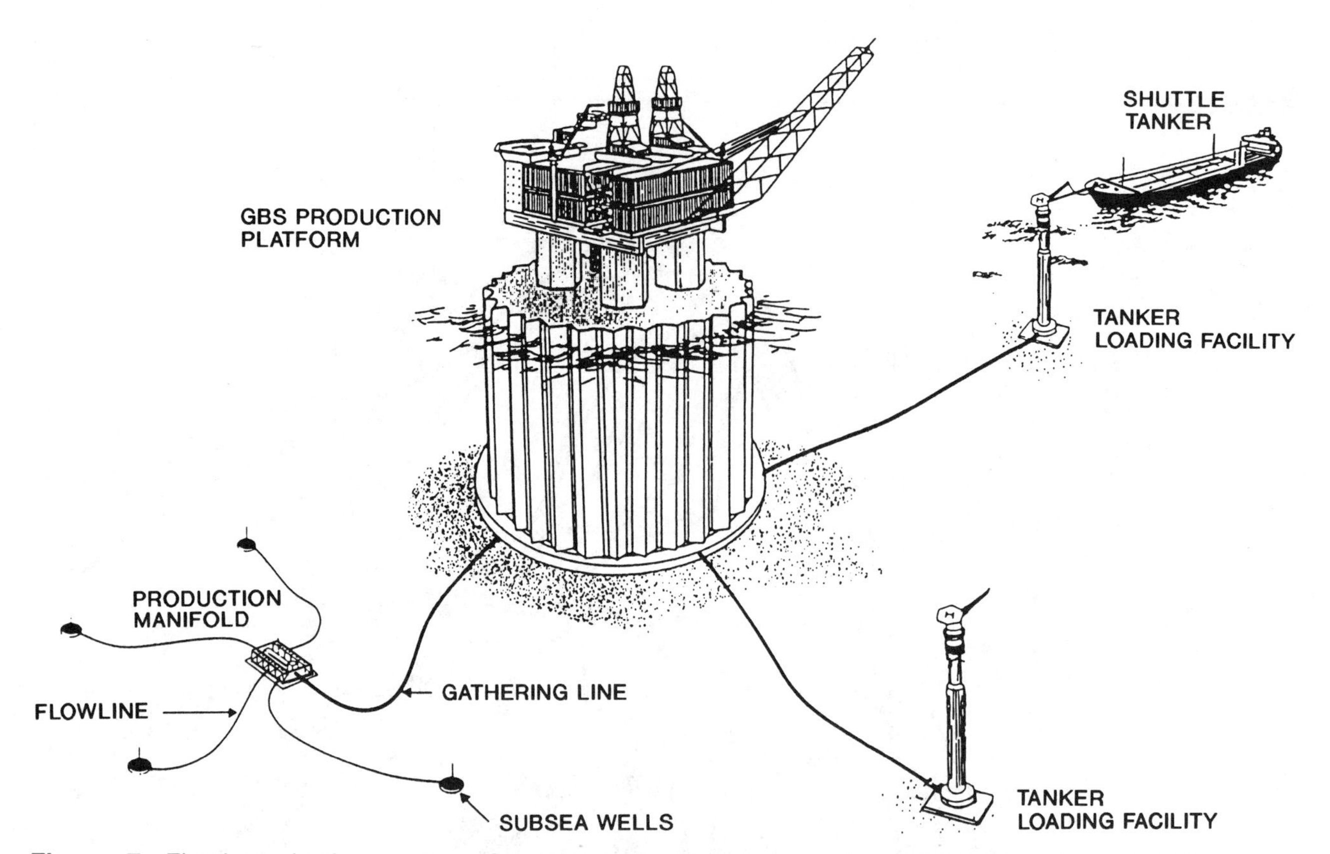

Figure 7. Fixed production system. Gravity-based structure envisaged for production of the Hibernia oil field, consisting of a concrete base slab, a caisson reaching from the seabed to approximately 5 m (16 ft) above mean low water level, and four shafts extending above the caisson to support the topside facilities. (Courtesy of Mobil Oil Canada Ltd.)

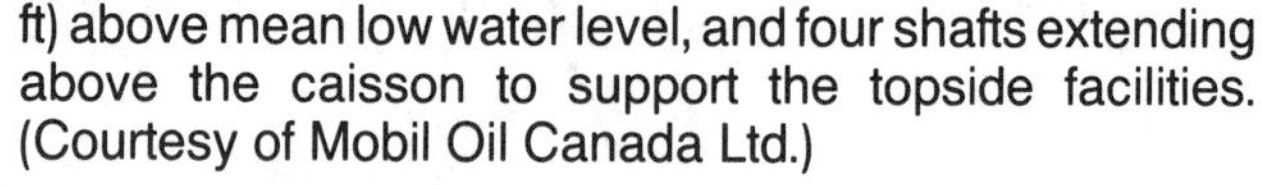

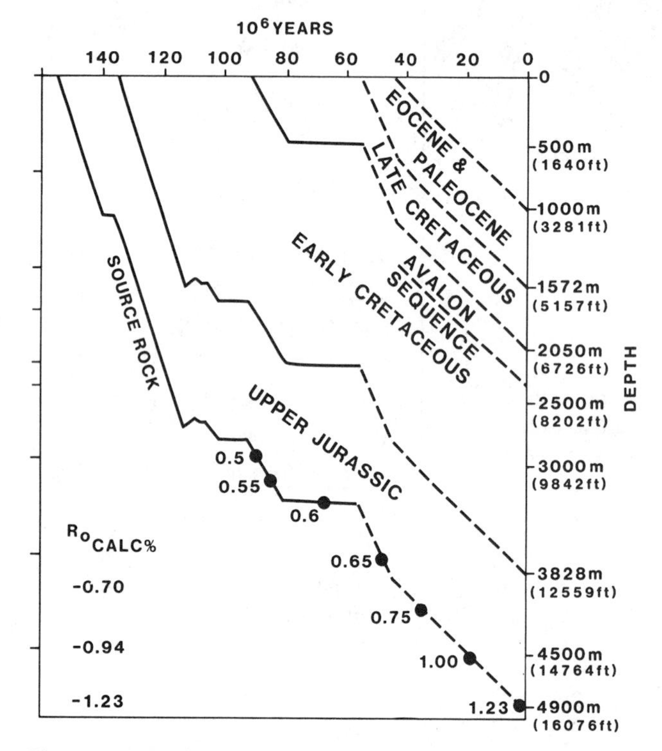

Figure 8. Burial history of Jeanne d'Arc basin shown for Hibernia K-18. (Modified after von der Dick et al., in press.)

interbedded mudstones and quartzose sandstones that are dissected by several unconformities. A prominent "break-up" unconformity of early Aptian age caps the Avalon sequence and marks the transition to post-rift subsidence and ocean opening between the Grand Banks and the Iberian peninsula (Tankard et al., 1989). The post-"break-up" Ben Nevis sequence consists of medium to very coarse grained, cross-bedded sandstones with mudclasts and coal spar on scour surfaces.

During the Late Cretaceous, relatively shallow seas dominated the Jeanne d'Arc shelf, resulting in deposition of prograding terrigenous wedges and blanketing limestones (Tankard and Welsink, 1987). The Petrel, a foraminiferal coccolithic limestone or chalk of Turonian to Coniacian age, is a prominent seismic reflector that drapes the Jeanne d'Arc basin and oversteps the Bonavista platform. Cretaceous and Tertiary sedimentation on the Grand Banks are separated by a base-Tertiary unconformity. In the Paleocene, prograding deltaic and submarine fans converged on the subtle depocenters above the Hibernia field, building thick sand-prone terraces.

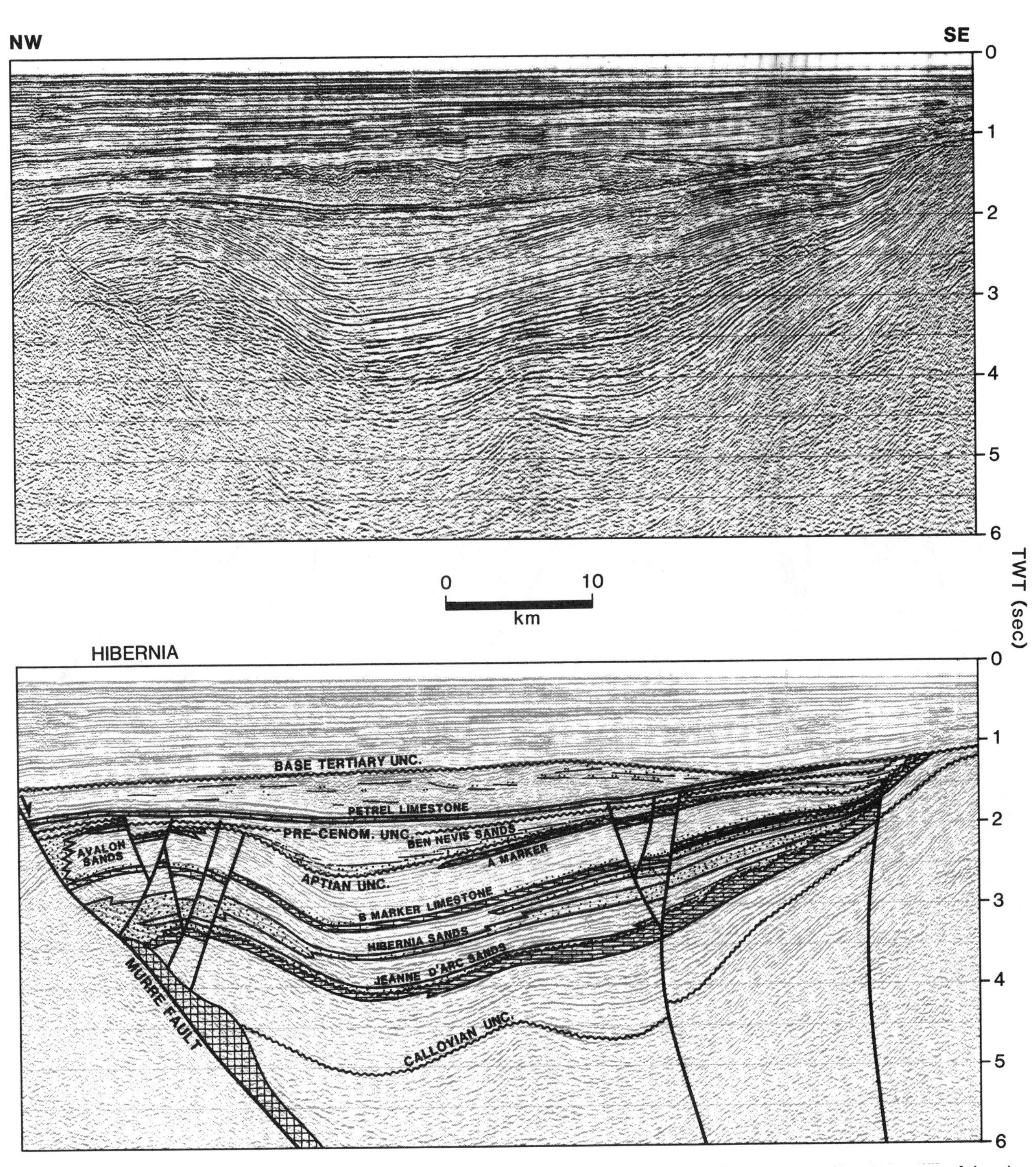

Figure 9. Seismic line 79-NF-103 (X–X′, Figure 3) showing half-graben asymmetry of Jeanne d'Arc basin and major unconformity-bounded sequences. Basin-bounding Murre fault was active during late Callovian–Aptian rifting, as shown by characteristic rollover of Hibernia structure. On eastern horst ramp of basin, merging unconformities indicate continuous uplift of Central ridge. (Seismic line courtesy Geophysical Service Inc.; after Tankard et al., 1989.)

The modern continental-terrace wedge was built mainly during the Tertiary and is dominated by mudstones with regional unconformities related to glacial-eustatic sea-level fluctuations (Tankard and Welsink, 1987).

TRAP

The anticlinal structure along the Murre fault provides the major trapping component for the

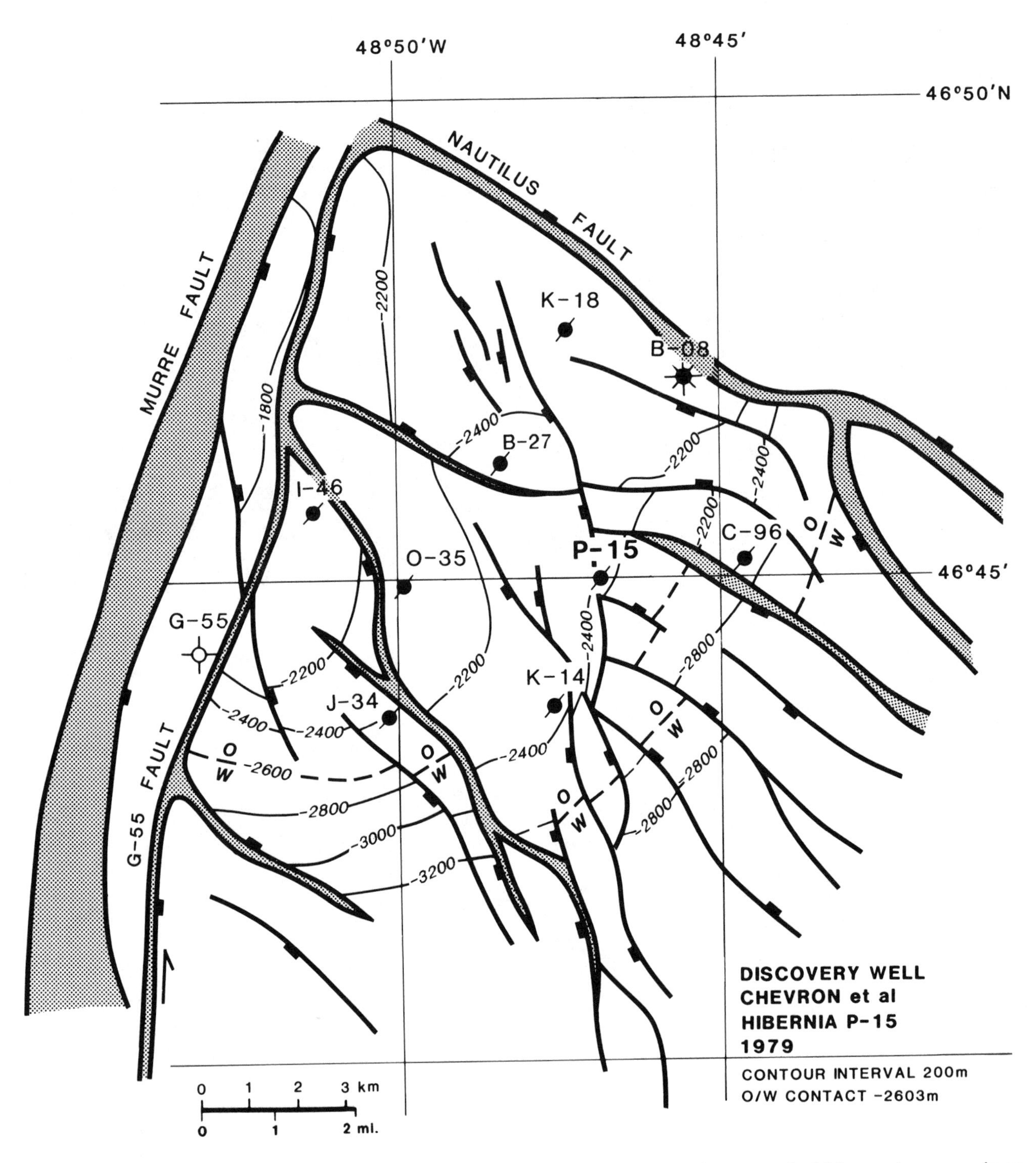

Figure 10. Depth-structure map at top of the Avalon reservoir. The structure is formed from the intersection of the Nautilus transfer fault with the listric normal Murre fault. The G-55 fault formed in post-rift time as a result of detachment and transport of Mesozoic cover down the axis of the basin, parallel to older basin-boundary faults.

primary reservoirs, Hibernia and Avalon sandstones, and secondary reservoirs, Jeanne d'Arc and Catalina sandstones (Figures 5 and 9). Conspicuous elements of this structural trap include extensional rollover that was enhanced by diapirism, offsetting transfer fault, and smaller scale Riedel shears and tension fractures. This trap evolved mainly during Callovian–Aptian time but was also disrupted structurally by later detachment and deformation of the cover sequence that formed the G-55 fault (Figure 10).

Each reservoir zone also contains a component of stratigraphic seal. The overpressured Jeanne d'Arc reservoir consists of discontinuous lenticular sandstones and conglomerates. Both facies and structural seals are recognized.

The primary trapping mechanism for the Hibernia reservoir is structural, while top seal is provided by overlying shales (Figures 5 and 11). An oil-water contact occurs at -3934 m (-12,907 ft); for example, in the Hibernia K-14 well. Although the sealing or conduit nature of the numerous faults dissecting the field cannot be determined with confidence until the field has been on production, the higher oil-water contact in the Hibernia P-15 well (-3862 m; -12,671 ft) indicates that faulting does create local variation in fluid contacts. A Hibernia sandstone gas-oil contact was encountered at -3544 m (-11,627 ft) in the Hibernia B-08 well.

The secondary Catalina reservoir is also structurally and stratigraphically sealed. The poor interconnectedness and interfingering relationships of the sandstone provide many separate oil-water contacts.

Trapping in the Avalon reservoir is dominated by structural (e.g., G-55 fault and Nautilus fault) and stratigraphic seals (Figures 10 and 12). The G-55 fault is attributed to post-rift detachment of the cover sequence (Tankard and Welsink, 1987). Movement along this fault in a strike-slip sense amounted to about 1.5 km (1 mi), resulting in substantial clay smearing. Hibernia G-55 was dry and abandoned, encountering wet sands west of the G-55 fault juxtaposed against oil-bearing reservoir east of the fault.

RESERVOIRS

Two principal hydrocarbon reservoirs, the Hibernia and Avalon sandstones, and two secondary reservoirs, the Jeanne d'Arc and Catalina sandstones, characterize the Hibernia field. Ten wells have been drilled on this structure to an average depth of 4250 m (13,944 ft); approximately 1300 m (4265 ft) of core have been recovered. Of the ten wells, seven were drilled to Jeanne d'Arc reservoir depths; the other three tested only the Avalon section.

Jeanne d'Arc Reservoir

The initial oil show in the Hibernia P-15 discovery well was at -4105 m (-13,468 ft) from a 120 m (394 ft) interval of overpressured Jeanne d'Arc sandstones and conglomerates of Kimmeridgian to Portlandian age. Four Jeanne d'Arc facies associations form an upward-coarsening sequence of mudstones, siltstones, and sandstones (Figure 13; Table 2). The coarseness, angularity, and immaturity of the matrix-supported conglomerate clasts (Figure 14) suggest substantial structural relief and short distances of transport. The association of matrix-supported conglomerate, clast-supported conglomerate, and sandstone facies have been attributed to braided alluvial plain deposition. Flood-basin deposition is suggested for the mudstone facies owing to its coal laminae, woody material, and pollen content. The rugged basin margin along the Murre fault supplied coarse detritus and debris flows (Tankard and Welsink, 1987). Porosity and permeability in the Jeanne d'Arc sandstones and conglomerates average 10% and 24 md, respectively. Porosity is generally secondary, developed from calcite dissolution (Figure 15). The barrier to fluid flow is mainly stratigraphic and is related to the discontinuous nature of the sands and conglomerates in the small braid deltas. Delineation drilling has shown that the Jeanne d'Arc does not contain significant reserves (Meneley, 1986).

Hibernia Reservoir

The prolific Hibernia sandstones were encountered at -3727 m (-12,228 ft) in the Hibernia P-15 well. Average gross thickness of the reservoir is 198 m (650 ft) and the sandstones are fine- to coarse-grained, poorly to moderately sorted quartz arenites and quartzose sublitharenites of Berriasian to early Valanginian age (Figures 16 and 17). The Hibernia interval consists of an upward-coarsening mudstone to fine-grained siltstone facies sequence that is overlain by a series of stacked, upward-fining, erosively based sandstones (Figure 16; Table 3). Lag deposits occur above the erosive bases, the sandstones are profusely cross-bedded, and each upward-fining sequence is overlain by bioturbated or rooted mudstones. The upward-coarsening mudstone-siltstone-sandstone facies tracts are attributed to delta front progradation. A sandy braid delta paleoenvironment is inferred from the association of alluvial channels with bayfill deposits that contain trace fossils and a marine invertebrate fauna (Tankard and Welsink, 1987). The burrowing of shale drapes suggests periodic marine encroachment during abandonment of these small sandy deltas.

In the Hibernia reservoir, the alluvial sequence forms a southward-thickening wedge attributed to structural control of the depocenter by relay and transfer faults, and increased stacking of channel sequences in the more rapidly subsiding low (Figure 18; Tankard et al., 1989). A rugged rift shoulder to the west shed terrigenous sediments, generating a suite of sandy deltas (Figures 16 and 19). The thinner Hibernia sequence at Nautilus C-92 was deposited across a structural culmination associated with a relay ramp and was only separated from Hibernia by the Nautilus fault in the late Barremian to Aptian (Tankard et al., 1989, their figures 6 and 18). The Hibernia alluvial sequence is overlain by a series of shales and thin sandstones that were deposited by transgressive onlap.

Porosity averages 16% and is generally secondary intergranular formed by dissolution of carbonate cement (Figure 20). Permeabilities average 700 md. Net pay in the wells averages 39 m (128 ft), with

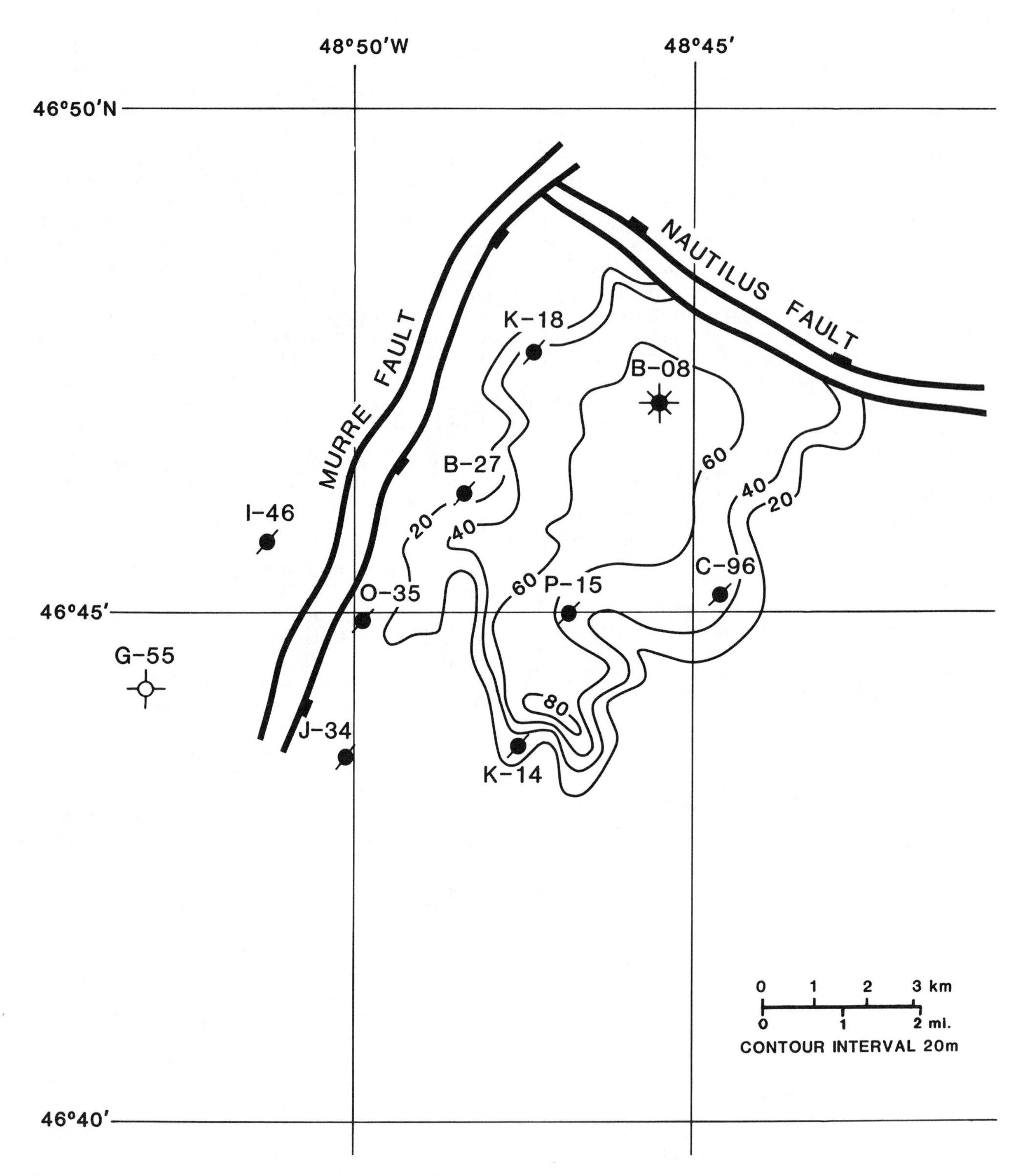

Figure 11. Hibernia sandstone net pay map. An oil-water contact is encountered to the south and east at -3934 m (-12,907 ft) and is also observed to the west as the structure dips into the Murre fault. The Nautilus fault seals the structure to the north. The thickest pay is to the south where the southward-thickening sandstone wedge is above the oil-water contact.

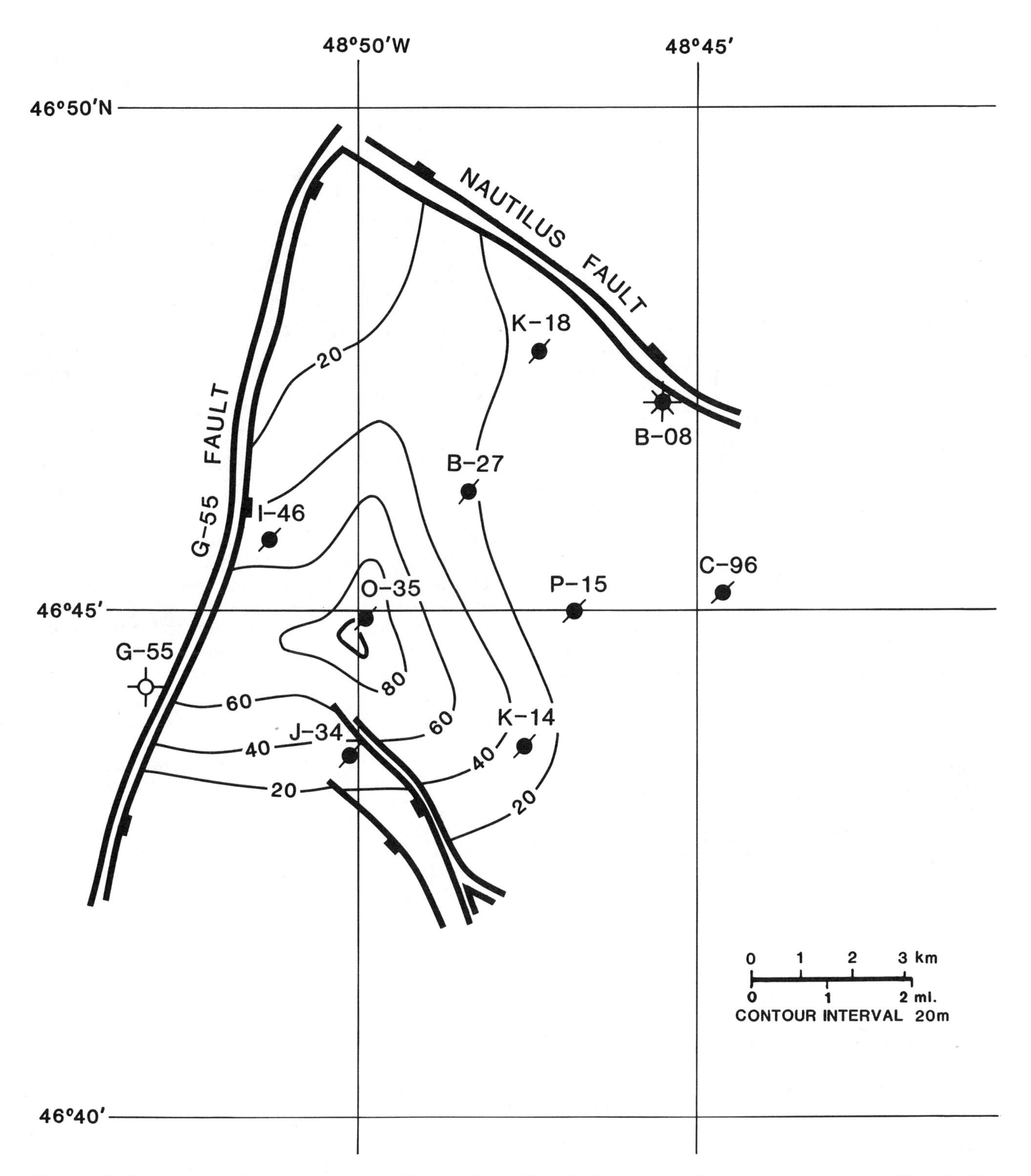

Figure 12. Avalon sandstone net pay map. The section dips below -2603 m (-8540 ft) oil-water contact to the south and east. Hydrocarbons are trapped to the north by the Nautilus fault and to the west by the G-55 fault. The thickest area of pay is centered on Hibernia O-35 where a thick section of reworked sandstones provides good reservoir.

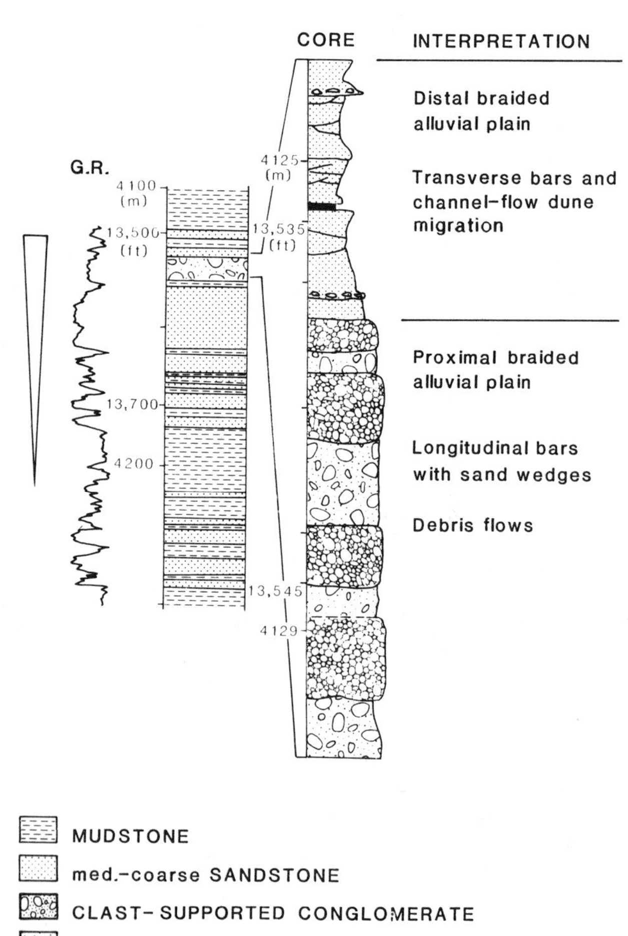

Figure 13. Jeanne d'Arc sandstone and conglomerate. Stratigraphic log of the Kimmeridgian-Portlandian Jeanne d'Arc sequence in Hibernia P-15. The overall upward-coarsening pattern resulted from basinward progradation with smaller-scale cyclicity attributed to rapid subsidence. (Modified after Tankard and Welsink, 1987.)

a maximum of 64 m (210 ft) (Figure 11). Table 4 summarizes the reservoir properties. Vertical communication results from erosive stacking of channel sandstones. Figure 18 shows the internal geometry of the Hibernia sandstones constructed on the basis of numerical models (cf. Allen, 1978) to predict sand-body geometry of a channel system (Tankard and Welsink, 1987).

An areal simulation study based on immiscible displacement indicates average recovery factors of 33%, and an average annual production rate of 17.5 × 10^3 m^3/d (110 × 10^3 BPD) oil is expected for the Hibernia field (CNOPB, 1986). The Hibernia project at present requires 58 development wells; the exact location and number will depend on progressively finer definition of reservoir characteristics as more wells are drilled.

Catalina Reservoir

The Catalina sandstones form a secondary reservoir interval at -3196 m (-10,486 ft) in the Hibernia P-15 well. These sandstones are generally very fine to fine-grained, poorly to moderately sorted sublitharenites. The Catalina sequence was deposited above the B-marker limestone during an episode of Valanginian regression. In core, the stacked upward-coarsening sequences are interpreted as bay margin, delta front, and shoreface deposits, with well-sorted quartz arenites associated with foreshore and tidal reworking (Figures 21 and 22; Table 5) (Tankard and Welsink, 1987).

Porosity in the sandstone reservoirs averages 16% and is mainly primary in nature; permeabilities average 100 md (Figure 23). Net pay averages 22 m (72 ft) with a maximum of 57 m (187 ft). Although sands in the Catalina sequence have flowed up to 574 m^3 OPD (3610 BOPD) 31° API oil with an average gas-oil ratio of 115 m^3/m^3 (646 ft^3/bbl), the zone

Table 2. Lithofacies of Jeanne d'Arc sequence, Hibernia oil field.

Facies	Description	Interpretation
Matrix-supported conglomerate	Angular clasts, nonresistant lithologies, poor sorting	Debris flows
Clast-supported conglomerate	Well-rounded cobble-size clasts, imbricated; fine-grained sandstone interbeds with ripple and ripple-drift cross-lamination	Longitudinal bar deposition with wedges of sandstone deposited during waning flow; proximal facies
Sandstone	Fine to coarse-grained, pebble lags on scour surfaces, plane-bedded and cross-bedded; shale partings; plant material	Channel-floor transverse bar deposition; distal facies
Mudstone	Mudstones and interbedded sandstones containing coal laminae, woody material, spores, pollens	Flood basin

Source: Tankard and Welsink (1987).

Figure 14. Core slab of Jeanne d'Arc alluvial facies, Hibernia P-15, illustrating the clast supported conglomeratic facies at -4128 m (-13,543 ft).

Figure 15A. Thin section photo showing Jeanne d'Arc porosity development, Hibernia P-15. Large dissolution pores, calcite cement, and quartz grains shown in thin section (-4127.8 m; -13,543 ft).

contributes little to overall field reserves. The unpredictable nature of the reservoir fluid contacts and pressure depletion in drill-stem tests indicates that hydrocarbon occurrence may be of limited extent in the Catalina zone (CNOPB, 1986). Development of the Catalina reservoir will depend on more rigorous analysis after drilling of production wells for the deeper Hibernia sandstones.

Avalon Reservoir

The Avalon sandstones form the youngest reservoirs in the field (Hauterivian–Barremian) and, like the Hibernia interval, are potentially prolific producers. They were encountered at -2191 m (-7188 ft) in the Hibernia P-15 well. The average gross thickness of the reservoir is 359 m (1178 ft) and consists of very fine to fine-grained, well-sorted, quartzose sublitharenites, subarkoses, and quartz arenites (Figures 5, 24, 25, and 26). The isopach distribution suggests that depositional thickening was controlled by hanging-wall rollover into the listric Murre fault and was only later disrupted by transfer faulting (Figure 26).

The Avalon sequence consists of a series of stacked, upward-coarsening bodies formed by progradation across the A-marker, a basin-wide, diachronous, arenaceous to argillaceous limestone (Table 6). Upward-coarsening facies of bioturbated mudstone and fine-grained sandstone dominated by dwelling burrows (*Paleophycus, Skolithos, Terebellina,* and *Ophiomorpha*) with a subordinate population of feeding burrows (*Teichichnus, Planolites, Chondrites,*

Figure 15B. SEM showing quartz grains with flat-faced overgrowths, rock fragments, and relatively open pore system (-4124.4 m; -13,535 ft).

Muensteria, Cylindrichnus, and *Thalassinoides*) indicate middle to lower shoreface paleoenvironments for the lower portion of the progradational packages (Tankard and Welsink, 1987; S. G. Pemberton, 1987, personal communication). More distal paleoenvironments are also indicated by abundant grazing (*Zoophycos*) and feeding (*Teichichnus, Chondrites, Cylindrichnus, Planolites, Asterosoma, Muensteria,* and *Rosselia*) ichnofacies (S. G. Pemberton, 1984, personal communication).

The upper parts or topset beds of each progradational package consist of channel sandstone and upward-coarsening mudstone to sandstone facies of shoreface origin (Tankard and Welsink, 1987). A low diversity assemblage of dwelling structures (vertical *Ophiomorpha, Skolithos, Terebellina,* and *Paleophycus*) reflect a predominantly foreshore suspension-feeding community (S. G. Pemberton, 1986, personal communication). These sandstones form the bulk of the Avalon reservoirs.

Each depositional unit is attributed to basinward progradation with periodic reworking of topset beds

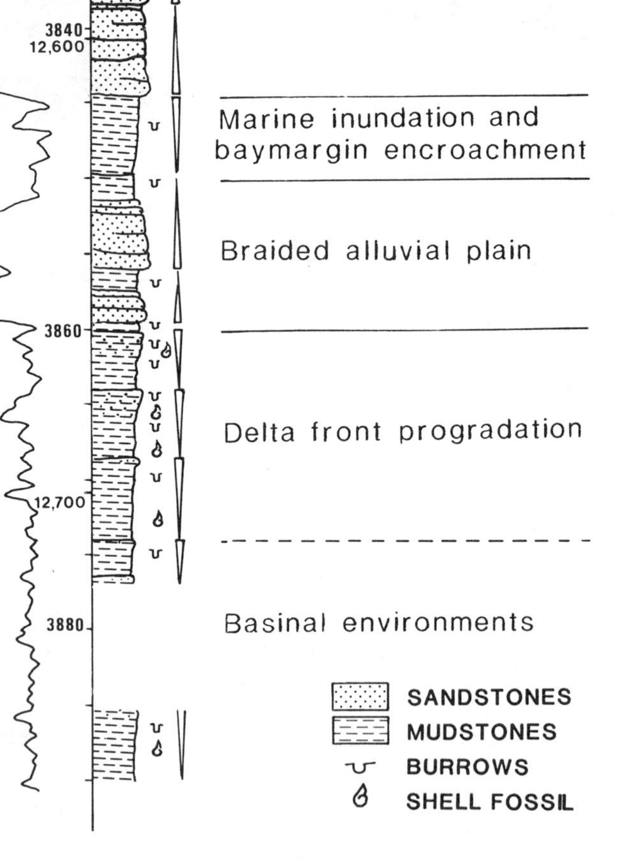

Figure 16. Hibernia sandstone. Stratigraphic log of Hibernia sequence in Hibernia K-18, built by bed-load river and flood-plain processes. (After Tankard and Welsink, 1987.)

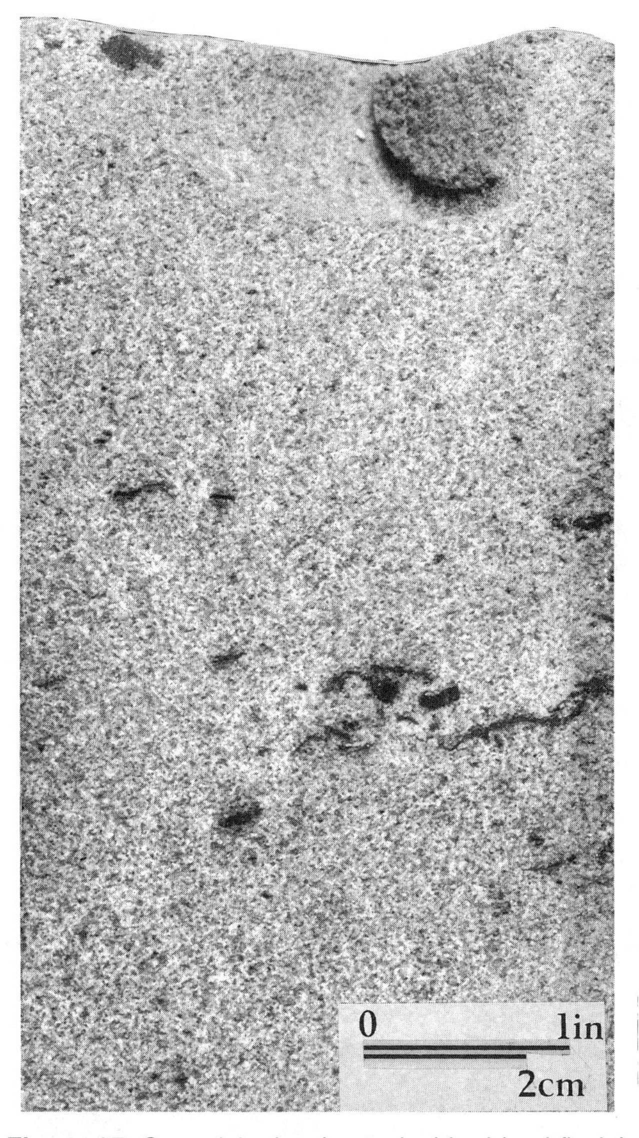

Figure 17. Core slab showing typical bed-load fluvial facies of Hibernia sequence at -3813.5 m (-12,511 ft) Hibernia K-18, generally finer grained than Jeanne d'Arc.

to form local disconformities. Basinward progradation is recorded in interbedded shoreface-foreshore deposits, dark mudstone, and red shale facies of tidal flat-lagoon origin (Table 6). Oxidizing conditions persisted above the structural culmination of the relay ramp (see Tankard et al., 1989).

Porosities and permeabilities in the Avalon sandstones average 20% and 220 md, respectively. Porosity is both primary and secondary, resulting from dissolution of ferroan calcite (Figure 27). The total net pay from wells averages 34 m (112 ft) with a maximum of 99 m (325 ft) (Figure 12). Table 7 summarizes the reservoir properties. Although sandstones in the Avalon have flowed up to 500 m^3 OPD (3150 BOPD), the reservoir is presently only considered for development as Hibernia reservoir production declines (CNOPB, 1986). A major barrier to fluid flow in the Avalon sequence is the discontinuous nature of the reservoir facies and impermeable calcite-cemented sandstones.

HYDROCARBON SOURCE

Von der Dick et al. (in press) described the source rock geology, generation, and expulsion of oil in the Hibernia field. Hydrocarbons accumulated in the Upper Jurassic–Lower Cretaceous synrift reservoirs. They exhibit homogeneous composition indicative of a local source rock and effective fault-controlled migration pathways. Geochemical well logging suggests a basin-wide occurrence of rich Kimmeridgian source beds developed in silled or semisilled depocenters (Swift and Williams, 1980; Powell, 1985; Tankard and Welsink, 1987). Sediment starvation resulted in concentration and preservation of marine organic matter (von der Dick, 1989). Increased sediment influx at the end of the Jurassic terminated source rock accumulation.

Total organic carbon values (TOC) average 3-4% but locally range up to 8%. In Hibernia K-18, the source rock interval is 190 m (623 ft) thick (4780–4970 m; 15,682–16,305 ft). The thickness of the source rock varies from more than 400 m (1312 ft) along the eastern flank of the basin to about 200 m (656 ft) in the Hibernia field and thins southward (von der Dick et al., in press).

Microscopic analysis shows that 75-90% of the organic matter in the source rock is liptinitic and oil prone. Land plant constituents are rare, suggesting low amounts of dilution by terrigenous material (von der Dick, 1989). The bulk organic fraction is dominated by dinoflagellates.

Figure 28 shows the excellent correlation, based on sterane fingerprinting, between the mature source rock (R_o 0.8%) and Hibernia oil (von der Dick et al., in press). During Tertiary subsidence, the source rock reached its present depth near 5000 m (16,400 ft). von der Dick et al. (in press) suggest that early mature conditions were reached in the Late Cretaceous. Subsidence during the Tertiary provided the thermal impetus for maturation and peak expulsion.

EXPLORATION AND DEVELOPMENT CONCEPTS

Twenty-five years ago the first well offshore Newfoundland was drilled in a pull-apart basin of the southern Grand Banks. This well and many subsequent wells focused on salt diapiric structures. In 1973, exploration activity switched to the Jeanne d'Arc half-graben; however, still in pursuit of diapiric structures. Hibernia was discovered when fault-

Table 3. Lithofacies of Hibernia sequence, Hibernia oil field.

Facies	Description	Interpretation
Channel sandstone	Fine- to medium-grained units average 3-6 m (10-20 ft) thick, erosively based, channel lag on scour surfaces, blocky to fining-upward pattern, upper contact abrupt or gradational; cross-bedded in sets up to 1 m (3 ft) (average 30 cm or 12 in.), shale drapes and woody material on bedding planes; internal erosion surfaces; shale partings (1-15 cm or 0.4-2 in.) burrowed	Sandy bed-load channels with channel-floor dune migration; fluctuating discharge and periodic incursions of salt-wedge and suspension sedimentation during low discharge
Mudstone-sandstone	Upward-coarsening sequence of mudstone, siltstone, and sandstone, 5-7 m (16-23 ft) thick; upper parts of sandstone small-scale cross-bedded; fossils include plants, burrows, and marine invertebrates	Broad embayments with bayhead delta and bay margin encroachment
Mudstone	Interlaminated mudstone, siltstone, fine-grained sandstone; 0.5-3 m (2-10 ft) thick; burrowed, plant debris, local seat earths	Brackish flood basin with vegetated margins
Dark mudstone-siltstone	Gray-black mudstone and fine-grained sandstones, gently coarsening upward, 2-6 m (7-20 ft) thick; sequences stacked, each progressively coarser; bioturbated, abundant pyritized mollusks and brachiopods	Delta-front progradation over basinal mudstones

Source: Tankard and Welsink (1987).

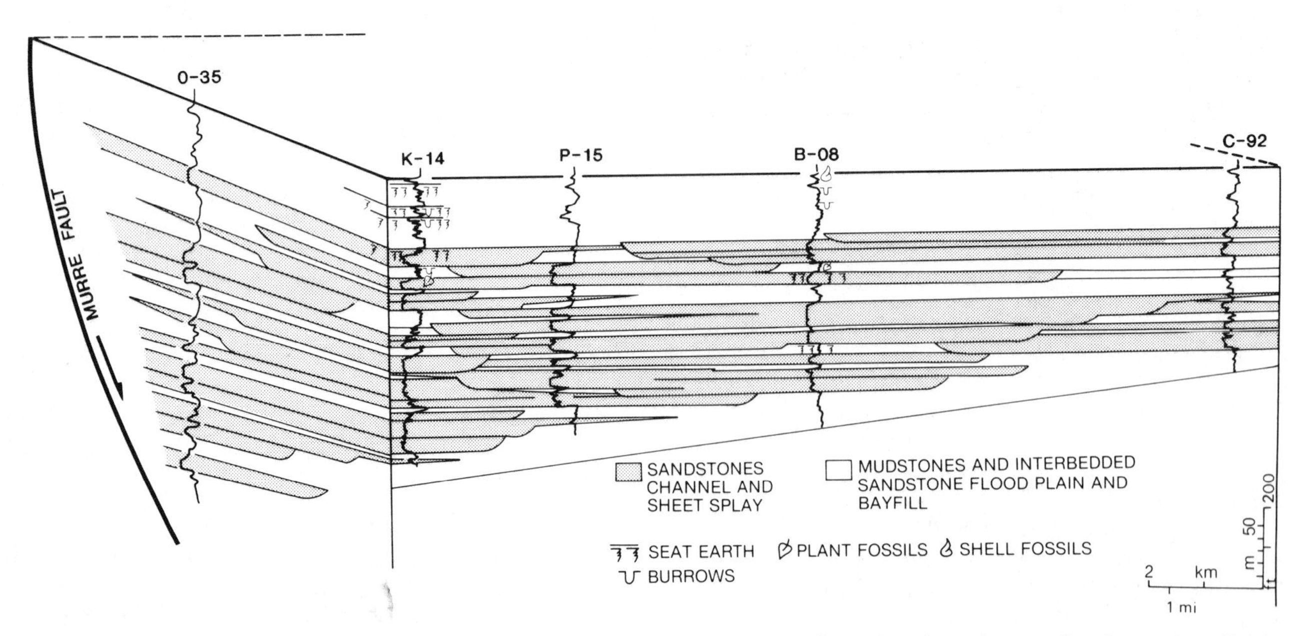

Figure 18. Stratigraphic cross section of Hibernia sequence showing interpretation of alluvial architecture based on quantitative models (cf. Allen, 1978). Wedge geometry attributed to hanging-wall rollover into listric fault.

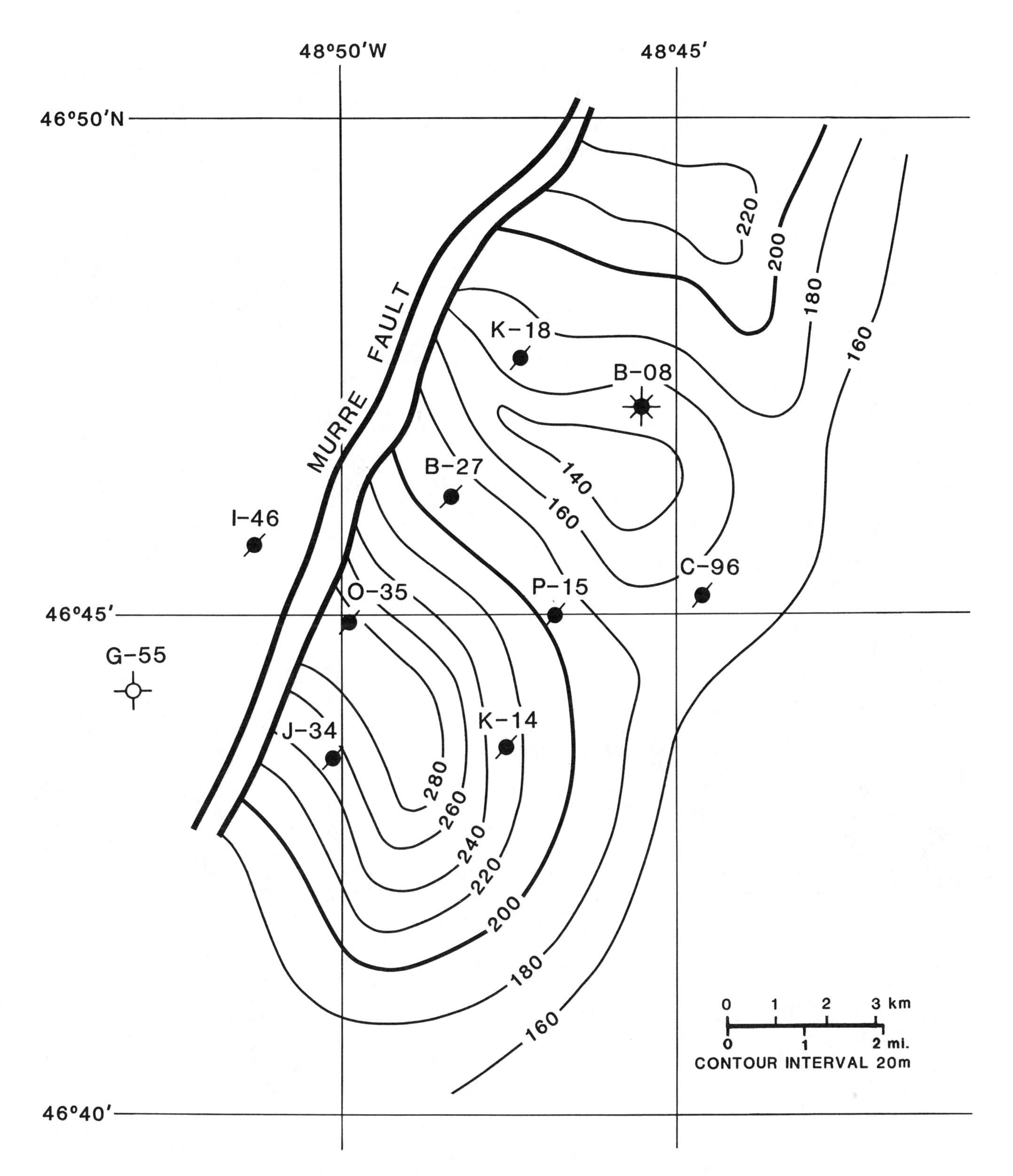

Figure 19. Hibernia sandstone gross isopach showing a southward-thickening wedge formed from increased stacking of channel sequences in the more rapidly subsiding low.

A

B

Figure 20. Thin section and SEM photos showing Hibernia porosity development, Hibernia K-18. (A) Thin section overview of moderately sorted, medium-grained quartzarenite with good porosity (-3815.4 m; -12,518 ft). (B) SEM showing good porosity reduced and modified by effects of compaction and small amounts of clay minerals. Primary porosity is further reduced by the presence of authigenic quartz overgrowths on some monocrystalline quartz grains (-3815.4 m; -12,518 ft). (Courtesy AGAT Consultants Ltd.)

Table 4. Hibernia sandstone reservoir properties.

Areal extent	6703 ha (16,563 ac)
Original oil in place	214×10^6 m^3 (1346×10^6 bbl)
API	34.6°
Gas/oil ratio	215 m^3/m^3 (1207 ft^3/bbl)
Original gas in place:	
Gas cap	18.8×10^9 m^3 (663.9×10^9 ft^3)
Solution gas	46.5×10^9 m^3 (1642.1×10^9 ft^3)
Oil reserves	71×10^6 m^3 (447×10^6 bbl)

Source: CNOPB (1986).

controlled structures became exploration targets in the Jeanne d'Arc basin. It took over 30 exploration wells before the Chevron et al Hibernia P-15 discovery well tested impressive pay zones in the Upper Jurassic and Lower Cretaceous succession. To date, 375,000 km of seismic line have been acquired, more than 100 wells drilled, and at least 2 billion barrels of recoverable reserves established in the Jeanne d'Arc basin.

Hindsight allows us to make several observations:

1. All significant discoveries have been made in the Upper Jurassic–Lower Cretaceous synrift succession of the Jeanne d'Arc basin. The pull-apart basins of the southern Grand Banks have not been revisited because uplift and erosion have breached and reduced synrift targets to small closures.
2. In the Jeanne d'Arc basin, significant discoveries of oil are generally related to intersecting normal and transfer fault trends.
3. In the Albian, the Mesozoic succession detached above basement and slumped northward up the

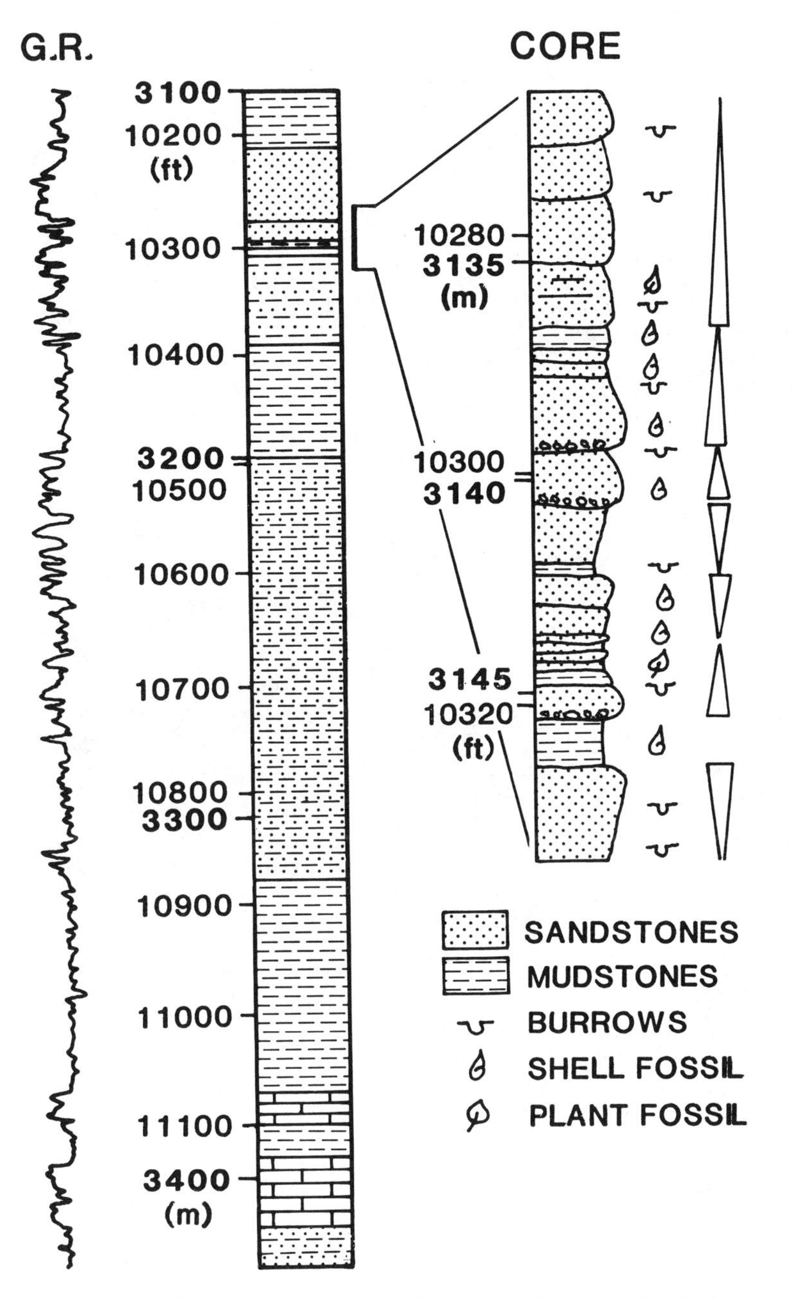

Figure 21. Catalina sandstone. Stratigraphic log of Catalina sequence in Hibernia K-18. The depositional facies are characteristic of shoreface environments dissected by tidal channels.

axis of the basin, creating the present structural complexity of Hibernia.

4. This supracrustal detachment induced strike-slip motion and probable clay smearing along the G-55 fault. The D & A Hibernia G-55 well was drilled on the wrong side (west) of this fault and encountered wet Avalon reservoir. It is interesting to speculate on how exploration would have proceeded had the G-55 well been the first Hibernia well drilled!
5. Hibernia reservoir geology reflects original depositional patterns and the way these responded to evolving structures. An example is the structural culmination that is attributed to increasing strain of a relay structure (Tankard et al., 1989, their figures 6 and 18). This culmination affected preservation (e.g., Avalon interval in the B-08 well is erosively stripped), reservoir quality (e.g., clean reworked Avalon sandstones are preserved in the O-35 well), isopachs, palynofacies trends, etc.
6. The pattern of normal faults and orthogonal transfer faults formed a compartmented basin that affected source rock accumulation.

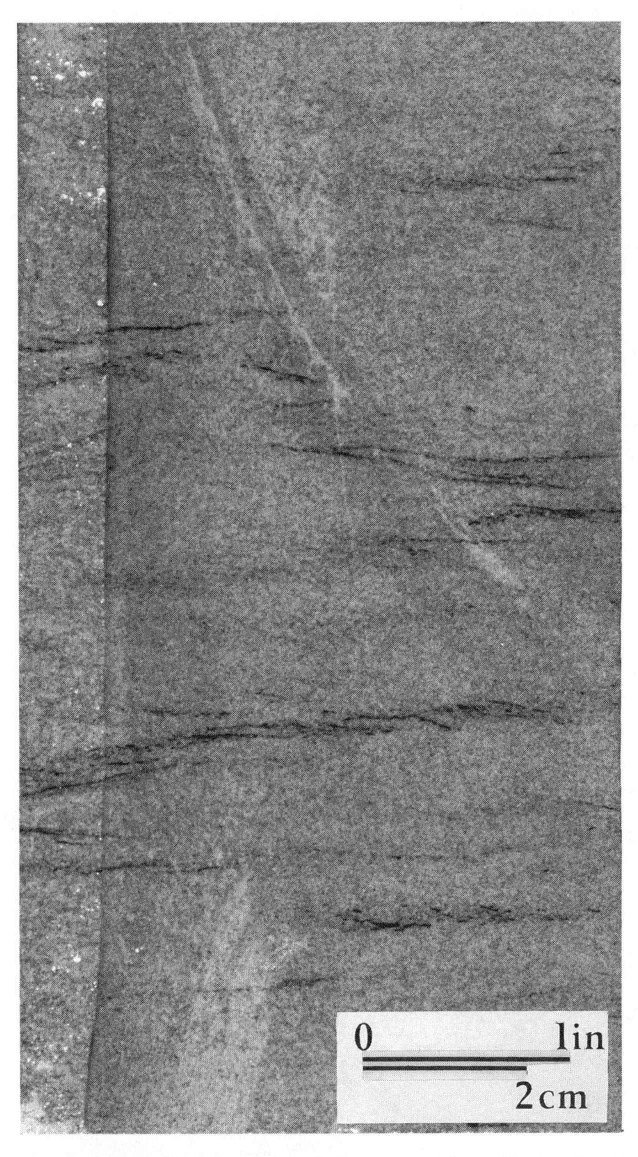

Figure 22. Core slab of Catalina sequence from -3133 m (-10,279 ft) Hibernia K-18, finer grained than the Hibernia with carbonaceous material marking some ripple laminations.

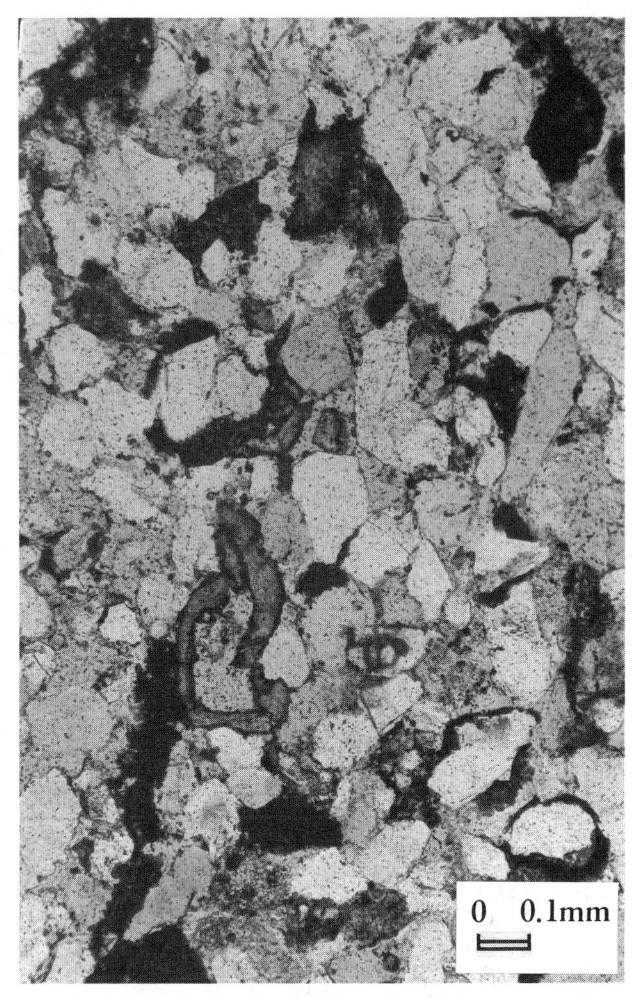

Figure 23A. Thin section photo showing Catalina porosity development, Hibernia K-18. Overview of moderately sorted, fine-grained quartzose sublitharenite with good porosity and bioclastic components (-3132.9 m; -10,279 ft).

Table 5. Lithofacies of Catalina sequence, Hibernia oil field.

Facies	Description	Interpretation
Channel sandstone	Fine- to medium-grained stacked sandstone units, cross-bedded, bioturbation, shell and coal lags	Tidal channel complex
Muddy sandstone	Fine- to medium-grained, upward-coarsening sandstone units, bioturbated, shell and plant fossils	Bay-margin-tidal flat complex

Figure 23B. SEM photo showing good intergranular porosity reduced and modified by compaction, clay minerals, irregularly distributed carbonates, and authigenic quartz (-3132.9 m; -10,279 ft). (Courtesy AGAT Consultants Ltd.).

7. Albian and younger detachment of the cover sequence resulted in *structural* deepening and maturation of the Kimmeridgian source rock.

The geology of the Jeanne d'Arc basin and Hibernia structure emphasizes the need for an integrated structural-stratigraphic approach, combined with dynamic reservoir analysis. It is also important that these studies should not be model driven. Honor the data!

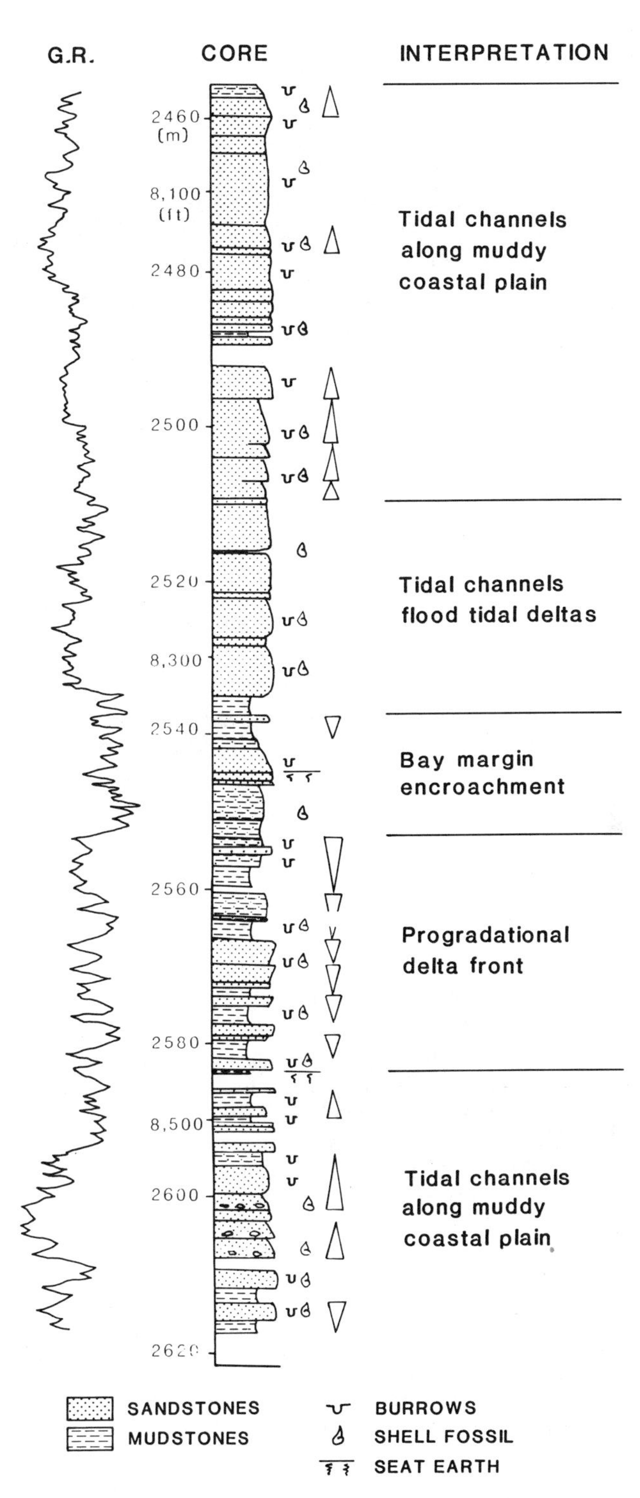

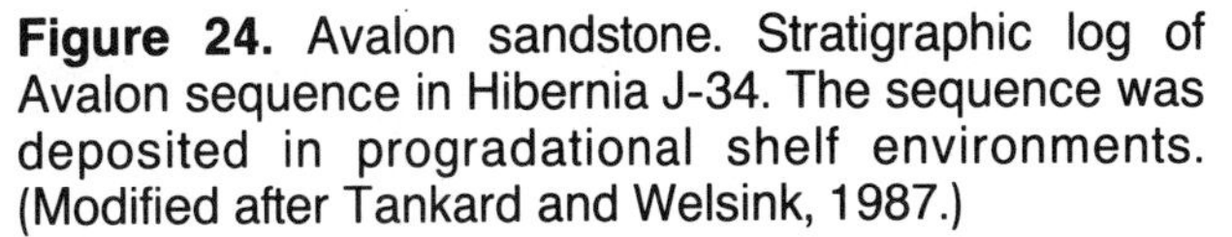

Figure 24. Avalon sandstone. Stratigraphic log of Avalon sequence in Hibernia J-34. The sequence was deposited in progradational shelf environments. (Modified after Tankard and Welsink, 1987.)

ACKNOWLEDGMENTS

We thank the staff at Petro-Canada, whose work provided the basis for this paper, as well as personnel from Mobil, Gulf, Chevron, and Columbia Gas for their contributions. We also thank E. Beaumont, E. Olynyk, L. Funkhowser, and reviewers at Chevron Canada and Mobil Canada for their comments. Typing was done by D. Holmes and O. Shearing. Finally, we thank Petro-Canada for permission to publish.

Figure 25. Core slab illustrating typical bioturbated, fossiliferous nature of the Avalon sequence at -2496 m (-8189 ft), generally fine- to very fine grained, Hibernia J-34.

REFERENCES

Allen, J. R. L., 1978, Studies in fluviatile sedimentation: an exploratory quantitative model for the architecture of avulsion-controlled alluvial sites: Sedimentary Geology, v. 21, p. 129-147.

Arthur, K. R., D. R. Cole, G. G. L. Henderson, and D. W. Kushnir, 1982, Geology of the Hibernia discovery, *in* M. T. Halbouty, ed., The deliberate search for the subtle trap: American Association of Petroleum Geologists Memoir 32, p. 181-195.

Benteau, R. I., and M. G. Sheppard, 1982, Hibernia—a petrophysical and geological review: Journal of Canadian Petroleum Technology, v. 21, p. 59-72.

Canada-Newfoundland Offshore Petroleum Board, 1986, Application for approval of Hibernia development plan—Decision 86.01: Canada-Newfoundland Offshore Petroleum Board, 143 pp.

Jansa, L. F., and J. A. Wade, 1975, Geology of the continental margin of Nova Scotia and Newfoundland: Geological Survey of Canada Paper 74-30, p. 51-105.

Leavitt, G. M., 1987, Canada's east coast offshore exploration and development: The Journal of Canadian Petroleum Technology, v. 26, p. 60-68.

Meneley, R. A., 1986, Oil and gas fields in the East Coast and Arctic basins of Canada, *in* M. T. Halbouty, ed., Future petroleum provinces of the world: American Association of Petroleum Geologists Memoir 40, p. 143-176.

Mobil Oil Canada, 1986, Hibernia development plan overview, prepared on behalf of, and in cooperation with, the joint venture participants: Gulf Canada Resources Incorporated, Petro-Canada Incorporated, Chevron Canada Resources Limited, and Columbia Gas Development of Canada Limited.

Powell, T. G., 1985, Paleogeographic implications for the distribution of Upper Jurassic source beds: offshore eastern Canada: Bulletin of Canadian Petroleum Geology, v. 33, p. 116-119.

St. John, B., A. W. Bally, and H. D. Klemme, 1984, Sedimentary provinces of the world, hydrocarbon productive and nonproductive: American Association of Petroleum Geologists Map Series, 35 p.

Swift, J. H., and J. A. Williams, 1980, Petroleum source rocks, Grand Banks area, *in* A. D. Miall, ed., Facts and principles of world petroleum occurrences: Canadian Society of Petroleum Geologists Memoir 6, p. 567-587.

Tankard, A. J., and H. J. Welsink, 1987, Extensional tectonics and stratigraphy of the Hibernia oil field, Grand Banks, Newfoundland: American Association of Petroleum Geologists Bulletin, v. 71, p. 1210-1232.

Tankard, A. J., and H. J. Welsink, 1989, Mesozoic extension and styles of basin formation in Atlantic Canada, *in* A. J. Tankard and H. R. Balkwill, eds., Extensional tectonics and stratigraphy of the North Atlantic margins: American Association of Petroleum Geologists Memoir 46, p. 175-196.

Tankard, A. J., H. J. Welsink, and W. A. M. Jenkins, 1989, Structure and stratigraphy of the Jeanne d'Arc basin, offshore Newfoundland: analysis of basin subsidence, *in* A. J. Tankard and H. R. Balkwill, eds., Extensional tectonics and stratigraphy of the North Atlantic margins: American Association of Petroleum Geologists Memoir 46, p. 265-282.

von der Dick, H., 1989, Environment of petroleum source rock deposition in the Jeanne d'Arc basin off Newfoundland, *in* A. J. Tankard and H. R. Balkwell, eds., Extensional tectonics and stratigraphy of the North Atlantic margins: American Association of Petroleum Geologists Memoir 46, p. 295-304.

von der Dick, H., J. D. Meloche, and F. J. Longstaffe, in press, Generation, expulsion and fault controlled migration of hydrocarbons in the Hibernia oil field, Grand Banks of Newfoundland: American Association of Petroleum Geologists Bulletin.

Welsink, H. J., S. P. Srivastava, and A. J. Tankard, 1989, Tectonic linkage between continental crust of the Grand Banks and oceanic crust during extension, *in* A. J. Tankard and H. R. Balkwill, eds., Extensional tectonics and stratigraphy of the North Atlantic margins: American Association of Petroleum Geologists Memoir, 46, p. 197-214.

Williams, G. L., L. R. Fyffe, R. J. Wardie, S. P. Colman-Sadd, R. C. Boehner, and J. A. Walt, eds., 1985, Lexicon of Canadian stratigraphy, Volume VI, Atlantic region: Canadian Society of Petroleum Geologists.

Williams, H., 1984, Miogeoclines and suspect terranes of the Caledonian-Appalachian orogen: tectonic patterns in the North Atlantic region: Canadian Journal of Earth Sciences, v. 21, p. 887-901.

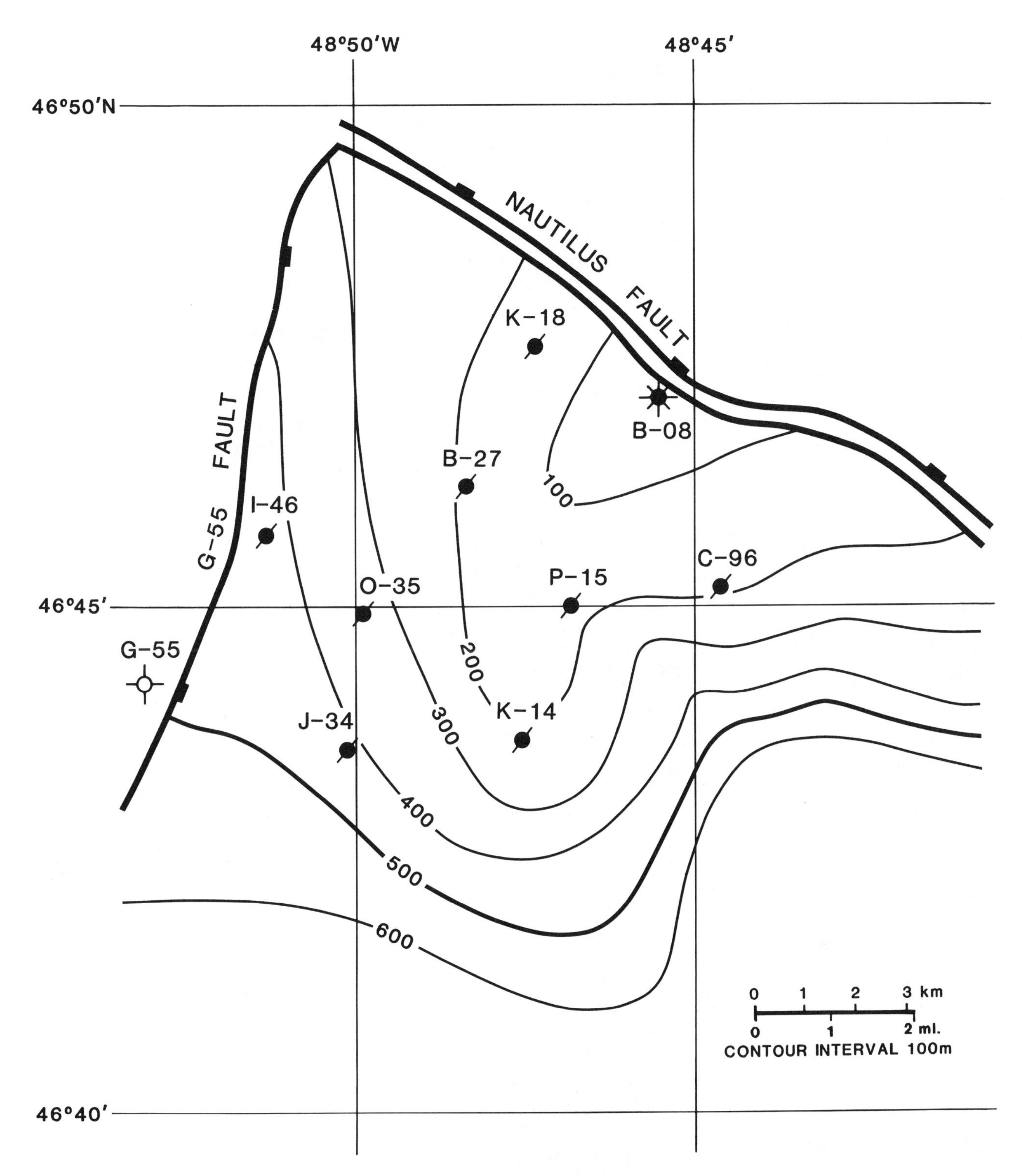

Figure 26. Avalon sandstone gross isopach showing a southwestward-thickening wedge. The Avalon sequence consists of interbedded mudstones and quartzose sandstones that were deposited in shelf, shoreface, and shore-zone environments.

Table 6. Lithofacies of Avalon sequence, Hibernia oil field.

Facies	Description	Interpretation
Channel sandstone	Fine- to medium-grained, quartzose, well sorted; upward-fining channels average 2.5 m (8 ft), shale clasts, coal spar, and woody material on scour surfaces, cross-beds 10-50 cm (4-20 in.), burrowed (*Ophiomorpha, Skolithos*), shelly fauna abundant (oysters, bivalves, gastropods), tops of some units rooted, coal laminae	Tidal channels draining shallow embayments Thicker sandstones (6-8 m; 20-26 ft) interpreted as flood tidal deltas, associated with lagoonal shales
Upward-coarsening mudstone-sandstone	Upward-coarsening sequences of silty mudstone and fine- to medium-grained sandstone 1-8 m (3-26 ft) thick; sandstone well-sorted, plane bedded, locally coquina-capped, burrowed (*Ophiomorpha, Planolites, Teichichnus, Paleophycos*)	Shoal-water sand bars with local reworking over crests; shoreface; water becoming more agitated upward
Silty sandstone	Very fine to fine-grained muddy and silty sandstones, lower contact abrupt or gradational, average 2 m (7 ft) thick comprising amalgamated units; horizontal bedding, rare cross-beds, hummocky cross-stratification, flasers, ripples, mudclast breccia, and bioclastic material on scour surfaces; bioturbated (*Skolithos, Chondrites, Planolites*), plant debris	Shelf-platform environment, deposition by isolated events such as storms; fluctuating energy conditions; sandbody amalgamation formed sheets
Silty mudstone	Gray mudstones, silty, average 2 m (7 ft) thick; lenticular bedding, flasers, ripples, horizontal bedding, load casts and deformation structures; burrowed, shell content increases upward (bivalves, oysters, gastropods), plant debris	Shelf-platform environment, fair-weather deposition; alternates with previous facies
Dark mudstone	Gray-black mudstones, average 1 m (3 ft) thick, bioturbated, shell fauna (*Turritella*, oyster), abundant plant debris and stems, coalified and pyritized, rooting of upper surface	Lagoonal or bay environments, periodic emergence, fringing marshes
Red shale	Red-brown shales and mudstones, green mottles, nodular, massive, black shell debris, seat earths	Tidal flat marsh-seat earth oxidized in intertidal zone

Source: Tankard and Welsink (1987).

Appendix 1. Field Description

Field name .... *Hibernia field*

Ultimate recoverable reserves .... *83 × 10⁶ m³ (522 × 10⁶ bbl) oil*

Field location:

- **Country** .... *Canada*
- **State** .... *Newfoundland*
- **Basin/Province** .... *Jeanne d'Arc basin Grand Banks*

Figure 27. Thin section and SEM photos showing Avalon porosity development, Hibernia J-34. (A) Very well sorted, fine-grained, quartzose sublitharenite in thin section with porosity from ferroan calcite dissolution (-2495 m; -8186 ft). (B) SEM showing porosity development (-2494 m; -8182 ft).

Field discovery:

Year first pay discovered .... *Upper Cretaceous Jeanne d'Arc, Lower Cretaceous Hibernia, Catalina, and Avalon formations 1979*

Year second pay discovered

Third pay

Discovery well name and general location:

First pay .... *Chevron et al Hibernia P-15 lat. 46°44′58.98″N, long. 48°46′51.18″W*

Second pay

Third pay

Discovery well operator .... *Chevron Canada Resources Ltd.*

Second pay

Third pay

IP in barrels per day and/or cubic feet or cubic meters per day:

First pay .... *Annual oil production rate of 17.5 × 10³ m³/d (110,000 bbl/d) is expected*

Second pay

Third pay

Table 7. Avalon sandstone reservoir properties.

Areal extent	1788 ha (4418 ac)
Original oil in place	104×10^6 m^3 (654×10^6 bbl)
API	31.8°
Gas/oil ratio	105 m^3/m^3 (590 ft^3/bbl)
Original gas in place:	
Solution gas	9.1×10^9 m^3 (321×10^9 ft^3)
Oil reserves	12×10^6 m^3 (75×10^6 bbl)

Source: CNOPB (1986).

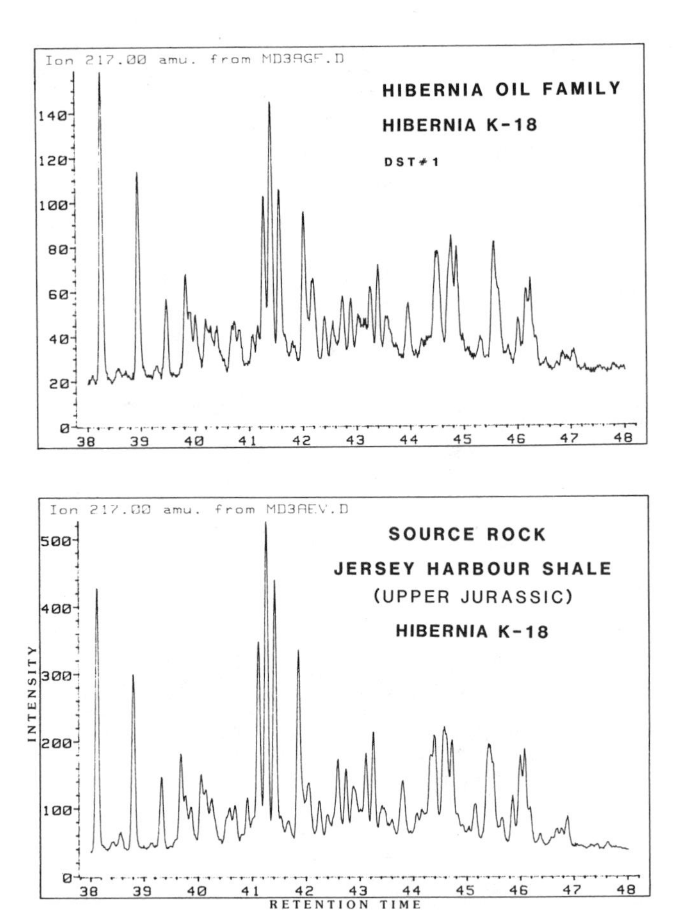

Figure 28. Oil-source rock correlation based on sterane finger-printing of Hibernia oils and fully mature source rock (after von der Dick et al., in press) .

All other zones with shows of oil and gas in the field:

Age	Formation	Type of Show
Jurassic	*Jeanne d'Arc*	*Oil (DST)*
Cretaceous	*Catalina*	*Oil (DST)*

Geologic concept leading to discovery and method or methods used to delineate prospect, e.g., surface geology, subsurface geology, seeps, magnetic data, gravity data, seismic data, seismic refraction, nontechnical:

(1) Exploration drilling had shown the occurrence of porous sandstones and oil shows in the Jeanne d'Arc basin. (2) Reflection seismic interpretation indicated a large structural closure. (3) Structure was successfully drilled.

Structure:

Province/basin type .... *Half-graben of Atlantic-type margin; Bally III, Klemme IIIA*

Tectonic history

The Grand Banks off Newfoundland was dissected into several rift basins by Mesozoic extensions and ocean opening. Hydrocarbon accumulation was associated with Upper Jurassic and Lower Cretaceous synrift sediments in the Jeanne d'Arc basin.

Regional structure

The Hibernia structure was formed by rollover into the listric normal basin-bounding fault and accentuated by salt diapirism. Smaller scale faulting of the Hibernia field is related kinematically to strike-slip motion of cross-basin transfer faults.

Local structure

Local structure is created by synrift subsidence and rollover into the listric Murre fault. The field is dissected by Riedel shears and tension gashes as a result of right-lateral shear couple on transfer faults.

Trap:

Trap type(s)

The Hibernia field is essentially a rollover structural trap with updip seals provided by the Murre fault (west) and Nautilus fault (north).

Basin stratigraphy (major stratigraphic intervals from surface to deepest penetration in field):

Chronostratigraphy	Formation	Depth to Top in m
Paleocene-Pleistocene	*Banqereau (P-15)*	
Coniacian-Maastrichtian	*Dawson Canyon (P-15)*	*-1451*
Turonian	*Petrel (P-15)*	*-1914*
Haut-Barr-Aptian-Albian	*Avalon (P-15)*	*-2191*
Barremian	*A-Marker (P-15)*	*-2489*
Valanginian-Hauterivian	*Catalina (P-15)*	*-3196*
Valanginian	*B-Marker (P-15)*	*-3383*
Berriasian-Valanginian	*Hibernia (P-15)*	*-3727*
Kimmeridgian-Tithonian	*Jeanne d'Arc (P-15)*	*-4105*
Kimmeridgian	*Egret (K-18)*	*-4382*

Location of well in field

Reservoir characteristics:

Number of reservoirs *2*
Formations .... *Avalon, Hibernia*
Ages .... *Avalon (Hauterivian-Albian); Hibernia (Berriasian-Valanginian)*
Depths to tops of reservoirs .... *Avalon, -2191 m; Hibernia, -3727 m (P-15 discovery well)*
Gross thickness (top to bottom of producing interval) .... *Avalon: max 547 m, avg 359 m; Hibernia: max 289 m, avg 198 m*

Net thickness—total thickness of producing zones

Average .... *Avalon, 34 m; Hibernia, 39 m*
Maximum .... *Avalon, 99 m; Hibernia, 64 m*
Average
Maximum

Lithology

Avalon sandstones are generally fine to very fine grained, well-sorted, quartzose sublitharenites, subarkoses and quartz arenites; Hibernia sandstones are generally fine- to coarse-grained, poorly to moderately sorted, quartz arenites, or quartzose sublitharenites

Porosity type

Hibernia Φ: avg 16% K: avg = 700 md; secondary intergranular porosity from dissolution of carbonate cements

Avalon Φ: avg 20% K: avg = 220 md; both primary and secondary from dissolution of ferroan calcite

Average porosity .... *Hibernia 16%; Avalon 20%*
Average permeability .... *Hibernia 700 md; Avalon 220 md*

Seals:

Upper

Formation, fault, or other feature .... *Transgressive shale*
Lithology

Lateral

Formation, fault, or other feature
Lithology

Source:

- **Formation and age** .... *Egret Formation Kimmeridgian*
- **Lithology** .... *Shale*
- **Average total organic carbon (TOC)** *3%*
- **Maximum TOC** *4%*
- **Kerogen type (I, II, or III)** .... *II*
- **Vitrinite reflectance (maturation)** .... $R_o = 1.0\text{–}1.1$
- **Time of hydrocarbon expulsion** .... *Late Cretaceous to Paleocene*
- **Present depth to top of source** .... *4900 m (K-18)*
- **Thickness** .... *300 m*
- **Potential yield**

Appendix 2. Production Data

Field name .... *Hibernia field*

Field size:

- **Proved acres** .... *Avalon, 1788 ha; Hibernia, 6703 ha*
- **Number of wells all years** .... *When under full production, 83 wells*
- **Current number of wells** *10*
- **Well spacing**
- **Ultimate recoverable** .... 83×10^6 m^3 *(*522×10^6 *bbl) oil*
- **Cumulative production**
- **Annual production** .... 17.5×10^3 m^3/d *(110,000 bbl/d) oil*
- **Present decline rate**
  - **Initial decline rate**
  - **Overall decline rate**
- **Annual water production**
- **In place, total reserves**
- **In place, per acre-foot**
- **Primary recovery**
- **Secondary recovery**
- **Enhanced recovery**
- **Cumulative water production**

Drilling and casing practices:

- **Amount of surface casing set** .... *141.1 m (P-15)*
- **Casing program**
  762 mm at 141.1 m; 508 mm at 345.6 m; 340 mm at 1077.4 m; 244 mm at 2897 m; 177 mm at 4113 m (P-15)
- **Drilling mud** .... *Gel chemical; gel seawater; Chrom lignosulfonate; KCl polymer*
- **Bit program**
- **High pressure zones** .... *Upper Jurassic Jeanne d'Arc Formation*

Completion practices:

- **Interval(s) perforated** ... *Wells were tested in Avalon and Hibernia by perforating casing, also in Catalina and Jeanne d'Arc if required*
- **Well treatment**

Formation evaluation:

- **Logging suites**
  (1) Combination dual induction-spherically focused log and long space sonic-gamma ray-caliper
  (2) Combination dual laterolog-microspherically focused log and synthetic microlog-gamma ray
  (3) Compensated neutron-formation density-gamma ray-caliper

(4) Four-arm dipmeter and direction survey
Repeat formation tester and various engineering logs were run as required.

Testing practices
Wells were usually tested in Avalon and Hibernia by perforated casing; Catalina and Jeanne d'Arc were also tested if hydrocarbons were indicated

Mud logging techniques
Mud density, temperature, and resistivity were monitored, in addition to using hydrocarbon detectors, cutting gas detectors, and chromatographs

Oil characteristics:

Type .... *Paraffinic*
API gravity .... *Hibernia, 34.6°; Catalina, 31°; Avalon, 31.8°*
Base
Initial GOR .... *Hibernia, 215 m^3/m^3; Catalina, 115 m^3/m^3; Avalon, 105 m^3/m^3*
Sulfur, wt% .... *Hibernia, 0.29%; Avalon, 0.89%*
Viscosity, SUS .... *Hibernia, 850 kg/m^3; Catalina, 865 kg/m^3; Avalon, 869 kg/m^3*
Pour point .... *Hibernia, +6°C; Avalon, +12°C*
Gas-oil distillate .... *Light and middle distillate*

Field characteristics:

Average elevation .... *Avalon, -2250 m; Catalina, -3100 m; Hibernia, -3750 m*
Initial pressure .... *Avalon, 24,800 kPa; Catalina, 32,000 kPa; Hibernia, 40,400 kPa*
Present pressure
Pressure gradient .... *Avalon, 11.02 kPa/m; Catalina, 10.32 kPa/m; Hibernia, 10.77 kPa/m*
Temperature .... *Avalon, 65°C; Catalina, 78°C; Hibernia, 95°C*
Geothermal gradient .... *0.023°C/m*
Drive .... *Water/miscible gas?*
Oil column thickness .... *2000 m*
Oil-water contact .... *Avalon O-W-2603 m; Hibernia O-W -3934 m, G-O -3544 m*
Connate water .... *Hibernia, SW = 16%; Avalon, SW = 26%; Catalina, SW = 22%*
Water salinity, TDS .... *Avalon, 60,000 ppm NaCl; Hibernia, 150,000 ppm NaCl*
Resistivity of water .... *Avalon, $R_w = 0.11$ at 25°C; Hibernia, $R_w = 0.055$ at 25°C*
Bulk volume water (%) *Hibernia, $\Phi \times SW = 0.16 \times 0.16 = 2.6\%$; Avalon, $\Phi \times SW$ $0.20 \times 0.26 = 5.2\%$; Catalina, $\Phi \times SW = 0.16 \times 0.22 = 3.5\%$*

Transportation method and market for oil and gas:
Gravity base structure producing oil that would be loaded on a tanker

Unity Field—Sudan
Muglad Rift Basin, Upper Nile Province

NORMAN R. GIEDT
Chevron Overseas Petroleum, Inc.
San Ramon, California

FIELD CLASSIFICATION

BASIN: Muglad
BASIN TYPE: Rift
RESERVOIR ROCK TYPE: Sandstone
RESERVOIR ENVIRONMENT OF DEPOSITION: Fluvial, Marginal Lacustrine, and Lacustrine Deltaic
RESERVOIR AGE: Cretaceous
PETROLEUM TYPE: Oil
TRAP TYPE: Faulted Anticline

LOCATION

Unity field is located in the Upper Nile province of south-central Sudan, on the northeast flank of the intracratonic Muglad rift basin (Figure 1). The northwest-southeast-trending Muglad basin complex is over 750 km (465 mi) long and in excess of 150 km (95 mi) wide. Khartoum, the capital of Sudan, is approximately 725 km (450 mi) to the northeast. The nearest port facility is Port Sudan on the Red Sea, approximately 1425 km (885 mi) to the northeast of the field. Twenty-five kilometers (15 mi) to the southeast of Unity is the village of Bentiu; it lies on a tributary of the White Nile River, which is an important transportation link for exploration operations in the Sudan interior.

Surface elevations in the field area average 1300 ft (400 m) above sea level. Nonmarine Cretaceous reservoirs underlie surficial Cenozoic deposits as shown on the geologic map of Sudan (Figure 2). Precambrian to Cambrian basement crops out at the surface about 100 km (60 mi) to the northeast of the field in the Nubian Mountains and over 150 km (90 mi) to the southwest. Shallow basement, less than 1 km (3300 ft) in depth, extends into parts of the Chevron Sudan Agreement area.

Other discoveries made by Chevron Overseas Petroleum in the Sudan interior include Heglig, approximately 68 km (42 mi) to the northwest of Unity; Kaikang, 60 km (37 mi) to the west; Abu Gabra and Sharaf, approximately 300 km (185 mi) to the northwest; and Adar-Yale, 370 km (230 mi) east of Unity in the Melut basin (Figure 2). Unity and Heglig fields have the greatest commercial potential; Unity is the larger and contains approximately 150 million bbl of recoverable oil of an estimated 600 million bbl original oil in place.

HISTORY

Southern Sudan was essentially unexplored for hydrocarbons prior to 1974, and no surface indications of oil or gas were known. The area had been considered to be a shallow intracratonic sag containing Tertiary and Cretaceous sediments overlying possible shallow basement. Chevron explorationists postulated that deep sedimentary basins related to Central and East Africa rifting might exist in the area, and further investigations were begun (Schull, 1988). The original concession agreement between the Democratic Republic of Sudan and Chevron Overseas Petroleum, signed in 1974, covered 516,000 km^2 (127,505,700 acres). It was changed to a Production Sharing Agreement in 1975, and, after a series of relinquishments and additions, the Chevron Sudan Agreement Area now totals 228,900 km^2 (56,562,109 acres). Royal Dutch Shell acquired a 25% interest in the entire concession in 1983 and relinquished that interest back to Chevron in late 1988.

The first well in the Unity area, Unity-1, was drilled to a depth of 14,483 ft (4417 m) in 1978 3 km (1.8 mi) north of the present limits of the field (Figure 3). It was the second well drilled by Chevron in the Sudan interior and it encountered numerous oil shows between 7500 and 13,000 ft (2290 and 3970 m) in thin and relatively impermeable Upper Cretaceous nonmarine sandstones. A test of one of the zones recovered 8 bbl of 38.2° API oil with a 10–12% wax content and a pour point of 80°F.

The discovery well, Unity-2, completed 28 April 1980 by Parker Rig No. 119, was the twelfth well to be drilled by Chevron in the Agreement Area. It was located on a seismically defined four-way dip closure 15 km (9.3 mi) southeast of Unity-1 on the

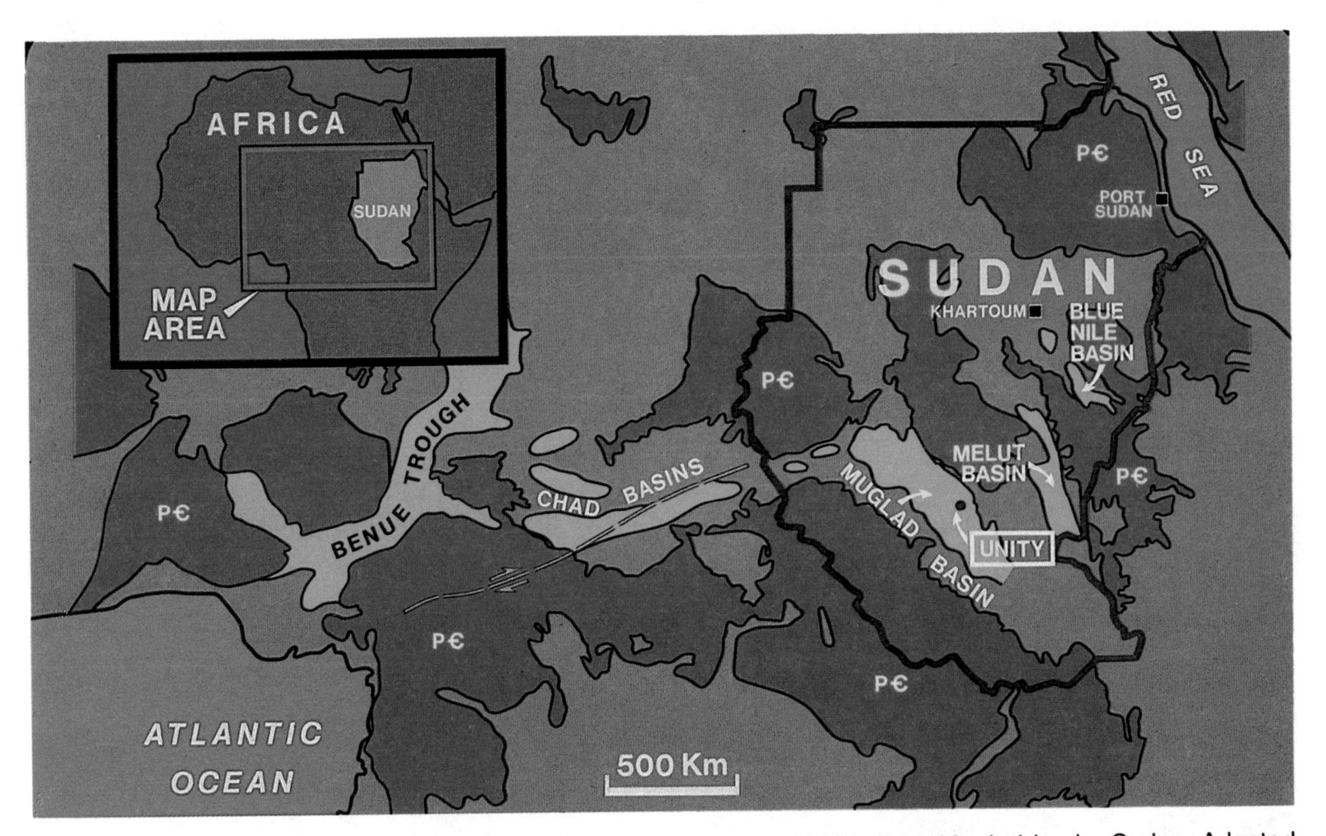

Figure 1. Generalized map of Central Africa showing Central Africa rift system, associated rift basins, and location of Unity field, Muglad basin, Sudan. Adapted from Browne and Fairhead (1983) and Fairhead (1988).

same structural trend. Drilled to 13,112 ft (4000 m), the well penetrated ten major reservoir intervals in a nonmarine sandstone-claystone sequence of Upper Cretaceous (Senonian to Cenomanian-upper Albian) age, and tested oil from six zones between 5900 and 8100 ft (1798 and 2469 m) at a composite flow rate of over 8000 BOPD. Individual sandstone intervals tested at rates ranging from 144 BOPD to 2939 BOPD of 31 to 38° API oil.

DISCOVERY METHOD

Initial studies of southern Sudan in 1974 consisted of LANDSAT and photogeologic analyses that revealed major regional lineaments, interpreted as faults bounding individual down-dropped blocks. An aeromagnetic survey conducted in 1975 identified the Muglad, Melut, and Blue Nile basins, each of which is tectonically divided into a number of subbasins. Maximum depth to basement in the Muglad basin in this early exploration period was estimated to be in excess of 22,000 ft (6700 m). With further geophysical data, depth to basement is now estimated to exceed 45,000 ft (13,700 m) in some areas.

A reconnaissance helicopter gravity survey was conducted in late 1975. The results of that survey further defined the configuration of the basins and the regional structural trends as interpreted from aeromagnetics. Observed linear gravity gradients indicated that, as suggested by early work, most of the defined anomalies were fault bounded. The Bouguer gravity map of the Muglad basin was used to guide the location of seismic programs and acquisition of gravity at shot points along seismic lines.

Between 1976 and 1980, conventional 2D seismic surveys in the Unity area outlined the structural trend on which the initial well and the discovery well were drilled. Follow-up 2D surveys and a 2500 km (1550 mi) 3D seismic survey in 1983 over the main Unity field area were conducted to further define the complex structure for the planned development of the field. The 3D survey covered a 9 by 14 km (5.6 by 8.7 mi) area, with 48 lines in the southwest-northeast direction (25 m [82 ft] depth point spacing) and 13 lines in a northwest-southeast direction (50 m [165 ft] depth point spacing). To date, 24 wells have been drilled by Chevron Overseas Petroleum in the Unity area, of which 18 are capable of production. Additional drilling will be required to define the limits of the field; however, exploration and development programs have been suspended since early 1984 following the onset of civil strife in the country.

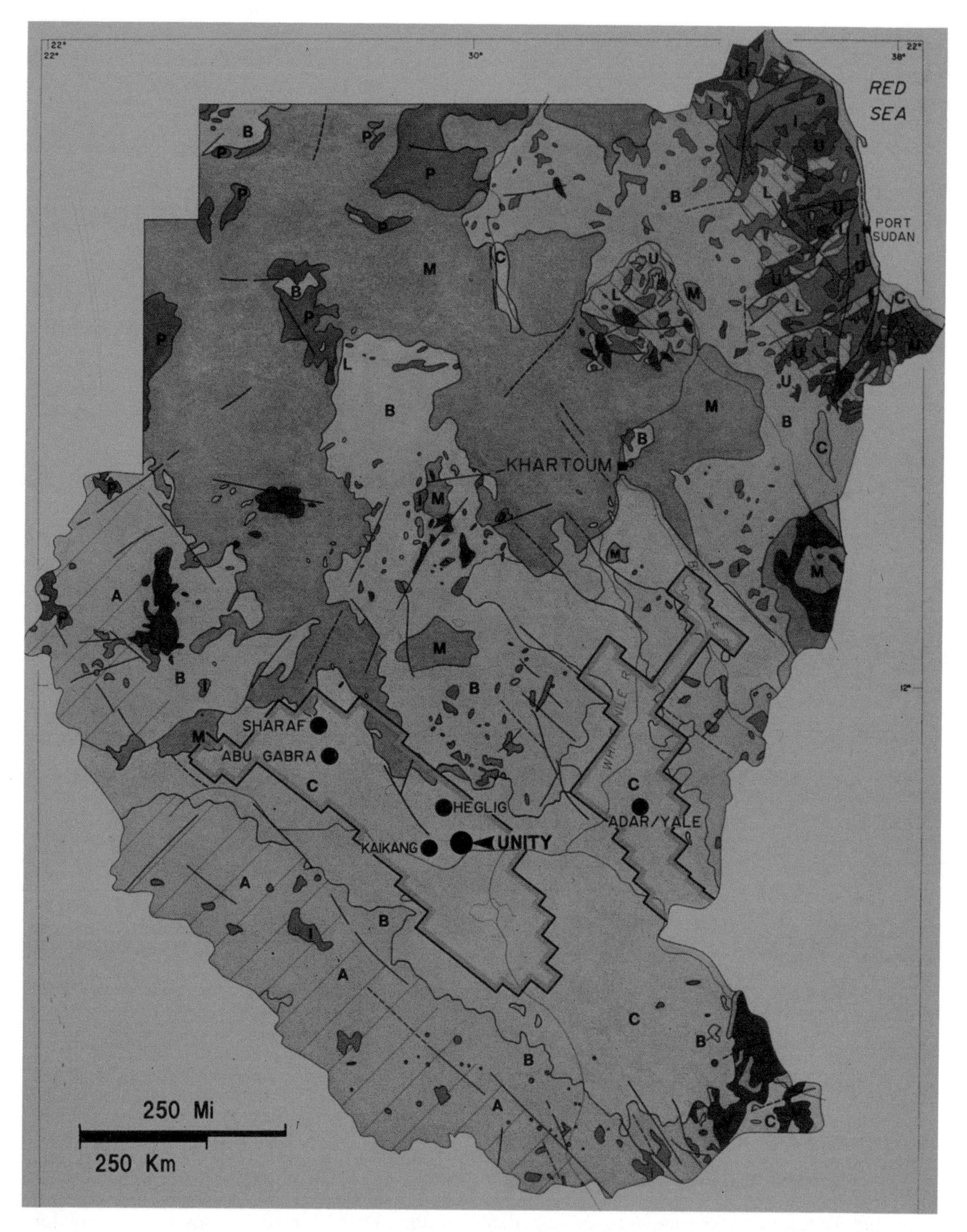

Figure 2. Generalized geologic map of Sudan with outlines of the Chevron Sudan Production Sharing Agreement area, the location of Unity field, other oil discoveries in the Muglad basin, and the Adar-Yale field in the Melut basin to the northeast.

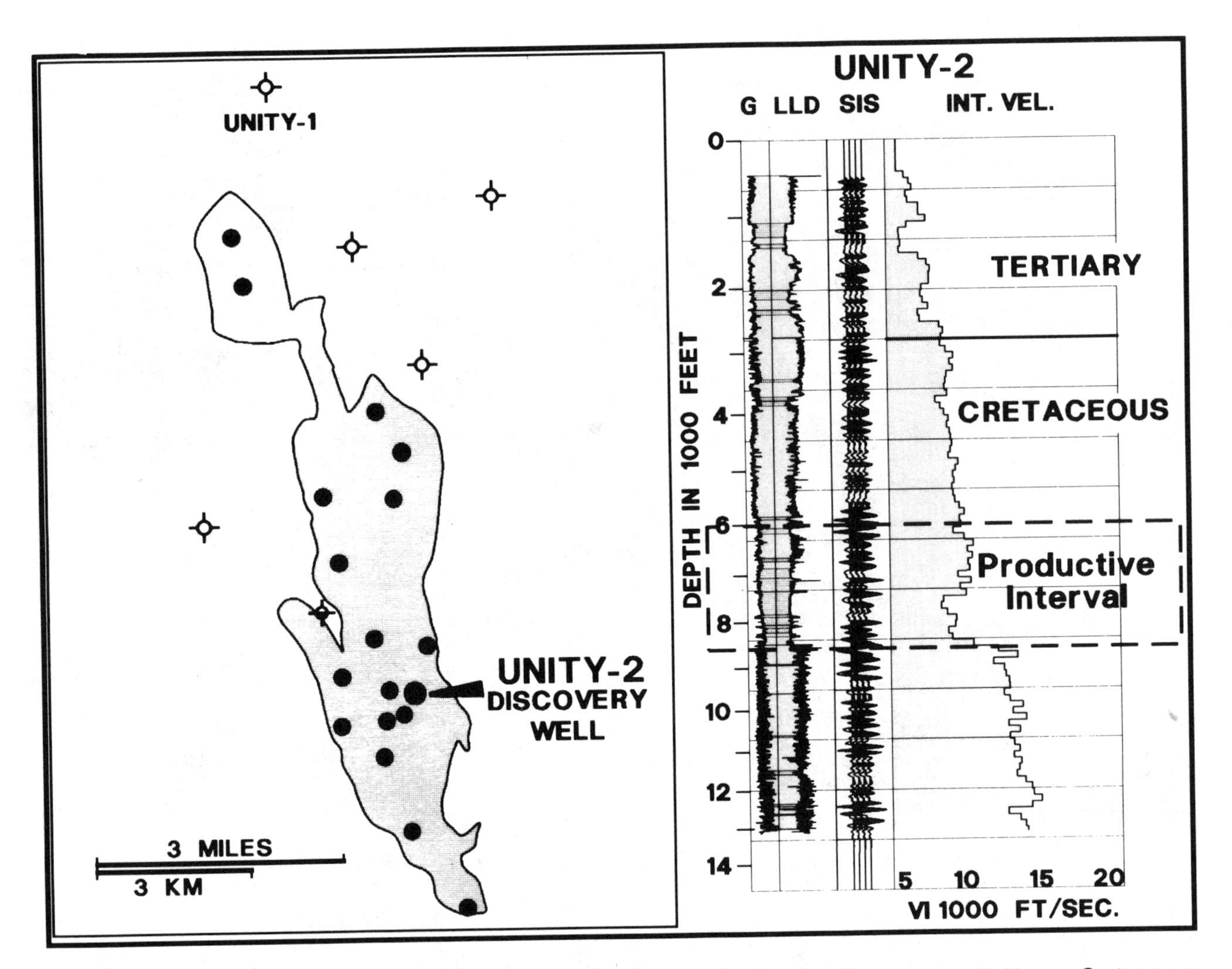

Figure 3. Map showing composite productive area of Unity field, with E log, synthetic seismogram, and interval velocity of Unity-2 discovery well. Upper Cretaceous productive interval is outlined.

REGIONAL STRUCTURE

The Muglad basin is one in a series of Cretaceous-Tertiary failed rifts that trend across the Central African craton from the Benue trough in Nigeria through Chad and the Central African Republic into Sudan (Figure 1). The right-lateral movement on the Central Africa rift system (Fairhead, 1988) is interpreted to have translated to northeast-southwest extension in Sudan, beginning by Barremian-Neocomian time, resulting in basins with an overall northwest-southeast trend that is nearly perpendicular to the shear zone. Rifting continued into the Tertiary, and seismic data indicate that approximately 35 km (22 mi) of tectonic extension have taken place in the Muglad basin. The Cretaceous structural grain of the central and southern basin including the Unity area trends north-northwest, whereas the strong Tertiary overprint of the deep basin trends more northwesterly. The northwestern part of the Muglad basin and the adjacent westernmost Bagarra basin are on trend with the northeast to easterly orientation of the south-central Chad basins along the Central African rift.

Unity field lies over a south-southeast "plunging" gravity anomaly on the northeast flank of the Muglad basin, shown on the Bouguer gravity map (Figure 4). The field is separated from shallow Precambrian to Cambrian basement to the northeast by a southeasterly plunging regional syncline, dating from the Upper Cretaceous, illustrated on structure section A-A′ (Figure 5). Based on geohistory reconstructions, initial hydrocarbons at Unity are thought to have been localized in a relatively simple anticline. Subsequent faulting complicated the structure with numerous listric and associated

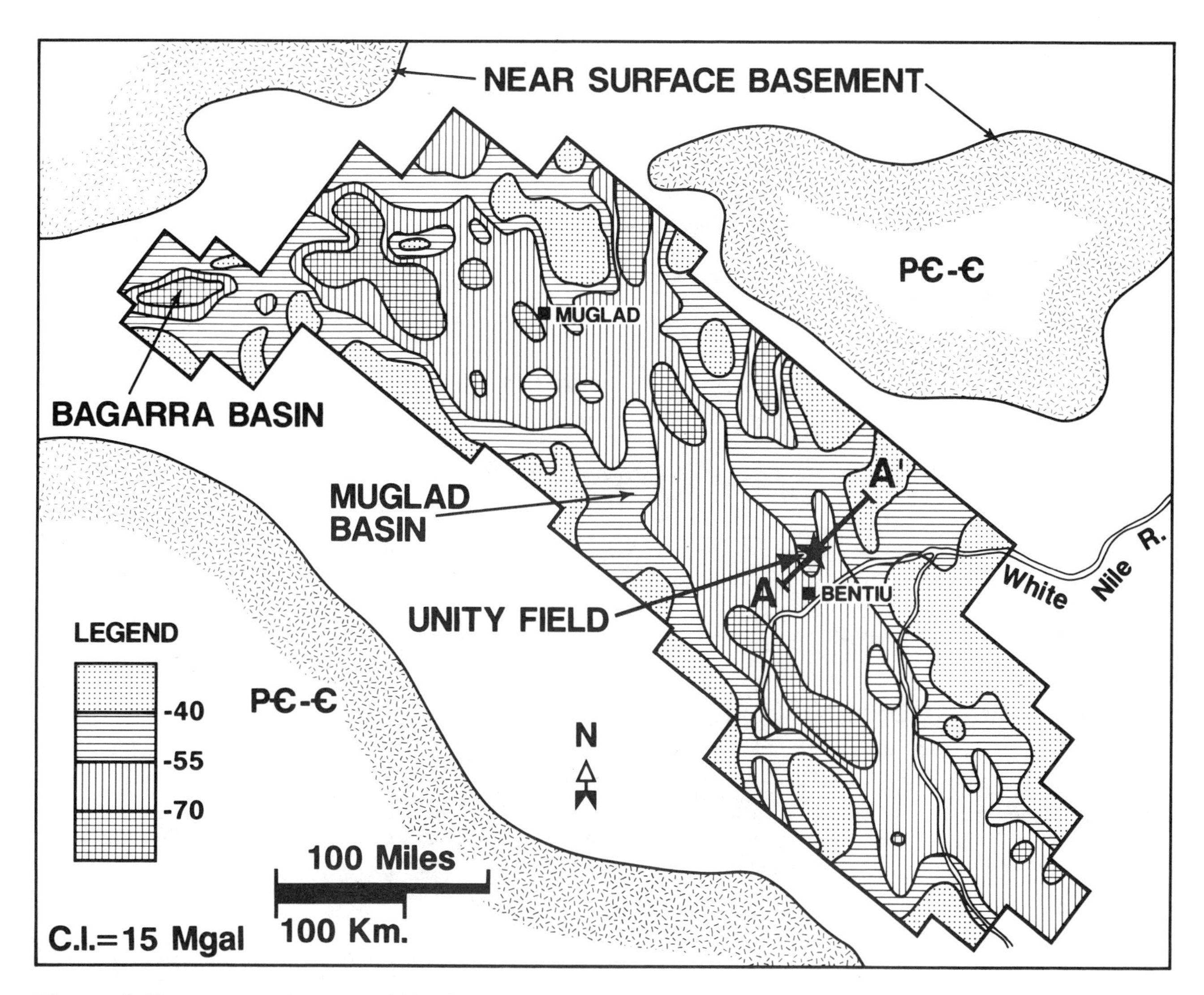

Figure 4. Bouguer gravity map of Muglad basin with location of Unity field.

antithetic faults, and additional migration and remigration from earlier traps probably occurred. Growth faults demonstrate deposition contemporaneous with faulting, particularly in the early Senonian section. Down-to-the-basin faults west of Unity field extend upward into the Tertiary, and displacements vary from under 200 ft to over 2000 ft. The result is a series of down-faulted rotated fault blocks that step down progressively toward the basin depocenter.

STRATIGRAPHY

The deepest part of the Muglad basin lies to the west of Unity field, as indicated on the Bouguer gravity map (Figure 4). It is estimated to contain over 15,000 ft (4572 m) of Tertiary and 20,000 to 30,000 ft (6100 to 9150 m) of Cretaceous nonmarine sediments derived from the surrounding basement complex. At Unity the total sedimentary section averages just over 20,000 ft (6100 m) thick, including a much thinner Tertiary section of 3000 to 4000 ft (915 to 1220 m). Generalized stratigraphic columns contrast the Unity section with the deeper basinal area to the west (Figure 6) and depict three rift cycles of sedimentation in the Muglad basin, each with a general coarsening-upward sequence of sediments (Schull, 1988). The ages of the cycles are: (1) Late Jurassic(?) to Cenomanian, (2) Turonian to Paleocene, and (3) early Tertiary.

The first rift cycle contains primarily lacustrine shales and claystones of the Sharaf and Abu Gabra formations in the deeper parts of the basin. These lacustrine deposits are overlain by the Bentiu Sandstone, which is the only formation of the first

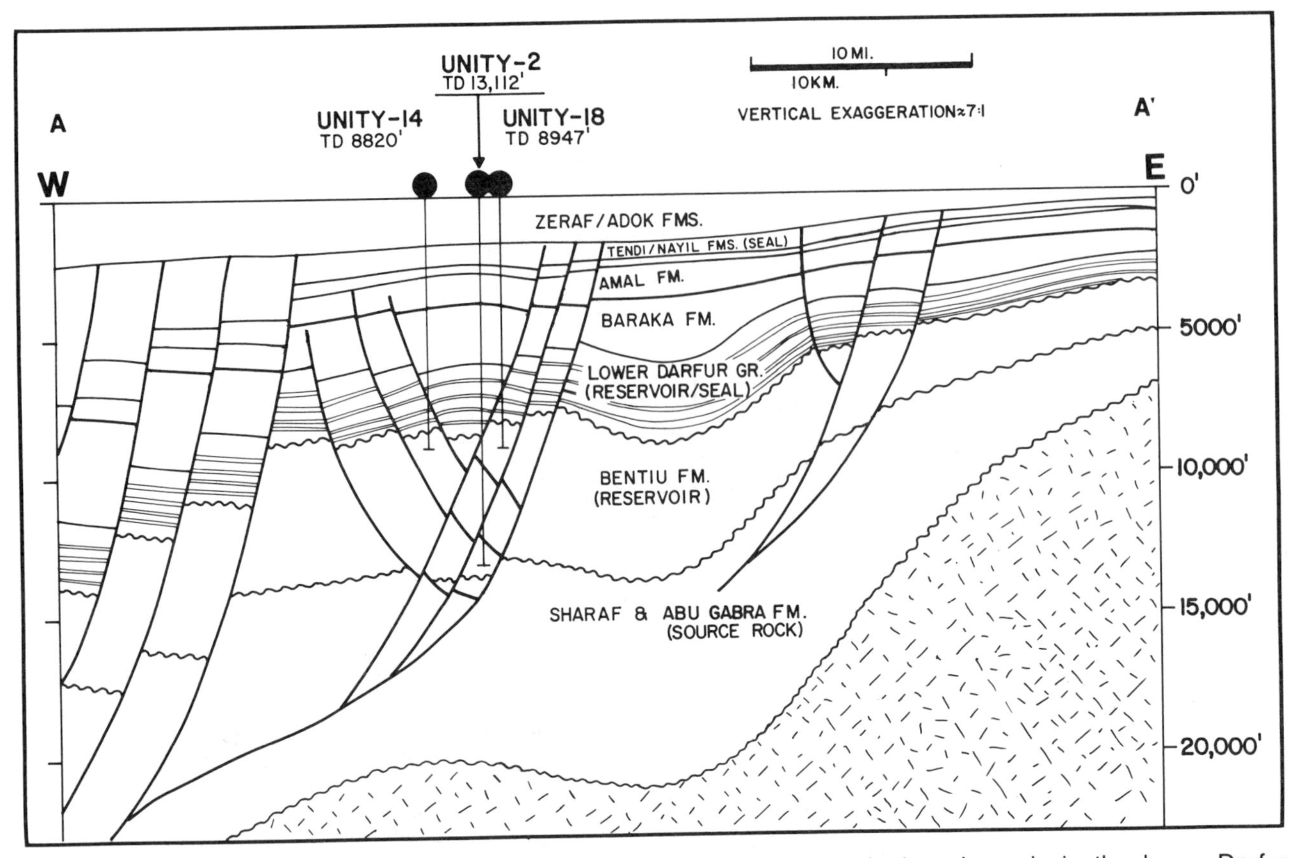

Figure 5. Schematic cross section A–A′, Unity field area (after S. E. Roe, Chevron, 1988). Productive reservoirs are the upper Bentiu Sandstone (upper Albian–Cenomanian) and sands in the lower Darfur Group (Senonian–Turonian). Refer to Figure 4 for location of profile.

rift cycle that has been completely penetrated at Unity, where it is about 4850 ft (1370 m) thick. It consists mostly of stacked channel and bar deposits, as suggested by the high sand-to-shale ratio characteristic of braided and meandering streams. Some of the finer grained Bentiu Sandstone intervals are calcite cemented in their upper parts, probably indicating periods of soil development. The Abu Gabra Formation contains the source rocks of the Muglad Basin; however, no rich source rocks have yet been encountered in limited penetration of this interval in the Unity area. It contains thin productive sands at the Abu Gabra and Sharaf fields in the northwestern part of the basin.

The second rift cycle began with deposition of the Darfur Group, a coarsening-upward interval consisting primarily of interbedded marginal lacustrine and fluvial-deltaic sandstones and claystones, 4850 ft (1370 m) thick at Unity. It contains the most important reservoirs in the field (the Aradeiba, Zarqa, and Ghazal formations) and includes all of the hydrocarbon seals. No significant organic content indicating source potential has been encountered in this unit to date. It is overlain by about 1000 ft (305 m) of the Paleocene Amal sandstone, representing the last phase of the second rift cycle. Despite the excellent reservoir characteristics of the Amal, it has not been found to contain oil.

The third rift cycle initiated deposition of the thick Eocene Nayil shale that grades upward into an increasingly sandy interval of the upper Nayil and younger Tertiary units. Although all of the rift cycles thin eastward toward Unity field, the Tertiary cycle thins dramatically, from up to 15,000 ft (4575 m) in the deep basin to an average of 2700 ft (825 m) in Unity field. No hydrocarbons have been found in the upper rift section at Unity.

Sediment types in the Unity section also include intraformational conglomerates, pebbly sandstones, and massive sandstones. Sedimentary structures are dominated by cross-bedding, ripple laminations, and parallel horizontal laminations. There is rare evidence of burrowing, boring, and other biological activity.

TRAP

Unity field oils are trapped in 11 major reservoirs in 77 separate oil accumulations. Trapping is in dip reversal closures and in upthrown rotated fault

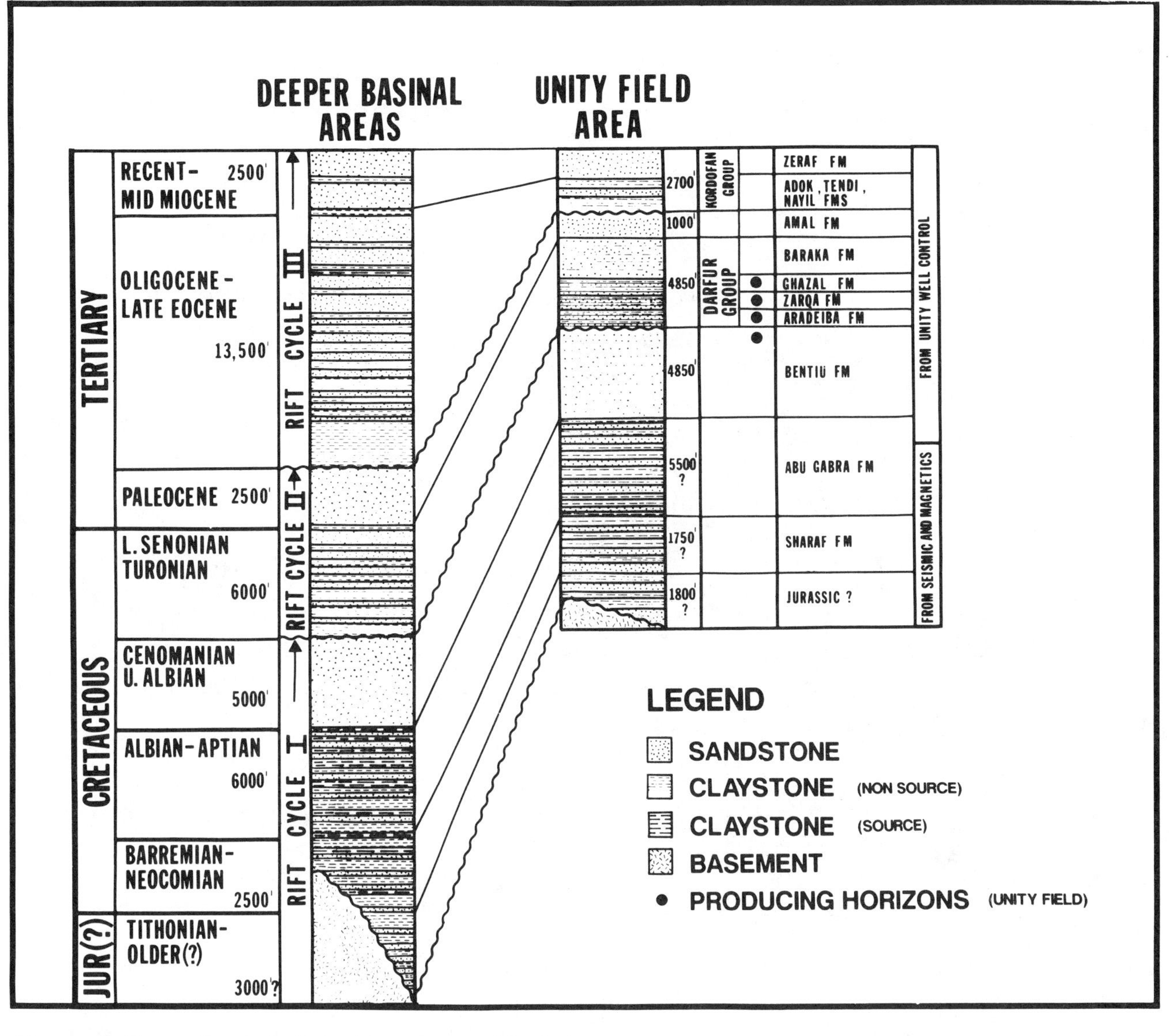

Figure 6. Generalized stratigraphic columns, Muglad basin and Unity field.

blocks between 5900 and 8900 ft (1800 and 2715 m) (Figure 7). They are sealed vertically by overlying shales and laterally by shales juxtaposed against reservoirs across block-bounding faults. Individual zones in the field have been tested at rates varying from about 150 to over 6000 BOPD.

The 10 reservoir sandstones of the Upper Cretaceous Ghazal, Zarqa, and Aradeiba formations range in thickness from approximately 20 ft (6 m) to over 200 ft (60 m), and they are generally continuous over the productive area of the field. There is some variation in the shaliness and thickness of the individual sand units, and some thin sands are not present in all of the wells. Interbedded claystones and shales range from 70 to 500 ft (21 to 150 m) thick. The underlying massive Bentiu Sandstone is productive only in its upper part.

A contour map on the top of the Aradeiba "A" reservoir, typical of Unity field structure, illustrates the distribution of the eight oil accumulations defined for this generally high productivity reservoir (Figure 8). The closures are separated from each other by basinward-dipping normal faults and associated antithetic faults dividing the field into the Main Unity, the northern Talih area, and the East Unity area. A major northwest-southeast-trending down-to-the-basin fault, with displacements ranging from 400 to about 1000 ft (122 to about 305 m), separates the East Unity area from the Talih and the Main Unity areas to the southwest.

The east-west 3D seismic section 2A, through the Unity-2 discovery, shows the down-to-the-basin normal faults to the northeast and the easterly dipping antithetic faults to the southwest (Figure 9).

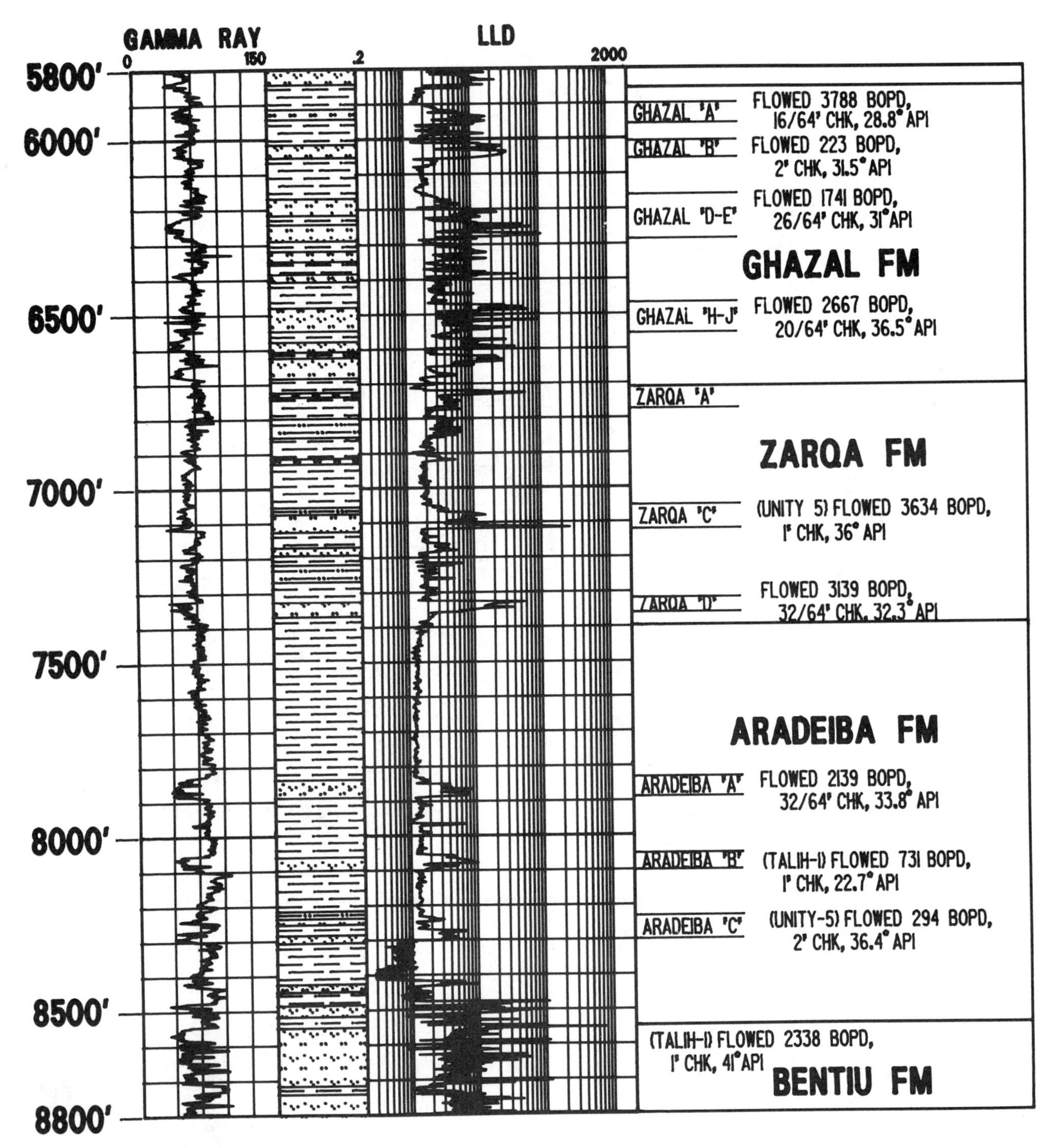

Figure 7. E log and lithologic log of productive interval, Unity-2 discovery well, with representative flow rates of individual Unity field reservoirs.

Indicated on the section are the productive horizons in the Unity-2 well in the Main Unity area, and in the Unity-14 well in a rotated fault block. Structure section B-B′ (Figure 10) shows the stratigraphic interval of the reservoirs and the structure of the central graben anticline in more detail. Typical dips of fault planes range from about 45° between 4500 and 5000 ft (1370 and 1525 m), to 25 to 30° between 6500 and 7500 ft (1980 and 2290 m). Oil-water contacts of individual reservoirs vary between fault blocks, ranging from -4689 to -7652 ft (-1430 to -2334 m) subsea in the greater Unity area. Seismic section 5 (Figure 11) illustrates the tilted fault block structure of the northern Unity area, and the upthrown block of East Unity where the Unity 8 and 12 wells produce from the Aradeiba and Bentiu

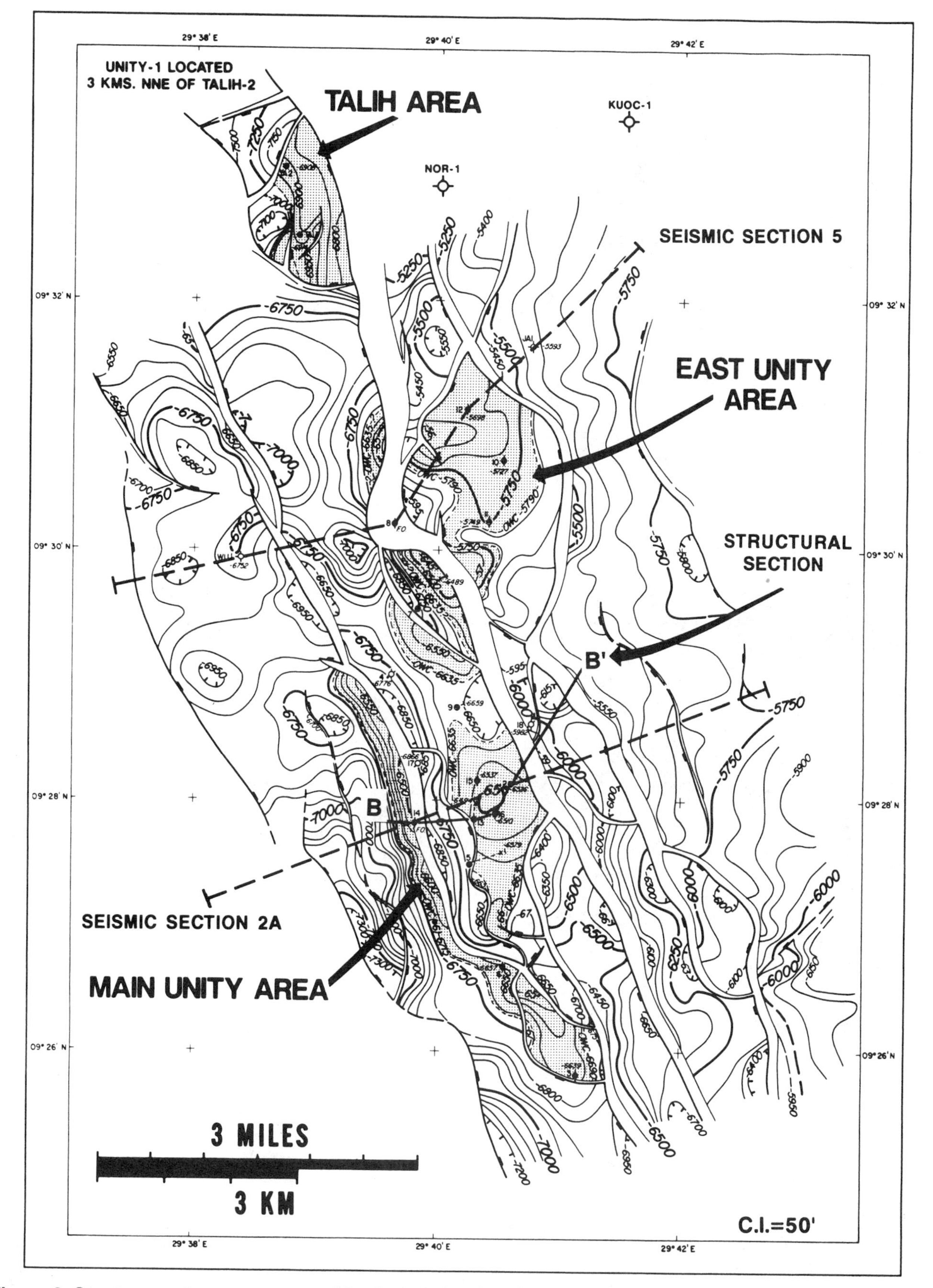

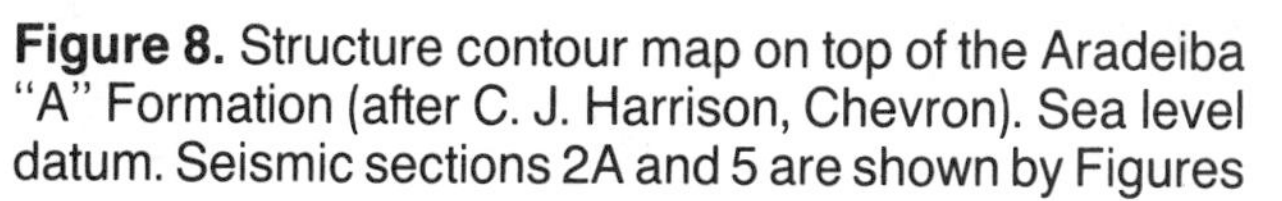

Figure 8. Structure contour map on top of the Aradeiba "A" Formation (after C. J. Harrison, Chevron). Sea level datum. Seismic sections 2A and 5 are shown by Figures 9 and 11, respectively. Structural section B–B′ is shown on Figure 10.

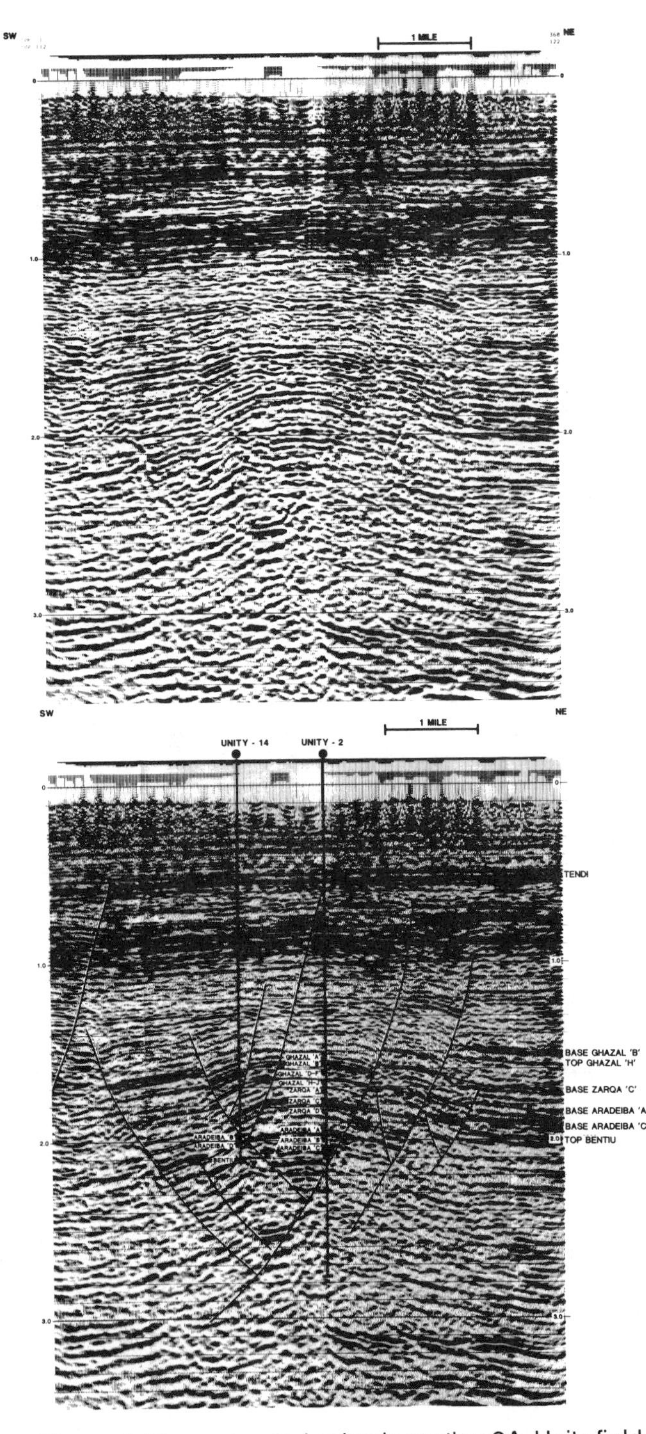

Figure 9. Time migrated seismic section 2A, Unity field. Times shown are two-way travel time in seconds. See Figure 8 for location of profile.

formations. Approximately 4500 BOPD of 30 to 33° API oil was tested from these formations in the East Unity-10 well, drilled in 1982 5 km (3 mi) north-northeast of Unity-2. The northernmost Unity area discovery to date is Talih-1. It tested at a composite rate of 8900 BOPD of 29 to 45° API oil from the Bentiu and six Zarqa zones.

Productive areas in Unity range from 4067 acres (1647 ha) for the Aradeiba Formation to 1128 acres (457 ha) for the Bentiu Formation. The composite productive area for all zones in the field, previously shown in Figure 3, totals approximately 7000 acres (2835 ha). Exploration potential remains on both flanks of the field and on trend to the northwest and to the southeast.

RESERVOIR CHARACTERISTICS

Most of the Unity reservoir sandstones generally are characterized by good porosity and permeability. Porosities range from 8 to 38%, averaging 27%. Permeabilities range from 40 to over 10,000 md, averaging 1600 md. Generally, the sandstones are very fine to medium-grained, moderately well-sorted, subarkosic arenites—quartz and feldspar being the most abundant minerals. Kaolinite is the most common authigenic clay, but minor amounts of smectite, chlorite, and illite also are present. Above 8000 ft (2440 m), sandstones show generally similar high porosity and permeability (Figure 12). Average values for both porosity and permeability show a steady decrease with increasing depth, reflecting largely a progressive change from friable to quartz overgrowth to pressure solution textures. However, depositional environment is an important factor in controlling porosity; some poor reservoir sands with carbonate cement or a muddy matrix occur throughout the section and their quality is not depth-dependent.

A Ghazal Formation arkosic arenite from a depth of 6315 ft (1926 m) in Unity-2 contains open primary pores and minor secondary porosity caused by partial dissolution of potassium feldspar (Figure 13). Minor decrease in porosity is caused by quartz overgrowths on detrital quartz grains and pore filling authigenic kaolinite (Figures 14A, 14B). An Aradeiba A reservoir from a depth of 7873 ft (2401 m) exhibits abundant open primary pores (Figure 15). The open pores often contain pore lining and bridging smectite, but reservoir quality is not affected significantly (Figure 16). A core of a fine-grained, oil stained Ghazal sandstone is shown in Figure 17, and cores showing some sedimentary structures of the Ghazal and Bentiu formations are shown in Figure 18.

Unity field crudes average 27 to 36° API and are highly undersaturated with average GORs of 4 to 171 scf/STB. Bubble points are very low and pour points are high (65 to 95°F). Wax content is high (about 12%) and viscosity ranges from 8 to 67 centistokes at reservoir conditions. Results of repeat formation tester analyses indicate the fluid gradients are normal, slightly less than a normal water gradient, with no indication of overpressured zones. Formation water is very fresh and is low in chlorides.

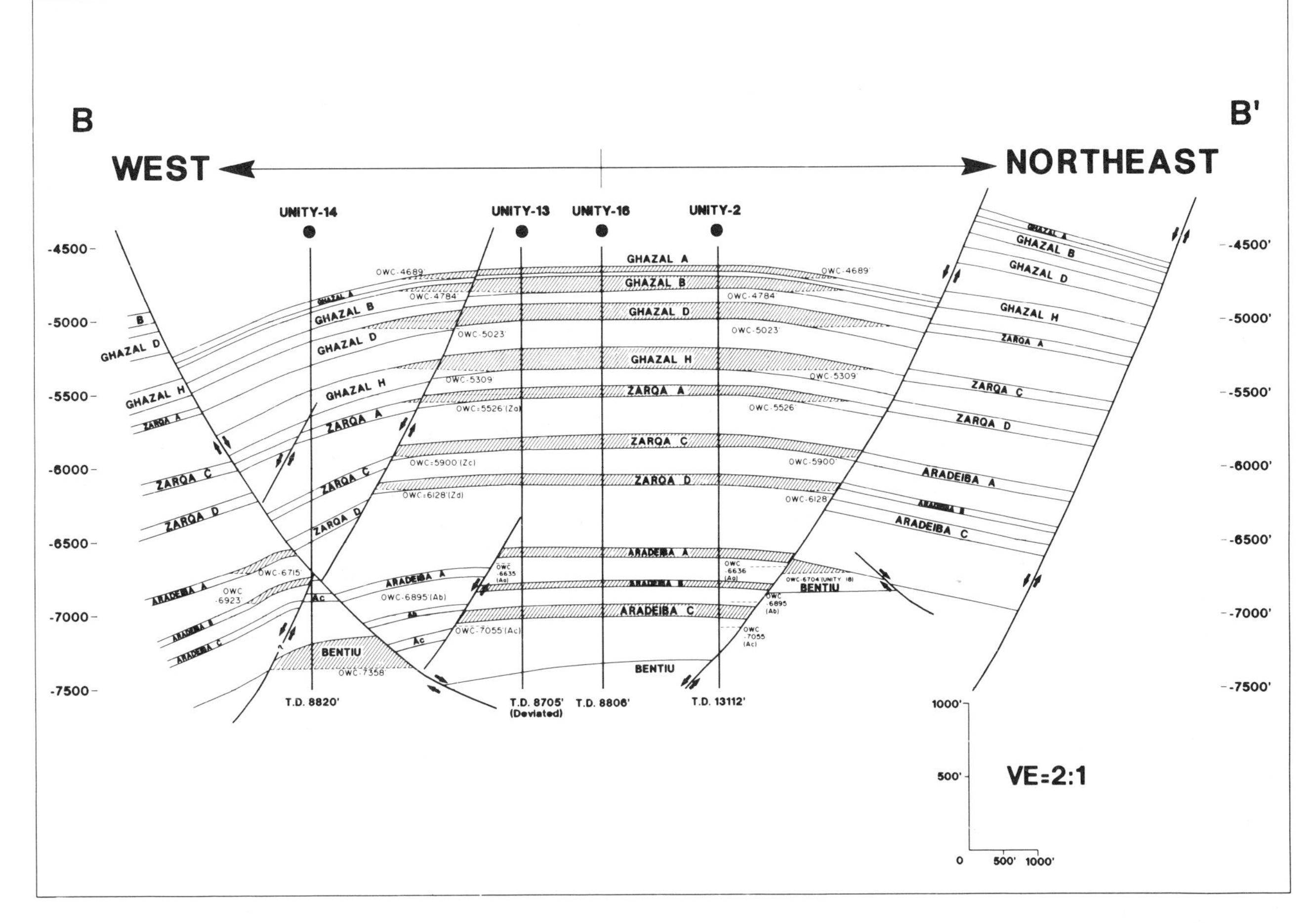

Figure 10. Structural section B–B′ through the main Unity field area (after L. S. Skander, Chevron). See Figure 8 for location of profile.

Geochemistry

The source rocks of the Muglad basin are organic-rich shales of the Abu Gabra Formation that were deposited in stratified lakes during Aptian to Albian time. In some parts of the basin the Barremian Sharaf Formation has also been found to contain some source intervals of appreciable volumes. Total organic carbon in the penetrated Abu Gabra section averages 1.3% but exceeds 7% in places in the northwestern part of the basin. Pyrolysis data indicate an average hydrogen index of 280 with some thin zones reaching 900. The kerogen is mostly of type I. Approximately 1500 ft (460 m) of the Abu Gabra Formation was penetrated by the Unity-1 well, where it was organic-poor to a depth of 13,000 ft (3960 m). Below that depth, organic richness increased slightly with the appearance of thin beds of black silty shales. The section reached the oil window at about 9500 ft (2900 m) (R_o = 0.7) and passed out of it near 14,300 ft (4360 m).

All Unity oils are mature, paraffinic, very low in sulfur, and chemically similar, suggesting that they were derived from the same or environmentally similar lacustrine source. They are low in cyclic gasoline-range (C_4–C_7) hydrocarbons. None of the oils have been heavily biodegraded, but slight biodegradation is seen in some of the Ghazal and Zarqa zones. Some oils in the Aradeiba and Bentiu formations are slightly more mature than others, probably because of repeated pulses of migration of increasingly more mature oils as the bounding growth faults in the field continued to move. There is no consistent relationship between API gravity and reservoir age or depth. A gas chromatograph of the C_{10+} saturated hydrocarbon fraction of an Aradeiba "A" oil from Unity-2 (Figure 19) shows a highly mature oil that has not been affected by biodegradation.

A Unity area geohistory diagram illustrates that the Abu Gabra Formation source section entered the oil window about 82 million years ago and passed

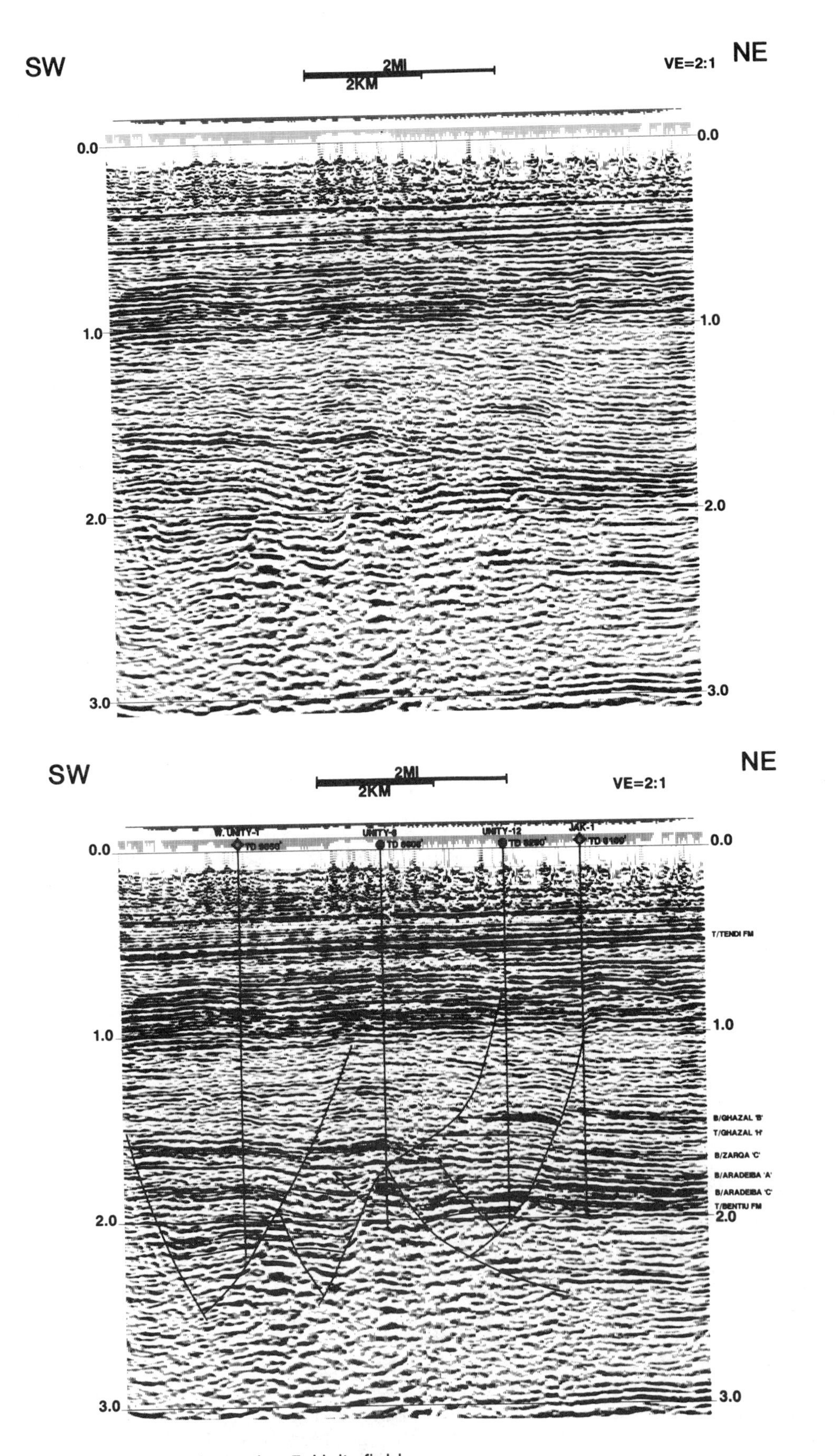

Figure 11. Time migrated seismic section 5, Unity field. Times shown are two-way travel time in seconds. See Figure 8 for location of profile.

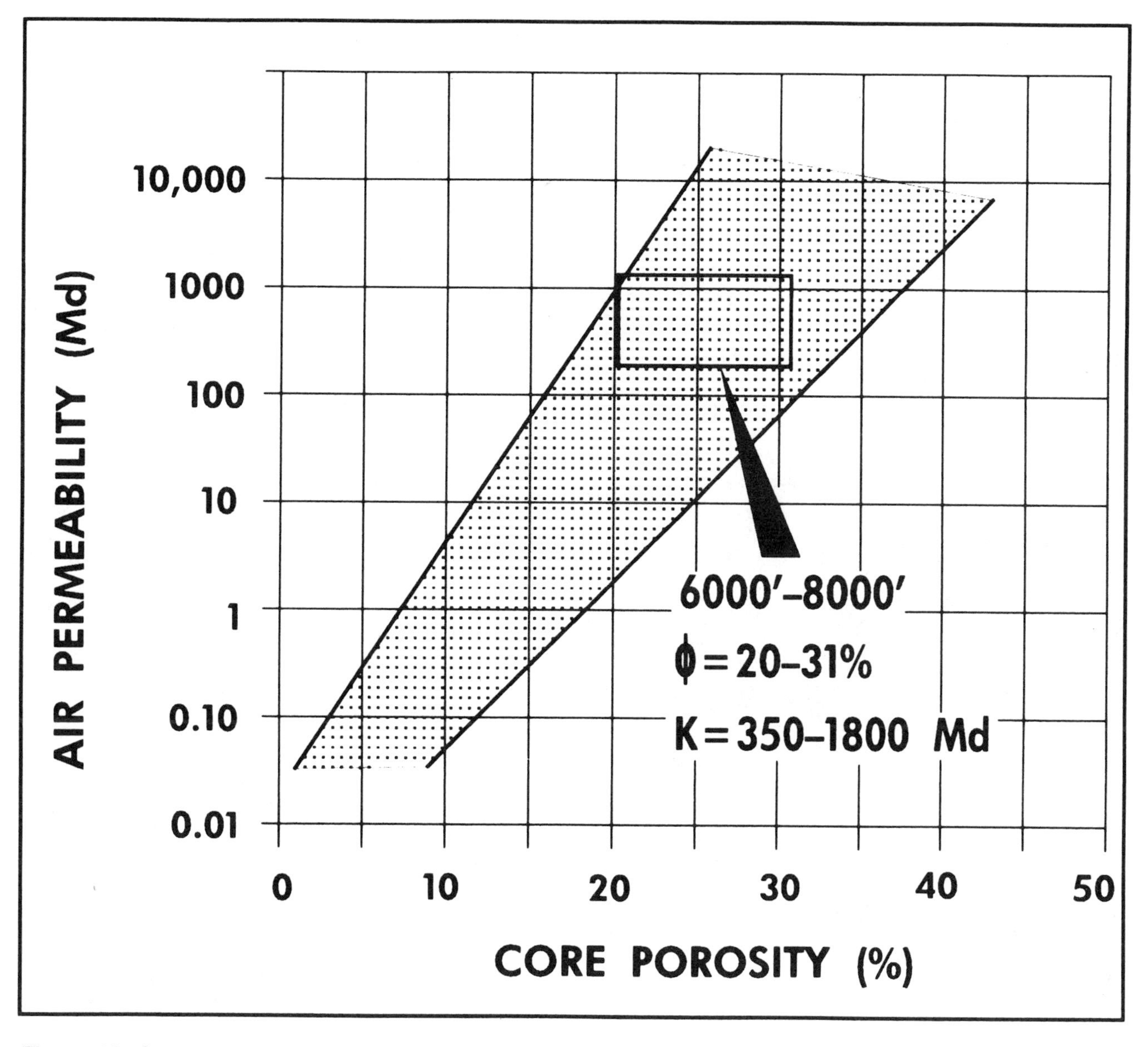

Figure 12. Core porosity/permeability plot, Unity field. Averages for values between 6000 and 8000 ft (1830 and 2440 m) are outlined.

through it approximately 42 million years ago, peaking after the deposition of the lower Darfur Group (Figure 20). Present-day peak oil generation is reached by about 12,000 ft (3660 m). Thinning of the Ghazal, Zarqa, and Aradeiba formations onto the Unity area demonstrates structural growth was in progress during the period that the Abu Gabra Formation passed through the generative window. Some down-flank generation would have predated this time, and initial generation and migration may have occurred prior to the formation of an effective trap.

The average temperature gradient at Unity is approximately 1.54°F/100 ft, at an ambient surface temperature of 85°F, somewhat higher than the average of 1.4°F/100 ft in other parts of the Muglad basin.

Reserves/Development Plan

Proven and probable recoverable reserves from the greater Unity area fields are estimated at 146 MMSTB. Reservoir simulation studies indicate that approximately 75% of these reserves will be recovered as the result of the waterflood that is planned to begin at the start-up of production. In the absence of this waterflood, recovery by primary depletion would be only 4 to 6% of an estimated 594 million

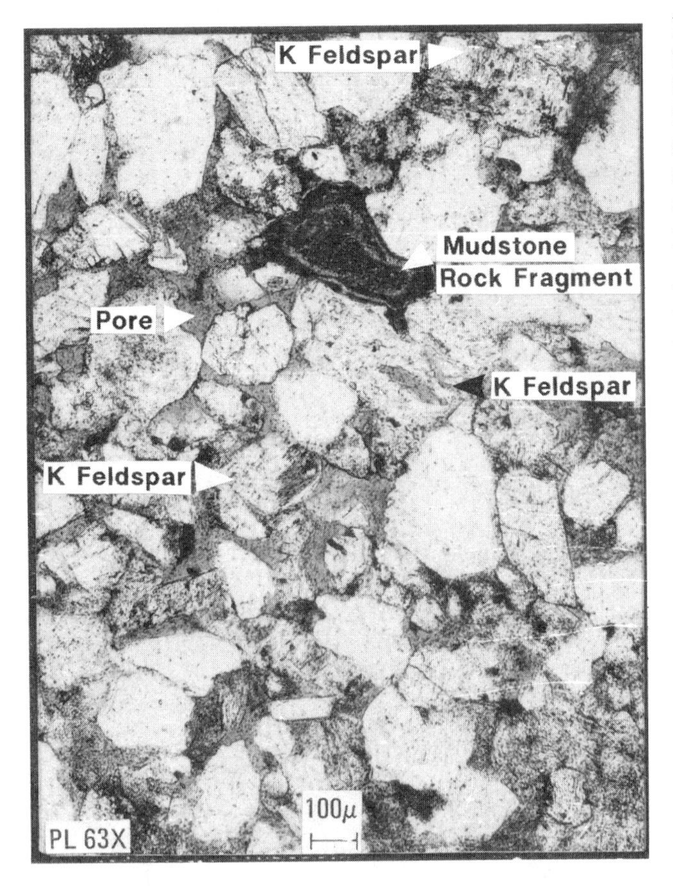

Figure 13. Photomicrograph of a Ghazal arkosic arenite (6315 ft [1926 m], Unity-2 well), with porosity of 25% and permeability of 333 md. Exhibits open pores and minor secondary porosity caused by dissolution of potassium feldspar. Sample contains about 45% quartz, 35% feldspar, 8% kaolinite, and lithic fragments.

bbl OOIP. In order to maximize water breakthrough time and minimize water production, it is planned that production wells will be placed, where possible, in high oil saturation columns and evenly spaced away from injection sources. Utilizing electrical submersible pumps, initial production rates are expected to vary from 1200 to 2600 BOPD per producer.

A minimum of 39 wells and a maximum of 55 wells will be required to develop the present Unity area. The maximum case would include 28 producers, 26 water injectors, and one water disposal well. A planned 50,000 to 200,000 BOPD export pipeline project from Unity and nearby Heglig fields to Port Sudan has been suspended.

EXPLORATION CONCEPTS

Unity field is one of a number of discoveries made by Chevron Overseas Petroleum in the Sudan interior as the result of a long-range evaluation program. At the start of exploration operations in 1975, there was essentially no data base other than from shallow water wells and limited surface geology. Geophysical, geologic, and geochemical data were acquired in a systematic manner to define areas of greater potential on which early drilling efforts were concentrated. The exploration program that was followed resulted in a high success ratio in the 1979 to 1984 period when all of the Sudan interior discoveries were made.

The first exploratory well in the Muglad basin was drilled in 1977 over 200 km (125 mi) to the northwest of Unity, and it established the presence of an encouraging reservoir and source rock section. The

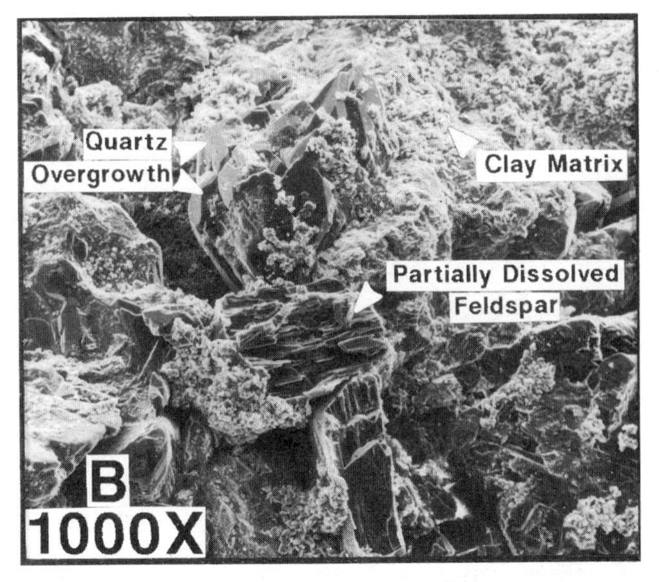

Figure 14. (A) SEM (300×) of Ghazal arkosic arenite (6325 ft [1929 m], Unity-2), shows open pores and a coating of kaolinite. (B) SEM (1000×) shows minor decrease in porosity caused by quartz overgrowths on detrital quartz grains and pore-filling authigenic kaolinite, and a slight increase in porosity caused by dissolved feldspar.

Figure 15. Photomicrograph of an Aradeiba "A" subarkosic arenite (7870 ft [2400 m], Unity-2) shows abundant open pores and measured values of 30.9% porosity and 2332 md permeability.

second well encountered hydrocarbon shows in Unity-1 in 1978. Almost two years after Unity-1, the second well in the Unity area (the twelfth well in the Sudan drilling program) made the first important discovery at Unity-2. Successive exploration and development wells and a detailed seismic survey have defined the present limits of Unity, a complex field in a new exploration province.

ACKNOWLEDGMENTS

The writer wishes to acknowledge that this paper reflects contributions of many employees of Chevron Overseas Petroleum Inc. and Chevron Oil Field Research who completed the extensive Unity field studies. Special recognition is due to Chris Harrison who interpreted the Unity 3D seismic data, and to Ernie Maxwell and Lillian Skander who completed the reservoir geology studies. Tim McHargue made helpful geologic and editorial suggestions, and Judy Bonilla prepared the illustrations.

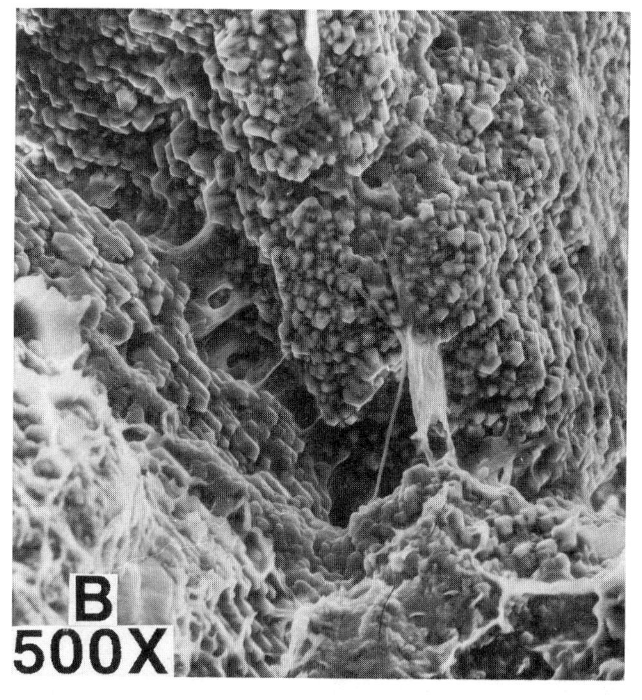

Figure 16. (A) SEM photo (50×) of Aradeiba "A" sandstone (7873 ft [2401 m], Unity-2) shows open pores with a thin coating of quartz overgrowths surrounding the detrital quartz grains. (B) SEM photo (500×) shows quartz overgrowths and pore lining and pore bridging smectite.

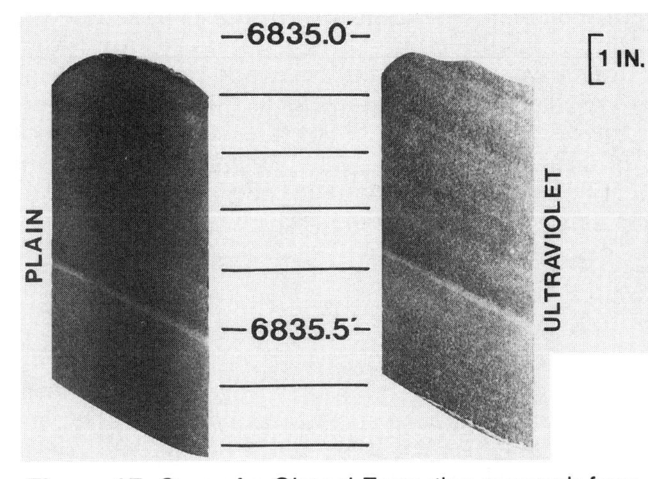

Figure 17. Core of a Ghazal Formation reservoir from a depth of 6835 ft (2085 m). This brown-gray, friable, subangular to subrounded, poorly sorted, evenly oil-stained sandstone has a measured porosity of 27.2% and a permeability of 1110 md.

REFERENCES

Browne, S. E., and J. D. Fairhead, 1983, Gravity study of the Central African Rift System, a model of continental disruption, 1. The Ngaoundere and Abu Gabra rifts: Tectonophysics, v. 94, p. 187-203.

Chevron proprietary reports.

Fairhead, J. D., 1988, Mesozoic plate tectonic reconstruction of the central South Atlantic Ocean: the role of the West and Central African rift system: Tectonophysics, v. 155, p. 181-191.

Geological & Mineral Resources Department, Khartoum, Sudan, Geologic Map of Sudan.

Schull, T. J., 1988, Rift basins of interior Sudan: petroleum exploration and discovery, American Association of Petroleum Geologists Bulletin, v. 72/10, p. 1128-1142.

SUGGESTED READINGS

Rosendahl, B. R., 1987, Architecture of continental rifts with special reference to East Africa: Annual Review of Earth and Planetary Science, v. 15, p. 445-503.

Rosendahl, B. R., D. J. Reynolds, P. M. Lorber, C. F. Burgess, J. McGill, D. Scott, J. J. Lambiase, and S. J. Derksen, 1986, Structural expressions of rifting: lessons from Lake Tangânyika, Africa, *in* L. E. Frostick et al., eds., Sedimentation in the African rifts: Geological Society Special Publication 25, p. 29-43.

Vail, J. R., 1978, Outline of the geology and mineral deposits of the Democratic Republic of Sudan and adjacent areas: Institute of Geological Sciences, Overseas Geology and Mineral Resouces 49, 68 p.

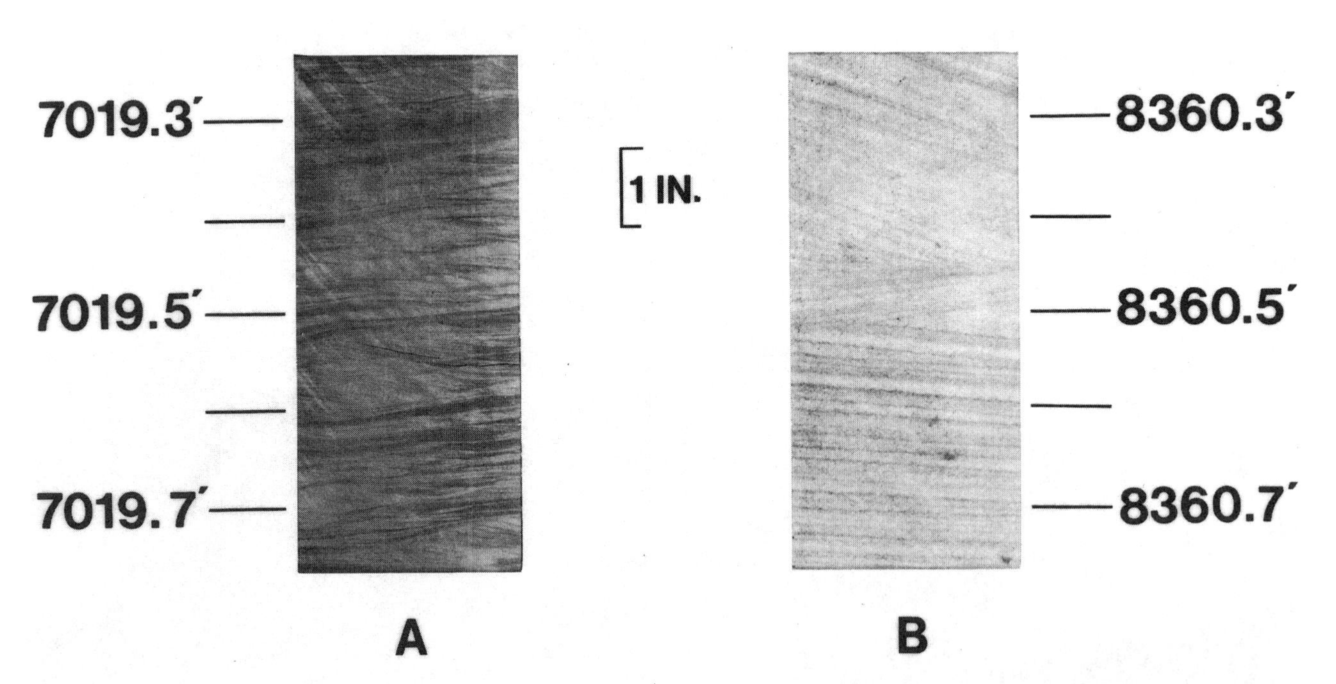

Figure 18. (A) Core photo of a Ghazal Formation ripple-laminated very fine sand, silt, and clay (7018 ft [2140 m] in Talih-1 well). Average porosity for this facies is 13% with low permeability. (B) Core photo of a Bentiu cross-bedded subarkosic to arkosic sandstone (8360 ft [2550 m] in Unity-8 well). The porosity of this facies ranges from 2.2 to 38.7% and the permeability range is 4 to 1870 md.

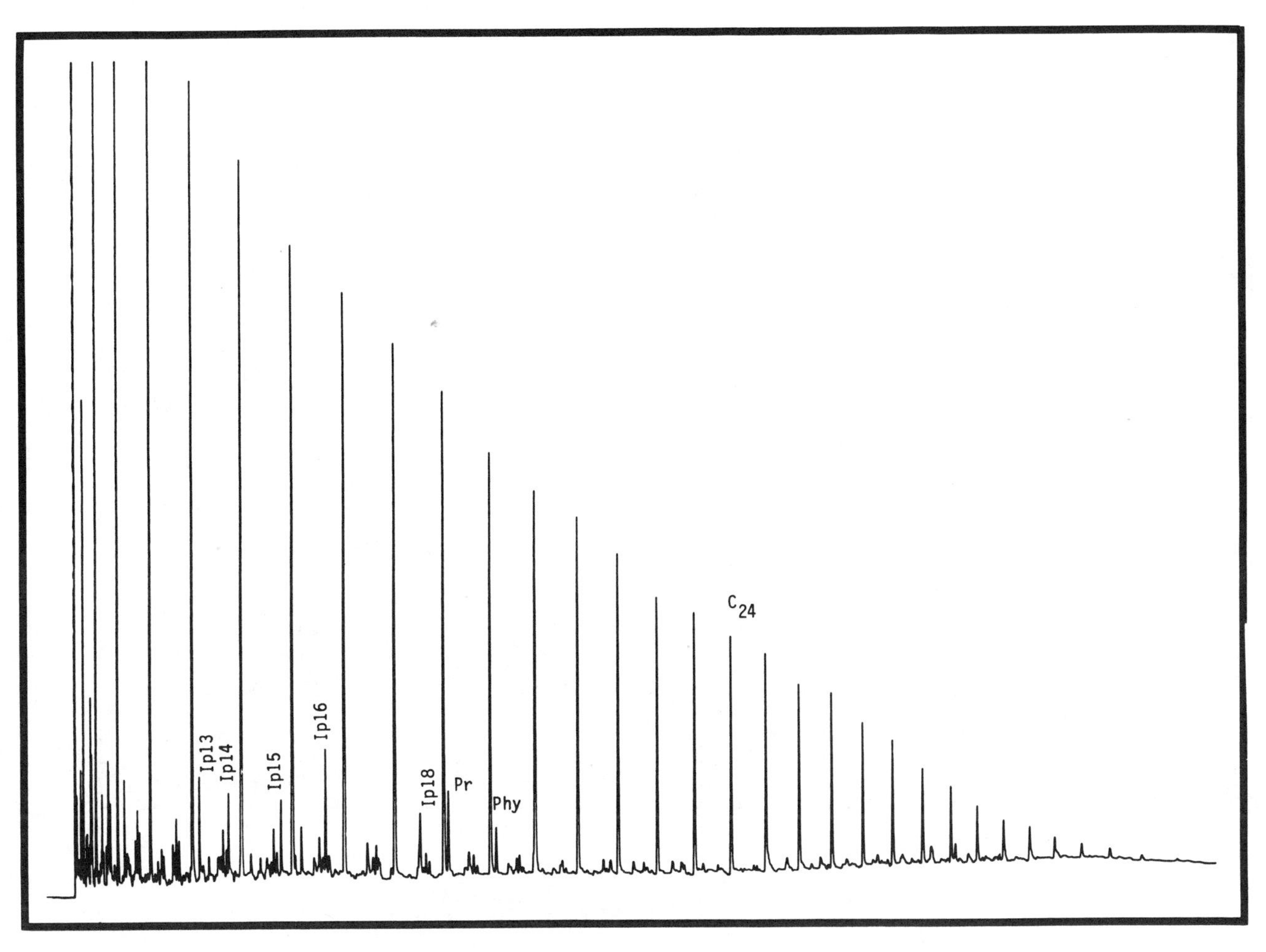

Figure 19. C_{10+} saturated hydrocarbon, Aradeiba "A" oil, Unity field.

Appendix 1. Field Description

Field name .... *Unity field*

Ultimate recoverable reserves .... *146 million bbl*

Field location:

- **Country** .... *Sudan*
- **State**
- **Basin/Province** .... *Muglad basin/Upper Nile province*

Field discovery:

- **Year first pay discovered** .... *10 major Upper Cretaceous reservoirs 1980*
- **Year second pay discovered** *1980*
- **Third pay** *1980*

Discovery well name and general location:

- **First pay** .... *Unity-2, lat. 9°28′11.5″N, long. 29°40′35″E*
- **Second pay** .... *Unity-2, lat. 9°28′11.5″N, long. 29°40′35″E*
- **Third pay** .... *Unity-2, lat. 9°28′11.5″N, long. 29°40′35″E*

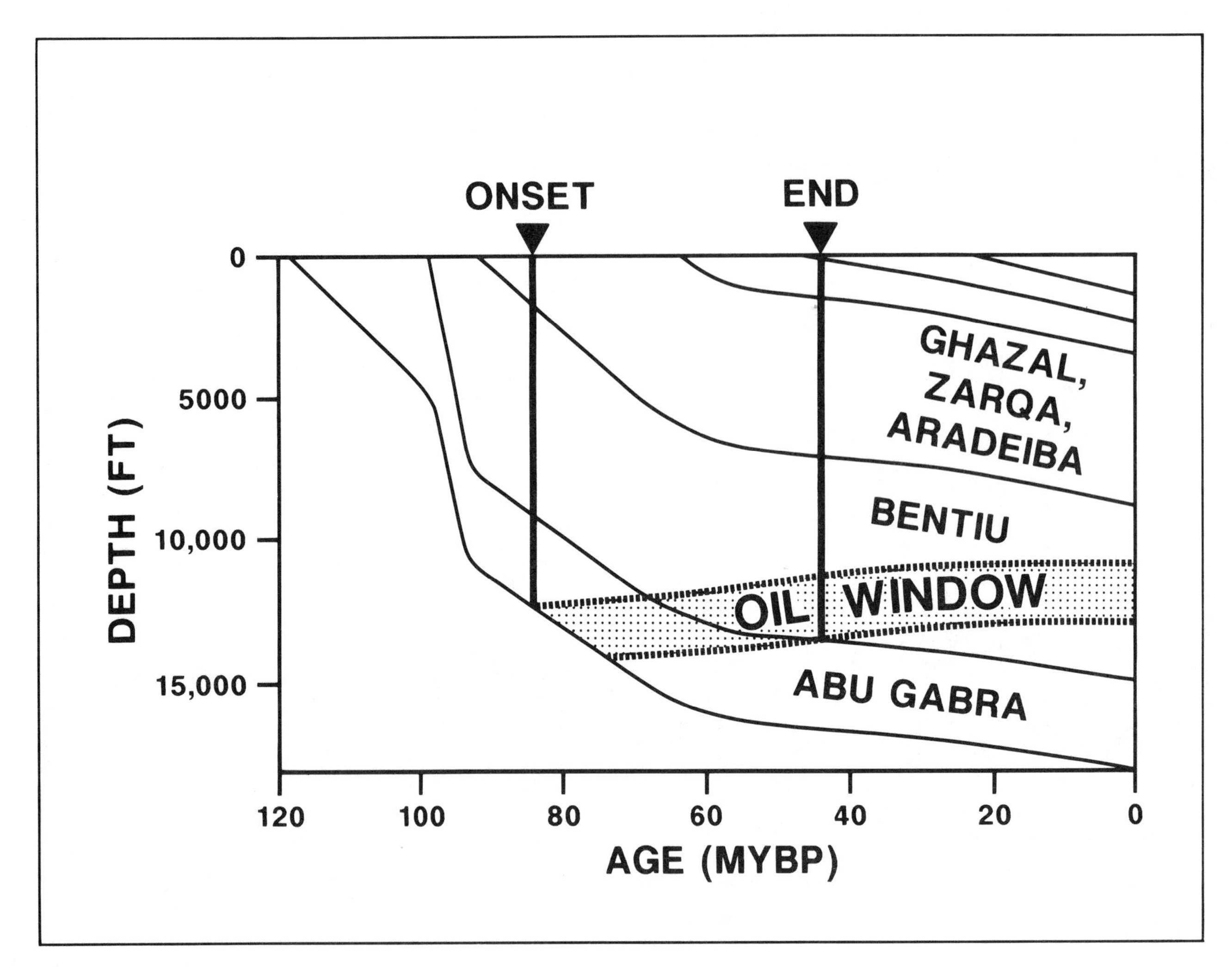

Figure 20. Geohistory diagram, Abu Gabra oil generation, Unity field.

Discovery well operator .... *Chevron*

Second pay

Third pay

IP in barrels per day and/or cubic feet or cubic meters per day:

First pay .... *2500–4800 bbl/day (Ghazal)*

Second pay .... *2527–3545 bbl/day (Zarqa)*

Third pay .... *2242–2690 bbl/day (Aradeiba)*

All other zones with shows of oil and gas in the field:

Age	Formation	Type of Show
Cenomanian	*Bentiu*	*2338 bbl oil*
Senonian	*Baraka*	*Oil stains*

Geologic concept leading to discovery and method or methods used to delineate prospect, e.g., surface geology, subsurface geology, seeps, magnetic data, gravity data, seismic data, seismic refraction, nontechnical:

Basins related to Central Africa rifting were postulated in the southern Sudan. Initial studies consisted of LANDSAT and photogeological analyses. This was followed by an aeromagnetic survey, helicopter gravity, reconnaissance and detailed seismic surveys with "shot point gravity," and a 3D seismic survey of the Unity field area.

Structure:

Province/basin type .... *Rift basin*

Tectonic history

The Muglad basin was formed as a result of Central Africa rifting which began in the Lower Cretaceous, followed by additional rifting in Upper Cretaceous and in the Tertiary. The reservoirs in Unity field were deposited late in the first rifting phase (Cenomanian Bentiu Formation) and early in the second rifting phase (Senonian Ghazal, Zarqa, and Aradeiba formations).

Regional structure

Unity field trends north-northwest and is located on the northeast flank of the Muglad basin over a south-southeast-plunging gravity high. The field is separated from shallow Precambrian basement to the northeast by a regional syncline. The basinal depocenter to the southwest of the field contains 35,000 to 45,000 ft (10,675 to 13,725 m) of Tertiary and Cretaceous fluvial sediments.

Local structure

The field is located within a complexly faulted graben over a broad regional high. Closures are anticlinal and fault bounded.

Trap:

Trap type(s)

(1) Anticlinal with multiple pays; (2) tilted fault blocks with multiple pays on the upthrown sides of normal faults

Basin stratigraphy (major stratigraphic intervals from surface to deepest penetration in field):

Chronostratigraphy	Formation	Depth to Top in ft (m)
Paleocene	*Amal*	*2700 (825)*
Senonian-Turonian	*Baraka*	*3700 (1130)*
Senonian-Turonian	*Ghazal*	*5850 (1785)*
Senonian-Turonian	*Zarqa*	*6650 (2030)*
Senonian-Turonian	*Aradeiba*	*7450 (2270)*
Cenomanian	*Bentiu*	*8550 (2610)*

Location of well in field

Reservoir characteristics:

Number of reservoirs *11*

Formations .... *Ghazal, Zarqa, Aradeiba, Bentiu*

Ages .... *Upper Cretaceous (Senonian-Cenomanian)*

Depths to tops of reservoirs .... *3700 to 8300 ft (1130-2530 m)*

Gross thickness (top to bottom of producing interval) .... *~2700 ft (825 m)*

Net thickness—total thickness of producing zones

Average .... *108 ft (33 m)*

Maximum .... *246 ft (75 m)*

Average

Maximum

Lithology .... *Fluvial-deltaic subarkosic arenites and wackes*

Porosity type .... *Primarily intergranular with minor secondary intergranular, intragranular and moldic due to partial or total dissolution of feldspars*

Average porosity *27%*

Average permeability .... *1600 md*

Seals:

Upper

Formation, fault, or other feature .... *Intrareservoir*

Lithology .... *Claystone/shale*

Lateral

Formation, fault, or other feature .... *Fault and dip closure*

Lithology .... *Claystone/shale*

Source:

Formation and age .... *Abu Gabra Formation (Albian-Aptian)*

Lithology .... *Lacustrine shales with interbedded sandstones*

Average total organic carbon (TOC) *1.3%*

Maximum TOC *7.5%*

Kerogen type (I, II, or III) .... *I and II*

Vitrinite reflectance (maturation) .... $R_o = 0.46$ *@ 6400 ft (1950 m) to 1.1 @ 14,300 ft (4360 m)*

Time of hydrocarbon expulsion .... *Upper Cretaceous (Senonian)*

Present depth to top of source .... *12,500 ft (3810 m)*

Thickness .... *1000-6000 ft (305-1830 m)*

Potential yield

Appendix 2. Production Data

Field name .... *Unity field*

Field size:

Proved acres .... *7000 (284 ha)*

Number of wells all years *18*

Current number of wells *18*

Well spacing .... *250 acres (101 ha) planned*

Ultimate recoverable .... *146 million bbl*

Cumulative production .... *None*

Annual production .... *Shut-in*

Present decline rate

Initial decline rate

Overall decline rate

Annual water production

In place, total reserves .... *594 million bbl*

In place, per acre-foot

Primary recovery .... *0 (4-6% of OOIP would be produced by primary depletion, but waterflooding will be initiated at commencement of production)*

Secondary recovery .... *146 million bbl (75% of proved and probable recoverable reserves will be recovered as the result of waterflooding)*

Enhanced recovery

Cumulative water production

Drilling and casing practices:

Amount of surface casing set .... *400 ft (122 m)*

Casing program

20-in. surface casing at 400 ft (122 m); 13⅜-in. casing at 2750 ft (839 m) (17½-in. hole); 9⅝-in. casing at 8800 ft (2684 m) (12¼-in. hole)

Drilling mud .... *Gel lignosulphonate, KCl Polymer*
Bit program
High pressure zones .... *None*

Completion practices:
Interval(s) perforated .... *14 intervals in Unity-2*
Well treatment .... *HCl*

Formation evaluation:
Logging suites .... *(#1) ISF, LSS, GR, FDC-CNL, DLL, MSFL, SWS; (@TD) LSS, GR, FDC-CNL, DLL, MSFL, PROX ML, DM, SWS, RFT*
Testing practices .... *DST*
Mud logging techniques .... *Samples collected at 10–30 ft intervals from 450 ft (139 m) to TD; samples and mud continuously monitored for shows*

Oil characteristics:
Type .... *Paraffinic*
API gravity .... *31–37°API*
Base
Initial GOR .... *3.6–15.6 scf/bbl*
Sulfur, wt% *0.04–0.07%*
Viscosity, SUS .... *8–67 cs @ reservoir conditions*
Pour point .... *65–95°F*
Gas-oil distillate

Field characteristics:
Average elevation .... *1300 ft (400 m)*
Initial pressure .... *3312 psi*
Present pressure
Pressure gradient .... *2500–3350 psi/ft*
Temperature .... *279°F @ 13,112 ft*
Geothermal gradient .... *1.54°F/ft*
Drive .... *Water*
Oil column thickness .... *2700 ft*
Oil-water contact .... *5989–8360 ft*
Connate water
Water salinity, TDS .... *Fresh water*
Resistivity of water
Bulk volume water (%)

Transportation method and market for oil and gas:
1425 km pipeline to export terminal on the Red Sea

Saticoy Oil Field—U.S.A.
Ventura Basin, California

ROBERT S. YEATS
Oregon State University,
Corvallis, Oregon
JAMES C. TAYLOR*

FIELD CLASSIFICATION

BASIN: Ventura
BASIN TYPE: Wrench
RESERVOIR ROCK TYPE: Sandstone
RESERVOIR ENVIRONMENT OF DEPOSITION: Submarine Fan
RESERVOIR AGE: Pliocene-Pleistocene
PETROLEUM TYPE: Oil
TRAP TYPE: Pinchout

LOCATION

The Saticoy oil field is located in the lower Santa Clara Valley of Ventura County, California. The field lies within the Ventura basin, a late Cenozoic east-west trough embedded in the western Transverse Ranges of southern California (Figure 1). The amount of oil plus energy-equivalent gas in place in the Ventura basin exceeds 8900 m^3 per cubic kilometer of total basin fill, with exploration of the offshore part of the basin still in progress (Bostick et al., 1978). Two reverse-fault systems form the boundaries of the thick central trough of the basin, the Oak Ridge fault on the south and the Red Mountain and San Cayetano faults on the north (Figure 1). In the Santa Clara syncline between these faults lies an extremely thick sequence of Pliocene and Pleistocene strata that thickens eastward from the coast to the vicinity of Fillmore, where the Pleistocene sequence is the thickest in the world (Yeats, 1977). These strata contain the Ventura Avenue oil field, with ultimate recovery close to a billion barrels of oil, and several smaller accumulations, including the Fillmore field, the McGrath pool of the West Montalvo field, and additional fields west of Ventura Avenue and offshore. The Santa Clara syncline is strongly asymmetric with its south limb locally overturned. The axis of the syncline is close to the Oak Ridge fault (Figure 1), and the Saticoy field produces oil trapped in Pliocene and early Pleistocene turbidites on the steeply dipping south limb adjacent to this fault. South of the fault is a second group of fields producing oil mainly from the nonmarine Sespe Formation of Oligocene age.

The Saticoy field merges to the east with the Bridge pool of the South Mountain field; the Bridge pool is part of the same oil accumulation. Ultimate recovery from the Saticoy field and Bridge pool is estimated to be 85.5 million barrels of oil.

HISTORY

Pre-Discovery

The Saticoy field is located in the footwall block of the Oak Ridge fault. Oil production was established in the nonmarine Sespe Formation of late Eocene and Oligocene age in the hanging-wall block of this fault as early as 1889 (Schultz, 1960), and by World War II, all but one of the major anticlines in the hanging-wall block had been found to be oil-productive.

By this time, the Ventura Avenue oil field had become recognized as a giant field producing from Pliocene sandstones folded into an east-west-trending anticline that extends offshore to the west. In the late 1940s, the Standard Maxwell and Superior Limoneira wells were drilled in the Santa Clara Valley south of the Ventura Avenue anticline, revealing that the sandstones productive at Ventura Avenue extend southward to the Santa Clara River

*Deceased. James C. Taylor died on 8 June 1976, less than a month after presenting a preliminary version of this paper in New Orleans, LA at the AAPG/NOGS short course on sedimentary environments and hydrocarbons. The paper was written by Robert S. Yeats based on a preliminary draft of Taylor's paper and on a more complete understanding of the Oak Ridge fault and of the timing of deformation of the Saticoy field, derived after Taylor's death. Yeats is a former development geologist for Saticoy field.

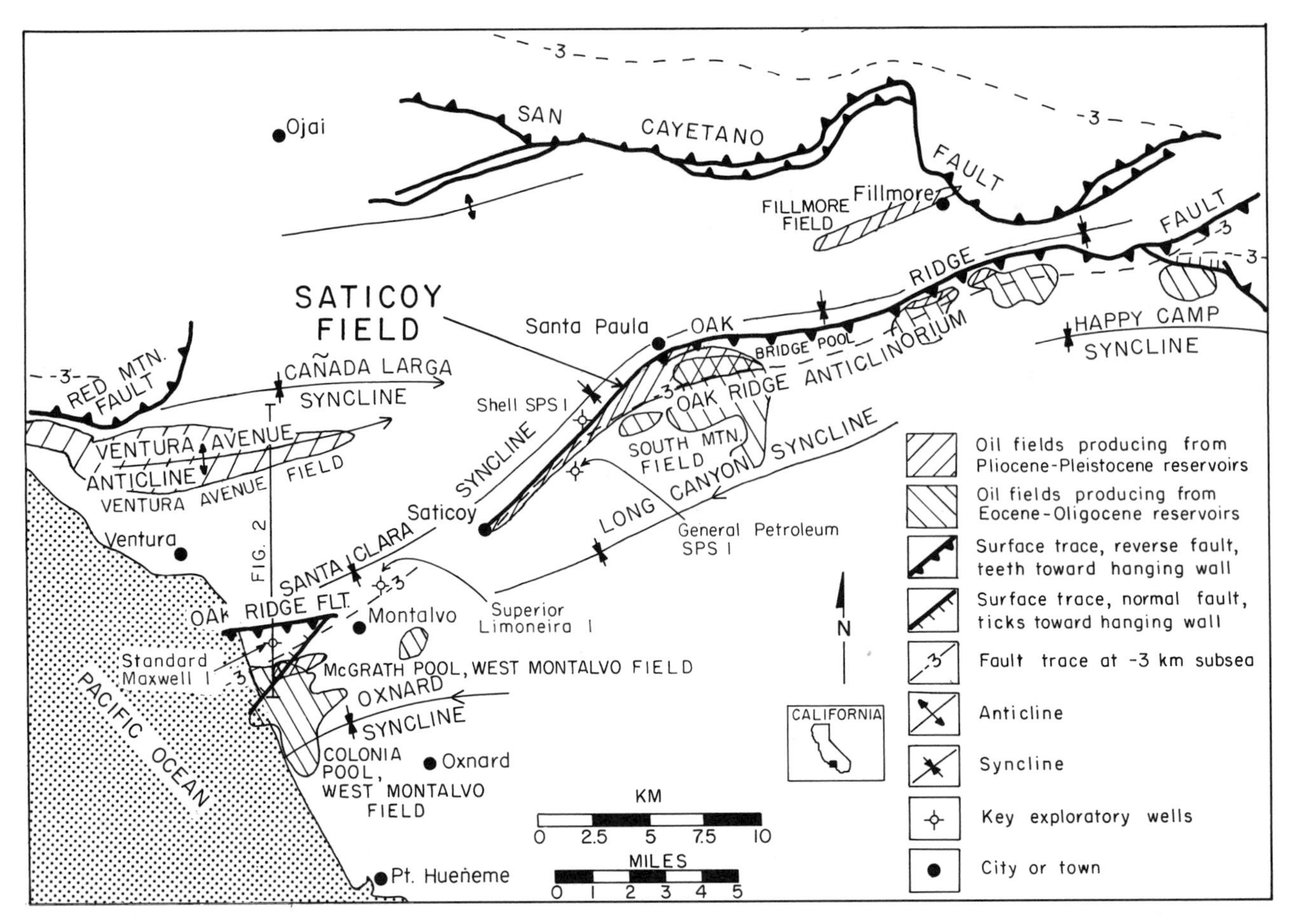

Figure 1. Index map of part of Ventura basin showing structural setting of Saticoy oil field. Line of cross section shown in Figure 2 is indicated.

(Figure 2). The Superior Limoneira well was drilled to a total depth of 18,734 ft (6807 m) in the lower Pliocene, whereas only a few miles away, south of the Oak Ridge anticlinorium at South Mountain field, the Pliocene was known to be relatively thin and underlain by Miocene Monterey Shale (Modelo Formation) and Eocene to Oligocene Sespe Formation. It became clear that the intervening Oak Ridge fault had very large displacement. Possibly the sandstones productive at Ventura Avenue could be found trapped against the Oak Ridge fault on the south side of the Santa Clara syncline. In 1942, Shell Oil Co. drilled the SPS (Santa Paula y Saticoy) no. 1 well to 5952 ft (2163 m) reaching, at total depth, steeply dipping to vertical upper Pliocene or lower Pleistocene beds north of the fault; no productive sandstones were found (Schultz, 1960). Farther south, General Petroleum SPS no. 1 was drilled in 1952 to 11,347 ft (4123 m) and abandoned in the Sespe Formation on the south side of the fault. A few years earlier, in 1947, oil had been found near the coast in lower Pliocene sandstones lensing out southward in the hanging-wall block of the Oak Ridge fault; this discovery was named the McGrath pool of the West Montalvo field (Figure 2).

Discovery

In 1954, guided by seismic-reflection lines and subsurface geology, Shell Oil Co. spudded SPS no. 2 and drilled a relatively straight hole to 12,020 ft (4367 m), reaching north-dipping upper Pliocene sandstones south of the Santa Clara syncline axis, but not finding oil (Figure 3). The well was redrilled southward and updip, but still the sandstones were found to be water-bearing. The third redrilled hole (Figure 3), still farther updip, found oil-saturated sandstone of the Pliocene and Pleistocene Pico Formation. The well was completed in May 1955 for 236 barrels per day of 32° API gravity oil (Jeffreys, 1958; Schultz, 1960). Seven months after the SPS no. 2 well was completed, Texaco and Union Oil Co. completed the Richardson Earl no. 1 well in sandstones of the Pico Formation for 205 barrels per day of oil as the discovery well for the Bridge pool of the South Mountain field (Ware and Stewart, 1958) (Figure 4). Unlike the SPS no. 2 well, which was in the footwall block of the Oak Ridge fault from top to bottom, the Richardson Earl well spudded in the hanging-wall block of the fault within the producing limits of the South Mountain oil field,

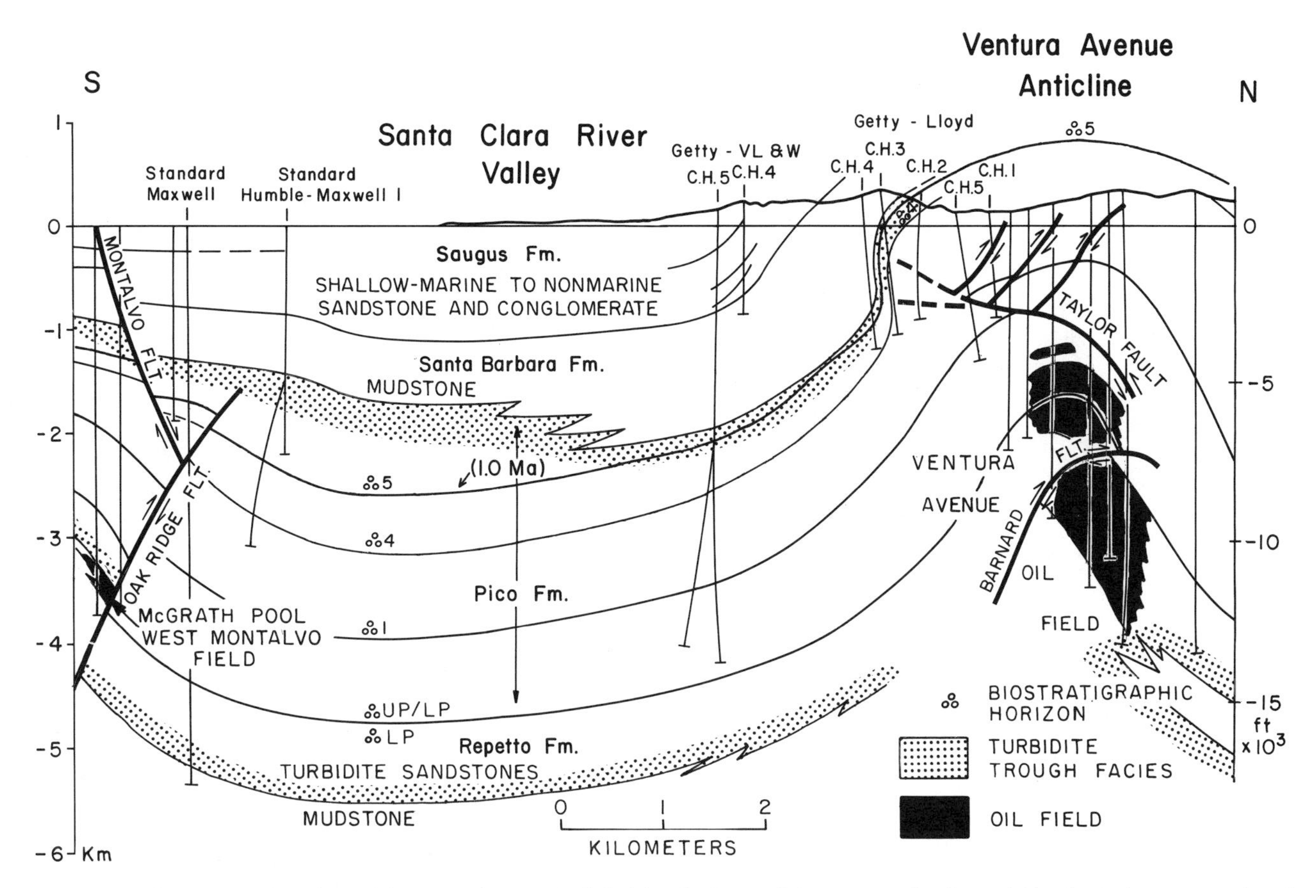

Figure 2. Cross section from Ventura Avenue oil field to Oak Ridge fault near coast, showing oil fields and turbidite sandstone facies. Standard-Maxwell well and McGrath pool of West Montalvo field provided evidence that sandstones productive at Ventura Avenue extend south to Oak Ridge fault and are productive where they lens out upstructure to mudstone. Cross section located on Figure 1. No vertical exaggeration.

producing from the Sespe Formation. Later development showed that the Saticoy field and Bridge pool are part of a single complex accumulation having little to do with the Sespe accumulation at South Mountain. Accordingly, this paper treats the Saticoy field and Bridge pool as a single oil field, although most of our experience is with the Shell portion of the Saticoy field.

Post-Discovery

The two discoveries led to an active exploratory drilling campaign along the entire length of the footwall block of the Oak Ridge fault, but no new major oil fields were discovered. Many wells found Pliocene and lower Pleistocene sandstones in trapping position close to the fault, but they are salt water-bearing. Further development at Saticoy showed that for many of the sandstones, the trap is stratigraphic; the sandstones lens out updip to mudstone below the fault (Figure 3). However, in the Bridge pool, the oil is trapped against the fault itself (Figure 4).

The field is now developed (Figure 5), although there have been attempts in recent years to extend Bridge pool production eastward. A total of 179 wells was completed on 1170 productive acres, and 118 were still producing in 1984. By the end of 1987, the Saticoy field, including the Bridge pool, had produced 64.2 million bbl of oil. The ultimate recovery is expected to be 85.5 million bbl of oil. The field is very narrow, limited by the Oak Ridge fault and updip pinchout of sandstones on the south and by downstructure water to the north. The west end of the field occurs where the pinchout lines and structure contours of the sandstones turn into the fault (Figure 6), and the Santa Clara syncline axis is so close to the fault that the syncline has virtually no south flank (Figure 6). The east end is more poorly defined. The sandstones are steeply dipping there, even locally overturned. The deepening of the basin to the east may result in individual sandstones being too far downstructure to produce, but poor correlation between sandstones in the Bridge pool and sandstones farther east makes this interpretation speculative.

Each producing sandstone has its own water table, and 20 individual reservoir sandstones have been identified (Figure 7). Two important observations were made early in the development of the field. First, the sandstones were deposited by multiple turbidity-current flows in deep water, a depositional regime

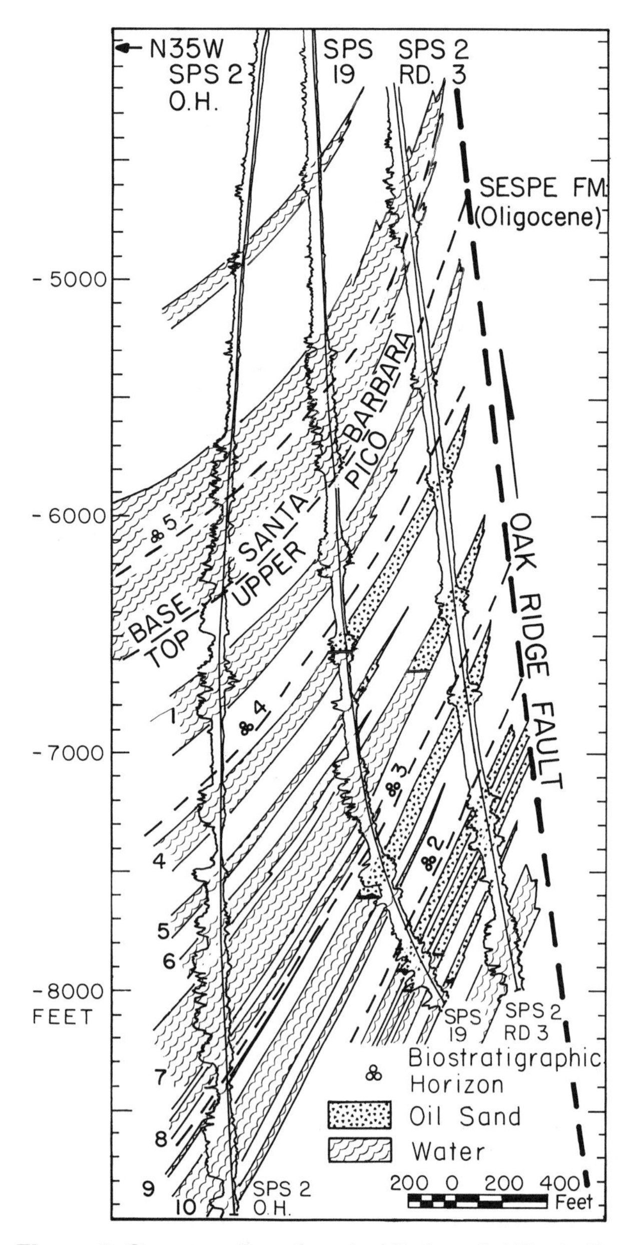

Figure 3. Cross section of central Saticoy field including discovery well, SPS no. 2, third redrill. Most turbidite sequences lens out below Oak Ridge fault, in contrast to the Bridge pool (Figure 4), in which turbidite sequences are truncated by fault. Fault dip is also steeper in Saticoy field than in Bridge pool. Cross section located on Figure 5.

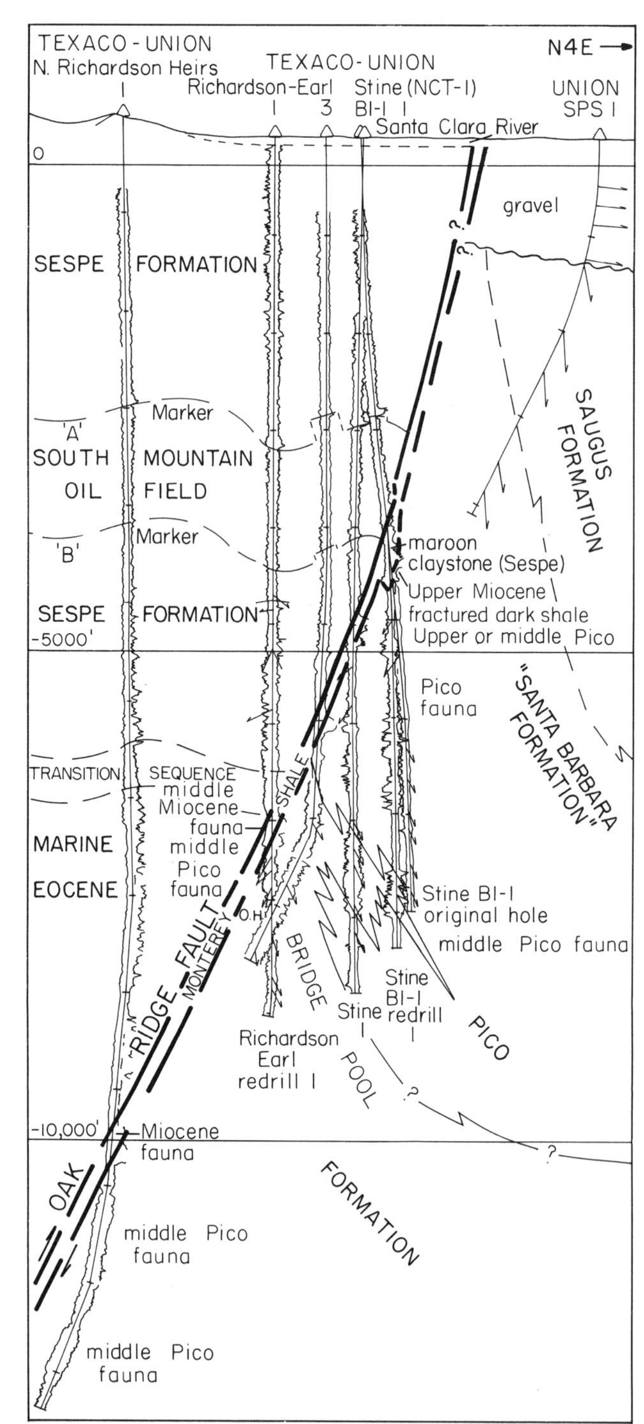

Figure 4. Cross section of Bridge pool including discovery well, Texaco-Union Richardson Earl no. 1, original hole. Wells spud in Oligocene Sespe Formation, major reservoir of the South Mountain oil field, cross a sliver of Miocene Monterey Shale at Oak Ridge fault, and reach total depth in steeply dipping Pico Formation. Turbidite reservoirs, more lensatic than those of central Saticoy field, are trapped directly by Oak Ridge fault. Cross section located on Figure 5.

that had been recognized only a few years before (Natland and Kuenen, 1951). This had a bearing in understanding reservoir characteristics; an individual sandstone reservoir would consist of many discrete turbidite sandstone bodies that are separated by hemipelagic mudstone near their updip pinchout. Second, most of the sandstones are of Pleistocene

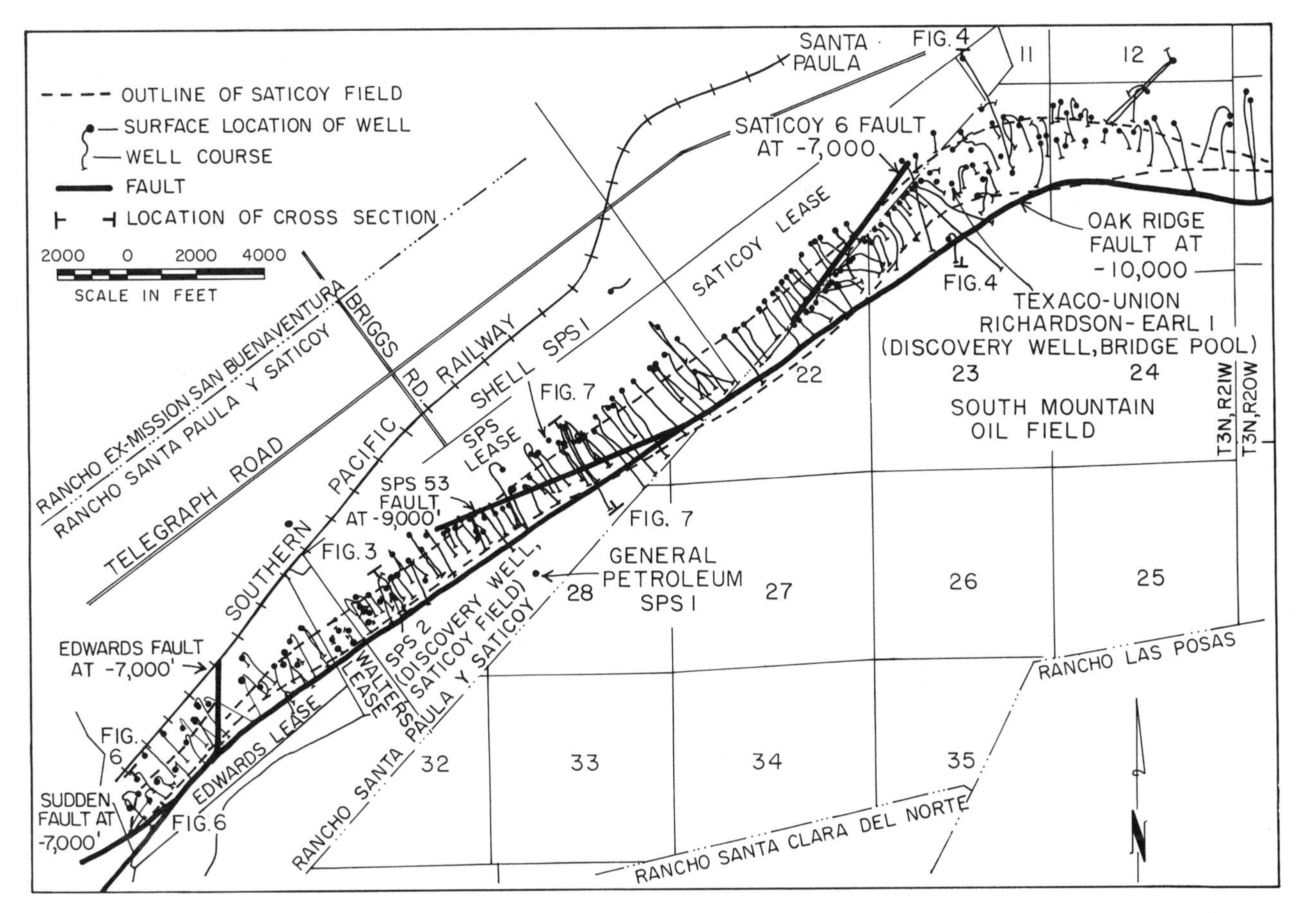

Figure 5. Map of Saticoy field, including Bridge pool, showing well locations (solid circles) and map projection of directionally drilled wells. Major faults shown in heavy solid lines. Cross section lines for Figures 3, 4, 6, and 7 are shown.

age, steeply dipping, and locally overturned. The Saticoy field demonstrated that deep-water sandstones of Pleistocene age can produce oil in commercial quantities.

DISCOVERY METHOD

The Saticoy field was discovered by a combination of subsurface geology and reflection seismic lines. The Standard Maxwell and Superior Limoneira wells showed that the sandstones productive at Ventura Avenue field are present on the downthrown block of the Oak Ridge reverse fault (Figure 2), and reflection seismic lines (of poor quality) suggested that these sandstones could be in trapping position beneath the fault.

It is now known that the primary trapping mechanism is stratigraphic; sandstones lens out to siltstone in the direction of a submarine high (see *Structure* below), or seaknoll, which was not recognized at the time of the discovery. Exploration for future prospects of this type should include the search for additional seaknolls that were positive during the time of deposition of the objective deep-water sandstones. Because these sandstones were deposited by turbidity currents, it is important to recognize that sandstones lens out to siltstone in the direction of the seaknoll, in contrast to subaerial positive features where sandstones may become coarser grained toward the feature. The trap should also be in updip drainage position from the deeply buried part of the basin where potential source rocks can generate hydrocarbons.

STRUCTURE

The Saticoy field occurs along the south flank of the Santa Clara syncline near its truncation by the Oak Ridge fault, constituting a continuous productive trend more than 7 mi (11 km) long. Bedding dips 60–90° in the main part of the field east of the Edwards fault (Figures 3, 7) and as low as 40° west of the Edwards fault (Figure 6). The main trap in the Saticoy field is stratigraphic. Turbidite sandstone reservoirs lens out updip southward to siltstone in the direction of the Oak Ridge fault. But in the eastern half of the field, including the entire Bridge pool, the trap is the Oak Ridge fault itself. Sandstone zero isopachs turn eastward into the fault, suggesting that

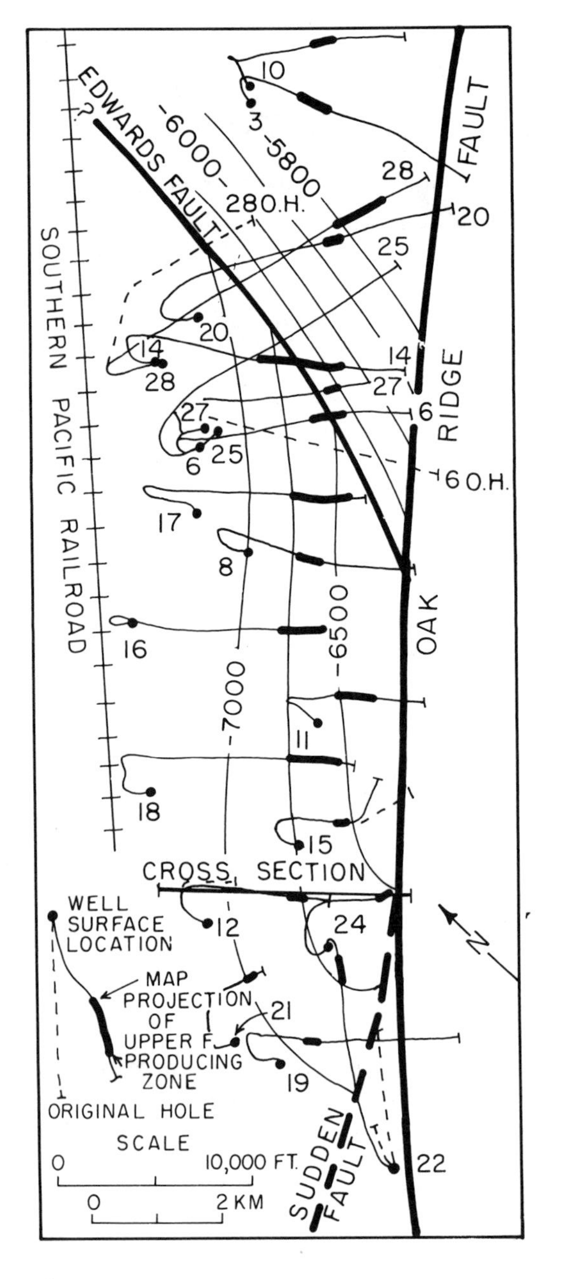

A

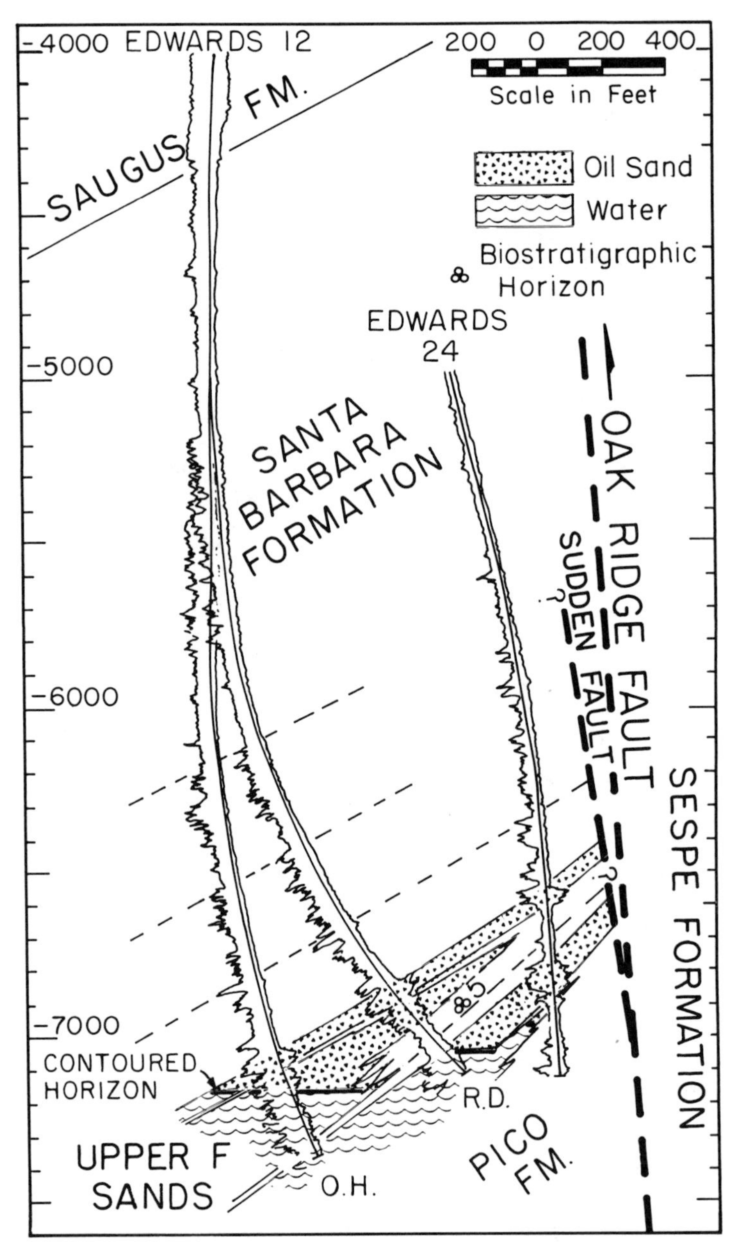

B

Figure 6. (A) Structure contour map and (B) cross section of Upper F pool at southwest end of Saticoy field, showing structural closure at end of field. Upper F pool extends across Edwards reverse fault to a west-dipping homocline, but cannot be correlated northeast of Edwards 10. All well numbers refer to Edwards lease wells. Contour interval 100 ft east of Edwards fault, 250 ft west of fault. O. H., original hole; R. D., redrilled hole.

the sand thicknesses were controlled by something other than the Oak Ridge fault.

Yeats (1965) showed that the South Mountain oil field, on the hanging-wall side of the Oak Ridge fault, is on the site of a seaknoll that was positive during deposition of Pliocene-Pleistocene turbidite sandstones. These sandstones occupied a trough south of South Mountain as well as north of it, and sandstones lens out to siltstone in the direction of South Mountain. Yeats (1965, p. 536) suggested that the pinchout of turbidite sandstones at Saticoy was produced by a positive feature on the seafloor, not the Oak Ridge fault. This positive feature may well have been influenced by an ancestral Oak Ridge fault. Yeats (1987) speculated that the fault had a Miocene normal-fault ancestor that influenced the location of Miocene structural highs along the Oak Ridge trend. However, the present-day Oak Ridge fault underwent nearly 2.5 km of displacement after the end of deposition of Pliocene and Pleistocene strata (Yeats,

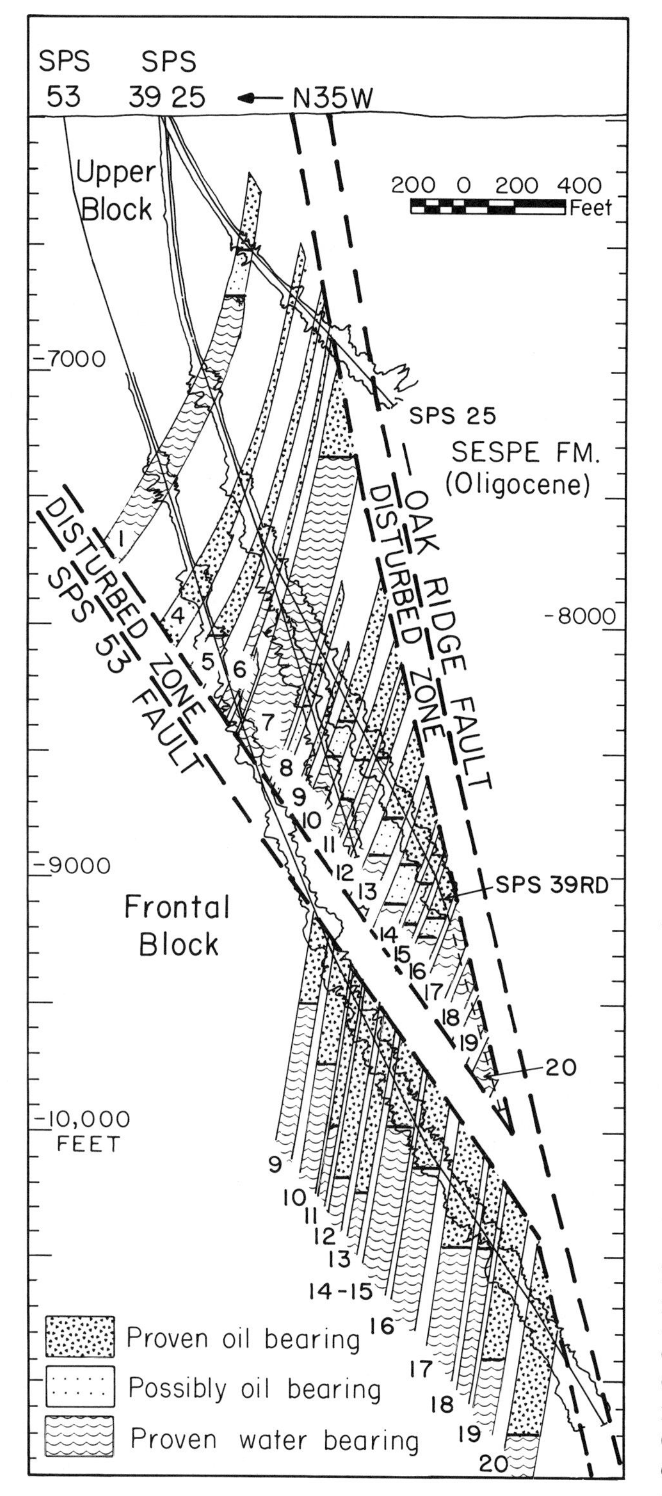

Figure 7. Cross section showing SPS 53 frontal fault and oil production in frontal block downstructure from water in SPS 53 well. Steeper dips in frontal block indicate that folding occurred prior to SPS 53 fault. In upper block, turbidite sandstone groups are truncated by Oak Ridge fault, whereas they lens out upstructure to mudstone farther west (compare Figure 3).

1988), and the difference in trend between this late Quaternary fault and the fault contemporaneous with Pliocene and Pleistocene deposition may have been enough to produce the discordance between isopachs of turbidite reservoirs and the present trace of the fault.

Figure 8 shows the vertical separation rates between various Pliocene and Pleistocene time-stratigraphic units along the entire onshore trace of the Oak Ridge fault. It is based on the difference in thickness of a given time-stratigraphic unit between the Santa Clara syncline on the north and the Oak Ridge anticline on the south, in turn based in part on thicknesses extrapolated north from the Long Canyon syncline and Happy Camp syncline farther south (Figure 1). Benthic foraminiferal assemblages indicate that at any given moment in turbidite depositional history, the troughs north and south of Oak Ridge were at the same depth. They acquired much of their vertical separation as the result of the greater subsidence of the northern trough relative to the southern trough. The vertical separation rates decrease gradually from east to west (Figure 8).

The vertical separation rate in the last few hundred thousand years, shown in Figure 8, is based on the estimated altitude of the top of the basin fill (top of the Saugus Formation as described under *Stratigraphy* below) in the trough of the Santa Clara syncline and the crest of the Oak Ridge anticline (see Yeats, 1988, for further details). In contrast to the gradual change in the older separation rates, the post-Saugus separation rate decreases abruptly from more than 10 mm/yr to zero near the coast. This shows that the post-Saugus tectonic style of the Oak Ridge fault, involving uplift of Oak Ridge, contrasts with the syndepositional Oak Ridge fault, which involved differential rates of subsidence of both hanging-wall and footwall blocks. The structure of Saticoy field and the timing of its oil accumulation can be understood best by considering syndepositional and postdepositional structures separately.

Yeats (1976) mapped the sandstone-siltstone contact of turbidite sandstones near his Microfaunal Horizon 5, dated as 1 million years old (Yeats, 1983), located above the Bailey ash horizon near the top of the turbidites in the main part of the Saticoy field (Figures 3 and 9). The sandstone-siltstone contact in the Saticoy field is based on isopachs of Sand 4 (Figure 10), although this sandstone group is slightly older than Horizon 5. Yeats (1976) correlated the eastward turn of the sandstone pinchout in the Saticoy field (point A, Figure 9; see also Figure 10) with the eastward turn of the Oak Ridge fault itself at the town of Santa Paula (point B, Figure 9). He indicated that the Oak Ridge fault had undergone 3.5 km of left slip and 3.75 km of vertical separation in the last million years, a net slip of 5.2 km in a northeast-southwest direction.

The Oak Ridge fault is relatively planar, separating Eocene and Oligocene strata on the south from Pliocene and Pleistocene strata on the north (Figures

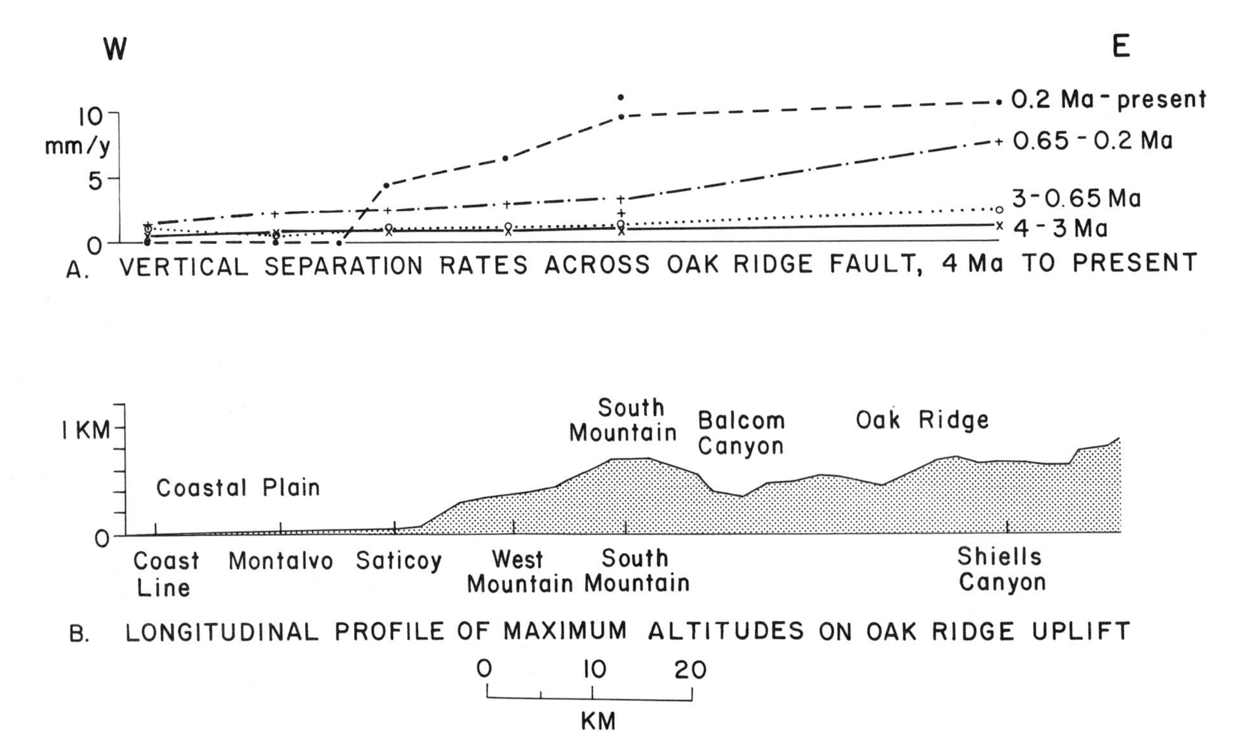

Figure 8. (A) Vertical separation rates across Oak Ridge fault, 4 Ma to present (Pliocene and Quaternary) plotted parallel to longitudinal profile of Oak Ridge and South Mountain. For location of place names in profile, see Figure 1. Saticoy producing measures deposited within the interval 3-0.65 Ma and Saugus during the interval 0.65-0.2 Ma. Displacement across Oak Ridge zone increases from west to east and upsection. (B) Longitudinal profile of maximum altitude on Oak Ridge uplift. Note sharp increase in displacement rates in last 200,000 years along Oak Ridge but not near coast.

3, 4, 5, and 6). The fault zone itself is characterized by highly fractured Monterey shale of Miocene age. As noted above, the fault changes in trend from east-west east of Santa Paula to northeast-southwest between the towns of Santa Paula and Saticoy. The offset of the shoulder of the South Mountain seaknoll suggests that the post-Saugus slip direction is north-northeast to northeast, and this is supported by focal-mechanism solutions of earthquakes and by borehole-breakout data based on four-arm, dual-caliper dipmeter logs (see also summary by Yeats et al., 1988). The fault is thus interpreted as largely dip slip east of Santa Paula and oblique slip southwest of Santa Paula with a large left-lateral component. The fault dip varies from 70° at the east end of the Bridge pool, where it is largely dip slip, to 85-87° at the southwest end of the Saticoy field, where it is in part strike slip.

The turbidite sandstones and siltstones of the Saticoy field are cut by microfractures, and the abundance of these microfractures increases toward the fault. Friedman (1969) noted that some of these fractures are filled with untwinned calcite, some are characterized by gouge and slickensides, and some are unfilled. The orientation of fractures suggests that they formed after folding and that they are related to the Oak Ridge fault. Friedman (1969) suggested that a study of microfracture orientation gives independent information about the subsurface fault pattern, including the orientation of frontal faults within the field.

Three subsidiary faults within the footwall block (Figure 5) were first described in an unpublished report by Hamann (1964). The SPS 53 fault was discovered by the SPS 53 well, which found oil-bearing sands downdip from water-bearing sands, apparently trapped against the SPS 53 fault (Figure 7). At first, the fault was difficult to detect because it has only about 100 ft of stratigraphic separation. Beds north of the fault dip more steeply than those to the south, and this fact, together with the presence of oil downstructure from water, clearly established the presence of the fault. Hamann (1964) proposed that the fault underwent about one-half mile of left-lateral strike slip with some reverse-slip component; faulting occurred after most of the folding of the south flank of the Santa Clara syncline and after the stratigraphic entrapment of oil. This, together with the observed stepwise displacement of the oil-water contacts downward to the northeast (discussed below), accounts for the presence of oil downdip from water and for the steeper dips in the northern block. Using this concept, several wells were completed successfully in the northern, frontal block.

At the southwest end of the field, the Edwards fault juxtaposes two different facies of Pleistocene

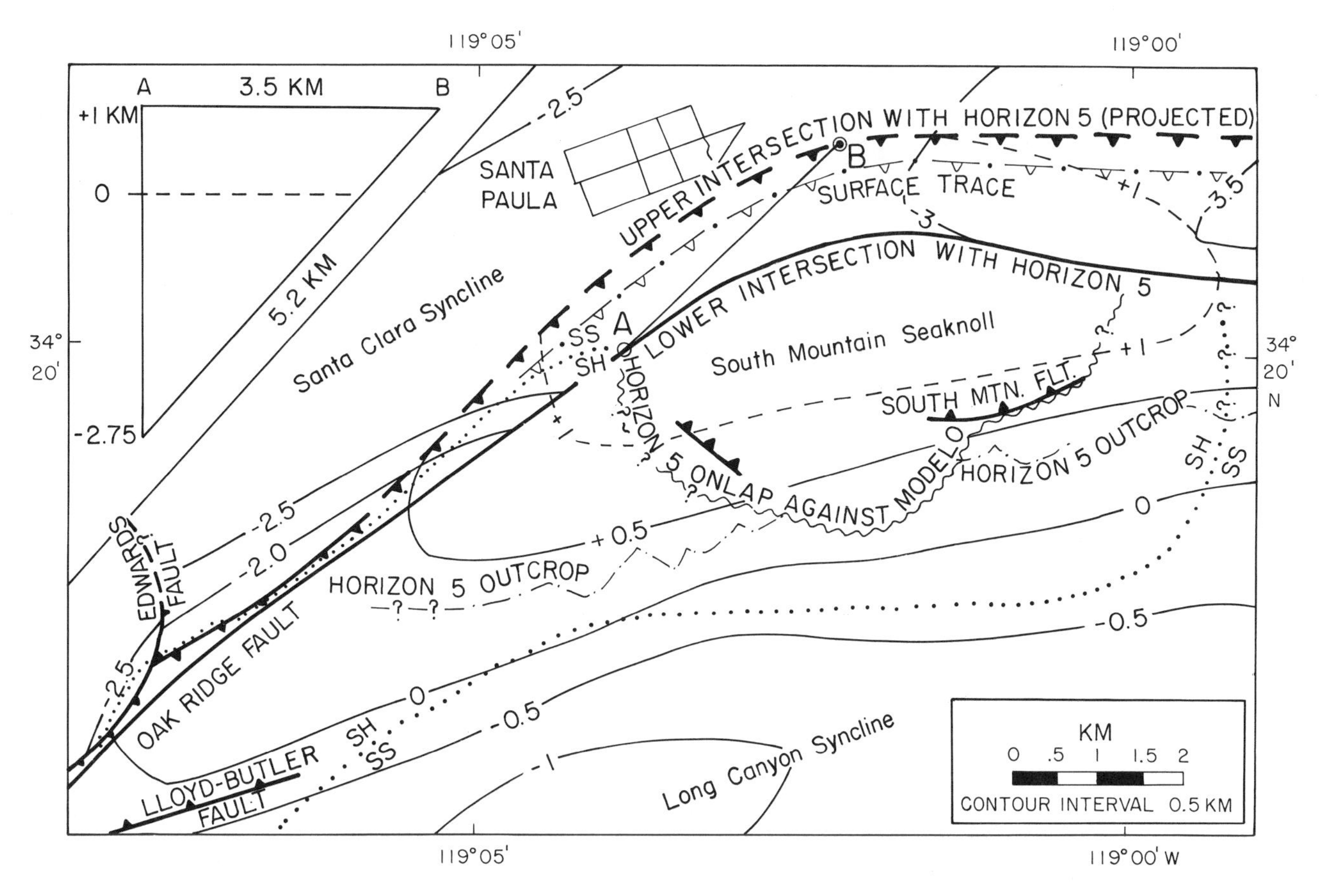

Figure 9. Structure contour map (in kilometers) of Microfaunal Horizon 5, about 1 Ma and above Sand 4, showing Saticoy field north and South Mountain seaknoll south of Oak Ridge fault. On south side, Horizon 5 onlaps against Modelo Formation on South Mountain and crops out. Contours projected into air based on curvature of older formations. Dotted line is approximate facies change of turbidite sandstone to siltstone at Horizon 5 on both sides of fault. Control on north side is from Saticoy field (see Figure 5) and on south side from Yeats (1965). Facies change turns toward the fault at A at Saticoy field, an apparent reflection of northwest shoulder of South Mountain seaknoll and bend in Oak Ridge fault projected into air at B. Net slip on Oak Ridge fault of 5.2 km reflects both reverse slip and left-lateral strike slip. Modified from Yeats (1976).

deep-water strata: thin-bedded, cross-laminated sandstones and siltstones on the west against coarse-grained channel sandstones and massive claystones on the east. Bedding dips are 40° in the western block and 65° in the eastern block. The Edwards fault diverges from the Oak Ridge fault in a northeasterly direction (Figure 6) and strikes northward into the Santa Clara syncline (Figure 1). Piercing-point offset of facies boundaries and sandstone isopachs suggest a displacement of about 3200 ft (1160 m), largely left-lateral.

The Saticoy 6 fault juxtaposes massive turbidite sandstones on the south against siltstone on the north (Figure 5). Hamann (1964) interpreted this fault as a left-lateral, strike-slip fault like the Edwards and SPS 53 faults. He proposed that the sandstones give way northward to siltstone, and left-lateral strike-slip faulting juxtaposed the sandstone-siltstone boundary. This explanation requires that the basin plain accommodating the turbidite sandstones be relatively narrow, since the northern edge of the turbidites would be near the north edge of the Saticoy field, and the southern edge would be truncated by the Oak Ridge fault.

STRATIGRAPHY

The stratigraphic nomenclature (Figure 11) follows generally accepted subsurface terminology (Vedder et al., 1969). Strata of early Pliocene age contain a Repettian Stage benthic microfauna according to the usage of Natland (1952) and are commonly called Repetto formation. Rocks of late Pliocene and early Pleistocene age contain Venturian Stage and Wheelerian Stage microfaunas (Natland, 1952) and are commonly called Pico formation; this formation is subdivided informally into a lower, middle, and upper member. The overlying Santa Barbara formation, also referred to as the Mudpit shale (Nagle and Parker, 1971, p. 277), contains Wheelerian and

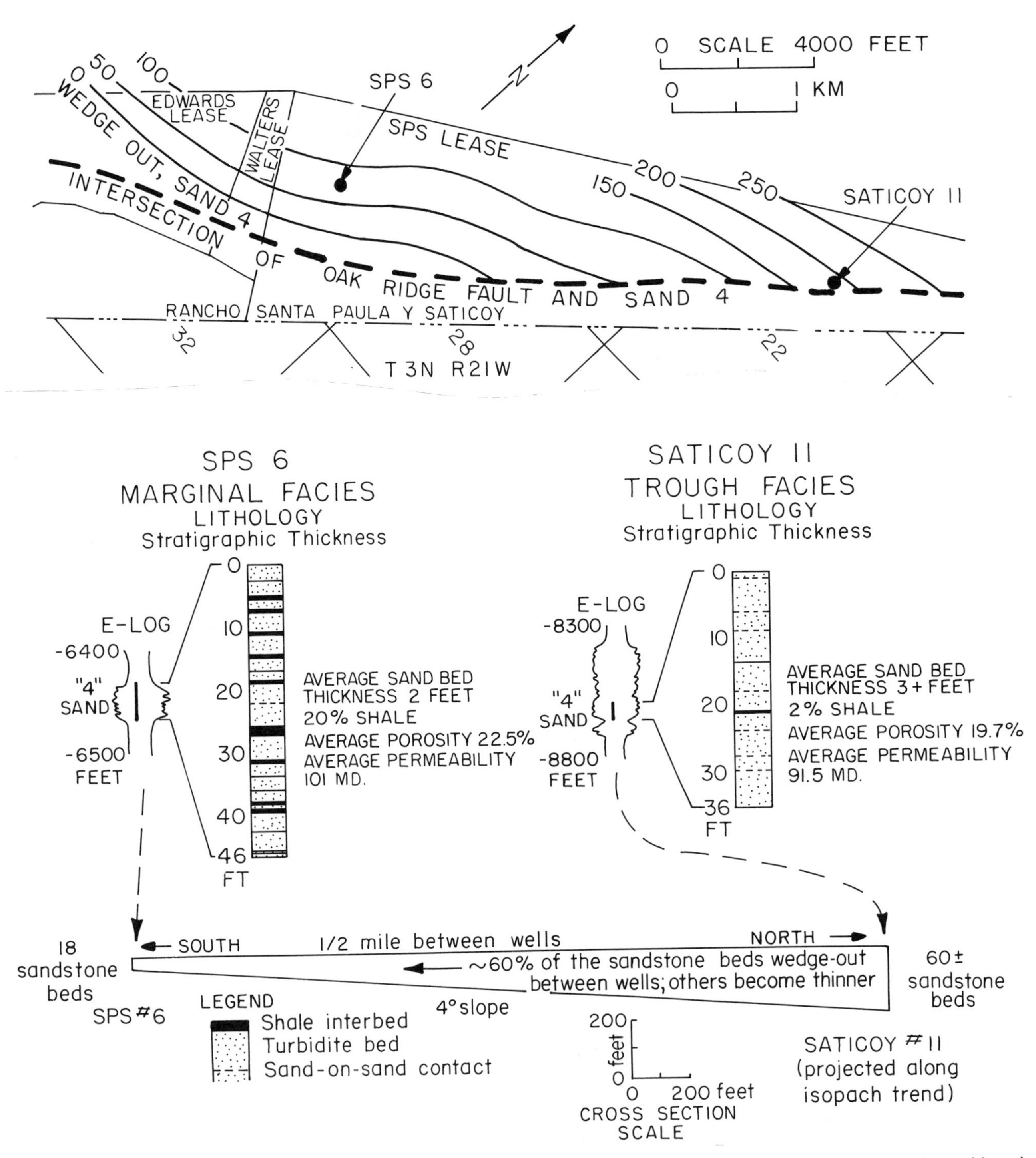

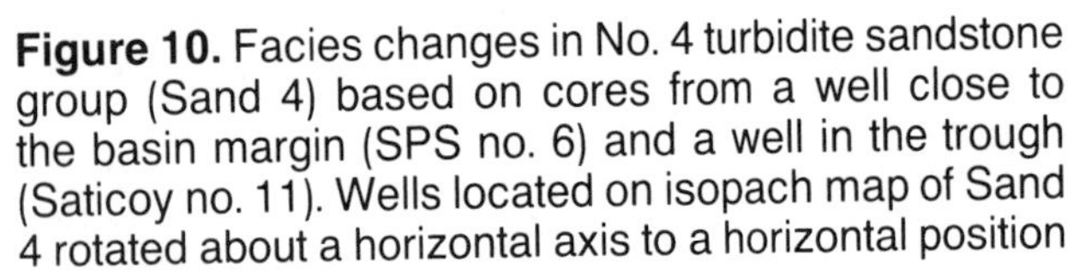

Figure 10. Facies changes in No. 4 turbidite sandstone group (Sand 4) based on cores from a well close to the basin margin (SPS no. 6) and a well in the trough (Saticoy no. 11). Wells located on isopach map of Sand 4 rotated about a horizontal axis to a horizontal position to show that Oak Ridge fault cuts across depositional strike. Isopach interval 50 ft. At bottom is a palinspastic cross section constructed at right angles to depositional strike, no vertical exaggeration, projecting Saticoy no. 11 parallel to isopachs downdip from SPS no. 6.

Hallian microfossils (Natland, 1952) and is of Pleistocene age. The Repetto, Pico, and Santa Barbara formations are equivalent to the Fernando Formation of Kew (1924). Overlying the Santa Barbara Formation in the Saticoy area are shallow-marine and nonmarine strata of Pleistocene age, variously referred to as the Saugus, San Pedro, or Las Posas Formation, and arbitrarily called Saugus Formation in this paper. Late Pleistocene and Holocene sediments are similar in lithology and alluvial facies

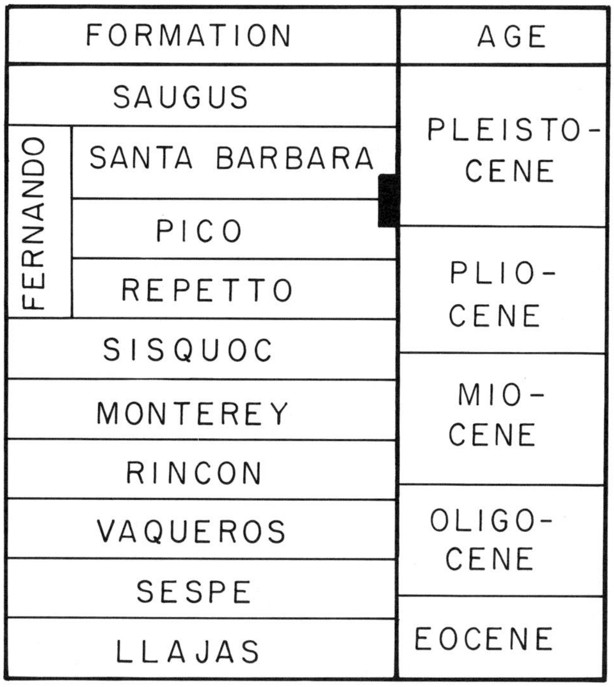

FORMATION		AGE
SAUGUS		PLEISTO-CENE
FERNANDO	SANTA BARBARA	PLEISTO-CENE
FERNANDO	PICO	PLEISTO-CENE / PLIO-CENE
FERNANDO	REPETTO	PLIO-CENE
SISQUOC		PLIO-CENE / MIO-CENE
MONTEREY		MIO-CENE
RINCON		MIO-CENE
VAQUEROS		OLIGO-CENE
SESPE		OLIGO-CENE / EOCENE
LLAJAS		EOCENE

Figure 11. Generalized stratigraphic column for the coastal Ventura basin. Heavy vertical bar indicates stratigraphic interval productive at Saticoy field.

to the Saugus Formation and are difficult to differentiate from the Saugus in the subsurface (Yeats, 1988).

The Saticoy sequence was correlated to the magnetic time scale by Blackie and Yeats (1976), who found that the producing sandstones were all deposited in the Matuyama Chron. Several ash beds in the Pico Formation and overlying strata have been correlated to their source areas where they are radiometrically dated or have been dated directly by the fission-track method (Yeats, 1983; Sarna-Wojcicki et al., 1987). The top of the Saugus Formation near the coast contains marine fossils with an estimated age of 0.4–0.2 Ma based on aminostratigraphy (Lajoie et al., 1982; K. R. Lajoie, written comm., 1987). Finally, the benthic microfaunal scale of Natland (1952) has been correlated to the planktic tropical oceanic scale based on pelagic foraminifers and coccoliths (for recent summary, see Lagoe, 1987).

The interbedded turbidite sandstones and hemipelagic siltstones were deposited in deep water in a continental-borderland trough similar to present-day basins off the southern California coast, based on benthic foraminiferal assemblages in the siltstones (Natland, 1952; Gorsline and Emery, 1959; Lagoe, 1987). The sandstones commonly contain shallow-water foraminifers, but these were transported downslope and redeposited by turbidity currents in the basin plain.

TRAP

The primary trapping mechanism is stratigraphic; the facies change from producing turbidite sandstone updip to siltstone in the direction of the South Mountain seaknoll. In the eastern part of the field, the oil is trapped against the Oak Ridge fault, but here we envision that the fault cut across an already existing stratigraphic trap. Lateral closure on the southwest end of the field occurs because the structural dips turn from northward to westward toward the Oak Ridge fault and are cut off by it. Closure on the east end is not clearly understood.

Reservoir Stratigraphy and Depositional Setting

Each reservoir group, from Sand 1 at the top of the Pico Formation to Sand 20 in the middle part of the Pico, is composed of many individual turbidite sandstones interbedded with hemipelagic siltstones. From electric logs, each group can be correlated for distances greater than 4 mi (6 km) along strike. Study of cores shows that sandstone and siltstone beds within these groups have similar sedimentary structures, textures, and facies changes. For this reason, the analysis of one group, Sand 4, is presented as representative of the entire sequence of Saticoy reservoir turbidites.

Depositional trends may be inferred from isopach maps, but such trends can be misleading where dips are as steep as they are in the Saticoy field. In Figure 10, Sand 4 has been rotated back to its position at the time of deposition. The isopach map shows a maximum thickness of 250 ft to the north and west and a pinchout of the sandstones to the south and southeast. The Oak Ridge fault cuts obliquely across the depositional strike of this group. Maximum sand deposition is inferred to have been in the deepest portion of the basin trough, and therefore the isopach lines should be approximately parallel to the paleobathymetric contours on the basin floor. Sedimentary structures measured in cores, including shale rip-ups and pull-overs, are shown in Figure 12 along with grain-orientation measurements. The structures indicate that the currents flowed from east to west.

Variations of the character and thickness of the sediments across the depositional trend can be compared by examining the cored sections from two wells shown in Figure 10. In the SPS no. 6 well, the entire 46 ft thickness of the Sand 4 group was cored. Eighteen sandstone turbidite beds are separated by siltstone interbeds several inches to several feet thick. The average sandstone bed thickness is 2 ft. These sandstones grade from Bouma type A (massive and graded sand; Bouma, 1962; Middleton and Hampton, 1973) at the base to type B, D, or E (parallel laminae, and/or fine-grained, pelagic, or low-density turbidite deposition) in the overlying siltstone interbed. Bouma type C layers,

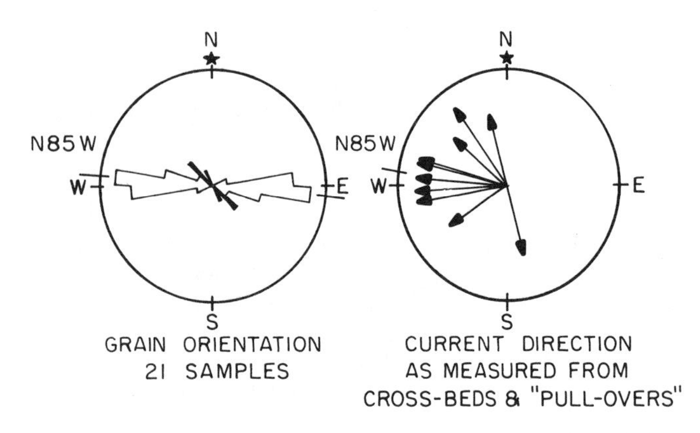

Figure 12. Directional features from cores of turbidite sandstone from well SPS no. 6. Well is located on Figure 10.

with ripples or wavy or convoluted laminae, are uncommon, and, where present, are poorly developed. Walker (1967) has observed that in proximal turbidites there is a strong tendency for type B and C layers to be absent.

In the second well, Saticoy no. 11, a 36 ft interval was cored near the base of the Sand 4 group, which at that locality has a thickness of 220 ft, nearly five times greater than at SPS no. 6 to the west. Most contacts between graded beds are erosional. Contacts of sand on sand, or "multiple amalgamated beds," of Walker (1967) are the rule, and siltstone interbeds the exception. Individual turbidite sandstone beds are thicker, averaging more than 3 ft. The average grain size is coarser, and the siltstones account for only 2% of the cored sequence. Texturally, beds are of Bouma type A, generally massive rather than graded, and many contacts between separate beds are difficult to distinguish where both are of massive type A and of similar grain size. Scouring and channeling are evident across the limited diameter of some of the cores.

The total number of sandstone beds in Sand 4 at the Saticoy no. 11 well is unknown because the entire interval was not cored. It is assumed that there are at least 60 beds by extrapolating the observations in the 36 ft cored interval to the entire sand group possessing similar electric-log characteristics. This contrasts with 18 sandstone turbidites cored in the SPS no. 6 well. Individual turbidite beds cannot be correlated from SPS no. 6 to Saticoy no. 11; these wells are 2.5 mi (4 km) apart, and they are located obliquely across depositional strike. A projection of Saticoy no. 11 parallel with depositional strike places it about 0.5 mi downslope from SPS no. 6, as shown diagrammatically in Figure 10. The 4° divergence across depositional strike between the base and top of Sand 4 must reflect basin differential subsidence between the base and top of this sand group rather than paleoslope. About 60% of the sandstone beds in the well near the axis of the depositional trough are not present in the more marginally-located well. The facies change between trough and marginal facies documented by the cored sections of these two wells is believed to be typical for all reservoir groups of the upper and middle parts of the Pico Formation.

In contrast to the turbidite sandstone groups described above, characterized by Sand 4, the sandstones at the top of the Pico are more irregular in distribution, with abrupt facies change. Sandstones commonly grade downdip to siltstone as well as updip toward the Oak Ridge fault. The sandstones contain slump structures and large siltstone clasts, but typical graded beds are uncommon. These sandstones are productive at the west end of the Saticoy field as the Upper F sands and Lower F sands. The Upper F sand group lenses out to siltstone toward the Oak Ridge fault (Figure 6); these strata could not be correlated with the main part of the Saticoy field. Farther east, the Lower F sand group is conglomeratic, containing clasts of Eocene sandstone derived from the Topatopa Mountains north of the San Cayetano fault. Blackie and Yeats (1976) interpreted these conglomeratic sandstones as part of a submarine channel fill resting with low-angle unconformity on the sandstones of the main part of the Saticoy field. Similar channel-fill conglomeratic sandstones of about the same age are exposed on the flanks of the Ventura Avenue anticline (Sexton-Lake conglomerate of Yeats, 1980) and in Santa Paula Creek (Crowell et al., 1966). The presence of the unconformity at the base of the Lower F sand group raises the possibility that the Lower F may be younger than the Upper F sands. This possibility cannot be tested directly because the two sand groups are not in direct contact.

Reservoir Characteristics

Production testing of the reservoirs soon after discovery demonstrated that the fluid pressure is hydrostatic, in contrast to the Ventura Avenue oil field, which is tectonically overpressured. At virgin pressures, reservoir temperatures were 160°F (71°C) at 7000 ft (2134 m) and 185°F (85°C) at 9000 feet (2743 m) (Blackie and Yeats, 1976).

The reservoir produces principally by solution gas drive aided by gravity drainage and, to a minor extent, water encroachment, particularly in Sand 7. The gas-oil ratio of the discovery well, 645 ft^3/bbl, is lower than that for the first 5 years of field production, 2045 ft^3/bbl (McCulloh, 1967), and McCulloh (1967) estimated that the original solution gas-oil ratio at virgin reservoir conditions was 550 ft^3/bbl. None of the reservoirs had an original gas cap, indicating that the reservoir fluid was not saturated with respect to gas (McCulloh, 1967). There is an absence of strong correlation of gas-oil ratio with structural position above the oil-water contact, a possible indication of poor reservoir continuity upstructure toward the Oak Ridge fault.

Oil-field water has an average salinity of 15,800 ppm (Schultz, 1960) and a total dissolved solids content of 16,200 ppm for water samples from the

Lower F sands. Oil gravity is 31° to 35° API, averaging 32.7°.

Reservoir porosities are 22% for the Upper F sands (Schultz, 1960), and, also in the western part of the field, 23.5% for Sand 4, 24% for Sand 7, 21% for Sands 8 and 10, and 23.6 to 17.5% for sands below Sand 10. In the eastern part of the field, porosities are lower, with values ranging from 11.4% to 21.3%. McCulloh (1967) noted that average porosities of sandstones 500 ft from the Oak Ridge fault decrease from 23% at -6500 ft (-2360 m) to 19% at -10,500 ft (-3815 m) whereas similar sandstones more than 1000 ft from the fault have porosities from 33 to 19% at those same respective depths. These low porosities occur in both the trough and marginal facies of turbidite reservoirs at Saticoy, so the low porosities are not explained only by facies. The explanation may be a combination of two factors: (1) The reservoirs were at a greater depth of burial and have been uplifted as they were tilted, and (2) the reservoirs have been altered, largely mechanically, accompanying movement on the adjacent Oak Ridge fault (McCulloh, 1967; Friedman, 1969).

Oil-Water Contacts

A penetration chart of Sands 4, 7, and 10, showing the original oil-water contacts (prior to production) and the updip pinchout of Sand 4 and its truncation by the Oak Ridge fault, is projected onto a vertical plane striking N 60° E in Figure 13. The updip shale-out line is preserved in the southwest but not in the northeast, where the Oak Ridge fault truncates the sand directly. The oil-water contacts along the field in this sand are progressively deeper to the northeast. In the southwest, the sands shale out a short distance above the oil-water contact, and the vertical oil column is only a few hundred feet high. In the east, the oil column is as much as 1200 ft (436 m) high. The oil-water contact in Sand 4 is at -6300 ft (-2289 m) in the southwest end of the field and -8750 ft (-3179 m) in the northeast end, a vertical difference of 2450 ft (890 m), or 0.5 mi in a horizontal distance of 4 mi. If the tilt were uniform and gradual, it would have a northeastward slope of 8°. But the change is stair-stepped with sharp changes in subsea position across short distances.

Table 1 shows the variation in subsea position of oil-water contacts from southwest to northeast for seven sand groups. The difference in altitude varies from 2250 ft (817 m) for Sand 5 to 2800 ft (1017 m) for Sand 10. There is no variation with stratigraphic position or with marginal vs. trough turbidite facies. The oil-water contacts change in a stair-stepped pattern in Sands 4, 7, and 10, where control is most complete; the stair-stepped pattern is also present for all reservoirs from Sands 1 to 10, although not documented in detail.

Except for one instance, the SPS 53 frontal fault, the stair-stepped pattern does not correspond to mapped faults. In most cases, the barriers for one sandstone group do not continue into the other sandstone groups (Figure 13). Explanations include capillary pressures resulting in a higher apparent oil-water contact closer to the shale-out line, but this appears unlikely because the vertical difference in oil-water contacts is the same for Sand 7, a thick-bedded sandstone group produced in part by water encroachment and characterized by somewhat higher porosity than adjacent, thinner bedded sandstones, as it is for thinner-bedded sandstone reservoirs. Furthermore, the original oil-water contact is higher for Sand 7 than it is for adjacent sandstones (Figure 7), whereas it should be lower if the water table were controlled by capillary pressure differences. A hydrodynamic origin is considered unlikely because the field is known to be cut by oil-sealing frontal faults that would prevent the northeastward flow required for a hydrodynamic origin. The explanation adopted here is that the sandstones are cut by calcite- and gouge-filled microfractures as described by Friedman (1969), and these microfractures are barriers to oil remigration. These microfractures may account in part for the relatively low porosities of reservoir sandstones at Saticoy noted by McCulloh (1967). Poor lateral continuity between the marginal-facies turbidites and those to the east may also affect oil remigration.

Timing of Folding, Faulting, and Oil Accumulation

When the effect of the SPS 53 frontal fault is removed, the northeast slope of the oil-water contacts in the upper block and frontal block varies from 3° to 8°, which may be compared to the 3° plunge of the Santa Clara syncline axis at the level of Horizon 5, a datum about 1 million years old near Sand 4. Most of this plunge occurred during the time of deposition of the Saugus Formation because the syncline axis is now at maximum burial. This leads us to conclude that the oil migrated into a primary stratigraphic trap against the South Mountain seaknoll during early stages of Saugus deposition. The Saticoy trap at that time resembled the shale-out of turbidite sandstones south of the South Mountain seaknoll documented by Yeats (1965). At the time of first accumulation of oil, oil-water contacts were horizontal, but these contacts were tilted northeastward as the remainder of the Saugus Formation was deposited. An implication of this interpretation is that if filled microfractures inhibited the readjustment of the oil-water contacts as tilting took place, these microfractures must have developed along the Oak Ridge fault during Saugus deposition and not during the faulting at higher rates that took place afterwards.

The broad-scale tilting of the turbidite reservoirs was followed by stronger folding, developing the asymmetric south flank of the Santa Clara syncline. The oil-producing reservoirs in the western end of the field may have been uplifted more than reservoirs

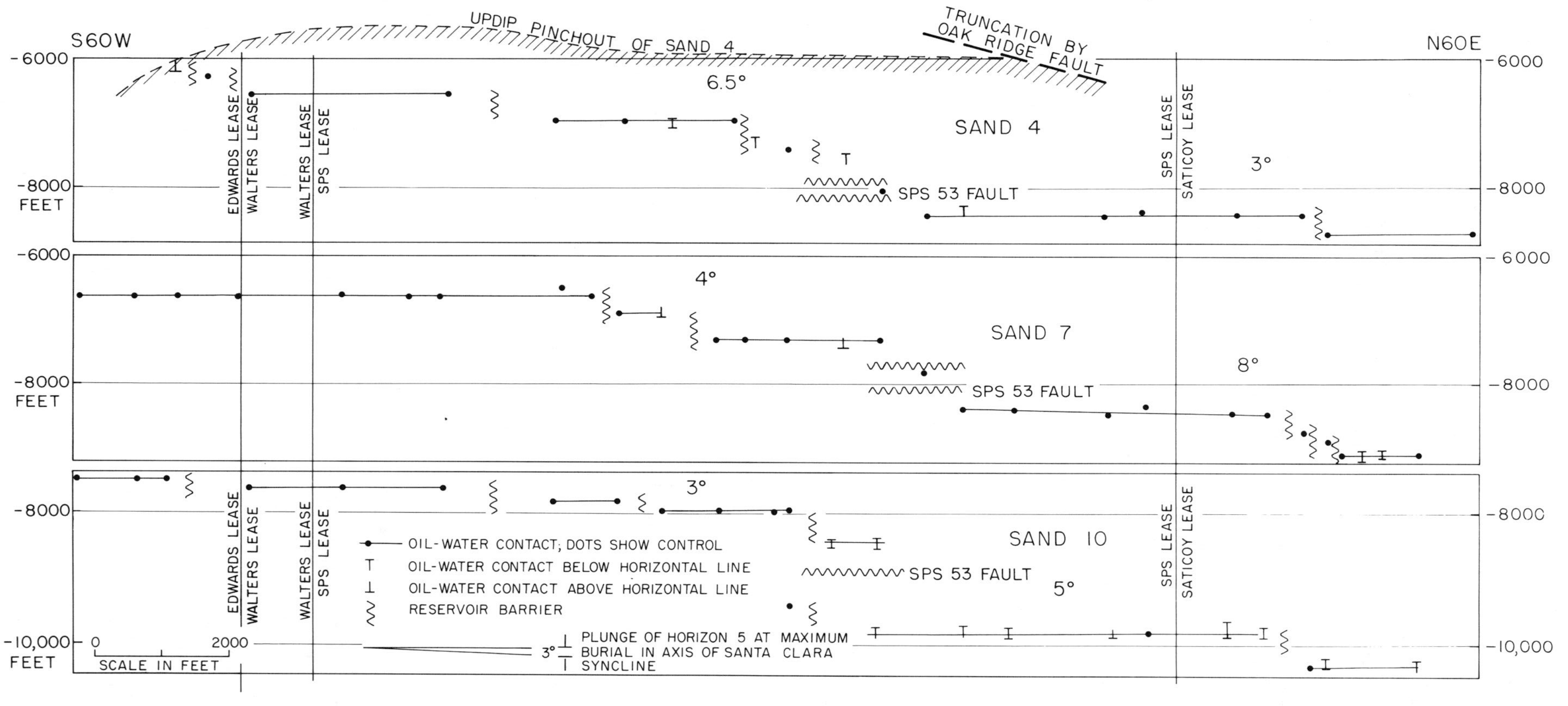

Figure 13. Variation in subsea depths of oil-water contacts in Sands 4, 7, and 10 from southwest to northeast end of Saticoy field, projected onto a vertical cross section parallel to Oak Ridge fault. No vertical exaggeration. Apparent dip of oil-water contact is 8° for Sand 4 and 9° for Sands 7 and 10. When the effects of the SPS 53 frontal fault are removed, the apparent dips are 6.5°, 4°, and 3° for Sands 4, 7, and 10, respectively, in the upper block and 3°, 8°, and 5° for these same sandstone groups in the lower block. Compare these with the 3° plunge of Horizon 5 (slightly younger than Sand 4) in the axis of the Santa Clara syncline. Change in oil-water contacts is not gradual; long segments with a common oil-water contact are separated by barriers that do not correspond to known faults. Individual barriers are not common to the three sandstone groups except for the SPS 53 fault and a barrier 2000 ft east of the SPS-Saticoy lease line. Updip pinchout and fault truncation of Sand 4 also shown.

Table 1. Variation in subsea depth of oil-water contacts from west to east ends of Saticoy field in different sand groups.*

Sand No.	Subsea Depth (ft) of Oil-Water Contact West End of Field	East End of Field	Differences in Depth (ft) of Oil-Water Contact
1	5370	7820	2450
4	6300	8750	2450
5	6710	8960	2250
6	6560	9030	2470
7	6650	9120	2450
8	7200	9700	2500
10	7500	10,300	2800

*See Figure 13 for penetration chart showing variation of oil-water contacts for Sands 4, 7, and 10. Only sand groups with positive electric-log correlation and oil-water contact control are listed; hence the Bridge Pool sand groups and the Upper F sands are excluded.

in the east end because the bed length from reservoir to syncline axis is greater in the west end. This differential uplift may account for the difference between the 3° plunge of the syncline axis at maximum burial and the 3° to 8° average slope of the oil-water contacts. The great difference in thickness of Saugus Formation in the Santa Clara syncline and around the South Mountain seaknoll suggests that some folding occurred during Saugus deposition, but much undoubtedly occurred later. Friedman (1969) observed that this is not drag folding related to the adjacent Oak Ridge fault, and Yeats (1988) suggested that both folding and faulting of basin strata may be a reflection of brittle faulting in subjacent basement rocks.

The SPS 53 frontal fault and the Edwards fault cut already folded strata, indicating that these faults were the latest structural feature to form. In addition, the observation that the Oak Ridge fault cuts across facies and across the oil-water contact suggests that the post-Saugus fault post-dates oil accumulation, and that post-Saugus faulting uplifted, exposed, and destroyed the updip portion of the eastern part of the Saticoy field on the hanging-wall side of the Oak Ridge fault.

This timing (early oil accumulation followed by folding followed by faulting) raises the question of why the tilted oil-water contacts did not reequilibrate. One explanation appears to be filled microfractures as described by Friedman (1969), but the relative simplicity of the fracture pattern suggested to Friedman that the fractures he studied formed after folding, whereas the fractures controlling the reequilibration of the oil-water contacts would have been in place before major folding. This paradox would strengthen the case for lateral stratigraphic discontinuities in the turbidites explaining the lack of reequilibration.

It should be noted here that tilted oil-water contacts are the rule, not the exception, not only in the Saticoy field, but also in the Ventura Avenue field and its western counterparts nearshore and offshore and the McGrath pool of the West Montalvo field. Yeats (1983) suggested that the anomalous oil-water contacts in the Ventura Avenue field may be a product of the very young age of folding, all in the last few hundred thousand years, and that the oil may not have had time to reequilibrate. Deformation is equally young at Saticoy, although nearly all deformation at West Montalvo had taken place by the end of Saugus deposition.

SOURCE

The source of the oil at Saticoy field is unknown. Siltstones interbedded with the reservoir turbidites are immature, and correlative fine-grained strata in the Fillmore field, near the point of maximum burial of Pliocene–Pleistocene strata in the Ventura basin, are also immature. Older Pliocene siltstones of the Repetto formation are immature in the Ventura Avenue field. They could be mature at Fillmore, but lower Pliocene strata have not been reached by drilling in that area. According to Bostick et al. (1978), gradients of vitrinite reflectance are unusually low in the Ventura basin, based on samples from the Standard-Maxwell and Superior-Limoneira wells (located on Figure 1). This is probably a response to the low geothermal gradient and the short time the strata have resided at maximum burial, where they are today. The source rocks may be the Monterey Shale of Miocene age, which is known to underlie the Pliocene strata in the Santa Clara syncline, although the Monterey has not been penetrated by drilling.

Migration occurred early, prior to strong folding and post-Saugus displacement along the Oak Ridge fault, because folding has distorted the original horizontal oil-water contacts, and late faulting has cut across the reservoir, uplifting and removing by erosion that part of the reservoir in the hanging-wall block of the fault. Because the source rocks must be older stratigraphically than the reservoir turbidites, oil migration had to cut across bedding, but the paths of migration are unknown.

EXPLORATION CONCEPTS

Saticoy field appeared at first to be trapped by the Oak Ridge fault, and exploration of the footwall block of the Oak Ridge fault was guided by this premise. Later, the field was shown to be stratigraphic, controlled by a seaknoll that caused turbidite sandstones to lens out toward it even though it was not itself a sediment source. No seaknoll, no oil accumulation! Oil could only accumulate north of the seaknoll where burial of source rocks in the Santa Clara syncline was sufficient to generate hydrocarbons. The turbidite reservoirs at Saticoy are Pleistocene in age, and so the source rocks could be Pleistocene as well, although this has not been substantiated.

Exploration for a Saticoy-type field involves three elements: (1) demonstration that a structural trap exists against an intrabasin or basin-margin fault, presumably using seismic-reflection profiles, (2) demonstration that the basin floor contains turbidite sands of reservoir quality and are either interbedded with or underlain by mature source rocks, and (3) demonstration that the stratigraphic equivalents of the turbidite objectives in the upthrown block are fine grained, indicating that a facies boundary between sandstone and siltstone or shale exists in the vicinity of the fault. In the case of Saticoy, the South Mountain seaknoll provided lateral closure for the initial stratigraphic trap, explaining why oil was not found against the Oak Ridge fault to the east or west.

ACKNOWLEDGMENTS

Much of the subsurface work was done while we were development geologists for Shell Oil Company; appreciation is expressed for the discussions with geologists and engineers concerned with Saticoy oil field development, especially S. H. Hamann, B. H. Mull, K. F. Pilgram, and D. C. Pontius. Bruce Hesson of the California Division of Oil and Gas provided updated production data for the Saticoy field and Bridge pool. A preliminary version of the paper was reviewed by S. H. Hamann, T. H. McCulloh, J. G. Vedder, and R. F. Yerkes, who made many useful suggestions for improvement. R. S. Yeats was supported by Grant 14-08-0001-G1194 from the U. S. Geological Survey and by the industry-supported Southern California Fault Studies Project at Oregon State University. The views expressed in this paper do not necessarily represent the views of any of these organizations and individuals.

REFERENCES CITED

Blackie, G. W., and R. S. Yeats, 1976, Magnetic-reversal stratigraphy of Pliocene-Pleistocene producing section of Saticoy oil field, Ventura basin, California: American Association of Petroleum Geologists Bulletin, v. 60, p. 1985–1992.

Bostick, N. H., S. M. Cashman, T. H. McCulloh, and C. T. Waddell, 1978, Gradients of vitrinite reflectance and present temperature in the Los Angeles and Ventura basins, California, *in* D. F. Oltz, ed., A symposium in geochemistry: low temperature metamorphism of kerogen and clay minerals: Pacific Section SEPM, p. 65–96.

Bouma, A. H., 1962, Sedimentology of some flysch deposits: Amsterdam, Elsevier, 168 p.

Crowell, J. C., R. A. Hope, J. E. Kahle, A. T. Ovenshine, and R. H. Sams, 1966, Deep-water sedimentary structures, Pliocene Pico Formation, Santa Paula Creek, Ventura basin, California: California Division of Mines and Geology Special Report 89, 40 p.

Friedman, M., 1969, Structural analysis of fractures in cores from Saticoy field, Ventura County, California: American Association of Petroleum Geologists Bulletin, v. 53, p. 367–389.

Gorsline, D. S., and K. O. Emery, 1959, Turbidity current deposits in San Pedro and Santa Monica basins off southern California: Geological Society of America Bulletin, v. 70, p. 279–290.

Hamann, S. H., 1964, Tectonic style of the Oak Ridge fault system in the Saticoy field: Unpublished report, 3p.

Hardoin, J. L., 1961, McGrath area of West Montalvo oil field: Summary of Operations, California Oil Fields, v. 47, n. 2, p. 43–53.

Jeffreys, S. R., 1958, Saticoy oil field, *in* J. W. Higgins, ed., A guide to the geology and oil fields of the Los Angeles and Ventura regions [California]: AAPG and SEPM Pacific Sections, p. 183–184.

Kew, W. S. W., 1924, Geology and oil resources of a part of Los Angeles and Ventura counties, California: U. S. Geological Survey Bulletin 753, 202 p.

Lagoe, M. B., 1987, Chronostratigraphic significance of late Cenozoic planktic foraminifera from the Wheeler Canyon and Balcom Canyon sections, Ventura basin, California, *in* T. L. Davis and J. S. Namson, eds., Structural evolution of the western Transverse Ranges: Pacific Section SEPM, v. 48A, p. 17–28.

Lajoie, K. R., A. M. Sarna-Wojcicki, and R. F. Yerkes, 1982, Quaternary chronology and rates of crustal deformation in the Ventura area, California, *in* J. D. Cooper, compiler, Neotectonics in southern California: Prepared for the 78th Annual Meeting of the Cordilleran Section of the Geological Society of America, Anaheim, California, April 19–21, 1982, p. 43–51.

McCulloh, T. H., 1967, Mass properties of sedimentary rocks and gravimetric effects of petroleum and natural-gas reservoirs: U. S. Geological Survey Professional Paper 528-A, 50 p.

Middleton, G. V., and M. A. Hampton, 1973, Sediment gravity flows: mechanics of flow and deposition, *in* Turbidites and deep-water sedimentation: Pacific Section SEPM, p. 1–38.

Nagle, H. E., and E. S. Parker, 1971, Future oil and gas potential of onshore Ventura basin, California, *in* I. H. Cram, ed., Future petroleum provinces of the United States—their geology and potential: American Association of Petroleum Geologists Memoir 15, p. 254–297.

Natland, M. L., 1952, Pleistocene and Pliocene stratigraphy of southern California: Ph.D. thesis, Univ. California, Los Angeles, 165 p.

Natland, M. L., and P. H. Kuenen, 1951, Sedimentary history of the Ventura basin, California and the action of turbidity currents: SEPM Special Publication 1, p. 76–107.

Sarna-Wojcicki, A. M., S. D. Morrison, C. E. Meyer, and J. W. Hillhouse, 1987, Correlation of upper Cenozoic tephra layers between sediments of the western United States and eastern Pacific Ocean and comparison with biostratigraphic and magnetostratigraphic age data: Geological Society of America Bulletin, v. 98, p. 207–223.

Schultz, C. H., 1960, Saticoy oil field: Summary of Operations, California Oil Fields, v. 46, n. 1, p. 58–69.

Vedder, J. G., H. C. Wagner, and J. E. Schoellhamer, 1969, Geologic framework of the Santa Barbara Channel region, *in* Geology, petroleum development, and seismicity of the Santa Barbara Channel region, California: U. S. Geological Survey Professional Paper 679-A, 11 p.

Walker, R. G., 1967, Turbidite sedimentary structures and their relationship to proximal and distal depositional environments: Journal of Sedimentary Petrology, v. 37, p. 25–43.

Ware, G. C., Jr., and R. D. Stewart, 1958, Bridge area, South Mountain oil field, *in* J. W. Higgins, ed., A guide to the geology and oil fields of the Los Angeles and Ventura regions [California]: AAPG and SEPM Pacific Sections, p. 180–182.

Yeats, R. S., 1965, Pliocene seaknoll at South Mountain, Ventura County, California: American Association of Petroleum Geologists Bulletin, v. 49, p. 526–546.

Yeats, R. S., 1976, Neogene tectonics of the central Ventura basin, California, *in* A. E. Fritsche, H. Terbest, Jr., and W. W. Wornardt, eds., The Neogene symposium: Pacific Section SEPM, p. 19–32.

Yeats, R. S., 1977, High rates of vertical crustal movement near Ventura, California: Science, v. 196, p. 295–298.

Yeats, R. S., 1980, Neotectonics of the Ventura Avenue anticline, *in* Subsurface geology of potentially active faults in the coastal region between Goleta and Ventura, California: Semiannual Technical Report to U. S. Geological Survey, Contract 14-08-0001-17730, 11 June 1979 to 10 December 1979, 28 p.

Yeats, R. S., 1983, Large-scale Quaternary detachments in Ventura basin, southern California: Journal of Geophysical Research, v. 88, n. B1, p. 569–583.

Yeats, R. S., 1987, Changing tectonic styles in Cenozoic basins of southern California, *in* R. V. Ingersoll and W. G. Ernst, eds., Cenozoic basin development of coastal California: Rubey Vol. VI, Englewood Cliffs, Prentice-Hall, Inc., p. 284–298.

Yeats, R. S., 1988, Late Quaternary slip rate on the Oak Ridge fault, Transverse Ranges, California: implications for seismic risk: Journal of Geophysical Research, v. 93, p. 12,137–12,149.

Yeats, R. S., G. J. Huftile, and F. B. Grigsby, 1988, Oak Ridge fault, Ventura fold belt, and the Sisar decollement, Ventura basin, California: Geology, v. 16, p. 1112–1116.

Appendix 1. Field Description

Field name .... *Saticoy (including Bridge Pool, South Mountain field)*

Ultimate recoverable reserves .... *85 × 10^6 BOE (64 × 10^6 BO)**

**BOE includes gas and condensates in oil equivalents.*

Field location:

- **Country** .... *United States*
- **State** .... *California*
- **Basin/Province** .... *Ventura*

Field discovery:

- **Year first pay discovered** .... *Pliocene-Pleistocene Pico Formation 1955*
- **Year second pay discovered** .... *Pliocene-Pleistocene Pico Formation 1956*
- **Third pay** .... *NA*

Discovery well name and general location
(i.e., Jones No. 1, Sec. 2T12NR5E; or Smith No. 1, 5 mi west of Sheridan, Wyoming):

- **First pay** .... *Shell Oil Co. S.P.S. 2, Sec. 29 T3N R 21W (formerly G-H-I-J-K sands; now Sands 1-20)*
- **Second pay** .... *Shell Oil Co. Edwards 3, Sec. 31, T3N R 21W (Upper and Lower F sands)*
- **Third pay** .... *NA*

Discovery well operator .... *Shell Oil Co.*
(if more than one pay in field, list operators of discovery well in other pays)

- **Second pay** .... *Shell Oil Co.*
- **Third pay** .... *NA*

IP in barrels per day and/or cubic feet or cubic meters per day:

- **First pay** .... *236 BO; 130 Mcf/gas*
- **Second pay** .... *883 BO; 440 Mcf/gas*
- **Third pay** .... *NA*

All other zones with shows of oil and gas in the field:

Age	Formation	Type of Show
None		

Geologic concept leading to discovery and method or methods used to delineate prospect, e.g., surface geology, subsurface geology, seeps, magnetic data, gravity data, seismic data, seismic refraction, nontechnical:

Deep wells established that thick Pliocene sands productive at Ventura Avenue oil field probably continued south to the Oak Ridge reverse fault. Possibly oil in the sands in the footwall were trapped against the fault. Oil in Oligocene sands at South Mountain oil field in the hanging wall suggested additional production below the fault. The prospect well was located using seismic reflection.

Structure:

Province/basin type (see St. John, Bally, and Klemme, 1984)

Bally 332; Klemme IIIBb

Tectonic history

The narrow Ventura trough was formed in the Pliocene. A thick sequence of turbidites grading upward to nonmarine sands and gravels accumulated until late Pleistocene (0.2 Ma); later folding and faulting of these beds continued to the present. The trough is flanked by the Oak Ridge, Red Mountain, and San Cayetano reverse faults that moved during Plio-Pleistocene deposition as well as afterward.

Regional structure

Turbidite sands on the steep south flank of the Santa Clara syncline are cut off by the Oak Ridge high-angle reverse fault.

Local structure

Dip of producing beds varies from 35°N to 80°S overturned. Frontal fault within field is oblique left and reverse slip.

Trap

Trap type(s)

Primary trap is stratigraphic: sands lensing to mudstone against the South Mountain seaknoll. Younger movement on the Oak Ridge fault has cut off the updip terminations of reservoir sands so that the present trap is now a fault trap for about half the reservoirs (20 unconnected reservoirs total).

Basin stratigraphy (major stratigraphic intervals from surface to deepest penetration in field):

Chronostratigraphy	Formation	Depth to Top in ft (m)
Pleistocene	*Saugus (San Pedro)*	*Surface*
Pleistocene	*Santa Barbara*	*2000-6700*
Plio-Pleistocene	*Pico*	*5000-10,200*

Location of well in field .... *NA*

Reservoir characteristics:

Number of reservoirs *20*

Formations .... *Pico; Lower Santa Barbara*

Ages .... *Late Pliocene to Pleistocene*

Depths to tops of reservoirs .... *5000-7000 ft*

Gross thickness (top to bottom of producing interval) .... *1400 ft*

Net thickness—total thickness of producing zones

Average .... *760 ft*

Maximum .... *Not determined*

Average

Maximum

Lithology

Coarse- to fine-grained, poorly sorted arkosic sandstone in repeated graded beds with characteristic Bouma A, B, D, and E bed types

Porosity type .... *Intergranular*

Average porosity *22%*

Average permeability .... *250 md*

Seals:

Upper

Formation, fault, or other feature .... *(1) Mudstone or (2) Oak Ridge fault zone*

Lithology .... *(1) Fine-grained siltstone and mudstone; (2) Monterey shale in fault zone*

Lateral

Formation, fault, or other feature .... *As above*

Lithology

Source:

Formation and age .... *Not determined, possibly Monterey Shale*

Lithology .... *Organic shale, dolomite*

Average total organic carbon (TOC) .... *Not known*

Maximum TOC .... *Not known*

Kerogen type (I, II, or III) .... *Not known*

Vitrinite reflectance (maturation) .... *Not known*

Time of hydrocarbon expulsion .... *Early stages of Plio-Pleistocene folding*

Present depth to top of source .... *7000 m*

Thickness .... *700 m*

Potential yield .... *Not known*

Appendix 2. Production Data

Field name .... *Saticoy (including Bridge Pool, South Mountain field)*

Field size:

- **Proved acres** *1170*
- **Number of wells all years** *179*
- **Current number of wells** *118*
- **Well spacing** .... *NA*
- **Ultimate recoverable** .... *85.5 million bbl or oil equiv.*
- **Cumulative production** .... *64.2 million bbl (Saticoy field, 21.4; Bridge Pool, 42.8); 127.158 bcf (Saticoy, 41.054; Bridge, 86.104)*
- **Annual production (1988)** .... *0.13 million bbl for Saticoy; not available for Saticoy and Bridge combined*
- **Present decline rate** .... *Not available*
  - **Initial decline rate** .... *Not available*
  - **Overall decline rate** .... *Not available*
- **Annual water production** .... *0.37 million bbl for Saticoy; not available for Saticoy and Bridge combined*
- **In place, total reserves** .... *Not determined*
- **In place, per acre-foot** .... *Not determined*
- **Primary recovery** .... *Not determined*
- **Secondary recovery** .... *Not determined*
- **Enhanced recovery** .... *Not determined*
- **Cumulative water production** .... *68.7 million bbl (Saticoy, 20.6; Bridge, 48.1)*

Drilling and casing practices:

- **Amount of surface casing set** .... *500–1000 ft of 13⅜ to 10¾-in.*
- **Casing program**
  Wells directionally drilled to TD, set water string 5½-in. or 7-in.) cemented at shoe, water shut-offs above and below producing zone, casing was gun-perforated
- **Drilling mud** .... *Water-base*
- **Bit program** .... *Not available*
- **High pressure zones** .... *None; all at hydrostatic pressure*

Completion practices:

- **Interval(s) perforated** .... *Gun-perforated after water shut-off above and below producing zone*
- **Well treatment** .... *None*

Formation evaluation:

- **Logging suites** .... *Mainly IES; occasionally dipmeter*
- **Testing practices** .... *DST*
- **Mud logging techniques** .... *Not available*

Oil characteristics:

- **Type** .... *Intermediate-intermediate (U.S. Bureau of Mines and USGS classification)*
  (Tissot and Welte Classification in "Petroleum Formation and Occurrence," 1984, Springer-Verlag, p. 419)
- **API gravity** .... *32.7°*
- **Base** .... *NA*
- **Initial GOR** *550*
- **Sulfur, wt%** *0.94*
- **Viscosity, SUS** .... *Not available*
- **Pour point** .... *Not available*
- **Gas-oil distillate** .... *Not available*

Field characteristics:

Average elevation .... *200 ft*
Initial pressure .... *Hydrostatic gradient*
Present pressure .... *Not available*
Pressure gradient .... *0.43 psi/ft (initial)*
Temperature .... *160°F at 7000 ft subsea; 185°F at 9000 ft subsea*
Geothermal gradient .... *0.012°F/ft*
Drive .... *Solution-gas drive, Sands 1-20; subordinate water drive, Sand 7; subordinate gravity drainage, Sands 1-20*
Oil column thickness .... *1400 ft*
Oil-water contact .... *Variable, deeper to northeast*
Connate water *40%*
Water salinity, TDS .... *18,000 mg/L*
Resistivity of water .... *0.428 ohm-m w/75°F*
Bulk volume water (%) .... *Not determined*

Transportation method and market for oil and gas:
Not available.

Midway-Sunset Field—U.S.A.
San Joaquin Basin, California

R. B. LENNON
Shell Western Exploration & Production Inc.
Houston, Texas

FIELD CLASSIFICATION

BASIN: San Joaquin
BASIN TYPE: Wrench
RESERVOIR ROCK TYPE: Sandstone
RESERVOIR ENVIRONMENT OF DEPOSITION: Alluvial, Shoreline, Shallow Marine, and Turbidite
RESERVOIR AGE: Miocene to Pleistocene
PETROLEUM TYPE: Oil
TRAP TYPE: Anticlinal and Stratigraphic

LOCATION

Midway-Sunset field is located in the southwest end of the San Joaquin Valley of California mainly in Kern County with a very small portion extending into San Luis Obispo County. The field, with an average width of 3 mi (4.8 km), extends from the vicinity of the town of McKittrick southeastward along the foothills of the Temblor Range over 25 mi (40 km) terminating approximately 6.5 mi (10.5 km) easterly beyond the town of Maricopa. It is easily accessible from State Highway 33, which traverses the field throughout its entire length (see Figure 1).

With over 10,000 producing wells and 21,830 proven productive acres, the field ranks fifth in U.S. oil and gas fields on the basis of ultimate recovery in equivalent barrels of oil with 2252 million barrels of oil (California State Division of Oil and Gas estimate). On the same basis, it ranks 107th in the world. It is bordered on all sides except the southwest by other oil fields including the large Buena Vista oil field, which was considered a part of Midway-Sunset until 1953.

The topography, with elevations ranging from 500 to over 1700 ft (150–520 m), varies from gently sloping alluvial fans to smoothly rounded hills occasionally dissected by gullies. Because of the low annual precipitation, the land not occupied by oil field installations or townsites is suitable only for grazing.

HISTORY

Pre-Discovery

The earliest use of petroleum in the region was by the Yokut Indians who used small balls of asphalt as an adhesive and caulking material for basketry and tool making. In historic time, early settlers of the region are known to have used crude oil from extensive seeps in the foothills at the base of the Temblor Range in the vicinity of the towns of McKittrick and Maricopa at the northwestern and southeastern extremities of the field as a lubricant for wagons. The settlers also utilized the asphalt deposits for roads in the area. As early as 1864, two refineries to process asphalt mined from tunnels and open pits around McKittrick were in operation, and hand-drilling of a well for oil was undertaken. The latter was abandoned when the bit was lost between 16 and 18 ft.

Discovery

Several early wells were drilled in the general area around the town of McKittrick between 1867 and 1894, but little is known about these wells. A well drilled by Solomon Jewett and Hugh Blodget in 1894 is known to have been located in Section 21-T11N-R23W in Kern County, southeast of Maricopa (see Figure 1). This well, the No. 1 Jewett and Blodget, is generally recognized as the discovery well of the Midway-Sunset field. The well produced a small amount of low gravity oil from a sandstone of the Tulare Formation at approximately 100 ft (30 m). The early development history in this area is quite obscure. One source reports that Jewett and Blodget completed 16 wells in Sections 21 and 28 which together produced 30 bbl/d of 8–6° gravity oil that was treated in a small refinery they erected to produce heavy distillates and refined asphaltum.

In 1901, the Belgian Oil Company found production in the Belgian Hills area of the field with their No. 1-A well in Section 34-T30S-R22E. Production was again from the Tulare with the depth and amount

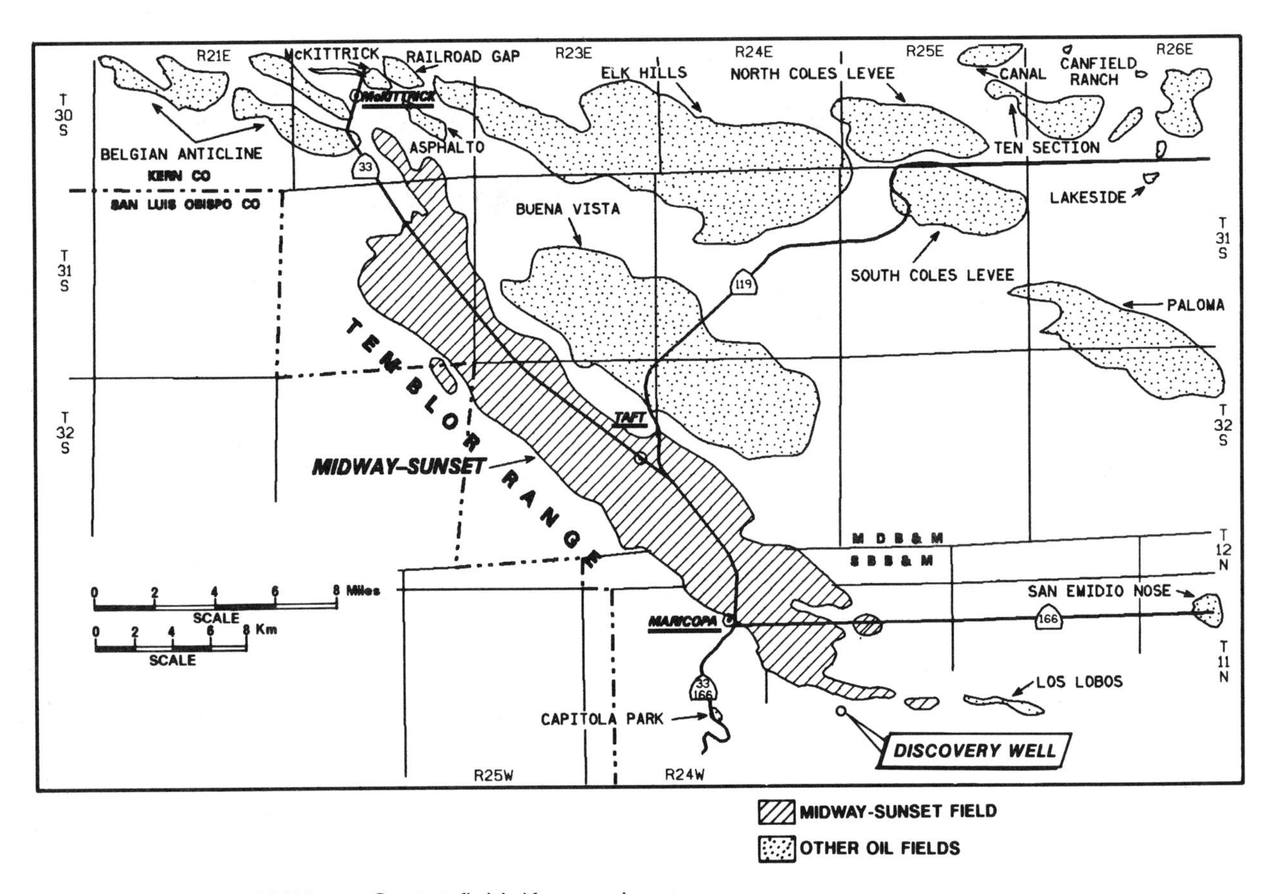

Figure 1. Location of Midway-Sunset field, Kern and San Luis Obispo Counties, California. (After Land and Anderson, 1965 Courtesy CDOG.)

of production unknown. Difficulties in obtaining water for drilling and a very low price for oil (10 cents a barrel in 1903) deferred development in both parts of the field and in a separate accumulation near McKittrick. However, with the completion of the Southern Pacific Railroad from Bakersfield to Maricopa in 1902 and the return of adequate oil prices in 1907, the field began to grow.

The development near Maricopa obtained the name of Old Sunset from the poetic descriptions of the sunsets in this area by Judge Lovejoy of Tulare, one of the early developers of this part of the field. For the other part of its name, Midway-Sunset has mule skinners to thank. A rest station on the hauling route midway between McKittrick and Maricopa, near present-day Fellows, was known by mule skinners as Midway, and the name was given to the field that developed there. When it became apparent that both accumulations were part of the same field, the field took the name of Midway-Sunset (Rintoul, 1984).

From 1901 to 1909, development of the field proceeded in cycles of exploration activity and new pool discoveries followed by periods of slow development. Drilling depths were generally less than 2000 ft (610 m) and production ranged from a few barrels to a few hundred barrels of oil per day. One well in 1909 was reported to have production of 2000 BOPD. By late 1910, more than 25 wells with initial production of at least 1000 BOPD were known, and the Webster area and Lakeview pools near Maricopa with production from Pliocene and Miocene sandstones had been discovered.

The most spectacular of the field's gushers was the Lakeview Oil Co.'s No. 1 in Section 25-T12N-R24W, which in March 1910 roared in flowing at rates estimated as high as 68,000 bbl/d. The well flowed uncontrolled for 544 days before the hole caved in, and the gusher died as suddenly as it started. Production was estimated at 9 million bbl of which about 3.5 million bbl were lost by evaporation and seepage.

Post-Discovery

From 1910 to 1928, exploration activity was relatively slow until 1928 when the Republic pool was discovered as the result of a "deep" test (2704 ft; 824 m) in the upper Miocene Antelope Formation. Exploration activity was again slow during the depression years of the early 1930s with the discovery of the Gibson Pool in 1936 stimulating development for a short time.

The period from 1940 to 1955 was marked by discovery of numerous pools that extended the field limits to about the present-day outline. A final discovery phase occurred about 1962, when the Olig, Monarch 10-10, and "1962" pools were discovered. No significant discoveries have been made in Midway-Sunset since 1962, but, as illustrated by Figure 2, the yearly production of oil has increased three and a half times since then. The reason for this increase has been the implementation of steam soak and steam drive recovery techniques. An in situ combustion project began in 1960 and some continue today; however, steam has claimed the major credit for the 55 million bbl of oil being produced annually. Steam soaking, which started about 1964, consists of injecting steam into a producing well periodically to stimulate its production while steam drives involve permanent steam injection wells in a pattern or line. Both processes have resulted in closer and closer well spacing.

The giant status of the Midway-Sunset field was recognized in 1910 with the flowing of the Lakeview gusher. The term "super-giant" was applied by the mid-1940s when the cumulative production reached more than 1 billion bbl of oil. However, this production included the adjacent Buena Vista field, which was determined to be a separate giant field in 1953. Midway-Sunset again entered the billion-barrel club in 1968, being the fourth U.S. field to qualify. For a tabulation of discovery data by zones as compiled by the California Division of Oil and Gas, see Table 1.

DISCOVERY METHODS

The discovery of Midway-Sunset field resulted from surface indications. The investigation of extensive oil seeps and asphalt deposits resulted in the discovery well being drilled near Maricopa in 1894 and the later discovery in 1901 in the Belgian Hills. The general areal extent of the field was recognized by 1902. Development at least up through the 1920s was based on random drilling, while since that time surface and subsurface geological studies have led to the drilling of deeper tests.

Had modern exploration techniques been available when the field was being discovered, they would have provided an earlier understanding of the structure and stratigraphy of this field. Geophysical data could have delineated most of the structural and unconformity traps, but it is doubtful such data could locate pools like the Gibson, which is a sand lens in a shale formation, or the Lakeview, which is located in a syncline with the trap produced by gravity drainage.

Surface evidence of hydrocarbons and structural closures are numerous at Midway-Sunset; i.e., oil seeps, tar deposits, and anticlinal structures. These attracted geologists in search of oil to the area nearly a century ago as they would certainly attract explorationists today to similar areas.

STRUCTURE

Tectonic History

The Midway-Sunset field is located in the southwest portion of the San Joaquin basin, a synclinal trough 250 mi (400 km) long and 50–60 mi (80–100 km) wide between the San Andreas fault in the Coast Range and the Sierra Nevada Mountains (see Figure 3). The basin is filled with over 25,000 ft (7620 m) of post-Jurassic sediments and is asymmetrical, with the west flank being the steeper. It originated as the southern portion of the Great Valley fore-arc basin when the Late Jurassic Nevadian orogeny resulted in intense folding of older sediments followed by intrusion and uplift of the Sierra Nevada batholith. At about the same time, uplift and slight folding of the ancestral Coast Ranges occurred. The subsiding basin between the two uplifts was occupied by a shallow inland sea until its withdrawal at the end of the Pliocene. During the early Tertiary, the San Joaquin basin was segmented from the rest of the Great Valley basin and probably more closely resembled a borderland style basin (Graham and Williams, 1985). By middle Tertiary time, plate interactions resulted in the northward propagation of the San Andreas transform fault system. Thus, the San Joaquin gradually was further isolated from the open Pacific by the northward translation of the Salinian block on the west side of the San Andreas.

Local and Regional Structure

As a result of compressional forces that were active periodically throughout the Tertiary and reached a climax in a major mid-Pleistocene orogeny, the tightly folded and locally overturned Temblor Range on the southwest flank of the San Joaquin Valley formed. Likewise the sediments along the west side of the basin were deformed into a series of anticlines and structural noses with axes oriented generally northwest-southeast. These are shown in Figure 4, a structure map contoured on the upper Miocene Antelope Shale "N" marker (Webb, 1981) (see correlation chart, Figure 7). As illustrated regionally in Figure 4 and in greater detail in Figures 5 and 6, Midway-Sunset field is made up of five or six of these structures.

During middle Pliocene time, the uplift of the Temblor Range and rifting along the San Andreas fault caused a northeast tilting of the Midway-Sunset basic structure. The Temblor Range is primarily a single anticline that is broken by numerous faults

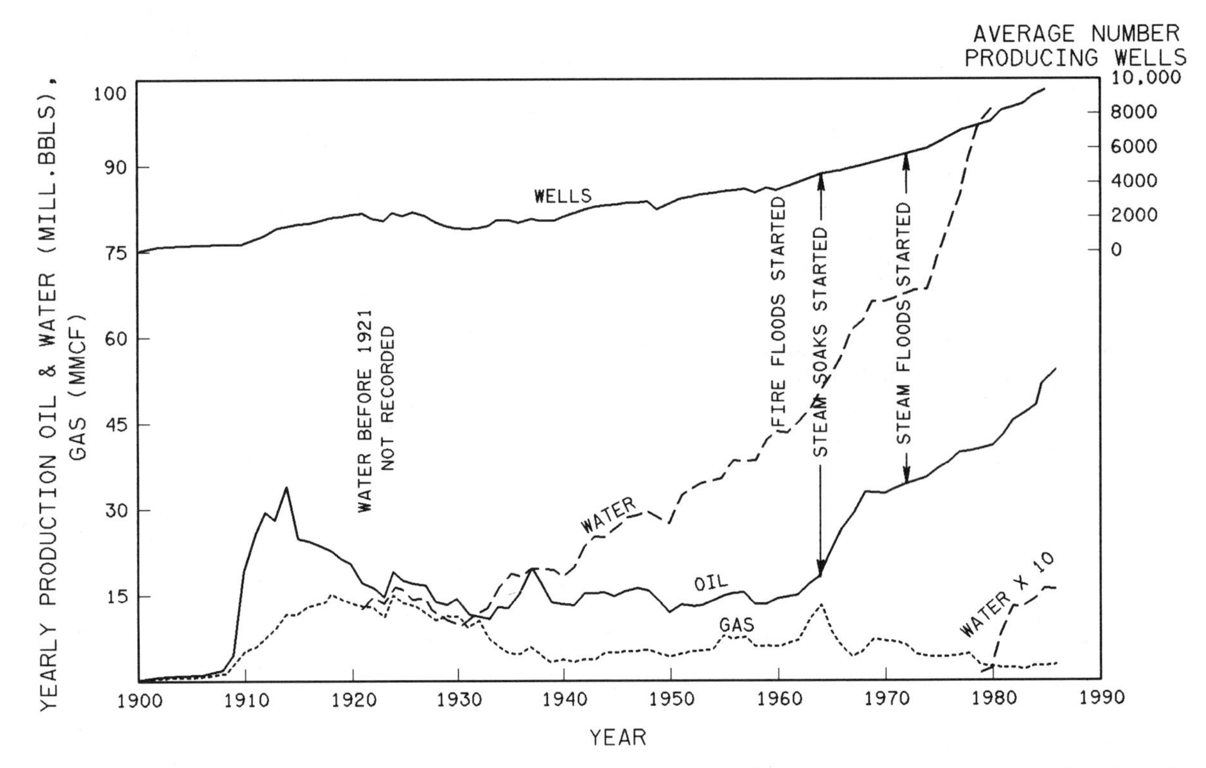

Figure 2. Graph of production history of the Midway-Sunset field, Kern and San Luis Obispo Counties, California, showing yearly production of water, oil, and gas as well as the number of producing wells.

Table 1. Zone discovery data, Midway-Sunset field

ZONE	PRESENT OPERATOR & WELL NAME	ORIGINAL OPERATOR & WELL NAME	SEC. T. & R.	* B&M	INITIAL DAILY PRODUCTION OIL (BBL)	INITIAL DAILY PRODUCTION GAS (MCF)	DATE OF COMPLETION
TULARE	OPERATOR NAME AND WELL NUMBER UNKNOWN	SAME AS PRESENT	N.A.	MD	N.A.	N.A.	PRIOR TO 1894
MYA TAR	GETTY OIL CO.NO.101	ASSOCIATED OIL CO. NO.101	2 31S 22E	MD	10	N.A.	JAN 1920
TOP OIL	OPERATOR NAME AND WELL NUMBER UNKNOWN	OPERATOR NAME AND WELL NUMBER UNKNOWN	N.A.	MD	N.A.	N.A.	N.A.
KINSEY	SAME AS ABOVE	SAME AS ABOVE	N.A.	MD	N.A.	N.A.	N.A.
WILHELM	SAME AS ABOVE	SAME AS ABOVE	N.A.	MD	N.A.	N.A.	N.A.
GUSHER	CHANSLOR-WESTERN OIL & DEV.CO.NO.2	CHANSLOR-CANFIELD MIDWAY OIL CO.NO.2	6 32S 23E	MD	3000	N.A.	NOV 1909
CALITROLEUM	OPERATOR NAME AND WELL NUMBER UNKNOWN	SAME AS PRESENT	N.A.	MD	N.A.	N.A.	N.A.
LAKEVIEW & SUB-LAKEVIEW	MOBIL OIL CORP. "LAKEVIEW" 1	LAKE VIEW OIL CO.NO.1	25 12N 24W	SB	68,000	N.A.	MAR 1910
POTTER	EXETER OIL CO.LTD. "EXETER-BAOC" 101-15	DOMINION OIL CO.NO.1	15 31S 22E	MD	100	N.A.	JAN 1910
MARVIC	MOBIL OIL CORP. "MARVIC" 1	MARVIC ASSOCIATES LTD.NO.1	16 31S 22E	MD	72	N.A.	MAY 1941
MONARCH	STANDARD OIL CO. OF CALIF."MONARCH" 28	SUNSET-MONARCH OIL CO.NO.1	2 11N 24W	SB	N.A.	N.A.	ABOUT 1902
WEBSTER	DIRECTORS OIL CO. NO.7	RUBY OIL CO.NO.7	2 11N 24W	SB	35	N.A.	DEC 1913
MOCO	MOBIL OIL CORP. "MOCO 35" WT504	GENERAL PETROLEUM CORP. "MOCO 35" 204	35 12N 24W	SB	188	20	JUL 1957
OBISPO	UNION OIL CO.OF CALIF."OBISPO" 6	OBISPO OIL CO.NO.6	32 12N 23W	SB	6000	N.A.	SEP 1925
PACIFIC	MOBIL OIL CORP. "PACIFIC" 4	GENERAL PETROLEUM CORP."PACIFIC"4	32 12N 23W	SB	1078	N.A.	JUN 1947
METSON	TENNECO OIL CO. "METSON" 47-24	BANKLINE OIL CO. "METSON" 47-24	24 11N 23W	SB	27	0	MAR 1953
LEUTHOLTZ	GULF OIL CORP.NO.2 - "I.M.WOODWARD USL"	WESTERN GULF OIL CO.NO. 2-"I.M.WOODWARD USL"	21 11N 23W	SB	1021	120	AUG 1945
REPUBLIC	SHELL OIL CO. "SEC.8" 25	REPUBLIC PETROLEUM CO. NO.25	8 32S 23E	MD	1114	350	MAR 1928

AFTER CALIFORNIA DIVISION OF OIL AND GAS

* B&M = BASE AND MEDIAN
MD = MT. DIABLO
SB = SAN BERNARDINO

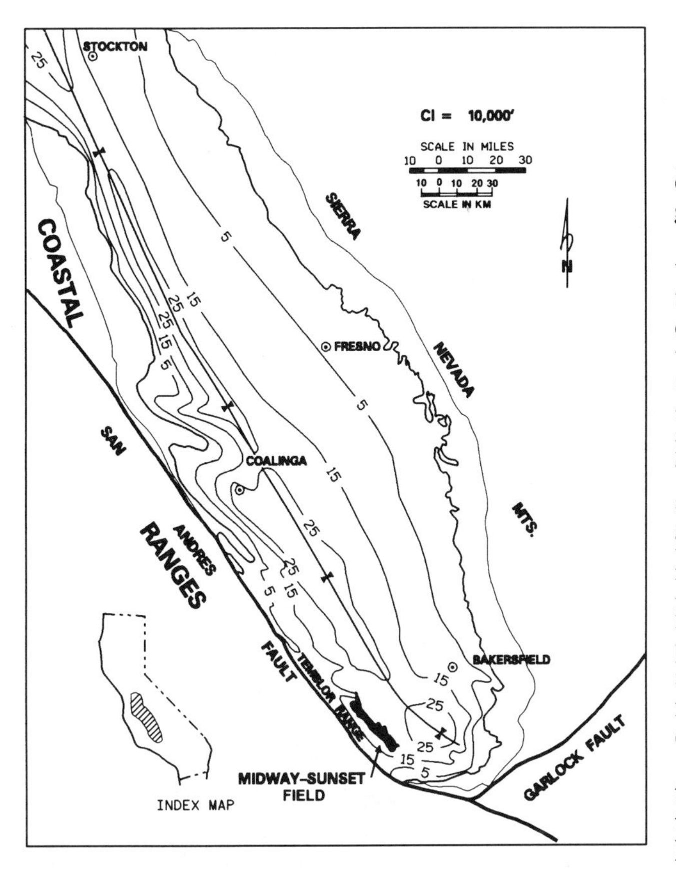

Figure 3. Isopach map of the total sediments in the San Joaquin Valley, California, showing the location of the Midway-Sunset field. (After Simonson, 1958.)

along the southwest flank while the southeastern flank adjacent to Midway-Sunset is relatively free of faulting. The field located in the foothills of the Temblor Range is a northeast-dipping regional homocline modified by minor folds and converging unconformities. A number of small anticlines and synclines are associated with major folds in the Temblor Range. These dominant structural features of the field often overlap each other such as the Globe, Midway, Spellacy, and Thirty-five anticlines. What faulting there is has little effect on hydrocarbon traps at Midway-Sunset.

STRATIGRAPHY

Stratigraphic sections along with composite electric logs for northern, central, and southern Midway-Sunset are illustrated in Figure 7. Only the geologic section through the upper Miocene is shown since no production has been obtained from older formations. This figure as well as most of the cross sections, some of the maps, and much of the description of the stratigraphy come from Land and Anderson (1965), as well as other publications of the California Division of Oil and Gas listed in the references. The deepest stratigraphic penetration within the field boundary was in Standard Oil Co. of California well No. 52, Sec. 33, T30S, R22E, which penetrated the Kreyenhagen formation of upper Eocene age. However, the section in this well is not considered typical of the Midway-Sunset stratigraphy because of its location. (See Figure 6.)

In the northern portion of the field, the deepest well, Shell Vedder USL 1, Sec. 9, T31S, R22E, reached the Santos shale of lower Miocene at a total depth of 10,319 ft (3145 m). The middle and lower Miocene was predominantly shale at this location. Superior C.W.O.D. 58-21 located in Sec. 21, T32S, R23E in the central portion of the field is the deepest well in the field with a total depth of 14,504 ft (4421 m). This well encountered a series of well-indurated sandstones and shales of middle and lower Miocene age and bottomed in the lower Miocene Media shale. Near the southern end of the field,Texaco General Petroleum 54X-20, Sec. 20, T11N, R23W bottomed in the upper Santos shale at 5650 ft (1722 m) after penetrating interbedded permeable sandstones and sandy siltstones of lower Miocene age. Uplift and faulting in this area are responsible for the shallow depth of the lower Miocene.

Middle and lower Miocene sediments are believed to have a combined maximum thickness of greater than 10,000 ft (3050 m) in the field while the upper Miocene sediments are approximately 8000 ft (2440 m) thick. Along the eastern edge of the field, the Pleistocene and Pliocene sediments attain a maximum thickness of 2000-2500 ft (610-762 m).

Pleistocene

The youngest sediments beneath the recent alluvium are the continental sands, silts, clays, and conglomerates of the Pleistocene Tulare Formation. These sediments, which range in thickness from 0 to 2500 ft (0-762 m) thick, unconformably overlie the upper Miocene or Pliocene formations. Tulare sediments consist of fluvial to lacustrine sandstones, lacustrine clays, siltstones, gypsum, and tuff. Beds contain a good number of mollusca representing freshwater snails and mussels and a few brackish-water clams. Freshwater diatoms are also present.

Pliocene

Sediments of Pliocene age at Midway-Sunset are divided into two formations: the older Etchegoin and the younger San Joaquin. In the Belgian Hills area at the northern end of the field, the Etchegoin is as much as 650 ft (200 m) thick while in the Thirty-five anticline area at the southwestern end, the San Joaquin reaches 1100 ft (335 m) in thickness. These sediments are truncated by the unconformity at the base of the Tulare along much of the west side of the field where the Tulare rests on the Miocene. Both formations are mainly marine clays and sandstones. However, some of the fine-grained beds appear to be

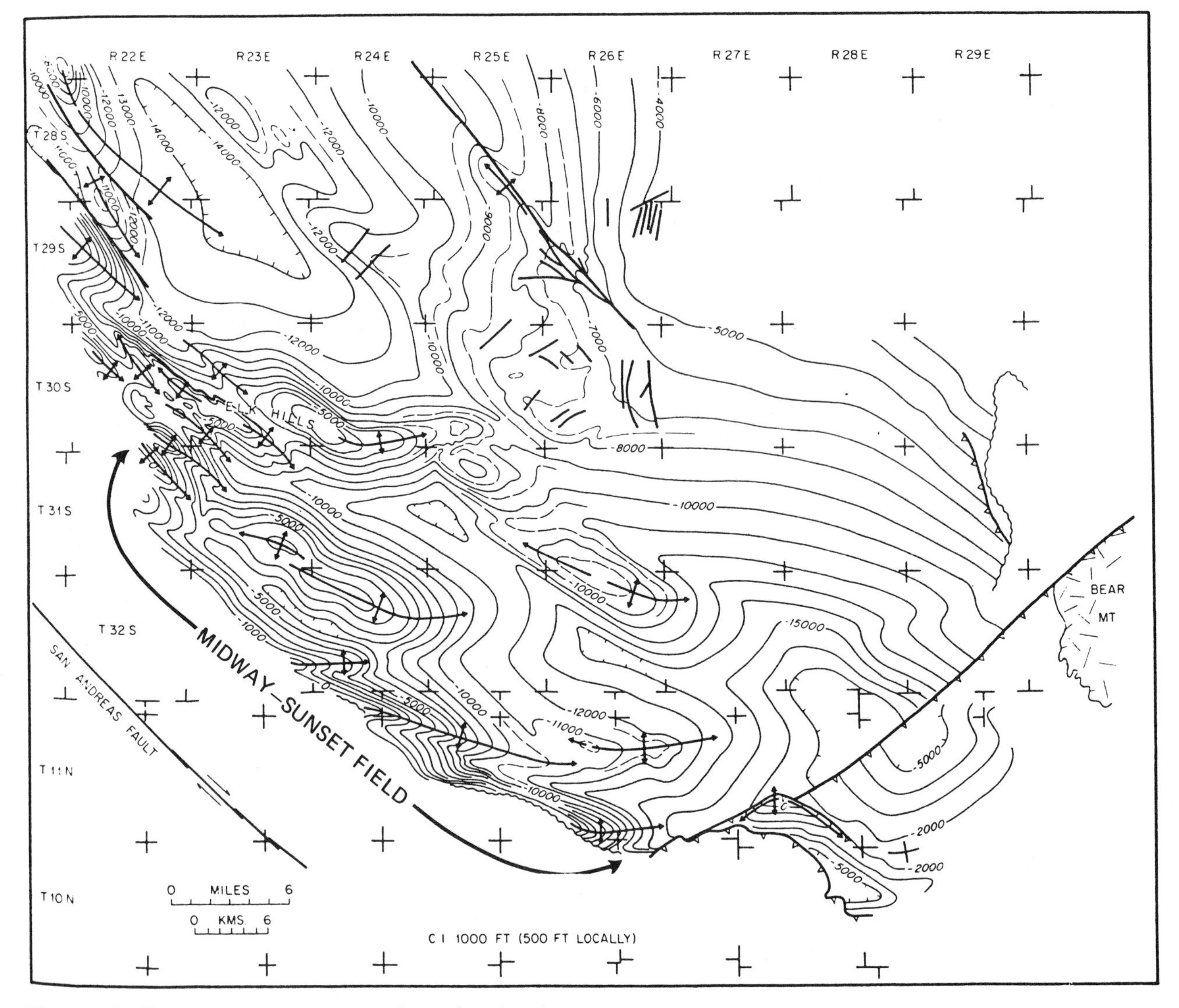

Figure 4. Structure map contoured on the Antelope Shale "N" Marker for the Southern San Joaquin Valley. (After Webb, 1981.)

nonmarine in origin: At least marine fossils have not been found in many of them and the remains of land plants and freshwater shells are present in some. Many of the sandstones produce oil.

Lithologically, the two formations are quite similar with the San Joaquin sandstones being in general slightly finer grained and containing green, blue, and brown clays. The division between the two is picked at the base of the "Top Oil" sandstone, which although thin is one of the few zones that can be correlated across the field. The Pliocene sediments lie on the Miocene unconformity.

Upper Miocene

The youngest formation of upper Miocene age present at Midway-Sunset is the Reef Ridge. It is a brown shale unit ranging from 0 to probably 2200 ft (670 m) in thickness, which may contain sandstone bodies up to 1000 ft (305 m) thick. These poorly sorted granitic sands and conglomerates are some of the main oil-producing horizons in the field. Generally, the Reef Ridge is absent along the Thirty-five anticline and in the Leutholtz and Metson pools at the south end of the field.

The Reef Ridge Shale is difficult to differentiate from the shale of the underlying Antelope Formation on the basis of lithology. Both are brown silty shales and diatomaceous with no fauna to aid in identifying the Reef Ridge. Thus, the contact between the two shales is usually selected arbitrarily.

The Antelope Formation, ranging in thickness from 0 to 4400 ft (0–1341 m), contains numerous thick coarse sandstones and conglomerates that produce oil. The shales, especially in the lower portion, are very diatomaceous and cherty. These are highly fractured and produce oil as well.

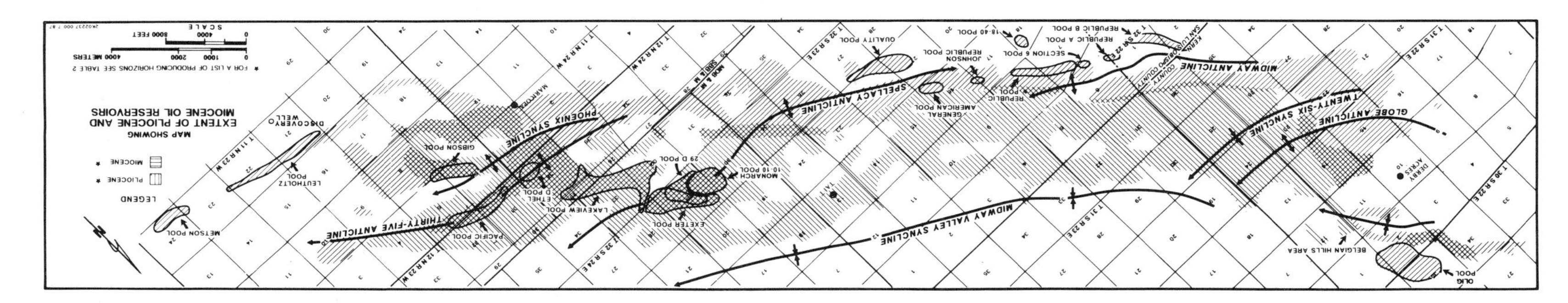

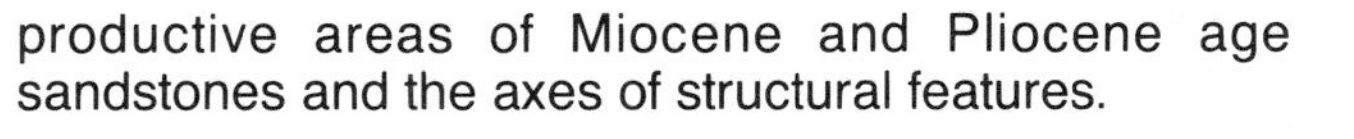

Figure 5. Map of the Midway-Sunset field, Kern and San Luis Obispo Counties, California, showing productive areas of Miocene and Pliocene age sandstones and the axes of structural features.

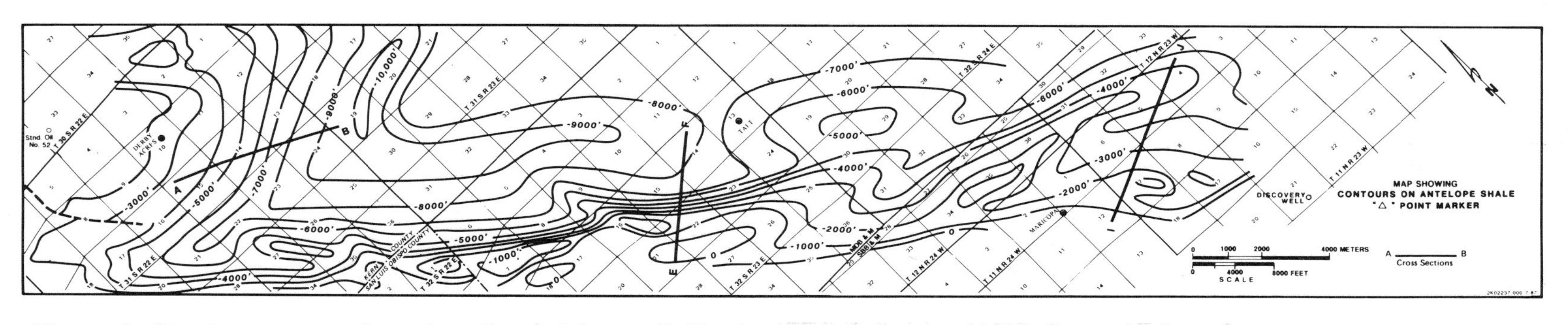

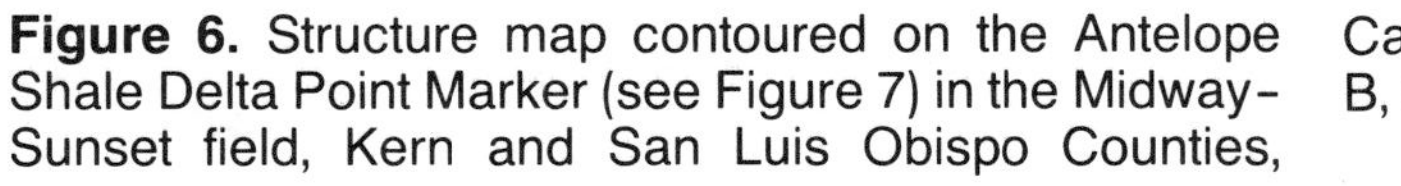

Figure 6. Structure map contoured on the Antelope Shale Delta Point Marker (see Figure 7) in the Midway-Sunset field, Kern and San Luis Obispo Counties, California. (After Callaway, 1967.) Cross sections A-B, E-F, and I-J are shown by Figures 8-10.

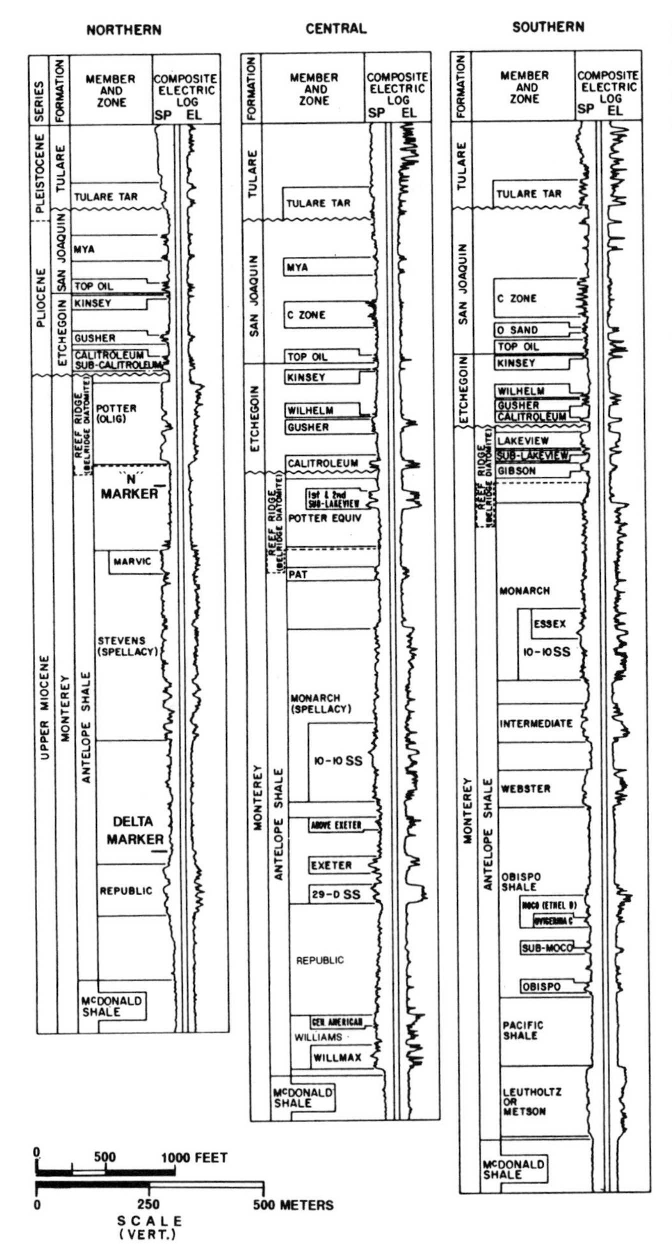

Figure 7. Type logs for Midway-Sunset field, Kern and San Luis Obispo Counties, California. (After Land and Anderson, 1965 Courtesy CDOG.)

The McDonald Formation, a massive chocolate brown silty shale, underlies the Antelope. Thickness of over 2000 ft (610 m) has been reported in the Thirty-five anticline area.

TRAPS

Numerous types of traps are present in the Midway-Sunset field. They include anticlinal closure, facies changes, sand pinchout, tar seals, onlap and overstepping unconformities, and porosity and permeability variations. Faulting is of minor importance in trapping but may have contributed to migration of hydrocarbons. Figures 8, 9, and 10 are cross sections through the field that illustrate many of the traps present.

Top and bottom seals are claystones, siltstones, and shale within the producing formations that range in age from Pleistocene to upper Miocene at depths from 500 to 5000 ft (152–1524 m).

The greatest amount of upper Miocene oil is found in overstepping unconformities where the Miocene subcrop is overlain by Pliocene or Pleistocene sediments to form the trap (Figure 8). In many cases, the oil does not extend all the way to the subcrop but deteriorates to a tar so that a tar seal is the trap. In the Republic, Quality, Ethyl D, Monarch 10-10, and Olig pools, Miocene oil is structurally trapped in anticlinal highs (Figures 9 and 10). (For location of pools, see Figure 5.) Stratigraphic traps with closure provided by facies changes on the regional homocline account for the Leutholtz, Metson, and Exeter/29-D pools. The Republic B pool has both stratigraphic (sand pinchouts) and structural (steeply dipping flank) traps. In the Belgian Hills area, the traps are due to variations of permeability and porosity with some fault control, and in the Lakeview pool the trap is a syncline with gravity drainage.

RESERVOIRS

All the producing zones in the field are listed in Table 2 and will be discussed by formation in this section.

Tulare

The Tulare Formation of Pleistocene age contains a number of fine to very fine grained moderately sorted sands. These sands are present over most of the field and often contain oil. They are considered subcommercial except on the east flank of Belgian Hills and Globe anticline and the south flank of the Thirty-five syncline. The depth to the sands ranges from 500 to 2500 ft (152–762 m). They are often 20-50 ft (6–15 m) thick separated by clays of similar thickness, but at a few locations the sands are a few hundred feet thick. Figure 11 is an isopach of net Tulare sand in the northern portion of the field (Burgdorf, 1986).

During the Pleistocene, a major portion of the central San Joaquin basin was covered by a large freshwater lake. The Tulare sands at Midway-Sunset were deposited along the shores of the lake as beaches and deltas and west of the lake as alluvial sands. (See Figure 12.) The sands are very poorly consolidated to unconsolidated with porosities as high as 30–37% and permeabilities in the darcys. Porosity is primary in origin with clay being the main destroyer of permeability.

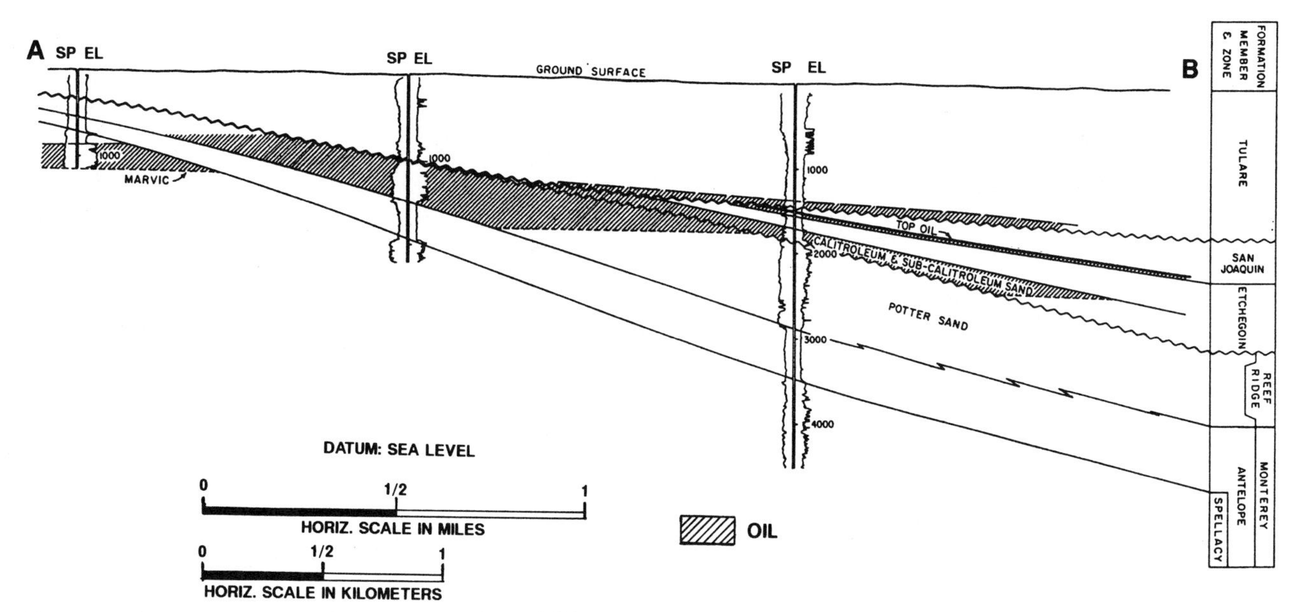

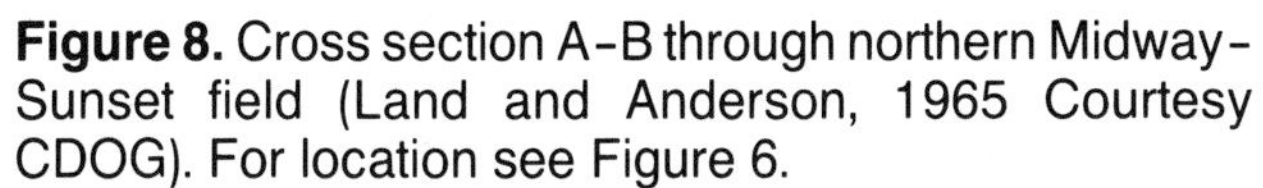

Figure 8. Cross section A-B through northern Midway-Sunset field (Land and Anderson, 1965 Courtesy CDOG). For location see Figure 6.

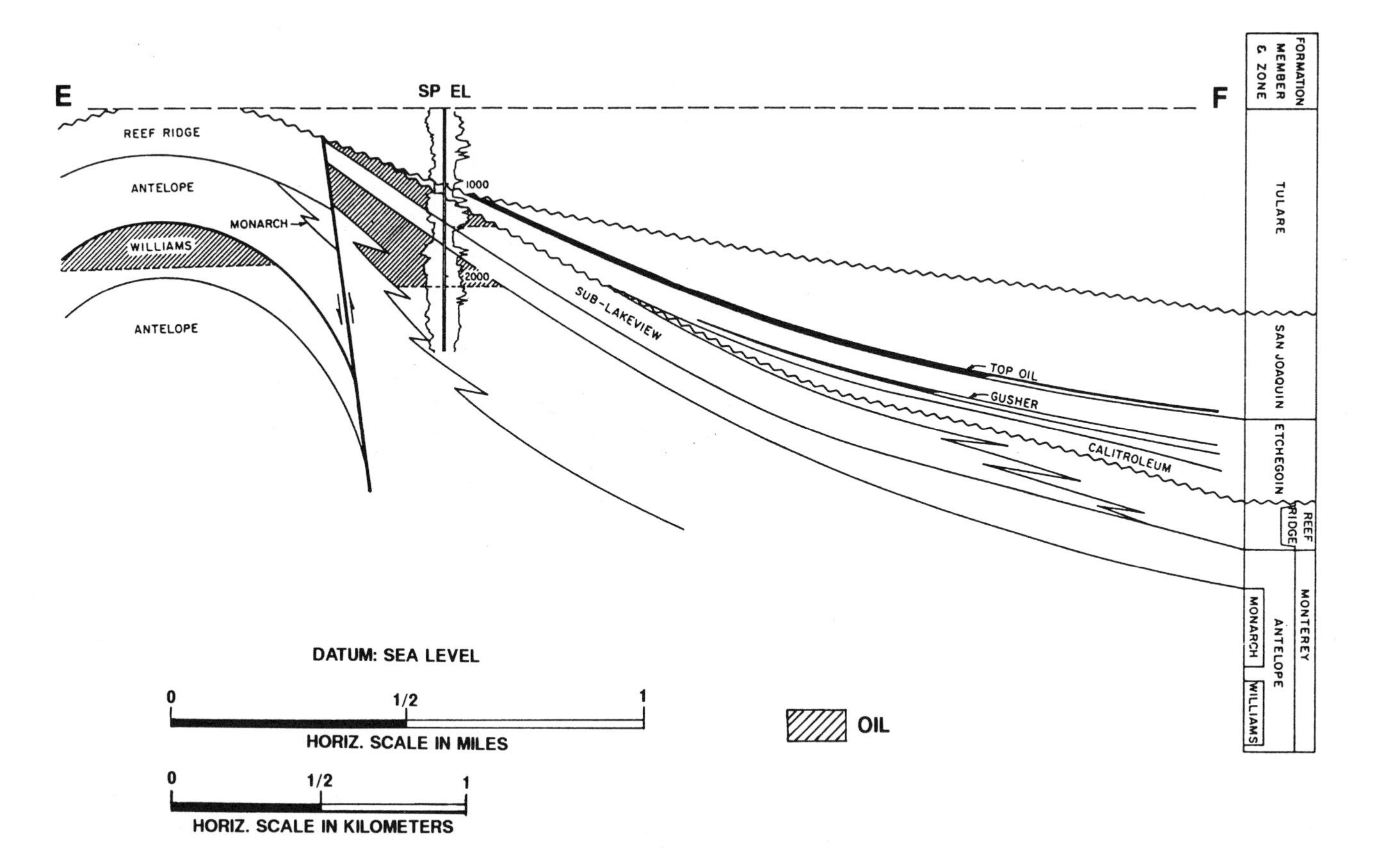

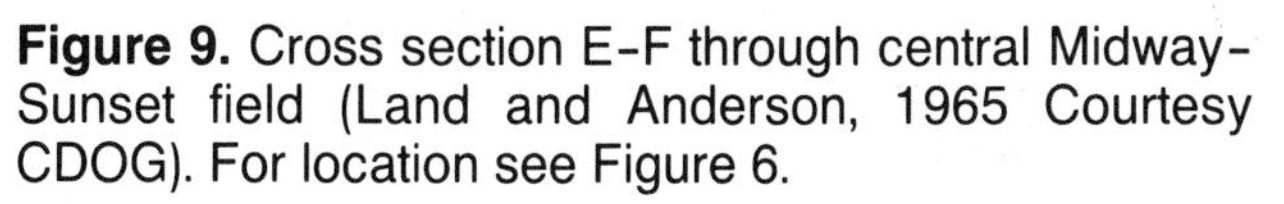

Figure 9. Cross section E-F through central Midway-Sunset field (Land and Anderson, 1965 Courtesy CDOG). For location see Figure 6.

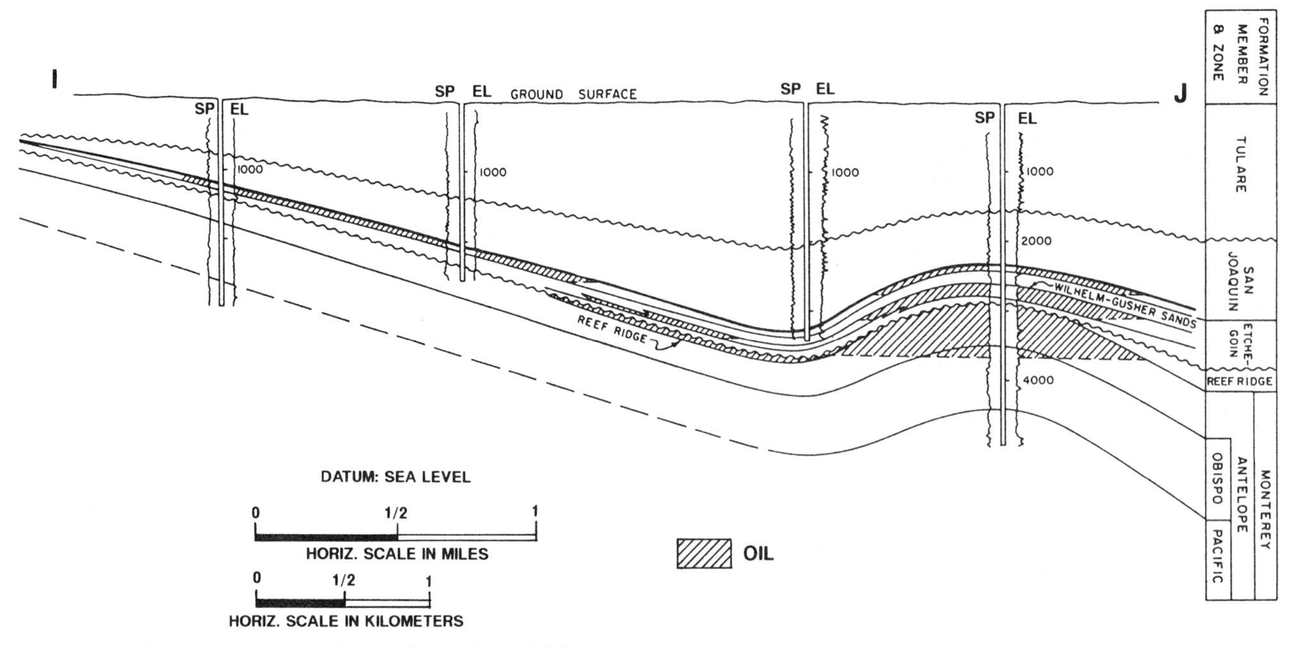

Figure 10. Cross section I-J through southern Midway-Sunset field (Land and Anderson, 1965 Courtesy CDOG). For location see Figure 6.

Table 2. Producing zones, Midway-Sunset field

ZONE	AVERAGE DEPTH (FEET)	AVG.NET THICKNESS (FEET)	GEOLOGIC		OIL GRAV. (°API) OR GAS (BTU)	SALINITY OF ZONE WATER GR/GAL
			AGE	FORMATION		
TULARE	200-1400	50-200	PLEISTOCENE	TULARE	13	200-1000
MYA TAR	1100	150	PLIOCENE	SAN JOAQUIN	12	260
TOP OIL	500-2500	20-50	PLIOCENE	SAN JOAQUIN	15-23	1490-2160
KINSEY	2000-3600	15-175	PLIOCENE	ETCHEGOIN	14-26	1500-1860
WILHELM	2000-3000	100	PLIOCENE	ETCHEGOIN	14-26	1700-2100
GUSHER	2000-3000	75	PLIOCENE	ETCHEGOIN	14-26	1440-1580
CALITROLEUM	1500-4500	80	PLIOCENE	ETCHEGOIN	14-26	1620-2040
LAKEVIEW	2600-3300	20-200	UPPER MIOCENE	MONTEREY	21	1670
SUB-LAKEVIEW	400-3100	10-300	UPPER MIOCENE	MONTEREY	22	440
POTTER	200-2500	60-500	UPPER MIOCENE	MONTEREY	14	5-400
MARVIC	1000	200	UPPER MIOCENE	MONTEREY	13	40
MONARCH	600-2000	50-400	UPPER MIOCENE	MONTEREY	13-17	50-1300
WEBSTER	1500-1800	50-250	UPPER MIOCENE	MONTEREY	14	N.A.
MOCO	2150	70-450	UPPER MIOCENE	MONTEREY	15	980
OBISPO	3600	50-1500	UPPER MIOCENE	MONTEREY	14-27	970
PACIFIC	3700	50-300	UPPER MIOCENE	MONTEREY	16	600
METSON	1250	400	UPPER MIOCENE	MONTEREY	8-12	790
LEUTHOLTZ	3200	40-400	UPPER MIOCENE	MONTEREY	15-24	550
REPUBLIC	1300-4900	150	UPPER MIOCENE	MONTEREY	12-24	70

The Tulare oil is very low gravity, ranging from 12° to 8° API. Thus, the reservoir has been called the Tulare tar, and the noncommerciality of the zone is often the result of low oil gravity. When it is found overlying the Miocene pays, it is often produced along with them, and thus it is difficult to determine the productive characteristic of the Tulare. At least three separate steam flood projects are being conducted in the Tulare tar in Midway-Sunset field at this time.

San Joaquin

The San Joaquin of Pliocene Age contains four productive sands in various areas of the Midway-Sunset field. These have been given the local names of Mya, C zone, O sand, and Top Oil (Figure 7). The latter is the most extensive, being found over most of the fields at the base of the San Joaquin. It is a fine to very fine grained sand usually 30 ft (9 m)

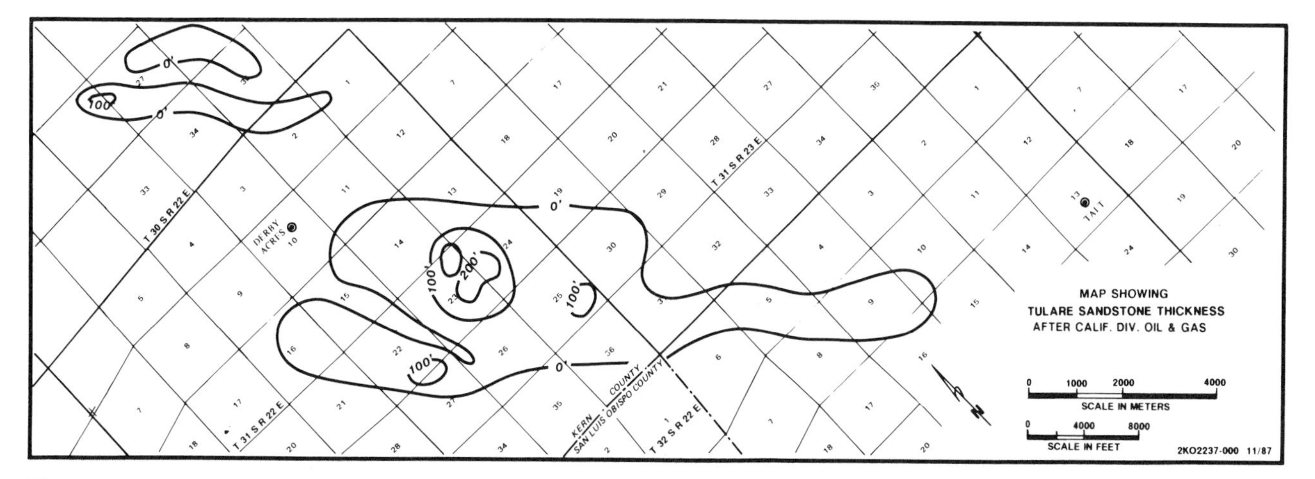

Figure 11. Map showing Tulare sandstone thickness in the Northern part of Midway-Sunset field, Kern and San Luis Obispo Counties, California. (After Burgdorf, 1986 Courtesy CDOG.)

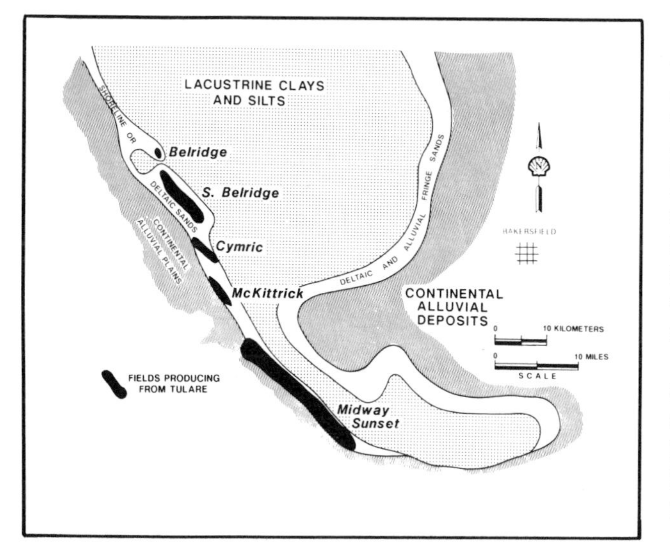

Figure 12. Tulare sands environments of deposition map for the San Joaquin Valley, California.

or less in thickness containing marine fossils. The Top Oil sand reservoirs are believed to have been deposited in a shallow-marine environment. Updip the sand shales out or is truncated below the Tulare unconformity (Figures 8 and 9). The shallower San Joaquin sands are probably nonmarine and are thicker but less continuous than the Top Oil.

As with the Tulare, it is often difficult to separate the production from these sands from that of the Miocene zones. However, it appears that the gravity of the oil produced from the San Joaquin sands is usually higher (18-22° API) than that from the shallow Miocene.

Etchegoin

The Etchegoin Formation is predominantly marine in origin and contains several sandstone reservoirs that are named in descending order: Kinsey, Wilhelm, Gusher, Calitroleum, and Sub-Calitroleum zones. They vary in thickness from a few tens of feet to 200 ft (61 m). Some were deposited onlapping the Miocene unconformity. The distribution of these sands is similar to that of the Top Oil sand. They are found productive on parts of the Thirty-five anticline and locally in other areas of the field with one exception—the Sub-Calitroleum, which is a well-developed sand only between Taft and Derby Acres.

The gravity of the oil produced from the Etchegoin zones is generally 25-28° API, although lower gravity oils (13-18° API) are found near their updip truncations. As with the other "shallow zones," it is not possible to determine how much oil has been produced out of the Etchegoin zones.

Upper Miocene Sands

By far the greatest amount of oil produced at Midway-Sunset has come from the upper Miocene reservoirs, and they also contain the lion's share of the remaining proven reserves. The reservoirs are moderate to very poorly sorted, medium-grained sandstones to conglomerates containing grains from silt to boulders in size. The larger grains can be identified as rounded igneous and metamorphic rock fragments. They are very poorly consolidated to unconsolidated, often being held together only by the black viscous oil. Because of the oil, sedimentary features are difficult to see. However, graded beds ranging from a few inches to several feet are readily apparent.

Figure 13 is a cross section through the majority of the pay section of the Potter sand, which is one of these upper Miocene sands. This detailed section, which is only about 1 mi (1.6 km) in length, is located on Figure 14. As illustrated, fairly thick diatomaceous silts separate the reservoir into sand packages a few hundred feet thick (labeled J, K, L, M, N, O on the

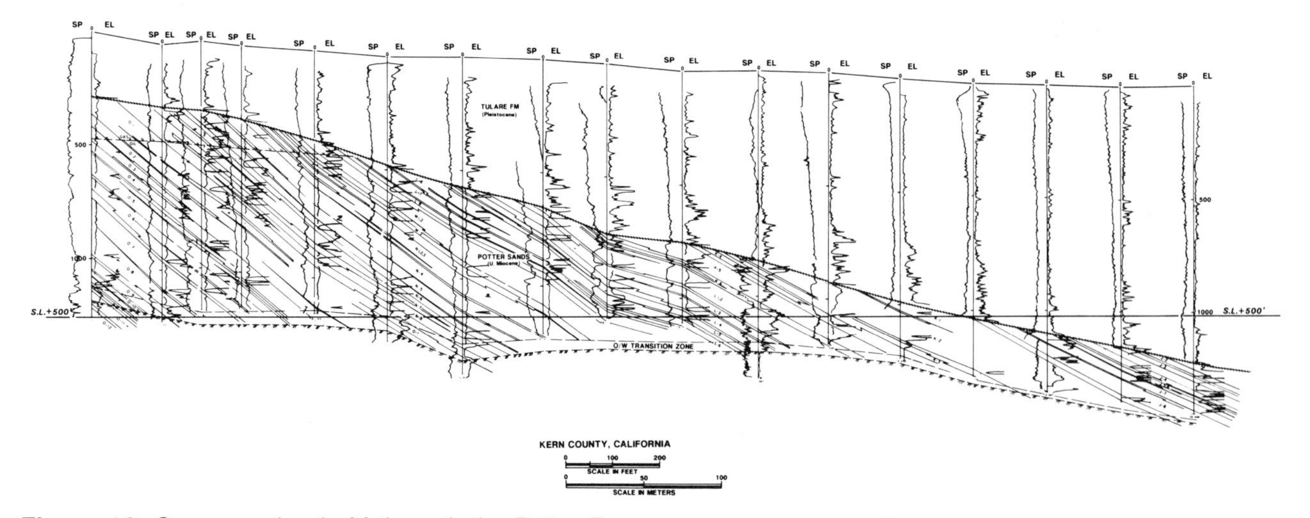

Figure 13. Cross section L-M through the Potter Pay Interval, Midway-Sunset field, Kern County, California. For location of section see Figure 14.

cross section) that are recognizable over as much as a half mile laterally. Individual sands (labeled K-1, K-2, etc.) are tens of feet thick separated by thin silts that are laterally discontinuous.

These sands were deposited as deep-water turbidite fans at the base of the slope near a coastline similar to the present-day California coast. Webb (1981) and Callaway (1962) have described the deposition and distribution of the sands in detail. Sediments came from the Salinian block to the west that was moving northward on the west side of the San Andreas fault. The Santa Margarita sands that outcrop on the west side of the Temblor Range are the shallow-marine equivalent of these turbidites. Submarine canyons provided pathways for the sands to be transported to the base of the slope where they were deposited in the form of submarine fans under at least 1000 ft (305 m) of water. The result was a series of overlapping and interfingering fans that gradually become younger going from south to north in the field. Figure 14 contains isopach maps of two such deposits, the Potter and Williams fans, with the Potter being the younger. As illustrated, these fans contain up to 1500 ft (457 m) of sand and cover 17-20 mi^2 (44-52 km^2). This is not all of the sediment originally deposited in these fans since with the uplift of the Temblor Range the fans were tilted to the northeast and a portion of the fans and the submarine canyon deposits were removed by erosion. As illustrated by Figure 13, in the Potter area the reservoirs dip at up to 45° and are truncated by the unconformity that dips at approximately 20°. An oil-water contact with a fairly thick transition zone limits the accumulation downdip so that not all the fan is filled with oil. Updip an air-oil contact limits the accumulation. This is not a gas-oil contact since only air fills the pores in this area. As illustrated, both oil contacts are tilted with a dip similar to the dip of the unconformity.

Figure 15 shows the logs of a typical 300 ft (91 m) section in the Potter reservoirs. The lithology of two 60 ft (18 m) cores in this well are illustrated as well as the lab porosities and permeabilities measured from these cores. The importance of measuring these properties under a stress similar to their depth of burial is illustrated by the porosity measurements that are given both stressed and unstressed. These differences are of course due to the poorly consolidated nature of these sands. The data shown by Figure 16 (after Neuman, 1980) show how the porosities from the density log are significantly less than routine unstressed core porosities for a 200 ft (61 m) section of another upper Miocene Midway-Sunset reservoir. Note that the porosities measured under stress are similar to the density log porosities. The effect on permeabilities is even more dramatic. Webb (1978) showed by an illustration similar to Figure 17 how the grain size, porosity, and permeability change in a typical graded bed in these turbidite sandstones. Detailed grain size distribution, capillary pressure, and relative permeability curves for two samples from the cores illustrated in Figure 15 are shown in Figure 18.

Upper Miocene Shale

Oil production has been reported from fractured diatomaceous shales of the Antelope Formation called the Obispo and Pacific shales along the Thirty-five anticline and in the Republic 18-40 pool. Little is known about the production from these shales, because they are usually produced along with adjacent oil sands, but it is presumed that the oil from the sands simply fills the fractures in the shale and may also exist in the diatomaceous matrix. Since 1985, a number of wells have been drilled for fractured diatomaceous shales in the Reef Ridge Formation in the Republic area.

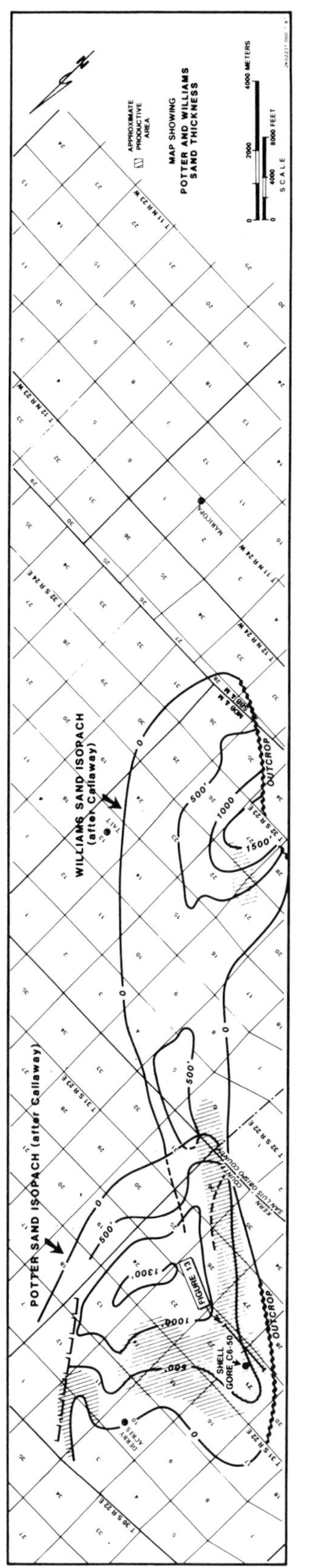

Figure 14. Isopach map of the upper Miocene Potter and Williams turbidite fans, Midway-Sunset field, Kern County, California. (After Callaway, 1982.)

SOURCE

Zieglar and Spotts (1977) stated that upper Eocene, Miocene, and Pliocene shaly beds of the San Joaquin Valley generally contain more than 1% by weight of organic material with some of the richer zones containing more than 5%. The organic material is largely sapropelic and should be considered oil-prone. Their isopach data indicate that prior to the Miocene, potential source bed in the Tejon depocenter adjacent to Midway-Sunset would not have been buried deep enough to be mature. With the addition of Miocene sediments, the entire lower Tertiary plus the lower part of the Miocene should have been generating hydrocarbons. With continued basin filling through Pliocene, Pleistocene, and up to the present, the Tejon depocenter received more than 10,000 additional feet (3050 m) of sediments, resulting in a cumulative thickness in excess of 25,000 ft (7620 m). Zieglar and Spotts believe the Eocene beds reached generation depths 15 million years ago and Miocene beds 5 to 6 million years ago.

Graham and Williams (1985) examined the tectonic, depositional, and diagenetic history of the Miocene Monterey Formation in the Central San Joaquin basin. Although their work was mainly concerned with the basin north of Midway-Sunset, they concluded that "the organic richness, kerogen type, maturation level and association with producing reservoirs argue that the Monterey Formation in the Central San Joaquin Valley likely was the source of major petroleum accumulations such as the Belridge, Cymric and Lost Hills fields as well as giant fields like Elk Hills farther south in the basin." The latter would presumably include Midway-Sunset. They showed that the Monterey contained two types of source beds: (1) a biogenic siliceous rock with an average total organic carbon of 3.43 ± 1.95% and (2) a clay shale and siltstone with a TOC of 2.06 ± 1.11%. The maximum TOC for these types were 9.16% and 4.87%, respectively. Kerogen types are I and II with II most abundant.

EXPLORATION CONCEPTS

As stated earlier, the evidence that led oilmen to explore in the area of Midway-Sunset was the presence of oil seeps. It was one of the first of numerous fields discovered along the west side of the San Joaquin basin that now can be recognized as a "regional play" even though at the time they were discovered many were the result of random

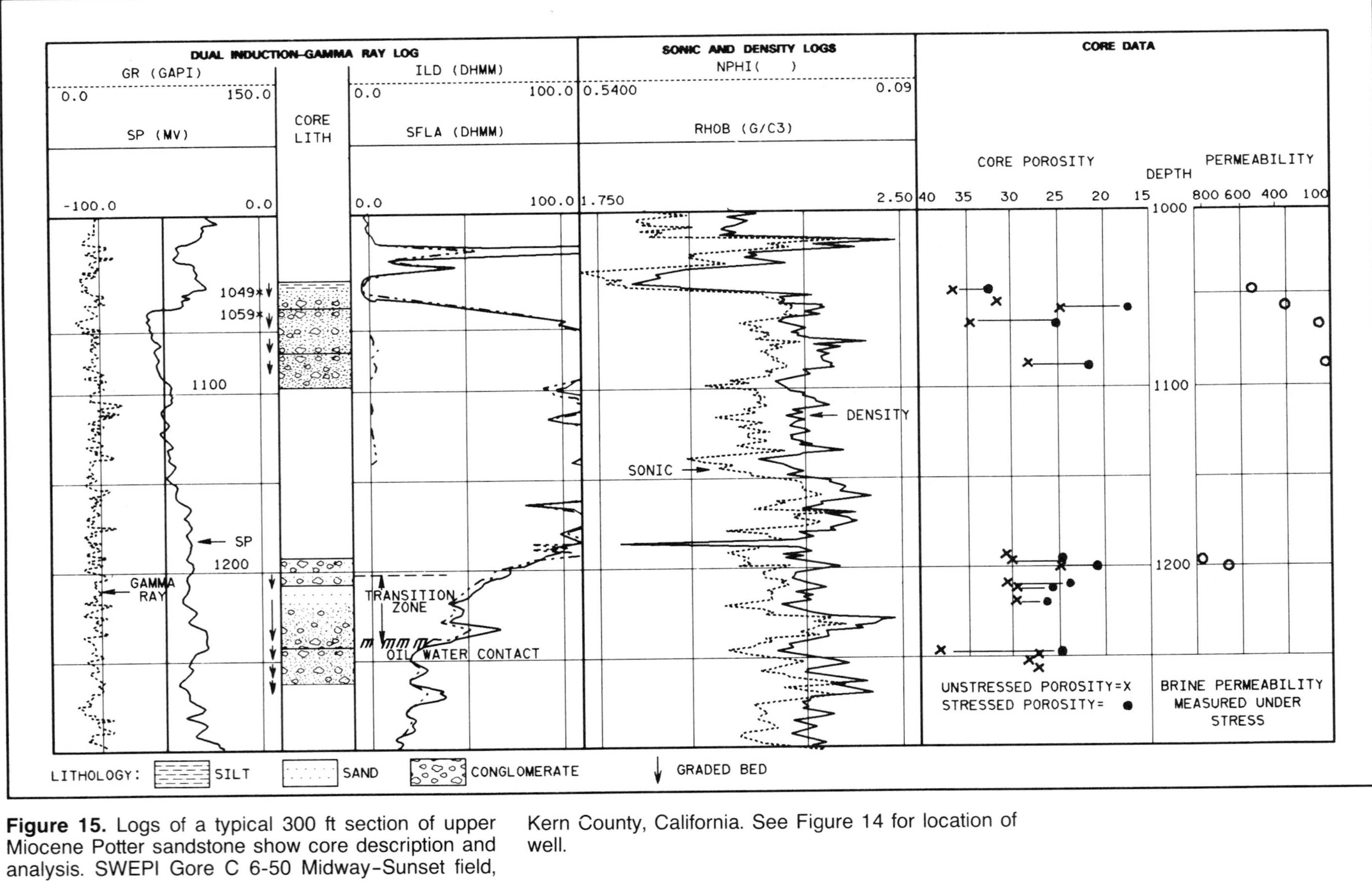

Figure 15. Logs of a typical 300 ft section of upper Miocene Potter sandstone show core description and analysis. SWEPI Gore C 6-50 Midway-Sunset field, Kern County, California. See Figure 14 for location of well.

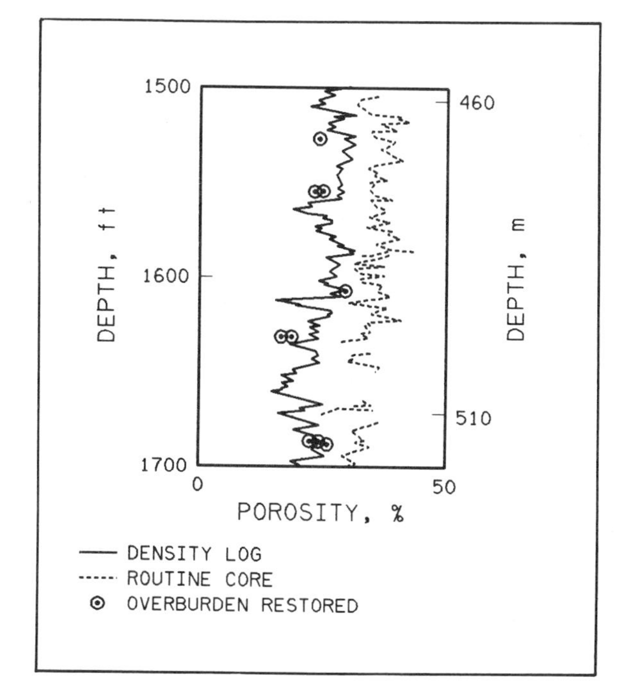

Figure 16. Density log porosities versus routine and stressed core porosities for a 200 ft section of upper Miocene Monarch sandstone, Midway-Sunset field, Kern County, California. Well not identified. (After Neuman, 1980 Courtesy SPE.)

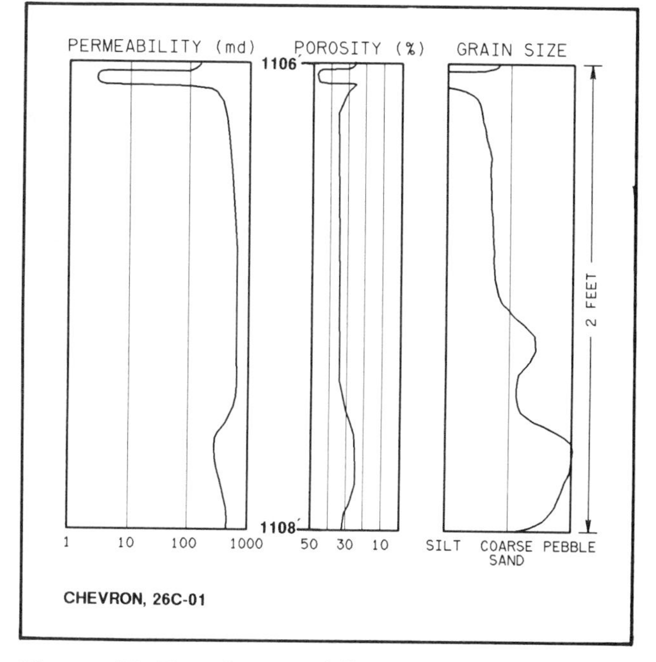

Figure 17. Plot of permeability, porosity, and median grain size versus depth for a 2 ft turbidite bed from the Monarch sandstone, Midway-Sunset field, Kern County, California. (After Webb, 1978 Courtesy CIM.)

drilling, trend following, seep drilling, etc. Harding (1976) has shown how the folds that trap hydrocarbons along the west side like those at Midway-Sunset and even those near the center of the basin are related to movement of the San Andreas fault. Therefore, in hindsight a structural analysis of the area would probably have led petroleum geologists to discover many of the accumulations. Modern seismic would certainly have been helpful in defining most of the structural traps.

In regard to stratigraphic trapping and reservoir parameters, the upper Miocene reservoirs at Midway-Sunset are similar to the time equivalent "Stevens" sandstones that produce in numerous fields in the southeast San Joaquin Valley. MacPherson (1978) has described these reservoirs in some detail in fields such as Strand, Ten Section, Greeley, etc.

From an exploration geologist's standpoint, the amazing thing about Midway-Sunset is that even though over 12,000 wells have been drilled in the field in over 90 years, probably fewer than a dozen wells have been drilled below 7000 ft (2134 m). Oil has been produced from the Oligocene and Eocene of Belgian anticline field (see Figure 1) a short distance northwest of the field for over 39 years, and yet no well within the main field has penetrated below the lower Miocene and only a few have gone below the upper Miocene. Many of the structures producing from the upper Miocene are expected to continue deeper. Early in the development of the field, the Maricopa shale was thought to be the source of the oil, which meant that the Pleistocene, Pliocene, and uppermost Miocene reservoirs are the only ones that could be charged. In 1928, the Republic pool was discovered with production at 2700 ft (823 m) in deeper Miocene sand, which changed this theory. Today it appears that the upper Miocene is the deepest section oilmen chose to explore for oil at Midway-Sunset.

It is also astonishing that for a field discovered in 1894, new pools such as the 29 Monarch and 10-10 (Gallear and Kistler, 1965) have been discovered as late as 1962. This certainly shows a lack of an aggressive exploration effort. One reason for the reluctance to test the deeper prospects may be the fact that no one company has a large continuous block of acreage on which to develop an attractive deep venture. It is also interesting from the development geologist's standpoint that more development wells have been drilled in the last 20 years than in the 54 years before 1968. The reason for this is of course due to the development of steam soak and steam drive technology since 1964 along with a proper economic atmosphere. For most of the accumulations at Midway-Sunset, the question has not been where the oil is but how to produce low gravity oil at economic rates. Recently, cogeneration has increased activity at Midway-Sunset by improving the economics and air pollution problems connected with heavy oil recovery. This is the generation of electrical power at the same time that steam is being generated by burning natural gas instead of crude oil. The

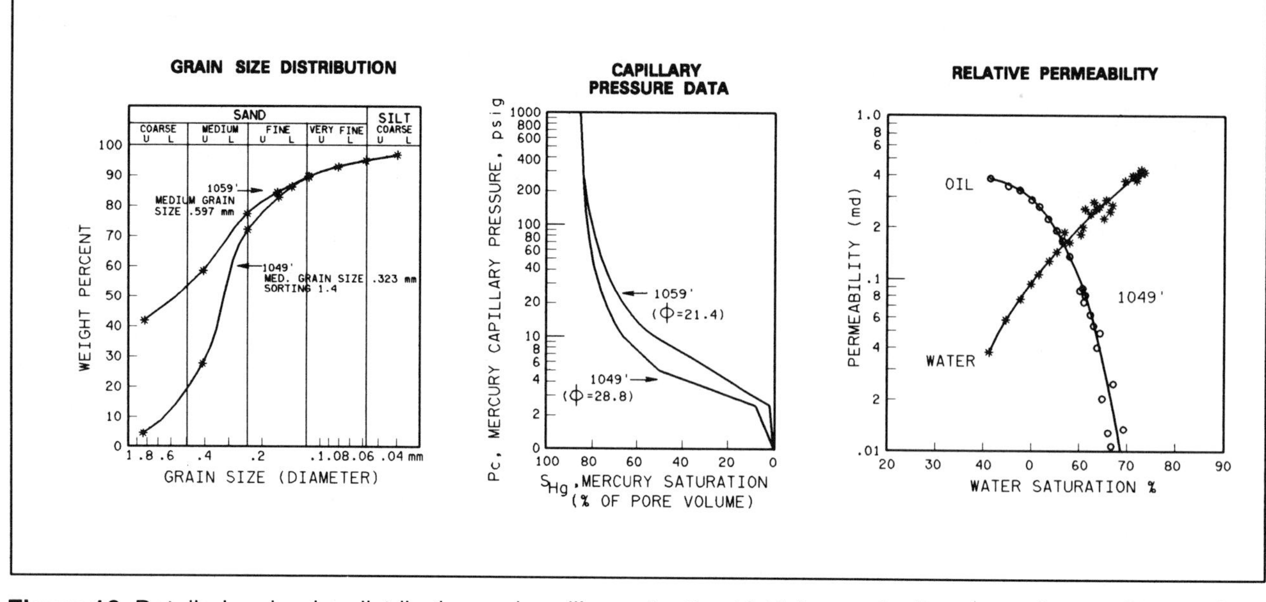

Figure 18. Detailed grain-size distribution and capillary pressure curves for sample from 1049 and 1059 ft of the core in Figure 15 and relative permeability curves for the 1049 ft sample. Potter sandstone, SWEPI Gore C 6-50 Midway-Sunset field, Kern County, California.

lesson here is that in unusual reservoirs all types of petroleum engineering technology must be developed early in order to expedite field development.

ACKNOWLEDGMENTS

The author is indebted to AMOCO who made available the Midway-Sunset field portion of their research department internal report, which was drawn from freely. They also provided a literature search for the field as well. I am also indebted to the drafting department of Shell Western Exploration and Production Inc. for their assistance with the illustrations.

REFERENCES

Burgdorf, G., 1986, Net Tulare oil sand, Midway-Sunset Field, Kern County, California: California Division of Oil and Gas, Special Map S-5.

Callaway, D. C., 1967, Midway-Sunset and Buena Vista Hills Map, contoured on Antelope Shale "Delta" Point Marker: Guidebook, Geology and Oil Fields, West Side Southern San Joaquin Valley, Pacific Sec. AAPG, SEG, SEPM.

Callaway, D. C., 1962, Distribution of upper Miocene sands and their relation to production in the North Midway area, Midway-Sunset Field, California, San Joaquin: Geological Society Selected Papers, v. 1, p. 47-55.

Gallear, D. C., and J. O. Kistler, 1965, The 29D Monarch and 10-10 Pool—a "sleeper" in the Old Midway-Sunset Oil Field, Kern County, California: San Joaquin, Geological Society Selected Papers, v. 3, p. 19-35.

Graham, S. A., and L. A. Williams, 1985, Tectonic, depositional and diagenetic history of Monterey Formation (Miocene), Central San Joaquin Basin, California: American Association of Petroleum Geologists Bulletin, v. 69, p. 386-411.

Harding, T. P., 1976, Tectonic significance and hydrocarbon trapping consequences of sequential folding synchronous with San Andreas faulting, San Joaquin Valley, California: American Association of Petroleum Geologists Bulletin, v. 60, p. 356-378.

Hillis, D., and W. T. Woodward 1943, Williams and Thirty-Five Hills area of the Midway-Sunset Oil Field: California Department of Natural Resources, Division of Mines Bulletin 118, p. 526-529.

Ingram, W. L., 1964, Olig Pool of Midway-Sunset Oil Field: California Oil Fields Summary of Operations, v. 50, n. 1, p. 47-52.

Land, P. E., and Anderson, D. N., 1965, Midway-Sunset Oil Field: California Oil Fields Summary of Operations, v. 51, n. 2, p. 21-29.

MacPherson, B. A., 1978, Sedimentation and trapping mechanism in upper Miocene Stevens and older turbidite fans of southeastern San Joaquin Valley: American Association of Petroleum Geologists Bulletin, v. 62, p. 2243-2274.

Neuman, C. H., 1980, Log core measurements of oil in place, San Joaquin Valley: Journal Petroleum Technology, August 1980, p. 1309-1315.

Rintoul, B., 1984, Midway-Sunset: on top again: Pacific Oil World, 1984, v. 77, n. 10, p. 24-26.

Simonson, R. R., 1958, Oil in the San Joaquin Valley, California, *in* L. G. Weeks, ed., Habitat of oil: American Association of Petroleum Geologists, p. 99-112.

Webb, G. W., 1981, Stevens and earlier Miocene turbidite sandstones: Southern San Joaquin Valley, California: American Association of Petroleum Geologists Bulletin, v. 65, p. 438-465.

Webb, M. G., 1978, Monarch Sandstone: reservoir description in support of steam flood, Section 26C, Midway-Sunset Field, California: Journal of Canadian Petroleum Technology, v. 17, n. 4, p. 31-40.

Woodward, W. T., 1941, Gibson area, Midway-Sunset Oil Field: Petroleum World, v. 38, n. 5, p. 40-42.

Woodward, W. T., 1943a, North Midway area of the Midway-Sunset Oil Field (California): California Department of Natural Resources, Division of Mines Bulletin 118, p. 519-521.

Woodward, W. T., 1943b, Gibson area of the Midway-Sunset Oil

Field (California): California Department of Natural Resources, Division of Mines Bulletin 118, p. 530-531.
Young, U., 1943, Republic area of the Midway-Sunset Oil Field: California Department of Natural Resources, Division of Mines Bulletin 118, p. 522-525.
Zieglar, D. L., and J. H. Spotts, 1977, Reservoir and source bed history in the Great Valley of California: Oil and Gas Journal, June 20, 1977, p. 116-119.
Zulberti, J. L., 1957, Republic sands of Midway-Sunset Field: California Oil Field Summary of Operations, v. 43, n. 2, p. 21-33.
Zulberti, J. L., 1958, Santiago area of Midway-Sunset Oil Field: California Oil Fields Summary of Operations, v. 44, n. 1, p. 65-73.
Zulberti, J. L., 1959, Thirty-five Anticline of Midway-Sunset Oil Field: California Oil Fields Summary of Operations, v. 45, n. 1, p. 36-43.
Zulberti, J. L., 1960, Exeter and 29-D Pools of Midway-Sunset Oil Field: California Oil Fields Summary of Operations, v. 46, n. 1, p. 40-50.
Zulberti, J. L., 1961, Lakeview Pool of Midway-Sunset Oil Field: California Oil Fields Summary of Operations, v. 45, n. 1, p. 29-38.

SUGGESTED READINGS

AAPG-SEPM-SEG Guidebook, Joint Annual Meeting, Los Angeles, 1952.
Arnold, R., and H. R. Johnson, 1910, Preliminary report on the McKittrick-Sunset Oil region, Kern and San Luis Obispo Counties, California: U.S. Geological Survey Bulletin 406.
Pacific Sections AAPG-SEG-SEPM, 1968 Guidebook, Geology and Oilfields, West Side Southern San Joaquin Valley, 43rd Annual Meeting.
Pack, R. W., 1920, The Sunset-Midway Oil Field, California, Part 1, Geology and Oil Resources: U.S. Geological Survey Professional Paper 116.

Appendix 1. Field Description

Field name .... *Midway-Sunset*

Ultimate recoverable reserves .... *2.2 billion bbl*

Field location:

- **Country** .... *United States*
- **State** .... *California*
- **Basin/Province** .... *San Joaquin basin/Pacific Border province*

Field discovery:

- **Year first pay discovered** .... *Pleistocene Tulare Formation 1894*
- **Year second pay discovered** .... *Upper Miocene Antelope Formation 1910*
- **Third pay** .... *Pliocene Lakeview Formation 1910*

Discovery well name and general location
(i.e., Jones No. 1, Sec. 2T12NR5E; or Smith No. 1, 5 mi west of Sheridan, Wyoming):

- **First pay** .... *Jewett & Blodget No. 1, Sec. 21, T11N, R23W*
- **Second pay** .... *Midway Oil Co., Webster No. 14, Sec. 35, T12N, R24W*
- **Third pay** .... *Lakeview Oil Co. No. 1, Sec. 25, T12N, R24W*

Discovery well operator .... *Solomon Jewett and Hugh A. Blodget*
(if more than one pay in field, list operators of discovery well in other pays)

- **Second pay** .... *Midway Oil Co.*
- **Third pay** .... *Lakeview Oil Co.*

IP in barrels per day and/or cubic feet or cubic meters per day:

- **First pay** .... *<10 BOPD*
- **Second pay** .... *Unknown*
- **Third pay** .... *68,000 BOPD*

All other zones with shows of oil and gas in the field:

Age	Formation	Zone	Type of Show
Pleistocene	*Tulare*	*Tar*	*Oil*
Pliocene	*San Joaquin*	*Mya*	*Oil*
Pliocene	*San Joaquin*	*Top Oil*	*Oil*
Pliocene	*Etchegoin*	*Kinsey*	*Oil*
Pliocene	*Etchegoin*	*Wilhelm*	*Oil*
Pliocene	*Etchegoin*	*Gusher*	*Oil*
Pliocene	*Etchegoin*	*Calitroleum*	*Oil*
Upper Miocene	*Reef Ridge*	*Lakeview*	*Oil*
Upper Miocene	*Reef Ridge*	*Gibson*	*Oil*
Upper Miocene	*Reef Ridge*	*Potter (Olig)*	*Oil*
Upper Miocene	*Reef Ridge*	*Marvic*	*Oil*
Upper Miocene	*Antelope Shale*	*Monarch (Spellacy)*	*Oil*
Upper Miocene	*Antelope Shale*	*Stevens (Spellacy)*	*Oil*
Upper Miocene	*Antelope Shale*	*Intermediate*	*Oil*
Upper Miocene	*Antelope Shale*	*Webster*	*Oil*
Upper Miocene	*Antelope Shale*	*Obispo*	*Oil*
Upper Miocene	*Antelope Shale*	*Exeter*	*Oil*
Upper Miocene	*Antelope Shale*	*29D*	*Oil*
Upper Miocene	*Antelope Shale*	*Metson or Leutholtz*	*Oil*
Upper Miocene	*Antelope Shale*	*Republic*	*Oil*

Geologic concept leading to discovery and method or methods used to delineate prospect, e.g., surface geology, subsurface geology, seeps, magnetic data, gravity data, seismic data, seismic refraction, nontechnical:

The discovery of the field is credited to surface indications. The discovery wells were drilled near asphalt deposits and oil seeps. Early development was based on random drilling and subsequent development on surface and subsurface geological studies.

Structure:

Province/basin type (see St. John, Bally, and Klemme, 1984)

Bally 332; Klemme IIIBb

Tectonic history

The San Joaquin basin began developing at the beginning of Tertiary time. The basin subsidence with sedimentation throughout Cenozoic was intermittently interrupted by uplift or orogeny that caused folding followed by erosion. The late Miocene orogeny produced northwest-southeast-trending folds and faults while mid-Pliocene movements caused tilting to the northeast.

Regional structure

Located southwest of the axes of a synclinorium between the Sierra Nevada Mountains and the San Andreas fault in the Coast Ranges being only 5 mi east of the fault.

Local structure

Regional homocline dipping at approximately 20° to the northeast at the Miocene unconformity, modified by minor folds and converging unconformities.

Trap

Trap type(s) .... *Anticlinal trap with multiple pays; truncation unconformity traps; pinchout traps; onlap traps*

Basin stratigraphy (major stratigraphic intervals from surface to deepest penetration in field):

Chronostratigraphy	Formation	Depth to Top in ft
Pleistocene	*Tulare*	*Surface*
Pliocene	*San Joaquin and Etchegoin*	*0–2000*
Upper Miocene	*Reef Ridge, Antelope, and McDonald*	*0–5000*
Middle Miocene	*Devilwater, Gould, Button Bed, Media, Carneros, Upper Santos*	*3000–11,000*

Location of well in field .... *NA*

Reservoir characteristics:

Number of reservoirs .... *13 major, 62 minor*

Formations .... *Tulare, San Joaquin, Etchegoin, Reef Ridge, Antelope Shale*

Ages .... *Pleistocene, Pliocene, upper Miocene*

Depths to tops of reservoirs .... *500–5000 ft (range); average 1000 ft*

Gross thickness (top to bottom of producing interval) .... *Variable*

Net thickness—total thickness of producing zones

Average .... *120 ft*

Maximum .... *1000 ft*

Lithology .... *Fine-grained, well-sorted to very coarse grained, poorly sorted sandstone to conglomerate plus fractured diatomaceous shale*

Porosity type .... *Primary intergranular in sands, fractures in shales*

Average porosity .... *Sands variable, 24–33%*

Average permeability .... *Sands variable, 10–3600 md*

Seals:

Upper

Formation, fault, or other feature .... *Overlying Recent, Pleistocene, Pliocene, or upper Miocene*

Lithology .... *Shales, siltstones, or claystones*

Lateral

Formation, fault, or other feature .... *Structural closure, or stratigraphic change*

Lithology .... *Shale, siltstone, or diatomites*

Source:

Formation and age .... *Monterey, middle to upper Miocene (Relizian to Mohnian)*

Lithology .... *(1) Biogenic siliceous rocks; (2) clay shale and siltstone*

Average total organic carbon (TOC) .... *(1) 3.43 ± 1.95%; (2) 2.06 ± 1.11%*

Maximum TOC .... *(1) 9.16; (2) 4.87%*

Kerogen type (I, II, or III) .... *I and II (mostly II)*

Vitrinite reflectance (maturation) .... $R_o = 0.7$ *average*

Time of hydrocarbon expulsion .... *Began in Miocene 5-6 million years ago; midpoint Pliocene 2 million years ago*

Present depth to top of source .... *10,000-13,000 ft*

Thickness .... *2000 ft net*

Potential yield .... *Unknown*

Appendix 2. Production Data

Field name .... *Midway-Sunset*

Field size:

Proved acres *21,830*

Number of wells all years .... *Unknown*

Current number of wells (as of year) .... *10,143 producing, 1790 shut in*

Well spacing .... *Variable; average 2 ac/well*

Ultimate recoverable .... *2252 million bbl*

Cumulative production .... *1.764 million bbl*

Annual production (1/1/1986) .... *57 million bbl*

Present decline rate .... *0%; increasing as a result of drilling*

Initial decline rate *3%*

Overall decline rate *0%*

Annual water production (1986) .... *150,000,000 bbl*

In place, total reserves .... *~4400 million bbl*

In place, per acre-foot .... *1680 average*

Primary recovery (1/1/1986) .... *895 million bbl*

Secondary recovery (1/1/1986) .... *869 million bbl*

Enhanced recovery *0*

Cumulative water production .... *2,565,681,000 bbl*

Drilling and casing practices:

Amount of surface casing set .... *0-500 ft*

Casing program

10¾-in., 11¾-in., or 13¾-in. surface casing; 7-in. or 8⅝-in. at top of pay; and 5-in., 5½-in., or 6⅝-in. slotted liner gravel packed through pay; or 7-in. or 8⅝-in. surface to TD and perforate pay zone

Drilling mud

Freshwater/gel mud to top of pay and convert to oil-base to under ream, hole below shoe if liner completion

Bit program

11-in. or 12-in. hole to TD, under ream to 15-in. through completion zone if liner completion

High pressure zones .... *None*

Completion practices:

- **Interval(s) perforated** .... *Variable*
- **Well treatment** .... *None*

Formation evaluation:

- **Logging suites** .... *Focused resistivity or induction log with SP; density and neutron logs with gamma ray and sidewall samples*
- **Testing practices** .... *Wireline formation tests presently being used to obtain pressure data*
- **Mud logging techniques** .... *None*

Oil characteristics:

- **Type** .... *Aromatic-naphthenic*
  (Tissot and Welte Classification in "Petroleum Formation and Occurrence," 1984, Springer-Verlag, p. 419)
- **API gravity** .... *8-30°, average 12°*
- **Base** .... *Asphalt*
- **Initial GOR** .... *Variable, 0-4300 ft^3/bbl*
- **Sulfur, wt%** .... *Generally <1.0%*
- **Viscosity, SUS** .... *18,000 to 400 at 100°F*
- **Pour point** .... *5-45°F*
- **Gas-oil distillate** .... *Variable, 10.8-26.9%*

Field characteristics:

- **Average elevation** .... *500-1700 ft*
- **Initial pressure** .... *75-2260 psi, depending on depth*
- **Present pressure** .... *Variable*
- **Pressure gradient** .... *0.452 psi/ft to much less for shallow zones*
- **Temperature** .... *80-100°F*
- **Geothermal gradient** .... *0.02°F/ft, western part; 0.011°F/ft, central part; 0.006°F/ft, eastern part*
- **Drive** .... *Variable; gas, water, and gravity*
- **Oil column thickness** .... *Variable*
- **Oil-water contact** .... *Variable*
- **Connate water** *25-30%*
- **Water salinity, TDS** .... *Variable; 1500-38,000 ppm*
- **Resistivity of water** .... *Variable; 5.46 to 0.168 ohm/m at 77°F*
- **Bulk volume water (%)** *6-10%*

Transportation method and market for oil and gas:

Pipeline to Los Anglos, Bakersfield, Santa Maria, and Bay Area.

Rio Vista Gas Field—U.S.A.
Sacramento Basin, California

DANE S. JOHNSON
Clovis, California

FIELD CLASSIFICATION

BASIN: Sacramento
BASIN TYPE: Forearc
RESERVOIR ROCK TYPE: Sandstone
RESERVOIR ENVIRONMENT OF DEPOSITION: Deltaic and Shelf
RESERVOIR AGE: Eocene to Upper Cretaceous
PETROLEUM TYPE: Gas
TRAP TYPE: Faulted Dome with Superimposed Stratigraphic Traps

LOCATION

The Rio Vista gas field, located in portions of T3 & 4N, R2 & 3E, M.D.B. & M. of Solano, Sacramento, and Contra Costa Counties, is the largest onshore natural gas field in California. The field is part of a major productive trend lying along the Midland fault zone in the southern portion of the Sacramento basin (Figure 1). Other significant fields lying along the trend and producing from Paleogene and upper Cretaceous strata are, from north to south, the Bunker, Maine Prairie, Lindsey Slough, Sherman Island, and Dutch Slough gas fields.

Cumulative production of gas and condensate from the Rio Vista field is 3,324,572 MMcfg (3.32 tcf) and 1,483,600 bbl of condensate through December 1988 (California Division of Oil and Gas, DOG, 1988). Estimated remaining gas reserves are 175,365 MMcfg. The field is ranked as the world's 396th largest field in the compilation by Carmalt and St. John (1986).

The California Division of Oil and Gas has established the total areal extent of the Rio Vista field at approximately 49,000 ac (76 mi^2, 198 km^2), with an estimated 25,000 productive acres. The field is 10 mi (16 km) long (north-south direction) and is up to 9 mi (14.5 km) wide. The Rio Vista Gas Unit (RVGU), operated by Amerada Hess Corporation (AHC), has an areal extent of approximately 30,000 ac (47 mi^2, 121 km^2) and covers the western portion of the field. The Isleton Gas Unit, in the northeastern portion of the field, is operated by Unocal.

HISTORY

Pre-Discovery

Hydrocarbon exploration began in the Sacramento basin near the turn of the century. The earliest recorded well in the Rio Vista area was drilled in 1901 by Rochester near a natural gas blowhole (Sec. 24, T5N-R1W) approximately 11 mi (17.7 km) northwest of the Rio Vista field discovery well. In 1919 Getty drilled the Suisun No. 1 well in Sec. 13, T5N-R1W, approximately half a mile north of the Rochester well. Getty followed this well with a 1921 test drilled in Sec. 9, T4N-R1W, approximately three-quarters of a mile west of the Portrero Hills field (discovered December 1938) and 2 mi southwest of a seeping natural gas vent. In 1922 P. G. & E. drilled the Montezuma Hills No. 1 in Sec. 2, T3N-R1E, approximately 5.5 mi (8.8 km) southwest of the Rio Vista field discovery. The P. G. & E. well is topographically higher than Rio Vista and lies in an area that is now recognized as a synclinal trough between the Rio Vista field and the Kirby Hill field. Exploration again moved northwest in 1935 where Trico drilled the Noonan No. 1 well in Sec. 10, T5N-R1W. These early exploratory wells were drilled on topographically high areas and many were drilled near natural gas seeps.

Hodgson (1980) has documented the reported and observed oil and gas seeps in California. Of 543 documented seeps, 65 occur within the Sacramento

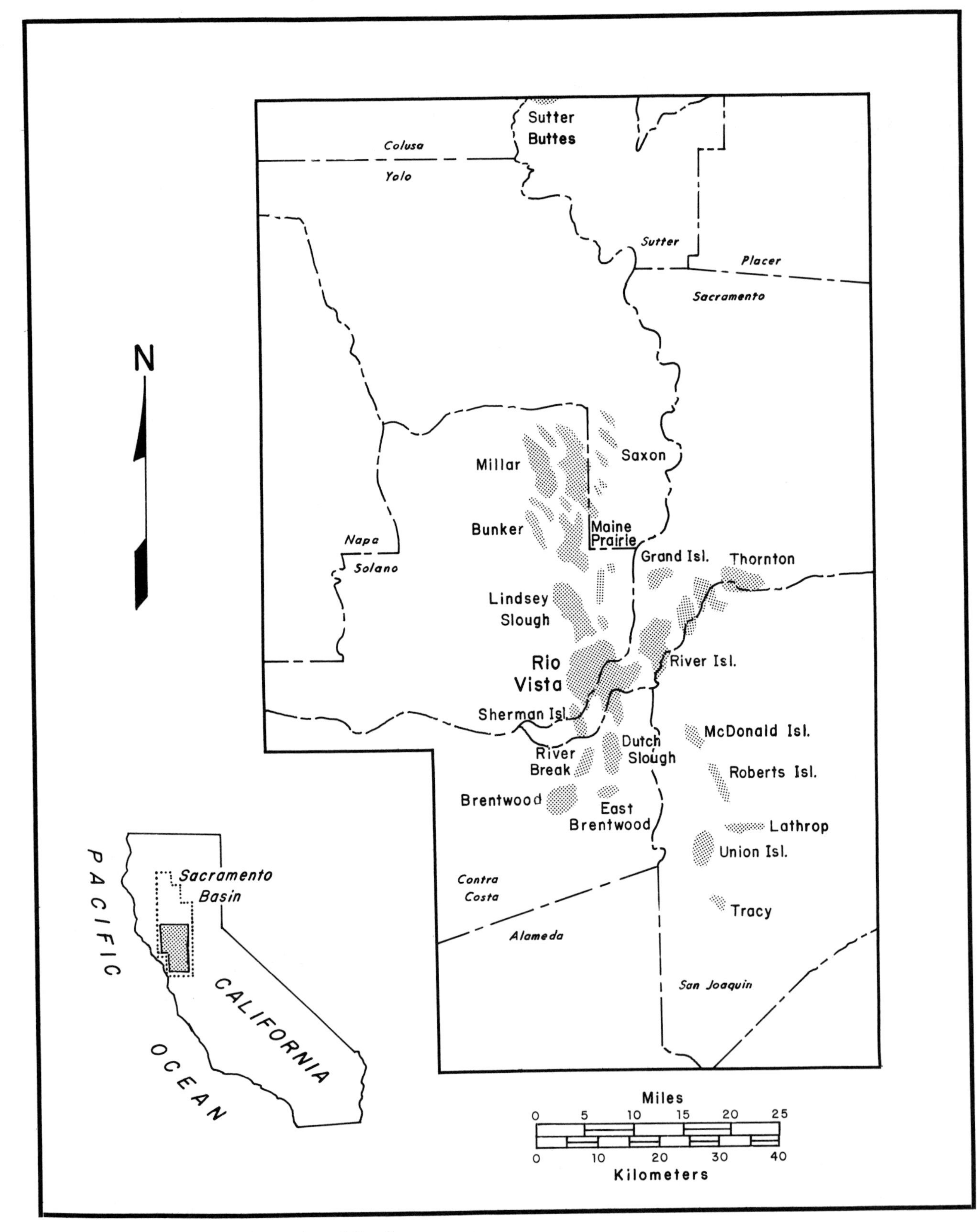

Figure 1. Regional map of a portion of the Sacramento basin, highlighting major producing fields along trend with the Rio Vista gas field.

basin. Although most of the seeps contain only natural gas, some also contain oil. To date, three Sacramento basin fields have produced oil ranging in gravity from 28° to 40° API.

Three fields were discovered in the Sacramento basin prior to the Rio Vista discovery. The initial discovery was the Sutter Buttes gas field by Buttes Resources Company in February 1933, approximately 67 mi (108 km) north of Rio Vista. In August 1935, Amerada Petroleum Company discovered the Tracey gas field about 25 mi (40 km) south-southeast of Rio Vista. Standard Oil of California followed with the discovery of the McDonald Island gas field, 15 mi (24 km) southeast of Rio Vista, in May 1936.

Discovery

The Rio Vista field was discovered by Amerada Petroleum Corporation. The discovery well, Emigh No. 1, was completed on 19 June 1936. Located in Sec. 26, T4N-R2E, in the Montezuma Hills of Solano County, the well is west of the town of Rio Vista and west of the Sacramento River, which bisects the field (Figure 2).

The Emigh No. 1 was completed flowing gas at the rate of 8.75 MMcfg/day through a ½-in. choke at a tubing pressure of 1375 psi (Burroughs, 1967; Burroughs et al., 1968). The well was drilled to a depth of 4485 ft (1366 m). Eighty-nine cores were recovered during the drilling of the well. The well was perforated between 4278 and 4484 ft (2669–1366 m) in the Eocene Domengine sand, which is locally termed the Emigh zone. (Although the lower Tertiary and upper Cretaceous sandstones of the Sacramento basin are semiconsolidated, they are customarily referred to as "sands" in both the literature and local terminology.) Open-hole flow potential was estimated at 81 MMcfg/day.

Post-Discovery

Development of the Rio Vista field began shortly after completion of the discovery well in June 1936. Within four years of the discovery, 30 wells had been drilled, 27 of which were completed. The three dry holes established the western and northern limits of the gas field. Twenty-three of the wells were completed west of the Sacramento River (Soper, 1943). Production was extended northeast, east, and southeast of the discovery well.

The third well in the field was drilled by Amerada Petroleum in 1936 (M. Hamilton No. 1, Sec. 26, T4N-R2E, RVGU #62). This deep test was drilled to 6791 ft (2068 m) and bottomed in the Paleocene McCormick sand. Gas was discovered in the sand below the Capay shale and the zone was named the Hamilton sand (Figure 3). Amerada plugged the well back for a Domengine sand completion following a formation test of the Hamilton sand, which produced gas at a rate of 4.16 MMcfg/day.

Until 1943 all wells drilled at Rio Vista were completed in the Domengine sand. In 1943 the Midland Fee No. 5 (RVGU #17) was drilled by Standard Oil Company of California in Sec. 4, T3N-R3E, east of the Sacramento River. Three separate productive zones were encountered—the Domengine sand, the Midland sand, and the Mokelumne River formation (M-5 zone)—with the latter two being new zone discoveries (Figure 3). Gas was subsequently discovered in the Anderson sand by Standard in 1944 in the P. Anderson No. 6 (Sec. 36, T4N-R2E).

Development through 1944 indicated that the northeast productive limits of the field had been reached. Producing intervals on the northeast part of the field were approaching the down-dip gas-water contact locations. In 1948, D. D. Feldman (later Brazos Oil & Gas Company and currently operated by Unocal) drilled the Gardiner No. 1 well (Sec. 35, T4N-R3E) and discovered gas in sand stringers in the Capay shale in an area east of the indicated limits of the Rio Vista field (Corwin, 1953). The Gardiner No. 1 discovered the Isleton area of the Rio Vista field. Subsequent development has demonstrated the structure in the Isleton area to be rising northeastward. The rising closed structure is truncated by the Isleton fault (Figure 4), which forms the northeast boundary of the Rio Vista field. By October 1949, six wells had been completed in the Isleton area.

Development in the Isleton area continued in the Capay shale and Midland sand. On 9 January 1951 the Isleton area was unitized with Brazos Oil & Gas as the operator. Currently, Unocal is the operator of the Isleton Gas unit.

Subsequent development through 1980 resulted in the discovery of seven additional productive horizons (Table 1). The most recent gas discovery was a stratigraphic trap in the upper Starkey sand of the Amerada Hess, Serpa No. 4 (Sec. 26, T4N-R2E). The entire Paleogene marine section at Rio Vista and the Upper Cretaceous Mokelumne River, Starkey, and Petersen sands are gas productive, for a total of 13 gas-producing zones.

In January 1965, the Rio Vista Gas Unit (RVGU) was established with Amerada Petroleum being named unit operator. The RVGU includes all stratigraphic units to a subsea depth of 5450 ft (1660 m). Excluded from the RVGU are the Isleton Gas Unit (previously unitized) and four wells operated by Tracey Drilling Company in the town of Rio Vista as well as all production established below a subsea depth of 5450 ft.

Recent development in the Rio Vista field has concentrated primarily on reservoirs below the RVGU and the Markley sand within the RVGU on the west side of the Sacramento River.

DISCOVERY METHOD

Exploration interest in the Sacramento basin was initiated by the presence of natural gas and oil seeps

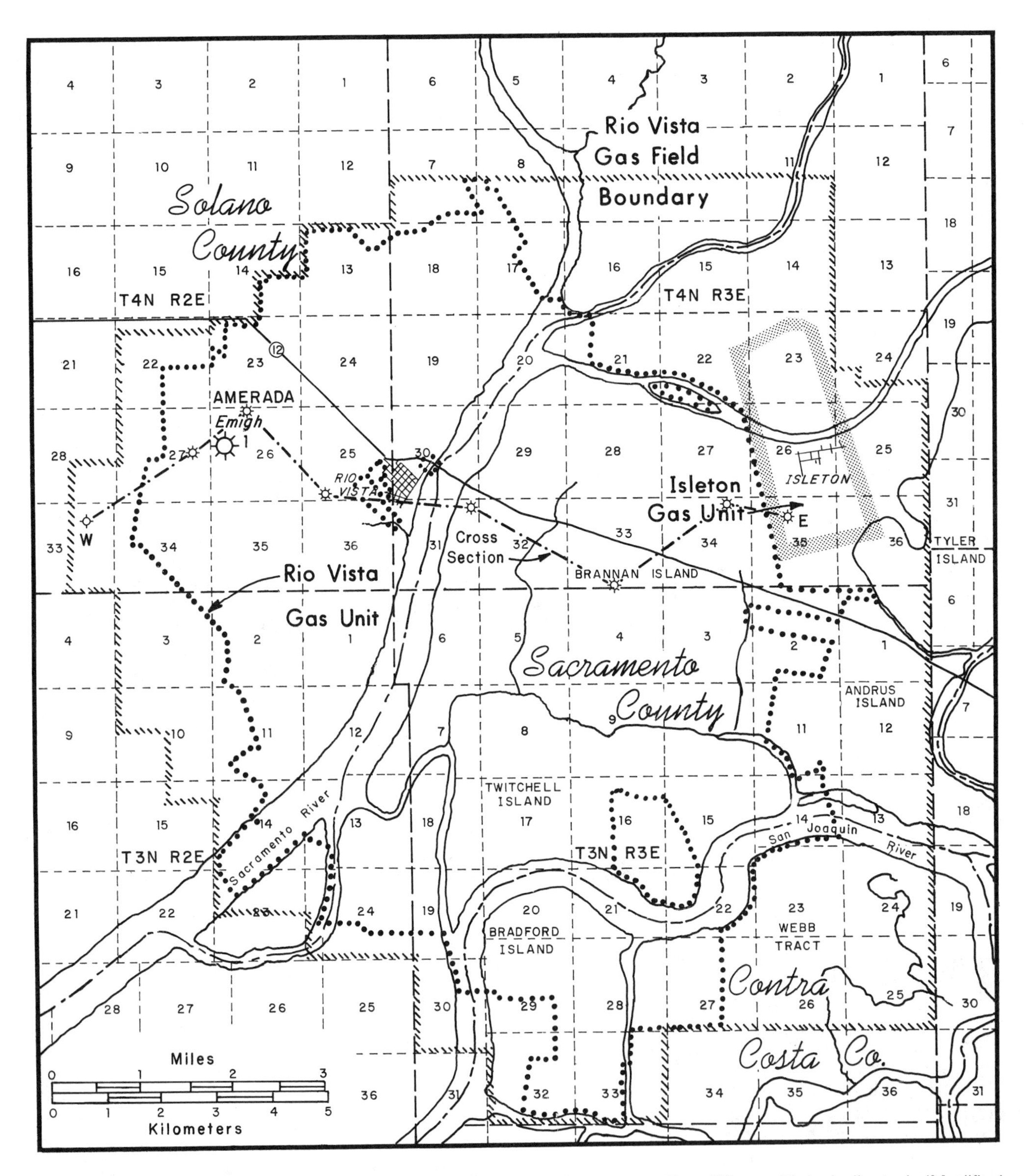

Figure 2. Plat of Rio Vista gas field showing Rio Vista Gas Unit, Isleton Gas Unit, field discovery well (Amerada Petroleum, Emigh No. 1), and Isleton area discovery well (D. D. Feldman, Gardiner No. 1). Line of the east-west cross section (Figure 5) is indicated. (Modified from California Division of Oil and Gas, Field Map No. 610.)

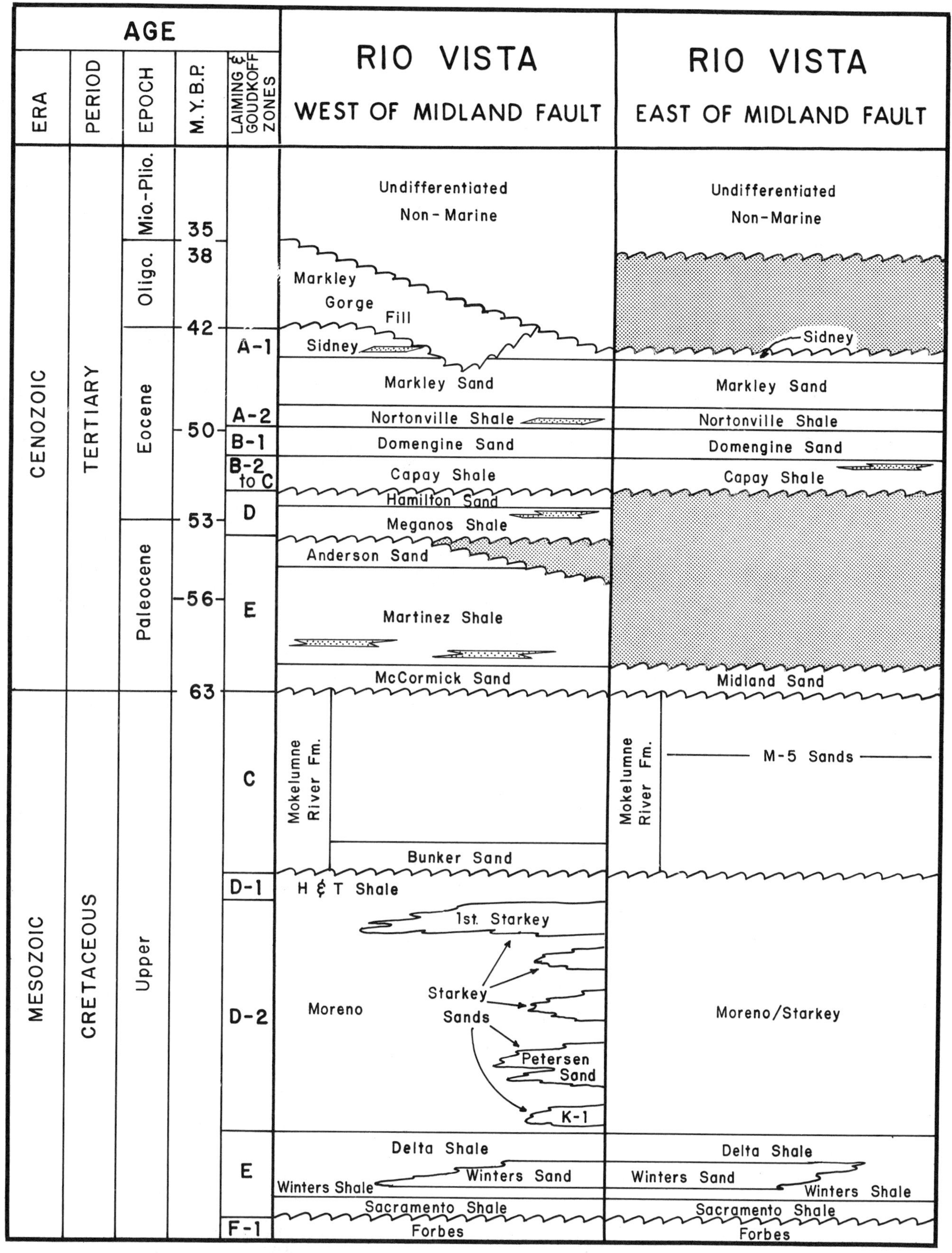

Figure 3. Generalized stratigraphic correlation chart for the Rio Vista gas field.

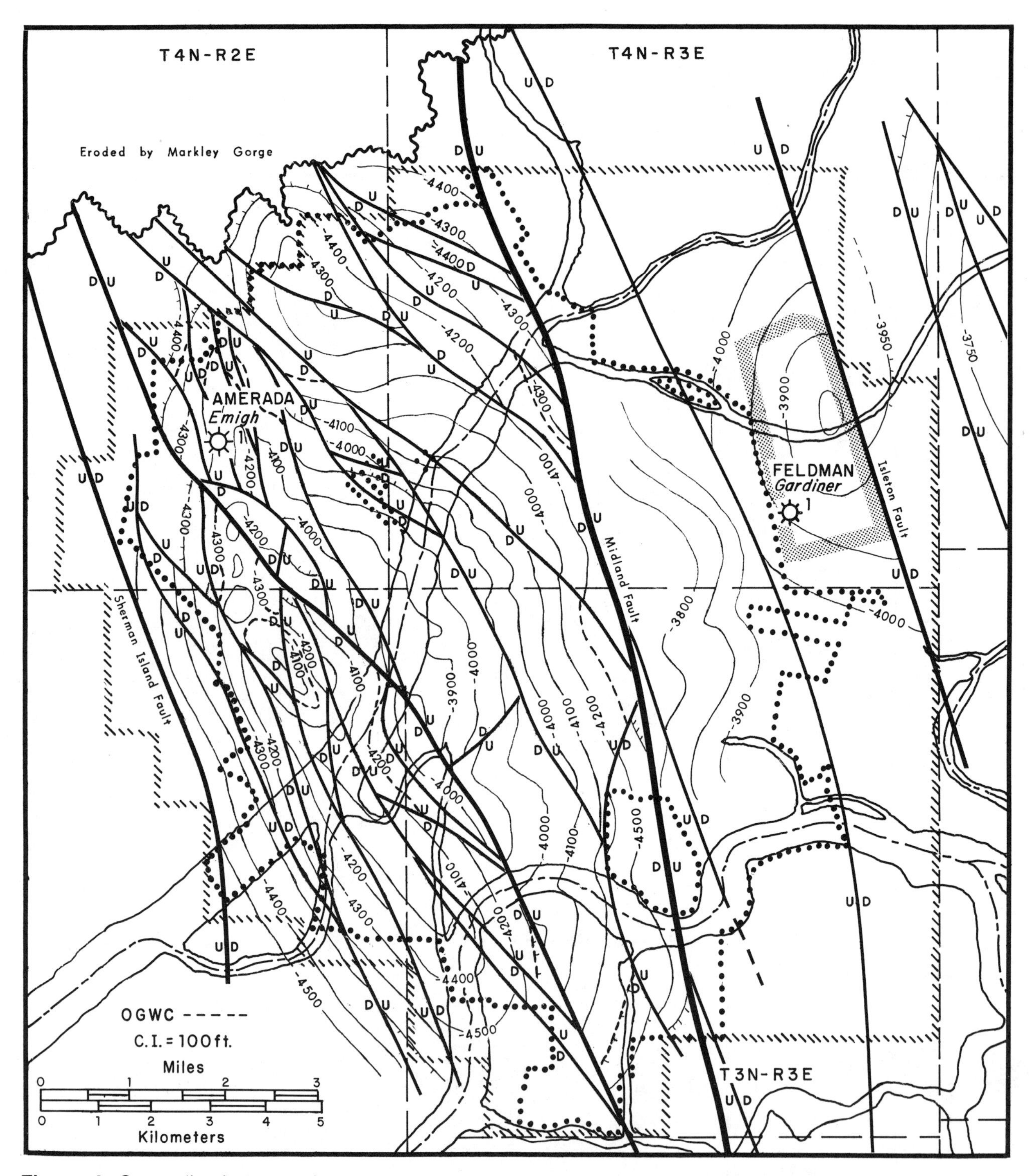

Figure 4. Generalized structural contour map of the Rio Vista gas field. Contours of top of Domengine sand. Contour interval 100 ft. Original gas-water contact (OGWC) indicated by dashed line.

in the western foothills of the valley. Many of the early exploration wells were drilled on topographic highs, expecting that they mimicked the subsurface structure in a fashion similar to that in the western San Joaquin basin to the south. This exploration concept generated interest in the Montezuma Hills area, which overlies a portion of the Rio Vista field west of the Sacramento River. The nonmarine strata exposed in the Montezuma Hills indicated a possible structural high. However, early exploration efforts

Table 1. Reservoir discovery data (modified from Burroughs, 1968).

Producing Zone	Operator	Well	Location Sec. T. R.	Perf. Depth	IP MCF/D	Disc. Date
Domengine	Amerada Petroleum	Emigh 1	26 4N 2E	4278-4484	8750	6-19-36
Hamilton	Amerada Petroleum	M. Hamilton 1	26 4N 2E	5522-5600 (open-hole test)	4160	11-22-36
Midland	Chevron	Midland Fee 5	4 3N 3E	4439-4460 (open-hole test)	5700	6-21-43
Mokelumne	Chevron	Midland Fee 5	4 3N 3E	5035-5043	13,337	8-3-43
Anderson	Chevron	P. Anderson 6	36 4N 2E	5735-5795	11,704	8-3-44
Capay	Feldman (now Unocal)	Gardiner 1	35 4N 3E	4494-4498 4500-4511	3013	5-18-48
Nortonville	Texaco	Brannan Isl. Un. 2	28 4N 3E	3690-3790 (at intervals)	189	9-15-50
Petersen	Amerada Petroleum	Drouin 8 (RVGU 188)	23 4N 2E	9762-9790 9810-9838	397	4-12-66
Martinez Sand Stringers	Chevron	Anderson 11	35 4N 2E	6126-6655 (at intervals)	4250	8-26-66
McCormick	Chevron	Anderson 11	35 4N 2E	6697-6707	5330	8-26-66
Markley Sand	Amerada Hess	RVGU 179	25 4N 2E	2620-2646	1818	8-19-77
Sidney	Amerada Hess	RVGU 181	36 4N 2E	2426-2471 (at intervals)	1490	10-13-77
Starkey	Amerada Hess	Serpa 4	26 4N 2E	8906-8920	2043	2-1-83

had not demonstrated the presence of the different structural configuration of the Paleogene and older rocks beneath the post-Eocene unconformity nor that the central portion of the Montezuma Hills overlies the synclinal axis of the Sacramento basin.

By 1929, J. C. Karcher of Amerada Petroleum had developed the use of the reflection seismograph (*Time*, 1952). This instrument permitted a much more accurate interpretation of subsurface structure than the refraction seismograph which was the standard industry tool until that time. Amerada Petroleum's staff developed the seismic reflection technique so well that Amerada soon dominated the U.S. field of petroleum geophysics.

Amerada employed the seismic reflection technique in the Montezuma Hills west of the Sacramento River searching for a structural trap. Interpretation indicated the presence of a dome under the eastern portion of the hills. The location for the discovery well was based on the reflection seismic interpretation.

STRUCTURE

The Sacramento basin, separated from the San Joaquin basin (to the south) by the Stockton arch, began to form in Late Jurassic to Early Cretaceous time as the Farallon plate began subducting beneath the North American plate. The result is an asymmetric, elongate, north-northwest-trending forearc basin with the structural depocenter lying at the southern end of the basin (Dickinson and Seely, 1979). The eastern limb of the forearc basin is characterized by gently dipping beds (4°–8° west) and undulating, low amplitude structures that are fractured and offset by north–northwest-trending normal faults. The western limb of the basin exhibits moderate eastward dips as the sediments of the Great Valley sequence form the eastern portion of the Coast Ranges. Structures on the west side of the Sacramento basin are complicated by thrust faults formed by imbrication of oceanic sediments that constitute the subduction complex of the Coast Ranges.

St. John et al. (1984) classified the Sacramento basin in combination with the San Joaquin basin. As such, the Bally classification is 332 (California-type basin) and the Klemme classification is IIIBb (continental rifted basin, rifted convergent margin, transform). These classifications are appropriate for the San Joaquin basin but are inappropriate for the Sacramento basin. The Sacramento basin has not been subjected to transform fault tectonics that have produced the structures on the west side of the San Joaquin basin. For the Sacramento basin, the proper Bally classification is 311 (forearc basin) and the proper Klemme classification is V (forearc basin).

The Rio Vista field lies at the southern end of the Sacramento basin on the east side of the Delta depocenter (Garcia, 1980, 1981). Sedimentation in the Rio Vista area was sourced from river systems originating in the arc system to the east that was forming the Sierra Nevada block. These rivers deposited their loads in delta systems along the eastern edge of the basin. Rapid sedimentation into the depocenter resulted in the overpressuring of basinal marine shales. This overpressuring resulted in the formation of the Midland fault zone, the dominant structural feature in the area.

Syndepositional faulting (growth-faulting) resulted in the formation of several roll-over structures on the basinward side of the Midland fault, the most notable of which forms the domal structure of the Rio Vista field. The domal structures of the Bunker and Lindsey Slough gas fields (along trend and to the north of the Rio Vista field) were similarly formed. Offset along the Midland fault increases with depth, with over 2600 ft (792 m) of offset being noted in the deeper portion of some wells (at a depth of approximately 9000 ft [2740 m]) on the north side of the Rio Vista field and the south side of Lindsey Slough field. Portions of the Paleocene and Eocene sections, which are present west of the Midland fault, either were not deposited or have been eroded from the stratigraphic section east of the fault.

The overall domal structure at Rio Vista (Figure 4) is cut by numerous longitudinal normal faults that trend roughly parallel to the Midland fault (north-northwest) and that dip either east or west (antithetic or synthetic, respectively). The major faults are connected by numerous smaller normal faults that trend approximately northwest. The strata on the east side of the Midland fault are offset by normal faults that are subparallel to the Midland fault. The structure at Rio Vista is an excellent example of a faulted dome produced by syndepositional faulting. Figure 5 presents a generalized cross section of the field oriented along the dip direction.

The Isleton area of the field appears to be a gently domed structure that is truncated by the Isleton fault to the northeast. The Isleton fault is a northeast-dipping, north–northwest-trending normal fault. It appears there may be a left-lateral component of strike-slip movement to the fault. The Isleton area appears to be separated from the main area of the Rio Vista field by faults.

Present-day relative movement along the Midland fault exhibits both normal dip-slip and reverse dip-slip (or thrust) displacement within the Markley stratigraphic section. Through Mesozoic and early Cenozoic time, the Midland fault was primarily a normal dip-slip growth fault. During Miocene and Pliocene time, displacement has probably been reverse. Reverse and wrench faults have not been identified in the Rio Vista area. However, there may be a strike-slip component to the offset of some of the normal faults that has been superimposed over prior growth-fault movement by later tectonic forces related to the wrench faulting of the San Andreas, Hayward, and Calaveras fault zones to the west.

STRATIGRAPHY

Stratigraphic relations within the Rio Vista field have been discussed by Soper (1943), Frame (1944), Corwin (1953), Burroughs (1967), Burroughs et al. (1968), Garcia (1980, 1981), and Cherven (1983). Figure 3 illustrates the generalized stratigraphic relations within the Rio Vista field, and Figure 6 is a composite type log with lithologic descriptions.

The stratigraphic section at Rio Vista consists of Cretaceous through Eocene marine rocks unconformably overlain by Miocene and Pliocene nonmarine sediments (Figure 3). The deepest well in the field, the Standard Oil Co., P. Cook No. 15, was drilled to 15,050 ft (4583 m) in Sec. 8, T4N-R3E and penetrated approximately 2200 ft (670 m) of the Upper Cretaceous Forbes Formation. The Forbes, a basinal to continental slope marine shale with interbedded sands, is the predominant producing horizon in the northern half of the Sacramento basin. In the Rio Vista area the Forbes sands are thin (5 to 20 ft [1.5–6 m] thick) and tend to produce an inverse deflection on the self-potential (SP) log. The inverse SP appears to be produced by an overbalanced drilling mud system that flushed the formation fluids back into the formation to the point where only the infiltrated mud is being analyzed by the wireline logs. An example is a Forbes sand at a depth of 14,000 ft, where normal hydrostatic pressure would be 6062 psia. The mud system used by Standard Oil produced a bottom hole pressure of 12,068 psia. The resulting overbalanced mud system was not conductive for an adequate evaluation of the gas-bearing potential of the Forbes sands.

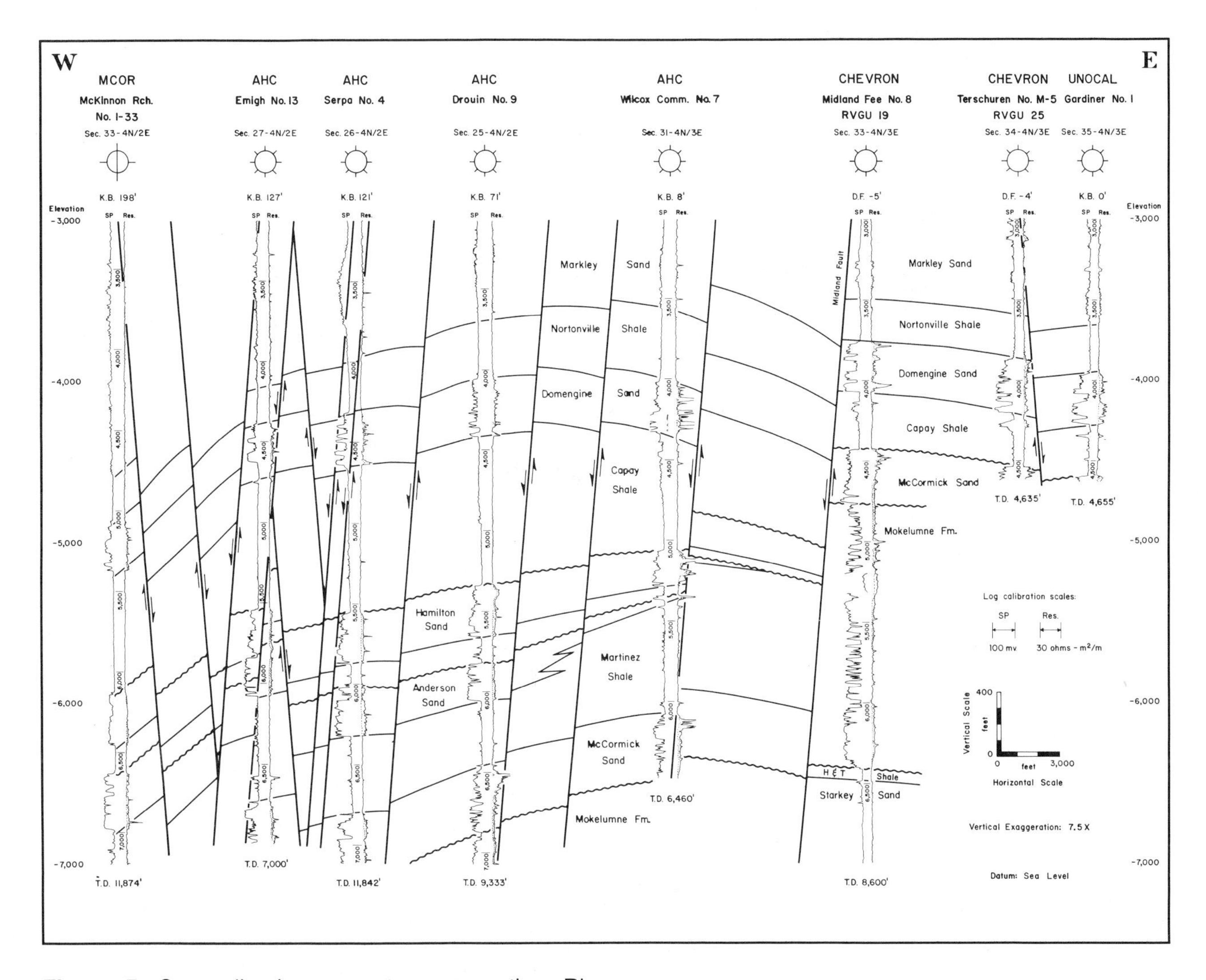

Figure 5. Generalized east-west cross section, Rio Vista gas field. Line of cross section is indicated on Figure 2.

The basinal marine Sacramento shale lies unconformably above the Forbes Formation. The Sacramento shale is characterized on wireline logs by minimal SP and resistivity deflection. Conformably overlying the Sacramento shale is the Winters shale. The Winters shale is a prodelta shale that progrades westward over the Sacramento shale.

Winters Sand

The Upper Cretaceous Winters sand overlies the Winters shale along the eastern side of the basin. The Winters is a series of overlapping submarine fans deposited by turbidity currents (Cherven, 1983; Garcia, 1980, 1981). This unit forms a north-northwest trend that pinches out to the east (proximal to source area) and west (distal from source area). The Winters sand is not currently productive in Rio Vista. However, this unit has produced gas in the Lindsey Slough gas field to the north and eastward in the River Island gas field. There have been very few tests of the Winters sand in the Rio Vista field. It is possible that the Rio Vista area is represented by a distal facies of the fan lobe, where porosity and permeability are reduced to a level where the Winters is noncommercial.

Delta Shale

Conformably above the Winters sand is the Delta shale, also occasionally referred to as the Sawtooth shale. The Delta shale is a prograding prodelta shale and is a probable source of some of the hydrocarbons present. The Delta shale is continuous with the Winters shale in areas where the Winters sand was not deposited. Above the Delta shale the formations

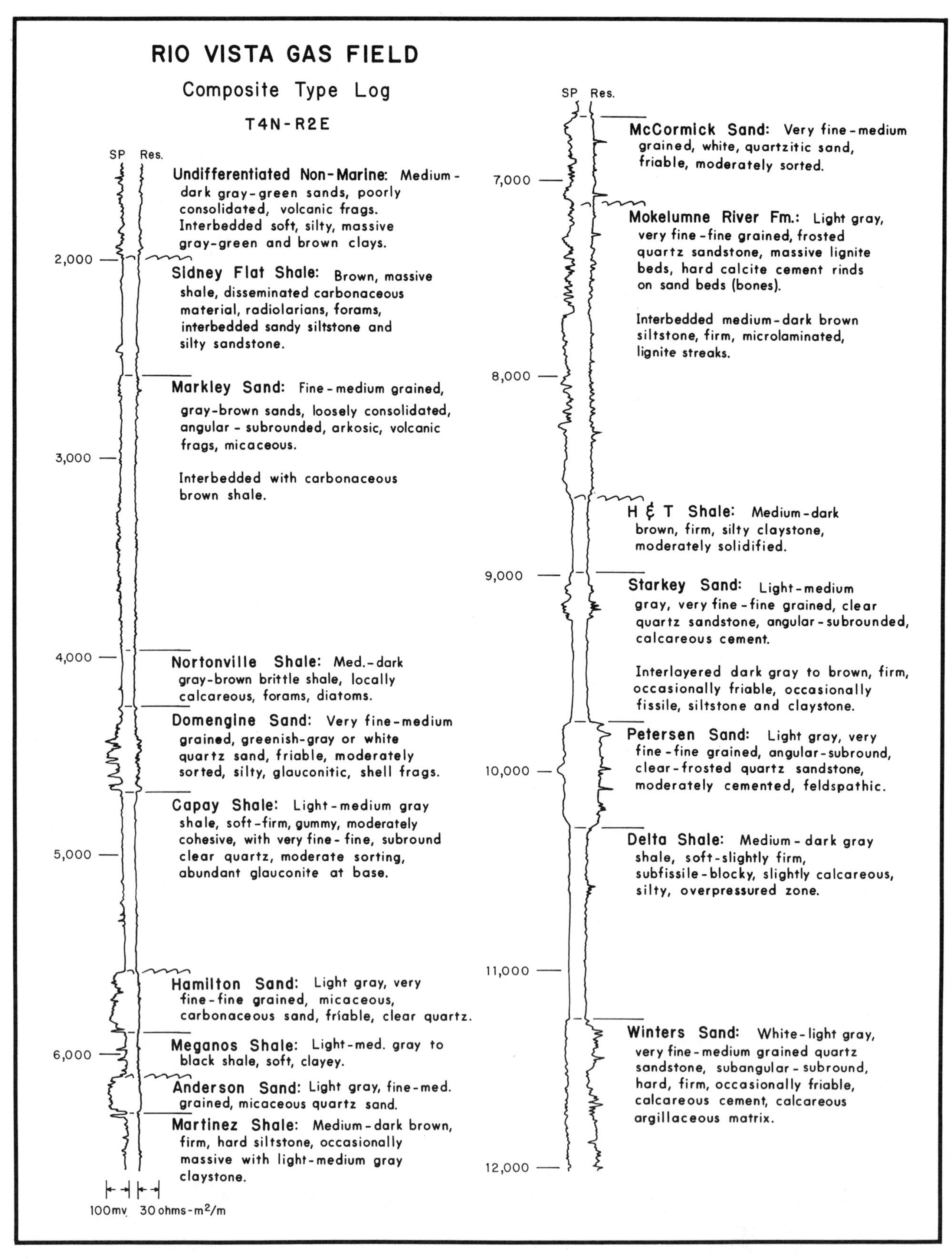

Figure 6. Composite type log (SP-resistivity) and generalized lithologic descriptions from Rio Vista gas field, west of the Midland fault.

exhibit normal hydrostatic pressure, whereas the Delta shale and older formations are overpressured approximately 5 to 10%.

Starkey Sand Sequence

The Starkey is a sequence of deltaic sands and intertonguing shales that conformably overly, and prograde westward over, the Delta shale. The Starkey sand, Delta shale, Winters sand, and Winters shale are different facies of the same depositional system and were contemporaneously deposited. In the Rio Vista area the Starkey sand intertongues with the Delta shale and has been subdivided with the lower units named to differentiate the productive intervals in the Lindsey Slough gas field. The shale intertonguing with the Starkey sands is referred to as the Moreno Shale. The basal sand, termed the "K-1" sand, is productive north of the Rio Vista field and is represented by a thin sand stringer at Rio Vista. The Petersen sand is separated from the "K-1" sand below and the Starkey sand above by the Moreno Shale.

Petersen Sand

The lower massive sand of the Upper Cretaceous Starkey sands is referred to as the Petersen sand. The Petersen is the lowermost productive unit in the Rio Vista field where only one well has produced a small amount of gas from the sand (Amerada Petroleum, Drouin No. 8 [RVGU #188] in Sec. 23, T4N-R2E, completed in 1966) prior to recompletion. Regional isopachs depict the Petersen as a series of offshore sand bars trending north-northwest, subparallel to the paleoshore-line and distal to the upper Starkey sands of deltaic origin.

Regionally, the Petersen becomes increasingly massive northward with increasing reservoir quality. Core data from the north end of the Maine Prairie field demonstrate a permeability of 4–5 darcys. Southward, in the Lindsey Slough field (5N/2E), where the Petersen is a primary producing horizon, increasing shale above the Petersen provides the updip seal in traps formed by faults. The amount of net sand in the Petersen decreases and the permeability also decreases.

Starkey Sand

The sands above the Petersen are represented by the uppermost Starkey sand. The unit has been penetrated, primarily, by a few wells in the northwest portion of the field. The Starkey is productive in a single well (Amerada Hess, Serpa No. 4, Sec. 26, T4N-R2E), in the northwestern part of the field, which was originally targeted to the Winters sand, in an updip stratigraphic pinchout of the unit. The reservoir sand occurs on the northern end of the Rio Vista structure and pinches out southward toward the crest of the structure. The deltaic Starkey sands are pinching out in the Rio Vista area and are absent over most of the western part of the field. It appears that a reentrant of the sea existed in the Rio Vista area during Starkey depositional time as evidenced by the shaling out of the Starkey section.

H & T Shale

A late Maastrichtian stage marine transgression conformably deposited the organic-rich H & T shale on top of the Starkey sands. The H & T shale is a middle- to outer-neritic deposit (water depth 20–200 m) with a foraminiferal fauna assigned to the uppermost D-1 and lower C zones of Goudkoff (1945). An unconformity separates the H & T shale from the overlying Mokelumne River formation.

Mokelumne River Formation

The Mokelumne River formation (Mokelumne) is the uppermost Cretaceous stratigraphic unit in the southern Sacramento basin. This unit is a series of interbedded sands and shales that were deposited in a delta system. The basal sand (Bunker sand to the north, "Third Massive" sand to the south) is an offshore bar sand and exhibits a coarsening-upward sequence on the wireline logs. The interbedded sands and shales of the Mokelumne above the Bunker sand are composed of channel, levee, and crevasse splay sands, overbank deposits, and interchannel silts and clays. Lignite beds commonly occur within the Mokelumne. Net sand content increases southward. South of the Rio Vista field, the upper portion of the Mokelumne (above the Bunker sand) is referred to as the "Second Massive" sand.

Few wells west of the Midland fault zone have penetrated the Mokelumne. Production is established east of the Midland fault in a section locally referred to as the M-5 zone. Sands in the zone are friable and poorly indurated and cemented. Traps are formed by updip faults with lateral structural closure. The Mokelumne is the uppermost producing Cretaceous horizon in the Rio Vista field.

McCormick Sand

The Tertiary section is separated from the Cretaceous strata by a regional unconformity. The basal Tertiary is represented by the Paleocene McCormick sand. Terminology in the area is complicated by the sand being given different names on either side of the Midland fault. The unit is termed McCormick sand west of the fault and Midland sand east of the fault. The unit is the lowermost producing Tertiary strata in the field. McCormick/Midland sand production has been established east and west of the Midland fault and is the primary target of most deep wells drilled within the field below the RVGU. Traps are formed by updip faults with lateral structural closure.

In the Isleton area of the Rio Vista gas field, the Midland sand is the lowermost productive formation. The upper portion of the McCormick sand was removed by Eocene erosion east of the Midland fault prior to deposition of the Capay shale. It is the oldest Tertiary strata showing the erosional effects related to movement along the Midland fault. South of the Rio Vista field the McCormick sand increases in thickness and is referred to as the "First Massive" sand.

The stratigraphic units that occur between the McCormick sand and the Eocene Capay shale (Anderson and Hamilton sands, and the Martinez and Meganos shales) are present west of the Midland fault only. This section was either removed by erosion or, probably in part, not deposited east of the Midland fault. Syndepositional tectonics, associated with growth-faulting along the Midland fault, may be partially responsible for the erosion/nondeposition of these strata. All Paleocene and Eocene strata are gas producing within the Rio Vista gas field, including the sands that occur within each of the shale units.

Martinez Shale

The Paleocene Martinez shale is a marine deposit that is only present west of the Midland fault and contains sand stringers 3 to 15 ft (1-5 m) thick. These stringers are frequently secondary targets in wells drilled into the McCormick sand. Production is from sand stringers 3 to 10 ft (1-3 m) thick in the lower portion of the unit and commonly produces gas with no water cut. Traps appear to be formed from updip fault truncation of the sand stringers, and laterally, possibly by stratigraphic pinchout.

Anderson Sand, Meganos Shale

The Paleocene Anderson sand conformably overlies the Martinez shale and thickens in a westward direction. The upper contact with the Paleocene-Eocene Meganos shale is an unconformity. A few sand stringers that occur in the Meganos are frequently productive.

Hamilton Sand

The Eocene Hamilton sand forms a sharp, conformable contact with the underlying Meganos shale. The Hamilton thickens westward and is a frequent producer within the RVGU.

Capay Shale

The Capay shale is the oldest Eocene unit to occur both east and west of the Midland fault. The Capay is separated from the Hamilton by an unconformity. East of the Midland fault the Capay is significantly thinner. Micropaleontologic data indicate that the lower portion of the Capay was deposited in an outer-neritic environment whereas the upper portion was deposited in an inner-neritic to brackish-water environment. Thus the Capay records a partial shoaling of the basin during Eocene time.

The Capay shale is productive east of the Sacramento River from an interbedded sand in the upper portion of the unit. Traps are formed by updip faults and lateral structural closure. The Capay was the discovery zone in the Isleton area of the field. All Capay production is part of the RVGU.

Domengine Sand

The Domengine sand, also referred to as the Emigh sand (pronounced "Amy") is the primary productive horizon both east and west of the Midland fault and is the discovery zone for the field. Contacts with the underlying Capay shale and overlying Nortonville shale are conformable. The unit consists of a series of interbedded marine sands and shales, with sand being the predominant lithology. The upper portion is referred to in the older literature as the green sand and the lower portion, as the white sand. The green sand contains a higher percentage of glauconite and is a dirtier, less mature, fine-grained sand. The white sand is more mature with better sorting and fewer interstitial clays than the green sand, with the grains being better rounded and containing fewer accessory mineral grains. The interlayered shales are also color differentiated. The shales in the upper part are a greenish gray with the lower shales being a light to medium gray. Some of the sands in the lower portion are coarse grained with vugular porosity and no cement to bind the sand grains.

The Domengine was the primary target within the field throughout the early development period. In deeper wells with producing zones below the Domengine, it was common practice to plug back the well and produce the Domengine.

Nortonville Shale, Markley Sand, Sidney Flat Shale

The Eocene section is completed, in ascending order, by the Nortonville Shale, Markley sand, and Sidney Flat Shale. These marine units are all conformable in sequence with and above the Domengine sand. The Nortonville Shale and Sidney Flat Shale produce from interbedded sand stringers.

The Markley sand is a poorly consolidated deltaic deposit of interbedded sands and shales. Formation waters in the Markley are not as saline as in the older formations, as evidenced by the suppressed deflection on SP logs. Traps appear to be formed by updip faults with lateral structural closure. It is possible that some stratigraphic traps also exist.

Development of the Markley has been undertaken during the past several years and the sand has been the target of many of the recent RVGU wells drilled by Amerada Hess. Most wells in the Rio Vista field were drilled in locations designed to test deeper strata on the upthrown side of potential trapping faults. Thus, most wells have tested the Markley on the downthrown side of potential trap forming faults and in structurally low positions within the encountered fault block. Additional development drilling will thus be necessary to fully evaluate Markley potential within the Rio Vista field.

Post-Eocene Strata

The Oligocene Markley gorge (Almgren, 1978; Kittridge and Wilson, 1984) cut a submarine canyon that truncates some of the Eocene strata in the extreme northwestern portion of the field. The gorge is filled in with a sticky ("gumbo") clay with occasional sand lenses. The Markley gorge fill acts as an updip seal for gas reservoirs in some fields north of Rio Vista. However, the gorge deposits do not appear to be associated with any of the traps in Rio Vista. The Markley gorge fill is not associated with the Markley sand and has completely eroded the Markley sand over a large area.

Unconformably overlying the Eocene and Oligocene strata are nonmarine alluvial sands and shales of Miocene age. The nonmarine sands are nonproductive. The structure and faults present in the Eocene section apparently do not, for the most part, extend into the nonmarine section. It is probable that the structures present in the Eocene and older strata, which were produced by syndepositional tectonics, had mostly stabilized by the end of the mid-Tertiary erosion. Regional uplift of the basin and Sierra Nevada block has produced a thick sequence (2200 ft, 670 m, more or less) of Miocene and Pliocene age alluvial sediments.

TRAP

Two trap types have been demonstrated at Rio Vista: structural closure against updip faults and stratigraphic barriers to fluid migration.

Faults are the most common trapping mechanism in the Rio Vista field. These are the result of normal faults that originated in the Eocene during the period of active growth-faulting and resulted in the domal structure west of the Midland fault and in the Isleton area. Traps are formed where the reservoir sand is offset across the fault by a shale. Lateral reservoir closure is commonly produced by structural dip beneath the gas-water contact, although lateral closure may also be formed by faulting or stratigraphic changes.

The stratigraphic traps, which are superimposed over the domal structure, are formed by pinchout of the reservoir sands and permeability reductions that inhibit fluid migration. Updip stratigraphic changes are the primary trapping mechanisms in the Starkey sand, in some producing sand stringers within the shales, and probably in some pools of the Domengine sand. Older structural contour maps of the Domengine sand indicate numerous gas pools with closures against updip faults. Detailed log analysis and seismic data examination cannot substantiate some of these faults, which also occur in areas of rapid stratigraphic change within the Domengine. It is probable that some of these pools are created by an updip facies change. It is possible that some stratigraphic traps have been previously misinterpreted as fault traps.

RESERVOIR

Different structures have been produced in the eastern and western portions of the Rio Vista field by growth-faulting along the Midland fault. A roll-over domal structure was produced west of the fault, and associated synthetic and antithetic normal faults formed east and west of the fault. The result is variable depths to the different reservoirs across the field. Depths to the reservoirs are generally deeper west of the fault (downthrown side) where the section thickens and the lower Eocene strata escaped the erosion that removed them from the area east of the fault. The average depth to the Eocene Domengine sand west of the Midland fault is 4300 ft (1309 m) and 3800 ft (1157 m) deep east of the fault. Average depth to the Paleocene McCormick sand reservoir is 6500 ft-7600 feet (1980-2315 m) west of the fault and 4500 ft (1370 m) east of the fault.

Table 2 lists the reservoir characteristics of each producing zone. Throughout the field there are 13 different stratigraphic units that are gas productive. This results in a wide variation in the gross thickness of the producing interval, depending upon location within the field and the number of stacked pay zones at any given well location. Net pay per producing horizon has varied up to 315 ft (96 m) and averages 40 ft (12 m). The majority of reserves in the field are trapped by updip fault closure, and the productive limits of the field are marked by an increase in water saturation. While the updip limits of some reservoirs are marked by stratigraphic pinchouts, there are no documented reservoirs that exhibit downdip pinchouts or permeability barriers.

The various producing fault blocks are hydraulically separated from each other. As such, each zone exhibits a unique gas-water contact in the different producing fault blocks across the field. Production is by active water drive in all reservoirs.

Table 2. Rio Vista gas field reservoir characteristics by producing zone (modified from California Division of Oil and Gas, 1981).

RIO VISTA GAS FIELD

ITEM	Sidney	Markley	Nortonville	Domengine	Capay	Hamilton	Anderson	Martinez	McCormick	Mokelumne	Starkey	Petersen
POOL DATA												
Initial production rates												
Gas (Mcf/day)	1,490	1,274	190	8,750	3,010	4,160	11,700	4,250	5,330-5,700	13,340	2,043	400
Flow pressure (psi)	731	918	–	1,375	1,670	290	2,145	1,810	1,925	1,635	2,230	125
Choke size (in.)	–	–	–	1/2	1/4	3/4	1/2	3/8	3/8	5/8	–	3/8
Initial reservoir pressure (psi)	1,110	1,190	1,230	1,715-1,915	1,930	2,415	2,550	2,550-3,000	2,060-2,930	2,210	3,790	4,860
Reservoir temperature (°F)	116	116	135	141-149	150	167	177	177-187	153-190	153	195	195
Initial gas content (MSCF/ac.-ft.)	500-690	560-740	700	1,200-1,300	800	1,100	1,600	920-1,400	1,100-1,400	1,100	930-1300	800-1,300
Average depth (ft.)	2,450	2,630	3,700-4,200	3,800-4,300	4,500-5,100	5,300	5,750	5,800-6,900	4,500-7,600	5,050	8,910	9,650
Average net thickness (ft.)	50	24	25	40-315	20-40	90	45	30-120	40-140	10	15	55
RESERVOIR ROCK PROPERTIES												
Porosity (%)	25-32	26-32	30	34	26	27	31	22-28	24-33	25	18-23	14-20
Sw (%)	35-40	35-40	35	30	45	40	25	35-40	35-40	30	45-50	45-50
Sg (%)	60-65	60-65	65	70	55	60	75	60-65	60-65	70	50-55	50-55
Permeability to air (md)	5-10	400-1,800	–	–	–	–	–	15-180	120	–	–	3-19
RESERVOIR FLUID PROPERTIES												
Gas:												
Specific gravity (air = 1.0)	.580	.580	.595	.580-.604	.599	.599	.601	.596	.599-.616	.583	.599	.608
Heating value (Btu/cu. ft.)	1,009	1,009	1,010	1,000-1,050	1,060	1,060	1,070	1,065	1,025-1,060	990	1,059	1,080
Water:												
Salinity, NaCl (ppm)	–	4,590	–	6,100-9,500	8,600-15,600	15,400-18,800	10,100-24,000	15,400	10,500-15,200	11,300	–	7,700
T.D.S. (ppm)	–	4,703	–	–	–	–	–	–	–	–	–	–
Rw (ohm/m) (77°F)	–	1.36	–	–	–	–	–	–	–	–	–	–

Petersen Sand

In the AHC, Serpa No. 4, in the northern portion of the field, permeability varies from 3 to 41 millidarcys in a gray, very fine to fine-grained silty sand. Porosity varies between 19 and 23%. At the northwest end of the Rio Vista field, reservoir quality in the Petersen is greatly reduced. Porosity in this fine-grained sand averages 16–18% and the permeability is reduced to 0.1 to 10 md. SEM and thin-section analysis reveal that authigenic chlorite occludes the pore throats, thus reducing the permeability.

This poor reservoir quality is evident in the Drouin No. 8. Initial gas flow during production testing was 14 Mcfg/day. Amerada Petroleum experimented with a sand frac in the Petersen during completion. At a perforation depth of 9762 to 9838 ft (2973–2996 m), 18,000 pounds of sand were squeezed away at pressures from 5700 to 6700 pounds. After cleaning out the hole, the well produced gas at the rate of 325 Mcfg/day. The Petersen is moderately friable and only semiconsolidated, which resulted in closing of the fractures around the sand propant. The Petersen zone was abandoned and the well recompleted in a shallower zone. Subsequent drilling to the west and northwest has demonstrated the same lithologic and reservoir characteristics as in the Drouin No. 8 but with an increase in chlorite content.

No other well has been completed in the Petersen sand within the field. Also, no other well within the Rio Vista field has been fractured during completion.

Starkey Sand

The Starkey reservoir is fine grained and exhibits 18 to 23% porosity by porosity log evaluation. A gas-water contact was not encountered in the Serpa No. 4. Water encroachment into the well had not occurred during the first few years of production and reservoir pressures had remained relatively constant.

Low permeability in the Starkey results in a relatively tight reservoir for this field. The absence of produced water and constant production pressures indicates that additional productive locations might exist down-dip from the existing well where reservoir quality is likely to improve. This reservoir is the best example of a stratigraphic trap within the field.

Martinez Zone Stringers

These 3 to 10 ft (1–3 m) thick stringers, when gas filled, may exhibit no more than 3 ohms deflection from background level on the deep-resistivity log curves. A 3 to 5 ft thick sand with 3 ohms resistivity deflection is capable of producing in excess of 1 MMcfg/day with no water cut. Minimal cross-over will be noted on neutron and density logs. It is therefore very important to examine the wireline logs closely in conjunction with the mudlog.

Domengine Sand

Reserves generally occur in the upper sands of the Domengine sand. Traps are normally formed by updip faults with lateral structural closure. Some reservoirs were previously mapped as fault traps, but the existence of the updip faults in the northwestern part of the field cannot be documented. The Domengine in the northwest area is stratigraphically more complex, and it is probable that some of these reservoirs are stratigraphic traps.

Produced Water

The produced-water characteristics from the formations in the field are shown by Table 2. Disposal of produced water is through Class II injection wells into depleted reservoirs, several of which are operating within the field. Class II wells are utilized for the disposal of oil field waste waters, permitted through the Underground Injection Control program, which is administered through the U.S. Environmental Protection Agency or a designated state agency. In California, the California Division of Oil and Gas permits and regulates the operation of Class II injection wells.

SOURCE

The origin and chemical properties of the natural gas at Rio Vista, and throughout the Sacramento basin, is discussed at length by Jenden and Kaplan (1989).

Total organic content (TOC) is variable with the Cretaceous strata containing a higher TOC percentage than the Tertiary strata. In the Tertiary section, the Paleocene Martinez shale contains 0.6–0.8% TOC. The Cretaceous H & T shale appears to contain the largest percentage of TOC with an average value of 1.8%. Analyses have recorded peak values of 5–8.3% TOC for the H & T shale. The Moreno Shale contains 1.0–1.2% TOC and the Delta shale contains 1.2–2.1% TOC. Vitrinite reflectance of less than 0.40% indicates that the organic matter is immature for the generation of hydrocarbons.

The organic material is dominated by higher plant debris and is thus prone to producing natural gas. Average composition of the organic matter is woody matter, 44%; inertinite, 8%; reworked woody material, 14%; herbaceous material, 16%; and amorphous material, 18%. Algal remains were largely absent from the samples examined by Jenden and Kaplan (1989). Their studies also indicated that much of the amorphous material could be type III, which is gas prone.

Although much of the natural gas in the Sacramento basin appears to be of microbial origin, Schoell

(1983) indicated that the gas at Rio Vista is probably of thermogenic origin. The gas present at Rio Vista may therefore be of mixed origin that is primarily derived from Cretaceous sediments.

EXPLORATION AND DEVELOPMENT CONCEPTS

The Rio Vista gas field was formed by faulting associated with syndepositional tectonics in a deltaic depositional environment. Movement on the downthrown side of a growth-fault produces a roll-over domal structure that can be identified with the aid of seismic data. Synthetic and antithetic faulting associated with the growth-fault produces numerous fault blocks. Traps are formed where a sand, at the updip side of the fault block, is offset opposite a shale, which prevents further updip migration of hydrocarbons. Faults with 35 ft of offset can form an economic reservoir. Consequentially, every fault block has the potential of producing hydrocarbons. Detailed subsurface geologic and seismic evaluations to delineate the location and extent of faults and possible stratigraphic traps that may be superimposed over the domal structure are imperative.

A major factor complicating seismic interpretation east of the Sacramento River at Rio Vista, where the land has been reclaimed from the delta and the surface elevation of some of the land is below sea level, is the presence of peat beds. The peat beds, which occur 10 to 15 ft beneath the surface, absorb the seismic energy and inhibit the acquisition of good quality seismic data. CGG American Services, Inc. (a geophysical service company), has recently evaluated the use of deep-seated shot holes and disposable downhole hydrophones as a means of achieving better resolution seismic data from beneath the peat beds (Vuillermoz et al., 1987). The improved quality of the seismic data thus acquired is documented in Vuillermoz et al. and the reader is referred to their paper for a discussion of the technique.

It is reasonable to expect that in other areas of deltaic deposition it may be difficult to obtain good quality seismic data owing to peat beds. Advanced seismic acquisition techniques can be used to overcome this problem.

Reservoir evaluation and completion techniques and subsurface geologic evaluations are equally important during the development of fields that occur in structural and sedimentologic environments similar to Rio Vista. Each reservoir presents unique characteristics that must be determined to achieve proper reservoir evaluation.

The Domengine sand is very friable and poorly consolidated. A primary problem in Domengine formation evaluation is caused by this lack of cohesion within the formation. Repeat formation testers (RFT) commonly plug with fine sand during tests, yielding inconclusive results. RFTs are thus used at Rio Vista primarily for pressure data to delineate virgin from depleted reservoirs. Another problem commonly encountered during testing with drill-stem test (DST) tools is differential pressure sticking of the DST tools. Most tests are thus conducted by running casing and perforating the interval(s) of interest.

Within the Markley sand, care must also be exercised when recompleting a well. Many of the early wells were drilled with over-weighted mud before being cased. Thus, when a well is recompleted in the Markley sand it may take several weeks for the formation to clean up and commence a steady flow. This phenomenon occurred in the J. C. Hamilton No. 3 (RVGU #61) well in Sec. 24, T4N-R2E, in 1984. The well was plugged back and perforated in the Markley sand. When there was no return to the surface, the well was shut in while the crew worked on another well. After about three weeks, the crew returned to the well to find that the perforations had cleaned out and pressure had increased. Patience is a virtue when working with this formation.

Several cased-hole logging tools have been tested in the past few years in an effort to better evaluate recompletion potential within the Markley sand. The carbon/oxygen log has been tested with inconclusive results. The thermal decay-time (TDT) log has been used with mixed results, yet has proven to be more useful than the carbon/oxygen log. The primary obstacle with Markley sand recompletions is that most of the recompleted wells were drilled in the 30s and 40s with SP and resistivity being the only wireline logs run and without the benefit of a mudlog. The high resistivity of the Markley sand formation water, owing to the relatively low total dissolved solids (TDS) value of the formation water (4500–4900 ppm), renders evaluation of the older resistivity logs difficult. Additionally, many of the older logs were not run through the entire Markley section.

Several operators have experimented with different borehole fluids during completion in the Petersen sand. The most effective completion fluid appears to be methanol. Methanol appears to avoid the swelling of interstitial clays, thus preserving the permeability. The clays, which occur as platelets and booklets, also tend to detach from the sand grains and can obstruct the pore throats. Caution must be exercised during testing and production not to overproduce and cause the physical plugging of pore throats by detached clays.

ACKNOWLEDGMENTS

I would like to thank George Sylvester, Ted Beaumont, Ted Bear, and Jim Dowdall for their critical review of the manuscript. I would also like to thank Larry Knauer of the California Core

Repository for his assistance in examining some of the cores from the Rio Vista field.

REFERENCES

Almgren, A. A., 1978, Timing of submarine canyons and marine cycles of deposition in the southern Sacramento Valley, California, *in* A. A. Almgren and P. D. Hacker, eds., Paleogene submarine canyons of the Sacramento Valley, California: Pacific Section American Association of Petroleum Geologists, S.V. 1, p. 1-16.

Amerada Hess Corp., 1985, Rio Vista at fifty: still going strong: Scope (Company magazine), Fall 1985, p. 1-7.

Burroughs, E., 1967, Rio Vista gas field: California Division of Oil and Gas, Summary of Operations—California Oil Fields, v. 53, n. 2, pt. 2, p. 25-33.

Burroughs, E., G. W. Beecroft, and R. M. Barger, 1968, Rio Vista gas field, *in* Natural gases of North America, v. I: American Association of Petroleum Geologists Memoir 9, p. 93-101.

California Division of Oil and Gas (DOG), 1981, California oil and gas fields, northern California: Pub. No. TR10, v. 3.

California Division of Oil and Gas (DOG), 1988, 73rd annual report of the state oil & gas supervisor, 1987: Pub. No. PR06, 156 p. (Plus monthly updates for 1988.)

Carmalt, S. W., and B. St. John, 1986, Giant oil and gas fields, *in* M. T. Halbouty, ed., Future petroleum provinces of the world: American Association of Petroleum Geologists Memoir 40, p. 11-54.

Cherven, V. B., 1983, A delta-slope-submarine fan model for Maastrichtian part of Great Valley sequence, Sacramento and San Joaquin Basins, California: American Association of Petroleum Geologists Bulletin, v. 67, p. 772-816.

Corwin, C. H., 1953, Rio Vista gas field, Isleton area: California Division of Oil and Gas, Summary of Operations—California Oil Fields, v. 39, n. 1, p. 13-16.

Dickinson, W. R., and D. R. Seely, 1979, Structure and stratigraphy of forearc regions: American Association of Petroleum Geologists Bulletin, v. 63, p. 2-31.

Frame, R. G., 1944, Rio Vista gas field: California Division of Oil and Gas, Summary of Operations—California Oil Fields, v. 30, n. 1, p. 5-14.

Garcia, R., 1980, How depositional systems relate to Sacramento Valley gas accumulation: Oil & Gas Journal, 19 May, p. 179-184.

Garcia, R., 1981, Depositional systems and their relation to gas accumulation in Sacramento Valley, California: American Association of Petroleum Geologists Bulletin, v. 65, p. 653-673.

Goudkoff, P. P., 1945, Stratigraphic relations of Upper Cretaceous in Great Valley, California: American Association of Petroleum Geologists Bulletin, v. 29, p. 956-1007.

Hodgson, S. F., 1980, Onshore oil and gas seeps in California: California Division of Oil and Gas, Pub. No. TR26, 97 p.

Jenden, P. D., and I. R. Kaplan, 1989, Origin of natural gas in Sacramento Basin, California: American Association of Petroleum Geologists Bulletin, v. 73, p. 431-453.

Kittridge, O., and M. Wilson, 1984, The Markley submarine valley and its stratigraphic relationships, *in* A. A. Almgren and P. D. Hacker, eds., 1984, Paleogene submarine canyons of the Sacramento Valley, California: Pacific Section American Association of Petroleum Geologists, S.V. 1, p. 81-99.

Rubey, W. W., and M. K. Hubbert, 1959, Role of fluid pressure in mechanics of overthrust faulting: Geological Society of America Bulletin, v. 70, p. 167-206.

Schoell, M., 1983, Genetic characterization of natural gases: American Association of Petroleum Geologists Bulletin, v. 67, p. 2225-2238.

Soper, E. K., 1943, Rio Vista gas field, *in* O. P. Jenkins, dir., Geologic formations and economic development of the oil and gas fields of California; Part Three, Descriptions of individual oil and gas fields: California Division of Mines, Bulletin No. 118, p. 591-594.

St. John, B., A. W. Bally, and H. D. Klemme, 1984, Sedimentary provinces of the world, hydrocarbon productive and non-productive: American Association of Petroleum Geologists Map Series, 35 p.

Time, 1952, The great hunter, Amerada's Jacobsen: Time Magazine Inc., 1 December, p. 68-76.

Vuillermoz, C., A. J. Bertagne, and R. Delzer, 1987, Sacramento delta: new approach to seismic exploration in area of near-surface problems: Oil & Gas Journal, 16 November, p. 63-66.

Zieglar, D. L., and J. H. Spotts, 1978, Reservoir and source-bed history of Great Valley, California: American Association of Petroleum Geologists Bulletin, v. 62, p. 813-826.

Appendix 1. Field Description

Field name .... *Rio Vista gas field*

Ultimate recoverable reserves .... *3,499,937 MMcfg*

Field location:

- **Country** .... *United States*
- **State** .... *California*
- **Basin/Province** .... *Sacramento basin, Delta Depocenter*

Field discovery:

- **Year first pay discovered** .... *Eocene Domengine sandstone 1936*
- **Year second pay discovered** .... *Eocene Hamilton sandstone 1936*
- **Third pay** .... *Paleocene Midland sandstone and Cretaceous Mokelumne River formation 1943*

Discovery well name and general location:

- **First pay** .... *Emigh No. 1, Sec. 26, T4N-R2E*
- **Second pay** .... *M. Hamilton No. 1, Sec. 26, T4N-R2E*
- **Third pay** .... *Midland Fee No. 5, Sec. 4, T3N-R3E*

Discovery well operator .... *Amerada Petroleum Company (now Amerada Hess Corp.)*

- **Second pay** .... *Amerada Petroleum Company*
- **Third pay** .... *Standard Oil Company (now Chevron)*

IP in barrels per day and/or cubic feet or cubic meters per day:

- **First pay** .... *8750 MMcfg/day*
- **Second pay** .... *4160 MMcfg/day*
- **Third pay** .... *5700 MMcfg/day*

All other zones with shows of oil and gas in the field:

Age	Formation	Type of Show
Eocene	*Sidney Flat Shale*	*Gas*
Eocene	*Markley Sand*	*Gas*
Eocene	*Nortonville Shale*	*Gas*
Eocene	*Capay Shale*	*Gas*
Eocene-Paleocene	*Meganos Shale*	*Gas*
Paleocene	*Anderson sand*	*Gas*
Paleocene	*Martinez shale*	*Gas*
Cretaceous	*Mokelumne River fm.*	*Gas*
Cretaceous	*Starkey sand*	*Gas*
Cretaceous	*Petersen sand*	*Gas*

Geologic concept leading to discovery and method or methods used to delineate prospect, e.g., surface geology, subsurface geology, seeps, magnetic data, gravity data, seismic data, seismic refraction, nontechnical:

Structure inferred and mapped using reflection seismic data, drilled high portion of time structure.

Structure:

Province/basin type (see St. John, Bally, and Klemme, 1984)

Bally 311; Klemme V (revised from St. John, Bally, and Klemme, 1984)

Tectonic history

Cretaceous through Eocene deltaic and submarine fans sediments deposited in a forearc basin. Resultant overpressuring of Cretaceous shales resulted in a north-northwest-trending growth fault system, producing several roll-over anticlinal structures. Rio Vista is largest and most prolific of these structures.

Regional structure

Gently westward-dipping beds with undulating, low-amplitude structures that are fractured and offset by north-northwest-trending normal faults.

Local structure

North-northwest-trending anticline formed by roll-over on down-thrown side of Midland fault (a growth fault).

Trap

Trap type(s)

Over 90% of the traps are updip fault closure with lateral structural/stratigraphic closure. Stratigraphic pinchout traps occur in the Domengine sand and the Starkey sand.

Basin stratigraphy (major stratigraphic intervals from surface to deepest penetration in field):

Chronostratigraphy	Formation	Depth to Top in ft
Pliocene-Miocene	*Undifferentiated nonmarine*	*Near surface to 2000*
Oligocene	*Markley Gorge fill*	*Present in N. part only*
Eocene	*Sidney Flat Shale*	*2000*
Eocene	*Markley sand*	*2580*
Eocene	*Nortonville Shale*	*4000*
Eocene	*Domengine sand*	*4345*
Eocene	*Capay shale*	*4685*
Eocene	*Hamilton sand*	*5595*
Eocene/Paleocene	*Meganos shale*	*5905*
Paleocene	*Anderson sand*	*6020*
Paleocene	*Martinez shale*	*6300*
Paleocene	*McCormick sand*	*6685*
Cretaceous	*Mokelumne River fm.*	*6895*
Cretaceous	*H and T shale*	*8625*
Cretaceous	*Starkey sand*	*8980*
Cretaceous	*Petersen sand*	*9745*
Cretaceous	*Delta shale*	*10,010*
Cretaceous	*Winters sand*	*10,970*
Cretaceous	*Winters shale*	*(12,390) est.*
Cretaceous	*Sacramento shale*	*(12,820) est.*
Cretaceous	*Forbes fm.*	*(13,000) est.*

Location of well in field .... *Well in northwest portion of field*

Reservoir characteristics:

Number of reservoirs *12*

Formations .... *See Table 2*

Ages .... *Cretaceous, Paleocene, and Eocene*

Depths to tops of reservoirs .... *2450 to 9650 ft*

Gross thickness (top to bottom of producing interval) .... *7200 ft*

Net thickness—total thickness of producing zones

Average .... *489 ft*

Maximum .... *974 ft*

Average

Maximum

Lithology

Light gray, fine- to coarse-grained silty sand, friable, quartzitic, subrounded, subangular, occasionally frosted, poorly to moderately indurated, marine shell frags., occasionally white or greenish gray

Porosity type .... *Intergranular*
Average porosity *20–34%*
Average permeability .... *5–1800 md*

Seals:

Upper

Formation, fault, or other feature .... *Fault*
Lithology .... *Reservoir sand offset across fault by shale*

Lateral

Formation, fault, or other feature .... *Structural closure or stratigraphic pinchout*
Lithology

Source:

Formation and age .... *Delta shale; Upper Cretaceous*
Lithology .... *Shale*
Average total organic carbon (TOC) *1.78*
Maximum TOC *2.15*
Kerogen type (I, II, or III) .... *II*
Vitrinite reflectance (maturation) .... $R_o = 0.34\%$
Time of hydrocarbon expulsion
Present depth to top of source .... *10,000 ft*
Thickness .... *960 ft*
Potential yield

Appendix 2. Production Data

Field name .... *Rio Vista gas field*

Field size:

Proved acres .... *±25,000 ac*
Number of wells all years *365*
Current number of wells *227*
Well spacing .... *Variable*
Ultimate recoverable .... *3500 bcfg*
Cumulative production .... *3277 bcfg plus 1,454,000 bbl condensate*
Annual production .... *29.25 bcfg plus 18,651 bbl condensate*
Present decline rate .... *13% est.*
Initial decline rate
Overall decline rate .... *15% est.*
Annual water production .... *210,115 bbl*
In place, total reserves .... *223 bcfg*
In place, per acre-foot
Primary recovery
Secondary recovery
Enhanced recovery
Cumulative water production

Drilling and casing practices:

Amount of surface casing set .... *Minimum 10% of Total Depth required*
Casing program .... *16-in. conductor pipe to 50 ft; 9⅝-in. surface casing to 1000 ft; 5½-in. production casing to TD; 2⅜-in. tubing landed in packer set above perfs.*

Drilling mud .... *Lignosulfonate*
Bit program .... *Variable*
High pressure zones .... *Delta shale and deeper horizons*

Completion practices:

Interval(s) perforated .... *Variable, upper portion of gas zone to avoid water coning*
Well treatment .... *The more friable sands require gravel packing*

Formation evaluation:

Logging suites .... *Older wells: SP and resistivity; modern wells: SP, resistivity, GR, sonic, neutron, density, dipmeter*
Testing practices .... *Typically production tested; occasionally wireline tested (RFT) for reservoir pressure data*
Mud logging techniques .... *Run on all wells and typically implemented near base of nonmarine section*

Oil characteristics:

Type
(Tissot and Welte Classification in "Petroleum Formation and Occurrence," 1984, Springer-Verlag, p. 419)
API gravity
Base
Initial GOR
Sulfur, wt%
Viscosity, SUS
Pour point
Gas-oil distillate

Field characteristics:

Average elevation .... *20 ft subsea to 180 ft*
Initial pressure .... *1110 to 4860 psi*
Present pressure .... *Variable*
Pressure gradient
Temperature .... *95 to 200°F*
Geothermal gradient .... *0.95°/100 ft*
Drive .... *Active water drive*
Oil column thickness
Oil-water contact
Connate water
Water salinity, TDS .... *4700-24,000 mg/L (ppm)*
Resistivity of water *1.36*
Bulk volume water (%) *30-60*

Transportation method and market for oil and gas:

Pipeline operated by Pacific Gas and Electric (P. G. & E.).

Badak Field—Indonesia
Kutai Basin/East Kalimantan (Borneo)

ROY M. HUFFINGTON and H. M. HELMIG
Roy M. Huffington Inc.
Houston, Texas

FIELD CLASSIFICATION

BASIN: Kutai/East Kalimantan
BASIN TYPE: Backarc
RESERVOIR ROCK TYPE: Sandstone
RESERVOIR ENVIRONMENT OF DEPOSITION: Deltaic
RESERVOIR AGE: Miocene
PETROLEUM TYPE: Oil and Gas
TRAP TYPE: Anticline

LOCATION

The Badak field is located at the northern edge of the Mahakam delta on the east coast of the island of Kalimantan (Borneo), Indonesia (Figure 1).

The Mahakam delta is part of the Tertiary Kutai basin, which contains a number of large oil and gas fields, notably Badak, Nilam, Sanga Sanga, Handil, Bekapai, Attaka, and Tunu, among others.

The eastern part of the Badak field lies in a coastal swamp characterized by mangrove and nipah vegetation. The western part of the field lies on higher ground, though still part of the coastal plain.

The Badak field has estimated total recoverable reserves of 6.5 tcf of gas, 96 million bbl of condensate, and 47 million bbl of oil. It ranks 178th in Carmalt and St. John's (1984) classification of giant fields.

HISTORY OF PETROLEUM EXPLORATION

Pre-Discovery

Exploration for oil in the Kutai basin started in the latter part of the nineteenth century, when numerous oil seeps were found during coal exploration operations. In 1897, the Sanga Sanga oil field was discovered. It has produced more than 260 million bbl of oil to date and is still producing about 3000 BOPD. Also the Samboja oil field, discovered in 1910, still is producing a few hundred barrels of oil per day. Little exploration took place in the Kutai basin in the period between 1942 and 1968, at the end of which a number of companies, including Huffco, began operating under new Production Sharing Contracts.

Roy M. Huffington, Inc. (Huffco) began its Indonesian operations in November 1968, with the arrival of its Exploration Manager, H. M. Helmig. During the course of 1969, he was joined by several geologists and a geophysicist.

During 1969, Huffco formed a consortium consisting of Ultramar, Union Texas Petroleum Company, Virginia International, and Austral Oil Company.

Exploration started with a review of existing literature, predominantly old B.P.M. (Shell) reports from Pertamina's (Indonesian National oil company) files, immediately followed by surface geological mapping, primarily aimed at obtaining fresh samples and accurate stratigraphic sections.

Pre-war aerial photographs, although in poor condition and offering only partial coverage of the contract area, were initially useful as an early reconnaissance tool. During the dry seasons of 1969 and 1970, a new aerial survey was carried out by KLM Aerocarto for Huffco, using false color infrared photography, which produced usable coverage over about 80% of Huffco's contract area. This new coverage was geologically interpreted and a set of new geological map quadrangles was produced that provided an excellent overview of the regional structural geology and that served as a base map and guide in the planning, execution, and interpretation of subsequent geologic and seismic surveys. It must be noted here that in 1968 there was no consistent topographic map coverage of East Kalimantan, so that Huffco's base maps had to be constructed "from scratch" and even now are periodically updated and improved with new data.

Figure 2 demonstrates the remarkable coincidence between the coastline morphology and the shallow structure of the Badak feature. This coincidence was suspected early and prompted special attention to the accurate positioning of some of the earliest seismic lines to ensure coverage of the Badak area

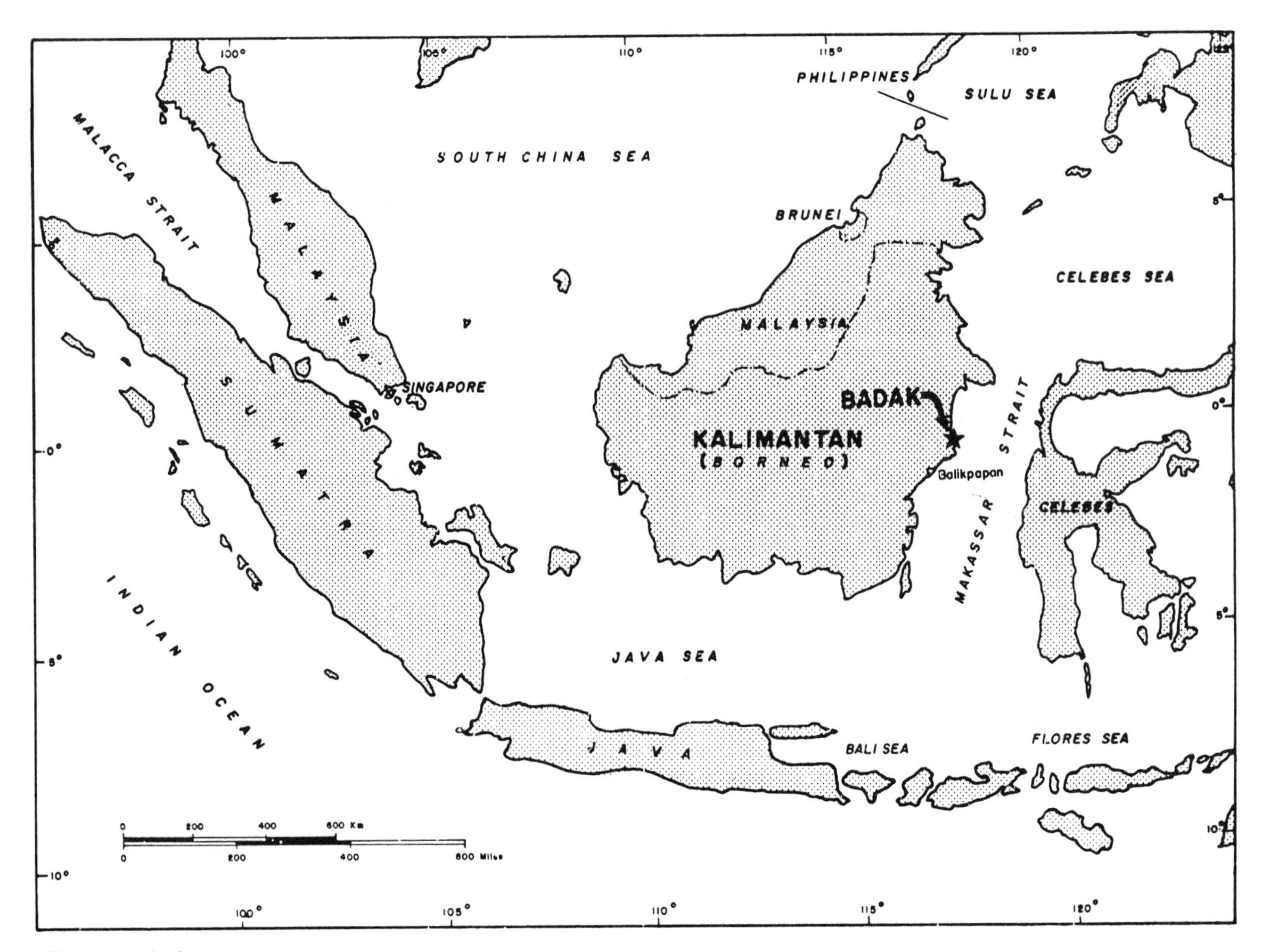

Figure 1. Index map.

where the presence of a structural anomaly was suspected.

In 1970, a joint aerial magnetometric survey was carried out by CGG, covering about two-thirds of the Kutai basin for Huffco, Pertamina, and Shell. The results of this survey contributed to the understanding of the basin's regional basement configuration.

In early 1970, a reconnaissance seismic survey was conducted in Balikpapan Bay and along the coast between Balikpapan and Samarinda, using a shallow-water barge and analog, single-fold methods. The quality of the results were mixed but gave an early indication of the seismic quality that could be expected and of some potential problem areas for seismic acquisition. Later in 1970, a full-scale survey was initiated with Seiscom Delta as contractor and initially using DSF III equipment, dynamite source, and multifold coverage. The first seismic lines shot across the Badak area, although of poor quality, clearly showed the structural rollover (Figures 3, 4, and 5), and early mapping indicated a sizeable closure at depths between 3500 and 12,000 feet (Figure 5). Later seismic coverage achieved much better resolution.

West of the Badak area, on the east flank of the Semberah surface anticline, the outcrops of a very sandy section, several thousand feet thick, can be projected eastward in the subsurface on seismic sections into the Badak area. Based on this information, the Badak No. 1 location was picked on a high point of the structure (Figure 6) and a Parker Drilling Company 1500 Helirig rig was mobilized, using barges and helicopters. Badak No. 1 was spudded on 27 November 1971 and completed on 11 February 1972 as an oil well with tested capacity of 2500 BOPD.

Post-Discovery

The Badak No. 2 and No. 3 wells were drilled in quick succession, respectively 2 km to the north and 4 km to the south of No. 1. Both these confirmation wells penetrated thick columns of gas reservoir but

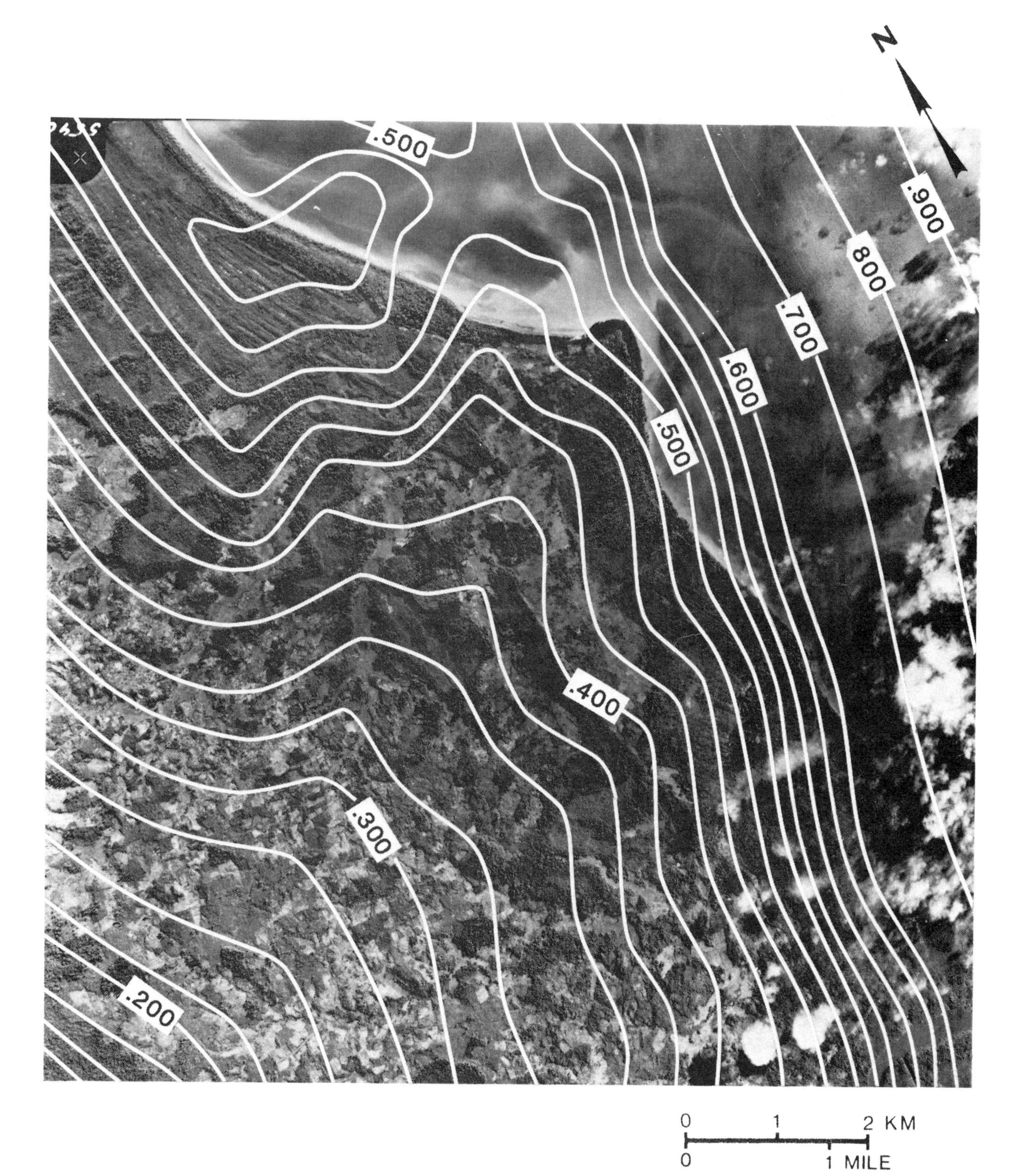

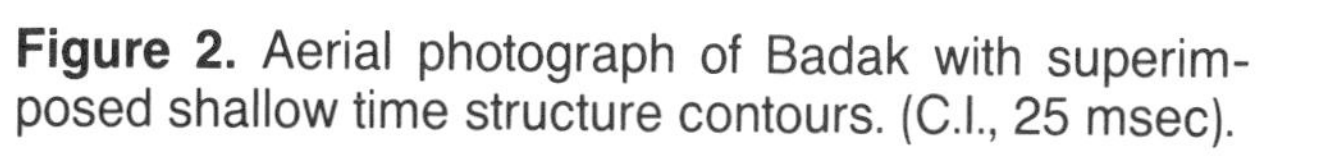

Figure 2. Aerial photograph of Badak with superimposed shallow time structure contours. (C.I., 25 msec).

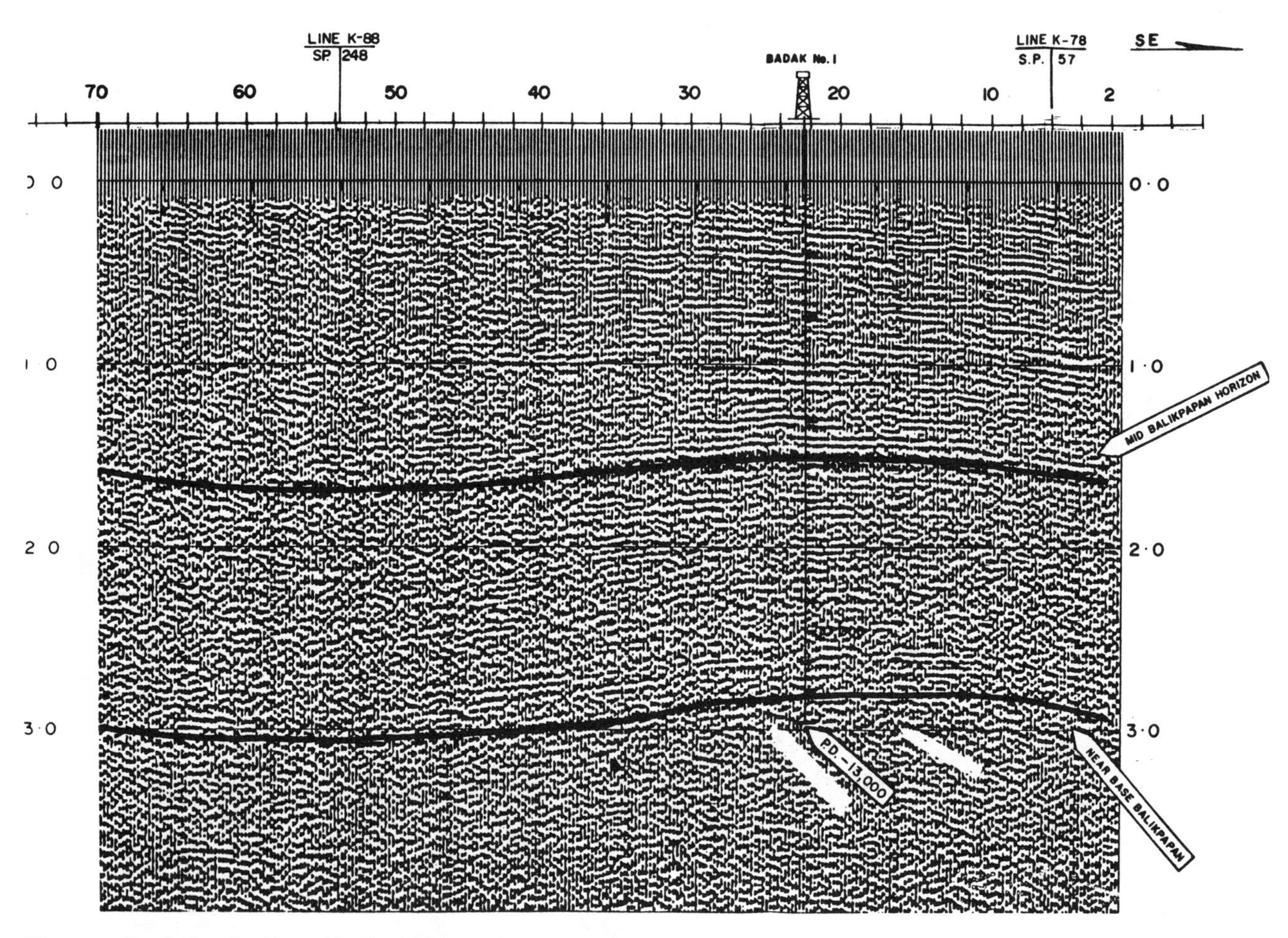

Figure 3. Seismic line K-70 with original (1971) interpretation (for location see Figure 6).

were completely void of oil reservoirs. This was unfortunate because Indonesia and, even less, East Kalimantan had no market for natural gas. On its own, the 120-ft oil column in Badak No. 1, in a reservoir of limited lateral extension, offered little justification for the considerable investment needed to fully explore the structure for oil, considering that world oil prices then were oscillating below U.S.$3.00/barrel.

However, in the succeeding six months the oil price began to rise sharply and the Arab oil embargo caused considerable discomfort and anxiety as to the future of energy supplies, especially in Japan. In addition, Shell had already proved the viability of the production and transport of LNG on a large scale in their plant in Brunei, which was being constructed about the time that Badak was discovered. The coincidence of these factors and others eventually led to the conclusion of a 20-year gas sales agreement with a consortium of Japanese buyers. Based on this agreement, the financing of the construction of the Bontang LNG plant, LNG tankers, and the receiving terminals was arranged.

A more complete account of the history of the LNG industry in Indonesia is provided in a Pertamina publication titled *Hands across the Sea* (Pertamina, 1985).

DISCOVERY METHOD

Although all conventional exploration tools except gravity mapping were employed in Huffco's operations preceding the discovery of the Badak field, it probably would not have been found without modern seismic methods. Indeed, the reason that Shell did not discover Badak previously was that prior to Shell's departure from Indonesia in 1965, seismic "state of the art" lacked the required deep resolution needed to explore the coastal plain east of the Sanga Sanga field.

The offshore Attaka oil field, discovered by Union Oil Company of California in 1970, had proved the existence of commercial oil accumulations in strata

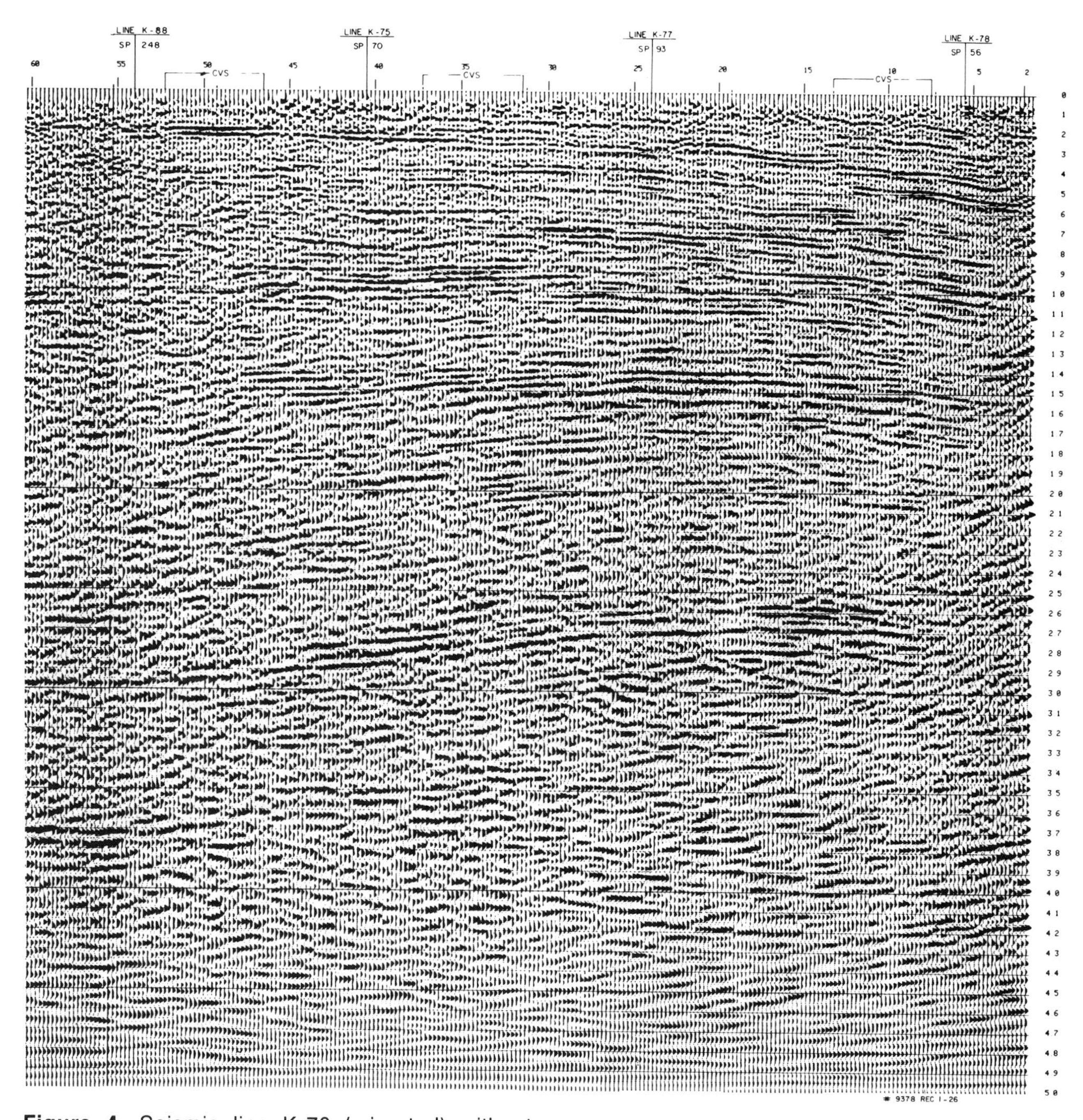

Figure 4. Seismic line K-70 (migrated) without interpretation.

younger than those in Sanga Sanga. Naturally, the area between Attaka and Sanga Sanga seemed highly prospective and the problem boiled down to finding a drillable anomaly, preferably a four-way structural closure within Huffco's contract boundary. Badak was the first and largest such anomaly outlined in Huffco's seismic structure maps.

STRUCTURE

Tectonic History

The Kutai basin apparently began as a subsidiary topographic depression in a major Eocene deposi-

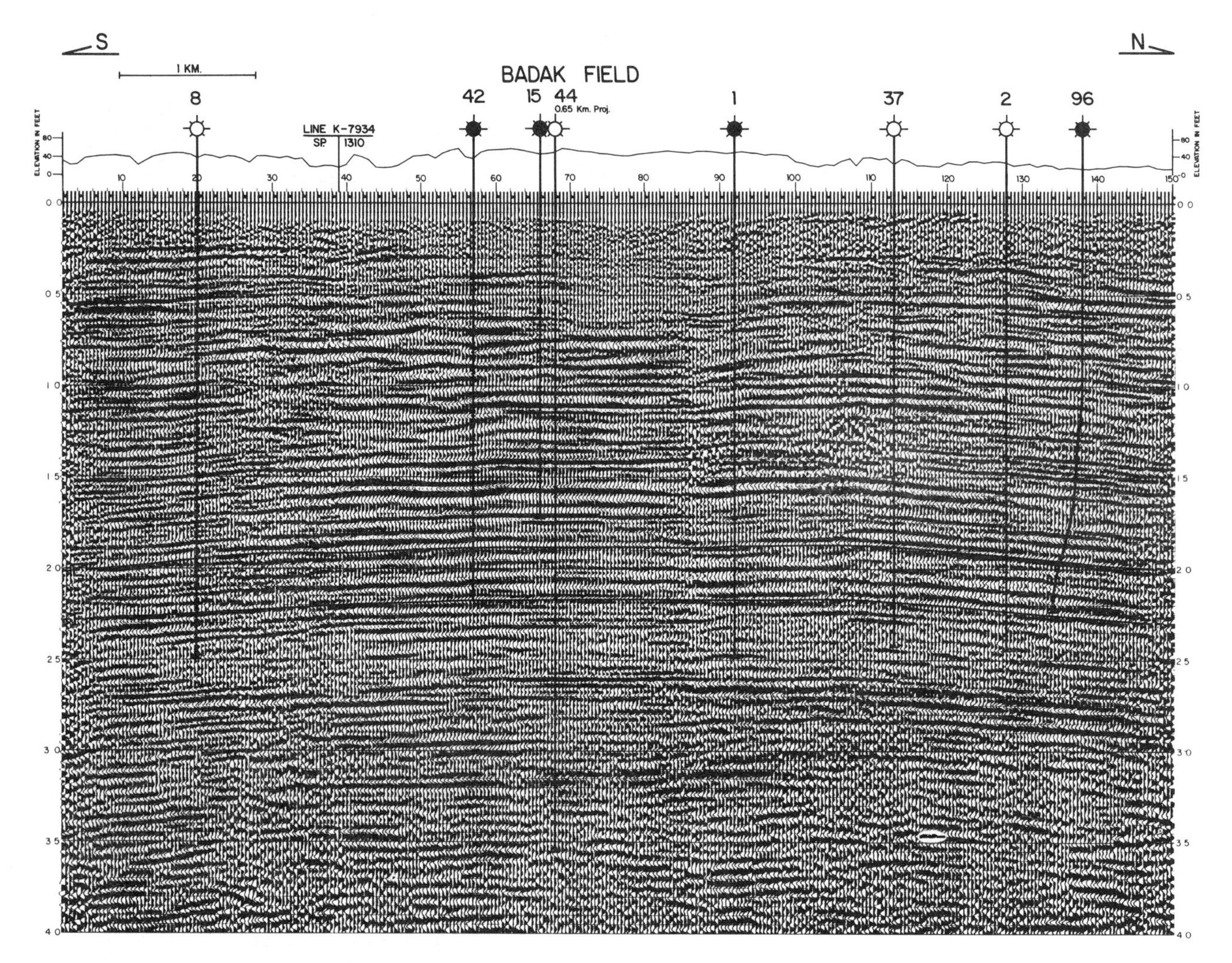

Figure 5. Seismic line K-77 (for location see Figures 6 and 12).

tional province that covered most of East and South Kalimantan (Figure 7). Thousands of feet of Eocene sediments are known in the northern, southern, and western parts of the basin, which are mainly shales and sandstones with occasional conglomerate beds.

Subsequently, during Oligocene, lower Miocene, middle Miocene, and upper Miocene to Recent periods, the depocenters shifted eastwards, so that the most recent depocenter occurs in the Makassar Strait.

The elongate north-northeast-trending structural anomalies, which are characteristic for the Kutai basin (Figure 8), seem to begin their growth during the middle Miocene. These linear anticlines, where drilled, all have cores of undercompacted, abnormally pressured shales, and their deformation decreases in intensity from west to east. The more westerly structures are affected by strong folding and faulting, and the existence of detached thrust sheets is suspected, although so far not proved, to occur as far east as the Sanga Sanga anticline.

The Badak structure forms the northernmost culmination on the Badak-Handil anticlinal trend, which is almost continuously productive from the northern tip of Badak to the southern tip of Handil, a distance of 75 km.

Badak Structure

The structure of the Badak field is that of a relatively simple unfaulted culmination on the northern end of a 75 km long anticline on the Badak-Handil trend (Figures 8 and 9).

From Figures 9 and 10 it can be seen that the Badak structure was formed before deposition of the "B" zone, probably during Pliocene times, since the critical west dip only exists below about 3000 ft (915 m) and the shallower strata form an east-dipping homocline with at best a structural terrace in the Badak area. This phenomenon explains the virtual

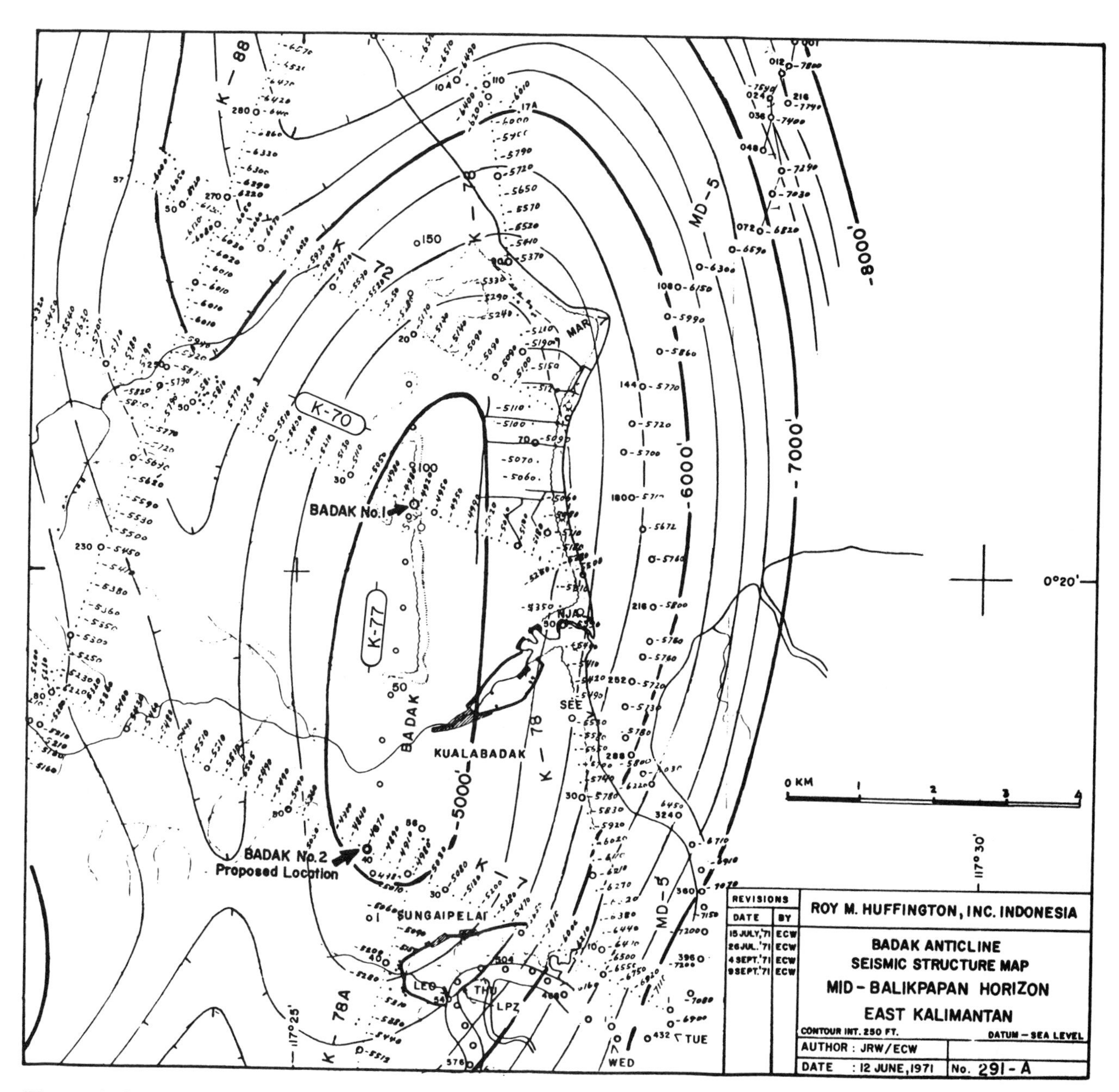

Figure 6. Structure map on "Mid-Balikpapan Horizon." Original map used in Badak No. 1 drilling proposal.

lack of hydrocarbons and abundance of fresh water in the upper 4000 ft (1220 m) of the Badak stratigraphic sequence.

Figure 11 shows the subsurface structure and the north and south plunge of the structure along seismic line K-77 (Figure 5). The shallowest closure in the Badak structure is depicted in Figure 12. Figure 13 illustrates the deeper structure in the Badak field. Figure 14 represents the structure on top of the "E" zone, based on subsurface (well) data. The areal closure on the "E" horizon is approximately 9500 acres (38 km^2) and the vertical closure is about 500 ft (152 m).

STRATIGRAPHY

Regional

The sedimentary sequence in the Kutai basin ranges in age from upper Eocene to Recent; however, in the Mahakam delta area, rocks older than lower Miocene (N4) are not known and will not be discussed in this paper. Figure 15 illustrates the general stratigraphic sequence in the Mahakam delta area.

More detailed and complete discussion of the regional stratigraphy is beyond the scope of this paper.

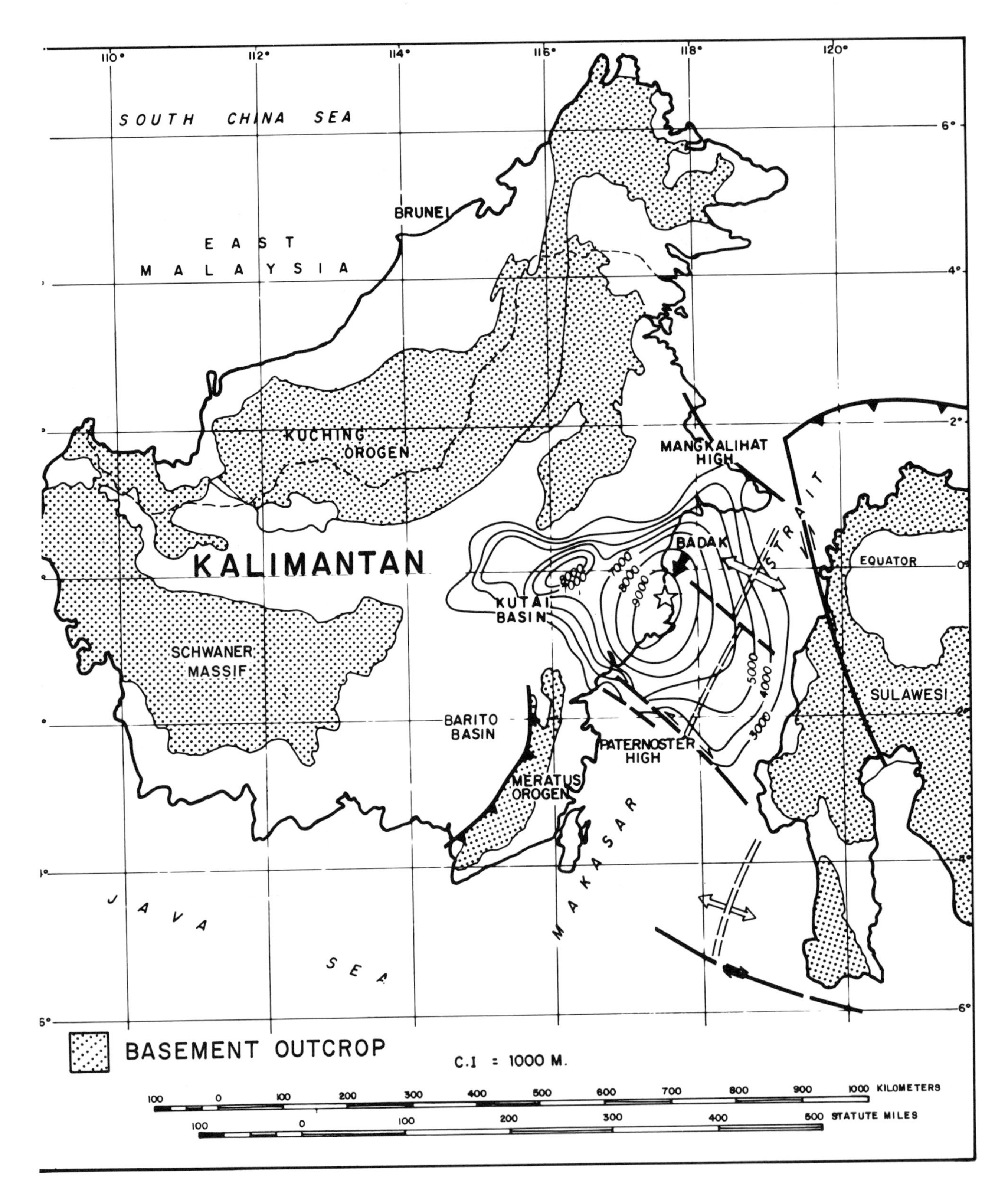

Figure 7. Kutai basin tectonic setting (after Hamilton, 1979).

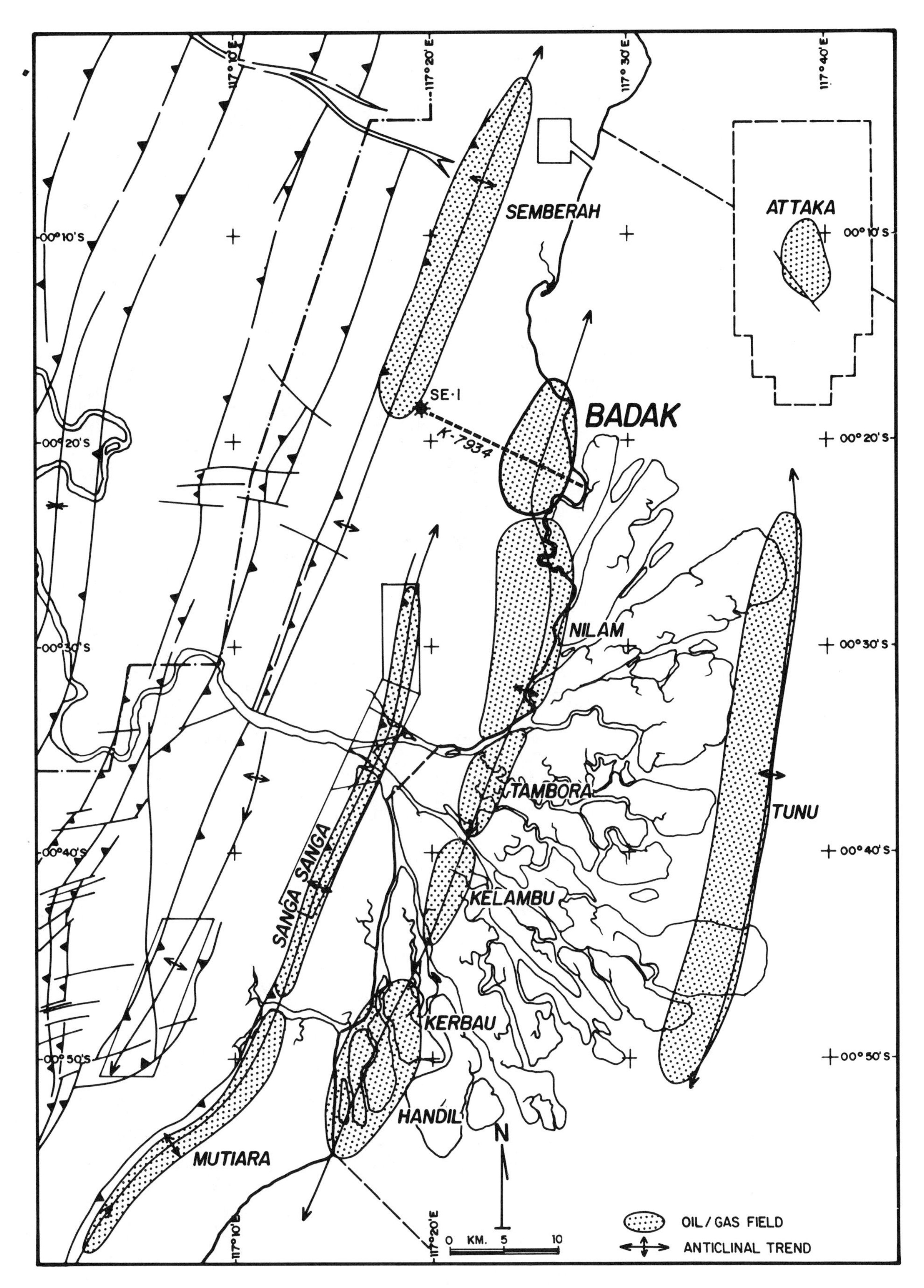

Figure 8. Structural sketch map and oil and gas fields, Mahakam delta.

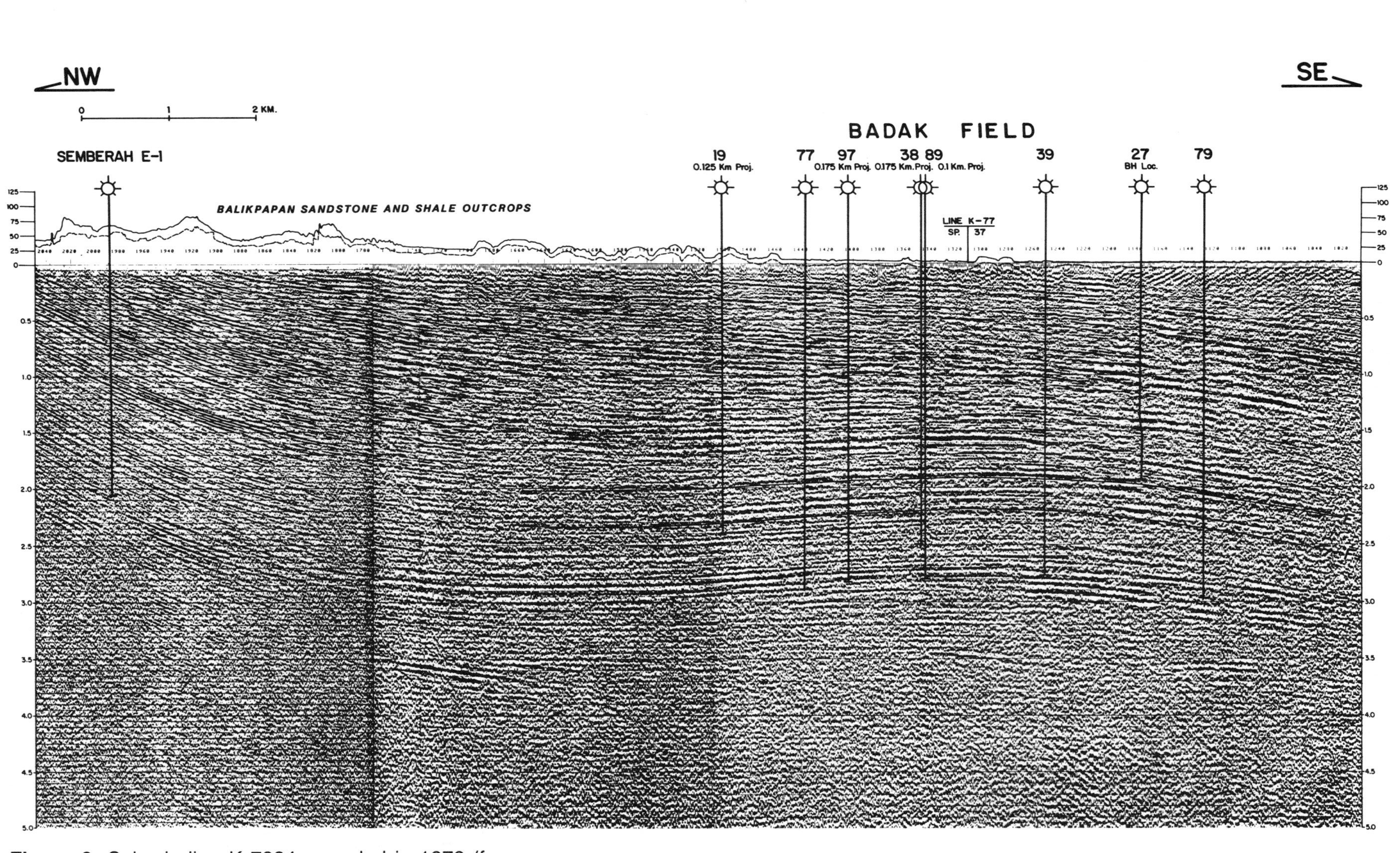

Figure 9. Seismic line K-7934 recorded in 1979 (for location see Figures 8 and 12).

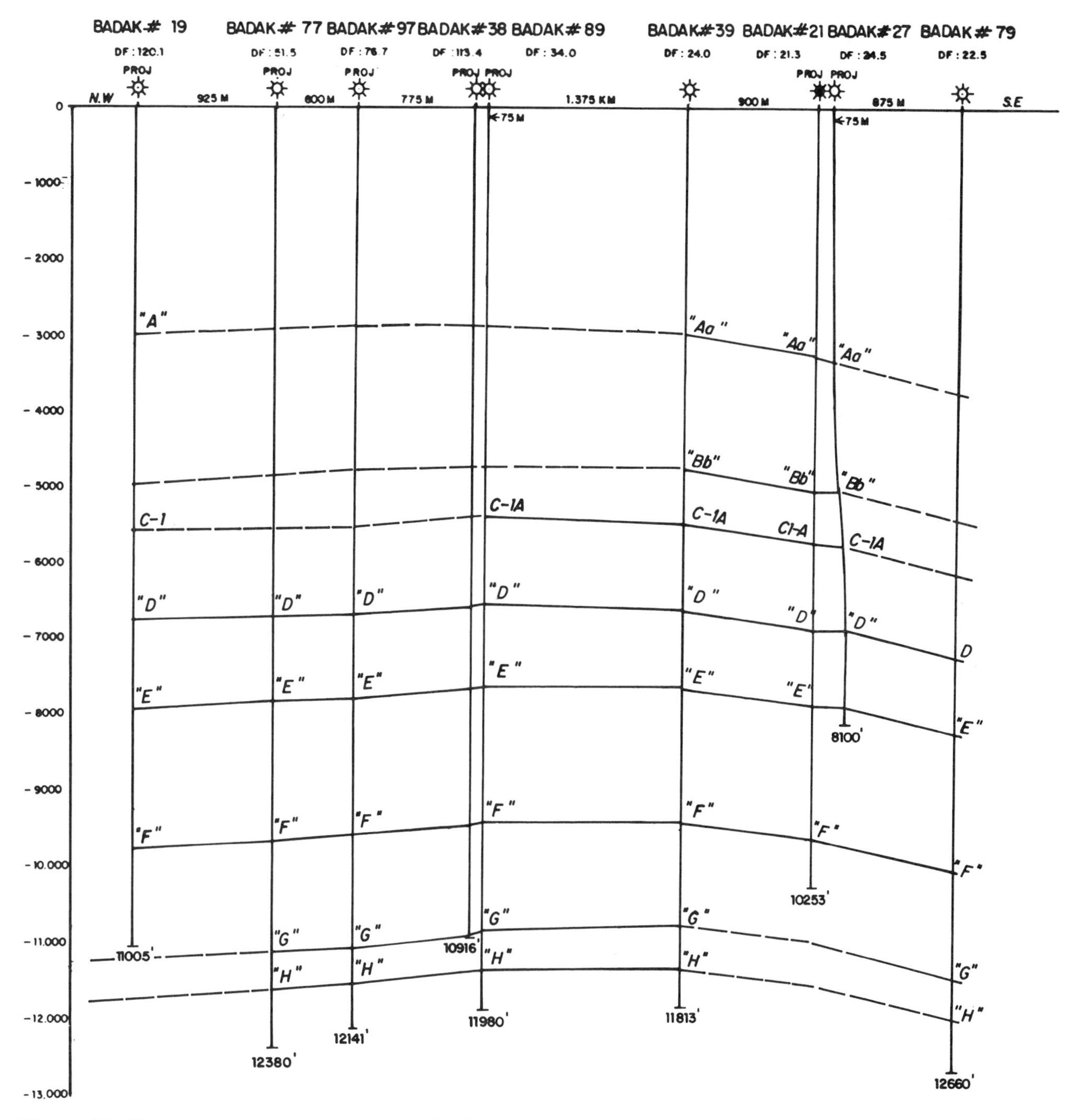

Figure 10. Structural cross section along seismic line K-7934 (location shown on Figure 12).

and has been documented by previous authors, such as Samuel and Muchsin (1975), Rose and Hartono (1978), and Marks et al. (1982).

Badak Field

The stratigraphic sequence in the Badak field area consists of a series of middle Miocene to Holocene clastics that were deposited in a deltaic environment, as presented in Figure 16. The total thickness of sediments penetrated by the drill in Badak is 16,408 ft (5002 m) in well No. 44. It is estimated from somewhat sparse gravity and magnetometry data that another 10,000–15,000 ft of Tertiary sediments (4573 m) may underlie the known sedimentary column in the Badak area. Figure 17 shows the increase in sand/shale ratio in the section from bottom to top, demonstrating the predominantly

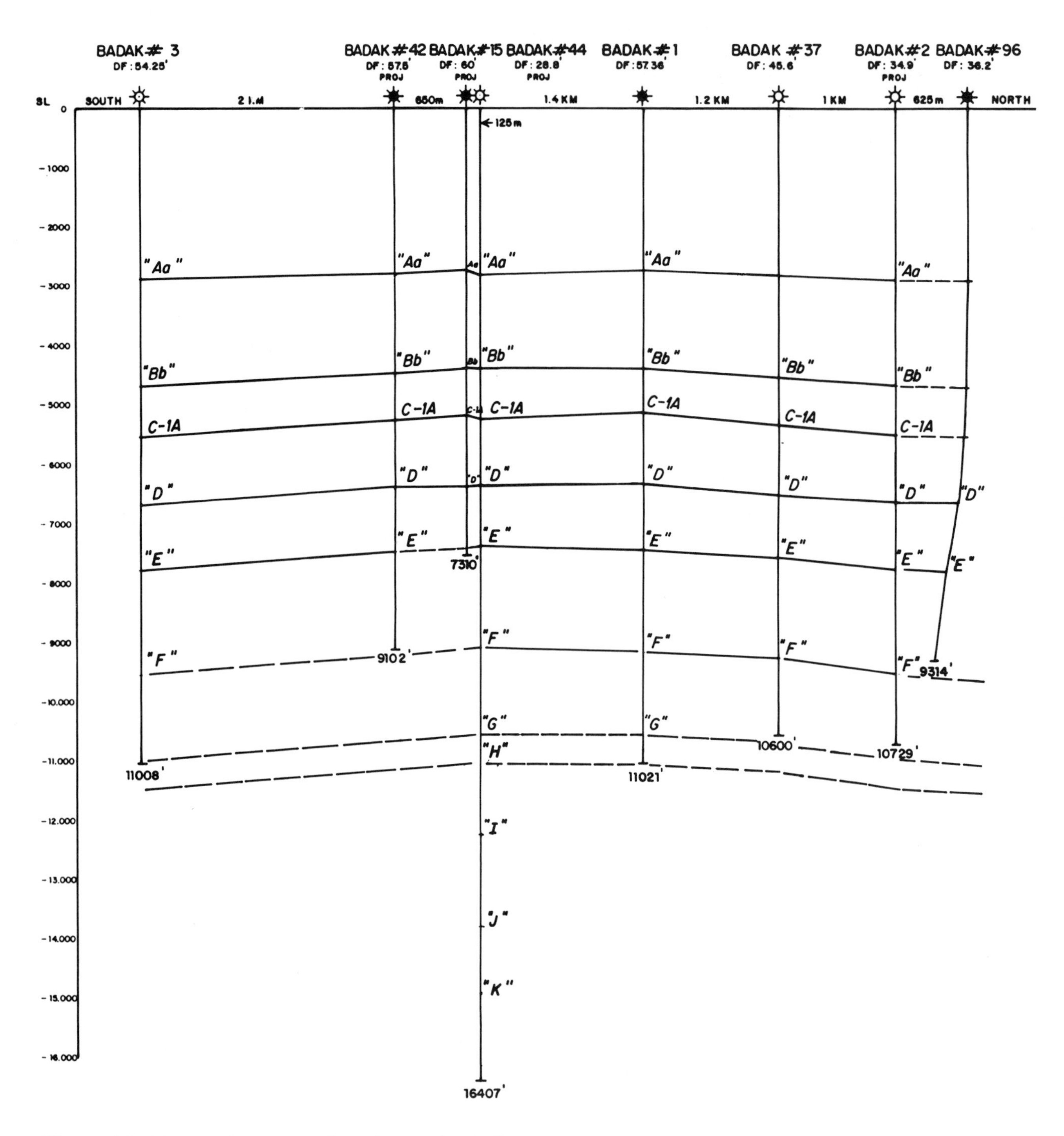

Figure 11. Structural cross section along seismic line K-77 (location shown on Figure 12).

regressive nature of the depositional regime of the Badak sequence.

The only stratigraphic subdivision in use in the Badak field is a reservoir zonation from "A" at the top to "K" at the bottom of the known (i.e., drilled) section, as shown in Figure 16. The basis for this subdivision, although apparently somewhat arbitrary, is the existence of a certain cyclicity in the sedimentary sequence in Badak. In the ideal case, such cycles begin with a shaly member with occasional limestone intercalations that gradually coarsens upwards, ending with a package of reservoir sands, and topped by a stratigraphic hiatus, evidenced by erosional phenomena or a carbonate

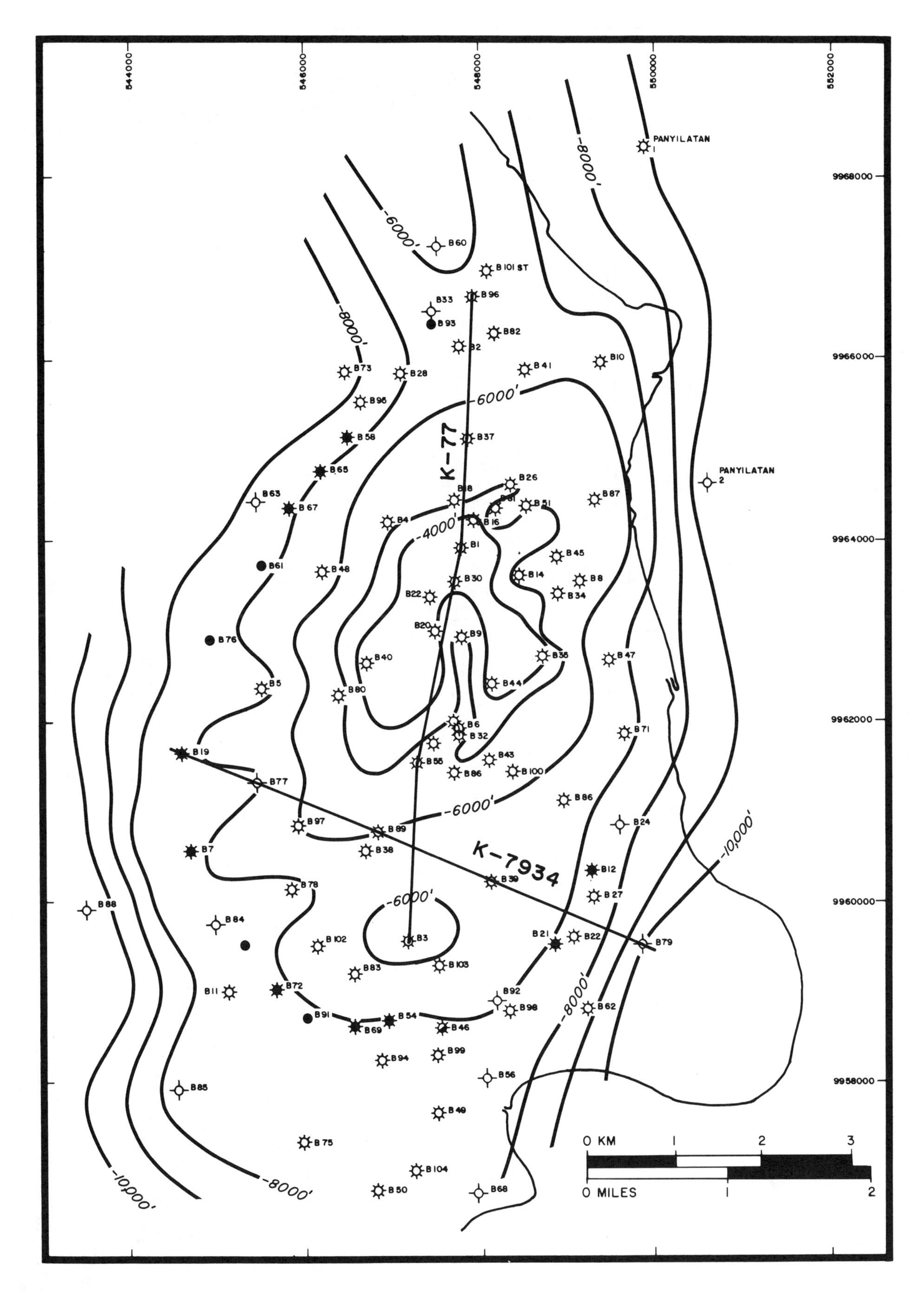

Figure 12. Structure map on first pay zone (Zone "A") encountered.

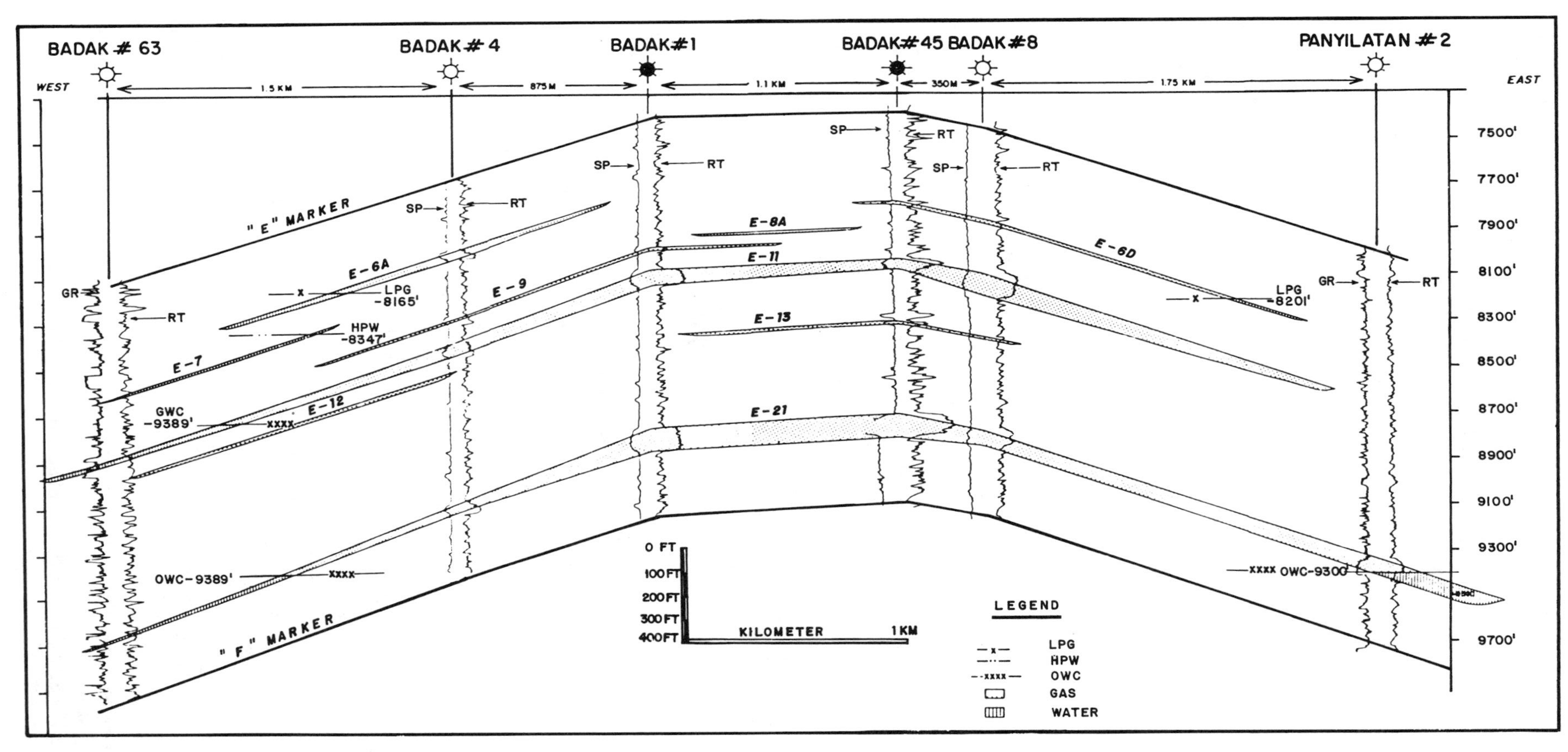

Figure 13. Structural cross section of Badak zone "E," showing fluid contacts (location shown on Figure 14).

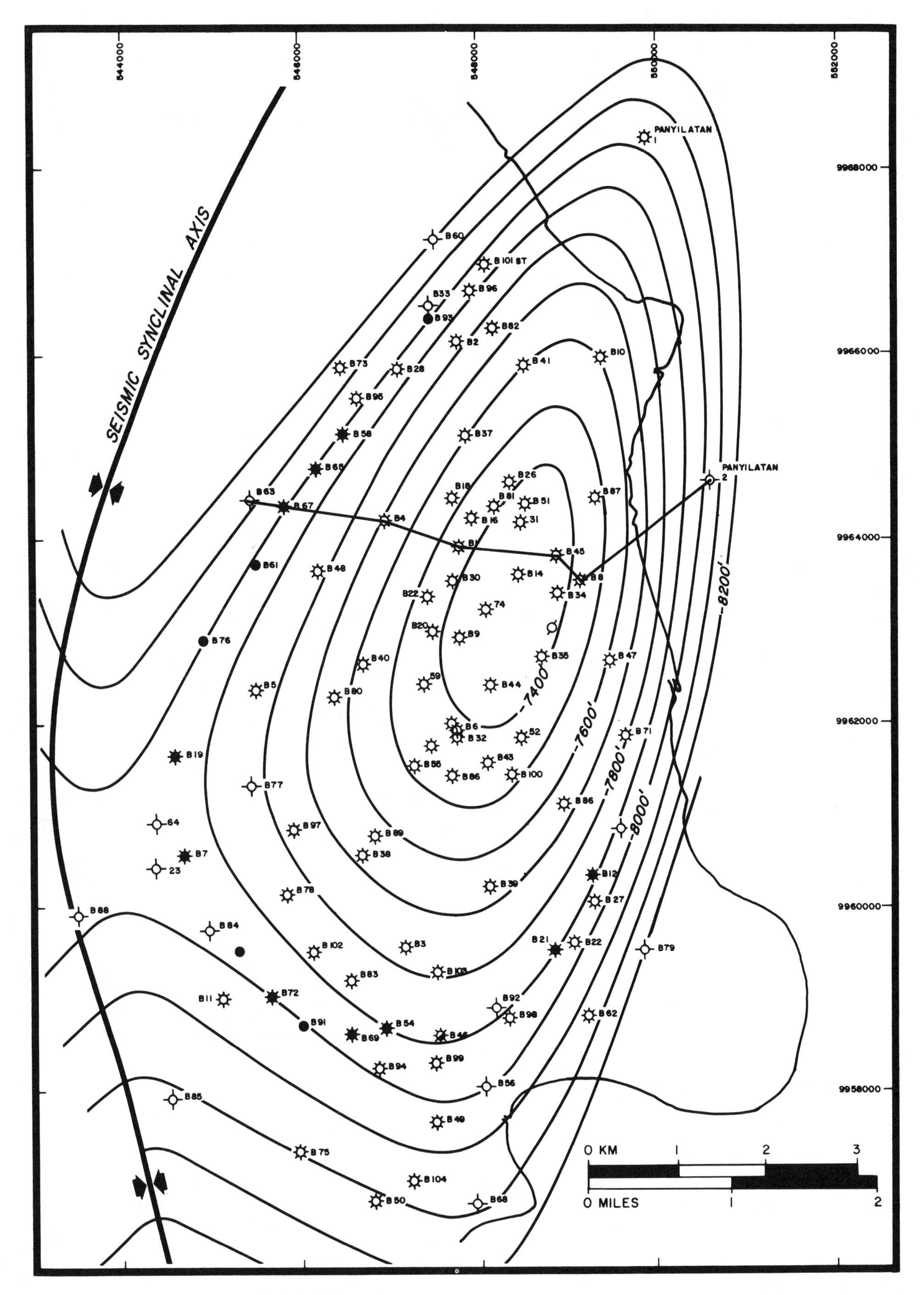

Figure 14. Structure map on Badak "E" marker.

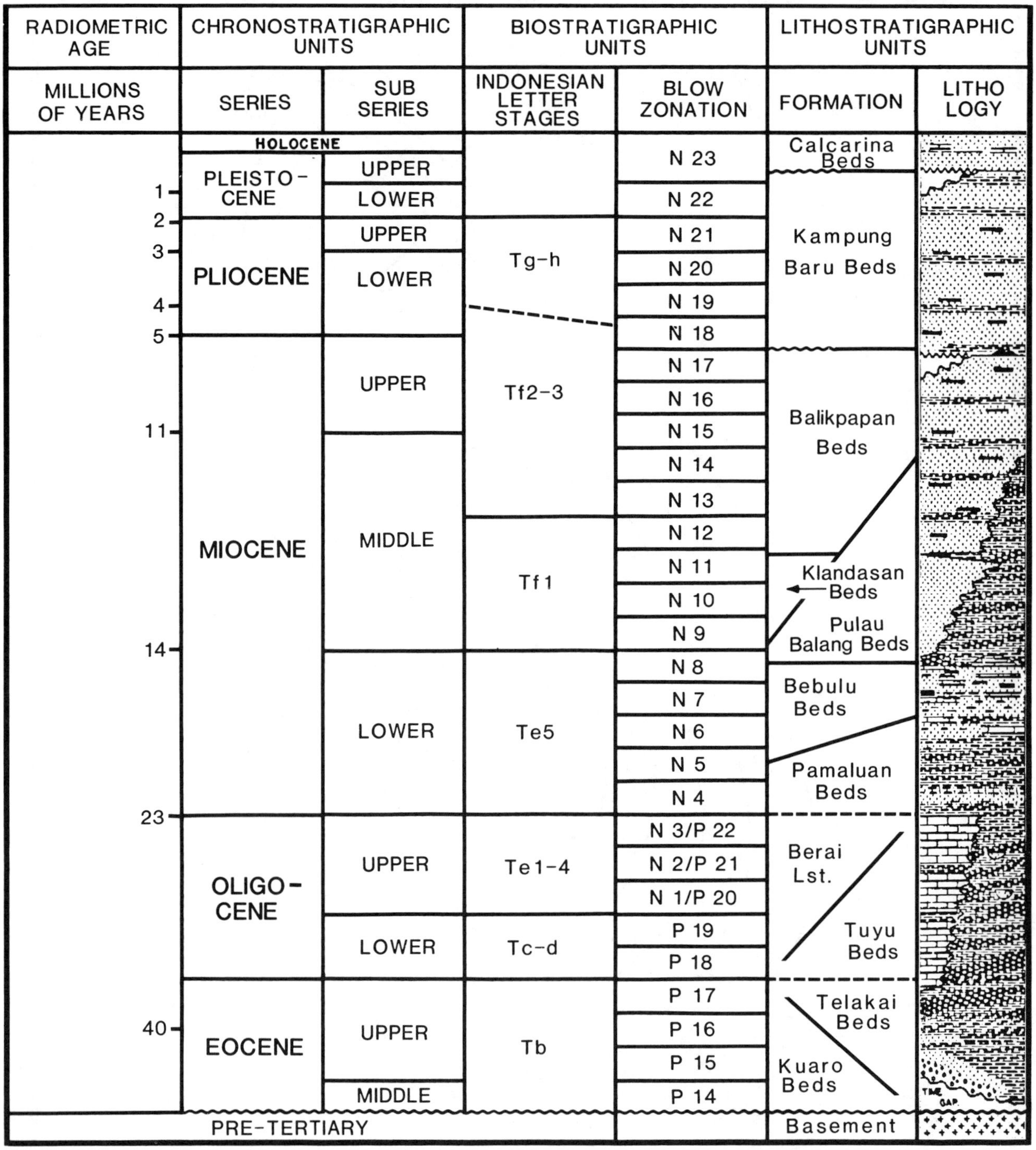

RADIOMETRIC AGE	CHRONOSTRATIGRAPHIC UNITS		BIOSTRATIGRAPHIC UNITS		LITHOSTRATIGRAPHIC UNITS	
MILLIONS OF YEARS	SERIES	SUB SERIES	INDONESIAN LETTER STAGES	BLOW ZONATION	FORMATION	LITHO LOGY
	HOLOCENE			N 23	Calcarina Beds	
	PLEISTO-CENE	UPPER				
1		LOWER		N 22		
2	PLIOCENE	UPPER		N 21	Kampung Baru Beds	
3		LOWER	Tg-h	N 20		
4				N 19		
5				N 18		
	MIOCENE	UPPER	Tf2-3	N 17	Balikpapan Beds	
				N 16		
11				N 15		
		MIDDLE		N 14		
				N 13		
			Tf1	N 12		
				N 11	Klandasan Beds	
				N 10		
14				N 9	Pulau Balang Beds	
		LOWER	Te5	N 8	Bebulu Beds	
				N 7		
				N 6		
				N 5	Pamaluan Beds	
				N 4		
23	OLIGO-CENE	UPPER	Te1-4	N 3/P 22	Berai Lst.	
				N 2/P 21		
				N 1/P 20		
		LOWER	Tc-d	P 19	Tuyu Beds	
				P 18		
	EOCENE	UPPER	Tb	P 17	Telakai Beds	
40				P 16		
				P 15	Kuaro Beds	
		MIDDLE		P 14		
	PRE-TERTIARY				Basement	

Figure 15. Generalized stratigraphic subdivision of the sedimentary sequence in the Mahakam delta area.

interval. The basal shale and limestone intervals indicate a more distal deltaic environment and are interpreted as representing transgressive pulses in the generally regressive depositional history of the Badak area.

The lateral stratigraphic development in the Badak field is illustrated in north-south direction in Figure 18 and in a northwest-southeast direction in Figure 19. Figure 20 shows the succession of paleoenvironments from faunal evidence in the Badak field.

More detailed information on the stratigraphy of Badak field is provided in the *Reservoir* section of this paper.

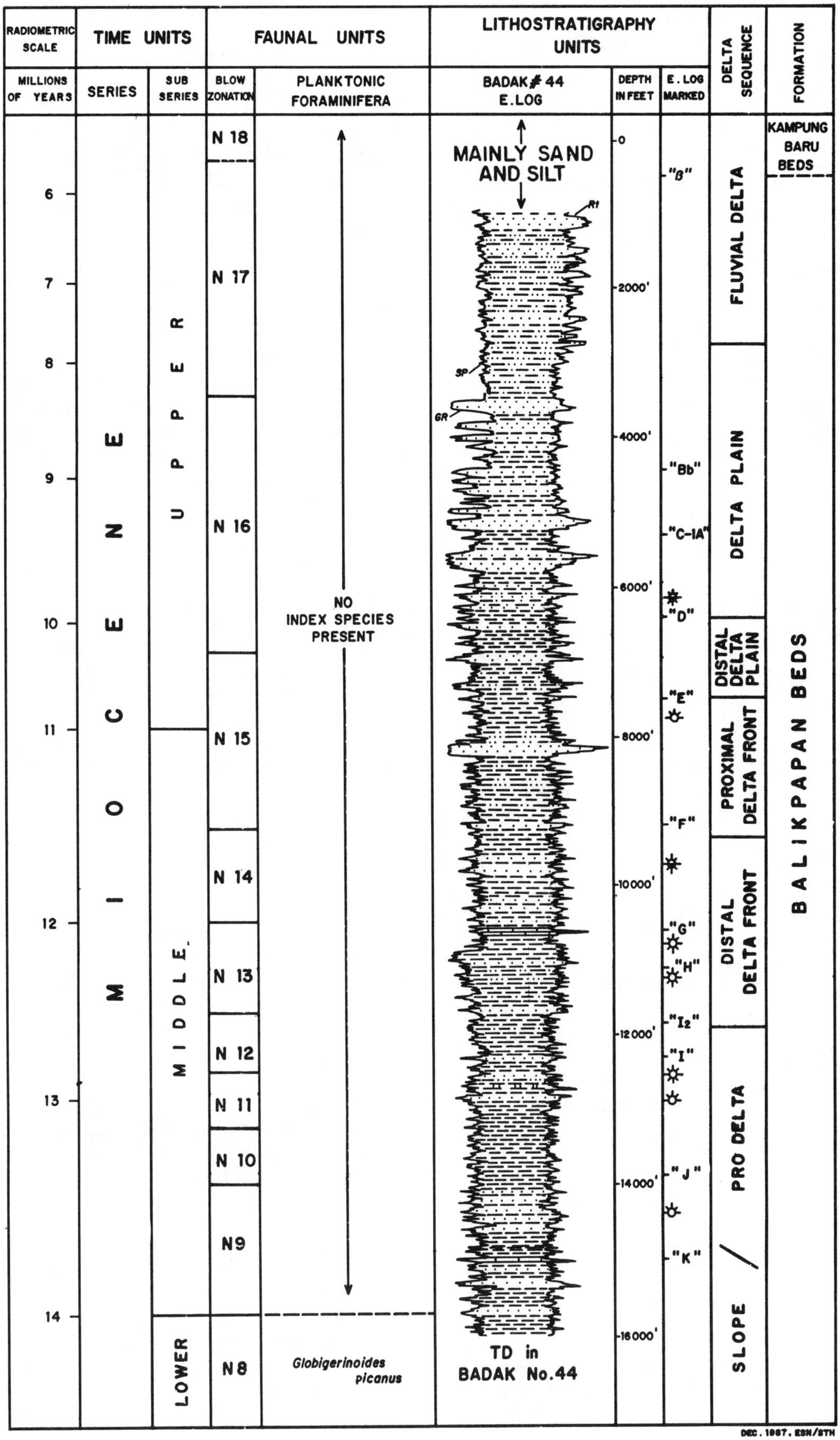

RADIOMETRIC SCALE	TIME UNITS		FAUNAL UNITS		LITHOSTRATIGRAPHY UNITS			DELTA SEQUENCE	FORMATION
MILLIONS OF YEARS	SERIES	SUB SERIES	BLOW ZONATION	PLANKTONIC FORAMINIFERA	BADAK # 44 E.LOG	DEPTH IN FEET	E. LOG MARKED		

6
7
8
9
10
11
12
13
14

M I O C E N E

U P P E R

M I D D L E

LOWER

N 18
N 17
N 16
N 15
N 14
N 13
N 12
N 11
N 10
N9
N8

NO INDEX SPECIES PRESENT

Globigerinoides picanus

MAINLY SAND AND SILT

Rt

SP

GR

TD in BADAK No.44

0
2000'
4000'
6000'
8000'
10000'
12000'
14000'
16000'

"B"
"Bb"
"C-1A"
"D"
"E"
"F"
"G"
"H"
"I2"
"I"
"J"
"K"

FLUVIAL DELTA

DELTA PLAIN

DISTAL DELTA PLAIN

PROXIMAL DELTA FRONT

DISTAL DELTA FRONT

PRO DELTA / SLOPE

KAMPUNG BARU BEDS

B A L I K P A P A N BEDS

DEC. 1987. ESN/STH

Figure 16. Stratigraphic column in Badak No. 44.

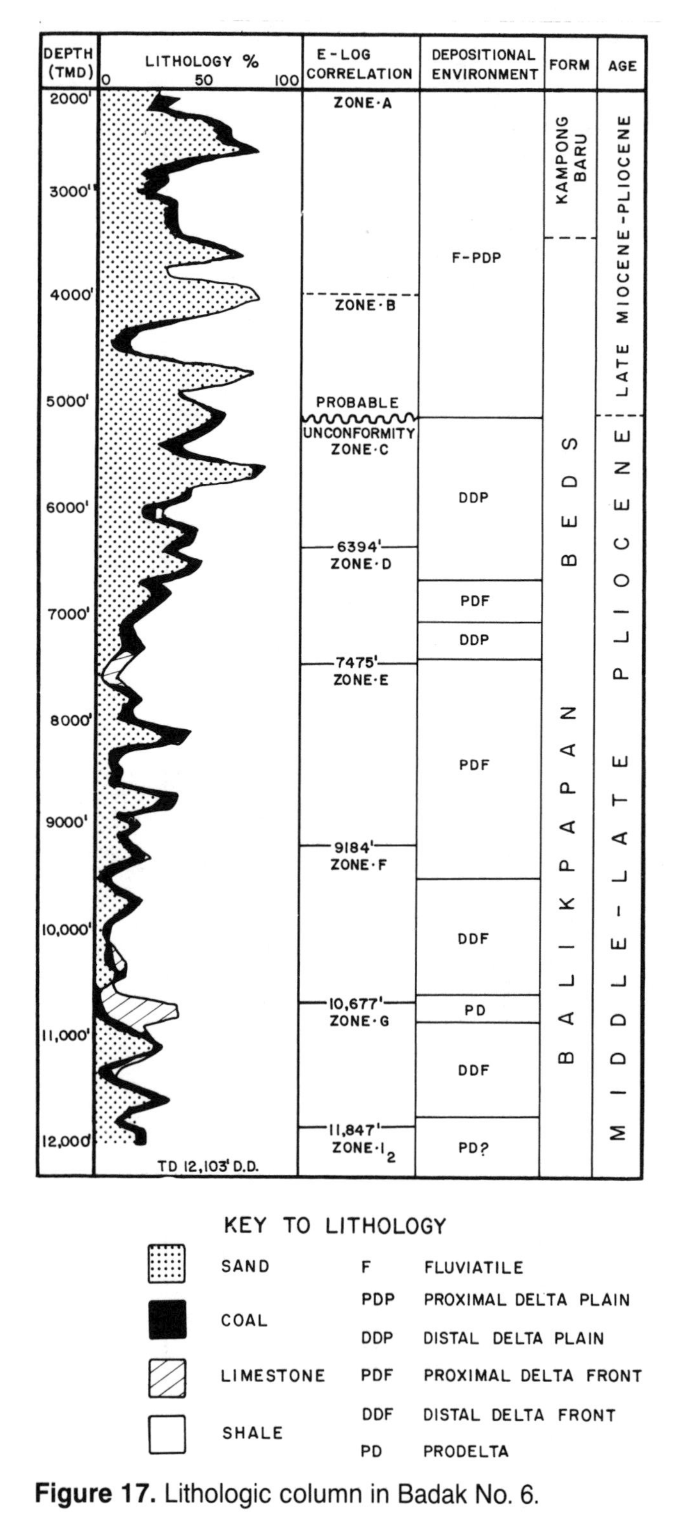

Figure 17. Lithologic column in Badak No. 6.

TRAP

The oil and gas in the Badak field are trapped in multiple deltaic sands in a simple, domal anticline. The maximum structural closure in Badak is about 11 km (6.9 mi) long and 5 km (3.1 mi) wide and covers an area of approximately 55 km² (13,585 acres).

Most of the structural growth in the Badak structure seemed to have occurred after deposition of the "F" zone and before deposition of the "B" zone. This timing may explain the relative paucity of oil in the "F" and lower zones, as the earlier and heavier hydrocarbons were only partially trapped in Badak, whereas the closure reached its maximum about the end of deposition of the "C" zone in which both early (oil) and later (gas) hydrocarbon products were trapped.

Reservoir

Reservoir Zones

A brief description of each of the Badak reservoir zones is given below, as developed in Badak No. 1 down to the base of the "F" zone at -10,597 ft (-3231 m) depth and in Badak No. 44 from the top of the "G" zone at -10,535 ft (-3212 m), to 16,400 ft (5000 m).

Zone "A," 0 to -4374 ft (-1333.5 m) SSL in Badak No. 1: Zone "A" consists of sandstones, siltstones, and lignite, which are interpreted as predominantly delta plain deposits (Figure 21). The sandstones were probably deposited in distributary channels and braided streams, whereas the siltstones and lignites were probably formed in lagoons and marshes between the stream channels. All porous intervals in zone "A" are filled with fresh or, in a few cases, brackish water. Wireline tests have indicated gas saturation in some of the lowermost "A"sands, which will be produced in later stages of field depletion.

Zone "B," -4374 to -5138 ft (-1566 m) SSL in Badak No. 1: Zone "B"predominantly consists of unconsolidated, clear, white, fine-grained, subrounded, well-sorted quartz sands and white, buff, or light brown, occasionally lignitic, siltstone. Brown, lignitic shales and lignites occur in subordinate amounts. As can be seen in Figure 22, zone "B" contains some thick reservoir sands, of which so far only the B-12 zone has been put on production and has produced about 38,000 bbl of condensate and 15 bcf gas. Zone "B" clearly corresponds with a fluvial deltaic to proximal delta plain environment.

Zone "C," -5138 to -6330 ft (-1930 m) SSL in Badak No. 1: Zone "C" also consists mainly of siltstone and sands, but the latter are increasingly consolidated downward in the section and become fairly well indurated sandstones (Figure 23). Some thick gas reservoirs occur in this zone, such as the C-1 and C-2 groups of sands and the C-8 sand. The C-22 sand contains 29° API oil. Some intercalations of dark brown to black, finely laminated, lignitic shale occur in the lower part of this zone in Badak No.

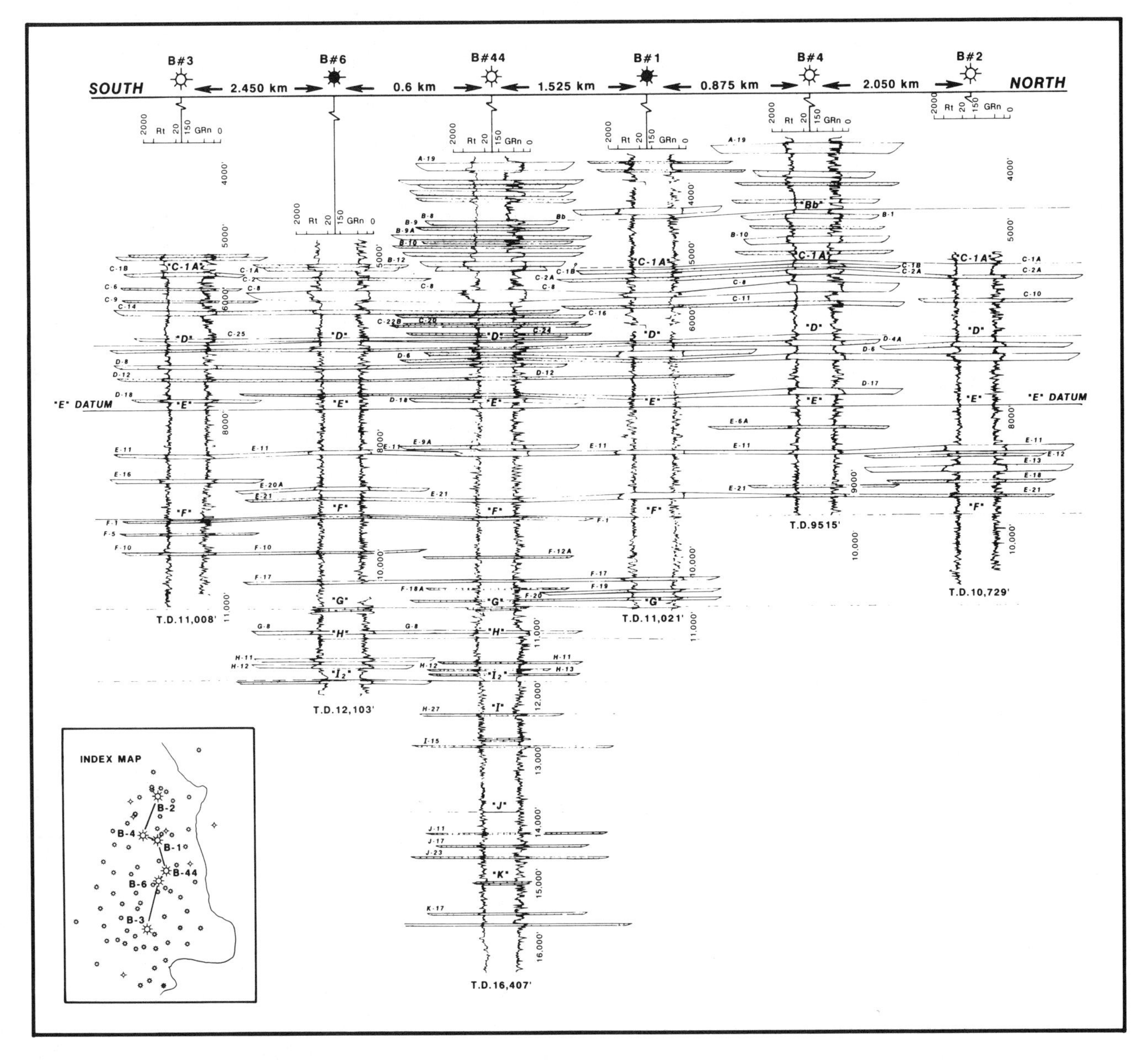

Figure 18. North-south stratigraphic cross section in Badak field. (Note that the gamma ray and resistivity curves are switched in this illustration.)

1. Near the base of zone "C" some thin streaks of a light brown to gray dolomite occur in Badak No. 1. Lignite is present throughout this zone, in both discrete beds and in finely disseminated form. The lithology suggests a delta plain environment, distal at the base and more proximal at the top. The average sand/shale ratio of zone "C" is approximately 30%.

Zone "D," -6330 to -1422 ft (-2263 m) SSL in Badak No. 1: At the top of zone "D" a number of well-developed sandstone reservoirs are found, the D-2 to D-6 sandstones (Figure 24). These are described as clear, opaque, white to red quartz sandstone, medium to coarse grained, hard, subrounded, and fairly well sorted. Further downward in zone "D," the rocks are predominantly shale, dark brown to black, and siltstone, with abundant lignite. Some thin dolomite intercalations are present. The lower part of zone "D" contains some thinner, but important, reservoir sandstones. Average sand/shale ratio is roughly 25%. As in the previous zone, the lithology suggests a distal delta plain environment at the base, becoming more proximal at the top.

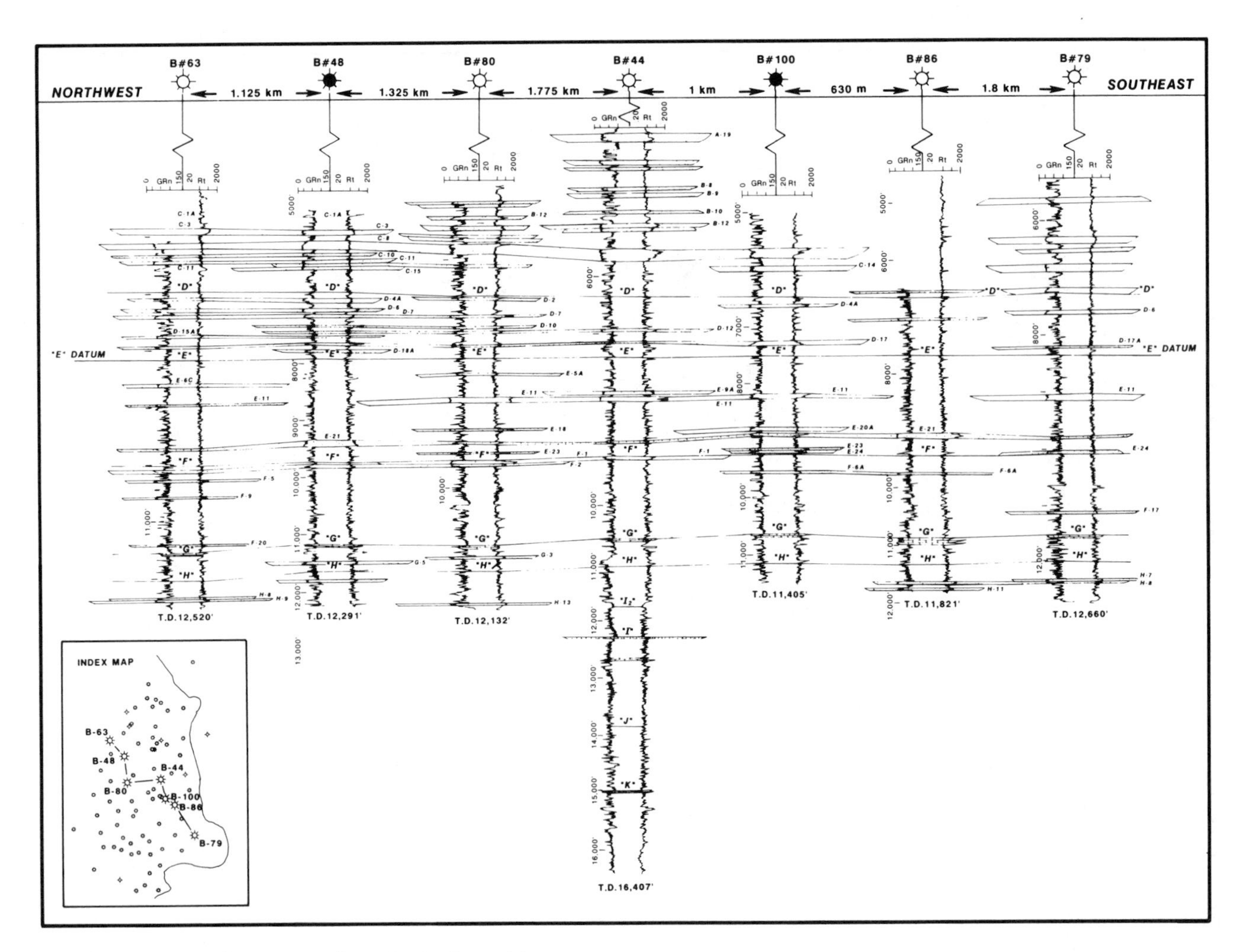

Figure 19. Northwest-southeast stratigraphic cross section in Badak field.

Zone "E," -7422 to -9168 ft (-2795 m) SSL in Badak No. 1: Zone "E"in Badak No. 1 starts at the top with a 140 ft (43 m) interval, which contains a number of dolomitic limestone beds with reefal fauna (Figure 25). This interval is interpreted as an indication of distal delta front environmental conditions and, therefore, a clear transgressive interval in the otherwise largely regressive Badak sequence. This is the second transgressive interval known in Badak as a preceding transgression occurred at the end of zone "F" deposition. Further downward in the section, zone "E" consists of more proximal delta front or perhaps distal delta plain siltstone and sandstones. Some exceptionally thick sandstones are present in zone "E," such as the E-11 (66 ft net gas pay in B-1) and E-21 (99 ft net gas pay in B-1), which are interpreted as being deposited in fluvial channels traversing the delta plain. These sandstones are stacked in places, such as in Badak No. 2, as illustrated in Figure 26.

Zone "F," -9168 to -10,597 ft (-3231 m) SSL in Badak No. 1: Zone "F" is generally characterized by a series of thin sandstones of lower reservoir quality than the "E" sandstones (Figure 27). All or most "F" sandstones are of a "coarsening-upward" character and are thought to represent delta front bar deposits. The average sand/shale ratio of zone "F" is less than 10%. Coal appears to be present in lesser quantity than in zone "E." The top of zone "F" is a distinct E-log shale interval that can be recognized throughout the Badak field.

Zone "G," -10,535 to -11,025 ft (-3361 m) SSL in Badak No. 44: The top of zone "G" is picked at the top of a limestone interval that is widely developed in the Badak area and is believed to represent a transgressive pulse in the otherwise largely regressive depositional record in the area (Figure 28). Several producible sandstone intervals occur in zone

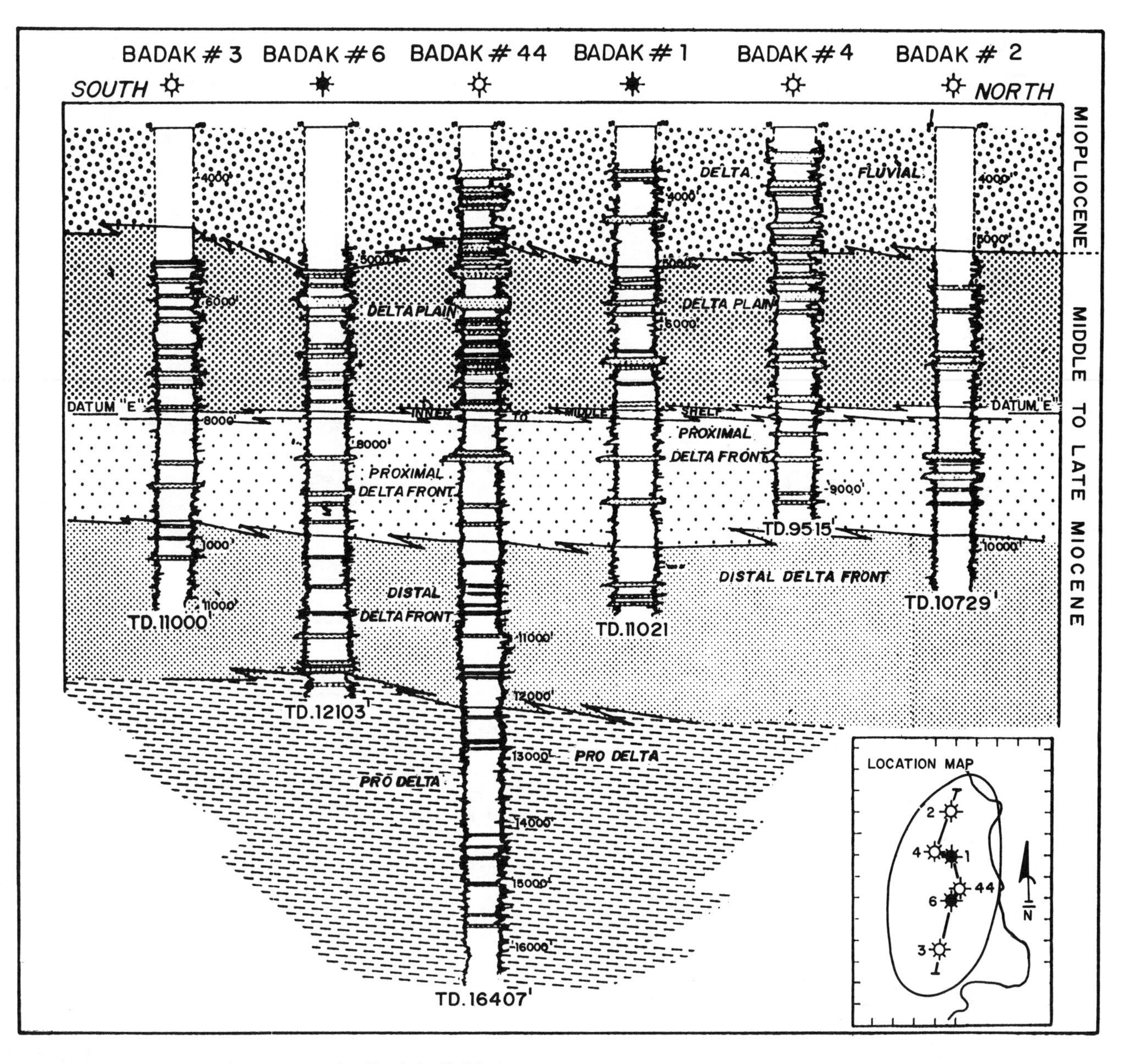

Figure 20. Paleoenvironments in Badak field, interpreted from faunal evidence.

"G," all of which, however, seem to be of somewhat limited areal extension. A few distinct coal beds are found in zone "G." As a whole, the lithology of zone "G" appears to indicate a more distal deltaic facies than either zone "F" or the underlying zone "H." Its average sand/shale ratio is about 7%. The lower boundary of zone "G" seems somewhat arbitrary and, at the time of writing, is less well known. It has only been penetrated by a few boreholes until recently and therefore has not been thoroughly studied.

Zone "H," -11,025 to -12,241 ft (-3732 m) SSL in Badak No. 44: In general, zone "H" appears to differ from zone "G" in that it forms a monotonous shale and siltstone sequence, interspersed in a seemingly cyclical manner with thin, coarsening-upward, shaly sandstone beds. "H" sandstones are generally rather poor reservoirs and their limited areal distribution makes their individual development difficult to predict. However, zone "H," as known from the 33 Badak wells that have so far partially or fully penetrated it (as of March 1987), has an average sand/shale ratio of 12% and contains several reservoirs with appreciable producible gas reserves. In Badak No. 44, between about -11,630 (-3546 m) and -11,731 (-3577 m), there is an interval of rather pure brown to gray shale, which is regionally recognizable as an E-log

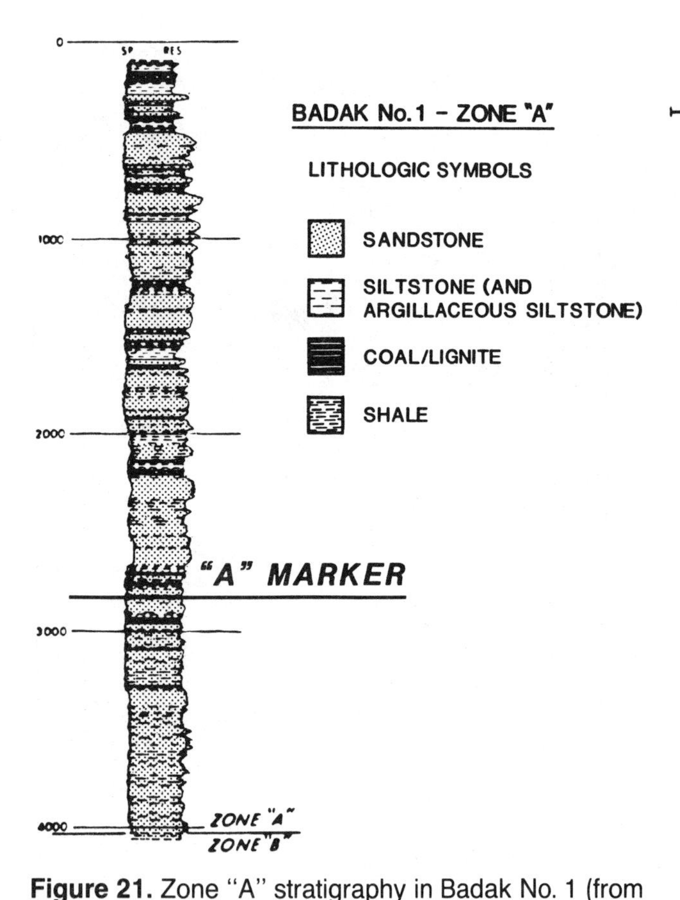

Figure 21. Zone "A" stratigraphy in Badak No. 1 (from Huffington and Helmig, 1980).

marker and is interpreted as representing a transgressive episode in the generally regressive deltaic regime (Figure 29). The base of this shale interval, at -11,731 ft (-3577 m) in Badak No. 44, is known as the "I_2" marker. The top of the I_2 shale interval generally lies below the top of the abnormal pressure transition zone in Badak, which is the reason that only a few wells have fully penetrated this horizon (Badak No. 6 and Badak No. 44).

Zone "I," -12,241 to -13,811 ft (-4211 m)

Zone "J," -13,811 to -14,921 ft (-4549 m)

Zone "K," -14,921 to ?

Only two Badak wells have penetrated the full I_2 shale interval and only Badak No. 44 reached the lower "H" and deeper zones (Figure 30). Therefore, only a few observations about this part of the Badak sequence, directly related to the latter well, can be made. First and most important, Badak No. 44 proved that sandstone porosity exists at this location to at least 16,400 ft (4877 m) depth. Down to almost 16,000 ft, coal beds occur although the lithology and fauna seem to indicate a prodeltaic depositional environment. The lower coal beds are probably of an

Figure 22. Zone "B" stratigraphy in Badak-1.

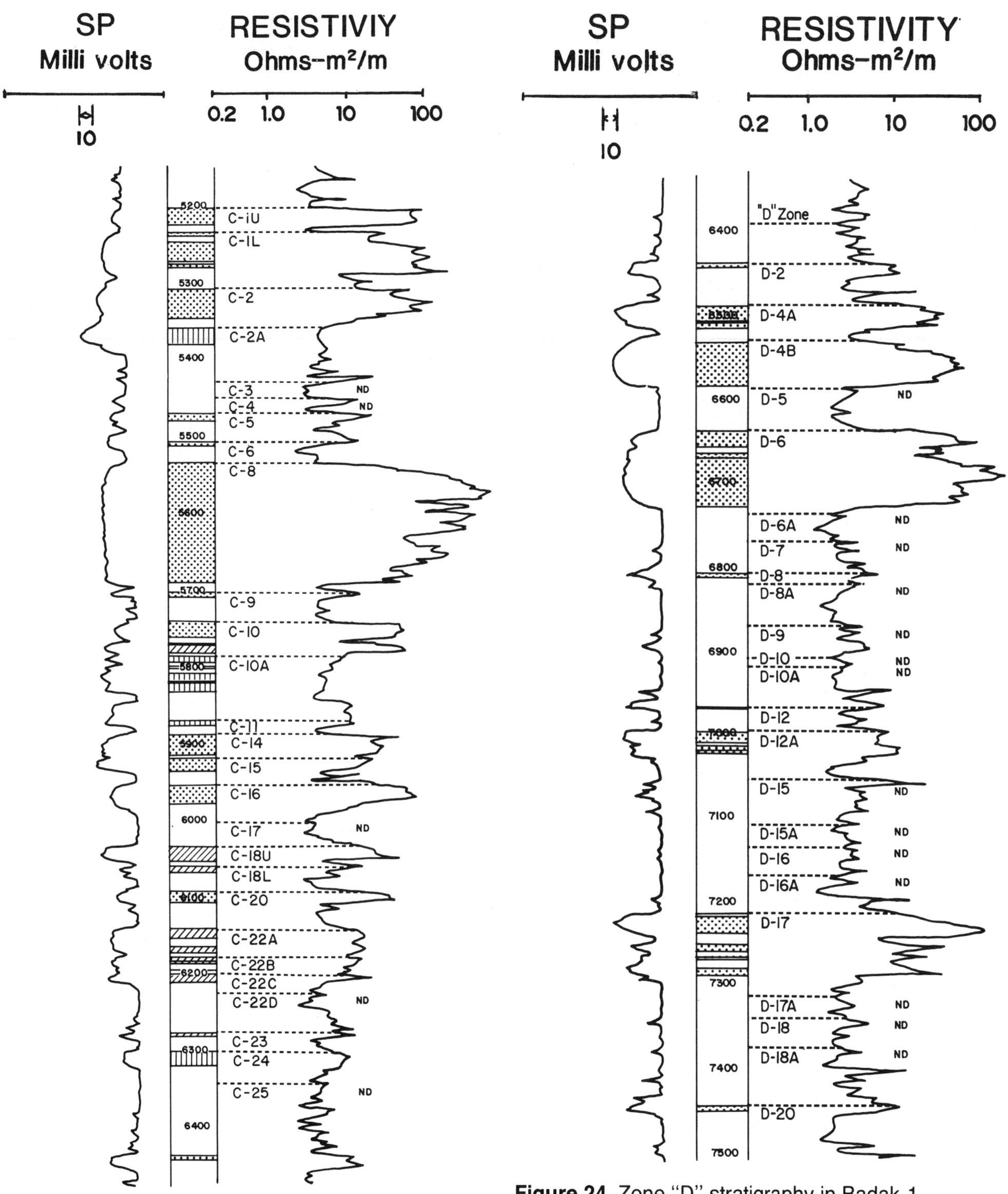

Figure 23. Zone "C" stratigraphy in Badak-1.

Figure 24. Zone "D" stratigraphy in Badak-1.

allochthonous nature, deposited in distal and prodeltaic zones.

Depositional Setting

The depositional setting of the Badak field is that of a major delta where clastics derived from a spasmodically rising hinterland to the west were deposited. The Badak reservoir sandstones were generally transported from southwest to northeast. The Mahakam delta, which may have started to build during lower Miocene (below zone N4) times, prograded from west to east in the Kutai basin. The

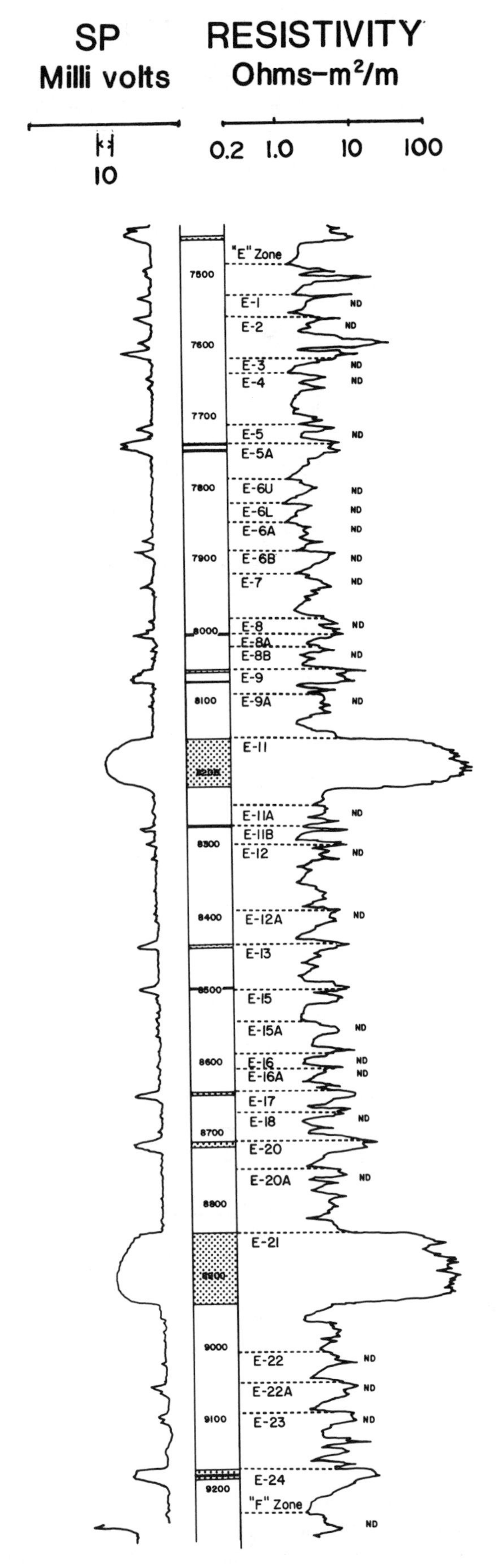

Figure 25. Zone "E" stratigraphy in Badak-1.

deepest Badak sands drilled so far are of middle Miocene age and indicate a prodeltaic environment. Younger Badak reservoirs show a gradual change upward from distal to proximal and finally fluvial deltaic facies, as shown in Figure 17.

Lithology*

The Badak and Nilam fields sandstone reservoirs consist of quartz arenites/wackes and feldspar arenites/wackes, decreasing in porosity and permeability from top to bottom, as illustrated in Figure 31. The type of porosity is primary interparticle porosity, which is partially filled with authigenic clay, detrital clay, and secondary quartz, as illustrated in the scanning electron micrograph (SEM) in Figure 32. Microporosity (i.e., pores with pore throat diameters of less than 1.0 micron) apparently contributes to the total porosity (effective for gas) of Badak reservoirs.

Authigenic clay minerals and interstitial quartz deposits are the main cementing agents in these reservoirs, although a small amount of carbonate and ferruginous cement is also observed. Kaolinite, illite, smectite, and chlorite clay minerals are distributed throughout all the reservoirs in variable amounts (Figure 33). Comparison of shallow and deeper reservoirs showed that diagenesis is the main factor controlling reservoir quality in these zones. A higher degree of postdepositional compaction and silicification, an increase in authigenic clay, and an increase in unstable minerals in the deeper reservoirs could be the most important factors causing porosity and permeability reduction with depth. The depositional recycling process may have improved reservoir quality in the upper zones.

Geopressures

The subsurface pressures in the Badak field roughly coincide with hydrostatic pressure down to the top of a pressure transition zone, where the pressure (and temperature) suddenly increases according to a steeper gradient. The top of the pressure transition zone does not conform to the stratigraphy; it occurs stratigraphically higher in the more distal deltaic parts of the field and stratigraphically lower in the proximal deltaic parts. It appears to follow roughly the top of the prodeltaic—i.e., shaly facies—at the bottom of the deltaic section, as illustrated by Figure 34. The normal pressure gradient in the Badak field is 0.433 psi/ft, but it increases to 0.898 psi/ft at the base of the pressure transition zone. The highest pressure recorded in the field was 11,978 psig, or 16.3 ppg equivalent mud weight, at a depth of 14,179 ft (4323 m) in Badak No. 44. (The maximum recorded temperature was 330°F [165.6°C] at 16,407 ft [5004 m] in Badak No. 44.)

The interfingering configuration of the pressure transition zone in Figure 34 is an interpretation of

*From Panigoro (1983).

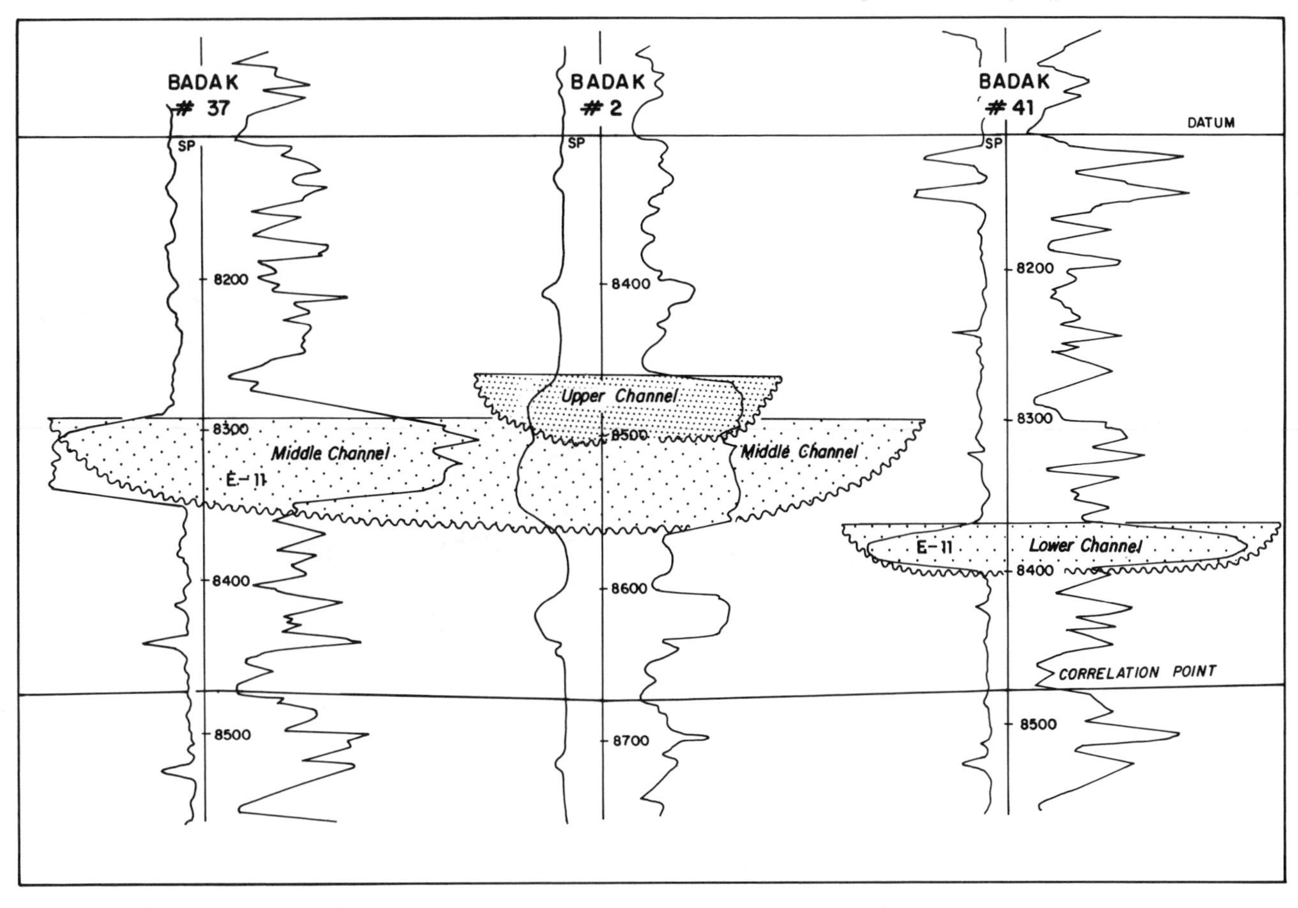

Figure 26. E-11 distributary channels illustrating stacked channels in Badak No. 2 well.

a phenomenon that occurred in a few Badak wells, whereby after the first pressure kick and setting of a protection string of casing, the bit would penetrate a normally pressured section before entering into overpressure again.

Source

In spite of intensive study of the geochemistry of the Mahakam delta, especially by the French Petroleum Institute, a convincing, comprehensive solution to the generation and migration of the hydrocarbons found in the Mahakam delta is still lacking. Figure 35 summarizes the results of the source rock analyses of nine samples from the Badak No. 1 well.

A major problem in developing a viable geochemical model for the Badak field is that measured and calculated maturities in a composite Badak well section are incompatible. In the deeper part of the section, down to approximately the zone of overpressuring, measured maturities are anomalously low compared to calculated maturities.

A solution proposed by researchers of the French Petroleum Institute was summarized as follows by Tissot and Welte (1984) for the Handil field:

> It is argued convincingly that gaseous and very light hydrocarbons migrate out of the overpressure zone more easily than heavier hydrocarbons and non-hydrocarbons. Then the light hydrocarbons on their way upward extract heavier compounds and carry them along. In the "hot" overpressured shales there seems to be a single supercritical hydrocarbon phase that, when moving upward to lower pressures and temperatures, progressively loses the heavier end of its hydrocarbons by retrograde condensation. In this manner the inverse specific gravity relationship of hydrocarbon occurrences in Handil can be explained, i.e. light products at the top and heavier at the bottom.

Vertical—i.e., cross-stratal—migration does not apply in the Badak field, where more than 100 gas reservoirs are stacked and separated from each other

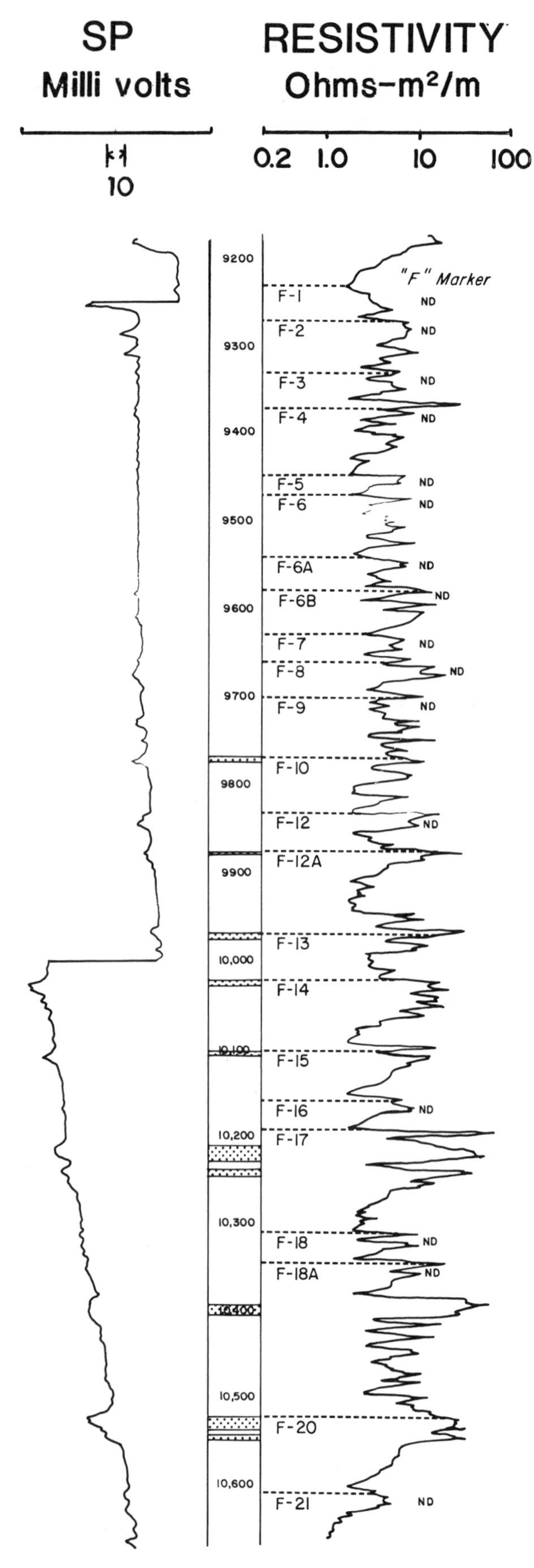

Figure 27. Zone "F" stratigraphy in Badak-1.

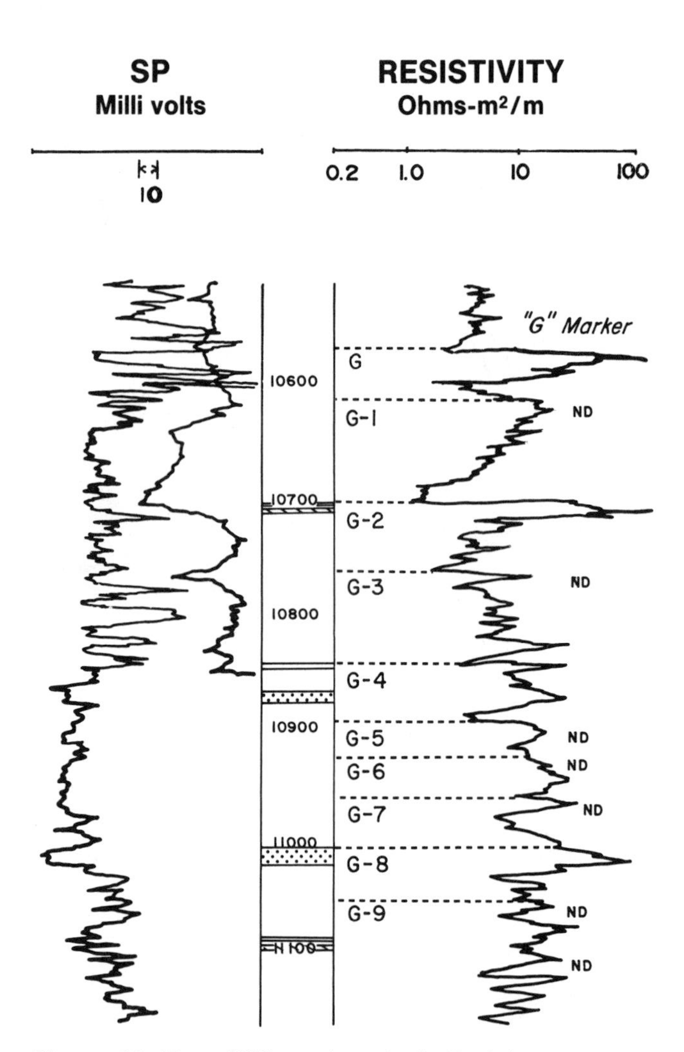

Figure 28. Zone "G" stratigraphy in Badak-1.

by cap rock that is apparently perfectly sealing, without any faults or fractures. Such a situation would preclude the continuous percolation of gases from a deep source gradually filling all reservoirs in the section. In addition, the oil found in Badak occurs in several reservoirs, which are more or less irregularly distributed throughout the overall pay section and which show no regular decrease in specific gravity of oil from the bottom of the section upward. However, some heavy, unproducible oil was found in a few deeper—i.e., more distal—deltaic sands in Badak. This may indicate that Tissot and Welte's hypothesis is applicable in a lateral sense and that the hydrocarbon gravity in a reservoir is related to its distance from its source, in this case the "hot" overpressured prodeltaic shales.

The present authors suggest a variation on the solution suggested above by the IFP researchers, which is as follows: Accepting that the generation of oil and gas takes place in the finer-grained, prodeltaic part of the sedimentary section that underlies the sandy delta and that is overpressured,

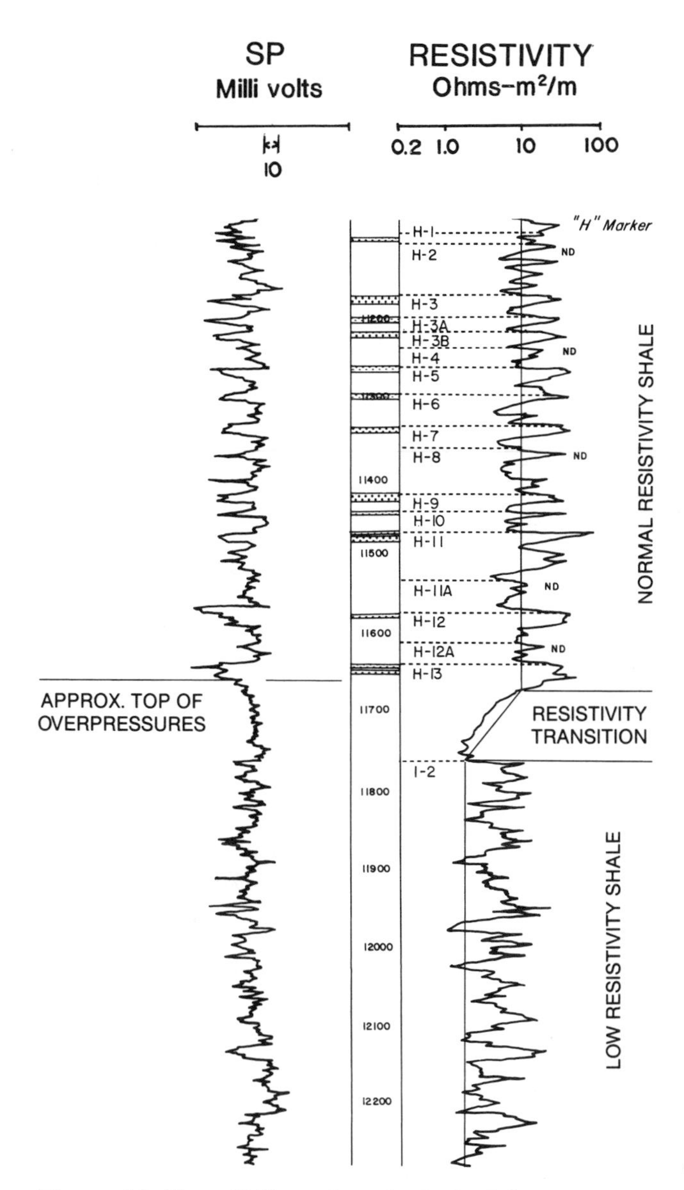

Figure 29. Zone "H" stratigraphy in Badak-44.

we suggest that the oil and gas generated is introduced through porous carrier stringers and laminae into the eventual reservoir bodies at the interfingering distal ends of prograding mega-foresets of each successive deltaic cycle as schematically illustrated in Figure 36. It appears that all oil and gas occurrences in the Mahakam delta could be generally explained by the above described model.

The fact that virtually all organic matter found in Tertiary rocks of the Mahakam delta is of humic origin (type III kerogen) may be thought by some to be incompatible with the abundance of oil in the basin. However, recent research, notably by the French Petroleum Institute (IFP), has found convincing evidence that coaly organic matter is a potential source for oil (Durand and Paratte, 1974; Durand and Oudin, 1980; Durand et al., 1983; Schoell et al., 1983, 1985; Tissot and Welte, 1984; Vandenbroucke et al., 1983).

Hydrocarbons

The stratigraphic distribution of the initial hydrocarbon reserves in place in the Badak field is illustrated by Figure 37.

Oil

The Badak field has produced a cumulative total of 37,645,317 BO as of 31 October 1987. An analysis of a representative oil sample from the C-22 reservoir in Badak No. 1 is given in Table 1. Figure 38 shows a decline curve for Badak oil production.

Most of Badak oil reservoirs exist as oil rims or legs under a typically large gas cap. Figures 39 and 40, respectively, illustrate the distribution of the C-20 and D-6 oil accumulations, the two major oil pools in Badak.

Condensate

In addition to the "black" oil, with the gas delivered as of 31 October 1987, a cumulative total of 30,613,910 barrels of 48° to 54° API condensate has been produced. This condensate is stabilized and blended with oil at the Badak field and subsequently pumped to the Santan Oil Terminal and sold.

Gas

Cumulative gas production as of 31 October 1987 from Badak field is 1986 billion ft^3 (56.74 billion m^3), which, after liquefaction, produced roughly 40 million metric tons of LNG, including gas used in the operation of the liquefaction facilities. The analysis of a representative gas sample from Badak field is given below:

Component	Mol%
CO_2	2.94
N_2	0.06
Methane	87.44
Ethane	4.51
Propane	2.84
Butane	1.29
Pentanes	0.43
Hexanes	0.17
Heptanes + heavier	0.32
Total	100.00

A decline curve for Badak gas production would be unrelated to its reservoir parameters because the production level is governed by the demand from the LNG plant, rather than by reservoir considerations.

The two major gas reservoirs in Badak are the E-11 and E-21 sandstones, with cumulative gas production as of 31 October 1987 of 224 and 297 bcf (6.4 and 8.5 billion m^3), with about 16 barrels of condensate per million cubic feet of gas. The distribution of the E-11 and E-21 gas reservoirs is shown in Figures 41 and 42, respectively.

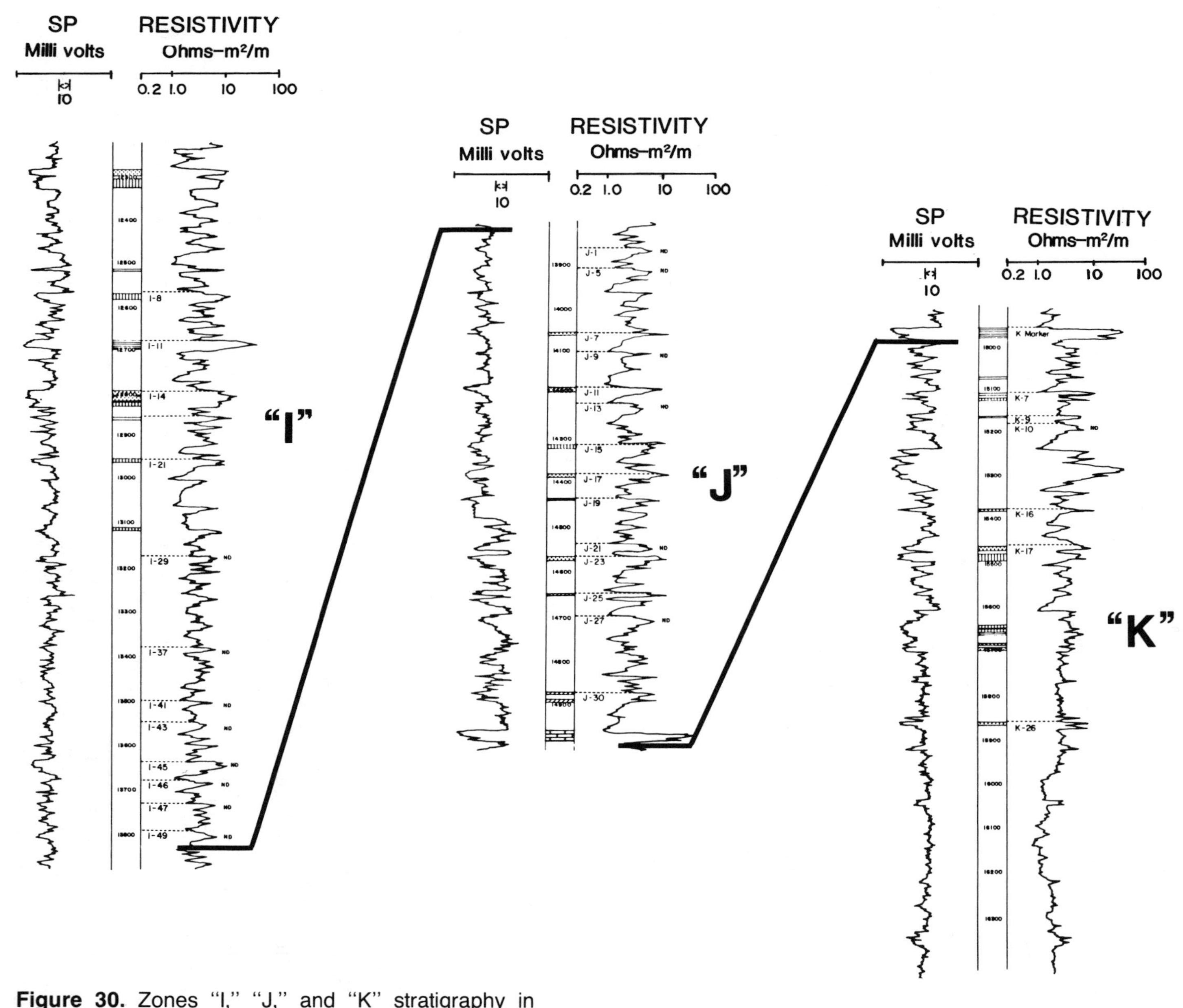

Figure 30. Zones "I," "J," and "K" stratigraphy in Badak-44.

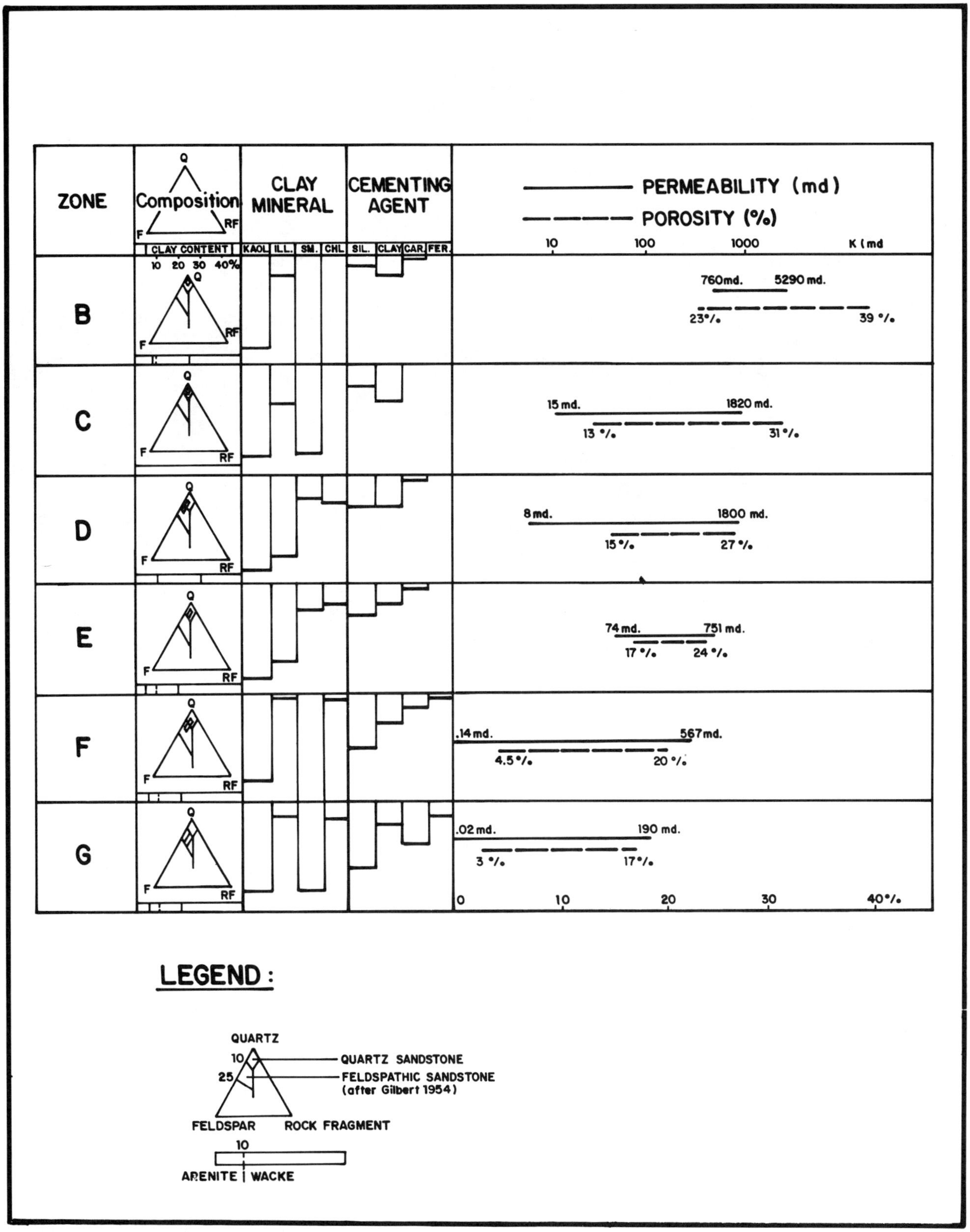

ZONE	Q / Composition / F RF, CLAY CONTENT 10 20 30 40%	CLAY MINERAL (KAOL. ILL. SM. CHL.)	CEMENTING AGENT (SIL. CLAY CAR. FER.)	—— PERMEABILITY (md) – – – POROSITY (%)
B				760md. 5290 md. / 23 %. 39 %.
C				15 md. 1820 md. / 13 %. 31 %.
D				8 md. 1800 md. / 15 %. 27 %.
E				74 md. 751 md. / 17 %. 24 %.
F				.14 md. 567md. / 4.5 %. 20 %.
G				.02 md. 190 md. / 3 %. 17 %.

K (md): 10 100 1000

Porosity scale: 0 10 20 30 40 %.

LEGEND:

QUARTZ
10 — QUARTZ SANDSTONE
25 — FELDSPATHIC SANDSTONE (after Gilbert 1954)
FELDSPAR ROCK FRAGMENT
10
ARENITE | WACKE

Figure 31. General composition, porosity, and permeability of sandstone reservoirs in Badak and Nilam fields.

Figure 32. Medium-grained sandstone from the B-12 zone showing scattered clay around the grains. Porosity 25%, permeability 2500 md. SEM photography 50 ×.

EXPLORATION CONCEPTS

The general concept that led to the exploration of the Kutai basin by Huffington was that of a major delta, comparable to the Mississippi and Niger deltas, which unlike the latter two had not been intensively explored. Oil had been found both in older (Sanga-Sanga) and younger (Attaka) deposits of the Mahakam delta and the area, and stratigraphy in between seemed highly prospective.

The Badak structure is an unfaulted, four-way anticlinal closure with more than 100 separate deltaic reservoir sandstones. These sandstones contain as much as 1000 ft (305 m) of net gas pay and up to 120 ft (37 m) of oil pay in an 8000 ft (2440 m) thick section. Before the discovery well was drilled, the existence of these sandstones in Badak was postulated on the basis of the occurrence of updip, sandier facies, correlative to the Badak section, in outcrops a few miles to the west.

The anticlinal structure is only developed in the subsurface, below about 3000 ft (915 m) in depth, although a structural terrace, which probably influenced the morphology of the present-day coastline, exists at shallow depths. Modern seismic methods were essential in mapping the structure properly and in locating the discovery and subsequent appraisal wells.

The key to the success in finding the Badak field was to first run a wide grid reconnaissance seismic program covering the entire area. If, instead, partial detail programs had been carried out first, as is often the practice in order to speed up the initiation of the drilling phase, the Badak field may well have been missed since it is located in an area with no discernible surface anomalies and for which piece-meal detail seismic lines probably would not have been programmed. Two or three initial dry holes might have led the Huffco consortium to farming out or relinquishment of the East Kalimantan area, and Badak would have been drilled much later, possibly too late for the 1974 opportunity window in the Japanese LNG market.

The deeper part of the sedimentary section in Badak has not yet been fully explored, and an evaluation of the deeper prospects is in progress.

The discovery of the Badak field in February 1972 gave the impetus to the rapid development of a gas

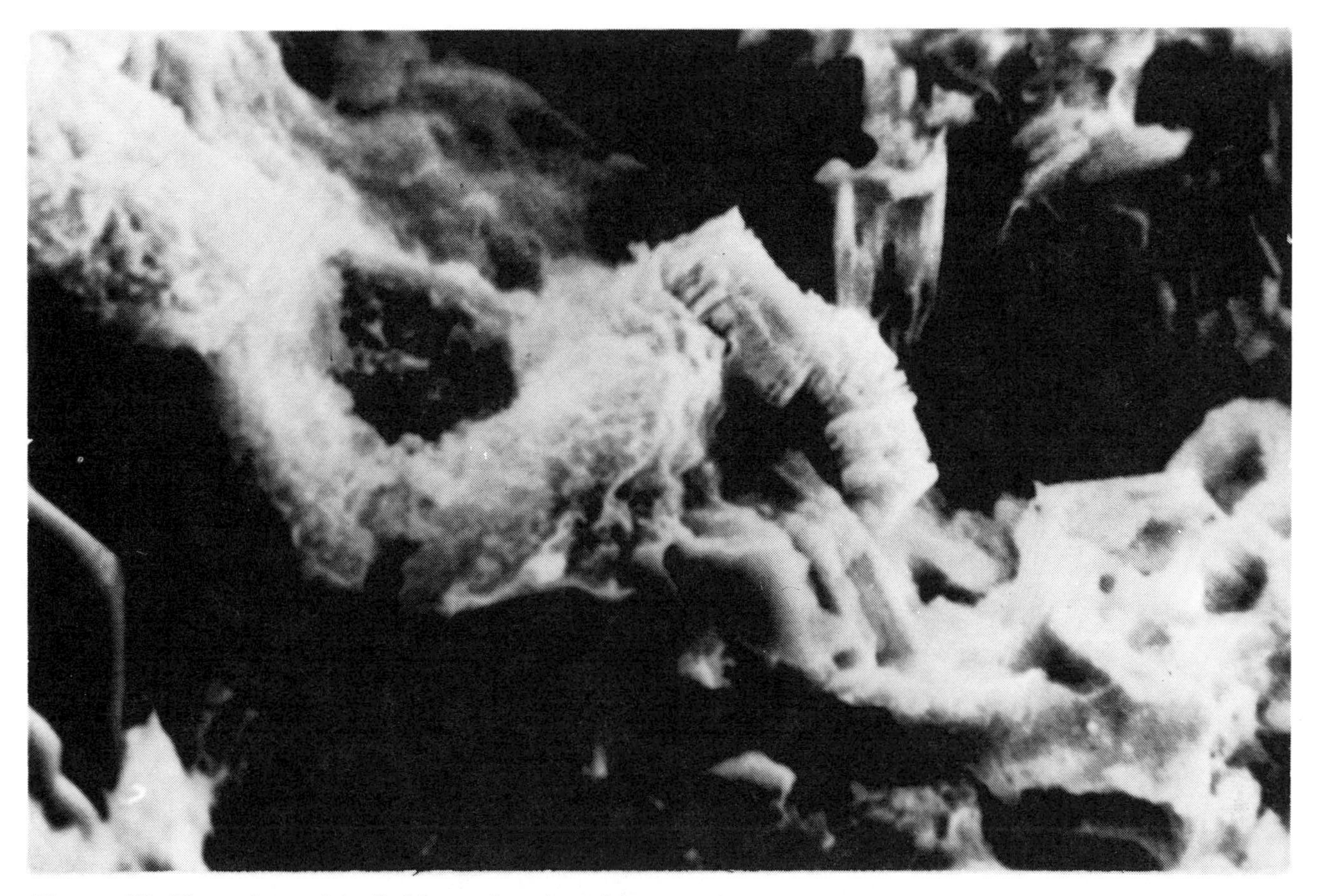

Figure 33. Flow channel in C-22 sand, reduced by kaolinite, illite, and quartz overgrowth. Porosity averages 20%, permeability 180 md. 2000 ×.

industry in Indonesia. Before that time, gas production and utilization had been limited to relatively minor quantities, mainly for feedstock for the initial PUSRI fertilizer plant in Sumatra. By 1980, Indonesia had become the largest producer of LNG in the world, mostly from gas production of the Badak and Arun (North Sumatra) fields.

ACKNOWLEDGMENTS

The authors thank the managements of Pertamina, American Ultramar Ltd., Union Texas Petroleum Company, Inc., and Universe Tankships, Inc., for their permission to publish this paper. Our gratitude goes out to the many members of the Huffco Indonesia Exploration Department who gathered the data, provided illustrations, and offered numerous suggestions and invaluable advice. Mr. Suradi and his staff in Jakarta and Mrs. Susan Bonora in Houston performed all the drafting work, in addition to their very busy normal workloads, and Mrs. Rita Barnett typed the manuscripts and provided general secretarial assistance, for which the authors thank all of them.

REFERENCES

Allen G. P., D. Laurier, and J. Thouvenin, 1979, Etude Sedimentologique du delta de la Mahakam: TOTAL/CFP Notes et Memoires 15.

Billman, H. G., and L. Witoelar Kartaadiputra, 1974, Proceedings of 3rd Annual Convention, Indonesian Petroleum Association, p. 301-310.

Carmalt, S. W., and B. St. John, 1984, Giant oil and gas fields, *in* M. T. Halbouty, ed., Future petroleum provinces of the world: American Association of Petroleum Geologists Memoir 40, p. 11-54.

Chapman, R. E., 1986, Geological reasoning for the petroleum source rocks of known fields: APEA Journal, p. 132-141.

Combaz, A., and M. de Matharel, 1978, Organic sedimentation and genesis of petroleum in Mahakam Delta, Borneo: American Association of Petroleum Geologists Bulletin, v. 62, n. 9, p. 1684-1695.

Durand, B., and J. L. Oudin, 1980, Exemple de migration des hydrocarbures dans une serie deltaique: le delta de la Mahakam, Kalimantan, Indonesie: Proceedings of 10th World Petroleum Congress, p. 3-11.

Durand, B., and M. Paratte, 1974, Oil potential of coals: a geochemical approach, *in* J. Brooks, ed., Petroleum geochem-

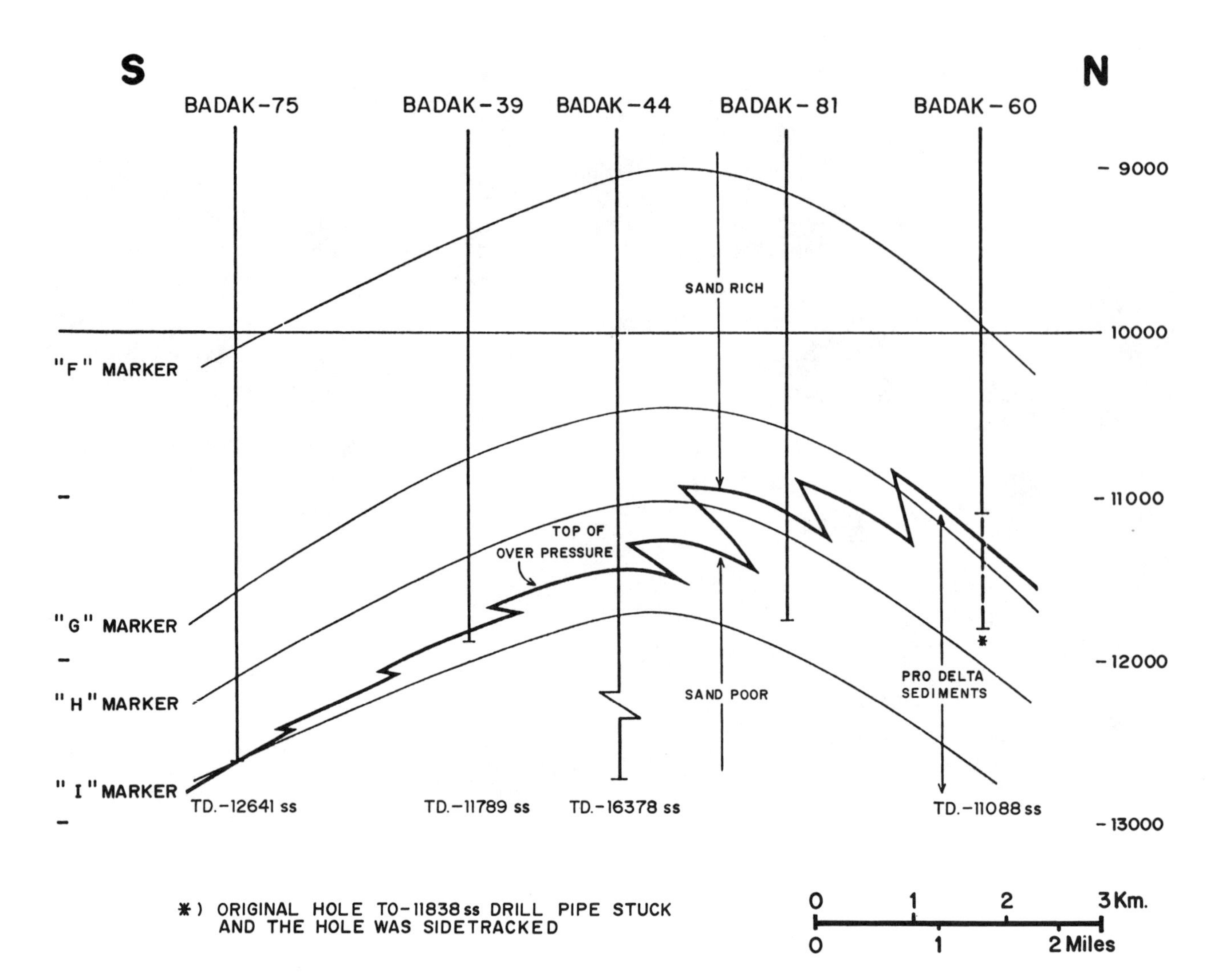

Figure 34. Cross section showing the discordance between the stratigraphy and the top of the overpressure in Badak field.

istry and exploration of Europe: Blackwell Scientific Publishers, p. 255-265.

Durand, B., P. Ungerer, A. Chiarelli, and J. L. Oudin, 1983, Modelisation de la migration de l'huile, applications a deux exemples de bassins sedimentaires: Proceedings of 11th World Petroleum Congress, p. 3-15.

Gerard, J., and H. Oesterle, 1973, Facies study of the offshore Mahakam area: Proceedings of 2nd Annual Convention, Indonesian Petroleum Association, p. 187-194.

Gwinn, J. W., H. M. Helmig, and L. Witoelar Kartaadiputra, 1974, Geology of the Badak Field: Proceedings of 3rd Annual Convention, Indonesian Petroleum Association, p. 311-331.

Hamilton, W., 1979, Tectonics of the Indonesian Region: U.S. Geological Survey Professional Paper No. 1078.

Huffington, R. M., and H. M. Helmig, 1980, The discovery and development of the Badak Field: American Association of Petroleum Geologists Memoir 30, p. 441-458.

Katili, J. A., 1978, Past and present geotectonic position of Sulawesi, Indonesia: Tectonophysics, v. 45, p. 289-322.

Magnier, P., T. K. Oki, and L. Witoelar Kartaadiputra, 1975, The Mahakam Delta, Kalimantan, Indonesia: 9th World Petroleum Congress, v. 2, p. 239-250.

Marks, E., H. Dhanutirto, T. Ismoyowati, B. Sidik, Sujatmiko, and L. Samuel, 1982, Cenozoic stratigraphic nomenclature in East Kutai Basin, Kalimantan: Proceedings of IPA 11th Annual Convention, p. 147-179.

Oudin, J. L., and P. F. Picard, 1982, Genesis of hydrocarbons in the Mahakam Delta and the relationship between their distribution and the overpressured zones: Proceedings of IPA 11th Annual Convention, p. 181-202.

Panigoro, H., 1983, Petrographic characteristics of Badak and Nilam Fields' sandstone reservoirs: Proceedings of IPA 12th Annual Convention, p. 191-206.

Pertamina, 1985, Hands across the sea.

Rose, R., and P. Hartono, 1978, Geological evolution of the Tertiary Kutai-Melawi Basin, Kalimantan, Indonesia: Proceedings of IPA 7th Annual Convention, p. 225-251.

Samuel, L., and S. Muchsin, 1975, Stratigraphy and sedimentation in the Kutai Basin, E. Kalimantan: Proceedings of IPA 4th Annual Convention, v. 2, p. 27-39.

Schoell, M., M. Teschner, H. Wehner, B. Durand, and J. L. Oudin, 1983, Maturity related biomarker and stable isotope variations and their application to oil/source rock correlation in the Mahakam, Kalimantan, *in* Advances in Organic Geochemistry 1981: John Wiley & Sons Ltd., p. 156-163.

Schoell, M., B. Durand, and J. L. Oudin, 1985, Migration of oil and gas in the Mahakam Delta, Kalimantan: evidence and quantitative estimate from isotope and biomarker studies:

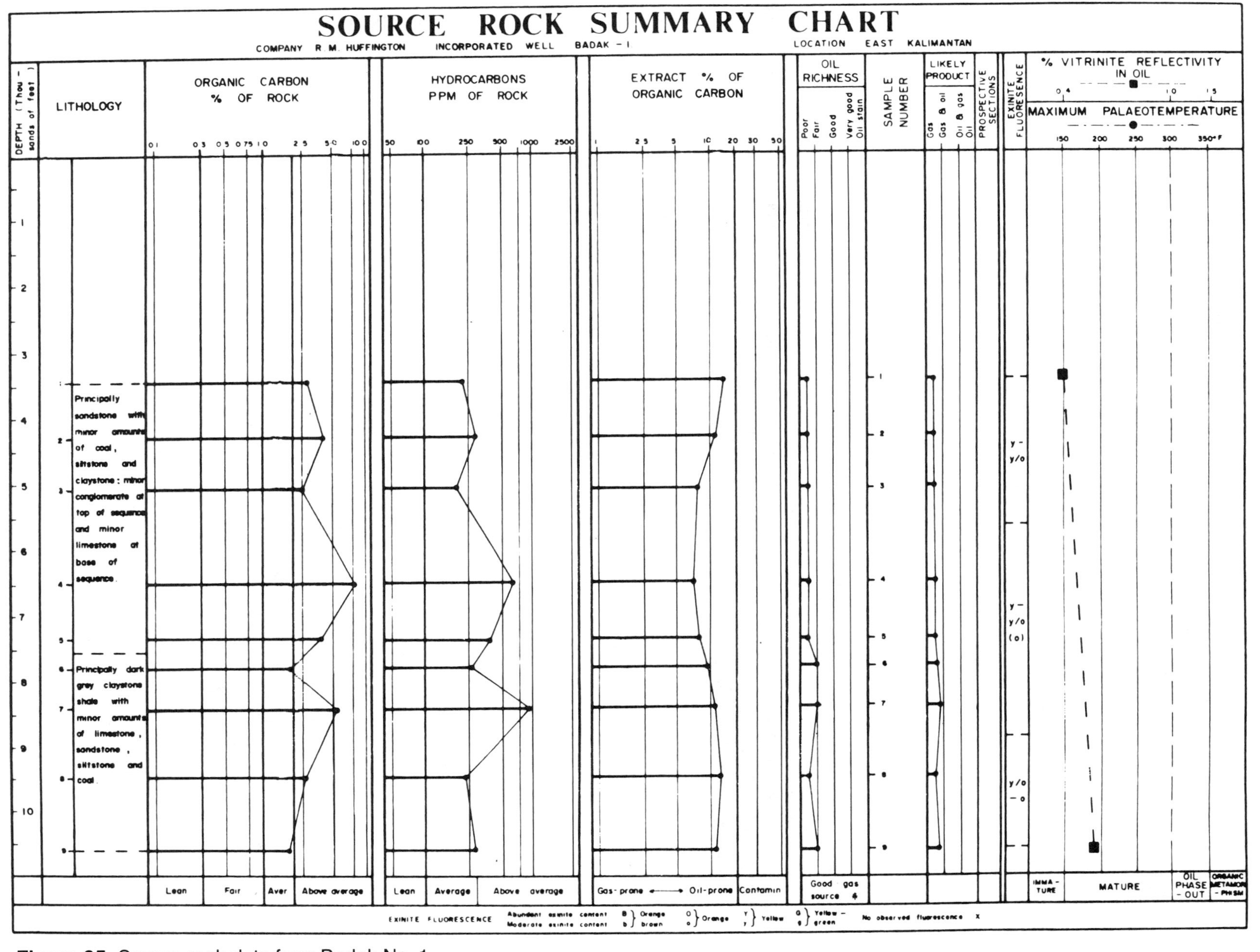

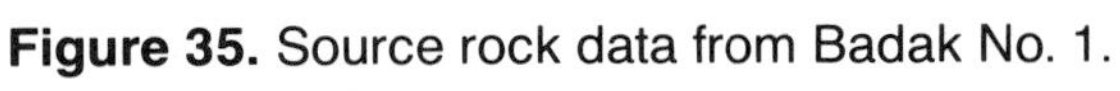
Figure 35. Source rock data from Badak No. 1.

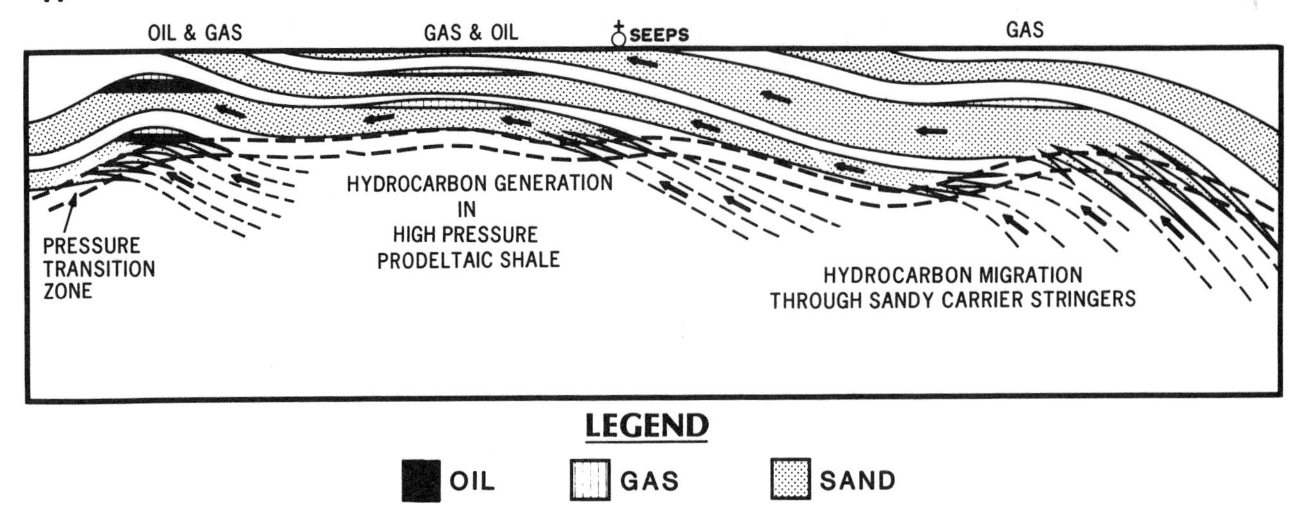

Figure 36. Schematic model of hydrocarbon migration in the Mahakam delta area.

Proceedings of 14th Annual Convention, Indonesian Petroleum Association, p. 49-56.

Schwartz, C. M., G. H. Laughbaum, B. S. Samsu, and J. D. Armstrong, 1973, Geology of the Attaka oil field, East Kalimantan, Indonesia: Proceedings of 2nd Annual Convention, Indonesian Petroleum Association, p. 195-215.

Tissot, B. P., and D. H. Welte, 1984, Petroleum formation and occurrence: Springer-Verlag.

Ungerer, P., F. Bessis, P. Y. Chenet, B. Durand, E. Nogaret, A. Chiarelli, J. L. Oudin, and J. F. Perrin, 1984, Geological and geochemical models in oil exploration; principles and practical examples. American Association of Petroleum Geologists Memoir 35, p. 53-77.

Vandenbroucke, M., B. Durand, and J. L. Oudin, 1983, Detecting migration phenomena in a geological series by means of C1-C35 hydrocarbon amounts and distributions, *in* Advances in Organic Geochemistry 1981: John Wiley & Sons Ltd., p. 147-155.

Weeda, J., 1958, Oil basin of East Borneo, *in* Habitat of oil: American Association of Petroleum Geologists Symposium, p. 1337-1346.

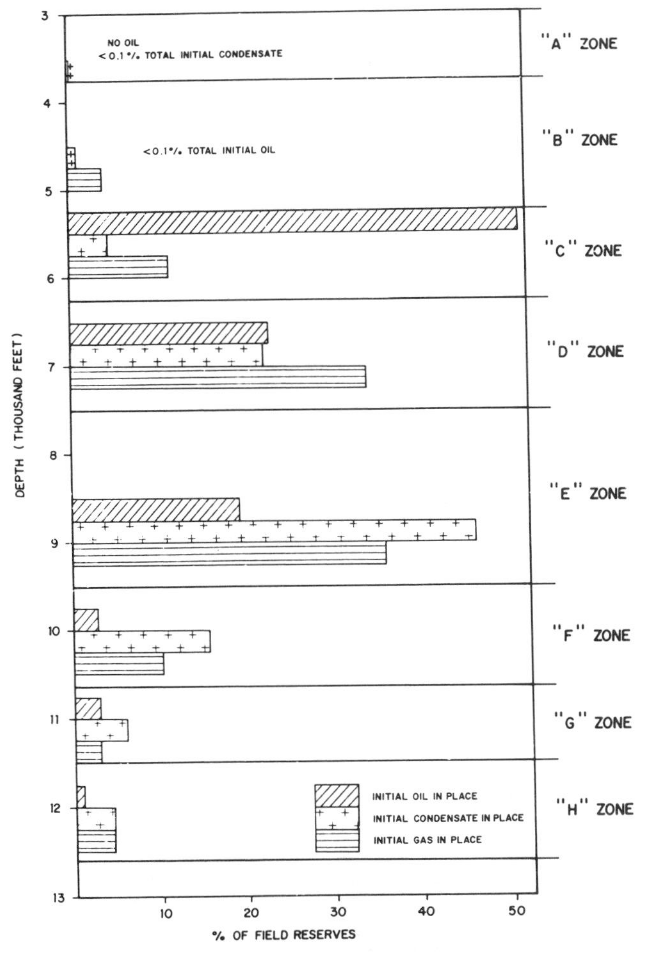

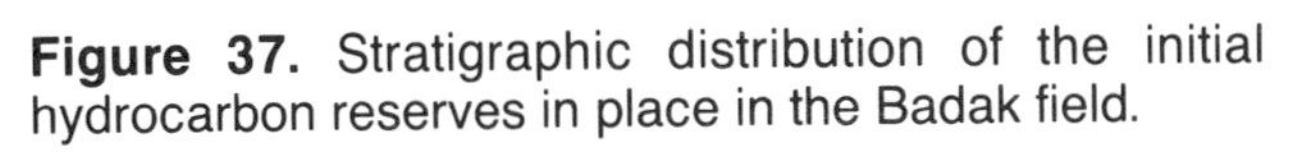

Figure 37. Stratigraphic distribution of the initial hydrocarbon reserves in place in the Badak field.

Table 1. Crude oil analysis

Specific gravity at 60°F		0.8318
A.P.I. gravity at 60°F		38.6
Pour point	°F	20
Kinematic viscosity at 100°F	cS	2.01
Kinematic viscosity at 122°F	cS	1.85
Redwood I viscosity at 100°F	secs	31.0
Engler viscosity at 50°C	°E	1.10
Water content	vol%	2.2
Flash point Abel	°F	<50
R.V.P. at 100°F	psi	2.2
Sulfur content	wt%	0.05
Wax content on crude	wt%	3.3
Congealing point of wax	°C	54.0
Water and sediment	vol%	2.7
A.S.T.M. Distillation		—
Initial boiling point	°C	68
10% vol. rec. at.	°C	124
20% vol. rec. at.	°C	149
30% vol. rec. at.	°C	180
40% vol. rec. at.	°C	212
50% vol. rec. at.	°C	241
60% vol. rec. at.	°C	264
70% vol. rec. at.	°C	286
80% vol. rec. at.	°C	320
90% vol. rec. at.	°C	
Recovery at 100°C	vol%	5
Recovery at 150°C	vol%	19
Recovery at 200°C	vol%	36
Recovery at 250°C	vol%	53
Recovery at 300°C	vol%	74
Recovery at 350°C	vol%	86.5
Recovery at 371°C	vol%	88
Distillate to 175°C	vol%.	28.0
Specific gravity 60/60°F		0.7673
Aromatic content	vol%	22.1
Distillate Fraction 175-300°C		47.0
Specific gravity 60/60°F		0.8418
Aromatic content	vol%	28
Smoke point	m.m.	14
Distillate Fraction 300-350°C	vol%	11.5
Specific gravity 60/60°F		0.8728
Pour point	°F	70
Diesel index		46
Kinematic viscosity at 100°F	cS	6.67
Redwood I viscosity at 100°F	secs	42.5
Residue above 350°C	vol%	13.5
Specific gravity 60/60°F		0.9286
Pour point	°F	115
Kinematic viscosity at 140°F	cS	79.3
Redwood I viscosity at 140°F	secs	321.2
Wax content on residue	wt%	22.2
Congealing point of wax	°C	56.0
Sulfur content	wt%	0.10

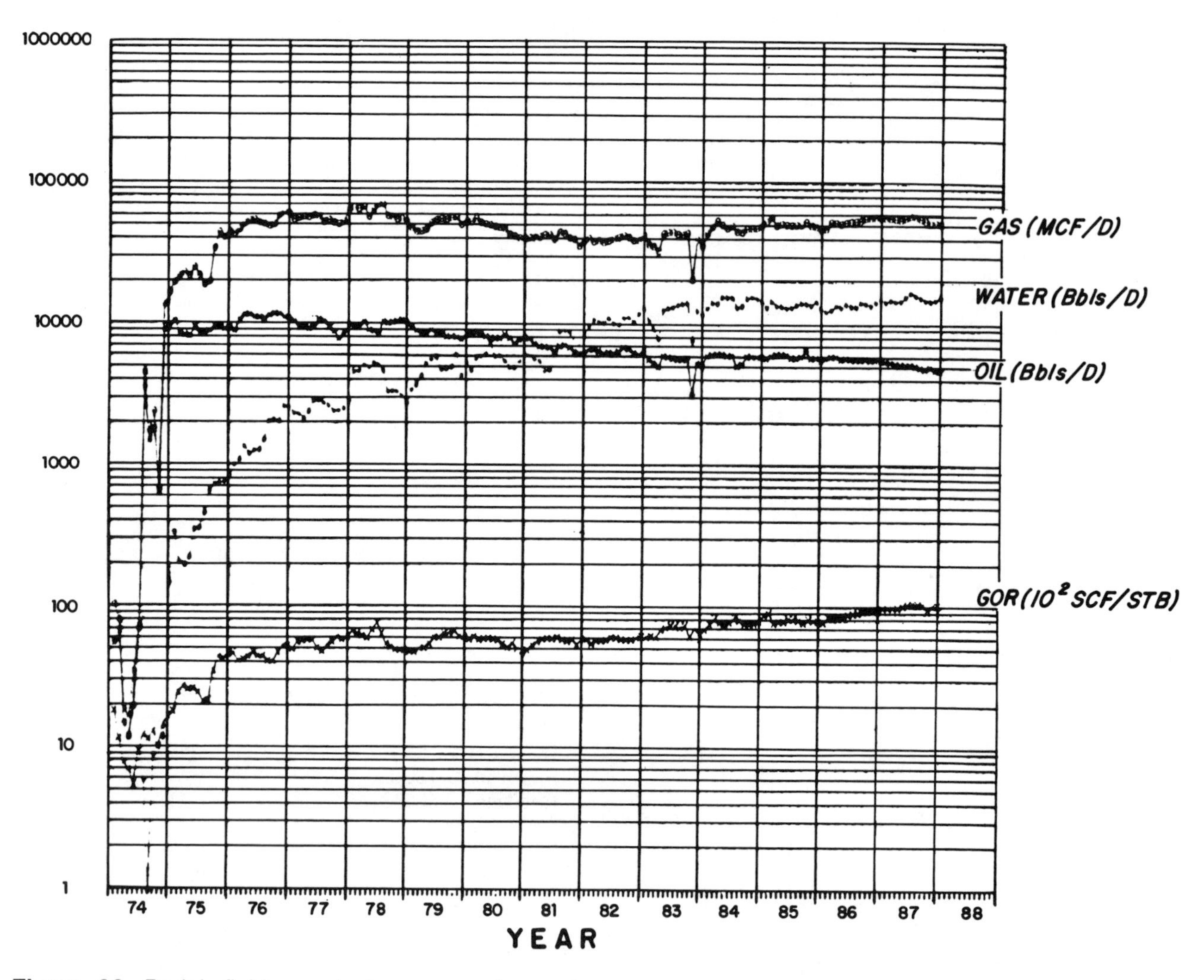

Figure 38. Badak fluids production curves from oil completions.

Appendix 1. Field Description

Field name .... *Badak field*

Ultimate recoverable reserves

Field location:

- **Country** .... *Indonesia*
- **State**
- **Basin/Province** .... *Kutai/East Kalimantan*

Field discovery:

- **Year first pay discovered** .... *Middle to late Miocene Balikpapan sandstones (A–F)* *1972*
- **Year second pay discovered** .... *Middle Miocene sandstones (G and H)* *1972*
- **Third pay** .... *Middle Miocene sandstones (I and J)* *1978*

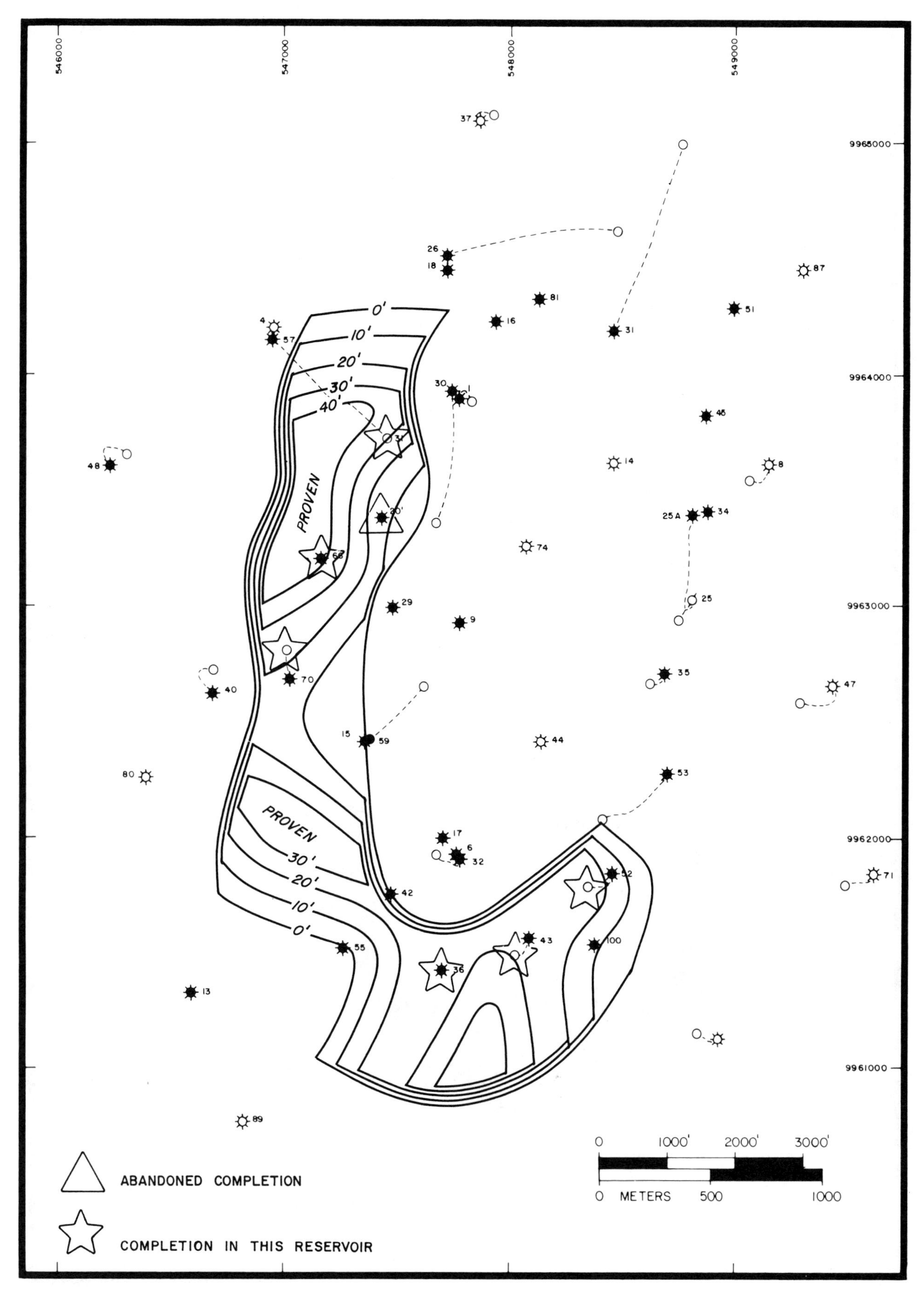

Figure 39. Distribution of the C-20 oil reservoir.

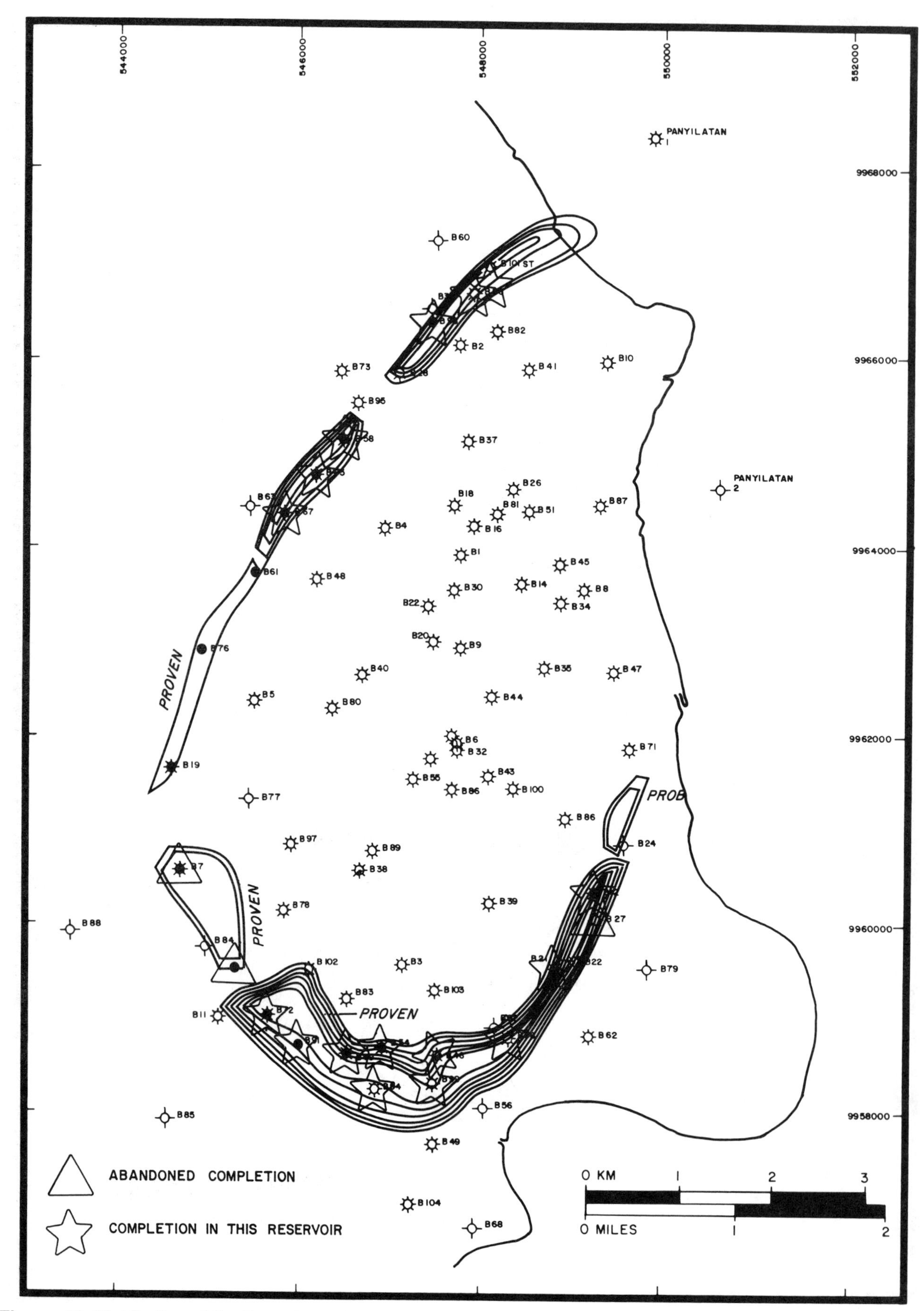

Figure 40. Distribution of the D-6 oil reservoir.

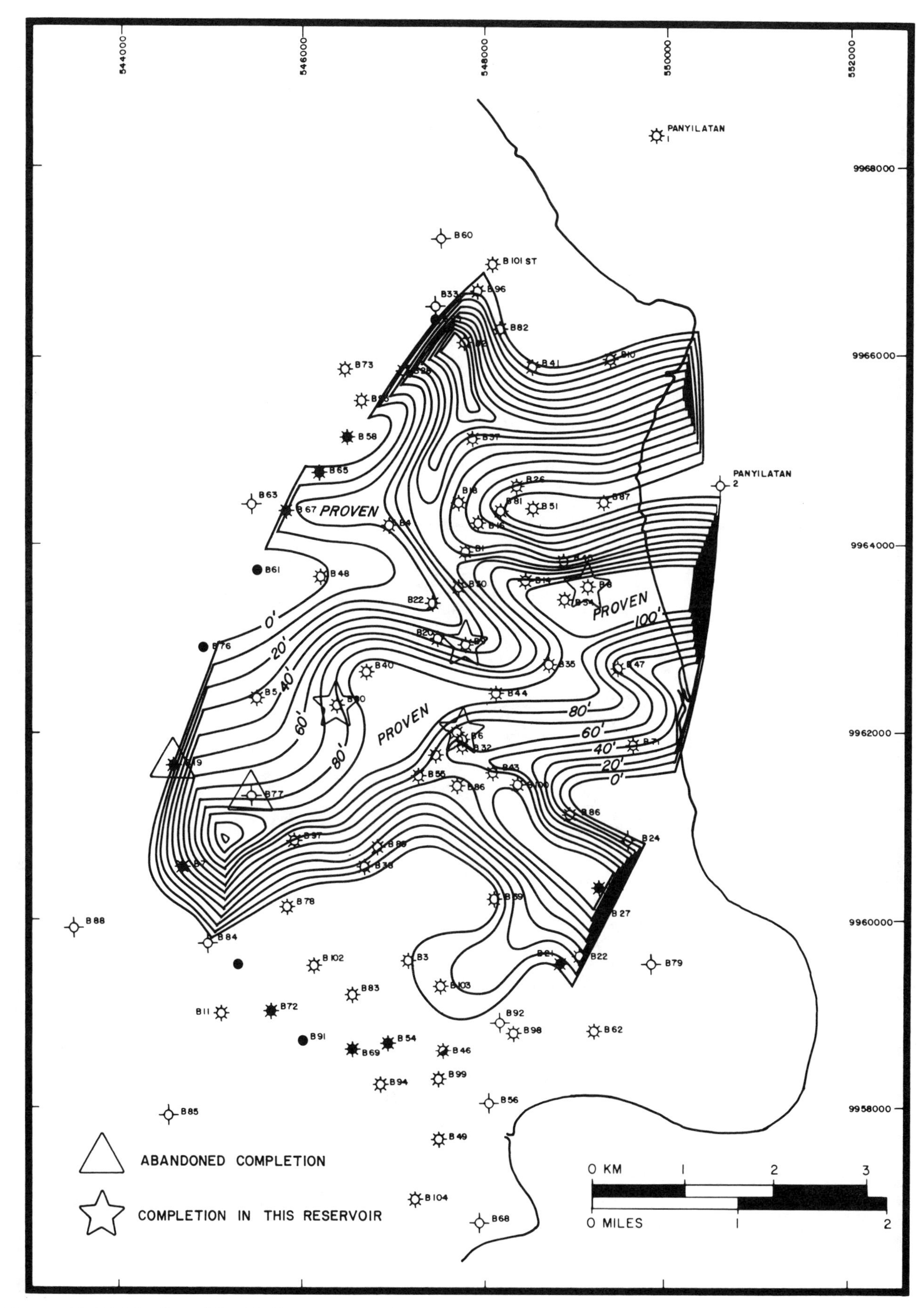

Figure 41. Distribution of the E-11 gas reservoir.

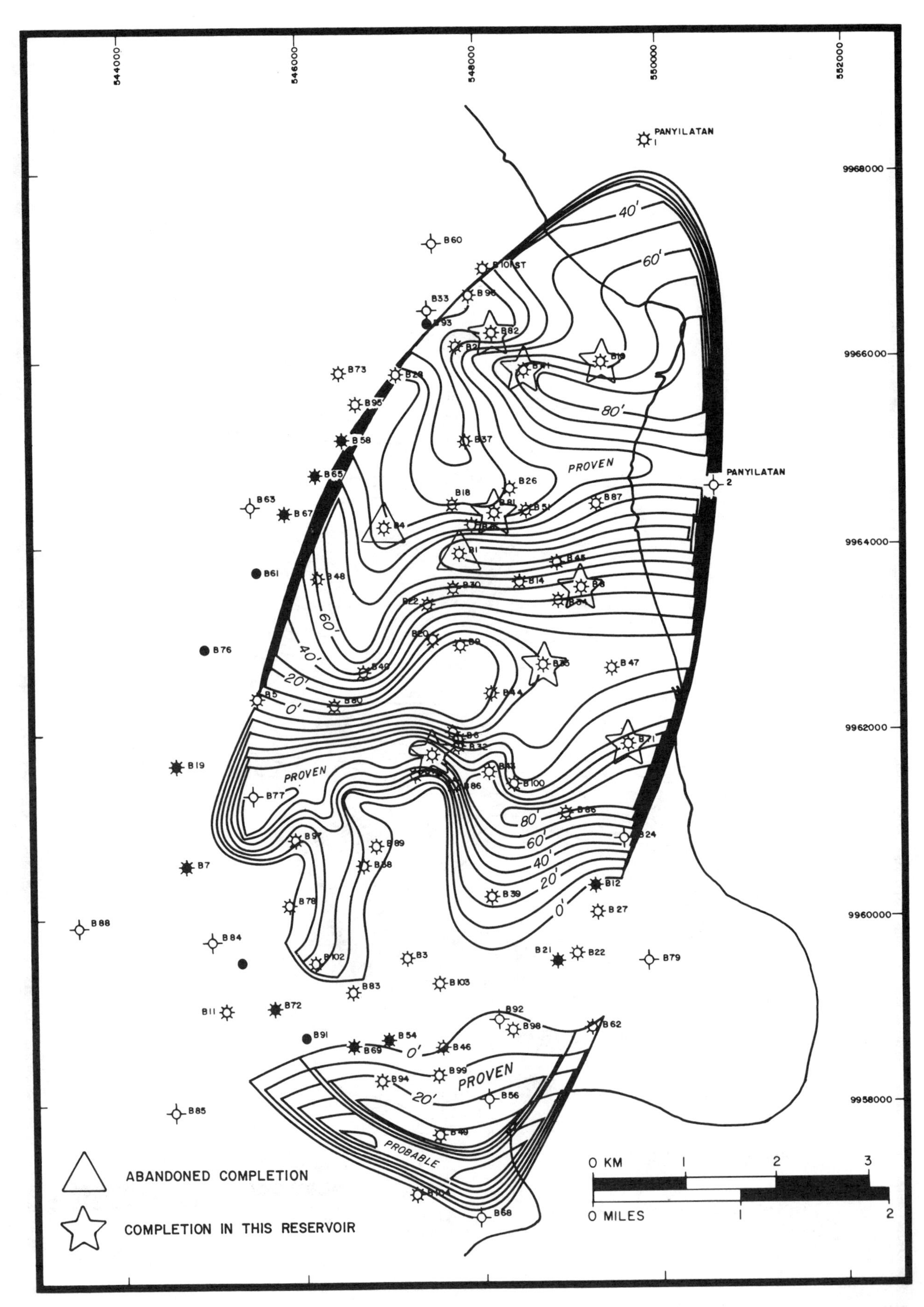

Figure 42. Distribution of the E-21 gas reservoir.

Discovery well name and general location:

First pay .... *Badak #1 coordinates: 117°25′45.7″E, 00°19′36.0″S*
Second pay .... *Badak #6 coordinates: 117°25′45.4″E, 00°20′40.3″S*
Third pay .... *Badak #44 coordinates: 117°25′57.3″E, 00°20′24.3″S*

Discovery well operator .... *Roy M. Huffington, Inc.*
Second pay .... *NA*
Third pay .... *NA*

IP in barrels per day and/or cubic feet or cubic meters per day:

First pay .... *Badak #6, DST, 7-12-81, 8148/8192, E-11, 25.396 MMscf, 566.8 BCPD*
Second pay .... *Badak #6, DST, 21-1-73, 11104/124, G-8, 10.151 MMscf, 324 BCPD*
Third pay .... *Not available*

All other zones with shows of oil and gas in the field:

Age	Formation	Type of Show
Middle and late Miocene	*Balikpapan*	*Oil and gas*

Geologic concept leading to discovery and method or methods used to delineate prospect, e.g., surface geology, subsurface geology, seeps, magnetic data, gravity data, seismic data, seismic refraction, nontechnical:

Neogene delta with previously discovered oil fields. Aerial photographic and magnetometric surveys, surface geology, and reflection seismic survey. Some geomorphologic evidence aided in accurately locating initial seismic lines.

Structure:

Province/basin type .... *Mahakam delta/continental margin*

Tectonic history

The Kutai basin was formed by the subsidence of a Late Cretaceous erosional surface. The original Eocene basin far exceeded the present-day limits of the Kutai basin, which latter were the result of more recent (late Pliocene) tectonic events. The formation of the Makassar Straits oceanic basin and the drifting apart of Sulawesi from Kalimantan probably started in the same late Pliocene period. There appears to have been a progradation of depocenters from west (Eocene) to east (Recent) in the Kutai basin so that the maximum thicknesses of the successive stratigraphic units are not stacked in any one place. The maximum thickness of the sedimentary column probably does not exceed 10,000 m (32,000 ft).

Regional structure

The regional structural geology around the Badak field is characterized by a series of north-northeast-trending linear anticlines. The Badak anticline is the northernmost culmination of an 85 km (53 mi) long anticlinal structure that might be called the Badak-Handil anticlinorium.

Local structure

Badak structure is a fairly symmetrical north-trending anticline, gently folded with dips up to 8°. Measured on the lowest closing structural contour, the anticline is about 15 km long in a north-northeast direction and 6 km wide in an east-southeast direction.

Trap:

Trap type(s)

The Badak oil and gas accumulation is generally trapped in a gentle, unfaulted anticline, although within the general trap there are a multitude of stratigraphic trap variations as usually found in deltaic environments. There are more than 100 individual reservoirs, which include anticlinal traps, updip pinchouts, sand bodies of various types forming lensoid traps, etc.

Basin stratigraphy (major stratigraphic intervals from surface to deepest penetration in field):

Chronostratigraphy	Formation	Depth to Top in ft (m)
Late Miocene	*Kampung Baru*	*Top eroded*
Middle-late Miocene	*Balikpapan*	*3200 (1000)*
Middle Miocene	*Pulau Balang*	*Probably not reached*

Location of well in field

Reservoir characteristics:

- **Number of reservoirs** ... *151*
- **Formations** ... *Balikpapan*
- **Ages** ... *Middle–late Miocene*
- **Depths to tops of reservoirs** ... *Shallowest: 3498 ft, A.19 in B #35; deepest: 12,704 ft, H.13 in B #75*
- **Gross thickness (top to bottom of producing interval)** ... *9206 ft*
- **Net thickness—total thickness of producing zones**
  - **Average** ... *515 ft (avg. from B #1, B #6, B #68, B #71, B #77)*
  - **Maximum** ... *1009 ft (in Badak #1)*
  - **Average**
  - **Maximum**
- **Lithology** ... *Fine- to medium-grained moderately well-sorted quartz sandstone*
- **Porosity type** ... *Interparticle porosity, partially filled with varying amounts of authigenic clay, detrital clay, and secondary quartz*
- **Average porosity** ... *25% (weighted average)*
- **Average permeability** ... *100 md in major reservoirs*

Seals:

- **Upper**
  - **Formation, fault, or other feature** ... *Balikpapan Formation*
  - **Lithology** ... *Shale*
- **Lateral**
  - **Formation, fault, or other feature** ... *Balikpapan*
  - **Lithology** ... *Shale*

Source:

- **Formation and age** ... *Balikpapan Formation*
- **Lithology** ... *Shale and coal*
- **Average total organic carbon (TOC)** ... *3.6%*
- **Maximum TOC** ... *7.6%*
- **Kerogen type (I, II, or III)** ... *Predominantly type III*
- **Vitrinite reflectance (maturation)** ... $R_o = 0.8$ *@ 14,400 ft (4390 m)*
- **Time of hydrocarbon expulsion**
- **Present depth to top of source** ... *3000 ft (915 m) (oil window below 12,000 ft)*
- **Thickness** ... *12,000 ft (3659 m)*
- **Potential yield**

Appendix 2. Production Data

Field name ... *Badak field*

Field size:

- **Proved acres** ... *15,100 acres*
- **Number of wells all years** ... *107*
- **Current number of wells** ... *107*
- **Well spacing** ... *160 to 640 acres*

Ultimate recoverable .... *6512 bcf*
Cumulative production .... *1981 bcf*
Annual production .... *220 bcf + 2.650 MMSTB condensate*
Present decline rate *17%*
Initial decline rate *20%*
Overall decline rate *18.5%*
Annual water production
In place, total reserves .... *7983 bcf*
In place, per acre-foot .... *1.026 MMCF + 12 bbl condensate/ac-ft*
Primary recovery .... *6500 bcf*
Secondary recovery .... *NA*
Enhanced recovery .... *NA*
Cumulative water production

Drilling and casing practices:

Amount of surface casing set .... *2000-3000 ft*
Casing program
13⅜-in. surface (2000-3000 ft); 9⅝-in. surface (10,000-12,000 ft); 7-in. (9⅝-in. casing - TD)
Drilling mud .... *Water-base mud (surface - 9⅝-in. casing); Invermul (9⅝-in. casing - TD)*
Bit program .... *26-in., 17½-in., 12¼-in., 8½-in.*
High pressure zones .... *Below productive zone (I_2), up to 0.908 psi/ft or 17.45 ppg (in H-27 RFT test at Badak #44)*

Completion practices:

Interval(s) perforated .... *Multiple zones (312 reservoirs)*
Well treatment .... *Fracturing for tight sand*

Formation evaluation:

Logging suites .... *Induction electrical log (ISF/MSFL/SR); FDC/CNL/GR*
Testing practices .... *FIT, RFT, DST, and production test*
Mud logging techniques .... *Drilling rate, gas detector/gas chromatography, sample analysis, temperature in/out, mud density in/out, d'exponent techniques, etc.*

Oil characteristics:

Type .... *Paraffinic-naphthenic*
API gravity .... *30-39°*
Base .... *Intermediate-intermediate*
Initial GOR *400-800*
Sulfur, wt% *0.09*
Viscosity, SUS .... *36 at 100°F*
Pour point .... *+40°F*
Gas-oil distillate

Field characteristics:

Average elevation .... *10.7 m*
Initial pressure .... *1600-6000 psi*
Present pressure .... *1200-5400 psi*
Pressure gradient
Temperature .... *160-280°F*
Geothermal gradient .... *0.015°F/ft in Badak #1*
Drive .... *Solution gas and water drive*
Oil column thickness .... *84 ft in Badak #1*
Oil-water contacts .... *5200-11,840 ft*
Connate water .... *(SW) 10.5-57.3%*
Water salinity, TDS .... *6000-14,000 ppm NaCl*

Resistivity of water .... *0.14–0.78 ohm-m*
Bulk volume water (%) .... *Water saturation variable*

Transportation method and market for oil and gas:
Gas—pipeline to Bontang LNG Plant; oil—pipeline to Santan oil terminal

Urengoy Gas Field—U.S.S.R.
West Siberian Basin, Tyumen District

JOHN D. GRACE
ARCO Oil and Gas Company
Plano, Texas

GEORGE F. HART
Louisiana State University
Baton Rouge, Louisiana

FIELD CLASSIFICATION

BASIN: West Siberia
BASIN TYPE: Complex
RESERVOIR ROCK TYPE: Sandstone
RESERVOIR ENVIRONMENT OF DEPOSITION: Fluvial and Nearshore Marine
RESERVOIR AGE: Cretaceous
PETROLEUM TYPE: Oil and Gas
TRAP TYPE: Anticline

LOCATION

Urengoy is the largest gas field in the world. It straddles the Arctic Circle in the northern part of the West Siberian basin of the U.S.S.R. The field has an estimated ultimate recovery of 335.44 trillion cubic feet (9.5 tcm) of natural gas, at least as much as 1.5 billion barrels (202 mm t) of condensate and 1.2 billion barrels (170 mm t) of oil, for a total of 58.6 billion barrels of oil equivalent (Zap. Sib. NIGNI, 1989).

The West Siberian basin is the largest sedimentary basin in the world, occupying a 1.31 million mi^2 (3.4 million km^2) lowland. It is bounded on the west by the Ural Mountains, on the east by the Siberian platform, on the south by the Kazakhistan highlands, and is open to the Kara Sea in the north (Figure 1). In discovered hydrocarbon resources, it is second in the world behind the Arabian platform. Oil fields are concentrated in the central and southern areas of the basin, primarily along the Ob and Irtysh rivers. Gas fields are found principally north of lat.64°N.

Urengoy is the largest of more than 70 gas and gas-condensate fields discovered since 1962 in the northern part of the West Siberian basin (Figure 2). The gas-bearing northern part of the basin has an area of 366,797 mi^2 (950,000 km^2). Collectively, the fields in this area contain approximately one-third of the world's discovered gas resources, 1137 tcf (32.2 tcm) of proved and probable reserves (Grace and Hart, 1986).[1] The fields share common hydrocarbon sources and character, have similar trapping mechanisms, and contain reservoirs within the same stratigraphic units.

The Urengoy field is remote, more than 400 mi (660 km) north of the main oil production center near the cities of Surgut and Nizhnevartovsk in the Middle Ob area. Winter temperatures average below -30°F (-34°C). Though the field area is overlain by thick permafrost, seasonal melting of the surface forms a tundra of swamps, marshes, and lakes.

HISTORY

Pre-Discovery

I. M. Gubkin, a leading Soviet geologist, first postulated the hydrocarbon potential of the West Siberian basin in the early 1930s (Vinogradov, 1972).

[1]Meyerhoff (1983) defined Soviet reserve categories in English. He stated that Soviet "A" and "B" category reserves are "proved" by Western standards. The difference between the two is that "A" is under production, and "B" is not. Categories "C_1" and "C_2" are "probable" to "inferred" in a Western sense. Here the word "reserves" refers to the sum of "A"+"B"+"C_1" and "resources" refers to reserves, plus the "C_2" category. Estimated ultimate recovery is the sum of cumulative production plus "resources." Gas volumes are measured at 68°F (20°C).

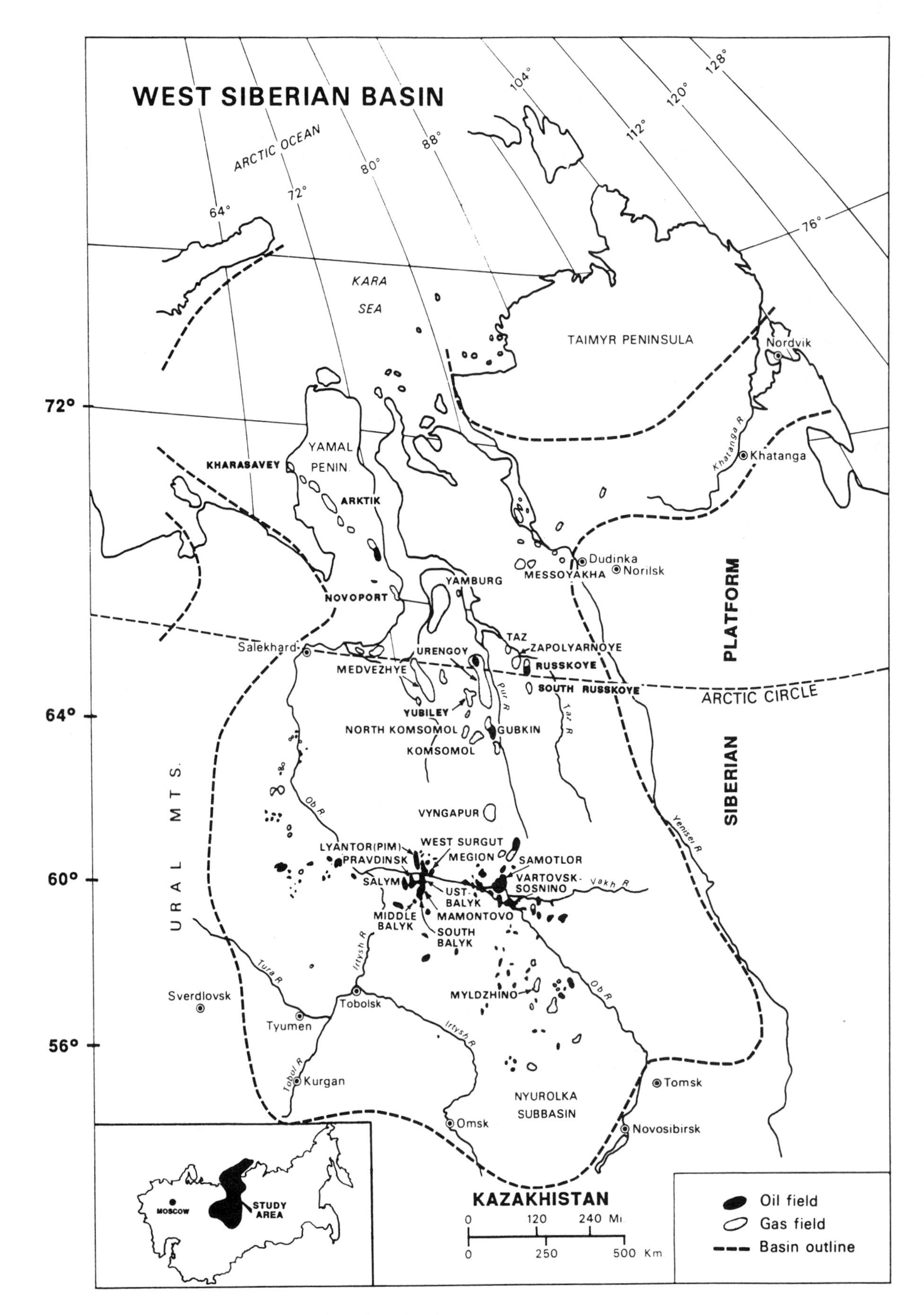

Figure 1. Location map of West Siberian basin. (Modified from Grace and Hart, 1986.)

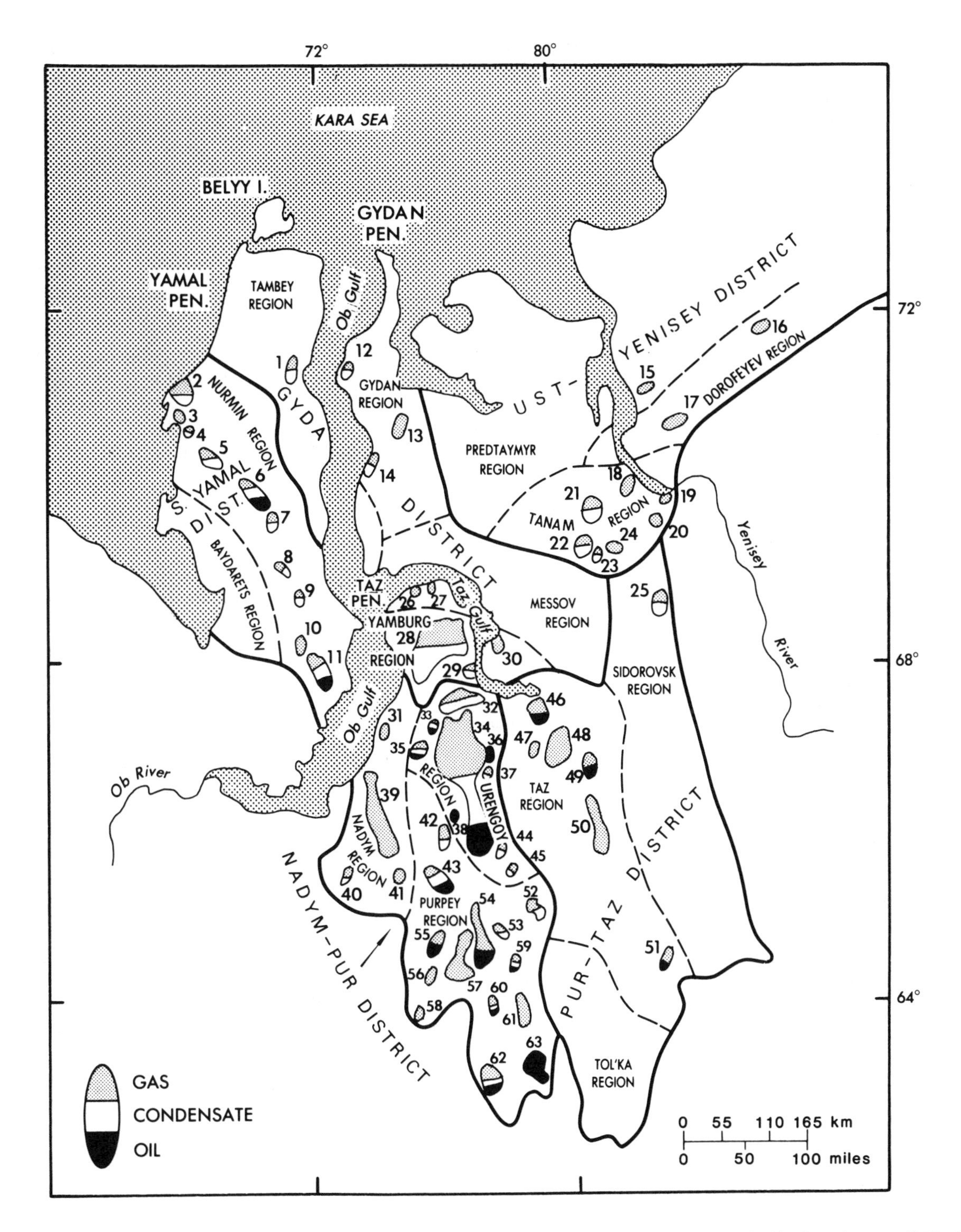

Figure 2. Fields of the northern West Siberian basin with boundaries of the Petroliferous Districts and Regions (Grace and Hart, 1986). Legend: (1) South Tambey, (2) Kharasavey, (3) Kruzenshten, (4) South Kruzenshten, (5) Bovanenko, (6) Neyto, (7) Arktik, (10) Maloyamal, (11) Novyy Port, (12) Utro, (13) Gydan, (14) Geofizika, (15) Deryabin, (16) Dzhangod, (17) Ozero, (18) Kazantsev, (19) Nizhnekheta, (20) Zima, (21) Pelyatka, (22) North Solenin, (23) South Solenin, (24) Messoyakha, (25) Suzun, (26) Semakov, (27) Antipayuta, (28) Yamburg, (29) Nakhodka, (30) Yurkharov, (31) Kharvuta, (32) North Urengoy, (33) Yanyakha, (34) Urengoy, (35) Pestsov, (36) Samburg, (37) South Samburg, (38) Noyabr', (39) Medvezh'ye, (40) Nadym, (41) Pangody, (42) Yubiley, (43) Yamsovey, (44) East Urengoy, (45) Pyrey, (46) Taz, (47) West Zapolyarnoye, (48) Zapolyarnoye, (49) Russkoye, (50) South Russkoye, (51) Ust-Chasel'ka, (52) East Tarkosalye, (53) West Tarkosalye, (54) Gubkin, (55) North Komsomol, (56) Verkhnepur, (57) Komsomol, (58) Muravlenko, (59) Tarasov (also known as Ayvasedopur?), (60) Vengayakha, (61) Vengapur, (63) Yarayner. Not shown are approximately ten fields discovered since 1985 for which no precise location data are available.

At that time, the major source of Soviet oil production was from the Baku area of Azerbaidzhan, on the southwest coast of the Caspian Sea. Most of the gas production came from the Dnepr-Donets basin in the Ukraine, but it was not exploited on a national basis.

World War II, distance from consumption centers, and the adequacy of supply from the new Volga-Ural discoveries delayed exploratory activity in West Siberia until 1948. The first oil discovery in the basin was in 1952, the first gas in 1953, both along the eastern margin of the Urals. The large oil discoveries in the Middle Ob area began with the Megion field in 1961 (Kontorovich et al., 1975). The 1962 discovery of the Taz field at the mouth of the Taz River demonstrated the gas potential of the northern West Siberian basin (Figures 1 and 2).

Discovery

Urengoy was the fourth major gas discovery in the Nadym-Pur and Pur-Taz petroliferous districts and the seventh field found in the northern part of the basin (Figure 2). It was discovered by an agency of the Ministry of Geology of the Russian Soviet Federated Socialist Republic, one of the 15 Union Republics of the U.S.S.R. This agency has become Glavtyumengeologiya, part of the national U.S.S.R. Ministry of Geology. Urengoy was named for a local village.

The field was discovered by the Urengoy #2R well, completed in early 1966 on the southern uplift of the field. It penetrated a 292 ft (89 m) gas column in the Cenomanian Pokur Formation. The following year, the Urengoy #1 well drilled the saddle between the southern and central uplifts. It demonstrated the productive potential of the Neocomian clastic section with successful tests of the BU_{14} reservoir (39 mmcf/day [1.1 mcm/day]) and an oil show in the BU_{16} reservoir, both in the Megion Formation (Tikhomirov et al., 1978).

Post-Discovery

After an extensive reflection seismic program, delineation drilling proceeded in two stages. The first stage, from 1969 to 1971, outlined the massive dry gas Pokur Formation $PK_{1\text{-}6}$ complex. The second, begun in 1972, delineated the Neocomian section and initially concentrated on the northern uplift (Tikhomirov et al., 1978). Commercial production from Urengoy began on 22 April 1978. The historical and projected production from Urengoy, in comparison with the totals for the Yamburg field, West Siberian basin, and total Soviet production is shown in Figure 3. The field provided the incremental increase in Soviet gas production in the 1980s and made that country the leading gas producer in the world in 1984 (Chernomyrdin, 1988).

First commercial production was from the $PK_{1\text{-}6}$ reservoir complex. By the end of 1978, there were 100 production wells in the field (Berman et al., 1979). In 1979, the Urengoy Gas Production Enterprise completed 37 more wells and in the following two years added approximately 100 more. Wells have typically been drilled vertically, four to a cluster, and have an average daily production of 35.3 mmcf/day (1 mmcm/day).

Commercial exploitation of the condensate-rich Neocomian reservoirs began in 1984 (Geresh, 1984). That year the U.S.S.R. Ministry of the Gas Industry announced plans for producing the Neocomian horizons involving several hundred wells, at least four major gas separation plants, and a 373 mi (600 km) liquid and gas pipeline collection system (Chernomyrdin, 1984; Yazik, 1984). By the end of 1988, there were probably more than 750 wells producing from both Neocomian and Cenomanian horizons. Annual peak production from the field may reach more than 11.72 tcf/yr (332 bcm/yr) in the 1990s (Oil and Gas Journal, 1987c). This will require more than 1000 production wells.

The condensate content of the Neocomian reservoirs of the Urengoy field is estimated to be at least 4.1 billion bbl in place.[2] Drilling crews completed 10 of the initial phase of 40 planned condensate production wells by the end of 1986. Between 1984 and 1987, commercial condensate production rose from 40,000 to 100,000 bbl/day (14 to 37 million bbl/year). Rather than processing the condensate at the field, it is piped 435 mi (700 km) south to Surgut for refining (Oil and Gas Journal, 1987a).

In addition to condensate, at least a third of the three to four dozen separate pools in Neocomian reservoirs have oil rims underlying the gas (Table 1). Oil production began in June 1987 and reached a total of 1.4 million bbl by the end of the year (Fueg, 1988). Soviet category C_1 plus C_2 oil reserves at Urengoy are 1.22 billion bbl (170 mm t) (Zap. Sib. NIGNI, 1989). Exploitation of Urengoy oil and condensate will play an increasing role in sustaining production of West Siberian liquids. Plans call for condensate and oil output to increase from 8.1 million bbl/day in 1987 to 10 million bbl/day by the year 2000 (Sagers, 1988; Oil and Gas Journal, 1987b).

The Urengoy gas trunk pipeline complex consists of six 56-in. (1420 mm) individual pipelines with an initial combined design capacity of 7.06 tcf/year (200 bcm/year) (Table 2). This was one of the largest construction projects in Soviet history, requiring 12 million tons of pipe and 175 compressor stations (Shabad, 1984). The system traverses more than 435 mi (700 km) of swamp, roughly 93 mi (150 km) of permafrost, and more than 500 large and small rivers, and cost more than 25 billion rubles ($40 billion) to build (Gurkov and Yevseyev, 1984).

[2]The stable condensate in gallons/mcf of dry gas was multiplied by the estimated ultimate recovery of dry gas (cumulative production + [A + B + C_1 + C_2 reserves]) for each reservoir reported in Zhabrev (1983). Data are given in Table 1.

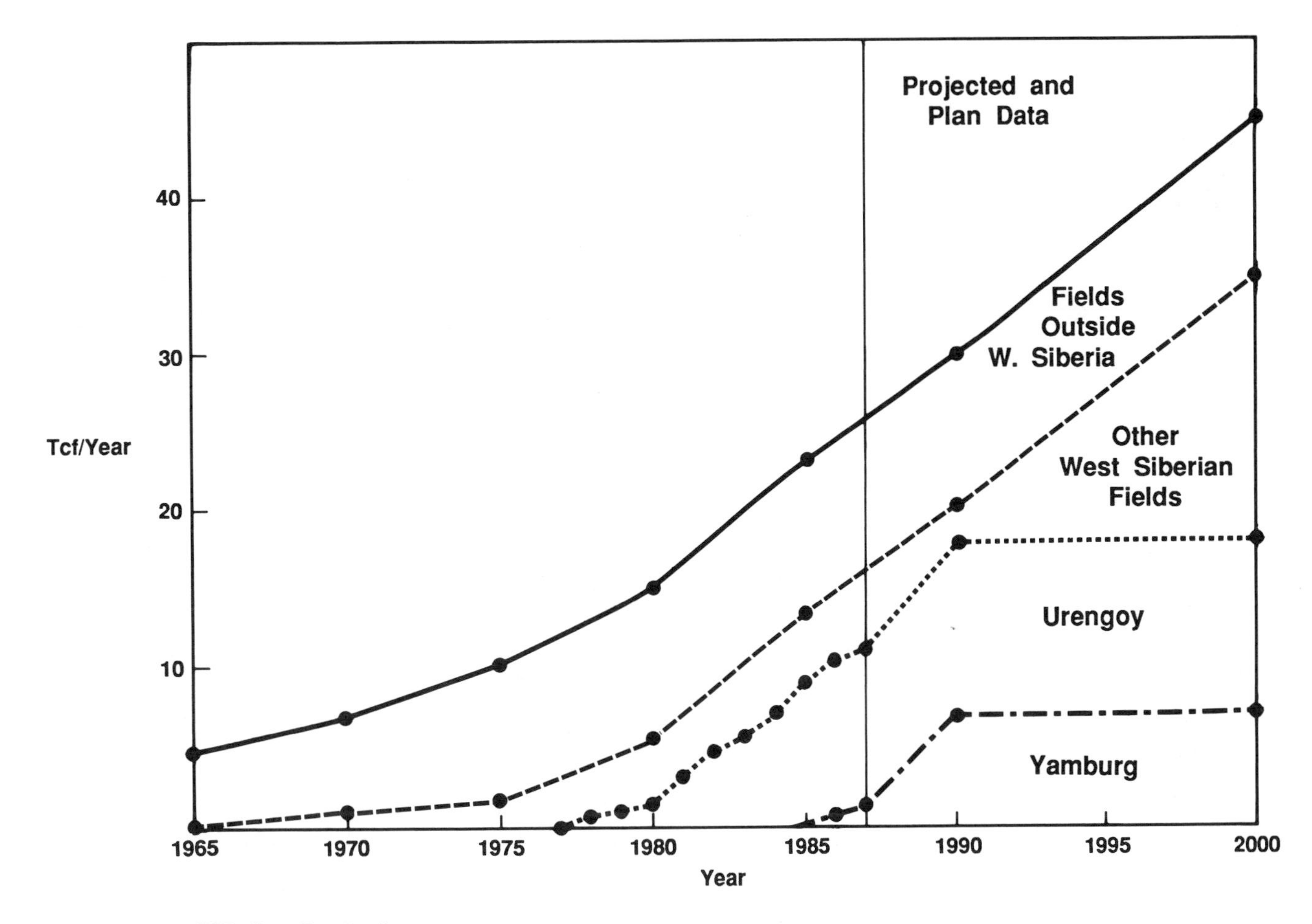

Figure 3. Historical and planned/projected gas production from the Urengoy and Yamburg fields, the West Siberian basin, and the total U.S.S.R. to the year 2000 (Oil and Gas Journal, 1984, 1987b, 1987c; Shabad, 1984; Sagers, 1988). Projections for both the Urengoy and Yamburg fields after 1990 are based on the assumption that peak production rates from these two fields can be maintained until the end of that decade.

DISCOVERY METHOD

Initial exploration results in the central basin in the early 1960s led to recognition of the role of major (more than 1544 to 1930 mi^2 [4000 to 5000 km^2] in area), shallow uplifts in the Mesozoic section in concentrating and trapping hydrocarbons. Moreover, the 1962 discovery of the Taz field demonstrated that the same type of large anticlines which are petroleum-bearing in the central part of the basin contained gas in the north. At the Taz field, dry gas was found in the Cenomanian Pokur Formation, and gas, gas-condensate, and oil rims were discovered in the Neocomian Megion and Vartov formations. This relationship between reservoir age and phase governs the distribution of hydrocarbons throughout the northern part of the West Siberian basin.

During 1964–1967, Soviet exploration teams conducted a major regional program of reconnaissance reflection seismic surveys across the Nadym-Pur and Pur-Taz petroliferous districts to identify anticlinal prospects in the Cenomanian and Neocomian sections (Figure 2) (Nesterov et al., 1971). More than three dozen of the structures identified at that time proved productive over the next ten years of exploration and were found to contain a total of more than 800 tcf (22 tcm) of gas resources. The same exploration concept applied in this region has led to the discovery of one-third of the world's natural gas reserves.

Table 1. Production data for Urengoy gas field.

Age	Formation Name	Reservoir Name	Uplift of Production	Hydro-carbon Phase	Year of Discovery	Depth to Top of Reservoir (ft)	Depth to O-W or G-W Contact (ft)	Area of ABC_1 Reserves (acres)	Area of C_2 Reserves (acres)	Average Porosity (%)	Average Permeability (md)	Liquid Gravity (glcc)	Sulfur Content (%)	Initial Formation Pressure (psi)	Average Reservoir Temperature (°F)	Average Condensate Content (gal/mcf)	Condensate In-Place (mmbbl)	Gas Resources $ABC_1 + C_2$ + Cum Prod (bcf)
Oxfordian	Vasyugan	Yu_2^1		Gas	1984	11,972	11,906	49	232	13								
Oxfordian	Vasyugan	Yu_1		Oil	1982	12,188	11,972	49	1501	10		0.83	0.06					
Oxfordian	Vasyugan	YuG_2		Oil	1984	12,421	12,202	146	1004	10		.083	0.08					
Valangian	Megion	$Achimov_5$		Gas	1988	12,103	11,939	279	1179	16		0.791						
Valangian	Megion	$Achimov_3$		Oil	1984	12,136	11,487		2349			.0843						
Valangian	Megion	$Achimov_2$	Middle	Gas	1979	11,939	11,752	802	941	16		0.781						
Valangian	Megion	$Achimov_2$	Middle	Oil	1979	12,169	11,883		311			0.804						
Valangian	Megion	$Achimov_2$	North	Gas	1975	11,644	11,391	20	237	20		0.737						
Valangian	Megion	$Achimov_1$		Gas	1980	11,382	11,283	86	64	20								
Valangian	Megion	Bu_{17}	Middle	Gas	1980	10,857	10,693	89	33	15								
Valangian	Megion	Bu_{17}		Gas	1987	11,152	10,995					0.768						
Valangian	Megion	Bu_{16}		Gas	1975	10,503	10,234	198		14								
Valangian	Megion	Bu_{16}		Oil	1975	10,801	10,627	37	372	14		0.800	0.09					
Valangian	Megion	Bu_{16}		Gas	1975	10,627	10,496	399	240	15								
Valangian	Megion	Bu_{15}		Gas	1981	10,375	10,168	166	161	15								
Valangian	Megion	Bu_{14}	North	Gas	1975	10,299	10,106	167		18		0.753						
Valangian	Megion	Bu_{14}	South	Oil	1975	10,217	9991	115	486	15		0.816	0.10					
Valangian	Megion	Bu_{14}	South	Gas	1975	10,109	9889	1653	27	15		0.753						
Valangian	Megion	Bu_{14}	South	Gas	1975	10,109	9889	1042		15		0.752	0.555					
Valangian	Megion	Bu_{13}^2		Gas	1983	10,358	10,168	148	15	18								
Valangian	Megion	Bu_{13}^1		Gas	1974	9676	9479	166	200	18		0.753						
Valangian	Megion	Bu_{13}^1		Gas	1974	9676	9479	179	192	15			0.023					
Valangian	Megion	Bu_{13}		Gas	1987	11,611	11,460					0.787						
Valangian	Megion	Bu_{13}		Gas	1987	11,611	11,460	40	111	14	42	0.787		4205	180	2.35	51	918
Valangian	Megion	Bu_{12}^2		Gas	1975	9728	9512	577	430	18		0.740						
Valangian	Megion	Bu_{12}^2		Gas	1975	9732	9512	613	183	16								
Valangian	Megion	Bu_{12}^{1-1}		Oil	1986	10,519	10,332	59		14		0.824						
Valangian	Megion	Bu_{12}^1	North	Gas	1978	9584	9299	177		19								
Valangian	Megion	Bu_{12}^1	Central	Gas	1980	9355	9532	173		19								
Valangian	Megion	Bu_{12}^1	South	Gas	1975	9381	9594	468		19		0.742						
Valangian	Megion	Bu_{12}^1	South	Gas	1975	9381	9594	580		15								
Valangian	Megion	Bu_{12}^1	South	Oil	1975	9512	9666		192			0.840	0.10					
Valangian	Megion	Bu_{12}^1	South	Oil	1975	9512	9666	80		18		0.828	0.10					
Valangian	Megion	Bu_{12}		Gas	1980	9683	9833	116		16	40			4060	178	1.85	138	3143
Valangian	Megion	Bu_{11}^2		Gas	1981	10,224	10,434	240	27	14				3915	176	2.45	810	13,912
Valangian	Megion	Bu_{11}^2		Oil	1984	0	10,257						0.082					
Valangian	Megion	Bu_{11}^1		Oil	1977	9250	9400		130			0.840	0.082					
Valangian	Megion	Bu_{11}	Middle	Gas	1977	9184	9381	118		16		0.755	0.010					
Valangian	Megion	Bu_{11}	Middle	Gas	1977	9184	9381	314		18		0.745	0.032					
Valangian	Megion	Bu_{11}	Middle	Oil	1977	9250	9400	249	61	16		0.836	0.032					
Valangian	Megion	Bu_{11}	North	Gas	1976	10,201	10,299	746		15		0.752	0.510					
Valangian	Megion	Bu_{10}^{2A}		Gas	1982	10,335	10,594	88	151	13			0.743					
Valangian	Megion	Bu_{10}^{2A}		Gas	1982	10,335	10,594	133		13			0.510					
Valangian	Megion	Bu_{10}^1		Oil	1985	0	10,824	30	141	13		0.810	0.080					
Valangian	Megion	Bu_{10}^1	North	Oil	1980	10,355	10,617	42		13		0.809	0.080					
Valangian	Megion	Bu_{10}^1	South	Gas	1982	10,496	10,680	138		12			0.755					
Valangian	Megion	Bu_{10}^1	South	Gas	1982	10,496	10,680	9	55	12	50		0.510					
Valangian	Megion	Bu_{10-11}		Gas	1972	9118	9387	2301		18		0.755						
Valangian	Megion	Bu_{10-11}		Oil	1972	9184	9430	1589		16		0.828	0.040					
Valangian	Megion	Bu_{10-11}		Gas	1972	9118	9387	1564		16		0.755	0.010					
Valangian	Megion	Bu_{10-11}		Oil	1972	9184	9430	643	320	16		0.850	0.082					
Hauterivian	Vartov	Bu_9^2		Gas	1982	9807	10,102	304	50	15	75		0.584					
Hauterivian	Vartov	Bu_9^2		Oil	1982	10,011	10,299	191	106	16	75	0.827	0.100					
Hauterivian	Vartov	Bu_9^2		Oil	1982	10,011	10,299	70		16	75	0.827	0.100					
Hauterivian	Vartov	Bu_9^1	North	Gas	1982	9912	10,155	11		16	75							
Hauterivian	Vartov	Bu_9^1		Gas	1982	9840	10,129	34	48	15	75		0.607					
Hauterivian	Vartov	Bu_9^1		Oil	1982	10,270	9938	99	224	15	75	0.813	0.010					
Hauterivian	Vartov	Bu_9^1	South	Gas	1985	10,070	9873	32	59	15	75			3770	172	1.81	225	5226
Hauterivian	Vartov	Bu_9	Central	Gas	1975	9007	8774	831		19		0.739						
Hauterivian	Vartov	Bu_9	South	Gas	1975	9017	8823	268		19		0.739						
Hauterivian	Vartov	Bu_9	South	Oil	1975	9053	8856	30		18		0.843	0.100					
Hauterivian	Vartov	Bu_9	South	Gas	1975	9017	8823	764		17			0.0123					
Hauterivian	Vartov	Bu_9	South	Oil	1975	9053	8856		627			0.840	0.080					

Table 1. (Continued)

Age	Formation Name	Reservoir Name	Uplift of Production	Hydro-carbon Phase	Year of Discovery	Depth to Top of Reservoir (ft)	Depth to O-W or G-W Contact (ft)	Area of ABC_1 Reserves (acres)	Area of C_2 Reserves (acres)	Average Porosity (%)	Average Permeability (md)	Liquid Gravity (glcc)	Sulfur Content (%)	Initial Formation Pressure (psi)	Average Reservoir Temperature (°F)	Average Condensate Content (gal/mcf)	Condensate In-Place (mmbbl)	Gas Resources $ABC_1 + C_2$ + Cum Prod (bcf)
Hauterivian	Vartov	Bu_8^3	North	Gas	1982	10,037	9774	49		14								
Hauterivian	Vartov	Bu_8^3	North	Oil	1982	10,299	9971	38	52	16		0.820	0.100					
Hauterivian	Vartov	Bu_8^3	South	Gas	1982	10,168	9906	39	114	13			0.617					
Hauterivian	Vartov	Bu_8^0		Gas	1974	8922	8659	2561		18								
Hauterivian	Vartov	Bu_8^0		Gas	1974	8922	8659	2085		15								
Hauterivian	Vartov	Bu_8^0		Oil	1974	8987	8758		258			0.840	0.053					
Hauterivian	Vartov	Bu_8^0		Oil	1966	8987	8758	182		18	8	0.840	0.060	3770	167	1.69	287	6921
Hauterivian	Vartov	$Bu_{8\text{-}9}$		Gas	1976	9531	9479	602		16		0.758	0.540					
Hauterivian	Vartov	$Bu_{8\text{-}9}$		Oil	1976	9722	8512	145		16		0.819	0.100					
Hauterivian	Vartov	$Bu_{8\text{-}9}$		Oil	1976	9722	8512	506	315	16		0.835	0.100					
Hauterivian	Vartov	Bu_8		Gas	1974	8843	8758	2332		18		0.739						
Hauterivian	Vartov	Bu_8		Oil	1974	9053	8836	136		18			0.090					
Hauterivian	Vartov	Bu_8		Gas	1974	8843	8758	2148		16								
Hauterivian	Vartov	Bu_8		Oil	1972	9053	8840	510	735	17		0.840	0.100					
Hauterivian	Vartov	Bu_5	North	Gas	1973	8151	7954	195		20		0.735		3335	158	0.95	31	1377
Hauterivian	Vartov	Bu_5	South	Gas	1973	8321	8118	146	75	20	167	0.735						
Hauterivian	Vartov	Bu_1	North	Gas	1975	7715	7518	414		21		0.734						
Hauterivian	Vartov	Bu_1	South	Gas	1975	7911	7708	104		21	166	0.734		3190	144	0.75	75	4237
Hauterivian	Vartov	Bu_0		Gas	1975	7551	0		358		166			3190	153	0.75	20	1130
Hauterivian	Vartov	Au_{10}		Gas	1980	7387	7183	77		21								
Albian	Pokur	Pk_{21}		Gas	1974	5950	5753	150		21								
Albian	Pokur	Pk_{21}		Gas	1974	5950	5753		115			0.732						
Albian	Pokur	Pk_{18}		Gas	1974	5838	5635	173		27			0.017					
Cenomanian	Pokur	$Pk_{1\text{-}6}$		Gas	1966	4182	3926	528		31								
Cenomanian	Pokur	$Pk_{1\text{-}6}$		Gas	1966	4182	3926	11,631		31	495			1740	88	<0.01	16	219,664

Table 2. Urengoy gas pipeline system.

Pipeline	Length (mi)	Year Completed	Market
Urengoy-Gryazovets-Moscow	1740	1981	European Russia
Urengoy-Petrovsk	1875	1982	European Russia
Urengoy-Novopskov	2217	1983	Ukraine
Urengoy-Uzhgorod	2764	1983	W.Europe (Export)
Urengoy-Center I	1875	1984	European Russia
Urengoy-Center II	1875	1985	European Russia

Sources: Shabad (1984) and Oil and Gas Journal (1984, 1985).

STRUCTURE

Tectonic Development of the Basin

The West Siberian basin is polycyclic, having three tectonostratigraphic divisions. As cited in St. John et al. (1984), using the basin classification of Bally and Snelson (1980), the basin is classed as 1212, occurring on rigid lithosphere, without associated megasutures. According to Klemme (1971), the class is IIB. These are continental, multicyclic complex basins (craton/accreted margin).

The lowest of the three tectonostratigraphic divisions is a crystalline Archean to Proterozoic basement. Reactivation of the major uplifts on the basement surface from the Paleozoic through Cenozoic eras produced the uplifts that trap most of the discovered hydrocarbons in the basin. The basement is overlain by a widely, but not completely, metamorphosed and folded Paleozoic division, which is the intermediate tectonostratigraphic division. Hercynian uplift in the Permian Period produced a major regional unconformity marking the top of the Paleozoic section. The top of the Paleozoic section is at a depth of 21,325 to 24,606 ft (6.5–7.5 km) at Urengoy (Bochkarev et al., 1978).[3]

East-west extension in the Late Permian through Early Triassic epochs created major north-south rifts in the erosional surface of the central and northern basin. The longest of these is the Koltogor-Urengoy rift system, which bounds the Urengoy megaswell to the east. It originates southeast of the city of Nizhnevartovsk, extends north for 745 to 807 mi (1200 to 1300 km) onshore, and then continues under the Taz Gulf (Bochkarev and Fedorov, 1977). Volcaniclastic deposits fill these rifts.

[3]All depths in this article are vertical distances below the surface of the earth.

More than 2.6 million mi^3 (10 million km^3) of Mesozoic to Cenozoic clastic sediments compose the uppermost of the three tectonostratigraphic divisions of the basin. Two types of activity produced the gross internal structure of this division: (1) Increased regional subsidence and sediment accumulation occurred, particularly in the northern basin, reaching maximum rates in the Early Cretaceous Epoch. (2) Substantial structural growth occurred on reactivated basement highs during the Early Cretaceous Epoch and again starting in Paleocene–Eocene time. These uplifts are generally elongated north-south in the north and are more circular in the central basin.

Local Structure

The Urengoy megaswell is a very large anticline in the central part of the northern West Siberian basin measuring 112 × 12–25 mi (180 × 20–40 km) on the top of the Jurassic section. Like most major uplifts of the region, the anticline is a prominent basement feature that has been periodically reactivated. Also typical of uplifts in the northern part of the basin, the Urengoy megaswell increases in structural complexity with depth. There is a single trap for the Cenomanian $PK_{1\text{-}6}$ reservoir complex, with a single gas-water contact, and nine separate structural uplifts on the top of the Jurassic System, within the confines of the field, that produced many hydrodynamically isolated reservoirs in the deeper Jurassic and Neocomian units (Konovalov and Glukhoyedov, 1977).

Closure on the Urengoy megaswell drops from 1312 ft (400 m) on the top of the Jurassic section, to 656 ft (200 m) on the top of Cenomanian Pokur Formation, and to 328 ft (100 m) on the top of the Paleocene Talitskoy Formation (Figure 4). The greatest growth in the Urengoy megaswell (e.g., on the central uplift, 17 ft/million years [5.2 m/million years]) occurred in the Berriasian through Valanginian ages during deposition of the Megion Formation and again in the Cenomanian Age during deposition of the Pokur Formation. Small-scale inversion occurred in the Aptian and Turonian ages and uplift began again in the Paleocene Epoch (Bochkarev et al., 1978).

STRATIGRAPHY

Three depositional megacycles comprise the sedimentary Mesozoic–Cenozoic stratigraphy of the basin. Each evolves from a basal continental section to marine at the top (Kontorovich et al., 1975). The first megacycle is from the Triassic System through the lower Aptian Series; the second is from the upper Aptian Series to the Oligocene Series; and the third encompasses the Oligocene Series through the Quaternary System (which includes only the basal section). Virtually all hydrocarbon production in the basin comes from the first two megacycles (Figure 5).

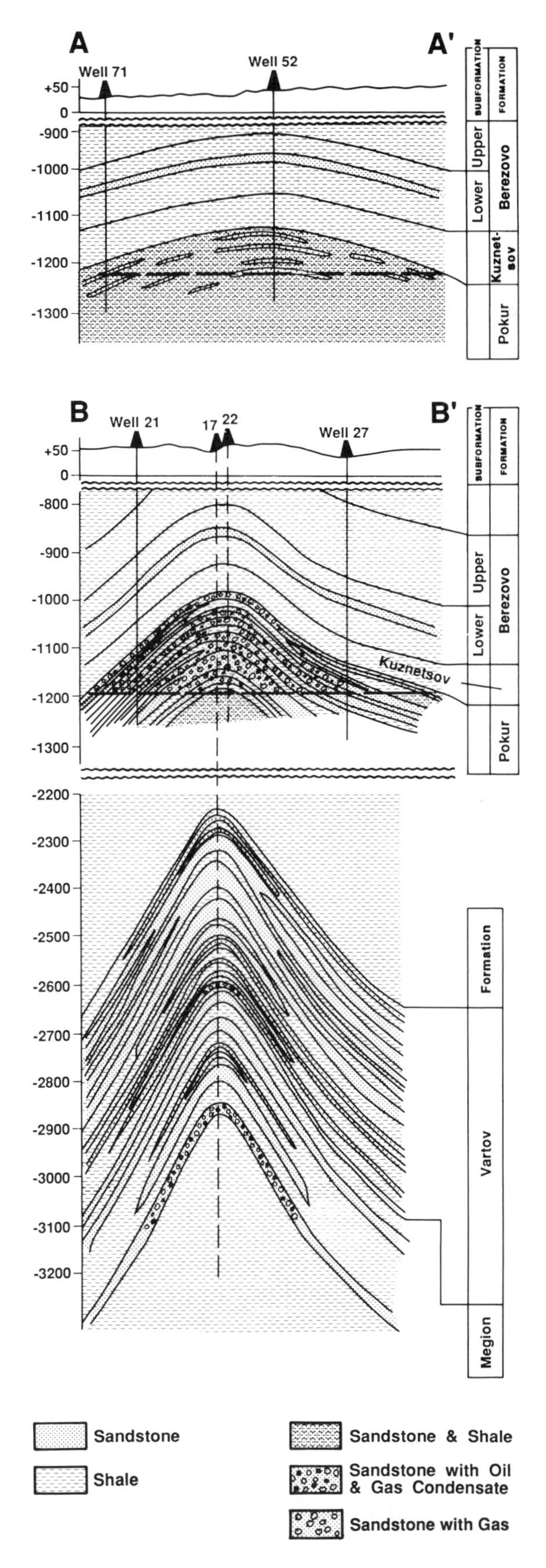

Figure 4. Cross sections of Urengoy field along lines A-A′ and B-B′ of Figure 7. Depths in meters. (Modified from Nesterov et al., 1971.)

Triassic-Neocomian Section

Basal continental Triassic sandstones and shales of the Tampey Formation unconformably overlie the Paleozoic surface and fill the major north-south rifts. The deposition of the Tyumen Formation (Lower and Middle Jurassic) represents a continuation of continental conditions in the southern and central basin, where it ranges from 820 to 1640 ft (250 to 500 m) in thickness. In the north, where its thickness reaches more than 8202 ft (2.5 km), it becomes progressively more marine (Rudkevich, 1988a).

The Upper Jurassic section is marked by three separate transgressive cycles of progressively increasing areal extent. The youngest cycle covers more than 849,000 mi^2 (2.2 million km^2), more than two-thirds of the basin (see "Area under water" graph in Figure 5). During the Callovian through Kimmeridgian ages, shallow-water facies occur extensively throughout the central basin, and deep-water shales were restricted to the northern basin.

The Volgian Bazhenov Formation was deposited during the last Late Jurassic cycle. The formation comprises bituminous deep-water facies over half of the area of marine deposition, ranging in thickness from 66 to 230 ft (20 to 70 m) (average 98 ft [30 m]). Paleobathymetry in the Urengoy area indicates water depths of more than 1640 ft (500 m) at the time of Bazhenov deposition (Filina et al., 1984). The Bazhenov Formation is a major hydrocarbon source rock in the basin.

The Bazhenov Formation is conformably overlain in the central and northern areas of the basin by marine transgressive shales of the lower Megion Formation (Berriasian). The thickness of this unit averages 656 ft (200 m) across the basin. Shale content decreases upward in the formation as a consequence of the localized reactivation of structural highs that accompanied regional regression. Sandstone reservoir porosities range from 18% to 26% and permeability ranges from 1 to 35 md.

The Vartov Formation (Hauterivian-Barremian) represents intensified regional regression. The basin area under water decreased to less than 20% of the total basin area by the end of the Barremian Age (Figure 5). This formation consists of more than 3280 ft (1000 m) of paralic sandstones interbedded with shales and coal. Sandstones in the Vartov Formation form traps in at least 18 stratigraphic units, which have reservoir porosities of 20% to 25% and permeabilities of 1 to 20 md. Interbedded shales from 98 to 164 ft (30 to 50 m) thick form the seals (Rudkevich, et al., 1988a).

Albian-Turonian Section

The Pokur Formation (Albian-Cenomanian) represents the basal unit of the second Mesozoic-Cenozoic depositional megacycle. The unit is from 2296 to 2624 ft (700 to 800 m) thick and is composed of friable sandstones to unconsolidated sands

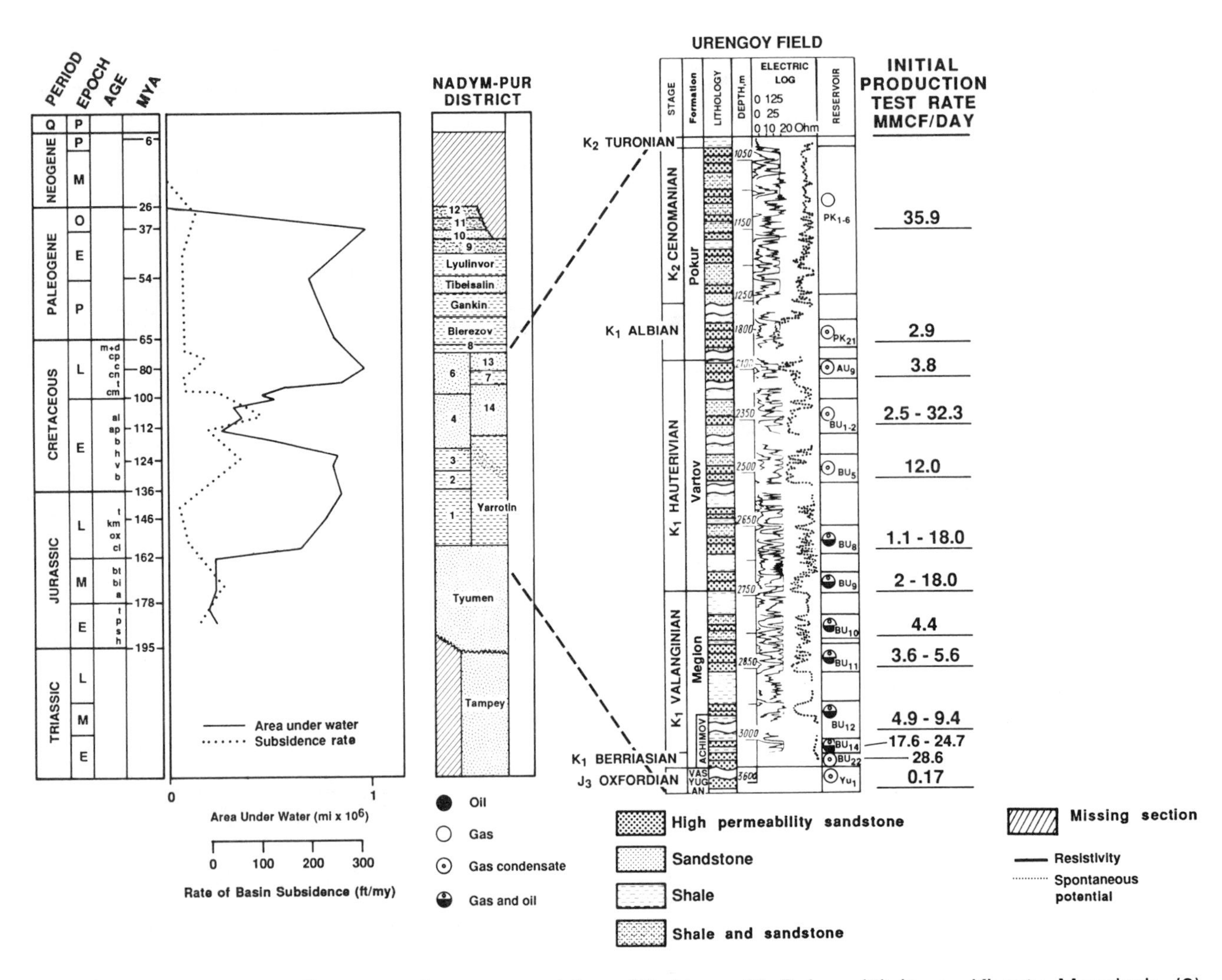

Figure 5. Left: Chart of the area under water and the rate of subsidence in the West Siberian basin. (Modified from Kontorovich et al., 1975, and Aleinikov et al., 1980.) Center: Regional stratigraphic column for the Nadym-Pur Petroliferous District (Grace and Hart, 1986). Legend: (1) Abalak, (2) Vasyugan, (3) Megion, (4) Vartov, (5) Alym, (6) Pokur, (7) lower Khanty Mansiysk, (8) Kuznetsov, (9) Chegan, (10) Atlym, (11) Novomikhailov, (12) Lagar'-Tomsk, (13) Marresalin, (14) North Sos'va. Right: Stratigraphic column for the Urengoy field. (Modified from Maksimov, 1987.) Production rates from Maksimov (1987) and Zhabrev (1983).

interbedded with siltstones and a few thin layers of shale (Sarkisyan and Komardinkina, 1971).

Extensive coal seams in the Pokur Formation contain an estimated 53 trillion tons (48.4 trillion metric tons) of predominantly humic organic matter (Kontorovich et al., 1975). There are at least 19 productive units in the Pokur Formation. The most important regionally and at Urengoy is the uppermost unit, called the PK_{1-6} complex. Pokur reservoirs contain two-thirds of the discovered gas of the northern part of the West Siberian basin, or nearly one-quarter of world gas resources.

A major marine transgression during the Turonian Age resulted in the deposition of tens to hundreds of feet of the Kuznetsov Formation shales that seal the reservoirs at the top of the Pokur Formation. Marine conditions dominated in the basin into the Eocene Epoch. Local and regional structural growth in the beginning of the Paleocene Epoch accompanied the final retreat of the sea to the north.

TRAP

Structure and Seals

The separate structural styles of traps in the Neocomian and Cenomanian sections of the gas-condensate fields of the northern basin arise from the differences in the nature of uplift between the

Early Cretaceous and Paleocene-Eocene epochs. In the earlier episode, highly localized structural growth occurred, typically producing several uplifts across the fields. Areally broader uplift across the larger structures starting in the Paleocene Epoch produced the singular massive traps in the Cenomanian Pokur Formation.

Syndepositional localized growth during Neocomian time produced the highly discontinuous sandstone/shale reservoir-seal packages of the Megion BU_{10} through BU_{22} reservoirs. These gas-condensate-oil accumulations are typically confined to sand packages on isolated uplifts within the larger four major uplifts of the Urengoy field: northern, middle, central, and southern (Figures 6 and 7). The sandstones generally thicken to the south, sealed vertically by interbedded shales of from 82 to 131 ft (25 to 40 m) thick, and are replaced laterally by shales in the saddles that separate the uplifts (Ognev et al., 1979).

Above the isolated sandstones of the Megion Formation, reservoirs of the lower Vartov Formation show a higher degree of continuity, have thinner shales seals (from 16 to 49 ft [5–15 m]) and thicker gas-oil columns (Table 1). The BU_8^0 is continuous across the entire field and a single gas-water contact is present (Figure 8). The three deeper productive units in the BU_8 and the two reservoirs of the BU_9 have isolated reservoirs confined to one or two of the four major uplifts within the limits of the field. The upper Vartov BU_0 and $BU_{1\text{-}2}$ reservoirs are productive only within the Northern Uplift, and the BU_5 on the northern and southern uplifts. Closure of Neocomian traps at Urengoy range from 43 to 558 ft (13 to 170 m), averaging 239 ft (73 m) (Zhabrev, 1983; Maksimov, 1987). They began to fill during the Cenomanian Age (Bochkarev et al., 1978).

Uplift beginning in the Paleocene Epoch, and intensifying in the Eocene Epoch, was generally uniform across the Urengoy megaswell. This created the singular massive trap for the Pokur $PK_{1\text{-}6}$ reservoir complex under shales of the Kuznetsov Formation. The maximum height of the gas column in the $PK_{1\text{-}6}$ trap, which is on the central and middle uplifts, is 649 ft (198 m) high (Figure 9). Growth of the traps at the Cenomanian level, starting in the Paleocene Epoch, accounted for an increase in the closure on the Cenomanian $PK_{1\text{-}6}$ trap of from 164 to 230 ft (50 to 70 m), or 20% to 30% of their present volume (Kuzin, 1983). The 197 ft (60 m) thick Kuznetsov shale, a basin-wide seal, overlies the $PK_{1\text{-}6}$ at Urengoy (Sarkisiyan and Komardinkina, 1971). The trap in the Cenomanian section is filled to the spill point.

Timing

In both Neocomian and Cenomanian reservoirs, periodic variations in the geometry of paleotraps suggest complex interaction between the volume of liberated gas available to charge the reservoir and the volume of trap available to accommodate the gas. Bochkarev et al. (1978) analyzed the development of 12 local uplifts at the Hauterivian Vartov Formation (BU_8^0 reservoir) level in the Urengoy region. Total trap area increased from 1639 mi^2 (4246 km^2) at the end of the Hauterivian Age to 1749 mi^2 (4530 km^2) at the end of the Aptian Age. After dropping to 1147 mi^2 (2970 km^2) during the Turonian inversion, total trap area reached a maximum of 1873 mi^2 (4852 km^2) by the end of the Paleocene Epoch, but lost one-fifth of its total volume by the end of the Eocene Epoch. This interpretation provides that maximum trap volume in the Neocomian section developed when the main hydrocarbon charge entered the Neocomian traps.

A similar pattern existed after deposition of the Cenomanian Pokur reservoirs and their associated Turonian Kuznetsov shale seal. The same uplift that increased the volume of the Cenomanian $PK_{1\text{-}6}$ trap starting in the Paleocene Epoch would have, with concomitant erosion, reduced formation pressures throughout the Pokur Formation, liberating gas dissolved in formation waters, thereby charging the trap. Kuzin (1983) estimated that uplift occurring during Neogene to Quaternary time alone totaled 656 ft (200 m) or more, reducing formation pressure by a cumulative total of from 294 to 367 psi (2.0 to 2.5 MPa). Akhiyarov (1982) established four zones in the $PK_{1\text{-}6}$ complex based on formation water salinity, resistivity, and amplitude of spontaneous potential, suggesting that the trap filled over four cycles (Figure 10).

The Role of Hydrates

In addition to the role of structural uplift in liberating and concentrating hydrocarbons, the cyclic formation of gas hydrates associated with Pleistocene glaciation may have focused regionally dispersed gas into the Urengoy structure. Kortsenshteyn (1970) and others proposed that glaciation produced gas hydrates as deep as 4921 ft (1500 m), which on glacial retreat liberated methane to migrate in the gas phase vertically behind the receding ice. Subsequent cycles of glacial advances and retreats produced successively larger accumulations of gas. The process terminated with the addition of that gas to the reservoirs in the top of the Pokur Formation on the large sealed uplifts across the northern basin, including the $PK_{1\text{-}6}$ complex on the Urengoy megaswell. For further discussion on the mechanism of gas concentration through cyclic hydrate formation, see Grace and Hart (1986).

Jurassic Reservoirs

There has been limited deep drilling (deeper than 14,763 ft [4.5 km]) at Urengoy. By 1978, only one well had penetrated the Jurassic section, the top of which is at a depth of between 11,500 and 12,000

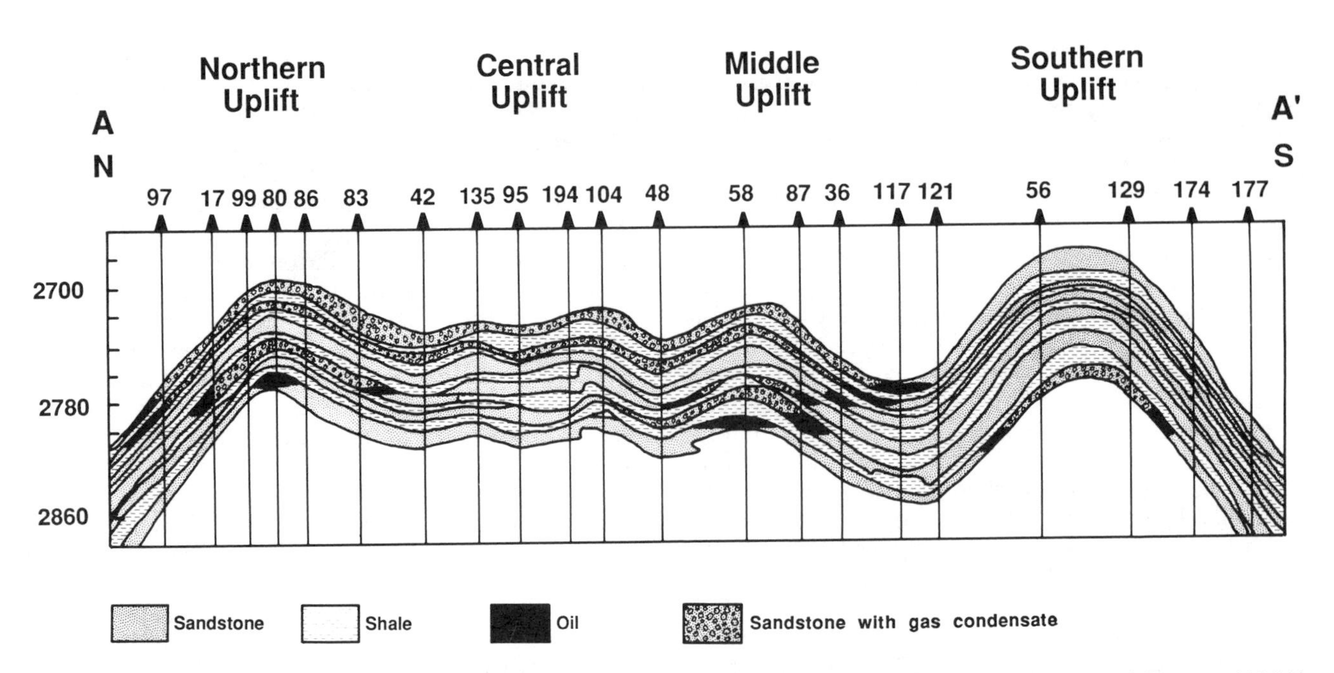

Figure 6. Cross section of the Valanginian Megion Formation BU_{10-11} reservoir complex along line A–A′ in Figure 11. Modified from Zykin and Tsarev (1982) and Maslennikov et al. (1980).

ft (3506 and 3641 m). By 1984, just seven had reached the Jurassic System (Borodkin, 1979; Nesterov et al., 1984). The deepest well, #266 (total depth of 16,469 ft [5021 m]), penetrated 4526 ft (1380 m) of Jurassic section, reaching total depth in the Lower and Middle Jurassic Tyumen Formation. In this area, the Tyumen Formation is 4920 ft (1500 m) thick. It is composed of alternating shallow-marine and paralic facies that have good source potential but poor reservoir quality in the absence of fractures. Rudkevich et al. (1988a) reported a recent gas-condensate discovery in the Tyumen Formation at Urengoy, perhaps as deep as 16,503 ft (5030 m).

The Callovian–Oxfordian formations, encountered at depths of from 11,808 to 11,972 ft (3600 to 3650 m), have a seismically determined thickness of from 4264 to 4920 ft (1300 to 1500 m). Deep drilling produced subcommercial oil accumulations in the Yu_1 reservoir of the Oxfordian Vasyugan Formation (Nesterov et al., 1984). Interbedded shales on subordinate uplifts at the Upper Jurassic level within the Southern and Central uplifts trap the hydrocarbons.

The Bazhenov Formation, which produces from fractures at the Salym field in the central part of the basin, is a dark gray to gray-black shale at Urengoy. However, its characteristic high bituminous content has not been found in all wells. The productive capacity of the unit has been demonstrated only in a subeconomic (13 bbl/day) flow on the northern uplift. Considering the depth of the Upper Jurassic section and the deeper Tyumen Formation, a commercial well requires production of at least 73 bbl/day (Nesterov et al., 1984).

Neocomian Reservoirs

Reservoirs of the 656 ft (200 m) thick Berriasian Achimov Member of the Megion Formation compose the lower part of this section. Deposition occurred in shallow water conditions with strong bottom currents producing narrow sand bodies elongated parallel to present structural strike, separated laterally by shales (Borodkin, 1979). The unit fines upward, is composed of fine-grained sandstone and siltstones, and in places is highly calcareous, with interbedded lenses of shale.

The Achimov BU_{22} reservoir is present across the field, though commercial flows have only occurred on the northern uplift of the field. This horizon contains the field's largest accumulation of condensate in place (1.2 billion bbl). It is geopressured. At well #95, pressure is 8787 psi (60.6 MPa), 3712 psi (25.6 MPa) above hydrostatic pressure (Nesterov et al., 1984).

Above the Berriasian Achimov Member, the Valanginian Megion and Hauterivian Vartov formations contain productive zones on the northern uplift spanning the BU_{14} to BU_0 reservoirs; gross thickness is 2820 ft (860 m). On the middle, central, and southern uplifts, the BU_{16} to BU_5 is productive, with a gross thickness of 1640 ft (500 m) (Table 1) (Ognev et al., 1979). The reservoirs of the Megion and Vartov formations are heterogeneous at two scales. Sandstone reservoirs in the same stratigraphic interval are divided by shale bodies that are distributed irrespective of present structure. Within individual reservoirs, continuity is broken by interbedded shale lenses (Zykin and Tsarev, 1982).

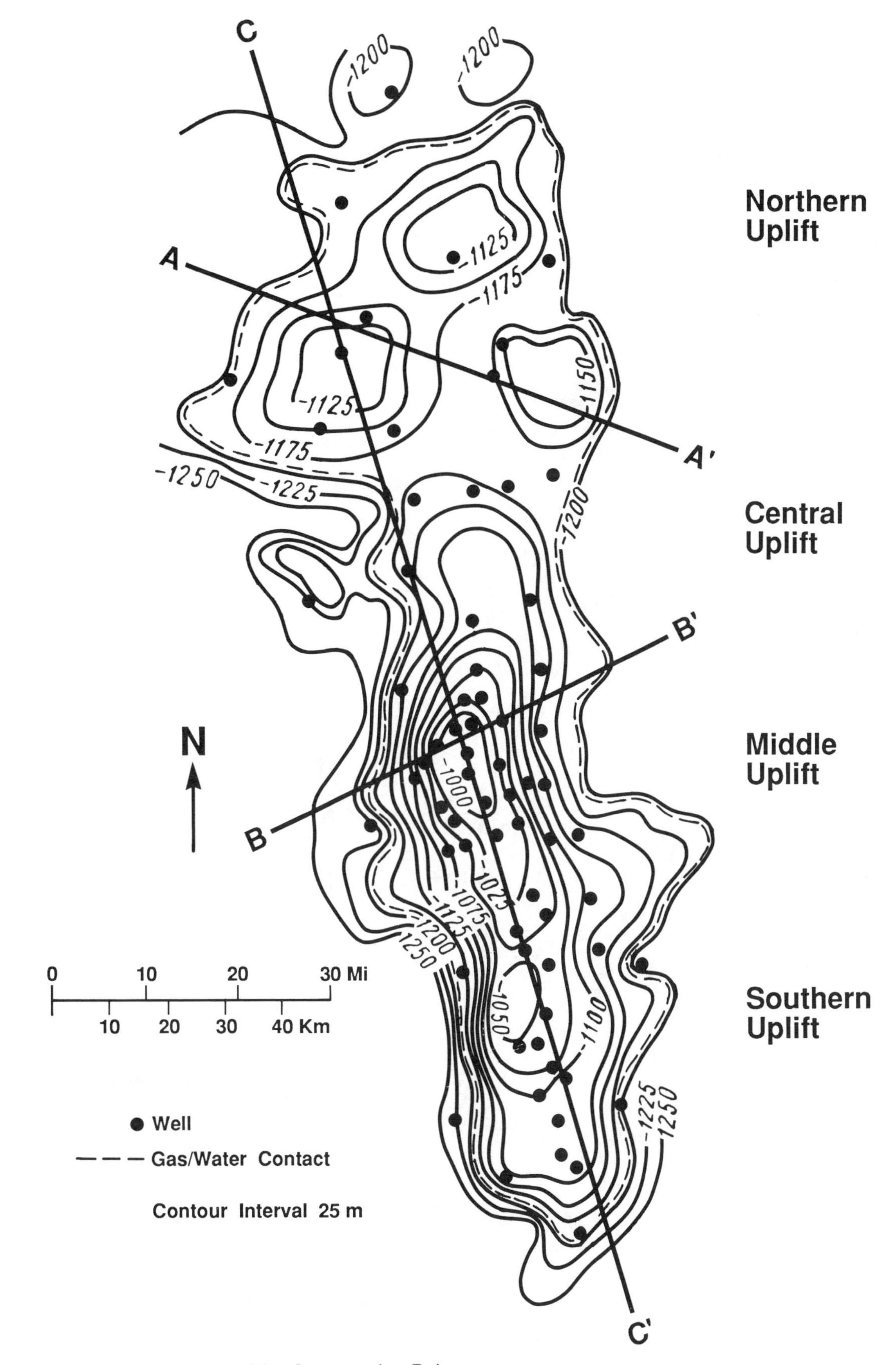

Figure 7. Structure map, top of the Cenomanian Pokur Formation PK_{1-6} reservoir complex. (Modified from Maksimov, 1987.)

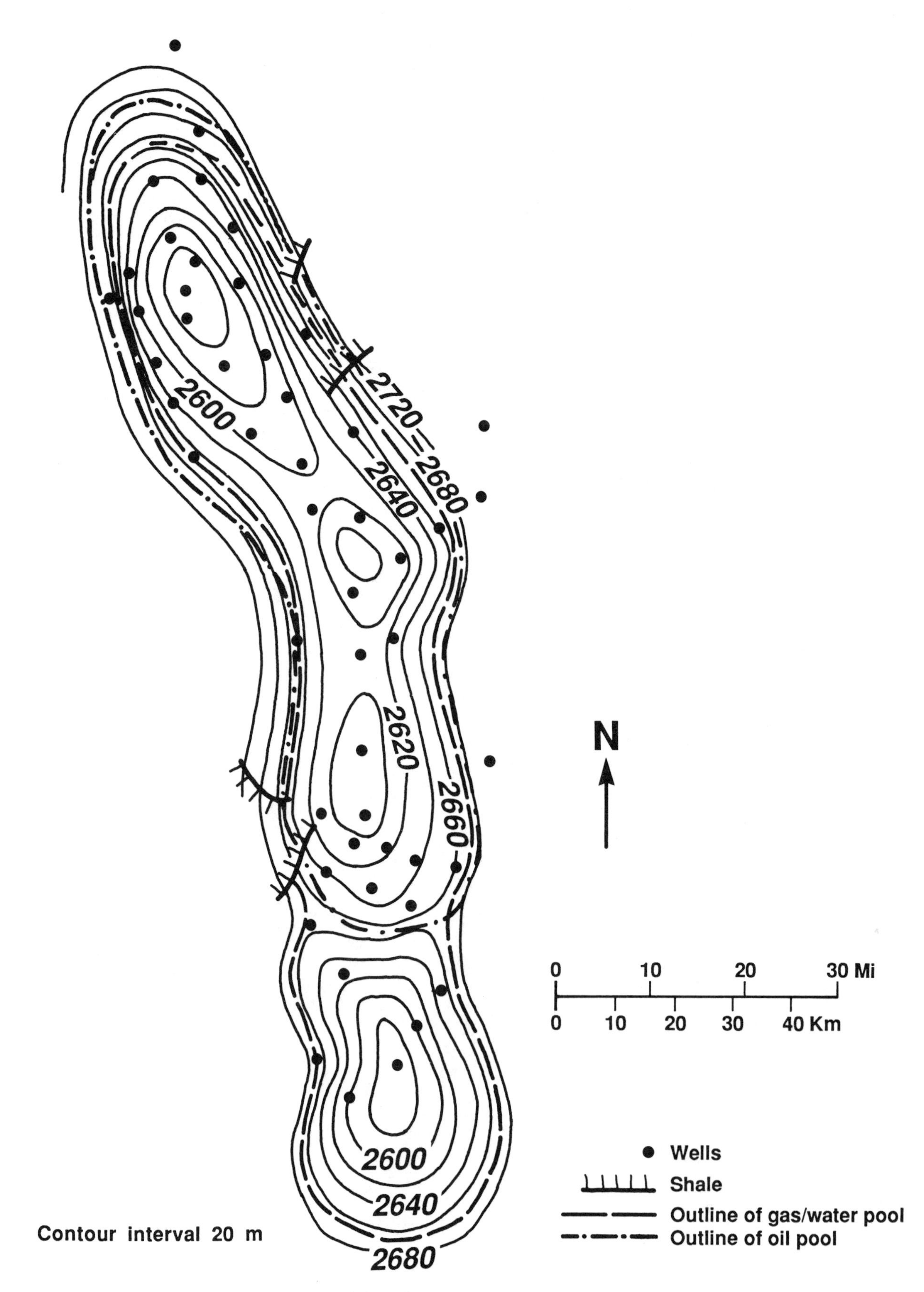

Figure 8. Structure map, top of the Hauterivian Vartov Formation BU_8^0 reservoir. (Modified from Tikhomirov et al., 1978.)

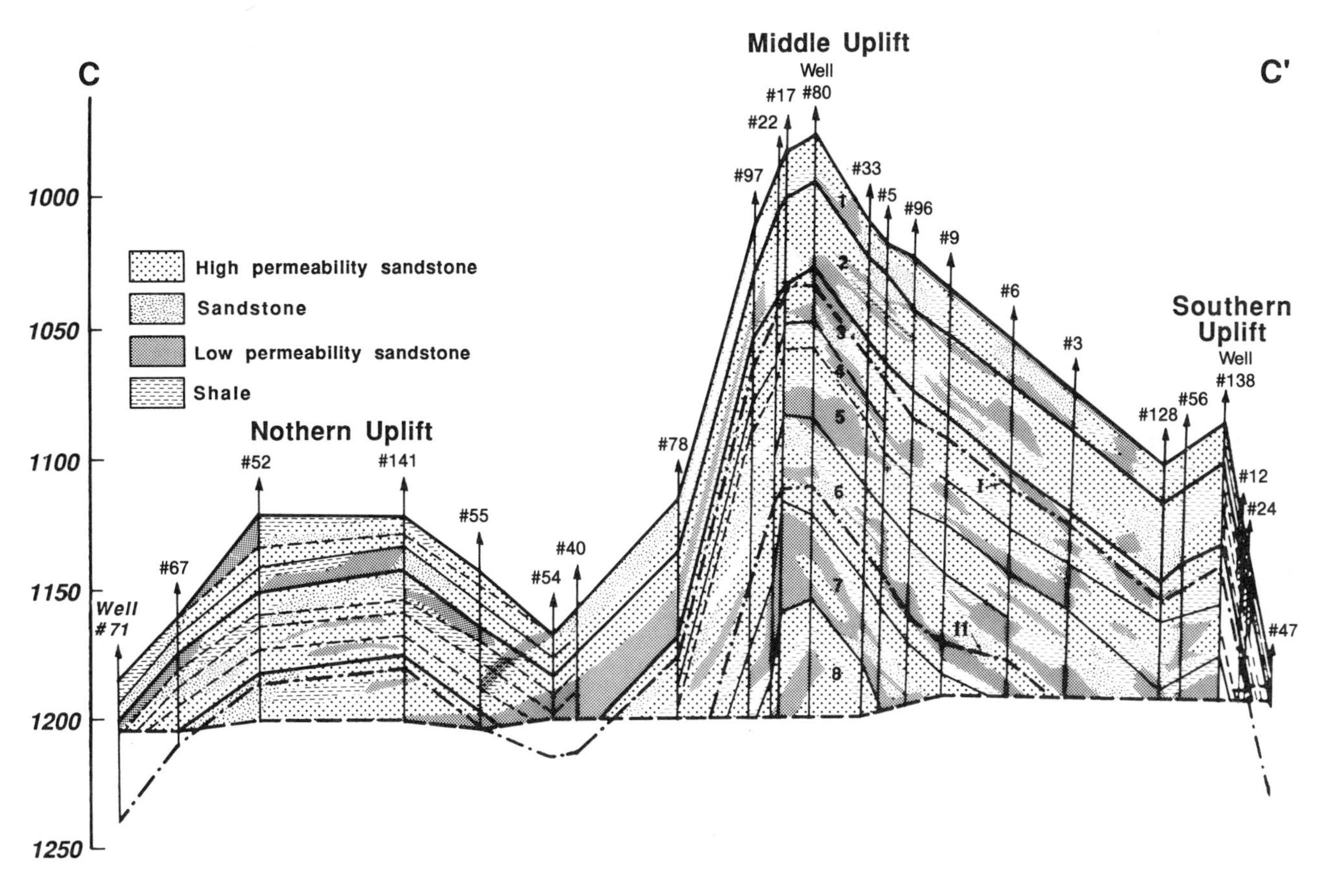

Figure 9. Schematic cross section of the Cenomanian Pokur Formation PK_{1-6} reservoir complex along line C-C′ in Figure 7. (Modified from Berman et al., 1978, and Yermakov and Shalya, 1982.) The eight internal fining-upward sequences are numbered and the positions of two of the intraformational unconformities are shown (I and II).

Conditions range from a single reservoir in hydrodynamic communication across the length of the field (e.g., the BU_8^0 reservoir, Figure 8) to two or more pools across all or part of each of the four major uplifts (e.g., the BU_{11} complex, Figures 6 and 11). Though buried at significant depths (from 8528 to 10,496 ft [2600 to 3200 m]), with resultant loss of porosity and permeability (Table 1), the wells have extraordinary productivity. Flows from Neocomian reservoirs exceed 35.3 mmcf/day (1 mmcm/day). In an open-hole test, the BU_8^0 flowed 49 mmcf/day (1.4 mmcm/day) of gas (Tikhomirov et al., 1978).

The Valanginian Megion Formation has a gross thickness of more than 1968 ft (600 m) at the Urengoy field, encompassing the BU_{16} through BU_{10} reservoirs (Galerkina, 1979). The deeper Megion BU_{16} through BU_{12}^1 sandstones are relatively thin and separated by thick shale units. The $BU_{15\text{-}16}$ reservoirs (at 9840 to 10,345 ft [3000 to 3155 m]) are composed of sand with interbedded mudstones and shale. Reservoir quality improves toward the crest of the uplift. The seal is a 262 to 344 ft (80 to 105 m) thick shale member of the Megion Formation.

In the BU_{14} through BU_{12} section, the sandstones are correlatable across the field, but hydrocarbon pools are generally isolated on individual uplifts or on several parts of a single uplift. Large single reservoirs spanning the central and middle uplifts also are present (Tikhomirov et al., 1978).

Deposition of the Megion BU_{12}^2 through BU_8 interval was in inner neritic to coastal environments during a period of regional uplift and the development of small local highs. The BU_{12} accumulations are confined to isolated traps on three of the four major uplifts in 66 to 98 ft (20 to 30 m) thick sandstone reservoirs sealed by interbedded (82 to 131 ft [25–40 m]) thick shales (Ognev et al., 1979).

The Valanginian upper Megion (Yuzhno-Balik Member) $BU_{10\text{-}11}$ reservoir complex has the second largest gas resources of the Neocomian section (13.9 tcf [394 bcm]) and third largest quantity of condensate in place (810 million bbl). The trapped gas, condensate, and oil are split between isolated accumulations on smaller uplifts within each of the major uplifts. The unit is composed of sandstone to siltstone with interbedded siltstone to shale lenses of varying thicknesses and lateral extent, but not exceeding a thickness of 16 ft (5 m). The thicker and more continuous of these siltstone-shales divide the complex into six subdivisions ($BU_{10}^{1\text{-}3}$ and $BU_{11}^{1\text{-}3}$) (Zykin and Tsarev, 1982).

Though there are separate gas-oil and oil-water contacts in the $BU_{10\text{-}11}$ complex on the four major uplifts, the distribution of sandstone bodies localizes

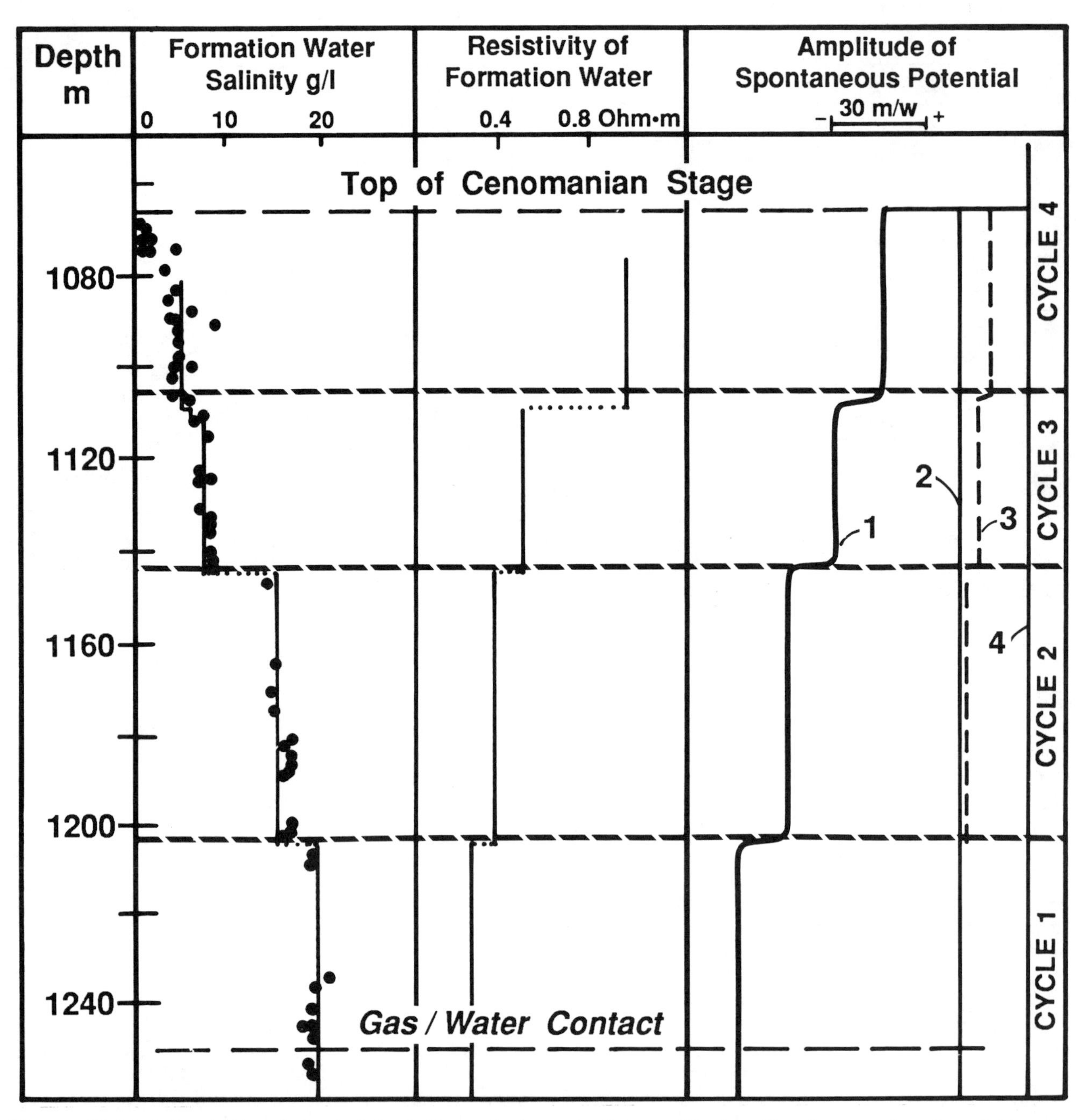

Figure 10. Variation in the characteristics of formation waters in the PK_{1-6} complex indicating the cyclic filling of the trap. Legend: 1, recorded amplitude of spontaneous potential; 2, limiting value of spontaneous potential amplitude without correction for change in salinity; 3, limiting value of spontaneous amplitude with correction for change in salinity; 4, shale line. (Modified from Akhiyamov, 1982.)

the hydrocarbons within parts of each uplift (Yermakov and Shalya, 1982a). The Yuzhno-Balyk Member thickens from south-southeast to north-northwest over the field, though the gross height of gas and oil columns increase in the opposite direction (Table 1). Net pay is generally 70 to 80% of gross sandstone (Maslennikov et al., 1980a). The liquid content of accumulations increases in lower units within the complex, as evidenced by oil pools without gas caps in the BU_{11}^{3} reservoir on both the northern and southern uplifts (Ognev et al., 1979).

The BU_9 reservoir is at a depth of from 8685 to 9092 ft (2648–2772 m) and marks the base of the Hauterivian Vartov Formation. The lower member of the Vartov, called the "chocolate" mudstone, encompasses the BU_9 and BU_8 reservoirs. Shales have a kaolinitic composition, and sandstones have weak zeolitic cements (Galerkina, 1979). In contrast to most

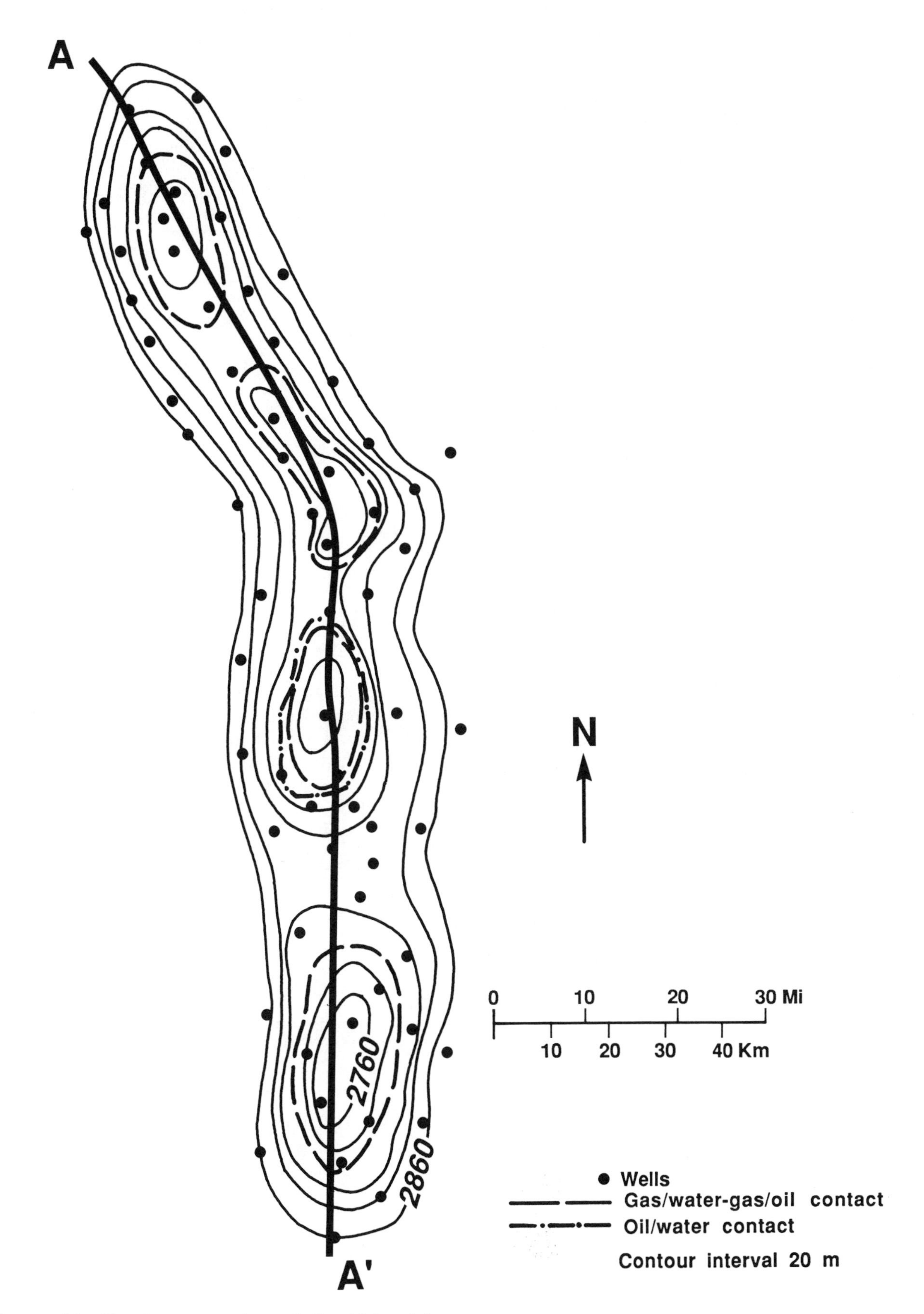

Figure 11. Structure map, top of the Valanginian Megion Formation BU_{11}^{2} reservoir. (Modified from Maslennikov et al., 1980.)

of the Megion reservoirs, the height of gross oil and gas columns increases to the north across the four major uplifts. Lower Vartov reservoirs also exhibit greater lateral continuity than either the underlying or overlying Neocomian horizons.

The BU_8 complex contains the largest gas resources in the Neocomian section (19.7 tcf [557 bcm]) and second largest accumulation of condensate in place (837 million bbl). There are four recognized stratigraphic subdivisions (BU_8^0 to BU_8^3), all of which produce gas and condensate on parts of all four major uplifts. All gas pools on the northern uplift, and most pools on the other three uplifts, have oil legs. Shale seals of from 13 to 33 ft (4 to 10 m) thick separate the units of the BU_8 reservoir vertically. The BU_8^0 reservoir is hydrodynamically connected across the entire field, though the high sand content on the northern uplift diminishes across the middle and central uplifts and is entirely replaced by shale in some wells on the southern uplift of the field (Figure 8) (Tikhomirov et al., 1978).

The upper Vartov Formation, which consists of the BU_5, $BU_{1\text{-}2}$, BU_0, and the AU_9 and AU_{10} reservoirs, is confined principally to the northern uplift and contains only gas and condensate. These reservoirs contain only a small share of the total resources of the Neocomian section and have low condensate content.

Albian to Cenomanian Reservoirs

The Albian-Cenomanian Pokur Formation is the continental basal unit of the second depositional megacycle in the Mesozoic-Cenozoic section of the basin. The formation is 6560 ft (2000 m) thick, composed of thick interfingered marine and nonmarine clastic members. Depositional environment of the lower Pokur Formation was fluvial (construed as the paleo-Pur River); these sediments were overlain by lower floodplain swamp and lake facies. The upper section shows higher frequency marine incursions with associated barrier and nearshore facies. The Cenomanian section of the Pokur Formation is 60 to 80% sandstone (Yermakov and Shalya, 1982b). The structure of the top of the Pokur Formation is shown in Figure 7 and in cross section in Figure 9.

The Cenomanian $PK_{1\text{-}6}$ reservoir complex marks the top of the Pokur Formation. It is a moderately heterogeneous unit both vertically and laterally, with a single gas-water contact. The structural plan of the field does not govern the distribution of the sandstone bodies. For example, the thickest sandstones occur in east-west belts, perpendicular to structural strike of the field (Yermakov and Shalya, 1982b).

Internally, there are eight fining-upward sequences, each from 49 to 164 ft (15 to 50 m) thick. The basal divisions of these cycles are fluvial facies, consisting of unconsolidated sands, sandstones, siltstones, and in the southern part of the field, gravels (Figure 9). Overlying rocks grade upward into lagoonal mudstones. Intraformational unconformities commonly divide the cycles but do not separate them hydrodynamically. Two of these unconformities are extensive, dividing the complex into three major productive intervals (Berman et al., 1979).

Berman et al. (1979) applied a classification algorithm to the reservoir characteristics of discrete lithologic units in the $PK_{1\text{-}6}$ complex. This analysis yielded eight classes of rock, which decrease in productive capacity as well as homogeneity of characteristics within groups down through the section (Table 3). Nevertheless, there are some regularities of reservoir quality that transcend these disparate lithologic subdivisions. Reservoir quality generally increases from the flanks of the structure to the crest:

1. Average grain diameter increases from 0.042–0.050 mm to 0.06–0.07 mm.
2. Average porosity increases from 25% to 30%.
3. Average permeability increases from 950–1000 md to 1300–1750 md, reaching a maximum in some areas of 7000 md (Zap. Sib. NIGNI, 1989).

Vertical and lateral heterogeneity and very high relative permeability of the $PK_{1\text{-}6}$ complex, to both gas and water, required strategic perforation to minimize coning of formation waters around production wells. In an analysis of well productivity in the Cenomanian Pokur reservoirs of Urengoy and Medvezh'ye fields, Temin (1987) recommended for four well clusters that only one be perforated over the entire productive interval, the second over the top two-thirds of the interval, the third over the top one-third of the interval, and the fourth only in the top part of the section. This produces the greatest pressure drawdown at the top of the reservoir, which has the maximum vertical distance from the gas-water contact, thereby minimizing the probability of localized water encroachment.

Though more than 99% of the discovered gas resources of the Pokur Formation are in the Cenomanian $PK_{1\text{-}6}$ complex, the Albian PK_{21} and PK_{18} reservoirs are productive (Zhabrev, 1983; Rudkevich et al., 1988a).

Sources of Jurassic-Neocomian Hydrocarbons

Basin-wide, the most prolific hydrocarbon sources are the Middle Jurassic to lower Neocomian transgressive marine shales. In the Urengoy region, these shales have an average total organic carbon content ranging from 2 to 3% in the Callovian to Oxfordian section and 6 to 9% in the Volgian Bazhenov Formation (Kontorovich et al., 1975; Filina et al., 1984). In the area of the Urengoy, Yamburg, and Medvezh'ye fields, the Upper Jurassic section is

Table 3. Reservoir characteristics of PK_{1-6} complex.

Unit	Porosity %	Permeability md	Gas Saturation %	Productivity Index*	Description
1	34.7	1000	89	1289	Sands; sandstone, predominantly 0.16 to 0.1 mm grains; medium to well sorted, weakly cemented
2	34.2	700	81	859	Sands; silty sandstone (silt 20%, fine sand 40%)
3	—	700	—	859	Same as 1 and 2 with interbedded shale and limestone
4	28.9	200	69	327	Siltstone, various grain sizes, low sand content; medium sorting, arkosic, weakly cemented
5	29.7	300	71	430	Same as 4
6	24.8	50	43	106	Sandstones, poorly sorted; grades to medium sorted siltstone
7	25.2	100	39	161	Interfingered subgroups 2, 4, 6, and 8 with shales
8	23.6	2	23	0	Siltstones, 39-47%, 0.05-0.01 mm grains, shale fraction, 38%

*mcf/day productive capacity per ft of pay for a 5689 psi drop in pressure.

Modified from Berman et al. (1979).

thermally mature for condensates and gas (Kontorovich, 1984).

Deep drilling in the field during in the 1980s allowed examination of the organic matter from the 4920 ft (1500 m) thick continental lake-swamp facies of the underlying Lower and Middle Jurassic Tyumen Formation. Based on pyrolysis data, both the gas and light liquid generation potential of the Tyumen Formation, particularly the top 1312 ft (400 m) of the section, exceeds that of both Upper Jurassic and Neocomian source rocks. Tyumen source rocks have also expelled a significant fraction of their hydrocarbons (Yemets et al., 1986).

Presently at a depth of from 11,677 to 13,776 ft (3560 to 4200 m), organic matter in the Tyumen Formation is mature for the generation of light liquids and gas. This raises the importance of the Lower and Middle Jurassic section, regionally and at Urengoy, as a source for Jurassic and Neocomian hydrocarbons and perhaps some share of the gas in the Pokur Formation.

Sources of Cenomanian Hydrocarbons

The origin of the gas in the Cenomanian PK_{1-6} complex has been highly controversial since the 1960s. Two alternative hypotheses on source mechanisms exist. The first hypothesis suggests that the same thermogenic Jurassic and Neocomian sources that charged Jurassic and Neocomian reservoirs produced the gas in the Pokur Formation. Kontorovich and Ragozina (1967) estimated that Upper Jurassic sources generated as much as 69 to 91 bcf/mi^2 (0.75 to 1 bcm/km^2) of wet gas in the Urengoy area. However, proposing a deep source relies on isotopic and molecular fractionation during migration to explain the 98 to 99% methane and isotopically light gas in the PK_{1-6} complex.

The second hypothesis suggests that biogenic alteration of humic organic matter in the Pokur Formation itself produced the gas. Yermakov et al. (1970) estimated biogenic gas generation of 17,665 tcf (500 tcm) from the Pokur coals—more than 20 times the volume of discovered gas resources basinwide.

The very high methane content suggests a biogenic origin, but analysis of $\delta^{13}C$ places the source in the transition zone between biogenic and thermogenic gas. This same combination of indicators is present in almost all Cenomanian gases in the northern West Siberian basin. Evidence reviewed by Grace and Hart (1986) favored a combination of *both* sources contributing to the gas in reservoirs in the Pokur Formation. Recently, speculation on the sources of the PK_{1-6} gases broadened further to encompass input from the overlying Kuznetsov Formation shale, the bottom 98 ft (30 m) of which are rich in humic organic matter (Rudkevich, 1988a; Clarke, 1989).

Hydrocarbon Characteristics

Two stratigraphic divisions can be made on the basis of differences in reservoired hydrocarbons: (1) the deep Jurassic, Neocomian, and Albian oil, condensate, and gas reservoirs; and (2) the shallow reservoirs of the Cenomanian Pokur PK_{1-6} complex, which contain 98 to 99% methane. Variations in the hydrocarbon characteristics provide additional information on the composition and thermal maturation of source material and the course of migration.

Oils and Condensates

Throughout the vertical extent of the field, there is a pronounced tendency toward increases downward of the liquid content and hydrocarbon density. Condensate is almost absent in the Cenomanian PK_{1-6} complex at a depth of 3608 ft (1100 m) (0.003 gal/mcf [0.3 g/cm³]). Of the 1 to 2% of C_{2+} methane homologs that do exist in the PK_{1-6} complex, medium density cyclanes compose almost all of the light, gasoline to naphtha, distillate fraction (distillation at less than 200°C [392°F]). Deeper, from a level of 0.07 gal/mcf (8 g/cm³) in the Albian PK_{18} pool at 5773 ft (1760 m), condensate content increases to a maximum of 6.1 gal/mcf (610 g/cm³) at 11,663 ft (3555 m) in the Achimov BU_{22} reservoir (Table 1 and Figure 12). Oil is restricted to the Megion Formation and the lower part of the Vartov Formation.

Condensate density also increases with depth, as does the percent arenes and alkanes in the light, gasoline to naphtha, fraction (Figure 12). The percent of cyclanes in the same distillate fraction decreases with depth. Associated with the increasing density, both the kerosene to lube oil distillate fraction (distillation greater than 200°C [392°F] and less than 420°C [788°F]) and the residual fraction (distillation at greater than 420°C [788°F]) of the condensate are higher in deeper pools.

In addition to differences between accumulations, these same parameters also vary within individual pools. The light, gasoline to naphtha, fraction of the condensates is highest and condensate density is lowest at the crest of the Neocomian reservoirs. Density increases toward the gas-oil contact (for example in the BU_8, from an average of 0.742 g/cm³ to 0.783 g/cm³) because of the reduction of light fractions (alkanes, isoalkanes, and light methane homologs) and a relative increase in the share of cyclanes and arenes (Rudkevich et al., 1988a).

Neocomian oil at Urengoy is light (0.8 to 0.856 g/cm³ [34 to 45° API]); has a low sulfur (0.1%) content with a light, gasoline to naphtha, fraction of 20 to 28%; and has 6% paraffins, 0.2% asphaltenes, and 2.5% residual (Goncharov, 1987). The volatile fractions of the oils have alkane, cyclo-alkane, and alkyl cycloalkane compositions. The composition of the oil in the BU_{12} (roughly in the middle of the oil-bearing section) is 50% iso- and n-paraffins, 47% naphthenes, and 3% aromatics. Though condensate density increases with depth, oil density diminishes from the lower Vartov Formation to deeper pools as heavier, more complex molecules are cracked.

In their study of northern West Siberian gas-condensate accumulations, Nemchenko and Rovenskaya (1987) differentiate "primary" and "secondary" condensates. The former is not associated with oil rims; the latter is. They suggest that the primary condensate accumulations in the northern part of the West Siberian basin arise directly from the organic matter of the Cretaceous System. Secondary condensates represent the "stripped" light, gasoline to naphtha, fraction of underlying naphthenic oils. In core, the BU_{8-9} and BU_{10-11} horizons both show residual oil saturation in the gas-bearing, oil-bearing, and water-bearing parts of the section (Maslennikov et al., 1980b). This supports the hypothesis that these were originally oil pools, with subsequent formation of the gas-condensate caps.

Shimanskiy et al. (1977) conducted chromatographic analysis of the Neocomian condensates and found significant differences between accumulations at different stratigraphic intervals, suggesting that the reservoirs were filled in stages. Lateral differences (from pool to pool in the same unit) indicate that the drainage area for the northern uplift of the field included the transformation products of organic matter of lower thermal maturity than that in the central and southern parts of the field. Separate analysis cited by Rudkevich et al. (1988a) suggested that lateral migration of hydrocarbons into Neocomian traps proceeded predominantly from north to south and from east to west.

Gases

The maximum methane content in the field is in the shallow PK_{1-6} complex. Methane decreases downward through the section, with corresponding increases in its heavier homologs (Table 4). Gases from all producing horizons have a low content of nonhydrocarbon gases. Hydrogen sulfide has been encountered in only one sample (Rudkevich et al., 1988a). The chemical composition of the hydrocarbon gases is given in Table 4.

The isotopic composition of the gases also varies vertically through the productive section. Data on $\delta^{13}C$ from Alekseyev (1974) and Yermakov et al. (1970) average -59 for the PK_{1-6} complex (at depths of from 3608 to 4100 ft [1100–1250 m]) and -43 for reservoirs of the Neocomian section (at depths of from 9889 to 9938 ft [3015–3030 m]). In his analysis of reservoir gases across the U.S.S.R., Alekseyev (1974) defined three diagnostic ranges of $\delta^{13}C$ for determining the method of formation: biogenic, from -55 to -95; transition, from -55 to -65; and thermogenic, from -35 to -58. With these criteria, the deeper gases are clearly thermogenic, but the methanes of the PK_{1-6} are in the transition zone, supporting the hypothesis of mixed sources.

EXPLORATION CONCEPT

The most important exploration question to ask about the largest gas field in the world is, why is

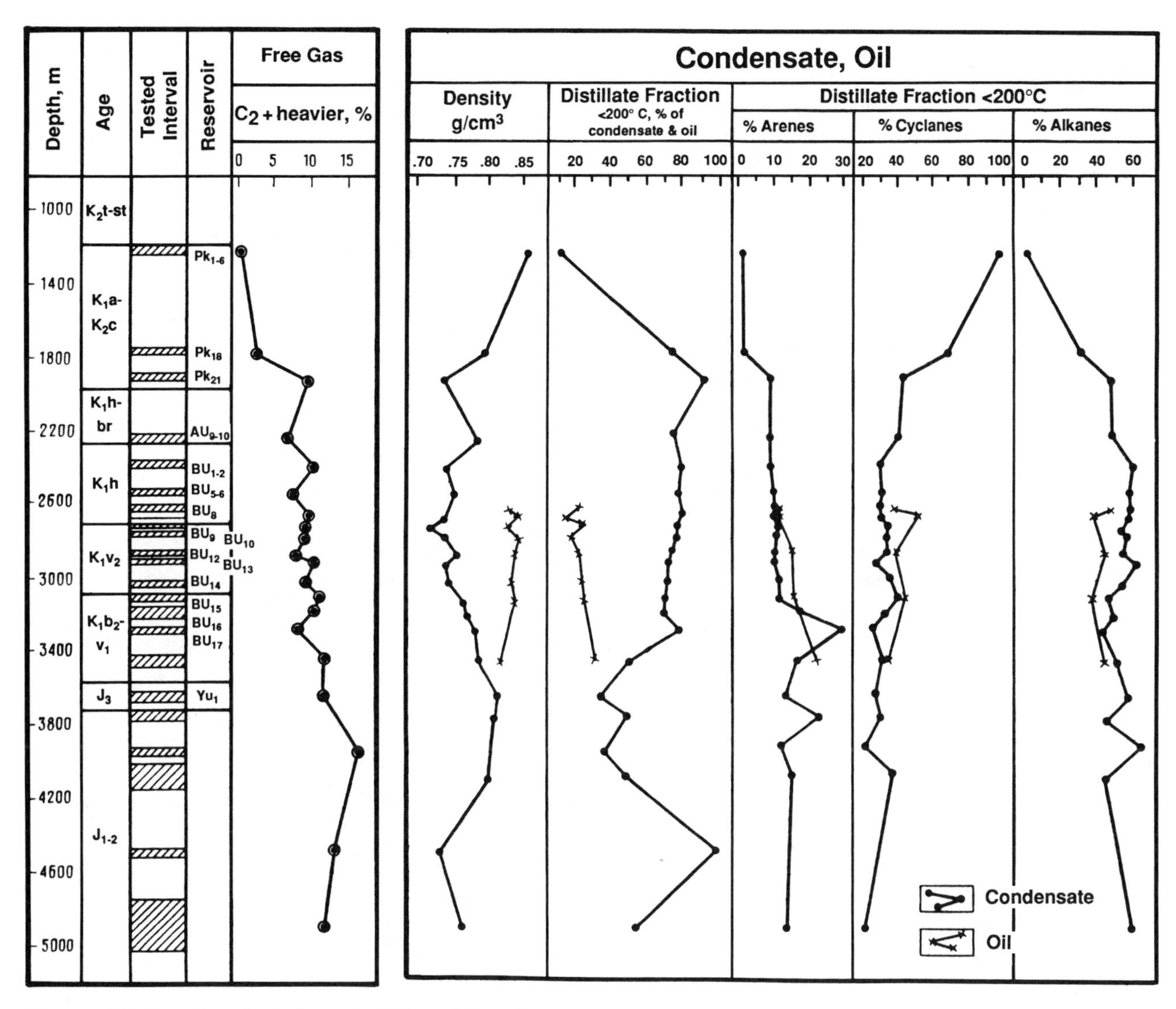

Figure 12. Geochemical characteristics of the oil, gas, and condensate of the Urengoy field. (Modified from Rudkevich et al., 1988a.)

Table 4. Chemical composition of hydrocarbon gases.

Reservoir	Density (g/cm³)	Percent Composition by Volume						
		CH_4	C_2H_6	C_3H_8	C_4H_{10}	C_5H_{12+}	CO_2	N_2
PK_{1-6}	0.560	99.05	0.06	0.01	—	—	0.08	0.80
PK_{21}	0.610	93.74	3.5	0.18	0.29	1.34	0.57	0.38
BU_{1-2}	0.653	89.00	5.15	2.33	1.08	1.44	0.19	0.81
BU_5	0.665	88.21	5.34	2.57	1.09	2.20	—	0.38
BU_8^0	0.706	86.20	5.71	2.25	0.91	4.20	0.30	0.43
BU_8^{1-3}	0.674	88.29	5.29	2.42	1.00	2.52	0.01	0.48
BU_9	0.685	88.49	4.64	2.23	0.91	3.18	0.46	0.09
BU_{10-11}	0.704	86.99	5.01	2.48	1.03	3.75	0.35	0.38
BU_{12}	0.670	89.88	4.46	2.11	0.90	2.52	0.17	0.46
BU_{13}	0.780	81.35	6.85	3.00	1.18	6.69	0.36	0.57
BU_{14}	0.747	82.60	6.21	2.91	1.00	5.39	0.25	1.59

Source: Zhabrev (1983).

it there? The answer is that *all* of the factors that control the size of oil and gas fields were maximized.

At the largest scale, two interrelated factors certainly played a fundamental role. First is the size of the West Siberian basin. The second is its low level of tectonic activity during the development of the regional hydrocarbon systems from the Jurassic Period to the present. Throughout the Mesozoic Era, stable low relief across the immense basin repeatedly fostered extensive shallow seas and very broad bands of deltaic plains and coastal environments. In this setting, several prolific regional hydrocarbon source rocks developed, which periodically introduced billions of barrels of oil and trillions of cubic feet of gas into the northern West Siberian basin. High frequency changes in sea level across the broad lowland created extensive interbedded high-quality reservoirs and competent local and regional shale seals.

On a local scale, the Urengoy megaswell, which includes the four major uplifts of the Urengoy field and neighboring uplifts that trap more than a half dozen smaller fields, was a key structural focus for migrating hydrocarbons. The timing of structural growth not only provided timely traps, but played a synchronous role in liberating as well as directing the gas into those traps. The role of glaciation in concentrating methane dissolved in formation waters also distinguishes mechanisms operating here from most other basins.

Considering the joint effect of the regional and local influences, it does not seem so unlikely that such a gigantic field developed where it did—nor is it surprising that such a propitious confluence of so many positive elements is an extremely rare event.

ACKNOWLEDGMENTS

The authors wish to express their deep appreciation to Drs. I. I. Nesterov, V. I. Shpil'man, and scientists at Zap. Sib. NIGNI for providing current information on the field. The authors also wish to thank Drs. R. M. Slatt, W. J. Ebanks, J. W. Clarke, and E. A. Beaumont for their comments on this paper and to the management of ARCO Exploration Research for permission to publish the paper.

REFERENCES

Akhiyarov, V. Kh., 1982, A method of delineation and evaluation of the Cenomanian reservoirs of the fields of northern Tyumen [in Russian]: Geologiya Nefti i Gaza, n. 5, p. 19-21.

Aleinikov, A. L., et al., 1980, Dynamics of the Russian and West Siberian platforms, *in* A. W. Bally, ed., Dynamics of plate interiors: Washington, D.C., American Geophysical Union, p. 53-71.

Alekseyev, F. A., 1974, Zonality in oil and gas formation in the Earth's crust based on isotope studies [in Russian]: Geologiya Nefti i Gaza, n. 4, p. 62-67.

Bally, A. W., and S. Snelson, 1980, Realms of subsidence, *in* A. D. Miall, ed., Facts and principles of world petroleum occurrence: Canadian Society of Petroleum Geologists Memoir 6, p. 9-94.

Berman, L. B., et al., 1979, Delineation of productive horizons within the limits of the Cenomanian pool at the Urengoy field [in Russian]: Gazovaya Promyshlennost', n. 1, p. 27-32.

Bochkarev, V. S., and Yu. I. Fedorov, 1977, Tectonics and the development of the Koltogor and Urengoy megatroughs (West Siberian plate) [in Russian], *in* V. S. Bochkarev, ed., Problemy tektoniki neftegazonosnykh oblastey Sibiri [Problems of tectonics of petroliferous districts of Siberia]: Tyumen, Trudy Zap. Sib. NIGNI, n. 125, p. 62-68.

Bochkarev, V. S., E. A. Latypova, and V. V. Tikhomirova, 1978, Paleotectonic analysis of the development of the Urengoy megaswell for the purpose of determining the time of formation of oil and gas pools in the Neocomian section [in Russian], *in* V. S. Bochkarev, Paleogeograficheskiye i paleotektonicheskiye kriterii razmeshcheniya zalezhey nefti i gaza v Zapadnoy Sibiri [Paleogeographic and paleotectonic criteria of the distribution of accumulations of oil and gas in West Siberia]: Tyumen, Trudy Zap. Sib. NIGNI, n. 133, p. 78-89.

Borodkin, V. N., 1979, Factors controlling the distribution of hydrocarbon pools in the Neocomian beds of the Urengoy oil-gas-condensate field [in Russian], *in* A. P. Sokovskii, ed., Kritirii poiskov i zakonomernosti razmeshcheniya zalezhey nefti i gaza v tsentral'nykh i severnykh raionakh Zapadnoy Sibiri [Criteria for exploration and the regularities of the distribution of oil and gas pools in the central and northern regions of West Siberia]: Tyumen, Trudy Zap. Sib. NIGNI, n. 145, p. 42-55.

Braduchan, Yu. V., F. G. Guari, V. A. Zakharov, et al., 1986, Bazhenovskiy gorizont Zapadnoy Sibiri (Stratigrafiya, paleogeografiya, ekosistema, neftegazonosnost') [The Bazhenov Formation of West Siberia (stratigraphy, paleogeography, ecosystem and oil and gas potential)]: Novosibirsk, Nauka, 217 p.

Chernomyrdin, V. S., 1984, Accelerated development of a northern giant [in Russian]: Gazovaya Promyshlennost', n. 1, p. 2-3.

Chernomyrdin, V. S., 1988, Present state and perspectives of the Soviet gas industry: Pipeline, August, p. 12, 13, 29.

Clarke, J. W., 1989, A possible source of the methane in the Cenomanian gas pools of northern West Siberia: Petroleum Geology, v. 23, n. 5/6, p. 206.

Clarke, J. W., and J. Rachlin, 1980, Salym: potential West Siberian giant: Oil and Gas Journal, v. 78 (June 16), p. 132-135.

Filina, S. I., M. V. Korzh, and M. S. Zonn, 1984, Paleogeographiya i neftenosnost' Bazhenovskoy svity Zapadnoy Sibiri [Paleogeography and oil potential of the Bazhenov Formation of West Siberia]: Moscow, Nauka, 35 p.

Fueg, J., 1988, The USSR sets another record: World Oil, August, p. 74-79.

Galerkina, S. G., 1979, Stroeniye nizhnemelovoy tolschahi i severa Zapadnoy-siberskoyplity [Structure of the Lower Cretaceous beds of the West Siberian plate] *in* Stratigrafiya Nizhnemelovykh Otlozheney Neftegazonosnykh Oblastey SSSR, Trudy: All-Union Scientific-Research Institute for Geologic Exploration, p. 1-30.

Geresh, P. A., 1984, Aspects of establishing production at the Urengoy field [in Russian]: Gazovaya Promyshlennost', n. 3, p. 30.

Goncharov, I. V., 1987, Geokhimiya neftey Zapadnoy Sibiri [Geochemistry of oils of West Siberia]: Moscow, Nedra, 177 p.

Grace, J. D., and G. F. Hart, 1986, Giant gas fields of northern West Siberia: American Association of Petroleum Geologists Bulletin, v. 70, n. 7, p. 830-852.

Gurkov, G., and V. Yevseyev, 1984, Tapping Siberian wealth: the Urengoy experience: Moscow, Progress Publishers, 194 p.

Klemme, H. D., 1971, What giants and their basins have in common: Oil and Gas Journal, v. 69, pt. I, n. 9, p. 85-90; pt. II, n. 10, p. 103-110; pt. III, n. 11, p. 96-100.

Konovalov, Yu. G., and Yu. M. Glukhoyedov, 1977, Characteristics of the subsurface and history of development of the Urengoy megaswell; Petroleum Geology, v. 18, p. 434-436. [Original Russian in Trudy Zap. Sib. NIGNI, n. 125, p. 74-79.]

Kontorovich, A. E., 1984, Geochemical methods for the quantitative evaluation of the petroleum potential of sedimentary basins, *in* G. Demaison and R. J. Murris, eds., Petroleum

geochemistry and basin evaluation: American Association of Petroleum Geologists Memoir 35, p. 79-109.
Kontorovich, A. E., and E. A. Ragozina, 1967, Masshtaby obrazovaniya uglevodorodnykh gazov v mezozoyskikh otlozheniyakh Zapdano-Sibirskoy nizhmennosti [Scales of formation of hydrocarbon gases in the Mesozoic formations of the West Siberian lowland]: Novosibirsk, Trudy SNIIG-GIMS, n. 65, 197 p.
Kontorovich, A. E., I. I. Nesterov, F. K. Salmanov, et al., 1975, Geologiya nefti i gaza Zapadnoy Sibiri [Oil and gas geology of West Siberia]: Moscow, Nedra, 678 p.
Kortsenshteyn, V. N., 1970, Effect of periodic glaciations on the formation of the enormously large gas fields of the northern part of the Tyumen Oblast' [in Russian]: Akademiya Nauk SSSR, Doklady, v. 191, p. 1366-1369.
Kuzin, I. L., 1983, Effect of recent tectonic movements, oscillations of sea level, and climate changes on the formation of oil and gas fields in West Siberia [in Russian], in Regional'naya neotektonika Sibirii [Regional neotectonics of Siberia]: Novosibirsk, Nauka, p. 26-31. [English in Petroleum Geology, v. 20, n. 8, p. 358-360.]
Maksimov, S. P., ed., 1987, Neftyanye i Gazovye Mestorozhdeniya SSSR: Spravochnik [Oil and gas fields of the USSR: A Handbook], volume 2: Moscow, Nedra, 303 p.
Maslennikov, V. V., Z. D. Khannanov, and Z. A. Akhmadeyeva, 1980a, Structure of the oil rims of the $BU_{10\text{-}11}$ reservoirs of the Urengoy oil-gas-condensate field [in Russian]: Neftegazovaya Geologiya i Geofizika, n. 10, p. 22-26.
Maslennikov, V. V., V. M. Zhilin, and T. I. Panfilova, 1980b, The results of study of the oil rims in oil-gas-condensate pools [in Russian]: Neftegazovaya Geologiya i Geofizika, n. 9, p. 12-16.
Meyerhoff, A. A., 1983, Soviet petroleum: history, technology, reserves, potential, and policy, *in* R. G. Jensen, T. Shabad, and A. W. Wright, eds., Soviet natural resources in the world economy: Chicago, University of Chicago Press, p. 302-362.
Nemchenko, N. N., and A. S. Rovenskaya, 1987, Genesis of the gas-condensate pools and forecasting the phase composition of hydrocarbons in the northern West Siberia [in Russian]: Geologiya Nefti i Gaza, n. 2, p. 25-31.
Nesterov, I. I., F. K. Salmanov, and K. A. Shpil'man, 1971, Neftyanye i gazovye mestorozhdeniya Zapadnoy Sibiri [Oil and gas fields of West Siberia]: Moscow, Nedra.
Nesterov, I. I., Yu. P. Tikhomirov, Yu. A. Stovbun, 1984, Oil and gas bearing potential of the Berriasian and Jurassic formations at the Urengoy field and the goals for their study [in Russian]: Geologiya Nefti i Gaza, n. 12, p. 9-13.
Ognev, A. F., Yu. A. Stovbun, and Yu. A. Trenin, 1979, Aspects of the geologic structure of the hydrocarbon pools of the Neocomian of the Urengoy field [in Russian], *in* N. Kh. Kulakhmetov, ed., Sovrmennye metody pri izuchenii geologii nedr Zapadno-Sibirskoy ravniny [Contemporary methods for the study of the geology of the West Siberian Plain]: Tyumen, Trudy Zap. Sib. NIGNI, n. 139, p. 3-7.
Oil and Gas Journal, 1984, Soviet gas industry hits fast clip, but exports likely to lag goals: v. 82 (October 1), p. 73-76.
Oil and Gas Journal, 1985, Soviets finish sixth big gas line from W. Siberia: v. 83 (March 18), p. 66.
Oil and Gas Journal, 1987a, Crude oil flowing from giant Urengoy: v. 84 (August 24), p. 24-25.
Oil and Gas Journal, 1987b, West Siberia will be key to plans to boost hydrocarbon production: v. 84 (December 7), p. 33-40.
Oil and Gas Journal, 1987c, Urengoi, Yamburg flows build, Soviet economists concerned: v. 84 (December 28), p. 28-29.
Rudkevich, M. Ya., L. S. Ozeranskaya, N. F. Chistyakova, et al., 1988a, Neftegazonosnyye kompleksy Zapadno-Sibirskogo basseyna [Oil and gas bearing complexes of the West Siberian basin]: Moscow, Nedra, 303 p.
Rudkevich, M. Ya., S. I. Shpil'man, and N. N. Shipovalova, 1988b, Questions on the oil and gas potential of the Lower-Middle Jurassic formations of the northern regions of the West Siberian oil-gas basin [in Russian]: Geologiya Nefti i Gaza, n. 7, p. 12-16.
Sagers, M. J., 1988, News notes—review of the Soviet oil industry in 1987: International Geological Review, v. 29, n. 1, p. 113-120.
Sarkisiyan, S. G., and G. N. Komardinkina, 1971, Litologo-fatsial'nye osobennosti senomanskikh gazonosnykh otlozhenii severa Zapadno-Sibirskoy nizhmenosti [Litho-facial characteristics of the Cenomanian gas-bearing beds of the northern West Siberian Lowland]: Moscow, Nauka.
Shabad, T., 1984, News notes: International Geologic Review, v. 26, n. 2, p. 242-246.
Shimanskiy, V. K., et al., 1977, Features of individual hydrocarbon composition of condensates of the Urengoy field [in Russian]: Geologiya Nefti i Gaza, n. 10, p. 64-68. [English in Petroleum Geology, v. 15, n. 10, p. 459-461.]
St. John, B., A. W. Bally, and H. D. Klemme, 1984, Sedimentary provinces of the world—hydrocarbon productive and nonproductive: Tulsa, American Association of Petroleum Geologists, 35 p.
Temin, L. S., 1987, A system for penetrating Cenomanian pools of northern fields [in Russian]: Gazovaya Promyshlennost', n. 7, p. 52-54.
Tikhomirov, Yu. P., V. I. Konyukhov, and A. F. Ognev, 1978, The geological structure and the results of delineation of the gas-oil-condensate and gas-condensate pools of the Urengoy field [in Russian] *in* Yu. P. Tikhomirov, ed., Metodika razvedki neftegazokondensatnykh mestorozhdeniy v Zapadnoy Sibiri [Methods of delineating oil-gas-condensate fields in West Siberia]: Tyumen, Trudy Zap. Sib. NIGNI, n. 134, p. 10-35.
Vinogradov, V. I., 1972, I. M. Gubkin—distinguished scholar, founder of the Soviet school of petroleum geology and petroleum formation, prominent government and social figure [in Russian], *in* Gubkin skiye chteniya, k 100-letiyu so dnya rozhdeniya [Gubkin reader, in honor of the 100th anniversary of his birth]: Moscow, Nedra, p. 13-21.
Yazik, A. V., 1984, An optimal system for the treatment of gas from the Urengoy field [in Russian]: Gazovaya Promyshlennost', n. 1, p. 8-9.
Yemets, T. P., N. V. Lopatin, and V. N. Litvinova, 1986, Catagenesis and hydrocarbon potential of the Jurassic formations of northern West Siberia [in Russian]: Geologiya Nefti i Gaza, n. 1, p. 53-58.
Yermakov, V. I., and A. A. Shalya, 1982a, The structure of the productive horizons of the Megion Formation of West Siberia [in Russian]: Geologiya Nefti i Gaza, n. 5, p. 13-18.
Yermakov, V. I., and A. A. Shalya, 1982b, Formation of the Cenomanian pay zone in the north of the Tyumen District: Geologiya Nefti i Gaza, n. 1, p. 40-45. [English in Petroleum Geology, v. 20, n. 1., p. 16-18.
Yermakov, V. I., V. S. Lebedev, N. N. Nemchenko, A. S. Rovenskaya, and A. V. Grachev, 1970, Isotopic composition of carbon in natural gases in the northern part of the West Siberian plain in relation to their origin [in Russian]: Akademiya Nauk SSSR Doklady, v. 190, p. 683-686.
Zap. Sib. NIGNI [The West Siberian Scientific-Research Institute of Petroleum Exploration Geology], 1989, personal communication.
Zhabrev, I. P., ed., 1983, Gazovye i gazokondensatnye mestorozhdeniia—Spravochnik [Gas and gas-condensate fields—a handbook]: Moscow, Moscow, 374 p.
Zykin, M. Ya., and V. V. Tsarev, 1982, Aspects of delineation of pools of the Lower Cretaceous complex in fields of northern West Siberia [in Russian]: Geologiya Nefti i Gaza, n. 5, p. 10-13.

Appendix 1. Field Description

Field name .... *Urengoy field*

Ultimate recoverable reserves .... *335 tcf gas; 1.5 billion bbl condensate; 1.2 billion bbl oil*

Field location:

Country .... *U.S.S.R.*
State .... *Russian Soviet Federated Socialist Republic*
Basin/Province .... *West Siberian basin*

Field discovery:

Year first pay discovered .... *Cenomanian Pokur formation zone 1966*
Year second pay discovered *1966*
Third pay *1966*

Discovery well name and general location:

First pay .... *Urengoy #2, 120 km southwest of village of Taz*
Second pay .... *Urengoy #1*
Third pay

Discovery well operator .... *Glavtyumengeologiya*

Second pay .... *Glavtyumengeologiya*
Third pay .... *Glavtyumengeologiya*

IP in barrels per day and/or cubic feet or cubic meters per day:

First pay .... *255 mcf/d (Pokur Formation)*
Second pay .... *28 mcf/d (Megion Formation)*
Third pay

All other zones with shows of oil and gas in the field:

Age	Formation	Type of Show
Neocomian	*Vartov*	*Oil/gas/condensate*
Upper Jurassic	*Vasyugan*	*Oil/gas/condensate*

Geologic concept leading to discovery and method or methods used to delineate prospect, e.g., surface geology, subsurface geology, seeps, magnetic data, gravity data, seismic data, seismic refraction, nontechnical:

Reconnaissance reflection seismic in early 1960s identified major structural features of northern West Siberian basin, including the major anticlines on which Urengoy and other gas giants were discovered.

Structure:

Province/basin type .... *Klemme 2A-complex*

Tectonic history

West Siberian basin formed after the collision of the Siberian and Russian plates in the Paleozoic. Basement is a composite of elements from the Russian, Siberian, and Kazakh plates. Mesozoic and Cenozoic sedimentary cover overlies the basin framework. Major structural features in the north are reactivated basement highs.

Regional structure

Urengoy megaswell is north-south-trending structure in line with similar structural uplifts in north-central part of the basin. Field occupies crest of structure, smaller fields on flanks.

Local structure

Urengoy megaswell is complicated by up to 9 separate uplifts at the Neocomian level that form unconnected reservoirs in the same stratigraphic units. Three subordinate uplifts exist at Cenomanian level, but reservoir hydrodynamically interconnected throughout.

Trap:

Trap type(s)

The anticlinal trap in the Cenomanian is sealed vertically by the Kuznetsov shale. Reservoirs in the Neocomian section are sealed by interbedded shale members.

Basin stratigraphy (major stratigraphic intervals from surface to deepest penetration in field):

Chronostratigraphy	Formation	Depth to Top in ft
Paleogene	*Several formations*	*Surface*
Turonian	*Kuznetsov*	*1214*
Aptian-Cenomanian	*Pokur*	*3444*
Neocomian	*Vartov*	*6888*
Neocomian	*Megion*	*9020*
Neocomian	*Achimov*	*9840*
Jurassic	*Vasyugan*	*11,152*
Volgian	*Bazhenov*	*11,480*

Location of well in field

Reservoir characteristics:

Number of reservoirs *88*

Formations .... *Vasyugan, Megion, Vartov, Pokur*

Ages .... *Upper Jurassic through Cenomanian*

Depths to tops of reservoirs .... *See above*

Gross thickness (top to bottom of producing interval) .... *8239 ft*

Net thickness—total thickness of producing zones

Average .... *82 ft in Cenomanian, 33 ft in Neocomian*

Maximum .... *328 ft in Cenomanian*

Average

Maximum

Lithology

The Cenomanian reservoir is a predominantly continental sandstone with interbedded paralic coals and shales. The Neocomian units are more heavily influenced by marine sedimentation and have a higher shale content.

Porosity type .... *Primary intergranular*

Average porosity .... *Cenomanian = 24%; Neocomian = 17%*

Average permeability .... *Cenomanian = 330 md; Neocomian = 73 md*

Seals:

Upper

Formation, fault, or other feature .... *Cenomanian reservoirs, Kuznetsov shale; Neocomian, interbedded Neocomian shales*

Lithology

Lateral

Formation, fault, or other feature .... *None*

Lithology

Source:

Formation and age .... *Middle-Upper Jurassic shales, Neocomian shales, Albian-Cenomanian coals*

Lithology .... *Shale and coals*

Average total organic carbon (TOC) .... *Middle-Upper Jurassic, 3%*

Maximum TOC *10%*

Kerogen type (I, II, or III) .... *Jurassic-Neocomian, II; Albian-Cenomanian, III*

Vitrinite reflectance (maturation) .... *R_o = immature to >1.2*
Time of hydrocarbon expulsion .... *Unknown*
Present depth to top of source .... *3–16,000 ft*
Thickness .... *Unknown*
Potential yield .... *Unknown*

Appendix 2. Production Data

Field name .... *Urengoy field*

Field size:

Proved acres .... *Approx. 1.7 million acres*
Number of wells all years .... *Approx. 1000*
Current number of wells .... *Unknown*
Well spacing .... *Unknown*
Ultimate recoverable .... *335 tcf*
Cumulative production .... *Approx. 77 tcf*
Annual production .... *Approx. 12 tcf*
Present decline rate .... *Not in decline*
Initial decline rate
Overall decline rate
Annual water production .... *Unknown*
In place, total reserves .... *Approx. 478 tcf*
In place, per acre-foot .... *Unknown*
Primary recovery .... *325 tcf*
Secondary recovery .... *Unknown*
Enhanced recovery .... *Unknown*
Cumulative water production .... *Unknown*

Drilling and casing practices:

Amount of surface casing set .... *Unknown*
Casing program .... *Unknown*
Drilling mud .... *Unknown*
Bit program .... *Unknown*
High pressure zones .... *Occur on east flank in Neocomian reservoirs*

Completion practices:

Interval(s) perforated .... *Unknown*
Well treatment .... *Unknown*

Formation evaluation:

Logging suites .... *Unknown*
Testing practices .... *Unknown*
Mud logging techniques .... *Unknown*

Oil characteristics:

Type .... *Light, low sulfur, 6% paraffins, 0.2% asphaltines*
API gravity .... *35–41°*
Base
Initial GOR .... *Cenomanian contains no liquids; Neocomian unknown*

Sulfur, wt% *0.1%*
Viscosity, SUS .... *Unknown*
Pour point .... *Unknown*
Gas-oil distillate .... *Unknown*

Field characteristics:

Average elevation .... *250 ft MSL*
Initial pressure .... *1784 psi*
Present pressure .... *Unknown*
Pressure gradient .... *0.45 psi/ft*
Temperature .... *93°F*
Geothermal gradient .... *0.025/100 ft*
Drive .... *Gas expansion, some water drive*
Oil column thickness .... *Variable up to 60 ft; see Table 1*
Oil-water contact
Connate water .... *Unknown*
Water salinity, TDS .... *1-18 g/L*
Resistivity of water .... *4-8 ohm-m*
Bulk volume water (%) *11-37%*

Transportation method and market for oil and gas:

Six 56-in. pipelines carry the gas to market within the U.S.S.R. and for export, one 12-in. liquid pipeline carries liquids to the city of Surgut for processing.

Bowdoin Field—U.S.A.
Bowdoin Dome, Williston Basin

DUDLEY D. RICE
U. S. Geological Survey
Denver, Colorado

GARY L. NYDEGGER
Consultant
Golden, Colorado

CHARLES A. BROWN
Snyder Oil Company
Denver, Colorado

FIELD CLASSIFICATION

BASIN: Williston
BASIN TYPE: Cratonic Sag
RESERVOIR ROCK TYPE: Sandstone
RESERVOIR ENVIRONMENT OF DEPOSITION: Shallow Shelf
RESERVOIR AGE: Cretaceous
PETROLEUM TYPE: Gas
TRAP TYPE: Combination Stratigraphic-Structural

LOCATION

Bowdoin field is located in eastern Phillips and western Valley Counties in north-central Montana. It is situated on the Bowdoin dome, a large structural uplift on the west flank of the Williston basin, a major oil-producing [petroleum province] in the Rocky Mountain region. The Bowdoin field is one of several fields in the northern Great Plains that produces microbial gas from low-permeability (tight), low-pressure reservoirs at depths less than 3000 ft (900 m) (Figure 1). The other major area of shallow gas production is in western Canada (Figure 1).

The Bowdoin field covers an area of almost 600 mi^2 (1550 km^2) and terminates abruptly at the Canadian border. The productive rocks probably extend into Canada where similar conditions exist. Ultimate recovery of natural gas is estimated to be 475 bcf (13×10^9 m^3).

HISTORY

Natural gas was discovered in the Bowdoin dome in 1913 in a shallow well (about 700 ft; 200 m) drilled for water on the Martin ranch (sec. 18, T.31N., R.35E.). The Bowdoin dome was discovered by surface geology mapped and described by Collier of the U.S. Geological Survey (1918). The central part of the Bowdoin dome was developed through the 1950s by Montana Dakota Utilities with open-hole completions in gas wells drilled where the pay zones are shallowest. The first gas production was reported in 1929. With the advent of improved completion techniques and increased gas prices in the 1970s, the field was expanded in size to almost 600 mi^2 (1550 km^2).

One major problem existed with the expansion of the field in the 1970s and 1980s. The development program was designed to be run concurrently with the construction of a large gas-gathering facility. The purpose was to increase economic viability and to develop a more efficient optimization of stimulation techniques and spacing criteria. However, delays in the approval of the pipeline facility in the 1970s forced the operators to use technology evolved in other areas. Specifically, the development of similar shallow gas accumulations in southeastern Alberta was used as an analog. Today, the Bowdoin field has more than 600 producing wells with the potential for many more if additional infill wells are drilled. However, the economic climate has reduced the desire to fully develop Bowdoin field at the present time.

DISCOVERY METHOD

The Bowdoin field is the result of serendipity. Natural gas was discovered in a shallow well (about

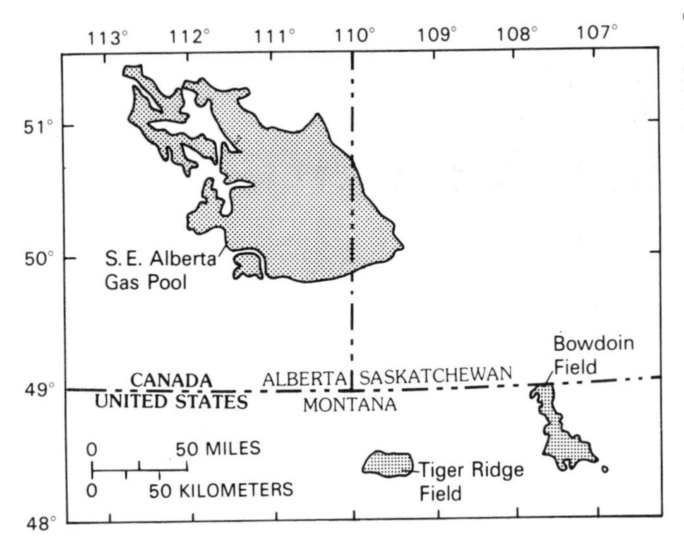

Figure 1. Map showing shallow gas fields (stipple) in north-central Montana, southeastern Alberta, and southwestern Saskatchewan. Bowdoin field and large area in western Canada produce from low-permeability reservoirs of Late Cretaceous age. Tiger Ridge field produces from structural traps of conventional reservoirs of Upper Cretaceous Eagle Sandstone.

700 ft; 200 m deep) drilled for water in 1913. Subsequent development of the natural gas resource followed the determination of the domal structure by surface mapping.

STRUCTURE

The Bowdoin dome is a very large (600 mi^2; 1550 km^2), nearly oval arch on the west flank of the structural-sedimentary, intracratonic Williston basin. The dome has structural closure of about 700 ft (200 m) at the top of the Upper Cretaceous Greenhorn Formation. Most recent field development is concentrated on a north-northwest-trending nose in north-central Montana that extends across the Canadian border. The symmetry of the dome is broken by local arches, saddles, noses, and by several northeast-southwest-trending faults (Figure 2).

STRATIGRAPHY

Natural gas in the Bowdoin field is trapped in shallow (less than 2000 ft; 610 m), low-permeability reservoirs of Late Cretaceous age (Cenomanian and Turonian). The carbonate and sandstone/siltstone reservoirs and surrounding shales were deposited on the shallow, western shelf of a north-south-trending Western Interior seaway that extended from the Gulf of Mexico to the Arctic Ocean. The producing intervals, from oldest to youngest, in the informal terminology of the drillers, are (1) the Phillips sandstone (Mosby Sandstone Member of the Belle Fourche Shale), which is as much as 200 ft (60 m) thick, (2) the Greenhorn lime (Greenhorn Formation), which averages 10 ft (3 m) in thickness, and (3) the Bowdoin sandstone (Carlile Shale), which is more than 250 ft (80 m) thick (Figures 3, 4, and 5). The major part of production is from the Phillips and Bowdoin sandstone reservoirs. The areal distribution of production from these units is shown in Figure 6.

Rice and Shurr (1980) showed the regional distribution of lithofacies that produce at Bowdoin field. Thin, discontinuous sandstones and siltstones typical of the Phillips and Bowdoin sandstones are referred to by them collectively as shelf sandstones. Sandstones are a minor part of the total shelf sequence that consists dominantly of shale. The Phillips sandstone is part of a southward-projecting lobe of shelf sandstones that covers much of central Montana and had a source area to the northwest in Canada (Rice, 1983) (Figure 7). The Bowdoin sandstone occurs in a northeast-trending belt of shelf sandstones that was derived from western Montana (Rice and Shurr, 1980).

The Greenhorn lime contains limestone, calcareous shale, and a widespread bentonite interval. It unconformably overlies the Phillips sandstone and was deposited during maximum transgression of the seaway. This unit and the underlying Phillips sandstone grade eastward into interbedded limestones and chalks typical of the Greenhorn Formation (Rice and Shurr, 1980).

Other potentially productive intervals include the Martin sandstone, a sandy zone of the Niobrara Formation, and the Shannon Sandstone Member of Gammon Shale (Figures 3 and 4). The Martin tested gas in the original discovery well but has not yet yielded commercial quantities of gas. On the southeast side of the Bowdoin dome near the Hinsdale fault (Figure 2), encouraging shows of gas have been encountered in the Gammon Shale. The Gammon correlates with the combined Telegraph Creek Formation and Eagle Sandstone to the west and with the Milk River Formation equivalent of southeastern Alberta. Shallow, thin laminae and beds of shaly sandstone and siltstone in the Milk River equivalent contain major reserves of gas in western Canada.

TRAP

Bowdoin field is a combination structural-stratigraphic trap. The gas is trapped in microstratigraphic traps consisting of thin, discontinuous sandstone and siltstone beds and laminae that are encased in shale with numerous bentonite beds that were deposited in a shallow shelf environment. The production is concentrated on a large domal feature, the Bowdoin dome, that experienced recurrent movement and influenced marine sedimentation during Late Cretaceous time. This interpretation is

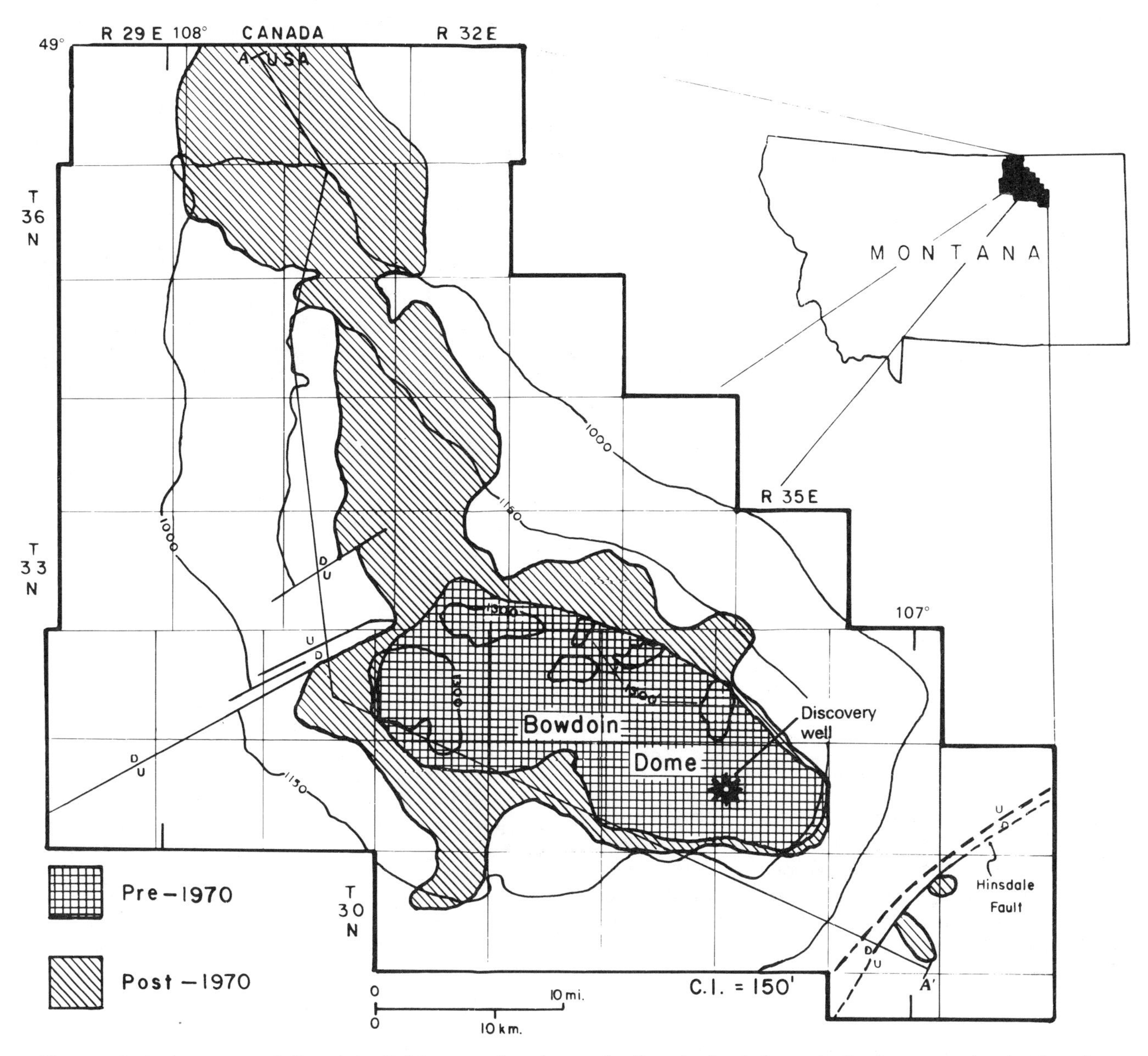

Figure 2. Index map of Bowdoin field area showing areas of pre- and post-1970 gas development and structure on top of the Greenhorn lime. Structure contours are in feet (1 ft = 0.3 m). Heavy lines indicate faults; dashed lines indicate approximate locations. Downthrown side, D. Upthrown side, U. Location of cross section A-A′ (Figures 4 and 5) is also shown.

particularly evident in the central part of the structure where the gas-productive shelf sandstones are better developed. The structure on which the Bowdoin field is located played an important role in trapping the gas in only the more porous intervals in the central part of the field. Elsewhere in the field, low permeability and discontinuity of the reservoirs have been the primary trapping mechanisms. The enclosing shales with numerous bentonite beds are not only the source of the gas, but also serve as the seal.

The exact age of the dome is difficult to establish since erosion has removed uppermost Upper Cretaceous and Tertiary strata. However, it is undoubtedly a product of the Late Cretaceous and early Tertiary Laramide orogeny, which occurred during the time represented by the eroded sediments. Stratigraphic evidence suggests that the general site of the dome must have existed as a shoal on the Late Cretaceous marine shelf that resulted in winnowing of the sediments.

RESERVOIRS

Thin, discontinuous, shaly siltstone and sandstone laminae and beds characterize the reservoirs of both the Phillips and Bowdoin sandstones (Figures 8 and

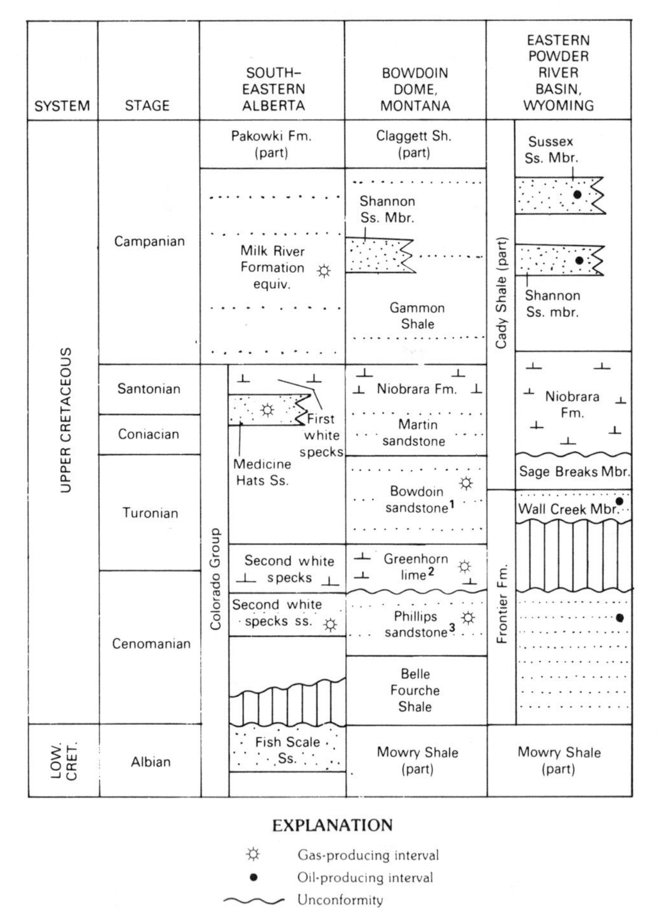

Figure 3. Chart showing Cretaceous hydrocarbon-productive zones in southeastern Alberta, Bowdoin dome, Montana, and Powder River basin, Wyoming. 1, informal unit equivalent to the Carlile Shale. 2, informal unit equivalent to the Greenhorn Formation. 3, informal unit equivalent to the Mosby Sandstone Member of the Belle Fourche Shale.

9). These reservoirs produce most of the gas in the Bowdoin field and occur at shallow depths (less than 2000 ft; 610 m). Individual beds, which are commonly less than 1 in. (3 cm) thick, cannot be correlated between wells and can be evaluated best with cores.

Examination of core samples by X-ray diffraction analysis, scanning electron microscope (SEM), and size fractionation analysis has defined the composition of the reservoir and the nature of the pore system. This information can be used to define the controlling factors of production and to design drilling and stimulation techniques. Part of the results of these studies is summarized by Starkey et al. (1978). Figure 10 illustrates some of the critical reservoir properties observed by SEM.

The Phillips sandstone consists of three coarsening-upward (progradational) sequences identified on gamma ray, density, and resistivity logs, and in cores (Figure 4). The sequences, as much as 75 ft (23 m) thick, grade upward from shale to very fine grained sandstone, which is commonly carbonate cemented at the top of each sequence. The sequences occur as elongate bodies, several hundred square miles in size, restricted to the main part of the domal structure and are referred to as sand ridges. All of the upper sequence and part of the middle sequence are progressively truncated at the north end of the dome by an unconformity below the Greenhorn lime (Figure 4). The sandstones at the top of the upper two sequences, where not removed by erosion, produce commercial quantities of natural gas and exhibit the best reservoir properties in the area.

Reservoirs in the Bowdoin sandstone are sandstones and siltstones that occur as clay-rich lenses and laminae and are interbedded with shale. Although individual reservoirs vary considerably from well to well, the overall interval with sandstone and siltstone maintains a fairly constant character over a large area. The Bowdoin sandstone is typically composed of 60% silt, 35% clay, and 5% sand. Furthermore, the Bowdoin sandstone is productive only where the silt-size material is concentrated as lenses and laminae usually less than 1 in. (3 cm) thick. Reservoirs in the Phillips usually have more sand, but in the productive intervals the coarser fraction occurs as lenses, laminae, or beds.

The silt- and sand-sized fractions of both the Phillips and Bowdoin sandstones consist dominantly of quartz with minor amounts of feldspar, gypsum, and pyrite. The quartz and feldspar are detrital in origin. The gypsum is probably a surface weathering phenomenon and does not occur in the subsurface reservoirs. Pyrite may be the end product of sulfate reduction that preceded microbial gas production. The Phillips sandstone also contains minor amounts of authigenic calcite and dolomite, which possibly precipitated early in conjunction with methane generation.

The clay-sized fraction of the Phillips and the Bowdoin represents the most important constituent of the rock in controlling production parameters. The clay-sized material is composed primarily of clay minerals: illite (mica) is the principal type, with minor amounts of kaolinite, mixed-layer smectite/illite, and chlorite in descending order of quantity. These clays are dispersed as a matrix between grains and are dominantly allogenic in origin—i.e., they were transported with the silt and sand grains and either deposited with them or mixed together by biological activity. Clays of authigenic origin, which formed by chemical precipitation, are of minor importance in the overall makeup of the reservoirs.

The clays are important because they have reduced greatly the permeability and hydrocarbon pore volume and have altered the geophysical log responses and calculations. In addition, they can greatly reduce recovery factors if not treated in the overall reservoir development. Most of the published research to date has dealt with reservoir problems created by authigenic clays. Almon and Davies (1978) discussed problems presented by each type of authigenic clay and the techniques used to offset these problems. The primary problem caused by

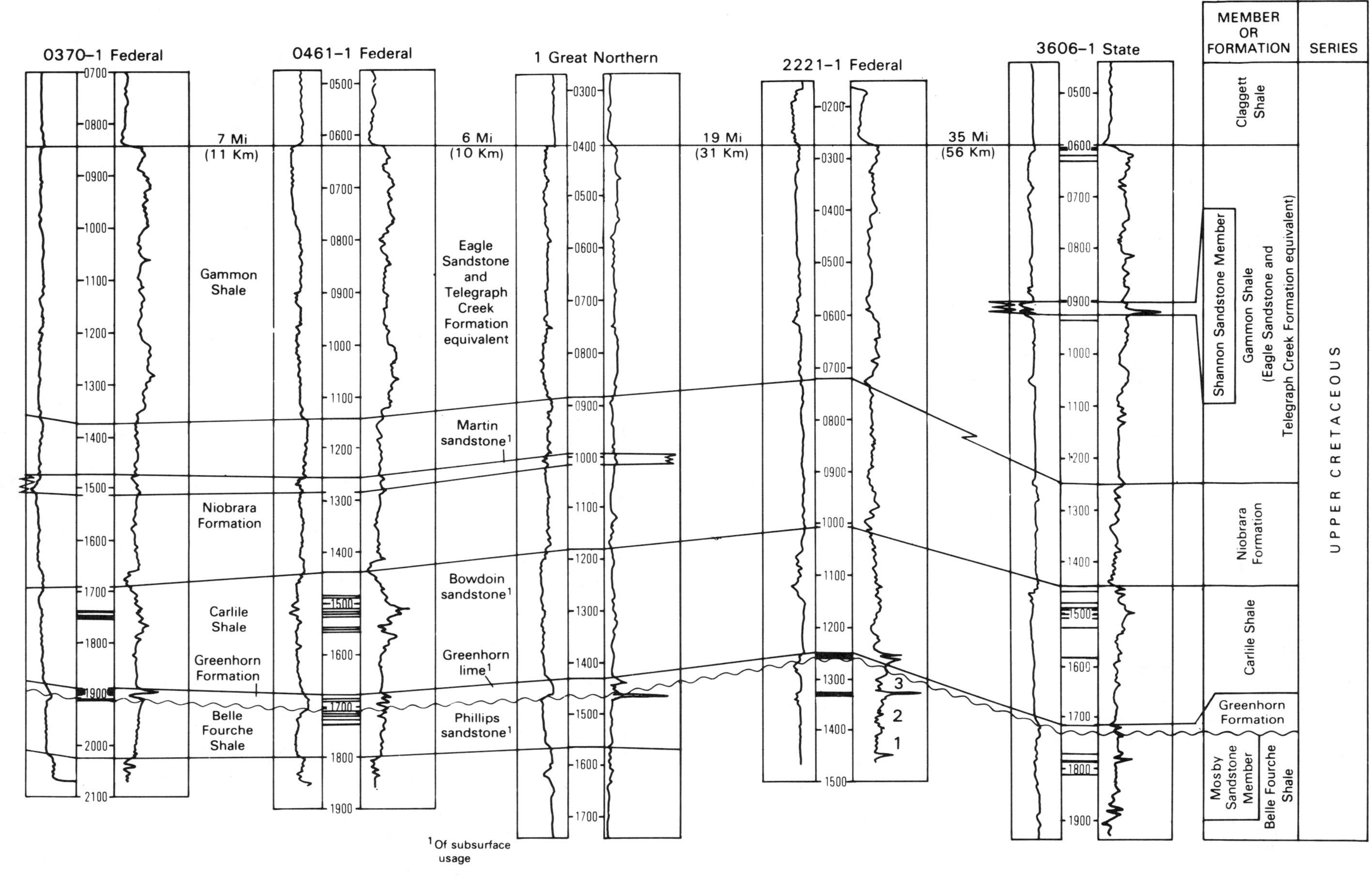

Figure 4. Electric log stratigraphic cross section across the Bowdoin field. Line of section shown on Figure 2.

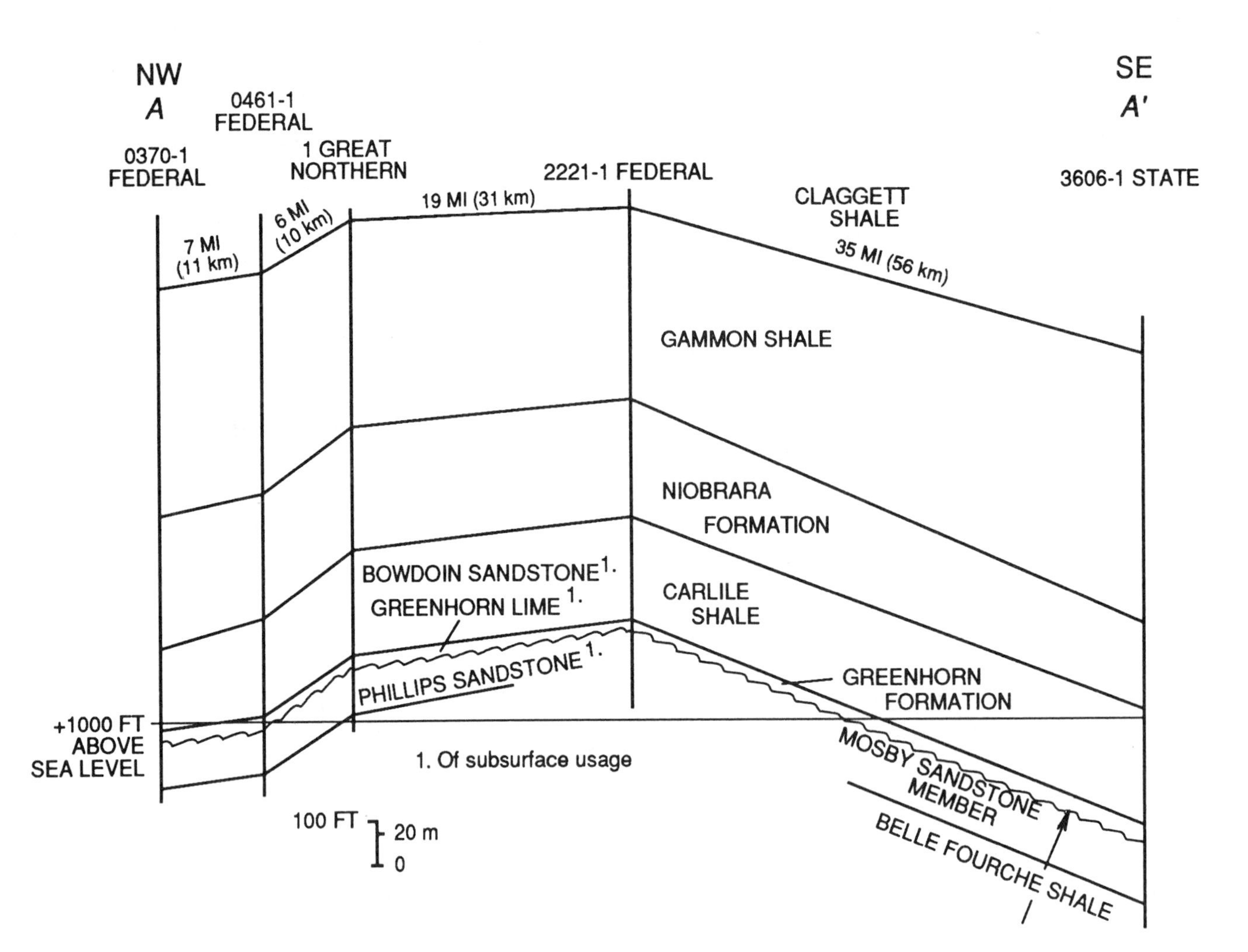

Figure 5. Structural cross section across the Bowdoin field. Line of section shown on Figure 2. Same line of section and wells with electric logs used in stratigraphic cross section (Figure 4).

allogenic clays is their high surface-area to pore-volume ratio, which increases their vulnerability to reaction with introduced fluids.

In contrast to the Phillips and the Bowdoin, the Greenhorn lime is dominantly a carbonate reservoir (Figures 10 and 11). The reservoir is a calcarenite with fragments of inoceramid shells and it contains discontinuous laminae of organic-rich shale that serve as a barrier to vertical fluid and gas movement. The unit is extremely brittle and readily fractures both naturally and by stimulation treatment.

A highly radioactive bentonite zone, which attains a maximum thickness of about 40 ft (12 m), makes up the basal part of the Greenhorn lime. The unit forms an impermeable barrier between reservoirs in the Phillips sandstone and the Greenhorn lime and acts as a seal for gas entrapment. The bentonite is composed of smectite (montmorillonite) and is highly susceptible to swelling.

Conventional measurements of porosity and permeability from cores alone are questionable in these clay-rich rocks because dehydration during and prior to analysis causes extensive cracking. However, a combination of core analysis, SEM, and transient pressure analysis has given some confidence to the estimated porosity and permeability ranges. Productive sandstones in the Phillips have an estimated average porosity from core of 15 to 17% and have permeability values as much as 2 to 3 md. Productive zones in the Bowdoin sandstone, which are not always the cleanest zones indicated on the logs, have porosity values ranging from 8 to 14% and permeability values ranging from less than 0.1 to 0.7 md. Wells with extremely high initial flow rates have probably resulted from localized natural fracturing. Porosity and permeability values in the Greenhorn lime are highly variable, as indicated by both core analysis and SEM; they range from 6 to 13% and from less than 0.1 to 6 md, respectively. Higher values have probably resulted from microfracturing induced by coring procedures. Reservoir simulation and pressure transient studies recently conducted on several wells indicate that in situ gas permeability is less than 0.1 md for all three units.

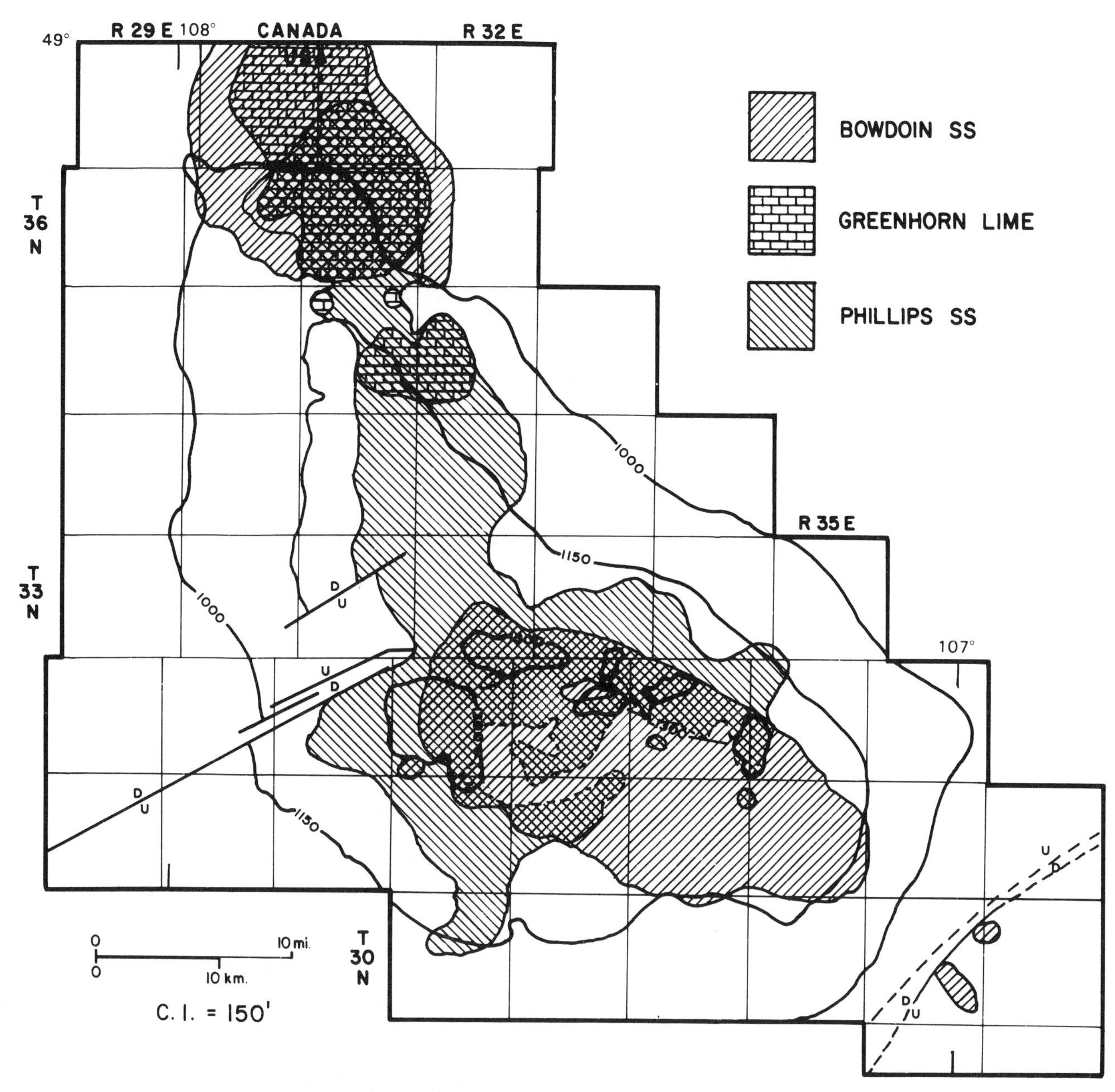

Figure 6. Map showing areal distribution of gas production from three main reservoirs in the Bowdoin field and structure on top of the Greenhorn lime.

Depositional Setting

The overall reservoir interval at Bowdoin dome was deposited on the western shelf of the Cretaceous Western Interior seaway. The shelf was probably storm dominated, and geostrophic storm flow was the main process responsible for transporting sediment (Swift and Rice, 1983). These storm flows were efficient at entraining sediment because of associated storm waves, and currents were capable of transporting sediment, especially during the winter storm season. The Western Interior shelf is interpreted to have had southward-flowing geostrophic currents induced by strong winds associated with the winter storms.

Reservoirs in the Bowdoin sandstone, which consist of thin beds and laminae of siltstone and sandstone, are widespread and can be explained by the "temporal acceleration" model (Swift and Rice, 1983) (Figure 12). This model is applicable under normal circumstances on the storm-dominated shelf during which currents accelerated and decelerated with time as storms waxed and waned. In this regime, subordinate amounts of sand were mixed with a silt and mud load without developing extensive sand deposits. The seafloor was eroded as the storm

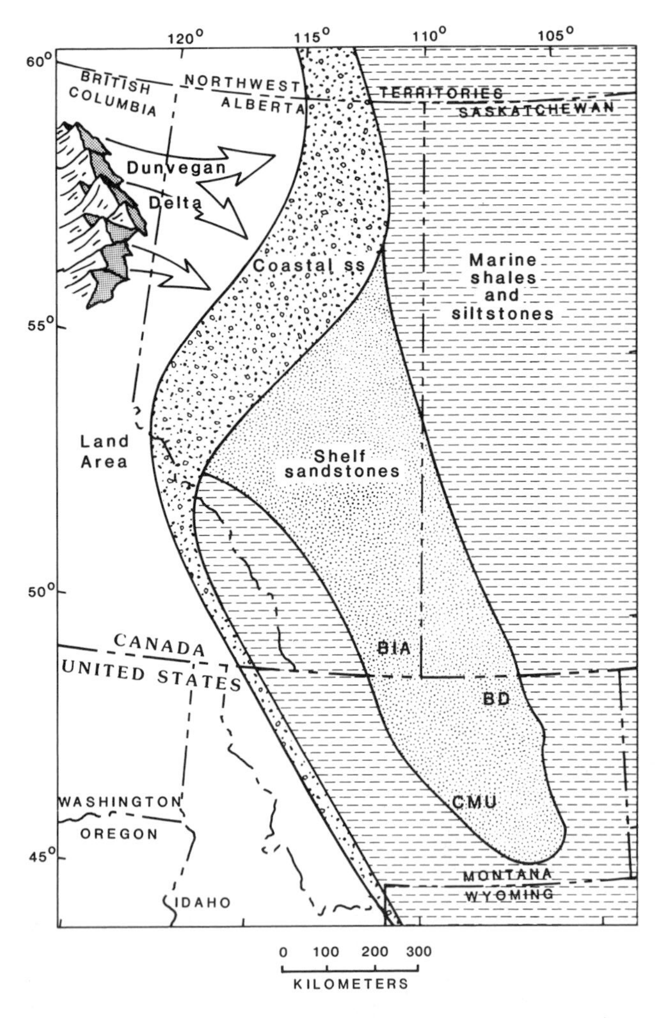

Figure 7. Paleogeographic map of northwestern United States and western Canada during time of deposition of Phillips sandstone. BIA, Bow Island arch; BD, Bowdoin dome; CMU, Central Montana Uplift.

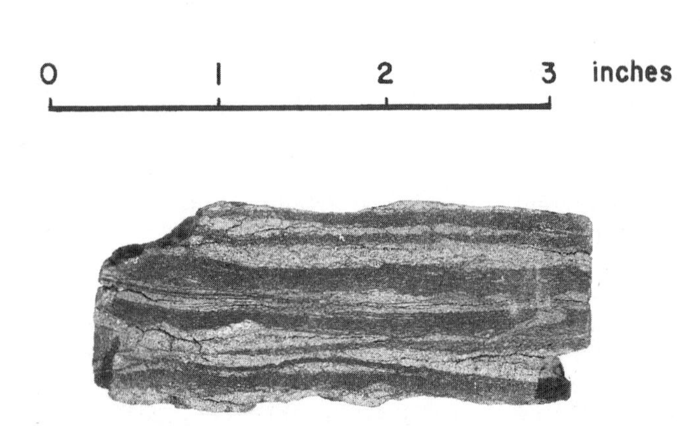

Figure 8. Core photo of lenses and laminae of very fine grained sandstone interlaminated with shale, Phillips sandstone. Federal No. 2-21 well, Sec. 21, T.33N., R.34E., 1169 ft (356 m). 1 in = 2.6 cm.

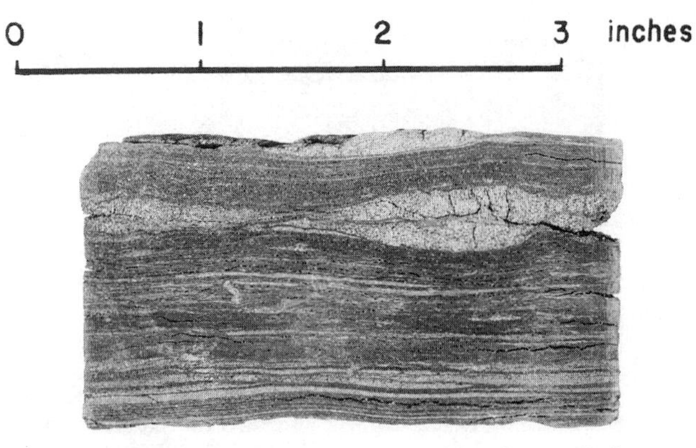

Figure 9. Core photo of lenses and laminae of siltstone and very fine grained sandstone interlaminated with shale, Bowdoin sandstone. Milk River No. 1-33 well, Sec. 33, T.32N., R.35E., 876 ft (267 m). 1 in = 2.6 cm.

currents accelerated, but the erosion ceased as soon as the flow reached capacity. Deposition of a graded bed took place as the storm waned and the storm deposit consisted of a sand and mud couple.

A different model, the "spatial acceleration" model (Swift and Rice, 1983) (Figure 13), was probably responsible for the development of thicker storm deposits and eventually sand ridges, which typify the Phillips reservoirs, particularly on the main part of the structure. Slight relief on the marine shelf, probably caused by recurrent movement on the ancestral Bowdoin dome, resulted in horizontal velocity changes. As a result, flow accelerated in space as well as time. Deceleration of flow occurred across the topographically high part of the dome as well as down its lee sides. The coarsest fraction, in this case very fine sand, was deposited in the zone of deceleration, and deposition occurred throughout the flow event. The process resulted in thicker storm beds in the Phillips sandstone than those in the Bowdoin sandstone. As the initial topographic relief of the dome was enhanced by the growth of the bed form, more sand was extracted from the transported load by each successive storm. Individual storm beds tend to fine upward, but each sequence as a whole coarsens upward because current perturbation and wave agitation were intensified as the amplitude of the bed form grew. The end result was the growth of sand ridges in the Phillips interval that are elongated in the direction of southward-directed flow.

In contrast, the dominantly carbonate reservoirs of the Greenhorn lime were deposited during a time of maximum transgression of the Western Interior seaway that coincides with a time of high global sea-level stand. Chemical precipitation of the carbonate facies resulted from a decrease in the supply of clastic detritus to the shelf areas.

Figure 10. SEMs illustrating properties of gas-producing intervals in the Bowdoin field. (A) Siltstone containing abundant laminae of allogenic clay, dominantly illite, which occurs as a matrix; individual silt grains are not distinguishable, Bowdoin sandstone. Federal No. 1 well 0370, Sec. 3, T.37N., R.30E., 1752 ft (534 m). (B) Close-up of A showing cubes of pyrite (P) and delicate needles of illite (I) occupying pore space. (C) Calcarenite with highly variable porosity and permeability, Greenhorn lime. Federal No. 1 well 1521, Sec. 15, T.32N., R.31E., 1902 (580 m). (D) Close-up of C showing coring-induced microfractures. (E) Very fine grained, silty sandstone with individual grains coated by dominantly allogenic clay, Phillips sandstone. Federal No. 1 well 1521, 1182 ft (360 m). (F) Close-up of E showing quartz grain (Q), authigenic calcite (C), and allogenic clay (A). Gypsum (G) is a surface weathering phenomenon.

Figure 11. Core photo of cross laminated calcarenite with shale laminae, Greenhorn lime. Federal No. 2-21 well, Sec. 21, T.33N., R.34E., 1163 ft (354 m). 1 in = 2.6 cm.

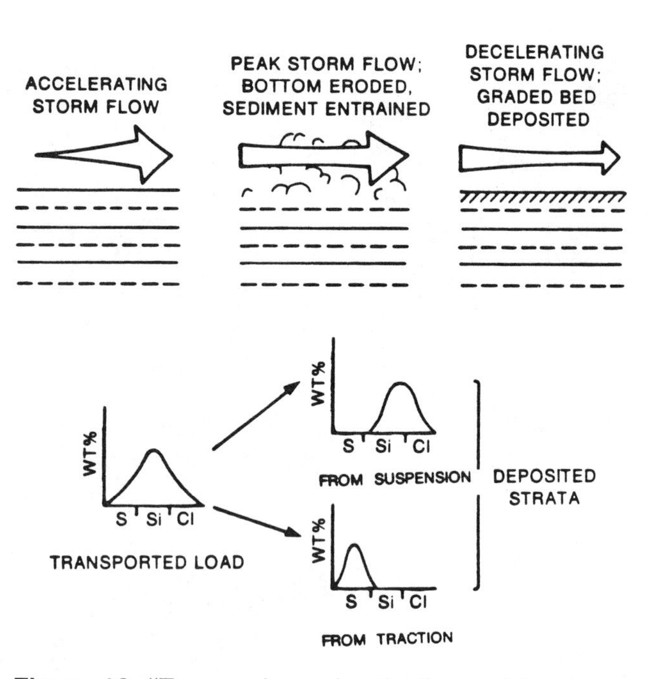

Figure 12. "Temporal acceleration" model for storm-dominated shelf sedimentation under normal conditions where velocity of flows varies primarily with time rather than in space. (From Swift and Rice, 1983.) The lower schematic shows the mode of transport according to particle size: S, sand; Si, silt; Cl, clay.

Reservoir Pressures

Original reservoir pressure of the Phillips, Greenhorn, and Bowdoin in the area of new development ranges from 400 to 630 psi (2760 to 4340 kPa). The reservoirs are slightly underpressured (0.3 to 0.4 psi/ft; 6.8 to 9.0 kPa/m) compared to a normal hydrostatic gradient (0.43 psi/ft; 9.7 kPa/m). In contrast, the reservoir of the Bowdoin sandstone in the old part of the field originally was more underpressured (0.23 psi/ft; 5.2 kPa/m). This difference indicates that the new development from the Bowdoin sandstone to the north is an entirely new reservoir, while the Phillips sandstone has a similar original pressure gradient in both the old and new portions of the field. Variations in pressure and pressure gradients can be expected in areas where stratigraphy, structure, and capillary pressures are major influences. Some pressure variations near the old part of the field are due to pressure depletion from 30 years of production.

Reservoir Evaluation

The evaluation of all wells has been difficult because of the thin, discontinuous, clay-rich reservoirs. During the initial phases of exploration, mud logs indicated gross pay intervals, which were subsequently drill-stem tested. The empirical cutoffs used for drillstem testing were 10 Mcf/d (280 m^3/d) for zones with little liquid recovery and 20 Mcf/d (570 m^3/d) for zones with more than 100 ft (30 m) of liquid recovery. If excessive amounts of water were recovered, the wells were plugged and abandoned.

Several open- and cased-hole logging devices have been used to evaluate the reservoirs. The initial logging program used in the 1974 expansion program included a traditional suite of induction, neutron, density, and gamma ray surveys. The spontaneous potential curve had very little definition, and the induction log could not detect thin beds. Neutron and density logs showed no gas effects. The gamma ray log was misleading because of the various types of clay. This suite was helpful only in the better Phillips wells with initial potentials greater than 450 Mcf/d (13×10^3 m^3/d), which included only one-third of the wells in the field.

Later, microresistivity and sonic logs were run in addition to the traditional suite. Application of crossplot analysis was attempted and no definitive results were obtained although slight anomalies were observed. Because a suitable logging program was not available, wells drilled in the new area during most of 1976 were developed empirically by correlating zones with good flow, either natural or after-stimulation tests.

The most successful logging technique has been a full suite of porosity logs (sonic, neutron, and density). Figure 14 shows typical log responses that are normalized in the Niobrara Formation and shales in the lower part of the Bowdoin sandstone. An

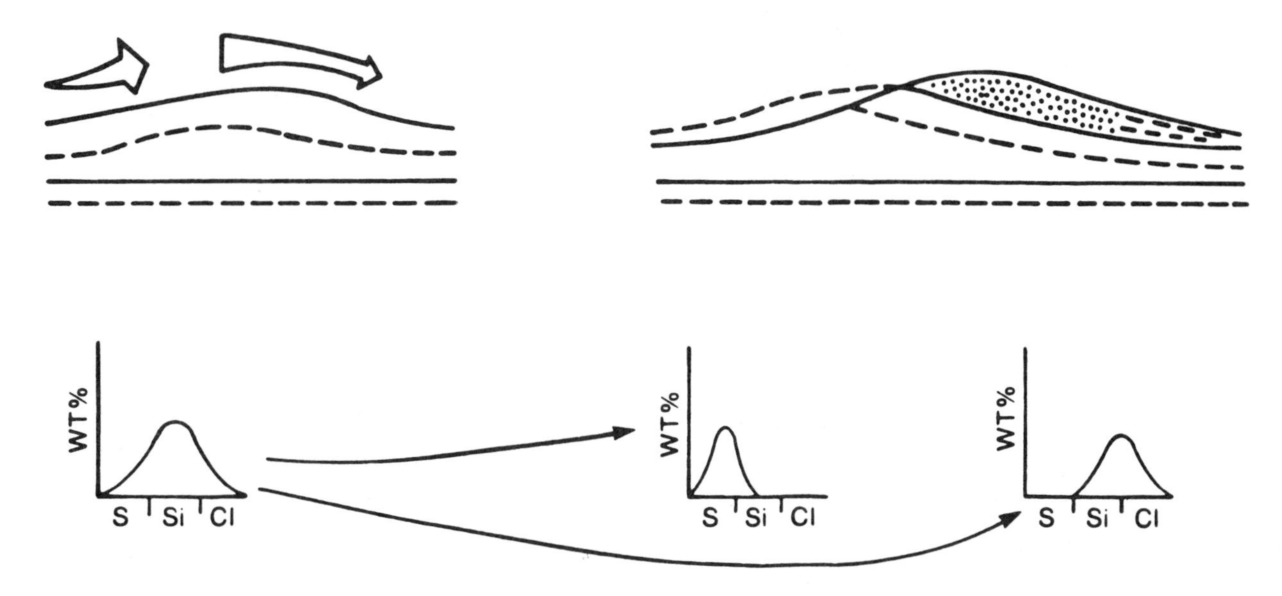

Figure 13. "Spatial acceleration" model for sand body formation when velocity of flows varies with space as well as time. (From Swift and Rice, 1983.) The lower schematic shows the mode of transport according to particle size: S, sand; Si, silt; Cl, clay.

overlay of interval transit time from the sonic log and porosity from the compensated neutron log is used as a gas indicator. A porosity overlay using sonic and density data is utilized for the determination of reservoir quality or clay content in a shaly sandstone. The crossover of the two overlays is interpreted as indicating a commercial gas-bearing reservoir. However, recent studies have indicated that this technique only identifies the better wells and potentially productive zones can be bypassed.

Pulsed-neutron and production (flow meter and temperature survey) surveys have been the primary cased-hole tools. No definitive results have been observed with the neutron logs owing to the high shale content. Production logs run on newly producing wells have yielded good results in wells with very little water production.

At this time, open- and cased-hole logging programs have been inconclusive in identifying commercially productive wells. Improvements must be made in distinguishing laminae and very thin beds and in determining effective porosity and gas saturation in shaly siltstone and sandstone reservoirs.

A research project funded by the U.S. Department of Energy's Western Tight Gas Sands Program evaluated the usefulness of the gamma ray spectral log for interpreting the geology and reservoir quality of the Bowdoin field. The log was successfully used to estimate organic-carbon content of source bed intervals and to identify subtle differences in clay composition (Gautier and Rice, 1983). These differences in clay composition affect the interpretation of values of irreducible water saturation, water sensitivity of the reservoirs, and selection of drilling and completion techniques.

SOURCE

Gas produced from the Bowdoin field contains 93% methane, 6% nitrogen, and minor amounts of ethane, propane, and CO_2. The gas has an average heating value of 950 Btu/ft^3 (35×10^3 kJ/m^3). The gases are considered of microbial origin because (1) the hydrocarbon composition consists primarily of methane, (2) the light isotope ^{12}C is enriched in the methane ($\delta^{13}C_1$ values of -68.3 to -72.3 ppt), and (3) the gases occur at shallow depths (1800 ft; 550 m), low temperatures (95°F; 35°C), and in immature (R_o <0.5%) sedimentary rocks (Rice and Claypool, 1981). They are typical of shallow Cretaceous gases throughout the northern Great Plains.

Microbial gas is the main product of the immature stage of hydrocarbon generation. The gas is generated by the breakdown of organic matter by anaerobic bacteria at shallow depths in rapidly accumulating sediments. In the Bowdoin dome area, this methane-rich gas generation probably took place in a relatively shallow-water, shelf environment at some depth below the sediment-water interface. In this marine environment, a succession of microbiological pro-

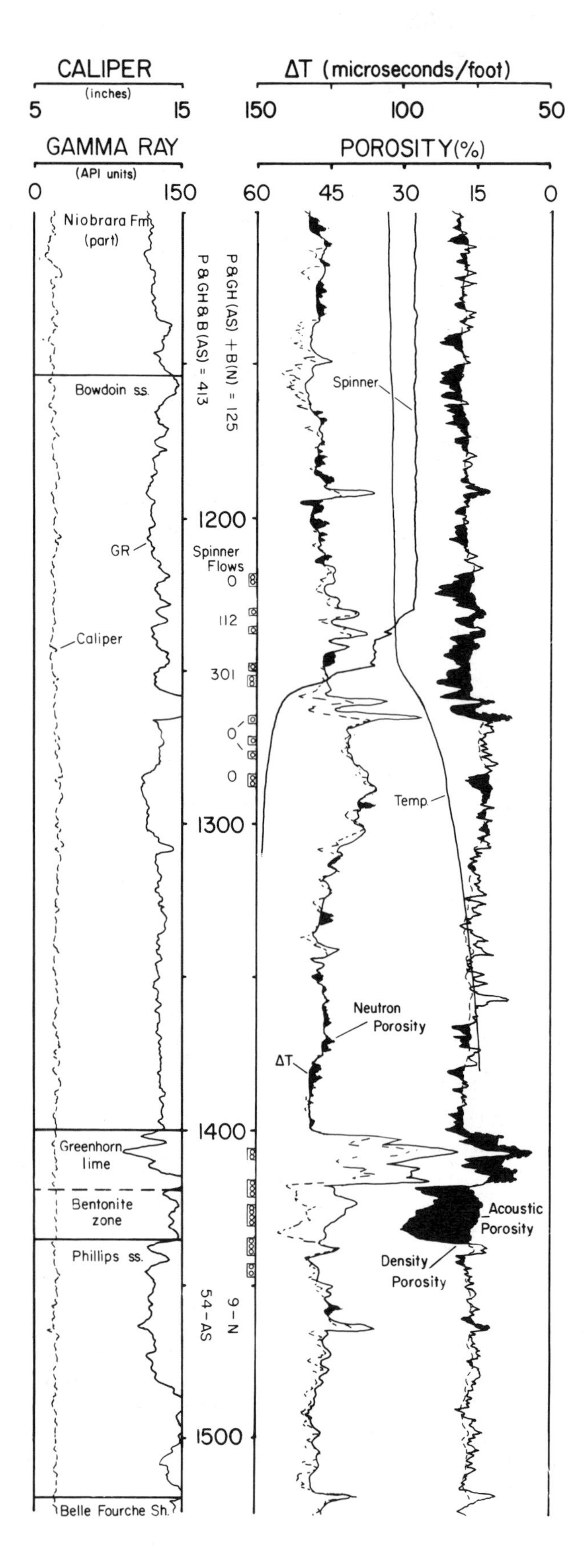

Figure 14. Suite of logs from Federal No. 1 well 1560 (Sec. 15, T.36N., R.30E.) showing natural flow (N) and after-stimulation flow (AS). All flow rates are given in thousand cubic feet per day (1 Mcf/d = 28 m^3/d), and depths are shown in feet (1 ft = 0.3 m). Shaded areas indicate crossover. Although crossover is indicated in the bentonite zone at the base of the Greenhorn lime, the bentonites serve as a seal and gas production has not been established from them.

cesses led to the generation of microbial gas (Figure 15). Generation began after the supply of oxygen was depleted and the sulfate in the sea water was greatly reduced. The gas is thought to have been generated in situ because it was trapped in discontinuous, low-permeability reservoirs, which prohibited migration. Also, the reservoirs are encased in organic-rich shales (average 2% TOC) with mostly type III kerogen that could have supported significant methane generation.

Most of the microbial gas was probably initially retained in solution in the pore water of surrounding shales because of higher methane solubility at higher hydrostatic pressures owing to the weight of the overlying water column (Rice and Claypool, 1981). The gas subsequently exsolved from the water during pressure release caused by uplift and erosion of the Bowdoin dome during Late Cretaceous and (or) early Tertiary time. Exsolution resulted in a free-gas state phase at depths less than about 2000 ft (610 m), which is the depth limit of free-gas phase gas in the Bowdoin field area. The free gas migrated to and accumulated in the discontinuous laminae and beds of siltstone and sandstone because of capillary pressure differentials between the shale and coarser-grained sediments.

DRILLING AND COMPLETION TECHNIQUES

A typical Bowdoin field well is drilled with native mud to the base of the glacial drift, which can be as much as 200 ft (61 m) thick. If necessary, lost circulation material is added to the mud system. A 7-in. (18 cm) surface casing is then set and cemented. A 6¼-in. (16 cm) hole is drilled with low-fluid-loss, polymer-based mud to the base of the Phillips sandstone. The polymer concentration is increased and the water loss is lowered through the pay zone. The well is logged and 4½-in. (11 cm) casing is set. The pipe is run with a float shoe, latch-down plug and baffle, centralizers to surface, and scratchers across the pay zones. In wells with multiple productive zones, multifracture baffles are also used. A 200 gal (0.757 m^3) mud-flush spacer is pumped, followed by 75 to 100 sacks of light cement and an equal amount of Class G cement. During the

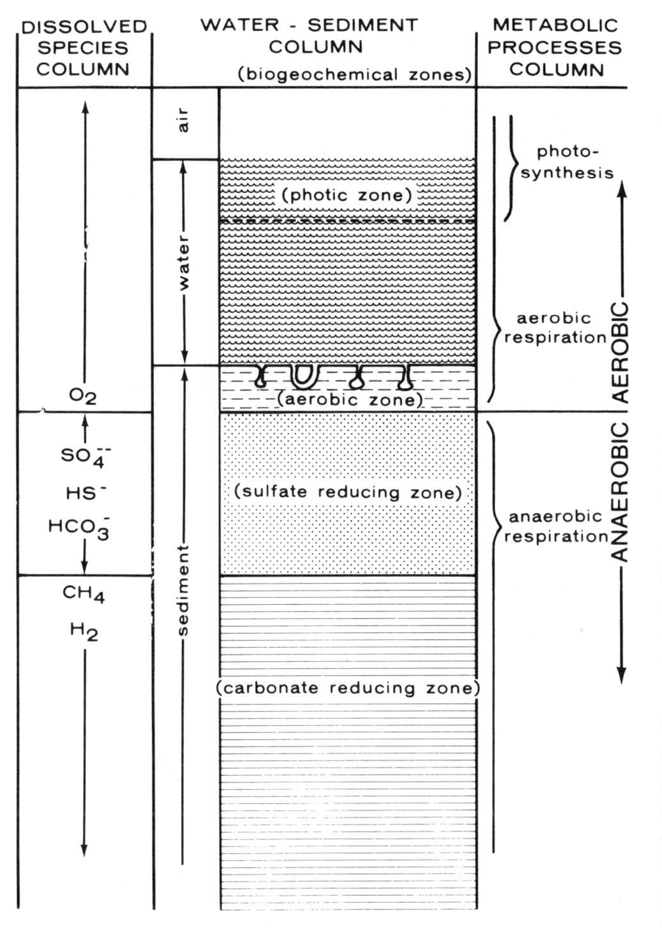

Figure 15. Diagrammatic cross section of organic-rich, open-marine setting showing succession of microbial ecosystems that led to microbial gas generation. (From Rice and Claypool, 1981.)

cementing procedure, the casing is reciprocated to facilitate bonding. Drilling time for these 1200 to 1900 ft (370–580 m) wells averages 2½ days from initial drilling to cementing the casing. The casing is swabbed dry and selectively perforated over an interval of 20 to 35 ft (6 to 11 m) with one shot per foot (0.3 m). Natural production rates usually range from 0 to 100 Mcf/d (0 to 2800 m^3/d).

A well usually is stimulated through 4½-in. (11 cm) casing with 40,000 lb (18,000 kg) of sand (half 8-12 mesh and half 10-20 mesh), 175 Mcf (5000 m^3) of CO_2, and approximately 325 bbl (50 m^3) of 4% KCl water. The fracture treatment begins with 60 bbl (9 m^3) of water to break down the formation followed by sand concentrations of 1 lb/gal (120 kg/m^3) increasing to 7 lb/gal (838 kg/m^3) at an injection rate of 45 bbl/min (0.12 m^3/s). High sand concentrations are an advantage because the fractures tend to heal on thin layers of sand proppant in these shallow, shaly reservoirs. The main disadvantage is the large amount of water used, which causes formation damage. The CO_2 is used to facilitate cleanup.

Other stimulation treatments that use less water have been investigated. These include nitrogen foam, nitrogen foam with alcohol, and nitrogen foam with carbon dioxide. However, not enough production data are available to evaluate these other treatments. Separate fracture treatments and tests are run on the Greenhorn-Phillips and the Bowdoin zones, and then production is commingled.

TESTING AND PRODUCTION

A typical Bowdoin dome area well is shut in for 30 minutes after the stimulation treatment and then flowed on choke to clean up for a period of 24 hours or more. After the flowback period, the well is shut in for 2 weeks, followed by a normal 24-hour initial potential test. A well's initial potential is the estimated hypothetical flow rate at a wellhead flowing pressure of 50 psi (345 kPa) at the end of the flow period. If justified, based on the test results, the well is connected to a pipeline.

The first wells from the recent expansion were connected to a pipeline at the end of 1975. Some concern developed over the rapid production falloff in the first few months. To counteract this sharp decline, additional compression was installed to reduce the flowing wellhead pressures. Although there was an immediate slight increase in producing rates, increased water production and the influx of formation clays and silts began to affect the well's production adversely. It was necessary to increase the wellhead flowing pressures (e.g., Figure 16). This procedure was very effective for most wells; however, some wells have continued to produce varying amounts of water. As a result, 1¼-in. (3.2 cm) tubing was installed in the wells to remove accumulated liquids. The technique is to produce the gas through the annulus with short, intermittent blows up the tubing to void the hole of liquid. Mechanical intermitters were installed. This method of using tubing has increased production significantly.

The present procedure is to maintain a flowing pressure at 70% of the well's initial shut-in pressure during the flush production period. The pressure then is decreased to 50% as the well's production becomes more stable. The higher pressure maintains the integrity of single phase flow, allowing the producer to maximize production.

The average initial potential of a Bowdoin field well is 662 Mcf/d (18.7×10^3 m^3/d). The cumulative production was 218 bcf (6.2×10^9 m^3) as of 31 December 1987.

RESERVOIR EVALUATION METHODS

Attempts were made to determine gas reserves using traditional methods. Since production histories

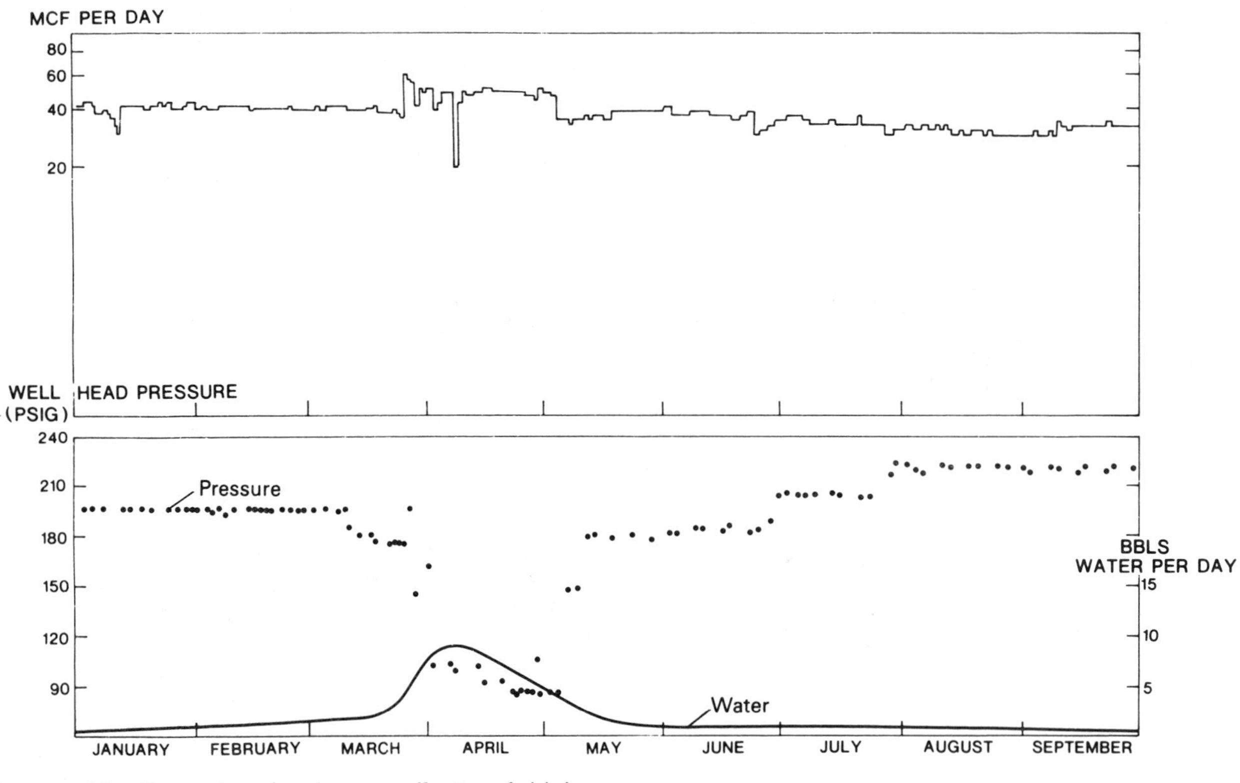

Figure 16. Example of adverse effects of higher pressure drawdown and increase in water production for Federal No. 1 well 2012 (Sec. 20, T.31N., R.32E.).

exist, standard application of extrapolating the empirical rate versus time plots and the material balance calculation using cumulative production versus shut-in pressure were two methods used. Both methods proved to provide inaccurate predictive models, especially in the early life of the Bowdoin field wells.

The production decline method, although useable with several years of data, was inaccurate when applied to short-term histories or where curtailment existed (Figure 17). Early reserve predictions were consistently pessimistic owing to the typical Bowdoin field well exhibiting a very strong curvature, an influence of the hydraulic fracture treatment. And as expected, curtailment further complicated the analysis with longer-life wells.

The material-balance method also was determined to be inappropriate. The actual reservoir pressure was not readily obtainable. Pressure measured during a short-term pressure buildup test in a Bowdoin field well reflects an unstabilized pressure much lower than the average reservoir pressure. Use of an approximation method to determine average reservoir pressure, such as the Horner plot (Horner, 1967), required an unacceptable amount of shut-in time and provided very questionable results. Estimates made with the material balance method have resulted in reserve predictions that are too pessimistic based on actual recoveries.

To accurately evaluate the low-permeability reservoirs of the Bowdoin field, a more rigorous approach was performed using numerical simulation of the reservoir behavior. The single well simulator (Crafton and Harris, 1976) included a history matching step to characterize the production and pressure behavior in both the induced fractures and the reservoir. After several satisfactory matches of the short-term production histories of individual wells and the simulation of long-term production were made, a projection of future production was made (Figure 18). The modeled curves for individual wells were quite similar in their shapes and an average curve was constructed to represent a typical Bowdoin field well decline curve (Figure 19). The curve represents an uncurtailed producing well in the better areas of production having an average 24-hour initial potential of 662 Mcf/d (18.7×10^3 m^3/d). This well can be expected to have a cumulative production of 330 MMcf (9×10^6 m^3) over a 10-year period. Depending on spacing density, the ultimate reserves could be as high as 500 MMcf (14×10^6 m^3) per well. It is important to note that if curtailment

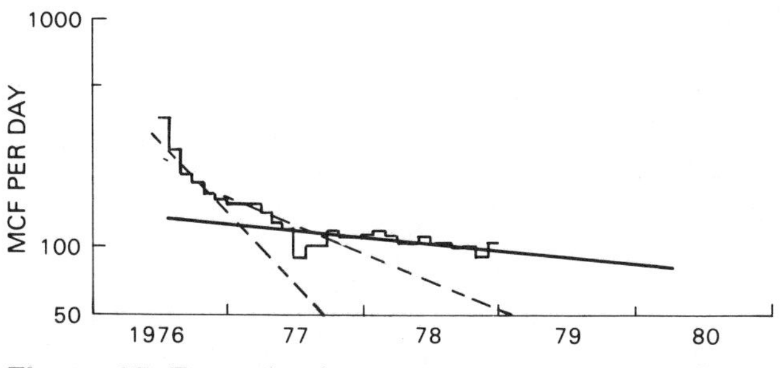

Figure 17. Example of conservative reserve prediction from production decline curves for Federal No. 1 well 1412 (Sec. 14, T.31N., R.32E.).

is anticipated, new simulation projections should be made using a new estimated pressure or production scenario for the deliverability projections.

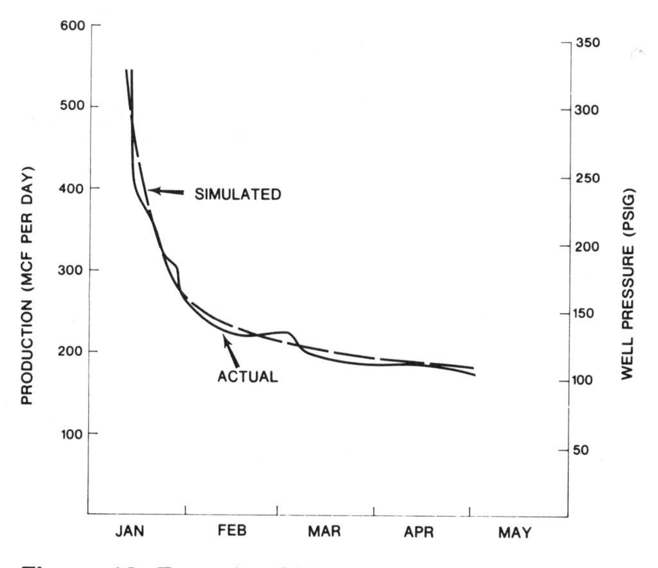

Figure 18. Example of history match for Federal No. 1 well 1512 (Sec. 15, T.31N., R.32E.).

EXPLORATION CONCEPTS

Large resources of natural gas are entrapped in shallow (less than 2000 ft; 610 m), low-permeability reservoirs of Late Cretaceous age in the northern Great Plains of the United States and Canada. The National Petroleum Council (1980) concluded that a study area of 120,000 mi^2 (311,000 km^2) in the U.S. portion contained recoverable gas reserves of about 100 tcf (3×10^{12} m^3). The reservoirs are thin, discontinuous laminae and beds of siltstone and sandstone, enclosed in organic-rich shale. Deposition of this type of sequence was widespread and took place in a shallow shelf environment.

The gas is microbial in origin and generation was widespread within the organic-rich laminae. Gas migration was generally limited to movement within the generating sequence itself. Entrapment of the gas occurred within each discontinuous laminae and bed of siltstone and sandstone encased in organic-rich shales that were the source. Although individual laminae and beds are highly discontinuous, the overall reservoir intervals are thick, widespread, and mappable.

To date, shallow microbial gas has been produced from "sweet spots" on structural highs, such as the Bowdoin dome, where the reservoirs are best developed. Recurrent tectonic movement has resulted in the winnowing of coarser grain sizes on these highs by geostrophic storm currents. In the future, exploration will occur off structure for lower-quality reservoirs. The free-gas phase will generally be at depths less than 2000 ft (610 m); at greater depths the gas will probably be in solution. Further development of this large resource will depend on higher gas prices and improved recovery technology.

In addition to the Bowdoin field, another important area of development of this type of gas accumulation is in southeastern Alberta and southwestern Saskatchewan (Figure 1). The production is widespread from shallow, low-permeability reservoirs of Late Cretaceous age and now covers an area of about 8000 mi^2 (20,700 km^2). In-place gas reserves in the Alberta portion of the accumulation are estimated to be about 15 tcf (420×10^9 m^3) (Energy Resources Conservation Board, 1986). The chief producing intervals in western Canada are the Second white specks sandstone, the Medicine Hat Sandstone, and the Milk River Formation equivalent as shown in Figure 3. All of the gas is produced from thin shelf siltstones and sandstones scattered throughout a thick shale sequence. The producing sequence is identical to that in the Bowdoin field. Rice (1981) showed in a cross section the correlation of Cretaceous rocks between the large producing area in western Canada and the Bowdoin dome and the gas-producing zones.

This type of geologic setting and the close association of low-permeability reservoirs and source rocks should have been widely developed and the resulting gas accumulations, particularly those of microbial origin, could be a potential source of energy for developing countries.

ACKNOWLEDGMENTS

This study was partially funded by the Department of Energy Western Tight Gas Sands Program.

REFERENCES

Almon, W. R., and D. K. Davies, 1978, Clay technology and well stimulation: Gulf Coast Association of Geological Societies Transactions, v. 28, pt. I, p. 1-6.

Collier, A. J., 1918, The Bowdoin dome, Montana, a possible reservoir of oil or gas: U.S. Geological Survey Bulletin 661-E, p. 193-208.

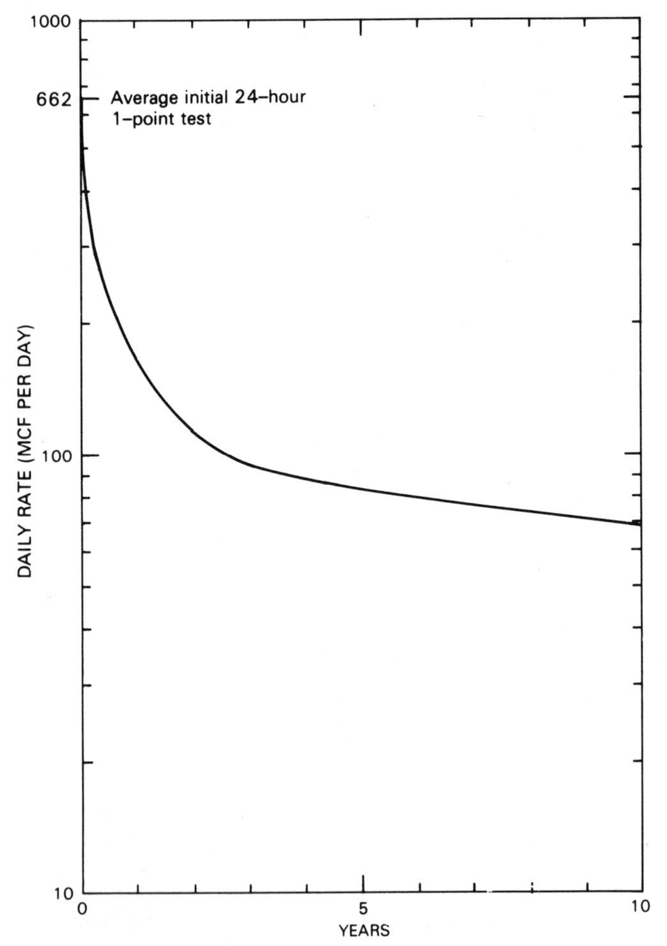

Figure 19. Simulated production decline curve for a typical Bowdoin field well (1 Mcf/d = 28 m^3/d).

Crafton, J. W., and C. D. Harris, 1976, Direct finite difference stimulation of a gas well with a finite capacity vertical fracture: Society of Petroleum Engineers Fourth Symposium on Numerical Simulation of Reservoir Performance, Los Angeles, Feb. 19-20, 1976, Paper No. 5736.

Energy Resources Conservation Board, 1986, Alberta's reserves of crude oil, oil sands, gas, natural gas liquids, and sulphur, 31 December 1986: Energy Resources Conservation Board Reserve Report Series ERCB-18, 379 p.

Gautier, D. L., and D. D. Rice, 1983, Significance of gamma-ray spectroscopy for evaluating shallow gas reservoirs from Bowdoin dome, Montana (abs.): American Association of Petroleum Geologists Bulletin, v. 67, p. 468.

Horner, D. R., 1967, Pressure buildup in wells: Pressure Analysis Methods, Society of Petroleum Engineers Reprint Series No. 9, p. 25-43.

National Petroleum Council, 1980, Tight gas reservoirs, Part II, *in* Unconventional gas sources, p. 10-1 to 170.

Rice, D. D., 1981, Subsurface cross section from southeastern Alberta, Canada, to Bowdoin dome area, north-central Montana, showing correlation of Cretaceous rocks and shallow, gas-productive zones in low-permeability reservoirs: U.S. Geological Survey Oil and Gas Investigations Chart OC-112.

Rice, D. D., 1983, Widespread, shallow-marine, storm-generated sandstone units in the Upper Cretaceous Mosby Sandstone, central Montana, *in* C. T. Siemers and R. W. Tillman, eds., Ancient shelf sediments: SEPM Special Publication No. 34, p. 145-163.

Rice, D. D., and G. E. Claypool, 1981, Generation, accumulation, and resource potential of biogenic gas: American Association of Petroleum Geologists Bulletin, v. 65, p. 5-25.

Rice, D. D., and G. W. Shurr, 1980, Shallow, low-permeability reservoirs of the northern Great Plains—an assessment of their natural resources: American Association of Petroleum Geologists Bulletin, v. 64, p. 969-987.

Starkey, H. C., P. D. Blackmon, and D. D. Rice, 1978, Mineralogical analyses of drill core samples from Midlands Gas Corporation wells—Federal 0370 No. 1 and Federal 2962 No. 1, Phillips County, Montana: U.S. Geological Survey Open-File Report No. 78-1001, 35 p.

Swift, D. J. P., and D. D. Rice, 1983, Sand bodies on muddy shelves—a model for sedimentation in the Western Interior seaway, North America, *in* C. T. Siemers and R. W. Tillman, eds., Ancient shelf sediments: SEPM Special Publication No. 34, p. 43-65.

SUGGESTED READING

Gautier, D. L., 1981, Lithology, reservoir properties, and burial history of the Gammon Shale (Cretaceous), southwestern North Dakota: American Association of Petroleum Geologists Bulletin, v. 65, p. 1146-1159.

Gautier, D. L., and D. D. Rice, 1982, Conventional and low-permeability reservoirs of natural gas in the northern Great Plains: Journal of Petroleum Technology, v. 34, p. 1600-1608.

Hankel, R. C., G. R. Davies, and H. D. Krouse, 1989, Eastern Medicine Hat gas field: a shallow, Upper Cretaceous, bacteriogenic gas reservoir of southeastern Alberta: Bulletin of Canadian Petroleum Geology, v. 37, p. 98-112.

Myhr, D. W., and N. C. Meijer-Drees, 1976, Geology of the southeastern Alberta Milk river gas pool, *in* M. M. Lerand, ed., The sedimentology of selected clastic oil and gas reservoirs in Alberta: Canadian Society of Petroleum Geology, p. 96-117.

Appendix 1. Field Description

Field name .... *Bowdoin field*

Ultimate recoverable reserves .... *475 bcf*

Field location:

- **Country** .... *U.S.A.*
- **State** .... *Montana*
- **Basin/Province** .... *Williston basin/Bowdoin dome*

Field discovery:

- **Year first pay discovered** .... *Upper Cretaceous Bowdoin and Phillips sandstones 1913*

Discovery well name and general location:

- **First pay** .... *Martin well, Sec. 18, T.31N., R.35E.*

Discovery well operator .... *Drilled by rancher for water*

IP in barrels per day and/or cubic feet or cubic meters per day:

- **First pay** .... *Not reported*
- **Second pay**
- **Third pay**

All other zones with shows of oil and gas in the field:

Age	Formation	Type of Show
Late Cretaceous	*Niobrara Formation*	*Gas*
Late Cretaceous	*Gammon Shale (Eagle Ss)*	*Gas*

Geologic concept leading to discovery and method or methods used to delineate prospect, e.g., surface geology, subsurface geology, seeps, magnetic data, gravity data, seismic data, seismic refraction, nontechnical:

Gas first discovered by well drilled for water. Initial development resulted from drilling of surface structure mapped by U.S. Geological Survey. Later development resulted from improved completion techniques and increased gas prices.

Structure:

Province/basin type .... *Bally 121; Klemme I*

Tectonic history

Williston basin is a structural-sedimentary intracratonic basin. The Bowdoin dome is a product of Laramide orogeny (Late Cretaceous), but was probably low relief feature in mid-Cretaceous time that influenced sedimentation.

Regional structure

Field is located on large arch on west flank of Williston basin.

Local structure

Very large oval arch with structural closure of 700 ft.

Trap:

Trap type(s)

Structural/stratigraphic—thin, discontinuous shelf sandstones that are well developed on large domal feature.

Basin stratigraphy (major stratigraphic intervals from surface to deepest penetration in field):

Chronostratigraphy	Formation	Depth to Top in ft
Cretaceous	*Muddy*	*1700*
Jurassic	*Swift*	*2300*
Jurassic	*Piper*	*2800*
Mississippian	*Madison*	*3100*

Mississippian-Devonian	*Bakken*	*4000*
Devonian	*Nisku*	*4100*
Silurian	*Interlake*	*5200*
Ordovician	*Red River*	*5400*
Ordovician	*Winnipeg*	*5600*
Cambrian	*Deadwood*	*5700*

Location of well in field

Reservoir characteristics:

Number of reservoirs *3*
Formations .... *Phillips and Bowdoin sandstones, Greenhorn lime*
Ages .... *Late Cretaceous*
Depths to tops of reservoirs .... *800-1800 ft*
Gross thickness (top to bottom of producing interval) .... *450 ft*
Net thickness—total thickness of producing zones
Average .... *30-50 ft*
Maximum .... *75 ft*
Lithology
Thin discontinuous shaly siltstone and sandstone laminae and beds, interbedded with shale (Bowdoin and Phillips); calcarenite (Greenhorn)
Porosity type .... *Intergranular*
Average porosity .... *Bowdoin, 8-14%; Greenhorn, 6-13%; Phillips, 15-17%*
Average permeability
Bowdoin, 0.1-0.7 md; Greenhorn, 0.1-6 md; Phillips, 2-3 md; in situ permeability indicated by reservoir stimulation and pressure transient studies is less than 0.1 md

Seals:

Upper
Formation, fault, or other feature .... *Niobrara Formation*
Lithology .... *Shale with bentonite beds*
Lateral
Formation, fault, or other feature .... *Pinchout of shelf sandstones and siltstones*
Lithology .... *Shale and bentonite*

Source:

Formation and age .... *Enclosing shales of Late Cretaceous age*
Lithology .... *Shale*
Average total organic carbon (TOC) *2%*
Maximum TOC *5%*
Kerogen type (I, II, or III) .... *Mostly III, some II*
Vitrinite reflectance (maturation) .... $R_o = 0.5\%$
Time of hydrocarbon expulsion .... *Late Cretaceous*
Present depth to top of source .... *Approx. 1000 ft*
Thickness .... *450 ft*

Appendix 2. Production Data

Field name .... *Bowdoin field*

Field size:

Proved acres *380,000*
Number of wells all years *693*
Current number of wells *547*

Well spacing .... *160 and 320 ac in different units*
Ultimate recoverable .... *475 bcf*
Cumulative production .... *218 bcf*
Annual production .... *As much as 10 bcf*
Present decline rate .... *By well, 8%*
Initial decline rate *65%*
Overall decline rate *8%*
Annual water production .... *13,000 bbl*
In place, total reserves .... *1120 bcf*
In place, per acre-foot .... *74 mcf*
Primary recovery .... *475 bcf*
Cumulative water production .... *740,000 est.*

Drilling and casing practices:

Amount of surface casing set .... *200 ft*
Casing program .... *7-in. surface casing; 4½-in. production casing to TD*
Drilling mud .... *Low fluid loss, polymer-based mud*
Bit program .... *6¼-in. standard tri-cone*
High pressure zones .... *All zones underpressured 0.23 to 0.4 psi/ft*

Completion practices:

Interval(s) perforated .... *Multiple intervals that vary between wells*
Well treatment .... *Hydraulic fracture using sand, CO_2, and KCl water*

Formation evaluation:

Logging suites .... *Older wells: induction, neutron, density, and GR; newer wells: microresistivity and sonic, in addition to those run in older wells*
Testing practices .... *Drill-stem tests used in initial phases of recent expansion*
Mud logging techniques .... *Used only in initial phases of development to define gross pay interval*

Oil characteristics:

Type .... *Microbial gas; $C_1/C_{1-5} > 0.99$, $\delta^{13}C_1 = -69$ to -72 ppt*
API gravity
Base
Initial GOR
Sulfur, wt%
Viscosity, SUS
Pour point
Gas-oil distillate

Field characteristics:

Average elevation .... *2300 ft*
Initial pressure .... *400–630 psi*
Present pressure .... *280–450 psi*
Pressure gradient .... *0.3–0.4 psi/ft*
Temperature .... *95°F*
Geothermal gradient .... *0.23°F/ft*
Drive .... *Pressure depletion (volumetric)*
Oil column thickness .... *Multiple pay zones; as much as 450 ft*
Oil-water contact .... *Poorly developed*
Connate water *40%*
Water salinity, TDS *9500–14,500*
Resistivity of water .... *0.4–0.7 ohm at 105°F BHT*

Transportation method and market for oil and gas:

Gas pipelines owned by KN Energy and Williston Basin Interstate.